AF658373

Advances in Endophytic Research

Vijay C. Verma • Alan C. Gange
Editors

Advances in Endophytic Research

Editors
Vijay C. Verma
Centre of Experimental Medicine
and Surgery (CEMS)
Institute of Medical Sciences
Banaras Hindu University
Varanasi, UP, India

Alan C. Gange
School of Biological Sciences
Royal Holloway University of London
Egham, Surrey, UK

ISBN 978-81-322-1574-5 ISBN 978-81-322-1575-2 (eBook)
DOI 10.1007/978-81-322-1575-2
Springer New Delhi Heidelberg New York Dordrecht London

Library of Congress Control Number: 2013953494

Printed on acid-free paper

Springer is part of Springer Science+Business Media (www.springer.com)

This book is dedicated to Pandit Madan Mohan Malaviya, founder of Banaras Hindu University, the largest residential university in Asia and one of the largest in the world.

India is not a country of the Hindus only. It is a country of the Muslims, the Christians and the Parsees too. The country can gain strength and develop itself only when the people of the different communities in India live in mutual goodwill and harmony. It is my earnest hope and prayer that this centre of life and light which is coming into existence, will produce students who will not only be intellectually equal to the best of their fellow students in other parts of the world, but will also live a noble life, love their country and be loyal to the Supreme ruler.

Pt. Madan Mohan Malaviya
(1861–1946)
Founder, Banaras Hindu University

Take up one idea. Make that one idea your life-think of it, dream of it, live on that idea. Let the brain, muscles, nerves, every part of your body, be full of that idea, and just leave every other idea alone. This is the way to success that is the way great spiritual giants are produced.

–Swami Vivekananda

World's most respected Vedanta thinker (1863–1902)
Vedanta Philosophy: Lectures by Swami Vivekananda, Kessinger Publishing, USA, 1996, Page 70

Foreword

I am delighted in writing a foreword on this very special achievement of Dr. Vijay Verma in the form of this book that contains innovative information about endophytic research. Microbial biodiversity is an ultimate source that may be utilized and applied to modern biology and biotechnology and has the potential to be developed as innovative and sustainable solutions to a wide range of problems of human beings. Microbes are omnipresent from normal to extreme to the tune of 10^{13} per mm^2 of our skin and natural to man-made environments, but what surprises me is their presence within the healthy internal tissues of higher plants. Being in medical science and working on plants of medicinal importance, I am well aware of the fascinating specifics of plant sciences but amazed by such vast information on endophytes and their biotechnological applications. Endophytes represent almost unlimited and sustainable sources of bioactive and chemically novel natural products with the potential for utilization in an array of medical, agricultural, and industrial applications. Since the discovery of the anticancer molecule "Taxol" from symbiotic endophytes as an alternative source to the host, the endophytes have become of core interest for drug discovery expedition. It is my pleasure to endorse Dr. Verma's long experience of working with endophytes and his collaborative interactions with the world leaders in this particular field. As a result of this effort as a book, not only Dr. Verma but the institution is also enriched and privileged to have opportunity of working with such international laureates. I am delighted to see that contributors of this book have several years of collaborative experience and have signatory status in endophytic research. I specially would like to mention Prof. Bacon, Prof. White Jr., Prof. Omacini, and Prof. Osono, who are international icons of this field. The information provided by them would be really beneficial to young researchers and will generate interest among them about these fascinating microbes. The potential importance of beneficial endophytes to plants and biotechnology really did not become clear until 1975, when Prof. Bacon discovered fungal endophytes in the family Clavicipitaceae growing systemically in pasture grasses. The contribution of Prof. Bacon about the future challenges of endophytic research is highlighted well in this book. As a book, I strongly feel that it is a gold mine of information about recent developments in the field of endophytic research and demonstrates a wealth of interesting details. This will surely enlighten the new minds and become a source of inspiration and information for those who wish to work in this fascinating area of research. In short, Dr. Verma's book is unique and surely a work to treasure for anyone who is interested in endophytic research. My heartiest congratulations and wishes are always with Dr. Verma for producing such a nice piece of work that will guide young minds for a long time to come.

Prof. Dr. R. G. Singh

MD, DM (Nep), MNAMS (Nep), FICA, FFIM
FIMAAMS, FIACM, FICN, FGSI, FMM, FICP, FIASM
FIAY, FISH, FICAAI, FAMS

Director
Institute of Medical Sciences
Banaras Hindu University

Varanasi, UP, India

Preface

Microbial biodiversity is a continuing problem in the estimates of overall biodiversity. A great extent of our knowledge about microbes results from the approximately 1 % of culturable microbes present on our planet. Thus, we are relatively unfamiliar with a vast magnitude of unculturable microbes that represent a significant part of microbial diversity. With recent revolutionary developments in different domains of "-omics" to which an appropriate term might be "Revolomics" that includes metagenomic and next-generation sequencing technologies, we are beginning to understand microbial diversity and the exploration of novel genes and metabolic products. Another aspect of diversity is to search for new microbial habitats that may possess hidden culturable microbes that may add to estimated and discovered microbial diversity. Many such unconventional habitats such as marine ecosystems, thermal vents, and ice caps are now being explored for novel microbes, and one result is that microbes in these habitats can provide new chemical diversity with the potential to be exploited as drug leads for many human pathogens. Since most of the planet is covered with marine ecosystems, it is reasonable to accept that a huge microbial diversity remains to be discovered from deep sea and from marine organisms. Regarding the land mass of our planet, 9.4 % is covered by the forests with a wealth of associated microbes. Forest vegetation not only has microbes on the surface (phyllosphere, epiphytes, rhizosphere, etc.) but also has symbiotic microbes within (endophytes, mycorrhizas, dark septate endophytes). All higher plants on this planet have a form of symbiotic association with microbes called "endophytic" symbiosis. A single plant may contain hundreds of microbes, and thus, the diversity of endophytes is likely to be many times greater than plant diversity. Mutualistic endophytic microbes with an emphasis on the relatively understudied fungal endophytes are the focus of this special book. Plants are associated with microorganisms, endophytic bacteria and fungi, which live inter- and intracellularly without inducing pathogenic symptoms, but have active biochemical and genetic interactions with their host. Endophytes play vital roles as plant growth promoters, biocontrol agents, biosurfactant producers, and enzymes and secondary metabolite producers, as well as in providing a new hidden repertoire of bioactive natural products with uses in pharmaceutical, agrochemical, and other biotechnological applications. Apart from these virtues, the microbial endophytes may be adapted to the complex metabolism of many desired molecules that can be of significant industrial applications. These microbes can be a useful alternative for sustainable solutions for ecological control of pests and diseases and can reduce the burden of excess of chemical fertilizers for this purpose.

This book is an attempt to review the recent development in the understanding of microbial endophytes and their potential biotechnological applications. We have tried to recognize several research domains of endophytic research in which significant progress has been made such as ecology and biodiversity, host-endophyte interactions, bioactive compounds from endophytes, and future challenges. Attempts have been made to summarize the development achieved so far and future prospects for further research in this fascinating area of research.

Varanasi, UP, India Vijay Chandra Verma
Egham, Surrey, UK Alan Christopher Gange

Contents

Part IV Bioactive Compounds from Endophytes

Part V Bio-control and Bioremediation

Contributors

Muhammad Afzal Environmental Biotechnology, National Institute for Biotechnology and Genetic Engineering (NIBGE), Faisalabad, Pakistan

Swechha Anand Centre of Experimental Medicine and Surgery, Institute of Medical Sciences, Banaras Hindu University, Varanasi, UP, India

Mariam Anees Department of Biochemistry, Quaid-i-Azam University, Islamabad, Pakistan

João Lúcio Azevedo Department of Genetics, Escola Superior de Agricultura "Luiz de Queiroz", University of São Paulo, Piracicaba, SP, Brazil

Charles W. Bacon Toxicology and Mycotoxin Research Unit, USDA, Agriculture Research Service, Russell Research Center, Athens, GA, USA

Kathrin Blumenstein Southern Swedish Forest Research Centre, Swedish University of Agricultural Sciences, Alnarp, Sweden

Sue Broughton School of Biological Sciences, Royal Holloway University of London, Egham, Surrey, England, UK

Andrea Campisano Sustainable Agro-ecosystems and Bioresources Department, IASMA Research and Innovation Centre, Fondazione Edmund Mach, San Michele all'Adige (TN), Italy

Amanda F. Currie School of Biological Sciences, Royal Holloway University of London, Egham, Surrey, England, UK

Mahmoud Fahmi Elsebai Department of Biology, University of Oulu, Oulu, Finland

Faculty of Pharmacy, Department of Pharmacognosy, Mansoura University, Mansoura, Egypt

David Ezra Department of Plant Pathology, ARO the Volcani Center, Bet Dagan, Israel

Tayaba Fatima National Bureau of Agriculturally Important Microorganisms, Kaithauli, Mau Nath Bhanjan, Uttar Pradesh, India

Christopher M.M. Franco Department of Medical Biotechnology, Flinders University, Bedford Park, SA, Australia

K.N. Ganeshaiah School of Ecology and Conservation, University of Agricultural Sciences, Bangalore, India

Ashoka Trust for Research in Ecology and the Environment, Bangalore, India

Department of Forestry and Environmental Sciences, University of Agricultural Sciences, Bangalore, India

Alan Christopher Gange School of Biological Sciences, Royal Holloway University of London, Egham, Surrey, UK

Mayank Gangwar Department of Microbiology, Institute of Medical Sciences, Banaras Hindu University, Varanasi, UP, India

Department of Pharmacology, Institute of Medical Sciences, Banaras Hindu University, Varanasi, UP, India

Pallavi Gaur National Bureau of Agriculturally Important Microorganisms, Kaithauli, Mau Nath Bhanjan, Uttar Pradesh, India

Venkadasamy Govindasamy Department of Medical Biotechnology, Flinders University, Bedford Park, SA, Australia

National Institute of Abiotic Stress Management, ICAR, Malegaon, Baramati, Maharashtra, India

Mary Ruth Griffin Department of Environmental and Plant Biology, Ohio University, Athens, OH, USA

Vadakattu V.S.R. Gupta Department of Medical Biotechnology, Flinders University, Bedford Park, SA, Australia

CSIRO Ecosystem Sciences, Glen Osmond, SA, Australia

Noemí Herrero Department of Plant Virology, Institute of Entomology Biology Centre of the Academy of Sciences of the Czech Republic, České Budějovice, Czech Republic

Dorothy M. Hinton Toxicology and Mycotoxin Research Unit, USDA, Agriculture Research Service, Russell Research Center, Athens, GA, USA

Su Hodgson School of Biological Sciences, Royal Holloway University of London, Egham, Surrey, England, UK

Ivelisse Irizarry Department of Plant Biology and Pathology, School of Environmental and Biological Sciences, Rutgers University, New Brunswick, NJ, USA

Liang Jin School of Biological Sciences, Royal Holloway University of London, Egham, Surrey, England, UK

Holly Johnson Department of Computer Science, Central Washington University, Ellensburg, WA, USA

Ari Jumpponen Division of Biology, Kansas State University, Manhattan, KS, USA

Janne J. Koskimäki Department of Biology, University of Oulu, Oulu, Finland

Vinod Kumar National Bureau of Agriculturally Important Microorganisms, Kaithauli, Mau Nath Bhanjan, Uttar Pradesh, India

Paulo Teixeira Lacava Centre of Biological Sciences and Health, Federal University of São Carlos, São Carlos, SP, Brazil

Orna Liarzi Department of Plant Pathology, ARO the Volcani Center, Bet Dagan, Israel

Riffat Naseem Malik Department of Environmental Sciences, Quaid-i-Azam University, Islamabad, Pakistan

Keerthi Mandyam Division of Biology, Kansas State University, Manhattan, KS, USA

B.L. Manjunatha Department of Crop Physiology, University of Agricultural Sciences, Bangalore, India

School of Ecology and Conservation, University of Agricultural Sciences, Bangalore, India

Juan A. Martín Departamento de Silvopascicultura, Universidad Politécnica de Madrid, Madrid, Spain

Pragya Mishra Ranjan Physiology and Biochemistry Laboratory, Department of Botany, Centre of Food Technology, University of Allahabad, Allahabad, India

Raghvendra Raman Mishra Department of Microbiology, Institute of Medical Sciences, Banaras Hindu University, Varanasi, Uttar Pradesh, India

P. Mohana Kumara Department of Crop Physiology, University of Agricultural Sciences, Bangalore, India

School of Ecology and Conservation, University of Agricultural Sciences, Bangalore, India

Gopal Nath Department of Microbiology, Institute of Medical Sciences, Banaras Hindu University, Varanasi, UP, India

Marina Omacini IFEVA, Facultad de Agronomía, Universidad de Buenos Aires, Buenos Aires, Argentina

Takashi Osono Center for Ecological Research, Kyoto University, Otsu, Shiga, Japan

Saurabh Kumar Patel Department of Microbiology, Institute of Medical Sciences, Banaras Hindu University, Varanasi, UP, India

Anna Maria Pirttilä Department of Biology, University of Oulu, Oulu, Finland

Johanna Pohjanen Department of Biology, University of Oulu, Oulu, Finland

Satya Prakash National Facility of Tribal and Herbal Medicine, Institute of Medical Sciences, Banaras Hindu University, Varanasi, UP, India

Chandra Bhan Pratap Department of Microbiology, Institute of Medical Sciences, Banaras Hindu University, Varanasi, UP, India

Enrique Quesada-Moraga Department of Agricultural and Forest Science and Resources, University of Córdoba (UCO), Córdoba, Spain

Shalini Rai National Bureau of Agriculturally Important Microorganisms, Kaithauli, Mau Nath Bhanjan, Uttar Pradesh, India

Amit Ranjan Centre of Experimental Medicine and Surgery, Institute of Medical Sciences, Banaras Hindu University, Varanasi, UP, India

G. Ravikanth School of Ecology and Conservation, University of Agricultural Sciences, Bangalore, India

Ashoka Trust for Research in Ecology and the Environment, Bangalore, India

N. Sachin Department of Crop Physiology, University of Agricultural Sciences, Bangalore, India

School of Ecology and Conservation, University of Agricultural Sciences, Bangalore, India

S. Sagar Jadhav Department of Crop Physiology, University of Agricultural Sciences, Bangalore, India

Hari S. Shukla Department of Surgical Oncology, Institute of Medical Sciences, Banaras Hindu University, Varanasi, Uttar Pradesh, India

Parjanya Shukla Department of Pharmaceutical Sciences, Faculty of Health Science, Sam Higginbottom Institute of Agriculture Technology and Sciences – Deemed University, Allahabad, India

S. Shweta Department of Crop Physiology, University of Agricultural Sciences, Bangalore, India

School of Ecology and Conservation, University of Agricultural Sciences, Bangalore, India

Archana Singh Department of Botany, Faculty of Science, Banaras Hindu University, Varanasi, Uttar Pradesh, India

Rana Gopal Singh Department of Nephrology, Institute of Medical Sciences, Banaras Hindu University, Varanasi, UP, India

Santosh Kumar Singh Centre of Experimental Medicine and Surgery, Institute of Medical Sciences, Banaras Hindu University, Varanasi, UP, India

Shriti Singh Centre of Experimental Medicine and Surgery, Institute of Medical Sciences, Banaras Hindu University, Varanasi, UP, India

Mariusz Tadych Department of Plant Biology and Pathology, School of Environmental and Biological Sciences, Rutgers University, New Brunswick, NJ, USA

Mysore V. Tejesvi Department of Biology, University of Oulu, Oulu, Finland

Adeline Su Yien Ting School of Science, Monash University Sunway Campus, Bandar Sunway, Darul Ehsan, Selangor, Malaysia

Mallika Tiwari Department of Surgical Oncology, Institute of Medical Sciences, Banaras Hindu University, Varanasi, Uttar Pradesh, India

Mónica S. Torres Department of Plant Biology and Pathology, School of Environmental and Biological Sciences, Rutgers University, New Brunswick, NJ, USA

R. Uma Shaanker Department of Crop Physiology, University of Agricultural Sciences, Bangalore, India

School of Ecology and Conservation, University of Agricultural Sciences, Bangalore, India

Ashoka Trust for Research in Ecology and the Environment, Bangalore, India

M.M. Vasanthakumari Department of Crop Physiology, University of Agricultural Sciences, Bangalore, India

School of Ecology and Conservation, University of Agricultural Sciences, Bangalore, India

Ajay Kumar Verma Department of Microbiology, Institute of Medical Sciences, Banaras Hindu University, Varanasi, UP, India

Vijay Chandra Verma DST-FAST Track Young Scientist, Centre of Experimental Medicine and Surgery (CEMS), Institute of Medical Sciences, Banaras Hindu University, Varanasi, UP, India

James Wearn Herbarium, Library, Art and Archives, Royal Botanic Gardens, Kew, Richmond, Surrey, UK

Hilary Wendt School of Biological Sciences, Royal Holloway University of London, Egham, Surrey, England, UK

James F. White Jr. Department of Plant Biology and Pathology, School of Environmental and Biological Sciences, Rutgers University, New Brunswick, NJ, USA

Johanna Witzell Southern Swedish Forest Research Centre, Swedish University of Agricultural Sciences, Alnarp, Sweden

Sohail Yousaf Department of Environmental Sciences, Quaid-i-Azam University, Islamabad, Pakistan

Iñigo Zabalgogeazcoa Department of Abiotic Stress, Instituto de Recursos Naturales y Agrobiología de Salamanca (IRNASA-CSIC), Salamanca, Spain

Part I

Ecology and Biodiversity

Diversity and Ecology of Endophytic and Epiphytic Fungi of Tree Leaves in Japan: A Review

1

Takashi Osono

Abstract

The phyllosphere is the living leaf as a whole and is colonized by endophytic and epiphytic fungi in the interior and on the surface of leaves, respectively. In this chapter, I summarize studies on the diversity and ecology of endophytic and epiphytic phyllosphere fungi on live leaves of trees in Japan. Studies to date have detected endophytes and epiphytes on leaves of at least 255 coniferous and broad-leaved tree species in 69 plant families, according to 45 papers published since 1990. These studies have recorded 24 endophytic and 22 epiphytic genera of fungi. Major trees used in the ecological studies of phyllosphere fungi include pines (*Pinus*), beech (*Fagus*), and dogwood (*Swida*). Focal topics include (1) the infection and colonization of leaves; (2) seasonal and leaf age-dependent patterns of temporal changes; (3) spatial distribution at various scales, from within-leaf, to within-canopy, to altitudinal and geographic distributions; (4) direct and indirect roles in decomposition of dead leaves; and (5) interaction with pathogens and herbivores and effects of simulated acid rain. Future research directions in Japan are suggested and discussed with reference to international literature on the ecology of endophytic and epiphytic phyllosphere fungi.

1 Introduction

The phyllosphere is the living leaf as a whole (including the interior and surface), which provides habitats for a variety of microorganisms, such as fungi, bacteria, and algae. Phyllosphere fungi include endophytes and epiphytes that colonize the interior or surface of leaves, respectively (Petrini 1991). Although the presence of phyllosphere fungi on tree leaves was known as early as the 1960s, studies of phyllosphere fungi increased in the 1980s, and a number of useful reviews have been published on their diversity and ecology (Hudson 1968; Carroll 1988, 1995; Petrini 1986, 1991; Boddy and Griffith 1989; Stone et al. 1996; Stone and Petrini 1997; Lindow and Brandl 2003; Arnold 2005, 2007; Saikkonen 2007), functional

T. Osono (✉)
Center for Ecological Research, Kyoto University, 2-509-3 Hirano, Otsu, Shiga 520-2113, Japan
e-mail: tosono@ecology.kyoto-u.ac.jp

V.C. Verma and A.C. Gange (eds.), *Advances in Endophytic Research*,
DOI 10.1007/978-81-322-1575-2_1, © Springer India 2014

roles (Rodriguez and Redman 1997; Saikkonen et al. 1998; Wilson 2000; Sieber 2007), ecophysiology (Petrini et al. 1992; Petrini 1996), and interactions with herbivores (Carroll 1991a, b).

Japan is an elongated island country located in the middle latitudes from 20 to 46°N, consisting of four main islands (Hokkaido, Honshu, Shikoku, and Kyushu) and a number of small islands. Approximately 66 % of the land is covered with a variety of forest types ranging from subboreal to temperate and subtropical forests (Fig. 1.1). More than 1,200 woody plants grow natively on the Japanese Archipel-ago (Biodiversity Center of Japan 2010) and merit the exploration of the diversity and ecology of endophytic and epiphytic fungi of their tree leaves. Accordingly, studies on the diversity and ecology of phyllosphere fungi of trees have been actively pursued in Japan since the 1990s. The purpose of this chapter is to review the major findings of these studies. First, I outline the richness of tree species examined and the genera of phyllosphere fungi reported in Japan. A variety of phyllosphere fungi have been detected on a variety of tree species, but ecological studies of phyllosphere fungi have concentrated on pine (*Pinus*, Pinaceae), beech (*Fagus*, Fagaceae), and dogwood (*Swida*, Cornaceae) as model systems. Focal topics of the ecological studies include (1) the infection and colonization of leaves; (2) seasonal and leaf age-dependent patterns of temporal changes; (3) spatial distribution at various scales, from within-leaf, to within-canopy, to altitudinal and geographic scales; (4) direct and indirect roles in decomposition of dead leaves; and (5) interaction with pathogens and herbivores and effects of simulated acid rain. Regarding topic (4), a large amount of information has already been included in the reviews of Osono (2006b, 2007) and Osono and Hirose (2009b). However, I realized that significant progress has been made in the last 3 years, especially on the roles of endophytic ascomycetes in lignin decomposition, which is summarized in this chapter. Finally, future research directions are suggested and discussed with reference to international literature about phyllosphere fungi. The potential applicability of molecular techniques, especially that of pyrosequencing using next-generation sequencers, is worth mentioning in this regard.

2 The Diversity of Endophytic and Epiphytic Fungi

2.1 Host Tree Species

To my knowledge, 45 papers have been published from 1990 to 2013 (as of January 2013) about the diversity and ecology of endophytic and epiphytic phyllosphere fungi in Japan.[1] Phyllosphere fungi have been detected on at least 255 tree species in 68 families in these studies (Table 1.1), including 202 broad-leaved and 53 coniferous species, 118 evergreen and 137 deciduous species, and 185 domestic and 70 exotic species. The majority (253 of 255) of tree species studied were examined for the detection of endophytic fungi, whereas only seven tree species were targeted for epiphytes: *Abies firma* (Pinaceae) (Aoki et al. 1990), *Bruguiera gymnorrhiza* (Rhizophoraceae) (Nakagiri et al. 1997), *Camellia japonica* (Theaceae) (Osono 2008), *Castanopsis sieboldii* (Osono et al. 2008), *Fagus crenata* (Osono 2002; Osono and Mori 2003), *Quercus myrsinaefolia* (Fagaceae) (Shirouzu et al. 2008), and *Swida controversa* (Cornaceae) (Osono and Mori 2004, 2005). Only four tree species were examined for both endophytes and epiphytes (see Sect. 2.2). The locations where these studies were conducted ranged from Hokkaido to Tohoku, Kanto, Chubu, Kinki, Kyushu, and Okinawa (Fig. 1.1). Reports from Kyoto and Mie Prefectures were especially notable in terms of the number of publications.

[1]Aoki et al. (1990), Asai et al. (1998), Hashizume et al. (2008, 2010), Hata and Futai (1993, 1995, 1996), Hata and Sone (2008), Hata et al. (1998, 2002), Ikebe et al. (2004), Ito et al. (2007), Kaneko and Kakishima (2001), Kaneko and Kaneko (2004), Kaneko et al. (2003), Koide et al. (2005a), Makisaka et al. (2005), Naito et al. (2002), Nakagiri et al. (1997), Nomura et al. (2003), Okane (2003), Okane et al. (1996, 1997, 1998, 2001a, b, 2003), Osono (2002, 2003, 2006a, b, 2008, 2012), Osono and Masuya (2012), Osono and Mori (2004, 2005), Osono et al. (2004a, 2008, 2013), Sahashi et al. (1999, 2000), Shirouzu et al. (2008), Suzuki et al. (2003), Tomita (2003), and Yoshihashi et al. (2000, 2001).

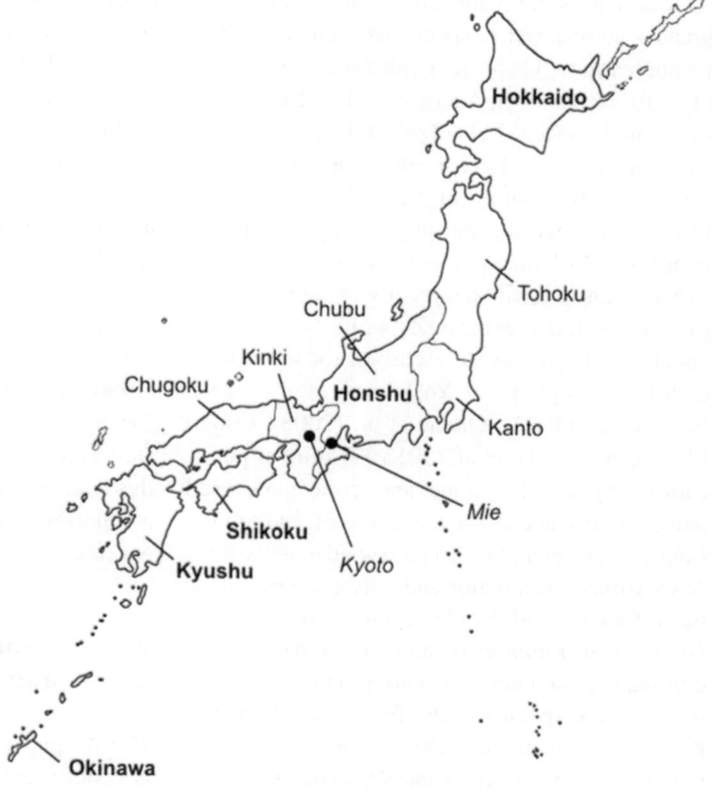

Fig. 1.1 Location of major islands and districts in Japan

Table 1.1 Family and number of tree species examined for endophytic and epiphytic phyllosphere fungi in Japan

Family	No. of tree species	Family	No. of tree species	Family	No. of tree species
Pinaceae	47	Theaceae	5	Araliaceae	2
Fagaceae	17	Anacardiaceae	4	Berberidaceae	2
Ericaceae	16	Magnoliaceae	4	Calophyllaceae	2
Aceraceae	12	Rutaceae	4	Cupressaceae	2
Betulaceae	12	Celastraceae	3	Daphniphyllaceae	2
Rosaceae	11	Cornaceae	3	Juglandaceae	2
Oleaceae	8	Euphorbiaceae	3	Pandanaceae	2
Fabaceae	7	Hamamelidaceae	3	Podocarpaceae	2
Lauraceae	7	Rhizophoraceae	3	Rubiaceae	2
Caprifoliaceae	6	Ulmaceae	3	Styracaceae	2
Moraceae	6	Verbenaceae	3	Symplocaceae	2
Aquifoliaceae	5	Actinidiaceae	2	Taxodiaceae	2
Saxifragaceae	5	Apocynaceae	2		

Families with one tree species: Boraginaceae, Cercidiphyllaceae, Clethraceae, Combretaceae, Coriariaceae, Davidiaceae, Ebenaceae, Ginkgoaceae, Goodeniaceae, Hippocastanaceae, Hypericaceae, Lardizabalaceae, Lecythidaceae, Liliaceae, Melastomataceae, Myricaceae, Pittosporaceae, Platanaceae, Punicaceae, Salicaceae, Sapindaceae, Scrophulariaceae, Simaroubaceae, Sonneratiaceae, Stachyuraceae, Sterculiaceae, Thymelaeaceae, Tiliaceae, Trochodendraceae, and Urticaceae

Pinaceae is the plant family that included the greatest number of tree species examined for phyllosphere fungi (47 species), followed by Fagaceae (17), Ericaceae (16), Aceraceae (12), Betulaceae (12), and Rosaceae (11) (Table 1.1). Less than 10 tree species each have been examined for the remaining 63 plant families. Hata and Futai (1996) compared endophytic mycobiota of needles of 45 *Pinus* species collected in a nursery in Kyoto and contributed greatly to the number of pinaceous trees examined. Four to five tree species in Fagaceae were chosen for the studies of foliar endophytes in Yoshihashi et al. (2000), Naito et al. (2002), Okane et al. (2003), Osono (2012), and Osono et al. (2013). A suite of publications by Dr. I. Okane and colleagues used leaves of Ericaceae for the study of endophytic fungi. Okane et al. (1996) described a new genus, *Discostroma*, as a major endophyte in ericaceous trees. Okane et al. (1998) found that species in *Guignardia*, *Phomopsis*, and *Colletotrichum* are dominant components of endophytic mycobiota of eight tree species in the Ericaceae planted in Kyoto. Okane et al. (2001a) then identified a *Guignardia* species (*G. endophyllicola*, the teleomorph of *Phyllosticta capitalensis*) frequently isolated from leaves of multiple ericaceous and other trees. Osono (2012), Osono and Masuya (2012), and Osono et al. (2013) contributed to the study of endophytic fungi associated with leaves of tree species in Aceraceae and Betulaceae.

Several studies have compared endophytic fungi between multiple tree species at single locations. For example, Yoshihashi et al. (2000) compared 29 tree species found on the campus of Mie University for endophytic fungi. Okane et al. (2003) tested leaves of 94 tree and herb species for the detection of *Guignardia* in a botanical garden in Kyoto. Okane et al. (1997) compared 21 evergreen tree species for endophytic mycobiota in Iriomote Island, located in the subtropical region in southern Japan. Okane et al. (2001b) described a new genus, new species, *Surculiseries rugispora*, isolated from leaves of a mangrove tree *Bruguiera gymnorrhiza*. Recently, Osono (2012) studied patterns of occurrence of endophytic fungal genera for leaves of 73 deciduous tree species in a cool temperate forest in Kyoto. In the same forest, Osono et al. (2013) examined 94 tree species (38 families) for the diversity and ubiquity of xylariaceous endophytes in live leaves.

The latent infection of endophytic fungi in healthy-looking tissues or the presence of epiphytic fungi (i.e., phylloplane fungi) on the surface (i.e., phylloplane) of tree leaves was already noted in Japan before 1990. For example, Soma and Saito (1979) observed fungi on live needles of *Pinus thunbergii*. Terashita (1973) detected *Colletotrichum* species on live leaves of 61 (88 %) out of 67 broad-leaved trees. Carroll (1990) detected endophytic fungi in needles of *Cryptomeria japonica* (Cupressaceae). However, studies published before 1990 are not included in this chapter, mainly because methods of isolation and presentation of results were somewhat different from those published after 1990.

2.2 Genera of Endophytic and Epiphytic Fungi

The 45 papers analyzed in this chapter include not only exhaustive surveys of phyllosphere mycobiota but also targeted studies that aimed at the detection of fungal taxa of particular interest. Fungi were identified to species in some papers but to genus in others. Moreover, methods of isolation of phyllosphere fungi differed among the studies (see below). These differences make it difficult to compare the richness and species composition of phyllosphere fungal assemblages reported from different papers. Instead, qualitative comparisons are made in this chapter to summarize the pattern of occurrence of fungal genera on tree leaves in Japan.

Twenty-three genera have been reported as endophytes of tree leaves in Japan (Table 1.2). These genera are divided into two groups in terms of the host specificity. The first group includes *Colletotrichum*, *Pestalotiopsis*, *Phomopsis*, *Phyllosticta*, and genera in the Xylariaceae (*Geniculosporium*, *Nodulisporium*, and *Xylaria*) that were isolated from a variety of host tree species and have low host specificity. It should be noted that host specificity at the level of fungal genus does not necessarily assure the specificity

Table 1.2 Genera of fungi isolated from tree leaves in Japan

Genus	Endophyte	Epiphyte	Genus	Endophyte	Epiphyte
Acremonium	–	+	*Periconiella*	+	–
*Alternaria**	+	+	*Pestalotiopsis**	+	+
*Arthrinium (Apiospora)**	+	+	*Phialocephala*	+	–
*Ascochyta**	+	+	*Phoma*	–	+
*Aureobasidium**	+	+	*Phomopsis**	+	+
Cenangium	+	–	*Phyllosticta* (*Guignardia*)*	+	+
Cladosporium	–	+	*Pseudocercospora* (*Mycosphaerella*)	+	–
Clonostachys	–	+	*Rhamichloridium*	–	+
Coccomyces	+	–	*Septonema*	–	+
*Colletotrichum**	+	+	*Sporobolomyces*	–	+
Coniothyrium	–	+	*Stachybotrys*	–	+
Discostroma	+	–	*Stenella*	–	+
Discula	+	–	*Surculiseries*	+	–
Epicoccum	–	+	*Trichoderma*	–	+
Geniculosporium (Nemania)	+	–	*Tritirachium**	+	+
Leptostroma (Lophodermium)	+	–	*Tubakia**	+	+
Nigrospora	+	–	*Xylocoremium* (*Xylaria*)	+	–
Nodulisporium (Biscogniauxia)	+	–			

*Indicates that the genus is known as both an endophyte and an epiphyte. Teleomorphs in parentheses.
+ present, – absent

at species level. However, molecular phylogenetic analyses demonstrated the identity as single species and the low host specificity of *Phyllosticta* and *Xylaria* isolated from multiple hosts (Okane et al. 2001a; Okane 2003; Wei et al. 2007; Osono et al. 2013). The second group includes fungal genera that show some degree of host specificity. Examples are *Lophodermium*, *Phialocephala*, and *Cenangium* in *Pinus* spp. (Kowalski 1982; Legault et al. 1989a; Hata and Futai 1996; Sieber et al. 1999), *Ascochyta* in *Fagus crenata* (Wei and Harada 1998; Sahashi et al. 1999, 2000; Kaneko and Kaneko 2004), and *Discula* and *Tubakia* in fagaceous trees (Sieber and Hugentobler 1987; Halmschlager et al. 1993; Wilson and Carroll 1994; Gennaro et al. 2003; Cohen 2004; Kaneko and Kaneko 2004; Shirouzu et al. 2008).

Twenty-two genera are reported as epiphytes of tree leaves in Japan (Table 1.2). These genera are generally known to have low host specificity and to be reported from a wide variety of tree species. Major epiphytes include *Alternaria alternata*, *Apiospora montagnei*, *Aureobasidium pullulans*, *Cladosporium cladosporioides*, *Clono-stachys rosea*, *Colletotrichum gloeosporioides*, *Pestalotiopsis* spp., *Phoma* spp., *Phomopsis* spp., *Trichoderma* spp., and *Tripospermum* spp. These fungi have been reported not only from Japan but also from distant continents, such as Europe and North and South America (e.g., Hudson 1968).

Ten genera are reported as both endophytes and epiphytes of tree leaves (asterisks in Table 1.2). *Ascochyta fagi* on leaves of *Fagus crenata* (Osono 2002), *A. montagnei* and *C. gloeosporioides* on leaves of *Swida controversa* (Osono and Mori 2004, 2005; Osono et al. 2004a), and *C. gloeosporioides* on leaves of *Camellia japonica* (Osono 2008) are examples of fungal species isolated frequently from both the interior and surface of leaves of single tree species. Previous studies showed a consistent trend that the number of species of endophytes is lower than that of epiphytes (Table 1.3). The similarity of species composition between endophytes and epiphytes varied among tree species,

Table 1.3 Comparison of endophytic and epiphytic mycobiota on leaves of tree species

		Number of species					
Tree species	Region	Total	Endophyte ($a+b$)	Epiphyte ($a+c$)	Common (a)	QS[a]	References
Nothofagus truncata	New Zealand	20	14	19	13	0.788	Ruscoe (1971)
Eucalyptus viminalis	Argentine	37	16	32	11	0.458	Cabral (1985)
Populus tremuloides	Canada	28	22	20	14	0.400	Wildman and Parkinson (1979)
Swida controversa	Kyoto, Japan	39	15	33	9	0.375	Osono and Mori (2005)
Camellia japonica	Kyoto, Japan	79	44	52	17	0.354	Osono (2008)
Castanopsis sieboldii	Okinawa, Japan	19	6	16	3	0.273	Osono et al. (2008)
Swida controversa	Kyoto, Japan	40	13	33	6	0.261	Osono et al. (2004a)
Fagus crenata	Kyoto, Japan	60	18	47	5	0.154	Osono (2002)
Pinus banksiana	Canada	31	8	25	2	0.121	Legault et al. (1989a, b)
Pinus resinosa	Canada	47	13	37	3	0.120	Legault et al. (1989a, b)

Modified from Osono and Mori (2004)
QS Sørensen's index of similarity
[a]$QS = 2a/(2a+b+c)$, where a is the number of common species and b and c are the number of species specifically isolated as endophytes or epiphytes, respectively

with Sørensen's index of similarity ranging from 0.12 to 0.79 (Table 1.3). The values for Japanese tree species are at the lower end to middle of this range (0.15–0.38).

2.3 Methods of Isolation and Incubation

The results of estimating fungal diversity using culture-dependent methods depend strongly on the methods of isolation because these are generally selective for some fungal species but not for others. Major factors affecting the selectivity are disinfectant (especially in the case of isolating endophytic fungi), type and composition of nutrient medium, incubation temperature, and incubation period. Methodological considerations are thus necessary to optimize the method of isolation and incubation for particular research purposes. Here, I summarize the methods of isolation and incubation adopted in the 45 papers.

Some authors isolated and incubated endophytic fungi using the method of Hata (1997), which basically follows the surface disinfection procedure of Kinkel and Andrews (1988) and utilizes 15 % hydrogen peroxide (v/v) as a disinfectant. Sodium hypochlorite solutions at various concentrations, and less frequently mercury (II) chloride ($HgCl_2$), are also used as disinfectants. Nutrient media used included potato dextrose agar (PDA), 1 or 2 % malt extract agar (MA) (w/v), and modified lignocellulose agar (LCA, Miura and Kudo 1970). The three media differ markedly in sugar content, resulting in differences in the isolation of endophytic (Hata 1997) and epiphytic (Osono and Takeda 1999) fungi. Incubation of leaf materials was done at constant temperature of 15, 17, 20, or 25 °C or variable room temperatures (18–25 °C), the commonest being constant temperature of 15, 17, or 20 °C. The duration of incubation varied from 0.5 to 3 months depending on the publication, the commonest being 1, 2, or 3 months.

The modified washing method of Tokumasu (1980) was consistently used to isolate epiphytic fungi from live leaves. Tokumasu (1980) described the procedure for washing leaf materials with 0.005 % Aerosol-OT (di-2-ethylhexyl sodium sulfosuccinate) solution (w/v) to isolate microfungi from dead pine needles and tested the effect of number of washes and incubation period. Osono and Takeda (1999) confirmed the applicability of the washing procedures of Tokumasu (1980) to dead leaves of *Fagus crenata*. The washing method developed for dead leaves is undoubtedly applicable to live leaves, which generally bear less microbial contaminants attached to the surface to be washed than dead leaves.

Micromorphological observations were commonly used to identify fungal isolates of endophytes and epiphytes. Several endophytes and epiphytes, however, cannot be thus identified due to the lack of sporulation despite efforts to promote sporulation. The proportions of non-sporulating endophytes can be as high as 54 % of total isolates (summarized in Promputtha et al. 2005). Recently, molecular methods are used to confirm the identity of not only non-sporulating "morphotaxa" but also sporulating fungal isolates (e.g., Okane et al. 2001a; Osono and Masuya 2012; Osono et al. 2013). Fungal isolates are then clustered into operational taxonomic units (OTUs) at given similarity thresholds of base sequences that are different among studies (Osono 2013).

3 Infection and Colonization of Leaves

Infection of leaves by endophytic and epiphytic phyllosphere fungi is generally thought to occur by (1) progressive growth of hyphae from buds and twigs, (2) attachment of airborne spores, and (3) transmission of propagules by insects (Petrini 1991). It is possible that phyllosphere fungi infect leaves by a combination of these routes. Successful colonization of leaves by phyllosphere fungi can then depend on microenvironmental conditions of the phyllosphere, such as light intensity and moisture, and structural and chemical properties of leaves, such as leaf thickness and nutrient contents. The infection and colonization of leaves by fungi were examined in Japan using the phyllosphere of pine, beech, and dogwood as model systems.

3.1 Infection from Twigs and Buds

Fungi frequently infect twigs and buds (e.g., Wildman and Parkinson 1979; Johnson and Whitney 1992; Toti et al. 1993). It is hence postulated that endophytic fungi are able to extend their mycelia from twigs and buds to infect leaves. This is especially true when endophytes occupy petioles of leaves and the basal part of needles proximate to twigs. For example, *Phialocephala* sp., an endophyte of pine needles, was detected more frequently on the basal than the apical portions and was also isolated frequently from current-year twigs, suggestive of infection via mycelia from current-year twigs to needles (Hata et al. 1998). Similarly, *Phomopsis* sp., an endophyte of beech (*Fagus crenata*), was isolated frequently from not only petioles but also current- and 1st- to 5th-year twigs, but not from lamina, suggesting the infection of petioles from twigs (Sahashi et al. 1999). Kaneko and Kaneko (2004) also demonstrated the mycelial colonization of beech leaves by two endophytic fungi in the Kanto district: *Periconiella* sp. from immature twigs within the winter buds and *Tritirachium* sp. from current- and 1st-year twigs. Osono and Mori (2003) in Kyoto reported results consistent with those of Sahashi et al. (1999) and Kaneko and Kaneko (2004), namely, that *Phomopsis* sp. was isolated from both lamina and current-year twigs of beech, and demonstrated a possibility that *Tritirachium oryzae* on leaves originated from twigs.

3.2 Airborne Spores and "Bagged" Leaves

No spore dispersal of *Discula* sp. was detected in the presence of snow cover in a beech forest in the Tohoku district, but the density of airborne spores of this fungus increased rapidly in late

May when the snow cover disappeared from the forest floor (Sahashi et al. 2000). The frequency of occurrence (FO) of *Discula* sp. on leaves just after budbreak on the canopy started to increase at the same time, suggesting that *Discula* sp. overwinters on fallen leaves under the snow and sporulates just after snow melt to infect expanded leaves via dispersed spores (Sahashi et al. 1999). Sahashi et al. (1999) confirmed that this fungus was not isolated from rolled-up leaves enclosed in winter buds. Another observation supports the idea that *Discula* sp. infects leaves via airborne spores. That is, Sahashi et al. (2000) found a delay of infection of live leaves by this fungus, probably because of the inhibition of spore discharge from the forest floor, in one study site (Asahimura) where the snowmelt was delayed compared to that at other sites. This type of lifecycle (sporulation on dead leaves and infection of live leaves via airborne spora) is also reported in some endophytic fungi, for example, *Mycosphaerella buna* on beech (Kaneko and Kakishima 2001; Kaneko et al. 2003) and *Lophodermium piceae* on *Picea abies* (Osorio and Stephan 1991a). Colonization by airborne spora was considered to be the major route of infection for epiphytes (e.g., Kinkel 1991; Levetin and Dorsey 2006).

"Branch-bagging" experiments have been performed in the field to manipulate the level of horizontal transfer of phyllosphere fungi via airborne spores. In manipulation experiments, winter buds were covered with well-ventilated vinyl bags before budbreak to exclude the infection of airborne spores onto expanded leaves and to examine the invasion of fungi from buds and twigs (Wilson 1996). In an experiment carried out in the Kanto district, Kaneko and Kaneko (2004) reported that *M. buna* and *Ascochyta fagi*, major endophytes of beech, were not isolated at all from bagged leaves, whereas these fungi were isolated frequently from leaves that were not bagged or leaves on saplings in pots experimentally placed in the study forest, suggesting that these endophytes infect leaves via airborne spores. This result is consistent with the report of Osono and Mori (2003) that in Kyoto the FOs of *A. fagi*, *Xylaria* sp., and *Geniculosporium* sp. on bagged beech leaves were lower than those on leaves that were not bagged. Infection of leaves by airborne spores was also probable in *Xylaria* endophytes of tropical trees (Bayman et al. 1998). Branch-bagging treatments resulted in a reduction in the FO of endophytic *Phoma* sp. in leaves of *Quercus serrata* (Ito et al. 2007).

Recognition and attachment of spores of phyllosphere fungi to the host leaf surface was examined for *Aureobasidium pullulans* and *Discula umbrinella* (Toti et al. 1992; Viret et al. 1993, 1994; Andrews et al. 1994; Viret and Petrini 1994). The mode of latent infection of endophytes within leaf tissues was investigated with micromorphological techniques (Suske and Acker 1987, 1989; Stone 1988; Johnson and Whitney 1989b; Viret et al. 1993; Deckert et al. 2001). No such studies have been performed for endophytic or epiphytic phyllosphere fungi of Japanese trees.

3.3 Effects of Microenvironments and Leaf Traits on Fungal Colonization

The tree canopy is heterogeneous in physical microenvironments, such as sunlight intensity and moisture, which can lead to phenotypic changes in physical and chemical traits of leaves. The heterogeneity of micro-microenvironments and properties of leaves is expected to influence the colonization of endophytic and epiphytic phyllosphere fungi and result in patterns of within-canopy distribution. Factors affecting the within-canopy distribution of phyllosphere fungi include height (Bernstein and Carroll 1977; Carroll 1979; Wildman and Parkinson 1979; Andrews et al. 1980; Johnson and Whitney 1989a), distance from the trunk (Andrews et al. 1980; Petrini and Carroll 1981), sun leaves versus shade leaves from different parts of the canopy (Wilson et al. 1997; Wilson and Faeth 2001), and compass direction (Johnson and Whitney 1989a; Petrini and Fisher 1990).

Osono and Mori (2003) compared the phyllosphere mycobiota of beech between sun leaves at the periphery of the canopy and shade leaves near the main trunk and suppressed by leaves in the upper canopy. The sun and shade leaves were collected at about 2 m height. Frequencies of

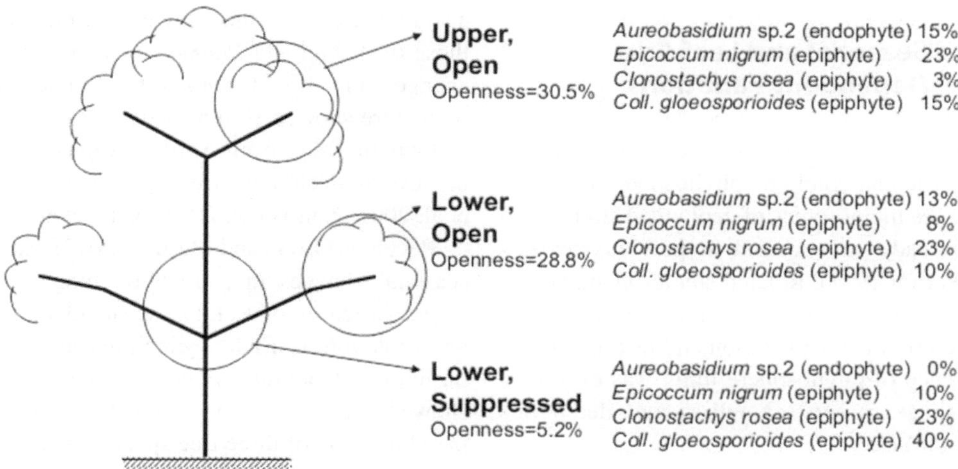

Fig. 1.2 Schematic diagram of canopy of giant dogwood (*Swida controversa*, Cornaceae) and frequency of occurrence of major phyllosphere fungi on leaves collected at different canopy positions (Data after Osono and Mori 2004)

occurrence of *Ascochyta fagi* and epiphytic *Apiospora montagnei* (anamorph: *Arthrinium*) were higher in shade than in sun leaves, whereas those of three epiphytic species, *Pestalotiopsis* sp., *Cladosporium cladosporioides*, and *Alternaria alternata*, were higher in sun leaves. Because sun and shade leaves of beech differed in the intensity of sunlight that leaf surfaces received, it is postulated that hyaline hyphae of *A. fagi* were sensitive to the higher sunlight intensity and possibly to severer desiccation on the surface of sun leaves. Conversely, the dematiaceous species *C. cladosporioides* and *A. alternata* were probably resistant to severe sunlight and desiccation and survived on the surface of sun leaves due to the dark pigmentation (melanization) of the hyphal wall (Butler and Day 1998). Hyphal tips of *C. cladosporioides* and *A. alternata* also showed high tolerance to desiccation (Park 1982).

Osono and Mori (2004) collected leaves from different positions of the canopy of giant dogwood (*Swida controversa*) to examine phyllosphere fungi and relate these to the height and openness (as an index of light environment) of the leaf positions and to the leaf properties, such as leaf mass per area (an index of leaf thickness) and contents of nitrogen and total polyphenols (Fig. 1.2). The canopy of giant dogwood in open sites was characterized by a multilayered distribution of leaves; that is, the distribution of leaves was discrete in relation to height, making the tree species suitable for the examination of the pattern of occurrence of fungi within the canopy. With respect to the openness, the FO of *Aureobasidium* sp.2 was higher on leaves collected at open positions at the periphery of canopy than on suppressed leaves near the trunk. In contrast, the FO of *Colletotrichum gloeosporioides* was higher on suppressed leaves. Alternatively, FOs of *Epicoccum nigrum*, *Clonostachys rosea*, and *C. gloeosporioides* were negatively correlated with leaf mass per area and/or contents of nitrogen and total polyphenol, suggesting possible effects of leaf properties.

The studies cited above indicated that the infection and colonization of leaves by endophytic phyllosphere fungi are affected by various factors, including physical microenvironments of the phyllosphere, physical and chemical properties of leaves, and possibly interactions with other microbes. More studies are needed to evaluate the ecological importance of individual factors and the relative importance of multiple factors. Useful in this regard is the sampling of leaves using "natural experiments," like that within the canopy of giant dogwood, as in the study of Osono and Mori (2004) (Fig. 1.2), and manipulation experiments that control individual factors of interest, such as bagging and shading of leaves.

4 Seasonal and Leaf Age-Dependent Variations

Seasonal and leaf age-dependent changes in endophytic and epiphytic phyllosphere fungi are among the major topics of ecological studies and were already summarized in previous reviews (e.g., Petrini 1991). Related studies using beech, birch, and pine leaves are available in Japan, and relative importance of season and leaf age on the occurrence of phyllosphere fungi was examined on both evergreen (camellia) and deciduous (dogwood) trees.

4.1 Seasonal and Leaf Age-Dependent Changes

Two patterns of temporal changes were recognized over the growing season for endophytic fungi associated with deciduous leaves of beech in the Tohoku district (Sahashi et al. 1999, 2000). Frequency of occurrence of *Discula* sp. was highest at leaf emergence in May to June, decreased temporarily in August and September, and then increased slightly at leaf senescence in October. In contrast, *Ascochyta fagi* was first isolated in late June and remained at a high FO (over 80 %) until the end of the growing season. Sahashi et al. (1999) postulated that the temporary decrease of *Discula* sp. during the summer might be caused by the hot, dry conditions, by defense responses of the host plant, and/or by interaction with another endophyte (*A. fagi*) that increased rapidly during that period. Differences in the phenology of spore discharge and/or of hyphal ingrowth from buds, as discussed in the previous section, may also account for the difference in temporal patterns between the endophytic fungal species. In another study conducted in the Kanto district (Kaneko et al. 2003; Kaneko and Kaneko 2004), the FOs of *Mycosphaerella buna*, *A. fagi*, *Periconiella* sp., and *Tritirachium* sp. in beech leaves increased from May and June to October, consistent with the results of Sahashi et al. (1999, 2000). In Kyoto, the FOs of *Geniculosporium* sp.1 and *Cladosporium cladosporioides* on beech leaves also increased over the growing season, whereas those of *A. fagi* and *Phoma* sp. showed no such changes (Fig. 1.3a; Osono 2002). Studies in Mie prefecture showed similar patterns of increasing endophytic fungi from April to August in leaves of several deciduous tree species (Yoshihashi et al. 2001; Naito et al. 2002; Ikebe et al. 2004).

Recently, Osono and Masuya (2012) examined seasonal changes in endophytic fungal assemblages in leaves of Betulaceae in subalpine and temperate forests. Endophytic fungal assemblages on leaves of *Betula ermanii* in subalpine forest showed relatively minor seasonal changes, compared to those of three tree species (*Alnus firma*, *Betula grossa*, and *Carpinus laxiflora*) in a cool temperate forest. The lower mean temperature and the lower variation in air temperature during the growing season in subalpine forest may partly account for the less marked changes in fungal composition than those in cool temperate forest (Osono and Masuya 2012).

Hata and Futai (1993) and Hata et al. (1998) studied seasonal and leaf age-dependent changes of endophytic mycobiota on middle and basal segments of pine needles in Kyoto. The longevity of needles used in these studies was 2 years for *Pinus densiflora* and 3 years for *P. thunbergii*. Virtually no endophytes were detected in needles just after emergence. The FO of *Leptostroma* sp. (an anamorph of *Lophodermium pinastri*) continuously increased with needle aging, with a rapid increase in current-year needles from September to December. Conversely, the FO of *Phialocephala* sp. on basal parts of needles slowly decreased with needle aging after a massive emergence in current-year needles. Another two major endophytes, *Cenangium ferruginosum* and unidentified sterile mycelium named BrS in Hata et al. (1998), were frequent on 1- and 2-year-old needles with negligible seasonal or leaf age-dependent variations. The continuous increase in FO of *Leptostroma* sp. with needle aging may be due to (1) cumulative infection with the time after needle flush, (2) improved habitat condition for the endophyte with the change in the physiology of needles with needle aging, and (3) increase in microscopic wounds or changes in the physical conditions of needles,

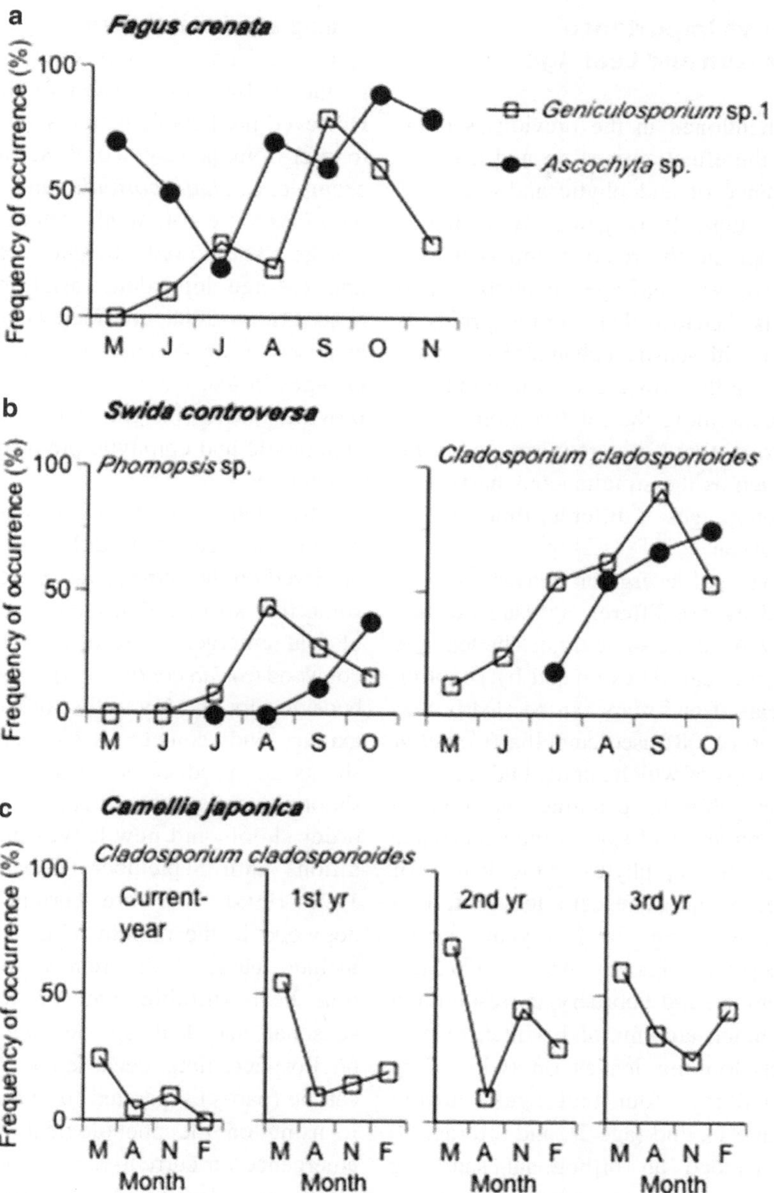

Fig. 1.3 Examples of seasonal and leaf age-dependent changes in frequency of occurrence of endophytic and epiphytic phyllosphere fungi. (**a**) Beech (*Fagus crenata*, Fagaceae; data after Osono 2002), (**b**) dogwood (*Swida controversa*, Cornaceae; modified from Osono and Mori 2005), and (**c**) camellia (*Camellia japonica*, Theaceae; modified from Osono 2008). *Open squares* and *filled circles* of (**b**) indicate the data of dogwood leaves on the first- and second-order shoots that produced leaves in May and June, respectively. Longevity of camellia leaves (**c**) was 4 years

which may facilitate fungal infection, according to Hata et al. (1998). Hata et al. (1998) suggested possible factors contributing to the decrease in the FO of *Phialocephala* sp. with needle aging to include (1) earlier fall of needles colonized by this endophyte, (2) aggravation of habitat conditions for the endophyte with the changes in the physiology of needles with needle aging, and (3) competition with other fungi, such as *Leptostroma* sp., whose FO increased with needle aging.

4.2 Relative Importance of Season and Leaf Age

The studies mentioned in the previous section demonstrated the effects of season and leaf age on the occurrence of endophytic and epiphytic phyllosphere fungi. It is generally difficult, however, to separate the relative importance of effects of season and leaf age on phyllosphere fungi. This is because leaf aging proceeds simultaneously with seasonal changes in climatic conditions during the growing season in temperate regions. Examining the relative importance of these factors, therefore, requires sampling procedures, such as the simultaneous harvest of leaves of different ages at different times during the growing season.

The longevity of evergreen leaves exceeds 1 year, so that leaves of different ages are attached on the same twigs at the same time. The leaf age of evergreen trees can be estimated by counting annual bud scars if bud scars can be clearly recognized. Osono (2008) used camellia (*Camellia japonica*) on twigs of which annual bud scars are readily recognizable to examine the relative effects of season and leaf age on the occurrence of endophytes and epiphytes. New leaves of camellia emerge and senescent leaves fall in May, and the leaf longevity is 4 years, at the study site in Kyoto. Twigs were harvested in May, August, November, and February, corresponding to spring, summer, autumn, and winter, respectively. Healthy-looking leaves on twigs were then classified into four leaf ages: current (age 0), 1st (age 1), 2nd (age 2), and 3rd (age 3) years. This yielded an orthogonal sampling design of four seasons and four leaf ages. Of the resulting eight major phyllosphere fungal species, six showed significant seasonal variations, one (*Geniculo-sporium* sp.) showed a leaf age-dependent variation, and one (*Cladosporium cladosporioides*) showed both seasonal and leaf age-dependent variations (Fig. 1.3c).

Studies were also conducted on evergreen trees *Nothofagus truncata* (Ruscoe 1971) and *Eucalyptus viminalis* (Cabral 1985) using the same sampling design as Osono (2008), demonstrating consistent patterns that endophytic and epiphytic fungi generally were influenced more frequently by seasonal variation than by leaf age. However, predictable patterns were still difficult to infer from the results of these case studies. For example, *C. cladosporioides* and *Aureobasidium pullulans* were commonly detected on the three host trees but showed different patterns of seasonal and leaf age-dependent variations on different hosts. Osono (2008) considered that more detailed analyses of the seasonal and leaf age-dependent changes in leaf environmental conditions might provide further insights into the dynamics of endophytic and epiphytic phyllosphere fungi on forest trees.

The longevity of deciduous leaves is less than 1 year, but leaves of different ages are often observed on the same twigs at the same time for some tree species that have a succeeding type of leaf emergence (Kikuzawa 1983). In giant dogwood (*Swida controversa*), for example, winter buds develop in May to first-order shoots, whose axillary buds sometimes elongate second-order shoots and produce new leaves in June. Such a shoot elongation pattern can give rise to higher-order shoots and new leaves under open conditions until September (Kodani and Togashi 1992, 1995). Therefore, current-year shoots of dogwood in the middle of the growing season include leaves of different ages. Dogwood can thus be a suitable material to examine the seasonal and leaf age-dependent changes of phyllosphere fungi because the effect of leaf age can be (partly) separated from that of season by focusing on the phenological patterns of leaf emergence on current-year shoots. Accordingly, Osono and Mori (2005) examined phyllosphere fungal assemblages on dogwood leaves of different age monthly over the growing season. The resulting seven major phyllosphere fungal species showed increasing patterns of occurrence from May to October. Osono and Mori (2005) then compared the FOs of these fungal species between leaves on the first-order shoots that elongated in May and those on the higher-order shoots that elongated from June to September. The results showed that the FOs of *Phomopsis* sp., *Pestalotiopsis* sp., and

Trichoderma viride were higher in first-order than in higher-order shoots, suggesting that their colonization was proportional to leaf age and/or that their potential to colonize leaves was lower later in the growing season. In contrast, the FOs of *Colletotrichum gloeosporioides*, *Clonostachys rosea*, *Cladosporium cladosporioides*, and *Phoma* sp. were not different between first- and higher-order shoots, suggesting the negligible effect of leaf age and the high potential to colonize leaves over the growing season. These patterns represented by *Phomopsis* sp. and *C. cladosporioides* are shown in Fig. 1.3b. Further studies are needed that compare the temporal patterns of phyllosphere fungi on deciduous trees with different types of leaf emergence to test the general validity of the findings of Osono and Mori (2005) on giant dogwood.

5 Spatial Variations

Assemblages of endophytic and epiphytic phyllosphere fungi show spatial patterns at various scales (Petrini 1991). Within-leaf, within-canopy, altitudinal, and geographic distributions have been demonstrated for phyllosphere fungi of tree species in Japan.

5.1 Within-Leaf Distribution

Frequencies of occurrence of three major endophytes, *Discula* sp., *Ascochyta fagi*, and *Phomopsis* sp. on beech leaves, were not significantly different between the edge, midrib, and lamina within single leaves (Sahashi et al. 1999). Similarly, the FO of *Mycosphaerella buna* on beech leaves was not significantly different between the top, middle, and base parts of midrib and lamina (Kaneko et al. 2003). It should be noted, however, that these results are not always the case, because a few studies performed outside Japan reported within-leaf differences in FOs of endophytes associated with broad-leaved tree species (e.g., Sieber and Hugentobler 1987; Wilson and Carroll 1994). The FO of major endophytes differed between the apical and basal portions of pine needles (Hata and Futai 1996; Hata et al. 1998), probably due to the difference in the route of infection (see Sect. 3.1). This is consistent with the finding of Dobranic et al. (1995) about the difference in FO of endophytes within positions of *Larix laricina* needles.

5.2 Within-Canopy Distribution

Trees of giant dogwood in open sites have canopies characterized by a multilayered distribution of leaves and are suitable for the examination of the within-canopy distribution of endophytic and epiphytic phyllosphere fungi (see Sect. 3.3). Major phyllosphere fungi of giant dogwood occurred either evenly or unevenly according to the positions within the canopy and the twigs examined. The latter included *Aureobasidium* sp. and *Epicoccum nigrum*, whose FOs on leaves were higher on upper than on lower canopy, and *Phoma* sp. and *Clonostachys rosea*, whose FOs were higher on lower canopy. The higher FOs in the lower canopy can be partly ascribed to the high density of airborne inoculum near the ground, because litter is another major habitat of *Phoma* sp. and *C. rosea* (Osono et al. 2004a). Moreover, *Aureobasidium* sp. had higher FO in leaves of the canopy periphery than near the main trunk, whereas *Colletotrichum gloeosporioides* had higher FO in leaves near the main trunk. Possible explanations for the observed differences in within-canopy distribution could include the different sensitivity of fungal species to microenvironments of the phyllosphere, as discussed in Sect. 3.3.

5.3 Altitudinal and Geographic Distributions

Hashizume et al. (2008) studied altitudinal distributions of endophytic fungi on leaves of an evergreen oak, *Quercus acuta*, on two mountains, Mt. Takao and Mt. Osuzu, located in eastern Honshu and Kyushu, respectively, 1,000 km apart. Fungi were isolated from healthy-looking leaves collected at 400 and 600 m on Mt. Takao and at 900, 1,100, and 1,300 m on Mt. Osuzu. The frequency of occurrence of *Tubakia* sp. was higher at lower altitude on both mountains. In contrast, FO of *Phomopsis* sp. on Mt. Takao and FO of *Discula* sp. on Mt. Osuzu were higher at higher altitude. Under pure culture conditions, the optimum temperature of hyphal extension was higher, and the lowest temperature at which hyphal growth ceased was higher for isolates of *Tubakia* sp. than for those of *Phomopsis* sp. and *Discula* sp. Based on the results of pure culture tests, Hashizume et al. (2008) attributed the higher FO of *Phomopsis* sp. and *Discula* sp. at higher altitudes to the adaptation to lower temperatures.

The species composition of major endophytic fungi was generally similar when a single host tree species from geographically separated forest stands was examined. *Tubakia* sp. from *Q. acuta* leaves is a typical example, as mentioned in the previous paragraph (Hashizume et al. 2008). Sahashi et al. (2000) examined endophytic mycobiota of beech leaves at five locations in the Tohoku district, approximately 250 km apart at the maximum, and found no significant difference in the composition of fungal species. Nevertheless, a delay of infection by *Discula* sp. was observed at one study site (Asahimura), probably due to the delayed snowmelt at that site compared to the other sites (see Sect. 3.2). On beech leaves, *Ascochyta fagi* and *Phomopsis* sp. were commonly isolated in Tohoku (Sahashi et al. 1999, 2000), Kanto (Kaneko and Kaneko 2004), Kyoto (Osono 2002; Osono and Mori 2003), and Kyushu (Hashizume et al. 2010), despite differences in the method of isolation. Hashizume et al. (2010) investigated endophytic fungi in beech leaves collected from pure beech stands at four locations with different summer temperatures, covering the range of natural distribution of the tree species. *Ascochyta fagi* was dominant at every site, and FO of this fungus was higher at sites with lower maximum air temperature, suggesting that its occurrence was influenced primarily by summer temperature rather than geographic distance.

Similarly, Suzuki et al. (2003) found the dominant occurrence of a single unidentified morphotype (coded as Ds) from leaves of a deciduous oak *Quercus serrata* collected at four locations in the Chubu and Kinki districts, approximately 200 km apart at the maximum. Nomura et al. (2003) studied endophytic fungi of *Pinus thunbergii* at coastal and inland stands in Mie prefecture and found similar species composition but differences in FOs of major fungal species. In a survey of endophytic fungi of *Pasania edulis* (Fagaceae), Hata et al. (2002) isolated *Phyllosticta* sp. frequently from leaves collected at a nursery (3 m asl), whereas FO of this fungus was lower, and FO of *Phomopsis* sp. was higher, on Mt. Takakuma (550 m asl), 30 km away from the nursery. Hata and Sone (2008) isolated endophytes from leaves of *Neolitsea sericea* (Lauraceae) in broad-leaved and coniferous forest stands located 200 m apart and found generally minor effects of canopy vegetation on the endophytic fungal assemblages. Exceptionally, *Cytosphaera* sp. in petiole segments was more frequently isolated from broad-leaved than coniferous stands, the reason for the difference being unclear.

A few studies have compared endophytic fungal assemblages across different climatic regions. Osono and Masuya (2012) found that the similarities in composition of endophytic fungal assemblages were generally low in 11 tree species in Betulaceae collected from subalpine, versus cool temperate, versus subtropical forests. The low similarity among the different climatic regions can be attributed to the difference in climatic conditions as well as difference in tree species. *Nemania diffusa* and *Xylaria cubensis* in the Xylariaceae, major components of endophytic fungal assemblages in tropical forests in Asia (Okane et al. 2008, 2012), were also common to subtropical and temperate forests in Japan (Osono et al. 2013; Ikeda et al. 2013).

6 Direct and Indirect Roles in Decomposition

As noted in the Sect. 1, the roles of endophytic and epiphytic phyllosphere fungi have been reviewed thoroughly in Osono (2006b, 2007) and Osono and Hirose (2009b). Here, recent progress in our understanding of the roles of endophytic ascomycetes in lignin decomposition is briefly summarized.

Some endophytic fungi on live leaves persist in dead leaves and grow saprobically. A few of these endophytes have ligninolytic and/or cellulolytic activity and participate in the decomposition of structural components, such as lignin and holocellulose (Osono and Hirose 2009b). Ligninolytic activity of endophytic fungi was demonstrated for species in the Xylariaceae associated with leaves of beech (Osono 2002) and dogwood (Osono et al. 2004a; Osono 2005) and in the Rhytismataceae associated with camellia leaves (Koide et al. 2005a, b). Osono et al. (2013) showed the ubiquity of xylariaceous endophytes on live leaves for more than 80 tree species in a cool temperate forest in Kyoto. In that study, *Xylaria* sp., detected as an endophyte on live leaves of 68 tree species, was shown to be associated with the bleached portions of dead leaves for 12 tree species, which were produced due to the selective decomposition of lignin by the fungal colonizer. Pure culture decomposition tests indicated that ligninolytic activity of *Xylaria* sp. was enhanced above 25 °C at the expense of cellulolytic activity (Osono et al. 2011). *Lophodermium pinastri* (Rhytismataceae), a dominant endophyte of pine needles, was shown to have ligninolytic activity in the pure culture and to reduce lignin content in needle portions colonized in the field (Osono and Hirose 2011; Hirose and Osono 2006). *Coccomyces sinensis* (Rhytismataceae) is the first to colonize recently dead leaves of camellia and takes part in selective delignification in leaf portions (Koide et al. 2005a). Osono and Hirose (2009a) recently demonstrated in pure culture tests that the prior decomposition of leaves by *C. sinensis* stimulated the subsequent decomposition by other fungi. This was partly owing to the selective delignification by *C. sinensis* and the concomitant increase in the availability of delignified holocellulose. The ligninolytic activity of *C. sinensis* appears to be sensitive to environmental changes: clear-cutting of a temperate secondary forest resulted in a decrease of the leaf area colonized and bleached by this fungus (Hagiwara et al. 2012). These recent findings suggest that ligninolytic endophytes are major components of decomposer fungal assemblages and play definite and unique roles in fungal succession and decomposition of dead leaves. The sensitivity of ligninolytic activity to temperature and environmental changes is of particular interest and deserves further analyses.

7 Interaction with Pathogens and Herbivores and Effects of Environmental Stress

7.1 Effects of Pathogens

Zonate leaf blight is a foliar disease of various evergreen and deciduous woody plant species caused by *Haradamyces foliicola* (Masuya et al. 2009). It is characterized by zonate necrosis, leading to defoliation during the early growing season (Osono et al. 2004b) and sometimes the death of the tree (Fig. 1.4). In particular, tree species in the Cornaceae, including the giant dogwood *Swida controversa*, are highly susceptible to the disease. Osono (2006a) studied endophytic fungi associated with leaves infested with zonate leaf blight. *Haradamyces foliicola* was exclusively isolated from zonate parts, and both endophytic fungi that were frequent on uninfected parts and *H. foliicola* were isolated from rim parts of symptomatic regions (Fig. 1.4). The FO of endophytic fungi was lower in leaf disks from which *H. foliicola* was detected than in those from which it was not detected, suggesting that endophytic fungi were excluded from leaf tissues as symptoms developed. Osono (2006a) then examined mycelial interactions between *H. foliicola* and nine species of endophytic and epiphytic phyllosphere fungi in dual culture tests on a nutrient medium. All phyllosphere fungi showed overgrowth, contact inhibition, or inhibition at a distance against isolates of *H. foliicola*. It is not likely, however, that the phyllosphere fungi have such antagonistic effects in vivo as observed in vitro, because the isolation experiment suggested the exclusion of phyllosphere fungi as symptom development proceeded.

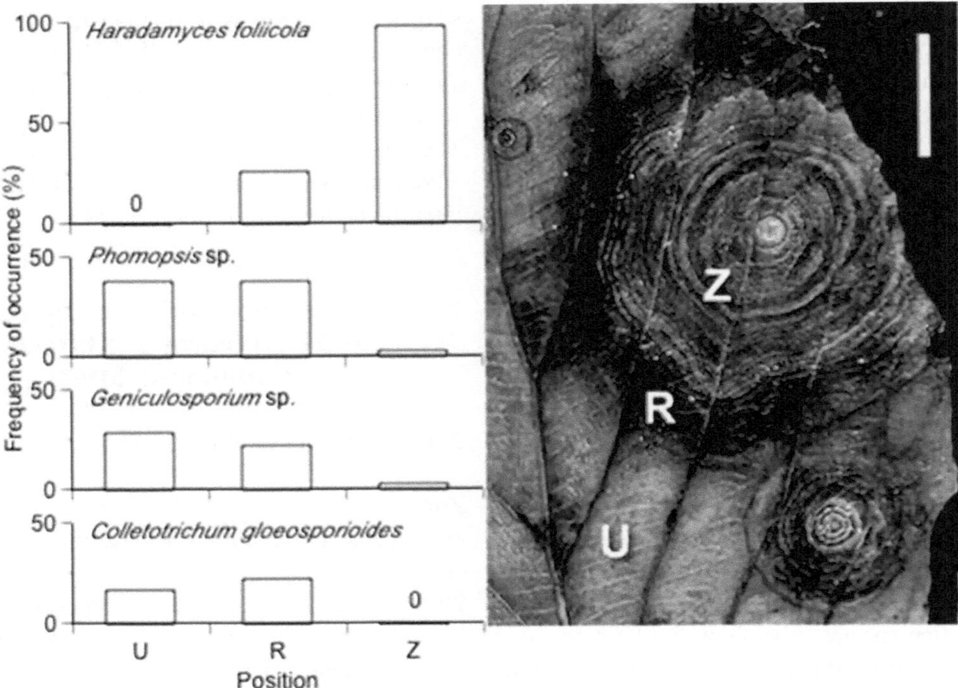

Fig. 1.4 Frequency of occurrence of fungi isolated from leaves infested by zonate leaf blight of giant dogwood (*Swida controversa*, Cornaceae). Fungi were isolated from zonate (Z), rim (R), and uninfected (U) positions of infested live leaves. Bar = 1 cm. *Haradamyces foliicola* is the causal agent of the disease (Data after Osono 2006a, b)

Induced resistance against pathogens in response to the infection by endophytic fungi was demonstrated for grass host (e.g., Clarke et al. 2006), but there are few known examples for tree hosts. Results of laboratory tests have shown inhibitory effects of epiphytic fungi on spore germination, germ tube elongation, and mycelial growth of pathogenic fungi (Dickinson and Skidmore 1976; Omar and Heather 1979; Fiss et al. 2000). Arnold et al. (2003) reported that the infection of leaves of cacao (*Theobroma cacao*, Malvaceae) seedlings by endophytic fungi reduced leaf mortality by a pathogen, *Phytophthora* sp. Future studies will be needed to explore the interaction between pathogens and endophytic and epiphytic phyllosphere fungi in the field.

7.2 Effects of Herbivores

The pine needle gall midge, *Thecodiplosis japonensis*, is an important forest pest of many pine species. Gall midges normally infest basal regions of needle fascicles that fuse to form galls. Hata and Futai (1995) isolated *Phialocephala* sp., *Phomopsis* sp., and *Pestalotiopsis* sp. as endophytes of the galls formed on *Pinus densiflora* and the F2 hybrid pine (*P. thunbergii* × (*P. thunbergii* × *P. densiflora*)). *Phialocephala* sp. was also isolated frequently from basal parts of healthy needles (see Sect. 3.1) but not from middle parts of healthy needles or galls formed at middle parts of needles, indicating that this fungus was position specific rather than gall specific. In contrast, *Phomopsis* sp. and *Pestalotiopsis* sp. were found to be gall specific, opportunistic colonizers of galls. No endophytic or opportunistic fungi were isolated in the early stages of gall formation, suggesting that no fungi are carried into the galls by the midge. Moreover, midge larvae collected from galls were almost free from fungi, indicating that larvae in the galls had probably not been in contact with endophytic fungi. Taking the quite low larval mortality of gall midges at the study site into account, Hata and Futai (1995) considered that the gall midge does not seem to have mutualistic or antagonistic associations with endophytic fungi. Faeth (2002) in his review of the three-way symbiosis of plant, herbivores, and endophytes showed that few reports are available demonstrating that fungal endophytes act as mutualistic symbionts of trees by functioning as antagonists against herbivores.

7.3 Effects of Simulated Acid Rain

Asai et al. (1998) experimentally treated living needles of 10-year-old *Pinus thunbergii* with simulated acid rain (SAR, pH 3) or tap water (control, pH 6.3) to examine the occurrence of an endophyte, *Lophodermium pinastri*. The SAR treatment resulted in increased needle fall, a decrease of FO of *L. pinastri* in live needles, and a decrease of FO of fruiting bodies on dead needles. These results were consistent with those of previous studies of SAR application (Helander and Rantio-Lehtimäki 1990; Helander et al. 1993a, b, 1994; Magan et al. 1995).

8 Future Directions

Most of our current knowledge on the diversity and ecology of endophytic and epiphytic phyllosphere fungi on tree leaves comes from original studies carried out in temperate regions of Europe and North America. Consequently, it is generally the case that a limited number conifers and broad-leaved trees (e.g., species in *Quercus* (oak) and *Betula* (birch)) have been used for the study of phyllosphere fungi, because a relatively limited number of tree species are available in these regions. In contrast, a rich number of tree species is distributed in Japan, and as many as 250 tree species have been explored for the diversity of phyllosphere fungi (see Sect. 2.1). Still, however, many tree species are yet to be examined for the occurrence and diversity of phyllosphere fungi. Besides, most of the ecological studies of phyllosphere fungi in Japan used as materials a few tree species (e.g., pines, beech, and dogwood) in temperate forests. Further studies are needed in less-studied climatic regions, that is, in subtropical, subalpine, and subboreal forests. This especially holds true for subtropical and tropical forests in southern Japan and Southeast Asia that are considered to harbor magnificent richness of phyllosphere fungi and are regarded as one of the "hotspots" of fungal diversity (Arnold et al. 2000; Arnold and Lutzoni 2007).

Studies on the diversity and ecology of phyllosphere fungi in Japan cover a wide range of topics, as is summarized in this chapter. Fields of research yet to be fully examined for phyllosphere fungi in Japan include trophic interactions. The interaction between trees, phyllosphere fungi, and herbivores and pathogens has been actively investigated for the grass-endophyte symbiosis (Cheplick and Faeth 2009) and for phyllosphere fungi of trees, relating to the activity of toxic substances derived from endophytes against herbivores (Miller et al. 1985, 2002; Miller 1986; Clark et al. 1989; Calhoun et al. 1992; Johnson and Whitney 1994), the association of foliar endophytes of oaks with gall-forming insects and leaf miners (Weis 1982; Lasota et al. 1983; Petrini et al. 1989; Butin 1992; Wilson 1995; Faeth and Hammon 1996, 1997a, b; Gaylord et al. 1996; Wilson and Carroll 1997; Wilson and Faeth 2001), the interaction between foliar endophytes and herbivores of birch (Lappalainen and Helander 1997; Lappalainen et al. 1999; Ahlholm et al. 2002a; Valkama et al. 2005), and the relationship between aphids and endophytes and epiphytes (Gange 1996; Stadler and Müller 1996, 2000). Regarding the physiological response of plants to fungal infection, phyllosphere fungi can accelerate the senescence of leaves (Skidmore and Dickinson 1973, 1976; Jachmann and Fehrmann 1989). Considering the record of over 1,200 species of trees and over 30,000 species of insects in Japan (Biodiversity Center of Japan 2010), future research will possibly reveal new and interesting interactions among trees, phyllosphere fungi, and insects. Another research field that remains to be investigated is the histochemical observation of latent infection by endophytic fungi (Suske and Acker 1987, 1989; Johnson and Whitney 1989b; Osorio and Stephan 1991b; Deckert et al. 2001).

Molecular biological methods are promising to identify, detect, and study the diversity of endophytic and epiphytic phyllosphere fungi. A growing number of publications have become available worldwide in this regard since around 2000. Isolates of phyllosphere fungi, especially endophytic ones, are often sterile and difficult to identify morphologically, which makes the evaluation of species richness difficult. Recent progress in molecular bar coding of fungi indicates its remarkable potential as a tool to facilitate identifying fungal species (Osono 2013). For example, molecular biological methods were used to characterize the taxa of isolates of endophytes and "sterile mycelia" in Japan (Okane et al. 2001a, 2003; Hashizume et al. 2008; Osono et al. 2013) and in other countries (e.g., Promputtha et al. 2005; Santamaria and Bayman 2005; Wang et al. 2005). Moreover, molecular tools are used in the evaluation of genetic diversity of endophytes (Guo et al. 2004; Lu et al. 2004; Cohen 2006) and in population genetics studies of endophytes (Burgess et al. 2004). More recently, environmental DNA was extracted directly from leaves and analyzed as clone libraries (Duong et al. 2006; Arnold et al. 2007; Yan et al. 2008). In Japan, fungal diversity on *Camellia japonica* leaves and the endophytic lifestyle of *Coccomyces sinensis* was successfully examined using an environmental clone library (Hirose et al. 2013). Such culture-independent methods will become popular and standard as a result of the development of next-generation sequencing technologies that have revolutionized large-scale sequencing of environmental fungal DNA (Jumpponen and Jones 2009, 2010; Osono 2013). This metagenomic approach will be essential for future explorations of yet-to-be-discovered hyper-diversity of phyllosphere fungi in subtropical and tropical forests (Sakaguchi et al. unpublished).

Molecular biological methods also throw light upon the functional aspects of endophytic fungi of trees. For example, the majority of endophytes of *Pinus monticola* (90 % of 2019 isolates) belonged to the Rhytismataceae, but not a single rhytismataceous endophyte was found to be most closely related to known rhytismataceous parasites of the host tree (Ganley et al. 2004). This study demonstrated that endophytes of *P. monticola* are merely cryptic or in a latent state of known parasites of the tree, and that endophytes are a unique functional group distinct from parasites of the same tree species. In contrast, Baayen et al. (2002) used molecular biological analysis and showed the identity of a cosmopolitan endophyte, *Guignardia mangiferae*, with nonpathogenic isolates of the citrus black spot fungus *G. citricarpa*. Similarly, the identity of endophytes in live leaves and saprobes in dead leaves of single tree species was confirmed by DNA sequence analyses (Müller et al. 2001; Deckert et al. 2002; Promputtha et al. 2007; Osono et al. 2013). Host (birch) genotypes influenced the probability of infection by, and the genotypes and genetic diversity of, an endophyte, *Venturia ditricha* (Ahlholm et al. 2002b). Undoubtedly, molecular biological methods will be a powerful tool for future studies on the diversity and ecology of endophytic and epiphytic phyllosphere fungi of trees in Japan.

Acknowledgments I thank Dr. Elizabeth Nakajima for her critical reading of the manuscript. This work has received partial financial support from the Ministry of Education, Culture, Sports, Science, and Technology of Japan (No. 23770083), the global COE program A06 to Kyoto University, and Grants for Excellent Graduate Schools, MEXT, Japan.

References

Ahlholm J, Helander M, Elamo P, Saloniemi I, Neuvonen S, Hanhimäki S, Sakkonen K (2002a) Micro-fungi and invertebrate herbivores on birch trees: fungal mediated plant-herbivore interactions or responses to host quality? Ecol Lett 5:648–655

Ahlholm J, Helander M, Henriksson J, Metzler M, Sakkonen K (2002b) Environmental conditions and host genotype direct genetic diversity of *Venturia ditricha*, a fungal endophyte of birch trees. Evolution 56:1566–1573

Andrews JH, Kenerley CM, Nordheim EV (1980) Positional variation in phylloplane microbial populations within an apple tree canopy. Microb Ecol 6:71–84

Andrews JH, Harris RF, Spear RN, Lau GW, Nordheim EV (1994) Morphogenesis and adhesion of *Aureobasidium pullulans*. Can J Microbiol 40:6–17

Aoki T, Tokumasu S, Tubaki K (1990) Fungal succession on momi fir needles. Trans Mycol Soc Jpn 31:355–374

Arnold AE (2005) Diversity and ecology of fungal endophytes in tropical forests. In: Deshmukh SK, Rai MK (eds) Biodiversity of fungi, their role in human life. Science Publishers, Enfield, pp 49–68

Arnold AE (2007) Understanding the diversity of foliar endophytic fungi: progress, challenges, and frontiers. Fungal Biol Rev 21:51–66

Arnold AE, Lutzoni F (2007) Diversity and host range of foliar fungal endophytes: are tropical leaves biodiversity hotspots? Ecology 88:541–549

Arnold AE, Maynard Z, Gilbert GS, Coley PD, Kursar TA (2000) Are tropical fungal endophytes hyperdiverse? Ecol Lett 3:267–274

Arnold AE, Mejia LC, Kyllo D, Rojas EI, Maynard Z, Robbins N, Herre EA (2003) Fungal endophytes limit pathogen damage in a tropical tree. Proc Natl Acad Sci USA 100:15649–15654

Arnold AE, Henk DA, Eells RL, Lutzoni F, Vilgalys R (2007) Diversity and phylogenetic affinities of foliar fungal endophytes in loblolly pine inferred by culturing and environmental PCR. Mycologia 99:185–206

Asai E, Hata K, Futai K (1998) Effect of simulated acid rain on the occurrence of *Lophodermium* on Japanese black pine needles. Mycol Res 102:1316–1318

Baayen RP, Bonants PJM, Verkley G, Carroll GC, van der Aa HA, de Weerdt M, van Brouwershaven IR, Schutte GC, Maccheroni W Jr, Glienke de Blanco C, Azevedo JL (2002) Nonpathogenic isolates of the citrus black spot fungus, *Guignardia citricarpa*, identified as a cosmopolitan endophyte of woody plants, *G. mangiferae* (*Phyllosticta capitalensis*). Phytopathology 92:464–477

Bayman P, Angulo-Sasndoval P, Baez-Ortiz Z, Lodge DJ (1998) Distribution and dispersal of *Xylaria* endophytes in two tree species in Puerto Rico. Mycol Res 102:944–948

Bernstein ME, Carroll GC (1977) Internal fungi in old-growth Douglas fir foliage. Can J Bot 55:644–653

Biodiversity Center of Japan (2010) Biodiversity of Japan: a harmonious coexistence between nature and humankind. Heibonsha, Tokyo

Boddy L, Griffith GS (1989) Role of endophytes and latent invasion in the development of decay communities in sapwood of angiospermous trees. Sydowia 41:41–73

Burgess TI, Gordon TR, Wingfield MJ, Wingfield BD (2004) Geographical isolation of *Diplodia scrobiculata* and its association with native *Pinus radiata*. Mycol Res 108:1399–1406

Butin H (1992) Effect of endophytic fungi from oak (*Quercus robur* L.) on mortality of leaf inhabiting gall insects. Eur J For Pathol 22:237–246

Butler MJ, Day AW (1998) Fungal melanins: a review. Can J Microbiol 44:1115–1136

Cabral D (1985) Phyllosphere of *Eucalyptus viminalis*: dynamics of fungal populations. Trans Br Mycol Soc 85:501–511

Calhoun LA, Findlay JA, Miller JD, Whitney NJ (1992) Metabolites toxic to spruce budworm from balsam fir needle endophytes. Mycol Res 96:281–286

Carroll GC (1979) Needle microepiphytes in a Douglas fir canopy: biomass and distribution patterns. Can J Bot 57:1000–1007

Carroll GC (1988) Fungal endophytes in stems and leaves: from latent pathogen to mutualistic symbiont. Ecology 69:2–9

Carroll GC (1990) Fungal endophytes in vascular plants: mycological research opportunities in Japan. Trans Mycol Soc Jpn 31:103–116

Carroll GC (1991a) Fungal associates of woody plants as insect antagonists in leaves and stems. In: Krischik VA, Jones CG (eds) Microbial mediation of plant-herbivore interactions. Wiley, New York, pp 253–271

Carroll GC (1991b) Beyond pest deterrence – alternative strategies and hidden costs of endophytic mutualisms in vascular plants. In: Andrews JH, Hirano SS (eds) Microbial ecology of leaves. Springer, New York, pp 358–375

Carroll GC (1995) Forest endophytes: pattern and process. Can J Bot 73(Suppl 1):S1316–S1324

Cheplick GP, Faeth SH (2009) Ecology and evolution of the grass-endophyte symbiosis. Oxford University Press, Oxford

Clark CL, Miller JD, Whitney NJ (1989) Toxicity of conifer needle endophytes to spruce budworm. Mycol Res 93:508–512

Clarke BB, White JF, Hurley RH, Torres MS, Sun S, Huff DR (2006) Endophyte-mediated suppression of dollar spot disease in fine fescues. Plant Dis 90:994–998

Cohen SD (2004) Endophytic-host selectivity of *Discula umbrinella* on *Quercus alba* and *Quercus rubra* characterized by infection, pathogenicity and mycelial compatibility. Eur J Plant Pathol 110:713–721

Cohen SD (2006) Host selectivity and genetic variation of *Discula umbrinella* isolates from two oak species: analyses of intergenic spacer region sequences of ribosomal DNA. Microb Ecol 52:463–469

Deckert RJ, Melville LH, Peterson RL (2001) Structural features of a *Lophodermium* endophyte during the cryptic life-cycle phase in the foliage of *Pinus strobus*. Mycol Res 105:991–997

Deckert RJ, Hsiang T, Peterson RL (2002) Genetic relationships of endophytic *Lophodermium nitens* isolates from needles of *Pinus strobus*. Mycol Res 106:305–313

Dickinson CH, Skidmore AM (1976) Interactions between germinating spore of *Septoria nodorum* and phylloplane fungi. Trans Br Mycol Soc 66:45–56

Dobranic JK, Johnson JA, Alikhan QR (1995) Isolation of endophytic fungi from eastern larch (*Larix laricina*) leaves from New Brunswick, Canada. Can J Microbiol 41:194–198

Duong LM, Jeewon R, Lumyong S, Hyde KD (2006) DGGE coupled with ribosomal DNA gene phylogenies reveal uncharacterized fungal phylotypes. Fungal Divers 23:121–138

Faeth SH (2002) Are endophytic fungi defensive plant mutualists? Oikos 98:25–36

Faeth SH, Hammon KE (1996) Fungal endophytes and phytochemistry of oak foliage: determinants of oviposition preference of leafminers? Oecologia 108:728–736

Faeth SH, Hammon KE (1997a) Fungal endophytes in oak trees: long-term patterns of abundance and associations with leafminers. Ecology 78:810–819

Faeth SH, Hammon KE (1997b) Fungal endophytes in oak trees: experimental analyses of interactions with leafminers. Ecology 78:820–827

Fiss M, Kucheryava N, Schonherr J, Arnold KG, Auling G (2000) Isolation and characterization of epiphytic fungi from the phyllosphere of apple as potential biocontrol agents against apple scab (*Venturia inaequalis*). Z PflKrankh PflSchutz 107:1–11

Gange AC (1996) Positive effects of endophyte infection on sycamore aphids. Oikos 75:500–510

Ganley RJ, Brunsfeld SJ, Newcombe G (2004) A community of unknown, endophytic fungi in western white pine. Proc Natl Acad Sci USA 101:10107–10112

Gaylord ES, Preszler RW, Boecklen WJ (1996) Interactions between host plants, endophytic fungi, and a phytophagous insect in an oak (*Quercus grisea* x *Q. gambelii*) hybrid zone. Oecologia 105:336–342

Gennaro M, Gonthier P, Nicolotti G (2003) Fungal endophytic communities in healthy and declining *Quercus robur* L. and *Q. cerris* L. trees in northern Italy. J Phytopathol 151:529–534

Guo LD, Xu L, Zheng WH, Hyde KD (2004) Genetic variation of *Alternaria alternata*, an endophytic fungus isolated from *Pinus tabulaeformis* as determined by random amplified microsatellites (RAMS). Fungal Divers 16:53–65

Hagiwara Y, Osono T, Takeda H (2012) Effects of clear-cutting on decomposition and bleaching of *Camellia japonica* leaf litter in a temperate secondary forest. Ap For Sci 21:1–6

Halmschlager VE, Butin H, Donaubauer E (1993) Endophytische Pilze in Blättern und Zweigen von *Quercus petraea*. Eur J For Pathol 23:51–63 (in German with English abstract)

Hashizume Y, Sahashi Y, Fukuda K (2008) The influence of altitude on endophytic mycobiota in *Quercus acuta* leaves collected in two areas 1000 km apart. For Pathol 38:218–226

Hashizume Y, Fukuda K, Sahashi N (2010) Effects of summer temperature on fungal endophyte assemblages in Japanese beech (*Fagus crenata*) leaves in pure beech stands. Botany 88:266–274

Hata K (1997) Collection, detection, and isolation of fungi: plant associated fungi (pathogens, symbionts, and saprobes): endophytes. Trans Mycol Soc Jpn 38:110–114 (in Japanese)

Hata K, Futai K (1993) Effect of needle aging on the total colonization rates of endophytic fungi on *Pinus thunbergii* and *Pinus densiflora* needles. J Jpn For Soc 75:338–341

Hata K, Futai K (1995) Endophytic fungi associated with healthy pine needles and needles infested by the pine needle gall midge, *Thecodiplosis japonensis*. Can J Bot 73:384–390

Hata K, Futai K (1996) Variation in fungal endophyte populations in needles of the genus *Pinus*. Can J Bot 74:103–114

Hata K, Sone K (2008) Isolation of endophytes from leaves of *Neolitsea sericea* in broadleaf and conifer stands. Mycoscience 49:229–232

Hata K, Futai K, Tsuda M (1998) Seasonal and needle age-dependent changes of the endophytic mycobiota in *Pinus thunbergii* and *Pinus densiflora* needles. Can J Bot 76:245–250

Hata K, Atari R, Sone K (2002) Isolation of endophytic fungi from leaves of *Pasania edulis* and their within-leaf distributions. Mycoscience 43:369–373

Helander ML, Rantio-Lehtimäki A (1990) Effects of watering and simulated acid rain on quantity of phyllosphere fungi of birch leaves. Microb Ecol 19:119–125

Helander ML, Neuvonen S, Sieber T, Petrini O (1993a) Simulated acid rain affects birch leaf endophyte populations. Microb Ecol 26:227–234

Helander ML, Ranta H, Neuvonen S (1993b) Responses of phyllosphere microfungi to simulated sulphuric and nitric acid deposition. Mycol Res 97:533–537

Helander ML, Sieber TN, Petrini O, Neuvonen S (1994) Endophytic fungi in Scots pine needles: spatial variation and consequences of simulated acid rain. Can J Bot 72:1108–1113

Hirose D, Osono T (2006) Development and seasonal variations of *Lophodermium* populations on *Pinus thunbergii* needle litter. Mycoscience 47:242–247

Hirose D, Matsuoka S, Osono T (2013) Assessment of the fungal diversity and succession of ligninolytic endophytes in *Camellia japonica* leaves using clone library analysis. Mycologia 105:837–843

Hudson HJ (1968) The ecology of fungi on plant remains above the soil. New Phytol 67:837–874

Ikebe M, Matsuda Y, Nakanishi K, Ito SI (2004) Temporal variations in isolation frequency of endophytic fungi in leaves of *Ginkgo biloba*. Chibu Shinrin Kenkyu 52:115–116 (in Japanese)

Ikeda A, Matsuoka S Masuya H, Mori AS, Hirose D, Osono T (2013) Comparison of the diversity and assemblages of xylariaceous endophytes on forest tree leaves in subtropical, cool temperate, and subboreal forests in Japan. Popul Ecol (in press)

Ito Y, Matsuda Y, Nakanishi K, Ito SI (2007) The effect of bagging branches on temporal infection levels of endophytic fungi in *Quercus serrata*. Chibu Shinrin Kenkyu 55:69–72 (in Japanese)

Jachmann HT, Fehrmann H (1989) Effects of phyllosphere microorganisms on the senescence of wheat leaves. Z PflKrankh PflSchutz 96:124–133

Johnson JA, Whitney NJ (1989a) An investigation of needle endophyte colonization patterns with respect to height and compass direction in a single of balsam fir (*Abies balsamea*). Can J Bot 67: 723–725

Johnson JA, Whitney NJ (1989b) A study of fungal endophytes of needles of balsam fir (*Abies balsamea*) and red spruce (*Picea rubens*) in New Brunswick, Canada, using cultural and electron microscope techniques. Can J Bot 67:3513–3516

Johnson JA, Whitney NJ (1992) Isolation of fungal endophytes from black spruce (*Picea mariana*) dormant buds and needles from New Brunswick, Canada. Can J Bot 70:1754–1757

Johnson JA, Whitney NJ (1994) Cytotoxicity and insecticidal activity of endophytic fungi from black spruce (*Picea mariana*) needles. Can J Microbiol 40:24–27

Jumpponen A, Jones KL (2009) Massively parallel 454 sequencing indicates hyperdiverse fungal communities in temperate *Quercus macrocarpa* phyllosphere. New Phytol 184:438–448

Jumpponen A, Jones KL (2010) Seasonally dynamic fungal communities in the *Quercus macrocarpa* phyllosphere differ between urban and nonurban environments. New Phytol 186:496–513

Kaneko R, Kakishima M (2001) *Mycosphaerella buna* sp. nov. with a *Pseudocercospora* anamorph isolated from the leaves of Japanese beech. Mycoscience 42:59–66

Kaneko R, Kaneko S (2004) The effect of bagging branches on levels of endophytic fungal infection in Japanese beech leaves. For Pathol 34:65–78

Kaneko R, Kakishima M, Tokumasu S (2003) The seasonal occurrence of endophytic fungus, *Mycosphaerella buna*, in Japanese beech, *Fagus crenata*. Mycoscience 44:277–281

Kikuzawa K (1983) Leaf survival of woody plants in deciduous broad-leaved forests. 1. Tall trees. Can J Bot 61:2133–2139

Kinkel LL (1991) Fungal community dynamics. In: Andrews JH, Hirano SS (eds) Microbial ecology of leaves. Springer, New York, pp 253–270

Kinkel LL, Andrews JH (1988) Disinfestation of living leaves by hydrogen peroxide. Trans Br Mycol Soc 91:523–528

Kodani J, Togashi K (1992) Leaf expansion and shoot elongation of *Cornus controversa* Hemsley. Jpn J Ecol 42:115–123 (in Japanese with English abstract)

Kodani J, Togashi K (1995) Foliage productivity and winter bud formation in relation to twig growth pattern in *Cornus controversa* Hemsley. Jpn J Ecol 45:237–245 (in Japanese with English abstract)

Koide K, Osono T, Takeda H (2005a) Colonization and lignin decomposition of *Camellia japonica* leaf litter by endophytic fungi. Mycoscience 46:280–286

Koide K, Osono T, Takeda H (2005b) Fungal succession and decomposition of *Camellia japonica* leaf litter. Ecol Res 20:599–609

Kowalski T (1982) Fungi infecting *Pinus sylvestris* needles of various ages. Eur J For Pathol 12:182–190

Lappalainen JH, Helander ML (1997) The role of foliar microfungi in mountain birch. Insect herbivore relationships. Ecography 20:116–122

Lappalainen JH, Koricheva J, Helander ML, Haukioja E (1999) Densities of endophytic fungi and performance of leafminers (Lepidoptera: Eriocraniidae) on birch along a pollution gradient. Environ Pollut 104:99–105

Lasota JA, Waldvogel MG, Shetlar DJ (1983) Fungus found in galls of *Adelges abietis* (L.) (Homoptera: Adelgidae): identification, within-tree distribution, and possible impact on insect survival. Environ Entomol 12:245–246

Legault D, Dessureault M, Laflamme G (1989a) Mycoflore des aiguilles de *Pinus banksiana* et *Pinus resinosa*. I. Champignons endophytes. Can J Bot 67:2052–2060 (in French with English abstract)

Legault D, Dessureault M, Laflamme G (1989b) Mycoflora of *Pinus banksiana* and *Pinus resinosa* needles. II. Epiphytic fungi. Can J Bot 67:2061–2065

Levetin E, Dorsey K (2006) Contribution of leaf surface fungi to the air spora. Aerobiologia 22:3–12

Lindow SE, Brandl MT (2003) Microbiology of the phyllosphere. Appl Environ Microbiol 69:1875–1883

Lu G, Cannon PF, Reid A, Simmons CM (2004) Diversity and molecular relationships of endophytic *Colletotrichum* isolates from the Iwokrama Forest Reserve, Guyana. Mycol Res 108:53–63

Magan N, Kirkwood IA, McLeod AR, Smith MK (1995) Effect of open-air fumigation with sulphur dioxide and ozone on phyllosphere and endophytic fungi of conifer needles. Plant Cell Environ 18:291–302

Makisaka K, Matsuda Y, Nakanishi K, Ito SI (2005) Decomposing ability of endophytic fungi in leaves of *Quercus serrata*. Chibu Shinrin Kenkyu 53:99–102 (in Japanese)

Masuya H, Kusunoki M, Kosaka H, Aikawa T (2009) *Haradamyces foliicola* anam. gen. et sp. nov., a cause of zonate leaf blight disease in *Cornus florida* in Japan. Mycol Res 113:173–181

Miller JD (1986) Toxic metabolites of epiphytic and endophytic fungi of conifer needles. In: Fokkema NJ, Heuvel J (eds) Microbiology of the phyllosphere. Cambridge University Press, New York, pp 223–231

Miller JD, Strongman D, Whitney NJ (1985) Observations on fungi associated with spruce budworm infested balsam fir needles. Can J For Res 15:896–901

Miller JD, Mackenzie S, Foto M, Adams GW, Findlay JA (2002) Needles of white spruce inoculated with rugulosin-producing endophytes contain rugulosin reducing spruce budworm growth rate. Mycol Res 106:471–479

Miura K, Kudo M (1970) An agar-medium for aquatic hyphomycetes. Trans Mycol Soc Jpn 11:116–118 (in Japanese)

Müller MM, Valjakka R, Suokko A, Hantula J (2001) Diversity of endophytic fungi of single Norway spruce needles and their role as pioneer decomposers. Mol Ecol 10:1801–1810

Naito S, Yoshihashi H, Matsuda Y, Ito SI (2002) Endophytic fungi in leaves of four species of the Fagaceae. Chibu Shinrin Kenkyu 50:113–114 (in Japanese)

Nakagiri A, Okane I, Ito T, Katumoto K (1997) *Lanceispora amphibia* gen. et sp. nov., a new amphisphaeriaceous ascomycete inhabiting senescent and fallen leaves of mangrove. Mycoscience 38:207–213

Nomura K, Naito S, Shimada H, Matsuda Y, Nakanishi K, Ito SI (2003) Comparison of the endophytic mycobiota in needles of *Pinus thunbergii* grown at seaside and inland districts. Chibu Shinrin Kenkyu 51:209–210 (in Japanese)

Okane I (2003) Diversity of xylariaceous fungi on natural substrates and its relationship with plant endophytes. In: Nakagiri A (ed) Clarification of subtropical microbial flora and collection and preservation of cultures. Report of Grants-in-Aid for Scientific Research (No. 11660326). Osaka, pp 31–45 (in Japanese)

Okane I, Nakagiri A, Ito T (1996) *Discostroma tricellulare*, a new endophytic ascomycete with a *Seimatosporium* anamorph isolated from Rhododendron. Can J Bot 74:1338–1344

Okane I, Nakagiri A, Ito T (1997) Preliminary study of endophytic fungi in evergreen plants from Ishigaki and Iriomote islands. IFO Res Commun 18:45–51

Okane I, Nakagiri A, Ito T (1998) Endophytic fungi in leaves of ericaceous plants. Can J Bot 76:657–663

Okane I, Nakagiri A, Ito T (2001a) Identity of *Guignardia* sp. inhabiting ericaceous plants. Can J Bot 79: 101–109

Okane I, Nakagiri A, Ito T (2001b) *Surculiseries rugispora* gen. et sp. nov., a new endophytic mitosporic fungus from leaves of *Bruguiera gymnorrhiza*. Mycoscience 42:115–122

Okane I, Lumyong S, Nakagiri A, Ito T (2003) Extensive host range of an endophytic fungus, *Guignardia endophyllicola* (anamorph: *Phyllosticta capitalensis*). Mycoscience 44:353–363

Okane I, Srikitikulchai P, Toyama K, Læssøe T, Sivichai S, Hywel-Jones N, Nakagiri A, Potacharoen W, Suzuki KI (2008) Study of endophytic Xylariaceae in Thailand: diversity and taxonomy inferred from rDNA sequence analyses with saprobes forming fruit bodies in the field. Mycoscience 49:359–372

Okane I, Srikitikulchai P, Tabuchi Y, Sivichai S, Nakagiri A (2012) Recognition and characterization of four Thai xylariaceous fungi inhabiting various tropical foliages as endophytes by DNA sequences and host plant preference. Mycoscience 53:122–132

Omar M, Heather WA (1979) Effect of saprophytic phylloplane fungi on germination and development of *Melampsora larici-populina*. Trans Br Mycol Soc 72:225–231

Osono T (2002) Phyllosphere fungi on leaf litter of *Fagus crenata*: occurrence, colonization, and succession. Can J Bot 80:460–469

Osono T (2003) Effects of prior decomposition of beech leaf litter by phyllosphere fungi on substrate utilization by fungal decomposers. Mycoscience 44:41–45

Osono T (2005) Colonization and succession of fungi during decomposition of *Swida controversa* leaf litter. Mycologia 97:589–597

Osono T (2006a) Fungi associated with *Swida controversa* leaves infested by zonate leaf blight and their antagonistic effects on the pathogenic fungus. Ap For Sci 15:7–12 (in Japanese with English abstract)

Osono T (2006b) Role of phyllosphere fungi of forest trees in the development of decomposer fungal communities and decomposition processes of leaf litter. Can J Microbiol 52:701–716

Osono T (2007) Role of endophytic fungi in grass litter decomposition. In: Popay A, Thom E (eds) Proceedings for the 6th international endophyte symposium, New Zealand Grassland Association, Christchurch, New Zealand, 25–28 March 2007, pp 103–105

Osono T (2008) Endophytic and epiphytic phyllosphere fungi of *Camellia japonica*: seasonal and leaf-age-dependent variations. Mycologia 100:387–391

Osono T (2012) Endophytic fungal assemblages on leaves of 73 deciduous tree species in a cool temperate forest. Ap For Sci 21:13–20

Osono T (2013) Metagenomic approach yields insights into fungal diversity and functioning. In: Sota S (ed) Species diversity and community structure, novel patterns and processes in plants, insects, and fungi. Springer, Tokyo

Osono T, Hirose D (2009a) Effects of prior decomposition of *Camellia japonica* leaf litter by an endophytic fungus on the subsequent decomposition by fungal colonizers. Mycoscience 50:52–55

Osono T, Hirose D (2009b) Ecology of endophytic fungi associated with leaf litter decomposition. In: Rai M, Bridge P (eds) Current advances in mycology. CAB International, Oxon, pp 92–109

Osono T, Hirose D (2011) Colonization and lignin decomposition of pine needle litter by *Lophodermium pinastri*. For Pathol 41:156–162

Osono T, Masuya H (2012) Endophytic fungi associated with leaves of Betulaceae in Japan. Can J Microbiol 58:507–515

Osono T, Mori A (2003) Colonization of Japanese beech leaves by phyllosphere fungi. Mycoscience 44: 437–441

Osono T, Mori A (2004) Distribution of phyllosphere fungi within the canopy of giant dogwood. Mycoscience 45:161–168

Osono T, Mori A (2005) Seasonal and leaf age-dependent changes in occurrence of phyllosphere fungi of giant dogwood. Mycoscience 46:273–279

Osono T, Takeda H (1999) A methodological survey on incubation of fungi on leaf litter of *Fagus crenata*. Ap For Sci Kansai 8:103–108 (in Japanese with English abstract)

Osono T, Bhatta BK, Takeda H (2004a) Phyllosphere fungi on living and decomposing leaves of giant dogwood. Mycoscience 45:35–41

Osono T, Mori A, Koide K (2004b) Defoliation of giant dogwood (*Swida controversa*) caused zonate leaf blight. Ap For Sci 13:161–164 (in Japanese with English abstract)

Osono T, Ishii Y, Hirose D (2008) Fungal colonization and decomposition of *Castanopsis sieboldii* leaves in a subtropical forest. Ecol Res 23:909–917

Osono T, Hagiwara Y, Masuya H (2011) Effects of temperature and litter type on fungal growth and decomposition of leaf litter. Mycoscience 52:327–332

Osono T, Tateno O, Masuya H (2013) Diversity and ubiquity of xylariaceous endophytes in live and dead leaves of temperate forest trees. Mycoscience 54:54–61

Osorio M, Stephan BR (1991a) Life cycle of *Lophoder-mium piceae* in Norway spruce needles. Eur J For Pathol 21:152–163

Osorio M, Stephan BR (1991b) Morphological studies of *Lophodermium piceae* (Fuckel) v. Höhnel on Norway spruce needles. Eur J For Pathol 21:389–403

Park D (1982) Phylloplane fungi: tolerance of hyphal tips to drying. Trans Br Mycol Soc 79:174–178

Petrini O (1986) Taxonomy of endophytic fungi of aerial plant tissues. In: Fokkema NJ, Van den Heuvel J (eds) Microbiology of the phyllosphere. Cambridge University Press, New York, pp 175–187

Petrini O (1991) Fungal endophytes of tree leaves. In: Andrews JH, Hirano SS (eds) Microbial ecology of leaves. Springer, New York, pp 179–197

Petrini O (1996) Ecological and physiological aspects of host specificity in endophytic fungi. In: Redlin SC, Carris LM (eds) Endophytic fungi in grasses and woody plants. APS Press, St. Paul, pp 87–100

Petrini O, Carroll GC (1981) Endophytic fungi in foliage of some Cupressaceae in Oregon. Can J Bot 59:629–636

Petrini O, Fisher PJ (1990) Occurrence of fungal endophytes in twigs of *Salix fragilis* and *Quercus robur*. Mycol Res 94:1077–1080

Petrini LE, Petrini O, Laflamme G (1989) Recovery of endophytes of *Abies balsamea* from needles and galls of *Paradiplosis tumifex*. Phytoprotection 70:97–103

Petrini O, Sieber TN, Toti L, Viret O (1992) Ecology, metabolite production, and substrate utilization in endophytic fungi. Nat Toxins 1:185–196

Promputtha I, Jeewon R, Lumyong S, McKenzie EHC, Hyde KD (2005) Ribosomal DNA fingerprinting in the identification of non sporulating endophytes from *Magnolia liliifera* (Magnoliaceae). Fungal Divers 20:167–186

Promputtha I, Lumyong S, Dhanasekaran V, McKenzie EHC, Hyde KD, Jeewon R (2007) A phylogenetic evaluation of whether endophytes become saprotrophs at host senescence. Microb Ecol 53:579–590

Rodriguez RJ, Redman RS (1997) Fungal life-style and ecosystem dynamics: biological aspects of plant pathogens, plant endophytes and saprophytes. Adv Bot Res 24:169–193

Ruscoe QW (1971) Mycoflora of living and dead leaves of *Nothofagus truncata*. Trans Br Mycol Soc 56: 463–474

Sahashi N, Kubono T, Miyasawa Y, Ito S (1999) Temporal variations in isolation frequency of endophytic fungi of Japanese beech. Can J Bot 77:197–202

Sahashi N, Miyasawa Y, Kubono T, Ito S (2000) Colonization of beech leaves by two endophytic fungi in northern Japan. For Pathol 30:77–86

Saikkonen K (2007) Forest structure and fungal endophytes. Fungal Biol Rev 21:67–74

Saikkonen K, Faeth SH, Helander M, Sullivan TJ (1998) Fungal endophytes: a continuum of interactions with host plants. Annu Rev Ecol Syst 29:319–343

Santamaria J, Bayman P (2005) Fungal epiphytes and endophytes of coffee leaves (*Coffea arabica*). Microb Ecol 50:1–8

Shirouzu T, Hirose D, Fukasawa Y, Tokumasu S (2008) Fungal succession associated with the decay of leaves of an evergreen oak *Quercus myrsinaefolia*. Fungal Divers 34:87–107

Sieber TN (2007) Endophytic fungi in forest trees: are they mutualists? Fungal Biol Rev 21:75–89

Sieber TN, Hugentobler C (1987) Endophytische Pilze in Blättern und Ästen gesunder und geschädigter Buchen (*Fagus sylvatica* L.). Eur J For Pathol 17:411–425 (in German with English abstract)

Sieber TN, Rys J, Holdenrieder O (1999) Mycobiota in symptomless needles of *Pinus mugo* ssp. *uncinata*. Mycol Res 103:306–310

Skidmore AM, Dickinson CH (1973) Effect of phylloplane fungi on the senescence of excised barley leaves. Trans Br Mycol Soc 60:107–116

Skidmore AM, Dickinson CH (1976) Colony interactions and hyphal interference between *Septoria nodorum* and phylloplane fungi. Trans Br Mycol Soc 66:57–64

Soma K, Saito T (1979) Ecological studies of soil organisms with references to the decomposition of pine needles. I. Soil macrofaunal and mycofloral surveys in coastal pine plantations. Rev Ecol Biol Sol 16:337–354

Stadler B, Müller T (1996) Aphid honeydew and its effect on the phyllosphere microflora of *Picea abies* (L.) karst. Oecologia 108:771–776

Stadler B, Müller T (2000) Effects of aphids and moth caterpillars on epiphytic microorganisms in canopies of forest trees. Can J For Res 30:631–638

Stone JK (1988) Fine structure of latent infections by *Rhabdocline parkeri* on Douglas-fir, with observations on uninfected epidermal cells. Can J Bot 66:45–54

Stone J, Petrini O (1997) Endophytes of forest trees: a model for fungus-plant interactions. In: Carroll GC, Tudzynski P (eds) The Mycota, V. Plant relationships, Part B. Springer, Berlin, pp 129–140

Stone JK, Sherwood MA, Carroll GC (1996) Canopy microfungi: function and diversity. Northwest Sci 70:37–45

Suske J, Acker G (1987) Internal hyphae in young, symptomless needles of *Picea abies*: electron microscopic and cultural investigation. Can J Bot 65:2098–2103

Suske J, Acker G (1989) Identification of endophytic hyphae of *Lophodermium piceae* in tissues of green, symptomless Norway spruce needles by immunoelectron microscopy. Can J Bot 67:1768–1774

Suzuki T, Naito S, Matsuda Y, Ito SI (2003) Differences of endophytic mycobiota in the leaves of *Quercus serrata* among sites and host trees. Chibu Shinrin Kenkyu 51:211–212 (in Japanese)

Terashita T (1973) Studies of an anthracnose fungus on broad-leaved trees in Japan, with special reference to the latency of the fungus. Bull Gov For Exp Sta 252:1–85 (in Japanese with English abstract)

Tokumasu S (1980) Observations on the fungal flora on pine needle litter. In: Biseibutsu no seitai 7. Gakkai Shuppan Center, Tokyo, pp 129–144 (in Japanese)

Tomita F (2003) Endophytes in Southeast Asia and Japan: their taxonomic diversity and potential applications. Fungal Divers 14:187–204

Toti L, Viret O, Chapela IH, Petrini O (1992) Differential attachment by conidia of the endophyte, *Discula umbrinella* (Berk. & Br.) Morelet, to host and non-host surfaces. New Phytol 121:469–475

Toti L, Viret O, Horat G, Petrini O (1993) Detection of the endophyte *Discula umbrinella* in buds and twigs of *Fagus sylvatica*. Eur J For Pathol 23:147–152

Valkama E, Koricheva J, Salminen JP, Helander M, Saloniemi I, Saikkonen K, Pihlaja K (2005) Leaf surface traits: overlooked determinants of birch resistance to herbivores and foliar micro-fungi? Trees 19:191–197

Viret O, Petrini O (1994) Colonization of beech leaves (*Fagus sylvatica*) by the endophyte *Discula umbrinella* (teleomorph: *Apiognomonia errabunda*). Mycol Res 98:423–432

Viret O, Scheidegger C, Petrini O (1993) Infection of beech leaves (*Fagus sylvatica*) by the endophyte *Discula umbrinella* (teleomorph: *Apiognomonia errabunda*): low-temperature scanning electron microscopy studies. Can J Bot 71:1520–1527

Viret O, Toti L, Chapela IH, Petrini O (1994) The role of the extracellular sheath in recognition and attachment of conidia of *Discula umbrinella* (Berk. & Br.) Morelet to the host surface. New Phytol 127:123–131

Wang Y, Guo LD, Hyde KD (2005) Taxonomic placement of sterile morphotypes of endophytic fungi from *Pinus tabulaeformis* (Pinaceae) in northeast China based on rDNA sequences. Fungal Divers 20:235–260

Wei CZ, Harada Y (1998) *Didymella fagi* sp. nov. and its anamorph *Ascochyta fagi*, causing the yellow leaf spot disease of *Fagus crenata* and *Quercus mongolica* var. *grosseserrata* in Japan. Mycoscience 39:63–69

Wei JG, Xu T, Guo LD, Liu AR, Zhang Y, Pan XH (2007) Endophytic *Pestalotiopsis* species associated with plants of Podocarpaceae, Theaceae, and Taxaceae in southern China. Fungal Divers 24:55–74

Weis AE (1982) Use of a symbiotic fungus by the gall maker *Asteromyia carbonifera* to inhibit attack by the parasitoid *Torymus capite*. Ecology 63:1602–1605

Wildman HG, Parkinson D (1979) Microfungal succession on living leaves of *Populus tremuloides*. Can J Bot 57:2800–2811

Wilson D (1995) Fungal endophytes which invade insect galls: insect pathogens, benign saprophytes, or fungal inquilines? Oecologia 103:255–260

Wilson D (1996) Manipulation of infection levels of horizontally transmitted fungal endophytes in the field. Mycol Res 100:827–830

Wilson D (2000) Ecology of woody plant endophytes. In: Bacon CW, White JF Jr (eds) Microbial endophytes. Marcel Dekker, New York, pp 389–420

Wilson D, Carroll GC (1994) Infection studies of *Discula quercina*, an endophyte of *Quercus garryana*. Mycologia 86:635–647

Wilson D, Carroll GC (1997) Avoidance of high-endophyte space by gall-forming insects. Ecology 78:2153–2163

Wilson D, Faeth SH (2001) Do fungal endophytes result in selection for leafminer ovipositional preference? Ecology 82:1097–1111

Wilson D, Barr ME, Faeth SH (1997) Ecology and description of a new species of *Ophiognomonia* endophytic in the leaves of *Quercus emoryi*. Mycologia 89:537–546

Yan L, Zhang C, Ding L, Ma Z (2008) Development of a real-time PCR assay for the detection of *Cladosporium fulvum* in tomato leaves. J Appl Microbiol 104:1417–1424

Yoshihashi H, Okuda A, Takeda A, Ito SI (2000) Endophytic fungi in leaves of the woody plants. Chibu Shinrin Kenkyu 48:155–156 (in Japanese)

Yoshihashi H, Okuda A, Takeda A, Matsuda Y, Ito SI (2001) Endophytic fungi in leaves of the woody plants. Temporal variations in isolation frequency of endophytic fungi. Chibu Shinrin Kenkyu 49:97–100 (in Japanese)

Endophytic Actinobacteria: Diversity and Ecology

2

Venkadasamy Govindasamy,
Christopher M.M. Franco,
and Vadakattu V.S.R. Gupta

Abstract

Actinobacteria are a group of Gram-positive microorganisms with a high G+C content in their DNA and belong to the phylum *Actinobacteria*, one of the largest phyla within bacteria. Some of these actinobacteria have an endophytic lifestyle which occurs abundantly in most plants. The abundance and diversity of endophytic actinobacterial colonisation depend on plant species, type of soils and other associated environmental conditions. *Streptomyces* spp. were reported as the most predominant species, and *Microbispora*, *Micromonospora*, *Nocardioides*, *Nocardia* and *Streptosporangium* are other common genera of endophytic actinobacteria isolated from a diverse range of plant species, including those found in estuarine/mangrove ecosystems and algae and seaweeds of marine ecosystems. Over the years, isolation media have been devised and numerous methods have been standardised for the isolation, identification and characterisation of these endophytic actinobacteria. Recent advances in molecular tools have revealed the 'not yet cultured' diversity within this group. Therefore, a combination of both culture-based and molecular techniques is essential to describe the diversity and ecology of endophytic actinobacteria. The quest for actinobacteria and their metabolic capabilities is ongoing, as they represent the largest ecological resource for secondary metabolites (plant hormones, antibiotics and other bioactive compounds), with potential biotechnological applications in agriculture, industry and medicine.

V. Govindasamy
Department of Medical Biotechnology, Flinders University, Bedford Park, SA 5042, Australia

National Institute of Abiotic Stress Management, ICAR, Malegaon, Baramati, Maharashtra 413115, India

C.M.M. Franco (✉)
Department of Medical Biotechnology, Flinders University, Bedford Park, SA 5042, Australia
e-mail: chris.franco@flinders.edu.au

V.V.S.R. Gupta
Department of Medical Biotechnology, Flinders University, Bedford Park, SA 5042, Australia

CSIRO Ecosystem Sciences, PMB No 2, Glen Osmond, SA 5064, Australia

V.C. Verma and A.C. Gange (eds.), *Advances in Endophytic Research*,
DOI 10.1007/978-81-322-1575-2_2, © Springer India 2014

1 Introduction

Plants are naturally associated with microorganisms both externally and internally in various ways. On the exterior surface of plants, diverse microbial interactions occur mostly in the root zone (rhizosphere) and on aerial parts, especially the leaves (phyllosphere) (Hiltner 1904; Yang et al. 2001; Lindow and Brandl 2003; Gray and Smith 2005). Some of the rhizosphere- and phyllosphere-derived microorganisms, which are either bacteria or fungi, are able to penetrate the interior of the plant and colonise intercellular spaces and vascular tissues, where they reside at least part of their lives showing beneficial/symbiotic, neutral or pathogenic interactions (Tervet and Hollis 1948; Hallman et al. 1997; Araujo et al. 2002; Rosenblueth and Martínez-Romero 2006). In the well-studied endosymbiotic beneficial interactions, like the root nodule symbiosis of legumes with rhizobia or the formation of arbuscular mycorrhiza with fungi, the formation of organised symbiotic structures is a common phenomenon, where the microsymbionts reside intracellularly surrounded by a host membrane (Fisher and Long 1992; Downie 1994; Wang and Qiu 2006). On the other hand, there are pathogenic interactions, in which bacteria or fungi often produce effector molecules/proteins inside plant host cells that elicit symptoms of plant disease, causing deleterious effects (Montesinos et al. 2002). In contrast to these interactions, another kind of beneficial interaction exists within the interior of the plant, which is poorly understood at the molecular level. The microorganisms involved in these interactions are commonly referred to as 'endophytes' (Wilson 1995). By definition, endophytes are bacteria or fungi that colonise the host tissues internally, sometimes in high numbers, without damaging the host or harming the host through symptoms of plant disease (Wilson 1995; Compant et al. 2005). Unlike endosymbionts, they do not reside inside the host cells or surrounded by a membrane compartment. Endophytes are distributed throughout the host in all plant organs roots, stems, leaves, flowers, fruits and seeds.

Plants are endophytically colonised by a variety of bacteria belonging to different phylogenetic groups (Chelius and Triplett 2001; Reiter and Sessitsch 2006; Berg et al. 2005). Among them, endophytic bacteria are mostly *Proteobacteria*, but also *Firmicutes*, *Actinobacteria* and *Bacteroidetes* (Rosenblueth and Martínez-Romero 2006). However, the structural composition of endophytic bacterial communities depends on the host plant genotype, the plant organ as well as on the vegetative stage, and may be significantly influenced by plant stress (Sturz et al. 1997; Sessitsch et al. 2002; Reiter et al. 2002; Rasche et al. 2006a, b) and soil type (Conn and Franco 2004a). The *Actinobacteria* are of interest as they are a primary source of secondary metabolites which include bioactive compounds with biotechnological significance. The actinobacteria mainly inhabit the soil, and a large number of actinobacteria have already been isolated and described. Recently, the rate of discovery of new actinobacteria isolated from soils has decreased. Therefore, researchers have examined other ecological niches, such as plant surfaces and interior tissues of plants, and also estuarine and marine ecosystems.

The actinobacteria represent a large portion of the rhizosphere microbial community (Lundberg et al. 2012). Early studies have demonstrated that some actinobacteria can form intimate associations with plants, such as the endosymbiotic association of *Frankia* species in nonleguminous plants and the pathogenic association of a narrow range of *Streptomyces* species on potato (Benson and Silvester 1993; Doumbou et al. 1998). Recent studies have revealed a diverse group of endophytic actinobacterial species with different functions from various plant species (Araujo et al. 2002; Coombs and Franco 2003a; Ryan et al. 2008; Bascom-Slack et al. 2009). Some of them can act as biological control agents (Coombs et al. 2004; Cao et al. 2005; Misk and Franco 2011), and some act as plant growth promoters (Igarashi et al. 2002; Hasegawa et al. 2006). However, the genotype, physiological status of the host plants and its surrounding environment (soil type, including its physicochemical properties, microbial load and diversity) have a major impact on species richness and diversity of endophytic actinobacterial populations and their related functions (Conn and Franco 2004b; Franco et al. 2007). Due to their ability to colonise the interior of plants coupled with their antimicrobial activities, many initial studies tested endophytic actinobacteria for biological control of plant diseases. In recent years, endophytic actinobacterial research has received special attention mainly as a result of their many other plant growth-promoting properties. In addition, actinobacteria cultured from different endophytic habitats are considered as a potential source for many novel secondary metabolites (Guo et al. 2008).

The aim of this chapter is to describe the recent taxonomy, ecology and diversity of endophytic actinobacteria and to summarise recent findings on isolation of novel endophytic actinobacteria from cultivated crops and also other unexplored plant sources from different ecosystems. Recent advances in the methods to study uncultured/not yet cultured endophytic actinobacterial diversity will also be covered.

2 Taxonomy and Molecular Phylogeny of Endophytic Actinobacteria

Taxonomically the endophytic actinobacteria are a group of Gram-positive bacteria belonging to the phylum *Actinobacteria.* With 6 classes, 25 orders, 52 families and 232 genera (Table 2.1), the phylum *Actinobacteria* represents one of the largest taxonomic units among the 18 major lineages currently recognised within the domain *Bacteria*, including 5 subclasses and 14 suborders (Stackebrandt and Schumann 2000). The phylum *Actinobacteria* comprises Gram stain-positive bacteria with a high G+C content in their DNA.

The species that constitute the *Actinobacteria* have morphologies that include a range of cell types, i.e. coccoid, rod-coccoid and hyphae, that fragment or are highly differentiated. In some genera the spores are formed from aerial mycelia, and may be motile, or may be contained in sporangia or other unusual spore-bearing structures. They have a diverse range of physiological properties and are sought after because of their production of extracellular enzymes but primarily for the production of secondary metabolites and increasingly for applications in agriculture.

Notably, many such secondary metabolites are antibiotics of medical importance (Lechevalier and Lechevalier 1967; Schrempf 2001). *Actinobacteria* play a crucial role in the recycling of biomaterials by organic matter decomposition and humus formation (Goodfellow and Williams 1983; Schrempf 2001; Stach and Bull 2005). This phylum includes human pathogens, e.g. *Mycobacterium* spp., *Nocardia* spp., *Tropheryma* spp., *Corynebacterium* spp. and *Propionibacterium* spp.; plant commensals, e.g. *Leifsonia* spp.; nitrogen-fixing plant symbionts, e.g. *Frankia* spp.; plant endophytes (many genera); plant pathogens, e.g. *Streptomyces* spp.; and inhabitants of the human gastrointestinal tract, e.g. *Bifidobacterium* spp.

Although *Actinobacteria* form a distinct cluster in the 16S rRNA phylogenetic trees, the only 'shared derived character' is a homologous insertion of ~100 nucleotides between helices 54 and 55 of the 23S rRNA gene (Ventura et al. 2007). Recent analysis has identified conserved indels and proteins that can be used to distinguish this important group of bacteria (Gao and Gupta 2005; Gao et al. 2006; Ventura et al. 2007; Hayward et al. 2009).

The initial genome sequencing results confirmed that, unlike most bacterial genomes, many *Streptomyces* genomes are linear (Dyson 2011) and so too are genomes of *Rhodococcus* spp., but the other genera have circular genomes (Bentley et al. 2002) with sizes ranging from 7.7 to 9.7 Mb (Redenbach et al. 2000) for the filamentous actinobacteria. In addition, large 'linear plasmids' typically possessing short inverted repeats at their termini and protein-bound 5′ends, are also reported to be present in the various genera of *Actinobacteria* (Kalkus et al. 1998; Redenbach et al. 2000). The first actinobacterial genome to be sequenced was that of the human tuberculosis agent, *M. tuberculosis* H37Rv (Cole et al. 1998). In the last few years, genomes of different *Actinobacteria* (including plant beneficial *Frankia*, *Leifsonia* and *Streptomyces* species) have been sequenced to completion

Table 2.1 Taxonomy of the phylum *Actinobacteria* and genera with endophytic life style as per *Bergey's Manual of Systematic Bacteriology* (Volume 5, Part A; 2nd edition, 2012) and 'List of Prokaryotic Names with Standing in Nomenclature' (Euzeby http://www.bacterio.cict.fr/)

Systematic position/taxonomic hierarchy	Orders	No. of families	No. of genera	Key genera reported to contain endophytes
Phylum XXVI. *Actinobacteria*				
Class I. *Actinobacteria*	Order I. *Actinomycetales*	1	5	*Actinomyces*
	Order II. *Actinopolysporales*	1	1	*Actinopolyspora*[c]
	Order III. *Bifidobacteriales*	1	7	ND
	Order IV. *Catenulisporales*	2	2	ND
	Order V. *Corynebacteriales*	6	13	*Corynebacterium*
				Dietzia[c]
				Gordonia[c]
				Mycobacterium
				Nocardia
				Rhodococcus
				Tsukamurella[c]
				Williamsia[c]
	Order VI. *Frankiales*	6	11[b]	*Blastococcus*[c]
				Frankia
				Jatrophihabitans[a]
				Modestobacter[c]
	Order VII. *Glycomycetales*	1	2	*Glycomyces*[c]
	Order VIII. *Jiangellales*	1	2	*Jiangella*[c]
	Order IX. *Kineosporiales*	1	3	*Kineococcus*[c]
	Order X. *Micrococcales*	15	84	*Arthrobacter*
				Brachybacterium[c]
				Citricoccus[c]
				Herbiconiux[a]
				Janibacter[c]
				Kocuria[c]
				Koreibacter[a]
				Leifsonia
				Microbacterium
				Micrococcus
				Oerskovia[c]
				Promicromonospora[c]
				Rathayibacter[c]
	Order XI. *Micromonosporales*	1	23	*Actinoplanes*[c]
				Dactylosporangium
				Jishengella[a]
				Micromonospora
				Phytohabitans[a]
				Phytomonospora[a]
				Planosporangium[c]
				Plantactinospora[c]
				Polymorphospora[c]
	Order XII. *Propionibacteriales*	2	18	*Actinopolymorpha*[c]
				Flindersiella[a]
				Kribbella[c]
				Nocardioides
	Order XIII. *Pseudonocardiales*	1	22	*Actinomycetospora*[c]
				Actinophytocola[a]
				Amycolatopsis[c]
				Kibdelosporangium[c]
				Pseudonocardia
				Saccharomonospora
				Saccharopolyspora[c]
				Saccharothrix

(continued)

Table 2.1 (continued)

Systematic position/taxonomic hierarchy	Orders	No. of families	No. of genera	Key genera reported to contain endophytes
	Order XIV. *Streptomycetales*	1	3[b]	*Kitasatospora*[c]
				Streptacidiphilus[c]
				Streptomyces
	Order XV. *Streptosporangiales*	3	22[b]	*Actinoallomurus*[c]
				Actinocorallia[c]
				Actinomadura[c]
				Allonocardiopsis[a]
				Microbispora
				Nocardiopsis
				Nonomuraea[c]
				Planotetraspora
				Streptomonospora
				Streptosporangium
	Order *Incertae sedis*[b]	0	1[b]	ND
Class II. Acidimicrobiia	Order I. *Acidimicrobiales*	2	5	ND
Class III. Coriobacteriia	Order I. *Coriobacteriales*	1	13	ND
Class IV. Nitriliruptoria	Order I. *Nitriliruptorales*	1	1	ND
	Order II. *Euzebyales*	1	1	ND
Class V. Rubrobacteria	Order I. *Rubrobacterales*	1	1	ND
Class VI. Thermoleophilia	Order I. *Thermoleophilales*	1	1	ND
	Order II. *Solirubrobacterales*	3	3	ND

[a] New genus discovered as an endophyte
[b] Includes genus *Incertae sedis*
[c] Contains recently identified/discovered endophytic species (after 2010); ND—no type strain identified as an endophyte

(Bentley et al. 2002; Monteiro-Vitorello et al. 2004; Normand et al. 2007), while sequencing of genomes from representatives of 43 or more actinobacteria are still in progress (http://www.ncbi.nlm.nih.gov/genomes/lproks.cgi).

In the recently published 2nd edition of *Bergey's Manual of Systematic Bacteriology* (Whitman et al. 2012), the polyphasic approach was followed for actinobacterial systematics. This taxonomic characterisation is inferred from many parameters, namely, its branching pattern in the 16S rRNA phylogenetic tree (Garrity and Holt 2001; Ludwig and Klenk 2005), taxon-specific 16S rRNA gene sequence signatures (Zhi et al. 2009), as well as chemotaxonomical, physiological and biochemical properties. The separation of this phylum from other bacterial taxa is supported by conserved indels in some proteins (e.g. cytochrome *c* oxidase subunit 1, CTP synthetase and glutamyl-tRNA synthetase), by the presence of a large insert in the 23S rRNA gene (Gao and Gupta 2005; Gao et al. 2006) and by distinctive gene arrangements (Kunisawa 2007).

3 Recent Advances in the Isolation and Characterisation of Endophytic Actinobacterial Diversity

3.1 Culture-Based Approaches

The method of isolation is one of the most crucial steps in obtaining pure cultures of endophytes; therefore, consideration should be given to the implementation of a plant-specific isolation protocol. Some detailed isolation methods and procedures, including plant sampling, surface sterilisation and media relevant for endophytic actinobacteria, were assessed by Hallmann et al. (2006), Qin et al. (2009) and recently by Kaewkla and Franco (2013a).

3.1.1 Plant Sampling, Surface Sterilisation and Processing

After the choice of host plant is made, the next decision is the age of the sample and the plant organ.

In most studies, sampling is a one-off event and the description of the endophytes obtained is provided with little or no acknowledgement of the possibility that the diversity can change with plant age and season or soil type (Conn and Franco 2004a). Very few studies are hypothesis driven, especially if the aim is to maximise the number and diversity of actinobacteria isolated. Sampling decisions should include the age or stage of the plant, the soil and climate and the parts of the plant (Zhang et al. 2006). In the case of trees, depending on the size, the location of the sample and the number of samples are likely to influence the outcomes. To date, there are no reports on the spatial diversity within a branch or root of a tree. However, as the abundance of endophytes is low (Kaewkla and Franco 2013a), it is recommended that a large amount of plant sample is collected to be able to increase the number and diversity of the strains cultivated.

Surface sterilisation of plant material is an obligatory step for endophytic actinobacterial isolation in order to kill all the surface microbes. It is usually accomplished by treating the plant tissues with an oxidising agent or general sterilant for a specific period, followed by repeated sterile water rinses. Commonly used surface sterilants include ethanol (70–95 %), sodium hypochlorite (3–10 %) and also hydrogen peroxide (3 %). Some surfactants such as Tween 20, Tween 80 and Triton X-100 can also be added to enhance the effectiveness of surface sterilisation (Sturz 1995; Hallmann et al. 2006). A general protocol involves a three-step procedure similar to that described by Coombs and Franco (2003a). It was recommend that a five-step procedure is optimum, and addition of sodium thiosulfate solution following the sodium hypochlorite treatment will improve cultivation efficiency because thiosulfate can neutralise the detrimental effects of residual NaOCl on the growth of microorganisms emerging from within the tissue (Qin et al. 2009). After this treatment, plant tissues can be soaked in 10 % $NaHCO_3$ solution to inhibit any endophytic fungi, which can outgrow the actinobacteria on isolation medium plates (Nimnoi et al. 2010a). The effectiveness of surface sterilisation should be checked to confirm the isolates are true endophytes. In general, the sterilisation procedure should be standardised for each plant type and tissue, especially the sterilisation time, as the sensitivity varies with plant species, age and tissue type. The concentration of the hypochlorite and the length of exposure should be adjusted to the type of plant tissue. For example, many leaves are more 'porous' than their root or stem surfaces and are prone to infiltration by the sterilant.

Samples containing extraneous material such as soil can be sonicated before sterilisation to remove any attached soil or microorganisms. Surface-sterilised plant samples are routinely air-dried or heated at 80 or 100 °C for 15–30 min to kill bacteria, resulting in a lowering of vegetative bacterial number if present. Commonly, plant materials are septically sectioned into small fragments of about 0.2 × 1.0 cm size (Coombs and Franco 2003a; Cao et al. 2004; Verma et al. 2009; Fialho de Oliveira et al. 2010) and then placed/distributed into various actinobacterial isolation media. In another method, surface-sterilised plant tissues can be aseptically crumbled into smaller fragments by commercial blender (Qin et al. 2008a, b; Li et al. 2009), to expose organisms from within the plant material and increase their recovery. These two preferred methods could recover a higher number of less commonly detected genera among the endophytic actinobacteria. One of the main objectives is to release the endophytes from the inner parts of plant tissue material and expose them to the growth medium. Some sterile samples were mixed in a mortar with 0.5 g of sterile powdered calcium carbonate and then placed in a Petri dish, and two millilitres of sterilised tap water was added to the sample to create a moist environment. After 2 weeks at 28 °C, the samples were air-dried at room temperature and placed in media plates, or samples were also placed in a glass dish and flooded with 50 ml of 10 mmol phosphate buffer containing 10 % plant or soil extract at 28 °C to liberate actinobacterial spores (Qin et al. 2009). Endophytes can also be separated from plant tissue using the method of Jiao et al. (2006) by grinding the plant material and subjecting it to enzymes that break down plant cell walls. The bacterial pellet is separated out by differential centrifugation, diluted and plated onto isolation media.

All the methods examined gave different populations, and none of them was recommended as being superior to any other.

3.1.2 Composition and Combination of Culture Media and Incubation Conditions

Successful culturing of microorganisms on laboratory media is dependent on the nutritional composition of the media and the incubation conditions. The use of a medium composition that mimics the micro-environments of inner part of the plants is a good strategy for isolation of endophytic actinobacteria. Some of the established media for isolation of actinobacteria from soil samples include humic acid vitamin B (HV) (Hayakawa 1990), International *Streptomyces* Project media 2 and 5 (Shirling and Gottlieb 1966), raffinose-histidine agar (Vickers et al. 1984) and starch casein agar (Küster and Williams 1964). Low-nutrient medium TWYE was found effective for isolation of endophytic actinobacteria from many plant species (Coombs and Franco 2003a; Qin et al. 2009; Li et al. 2009), due to the fact that high nutrient concentration allowed fast-growing bacteria to overgrow slower growing actinobacteria. Inside the plant, amino acids are the major source of nitrogen, and cellulose and xylan are the primary sources of carbon. Media containing amino acids (proline, arginine and asparagine) as nitrogen sources and cellulose, xylan, sodium propionate and sodium succinate as carbon sources improved isolation effectiveness and yielded uncommon and rare endophytic actinobacterial genera (Qin et al. 2009). Similarly, addition of plant or soil extracts into the isolation medium could help meet specific requirements of actinobacteria from plant tissues and soil environments (Okazaki 2003). Janso and Carter (2010) used arginine vitamin agar supplemented with 3 % soil extract to isolate several phylogenetically unique endophytic actinobacteria such as *Sphaerisporangium* and *Planotetraspora* from tropical plants of Papua New Guinea and Mborokua Island, Solomon Islands. In another example, the use of media with low concentrations of plant polymers (gellan gum, xylan and pectin), their constituent sugars (glucose, galactose, xylose, arabinose, glucuronate, galacturonate, ascorbate, gluconate and carboxymethylcellulose), and 17 amino acids improved the isolation of 16 rare actinobacterial genera including a new genus *Flindersiella* in the family *Nocardioides*, while other 11 strains were accepted as new species of endophytic actinobacteria (Kaewkla and Franco 2013a).

Kaewkla and Franco (2013a) recommend incubation of isolation plates under moist conditions for up to 16 weeks with removal of colonies every week, as they found that the majority of non-streptomycetes emerged after 6 weeks of incubation.

A list of isolation protocols and media used to study the diversity of endophytic actinobacteria is shown in Table 2.2.

4 Diversity of Endophytic Actinobacteria in Plants of Terrestrial Ecosystems

4.1 Agricultural/Field Crops

Early studies on endophytic actinobacterial associations in agricultural crop plants were reported from Italy by Sardi et al. (1992) who isolated 499 strains from surface-sterilised root samples of 28 plant species including different field crops such as barley, rye, oats and soybean, with the majority of the isolates belonging to the genus *Streptomyces*. Okazaki et al. (1995) isolated endophytic actinobacteria from other part of crop plants, e.g. leaves and leaf litter, with the majority belonging to the genera *Streptomyces* and *Microbispora*. *Microbispora* spp. was the most common actinobacteria isolated from the surface-sterilised roots and leaves of field-grown maize plants

Table 2.2 Methodology used in culture-based studies for the isolation of endophytic actinobacteria from different plant species

Plant type	Methods	Media used	List of reported/cultured genera	References
Australian endemic trees (*Callitris preissii*, *Eucalyptus camaldulensis*, *Eucalyptus microcarpa*, *Pittosporum phylliraeoides*)	Surface sterilisation and prolonged incubation at 27 °C up to 16 weeks	Mannitol mung bean yeast extract mineral salt agar (MMYA), yeast extract casamino acid glucose agar (YECG), humic acid vitamin B agar (HVA), HVA with gellan gum (HVG), VL 70 gellan gum with different combinations of sugar, amino acid mixtures	*Actinomadura*, *Actinomycetospora*, *Actinopolymorpha*, *Amycolatopsis*, *Flindersiella*, *Gordonia*, *Kribbella*, *Micromonospora*, *Nocardia*, *Nocardioides*, *Nocardiopsis*, *Nonomuraea*, *Polymorphospora*, *Promicromonospora*, *Pseudonocardia* and *Williamsia*	Kaewkla and Franco (2013a)
Cabbage (*Brassica campestris*, China)	Surface sterilisation and incubation at 30 °C up to 3 weeks	Humic acid vitamin B agar (HV) and corn meal agar (CMA)	*Microbispora*, *Streptomyces*, *Micromonospora*, *Nocardia*, *Verrucosispora*, *Nonomuraea*, *Actinomadura* and *Thermonospora*	Lee et al. (2008b)
Ethanobotanical trees (*Cinnamomum zeylanicum*, *Zingiber spectabile*, *Elettariopsis curtisii* and *Labisia pumila*) Thailand	Four different surface sterilisation procedures and incubation at 28 °C up to 3 weeks	Starch yeast casein agar (SYCA), actinomycetes isolation agar (AIA), HV agar, tap water yeast extract agar (TWYA) and coal vitamin agar	*Streptomyces* and one unknown genus	Zin et al. (2010)
Lentil, chickpea, pea, faba bean and rye (Parksville, South Australia)	Surface sterilisation and incubation at 27 and 37 °C up to 4 weeks	HV agar, starch casein medium and TWYA	*Streptomyces* and *Microbispora*	Misk and Franco (2011)
Medicinal plants (Hainan, China)	Surface sterilisation and incubation at 28 °C up to 3 weeks	ATCC 172 agar, Gauze's No. 2 agar, glucose-asparagine agar, HV agar and starch-casein-mineral salts agar	*Amycolatopsis*, *Micromonospora*, *Nocardia*, *Nonomuraea* and *Streptomyces*	Huang et al. (2012)
Medicinal plants (Xishuangbanna, China)	Surface sterilisation followed by four different selective isolation procedures and incubation at 28 °C for 2–8 weeks	TWYE, modified TWYE with plant extract, glycerol-asparagine agar (ISP 5), HV agar, M5 inorganic salts-starch agar (ISP 4), YIM 38 medium, raffinose-histidine agar, sodium propionate agar, cellulose-proline agar, trehalose-proline medium, xylan-arginine agar	*Actinocorallia*, *Blastococcus*, *Dactylosporangium*, *Dietzia*, *Jiangella*, *Oerskovia*, *Promicromonospora* and *Saccharopolyspora*	Qin et al. (2009)

(continued)

Table 2.2 (continued)

Plant type	Methods	Media used	List of reported/cultured genera	References
Medicinal tree (*Maytenus austroyunnanensis*) (Xishuangbanna, China)	Surface sterilisation followed by enzymatic homogenisation, diluted supernatant used for isolation and incubation at 28 °C for 2–8 weeks	Same as above	*Amycolatopsis*, *Cellulosimicrobium*, *Glycomyces*, *Jiangella*, *Micromonospora*, *Mycobacterium*, *Nocardia*, *Nocardiopsis*, *Polymorphospora*, *Pseudonocardia*, *Saccharopolyspora* and *Streptosporangium*	Qin et al. (2012a, b, 2013a)
Native herbaceous plants (South Korea)	Surface sterilisation followed by isolation from homogenised solution of plant materials and incubation at 30 °C for 2 weeks	Starch casein agar	*Arthrobacter*, *Dietzia*, *Herbiconiux*, *Kitasatospora*, *Microbacterium*, *Microbispora*, *Micrococcus*, *Micromonospora*, *Mycobacterium*, *Nocardia*, *Rathayibacter*, *Rhodococcus*, *Streptacidiphilus*, *Streptomyces* and *Tsukamurella*	Kim et al. (2012)
Neem tree (*Azadirachta indica*) (India)	Surface sterilisation and incubation at 28 °C for 3–4 weeks	S-agar and water agar	*Microbispora*, *Nocardia*, *Streptomyces*, *Streptosporangium*, *Streptoverticillium* and *Saccharomonospora*	Verma et al. (2009)
Rice (*Oryza sativa*) (China)	Surface sterilisation and incubation at 26 °C for 1 week	S (*Streptomyces*) medium	*Streptomyces* and *Nocardioides*	Tian et al. (2007)
Tomato (*Lycopersicon esculentum*) (Murray Bridge, South Australia)	Surface sterilisation and incubation at 27 °C up to 4 weeks	TWYE agar, HV agar and yeast extract, casamino acid medium	*Microbispora*, *Nonomurae* and *Streptomyces*	Inderiati and Franco (2008)
Tropical native plants (Papua New Guinea, Mborokua and Solomon Islands)	Surface sterilisation and incubation at 23–25 °C up to 8 weeks	Arginine vitamin agar supplemented with soil extract from organic humus	*Actinoplanes*, *Amycolatopsis*, *Dactylosporangium*, *Kibdelosporangium*, *Kitasatospora*, *Lechevalieria*, *Lentzea*, *Microbispora*, *Nonomuraea*, *Planotetraspora*, *Pseudonocardia*, *Sphaerisporangium*, *Streptomyces* and *Streptosporangium*	Janso and Carter (2010)
Wattle tree (*Acacia auriculiformis*) (Thailand)	Surface sterilisation followed by isolation from solution of crushed plant materials and incubation at 28 °C up to 4 weeks	Starch Casein agar containing 100 g/ml ampicillin, 2.5 U/ml penicillin G, 50 g/ml amphotericin B and 50 g/ml cyclohexamide	*Actinoallomurus*, *Amycolatopsis*, *Kribbella*, *Microbispora* and *Streptomyces*	Bunyoo et al. (2009)
Wheat (*Triticum aestivum*) (South Australia)	Sonication followed by surface sterilisation and incubation at 27 °C up to 4 weeks	TWYE agar, HV agar, flour-yeast extract-sucrose-casein hydrolysate agar, flour-calcium carbonate agar	*Microbispora*, *Micromonospora*, *Nocardioides* and *Streptomyces*	Coombs and Franco (2003a)

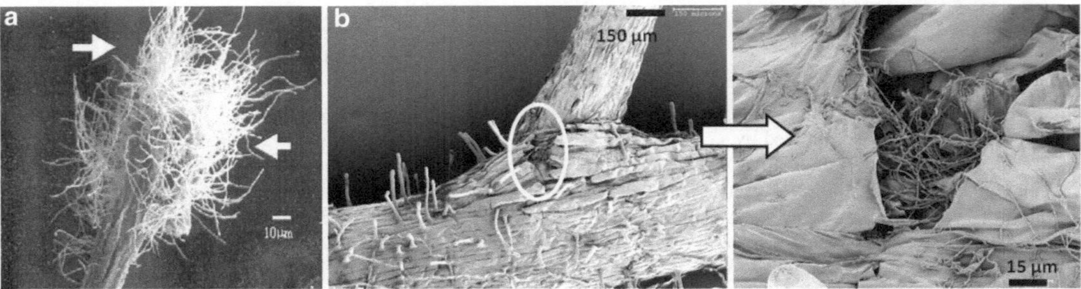

Fig. 2.1 Identification of endophytic actinobacterial colonisation in surface-sterilised wheat plants. (**a**) SEM image of *Streptomyces* aerial hyphal growth on a surface-sterilised root fragment from an isolation agar plate (Coombs and Franco 2003a). (**b**) SEM image showing the endophytic colonisation in a lateral root junction of a wheat plant by *Streptomyces* sp. EN27 (Courtesy V Conn and C Franco)

(*Zea mays* L.) (de Araujo et al. 2000), although *Streptomyces* and *Streptosporangium* spp. were also represented and some of them showed antimicrobial activity against one or more tested bacteria and yeast.

Coombs and Franco (2003a) reported the isolation of filamentous actinobacteria from surface-sterilised root tissues of healthy wheat plants (*Triticum aestivum* L.) (Fig. 2.1). Of the 49 endophytic isolates that belonged to *Streptomyces*, *Microbispora*, *Micromonospora* and *Nocardioides* were strains found to be similar to *S. caviscabies* and *S. setonii* that had been isolated originally from potato scabs. Therefore, detection of pathogenicity was required as the endophytic isolates were potential biocontrol agents. The isolates were found to be nonpathogenic, as they neither had *nec1*, a pathogenicity-associated gene, nor produced the toxin thaxtomin. In other studies, they visually demonstrated the colonisation of germinating wheat seed embryo, endosperm and emerging radicle with one of these endophytic actinobacteria, *Streptomyces* sp. strain EN27, tagged with the *egfp* gene. These observations show that the endophytic actinobacterium was able to associate with its host at a very early stage in the development of the plant (Coombs and Franco 2003b). Similarly, in pea plants, Tokala et al. (2002) showed a remarkable degree of preferential colonisation of pea nodules relative to roots by *Streptomyces lydicus* strain WYEC108 that was isolated from a rhizosphere soil. This observation and other studies indicated that actinobacteria isolated from soil could be capable of endophytic colonisation.

Tian et al. (2007) identified actinobacterial strains from the surface-sterilised stems and roots of rice and described differences in endophytic populations from these plant parts. Strains similar to *Streptomyces cyaneus*, *S. aurantiacus* and *S. paresii* were also isolated from roots and stems, whereas *Nocardioides thermolilacinus*, *S. exfoliates*, *S. glauciniger* and *S. kathirae* were only isolated from roots and *S. caviscabies* and *S. scabies* were isolated from stems only, indicating that more diverse actinobacteria were isolated from roots than stems. Their results also suggest the presence of more diverse communities of uncultured actinobacteria within stems and roots of rice. Velazquez et al. (2008) selected the apoplastic sap of the medullary parenchyma of the stem of healthy sugarcane plants to identify endophytic isolates belonging to the genera *Microbacterium, Micrococcus* and *Kokuria*. Root nodules of the grain legume *Lupinus angustifolius* yielded 136 different orange-pigmented actinobacterial colonies from surface-sterilised nodules which belonged to the genus *Micromonospora,* and a detailed taxonomic study on six of these isolates identified two novel species, *Micromonospora lupini* and *M. saelicesensis* (Trujillo et al. 2007). Misk and Franco (2011) found a physiologically diverse group of endophytic actinobacteria from grain legume plants such as lentil, chickpea, pea and faba bean. Some of the biotic activities observed included siderophore and cyanogen production, antifungal activity and phosphate solubilisation. These studies exemplify the value of using different approaches to characterise the culturable diversity of endophytic isolates obtained from a few crop plants. A large number of studies have since been reported from most crop plants confirming their ubiquitous presence. This group of microbes can colonise the internal tissue of crop plants and are capable of producing plant growth-promoting chemicals, enhancing nutrient uptake as well as producing secondary metabolites that can inhibit microbial pathogens and induce systemic resistance. Therefore, their functions have been a major factor for their isolation as they promise to offer an advantage in terms of reliability and efficacy as inoculants due their endophytic nature. A summary of these beneficial functions is shown in Table 2.3.

4.2 Horticultural Crops

Cao et al. (2004) compared the endophytic actinobacteria from roots and leaves of healthy and wilting banana plants. Community analysis of the 242 isolates demonstrated increased actinobacterial diversity in wilting leaves compared to that in healthy leaves, although actinobacterial communities in roots were similar. The same laboratory tested a total of 131 strains, identified as *Streptomyces*, *Streptoverticillium* and *Streptosporangium* spp., that were successfully isolated from surface-sterilised banana roots (Cao et al. 2005). About 18.3 % of these isolates inhibited the growth of pathogenic *Fusarium oxysporum* f. sp. *cubense*, the causal organism of Panama wilt disease of banana, on banana tissue extract medium. About 37.5 % of the most frequently isolated *S. griseorubiginosus* strains were antagonistic to this pathogen, but the antagonism was lost when $FeCl_3$ was introduced into the inhibition zone. These findings indicate the potential of developing siderophore-producing *Streptomyces* endophytes for the biological control of *Fusarium* wilt (Panama) disease of banana (Cao et al. 2005).

Actinobacteria were reported for the first time as endophytes of grapevines, with a number of other isolates identified as *Streptomyces* spp. and also the rare actinobacterium *Curtobacterium* spp. (Bulgari et al. 2009; West et al. 2010).

In a survey of endophytic bacteria colonising roots of processing carrot cultivars (Carochoice, Red Core Chantenay) grown at two locations in Nova Scotia, Surette et al. (2003) reported the association of *Arthrobacter*, *Kokuria* and *Microbacterium* as endophytes. In a similar study on potato-associated bacteria, the *Streptomyces* spp. had the highest antagonistic activity among endophytic actinobacteria against most of the fungal as well as bacterial pathogens (Sessitsch et al. 2004). A total of 619 actinobacteria, all *Streptomyces* spp., were isolated from different cultivars of tomato. The *aureus* group of *Streptomyces* was the most frequent isolate group, but the population composition of *Streptomyces* varied according to tomato cultivars, physiological status and soil types (Tan et al. 2006). *Microbispora* spp. (67 %) were the most common isolates of the 81 endophytic actinobacteria from Chinese cabbage roots (Lee et al. 2008b), followed by *Streptomyces* spp. (12 %) and *Micromonospora* spp. (11 %). The three antagonistic isolates were identified as *Microbispora rosea* subsp. *rosea* (A004 and A011) and *Streptomyces olivochromogenes* (A018), which effectively suppressed the disease club root of cabbage caused by *Plasmodiophora brassicae*. Recently, Khan and Doty (2009) reported a diverse array of endophytic bacteria associated with sweet potato plants (*Ipomoea batatas* L.) which included the actinobacterial genus *Arthrobacter*.

Shimizu et al. (2000) explored endophytic actinobacteria from the flowering plant Rhododendron. Nine, six and two isolates, with distinguishing characteristics based on the macroscopic appearance of colonies, were obtained from roots, stems and leaves, respectively, and shown to have antagonism against two major fungal pathogens of rhododendron, *Phytophthora cinnamomi* and *Pestalotiopsis sydowiana*. Similarly, Nishimura et al. (2002) isolated a total of 73 actinobacteria from leaves, stems and roots of the other *Ericaceae* plant called mountain laurel (*Kalmia latifolia* L.), and most of them were *Streptomyces* spp. with a broad and intense antimicrobial spectrum against various yeasts and

Table 2.3 Functional aspects of endophytic actinobacteria isolated from different plant species and habitats

Plant type	Endophytic actinobacterial genera	Functional role established	References
Arabidopsis	*Micromonospora* sp. strain EN43 and *Streptomyces* sp. strain EN27	Induction of defence through SAR and JA/ET pathways	Conn et al. (2008)
Banana	*Streptomyces*	Siderophore production and antibiosis	Cao et al. (2004, 2005)
Cabbage	*Microbispora* and *Streptomyces*	Antibiosis	Lee et al. (2008a, b)
Cucumber	*Actinoplanes campanulatus, Micromonospora chalcea* and *Streptomyces spiralis*	Antibiosis and glucanolytic activity	El-Tarabily et al. (2009)
Eaglewood tree	*Actinomadura, Nocardia, Nonomuraea, Pseudonocardia* and *Streptomyces*	Ammonia, indole acetic acid (IAA) and siderophore production	Nimnoi et al. (2010a)
Epiphytic vine	*Streptomyces*	Antibiosis	Ezra et al. (2004)
Foliose lichens	*Nocardia, Nocardiopsis* and *Streptomyces*	Antibiosis	da Silva et al. (2011)
Herbaceous and woody plants	*Microbispora, Micromonospora, Nocardia* and *Streptomyces*	Antibiosis	Taechowisan et al. (2003)
Lentil, chickpea, pea, faba bean and rye (South Australia)	*Microbispora* and *Streptomyces*	Siderophore and cyanogen production; phosphate solubilisation and antibiosis	Misk and Franco (2011)
Lichens	*Amycolatopsis, Actinomadura, Micromonospora, Streptomyces* and *Streptosporangium*	Antibiotic biosynthetic genes detected and antibiosis	González et al. (2005)
Madagascar periwinkle	*Streptomyces*	Antibiosis	Kafur and Khan (2011)
Mangrove plants in China	*Micromonospora* and *Streptomyces*	Antibiosis and inhibition of anticancer protein synthesis	Hong et al. (2009)
Marine sponges and soft corals	*Streptomyces*	Antibiosis	EI-Bondkly et al. (2012)
Medicinal plants	*Amycolatopsis, Micromonospora, Nocardia, Nonomuraea* and *Streptomyces*	Antitumour activity and antibiosis	Huang et al. (2012)
Medicinal plants in Panxi plateau, China	560 isolates belonging to different genera	Antibiotic biosynthetic genes detected and antibiosis	Zhao et al. (2010b)
Medicinal plants in Xishuangbanna, China	2174 isolates belonging to different genera	Antibiosis	Qin et al. (2009, 2012a, b, 2013a)
Native herbaceous plants in Korea	21 straining belong to different genera	Antibiosis, IAA and hydrolytic enzyme production; phosphatase activity	Kim et al. (2012)
Neem tree	*Nocardia, Streptomyces* and *Streptosporangium*	Antibiosis	Verma et al. (2009)
Rhododendron	*Streptomyces*	Antibiosis	Shimizu et al. (2000)
Snakevine	*Streptomyces*	Antibiosis	Castillo et al. (2006)
Tomato	*Microbispora, Nonomuraea* and *Streptomyces*	Siderophore production and antibiosis	Tan et al. (2006), Inderiati and Franco (2008)
Tropical native plants in Papua New Guinea, Mborokua and Solomon Islands	*Micromonospora, Nonomuraea, Pseudonocardia, Sphaerisporangium, Streptomyces, Streptosporangium* and *Thermomonospora*	Detection of bioactive extracts and biosynthetic genes for PKS-I, PKS-II and NRPS	Janso and Carter (2010)
Wattle tree	*Amycolatopsis* and *Streptomyces*	Antibiosis	Bunyoo et al. (2009)
Wheat	*Microbispora, Nocardioides* and *Streptomyces*	Antibiosis and plant growth promotion	Coombs and Franco (2003a), Coombs et al. (2004)

fungal pathogens of *Ericaceae*. In recent years, members of the genus *Micromonospora* have also been recovered from diverse plant tissues, especially nitrogen-fixing root nodules (Valdes et al. 2005; Trujillo et al. 2010). A new species *Streptosporangium oxazolinicum* sp. nov. in the genus *Streptosporangium* was isolated from the roots of a variety of orchids collected in the subtropical Okinawa prefecture by Inahashi et al. (2011) which was shown to produce a new group of antitrypanosomal antibiotics, spoxazomicins.

4.3 Medicinal Plants

It is believed that the greatest diversity of bacterial endophytes is likely to occur in the plant species of tropical and temperate regions (Strobel and Daisy 2003). From 36 medicinal plant species in Thailand, Taechowisan et al. (2003) isolated 330 strains belonging to four genera of endophytic actinobacteria, namely, *Streptomyces*, *Microbispora*, *Nocardia* and *Micromonospora*. Medicinal plants in Xishuangbanna tropical rainforest of China were subjected to diverse pretreatment methods and selective media, resulting in an unexpected variety of 10 different suborders and 32 genera, including at least 19 new taxa (Qin et al. 2009, 2010b). Huang et al. (2012) carried out the isolation of endophytic actinobacteria from the surface-sterilised tissues of 12 medicinal plants in Hainan, China, using different media. Of the 280 isolates recovered, 154 were from roots, 73 from stems and 53 from leaves, and they were identified as *Streptomyces, Micromonospora, Nocardia, Nonomuraea* and *Amycolatopsis* spp.

A total of 38 endophytic actinobacteria were isolated from surface-sterilised leaves of *Catharanthus roseus* (L.) (Kafur and Khan 2011). Similarly, from the medicinal plant *Artemisia annua*, a total of 228 isolates representing at least 19 different genera of actinobacteria were obtained and several of them were novel taxa (Li et al. 2012). From the plant *Maytenus austroyunnanensis* alone, a total of 312 endophytic actinobacteria were obtained and they were affiliated with the order *Actinomycetales* (distributed into 21 genera). Notably, a new genus *Polymorphospora* and seven new species were also isolated (Qin et al. 2012a).

Similarly, Kim et al. (2012) reported on the diversity of endophytic actinobacteria and their physiological properties in various Korean native plant species. Using a culture-based approach, the members of the genus *Rhodococcus* and the family *Streptomycetaceae* were found to be the main constituents of the endophytic actinobacterial community. In addition, *Arthrobacter, Dietzia, Herbiconiux, Mycobacterium, Nocardia, Rathayibacter, Tsukamurella, Streptacidiphilus* and *Kitasatospora* were reported for the first time as endophytes.

Higashide et al. (1977) isolated an actinomycete *Actinosynnema pretiosum* that produces maytasinoid compounds. These compounds are usually found in the Chinese medicinal tree *M. austroyunnanensis,* but no endophytic actinobacteria producing this compound were isolated from this plant (Qin et al. 2012a). Similar is the case with *Artemisia annua*, where many endophytic actinobacteria were reported, but none of them produced the compound artemisinin, an antimalarial drug.

4.4 Perennial Trees

Recent studies suggest that many of the perennial trees are an untapped source of endophytic actinobacteria of the non-*Frankia* type. Eleven strains of endophytic actinobacteria were isolated from the healthy roots of wattle trees *Acacia auriculiformis*, collected from Bangkok and Nakhonpathom, Thailand. Analysis of 16S rRNA sequences of those strains revealed that they belong to the genera *Streptomyces*, *Actinoallomurus*, *Amycolatopsis*, *Kribbella* and *Microbispora* (Bunyoo et al. 2009). Similarly, Verma et al. (2009) reported the isolation of endophytic actinobacteria from a neem tree *Azadirachta indica.* A total of 55 separate isolates were obtained from 20 plants, and 60 % of these showed inhibitory activity against one or more pathogenic fungi and bacteria. Actinobacteria were most commonly recovered from roots (54.5 % of all isolates), followed by stems (23.6 %) and leaves (21.8 %). The dominant genus was *Streptomyces* (49.09 % of all isolates), while *Streptosporangium* (14.5 %),

Microbispora (10.9 %), *Streptoverticillium* (5.5 %), *Saccharomonospora* (5.5 %) and *Nocardia* (3.6 %) were also recovered.

Zin et al. (2010) carried out the isolation of endophytic actinobacteria from the root and stem samples of ethanobotanical trees, namely, *Cinnamomum zeylanicum*, *Zingiber spectabile*, *Elettariopsis curtisii* and *Labisia pumila*, in the northern part of the Malay Peninsula. Sixty six *Streptomyces* spp., and one unidentified isolate were successfully isolated. Of the total isolates obtained, 61.2 % were isolated from root and 38.8 % from the stem. Of these 56.7 % of the endophytic actinobacteria were isolated from the outermost parts of the surface-sterilised plants and 43.3 % were from the internal part of the plants.

Chen et al. (2011) revealed species diversity of endophytic actinobacteria from cinnamon trees *Elaeagnus angustifolia*, mainly distributed in northwest of China and western inner parts of Mongolia. Eight strains of endophytic actinobacteria were successfully isolated from root nodules of *Elaeagnus angustifolia* by the method of nodule slicing, and the result showed that five of these strains belonged to *Micromonospora* and the other three strains were *Nonomuraea*, *Pseudonocardia* and *Planotetraspora*, respectively.

Recently, Kaewkla and Franco (2013a) reported the presence of a wide range of actinobacterial genera as endophytes by incubating plates for up to 16 weeks, but removing emerging colonies as soon as they were 1 mm in diameter. The majority of 576 actinobacterial isolates from leaf, stem and root samples of four Australian endemic trees—*Callitris preissii* (native pine tree), *Eucalyptus camaldulensis* (red gum), *Eucalyptus microcarpa* (Grey Box) and *Pittosporum phylliraeoides* (native apricot tree)—were *Streptomyces* spp., and the others belonged to 16 other actinobacterial genera, namely, *Actinomadura*, *Actinomycetospora*, *Actinopolymorpha*, *Amycolatopsis*, *Gordonia*, *Kribbella*, *Micromonospora*, *Nocardia*, *Nocardioides*, *Nocardiopsis*, *Nonomuraea*, *Polymorphospora*, *Promicromonospora*, *Pseudonocardia*, *Williamsia* and a novel genus *Flindersiella*. One of the strains represented a novel genus in the family Nocardioides and the other 11 strains were accepted as novel species. The literature from the limited number of studies with a limited number of trees has indicated the need for more research and the strong prospect for the culturing of diverse endophytic actinobacteria, including novel and rare genera residing in perennial trees.

The majority of agricultural crops, or other small medicinal, herbaceous weeds, are mostly seasonal, annual or biennial plants. In comparison, trees are perennial and growing for many years and exposed to varying soil conditions (with depth) and changing environmental conditions over many growth cycles. Both belowground and above-ground parts of perennial trees are exposed to continuous changes which occur with respect to climatic and environmental conditions. These spatio-temporal interactions may lead to the enrichment of many rare bacterial groups or more fastidious actinobacteria in their interior as endophytes.

5 Diversity of Endophytic Actinobacteria in Mangrove Ecosystems, Lichens and Mosses

5.1 Mangrove Ecosystems

Mangroves are the coastal wetland forests mainly found in the intertidal zone of estuaries, backwaters, deltas, creeks, lagoons, marshes and also mudflats of the tropical and subtropical latitudes. It is estimated that mangrove forests cover a total area of over one fourth of the world's coastline (Spalding et al. 1997; Alongi 2002). Mangroves are highly productive ecosystems, and little is known about the microbial communities living therein. Mangrove sediments contain populations of *Streptomyces*, *Micromonospora* (Eccleston et al. 2008) and other novel actinobacteria, as illustrated by the isolation of *Asanoa iriomotensis* (Han et al. 2007), *Nonomuraea maheshkhaliensis* (Ara et al. 2007) and *Micromonospora rifamycinica* (Huang et al. 2008). Hong et al. (2009) isolated over 2,000 bioactive actinobacteria from both rhizosphere soil and plant materials (including endophytes) of 23 plant species collected from 8 mangrove sites in China. The highest number of bioactive strains was observed from the plant tissues of *Bruguiera*. Taxonomic diversity of these bioactive actinobacteria assigned most of them to the genera *Micromonospora* and *Streptomyces* and less to the other genera *Actinomadura, Nocardia, Nonomuraea, Rhodococcus* and *Verrucosispora*.

A study of 19 different mangrove plant species in Bhitarkanika, Orissa, India, revealed that three species of *Streptomyces*, namely, *S. halstedii, S. longisproflavus* and *S. albidoflavus*, were found to be associated with *Kandelia candel*. Similarly, *S. atroolivaceous* was found in phyllosphere of *Sonneratia apetala* and *S. caseolaris* of Dangmal and Khola region respectively. Two species *S. exfoliates* and *S. aurantiacus* were found to be associated with almost all mangrove plants studied (Gupta et al. 2009). An endophytic actinobacterial strain *Nocardiopsis* sp. A00203 isolated from the leaves of mangrove plant *Aegiceras corniculatum* collected from Jimei, Fujian Province, China, was shown to produce three biologically active 2-pyranone compounds (Lin et al. 2010). In another study, Mangamuri et al. (2012) isolated a rare actinobacterium closely related to *Pseudonocardia endophytica* from a mangrove ecosystem of Nizampatnam, India, which produced bioactive metabolites with broad-spectrum inhibitory effects on Gram-positive, Gram-negative bacteria and fungi. Baskaran et al. (2012) reported a higher proportion of actinobacterial endophytes in the mangrove plant *B. gymnorrhiza* of the Andaman Islands. However, the ecto- and endorhizosphere of plants in the mangrove ecosystems are still largely an unexplored source for screening and isolation of novel endophytic actinobacteria with rich potential to produce active secondary metabolites.

5.2 Lichens and Mosses

As pioneers of the colonisation of terrestrial habitats, lichens are found from the Arctic to tropical regions and are present on stones, in arid soils or as epiphytes on plants (Ahmadjian 1993). About 10 % of lichen-forming fungi are associated with nitrogen-fixing cyanobacteria (e.g. *Peltigerales* and *Lichinomycetes*); however, the remaining 90 % of lichen-forming fungi are not known for their intimate association with many other bacteria (Richardson and Cameron 2004; Liba et al. 2006). Studies have described the isolation of different species of the actinobacteria of the genera *Micromonospora* and *Streptomyces* from this environment (Hirsch et al. 2004). González et al. (2005) reported on the diversity in actinobacterial population from three regions: Within tropical lichens studied, *Micromonospora* strains were isolated with similar frequencies from different types of lichens, whereas arboricolous lichens from Hawaii were richer in *Streptomyces* than saxicolous samples. In addition, members tentatively assigned to the order Pseudonocardiaceae and the genera *Actinoplanes* and *Actinomadura* were isolated. Other genera isolated from lichens collected in Alaska belonged to *Rhodococcus* spp., from Hawaii belonged to *Saccharopolyspora* spp. and *Geodermatophilus* spp. and from Reunion Island belonged to *Planobispora* spp. and *Streptosporangium* sp. Two lichen-derived actinobacteria identified as new species of *Streptomyces* produced novel angucycline and butenolide compounds having cytotoxic activities against cancer cells and antibacterial activity. Two novel actinobacterial strains *Actinomycetospora iriomotensis* and *Actinomycetospora rishiriensis* were isolated from a lichen sample from Iriomote Island and Rishiri Island of Japan, respectively (Yamamura et al. 2011a, b). Recently, da Silva et al. (2011) isolated 71 isolates of actinobacteria associated with the foliose lichens from an Amazonian ecosystem in Brazil. The morphological characteristics and characterisation of cell wall amino acid of actinobacteria isolated from foliose lichens indicated that from the total of 71 actinobacteria, 91.5 % were *Streptomyces*, 4 % *Nocardia* and 1.5 % *Nocardiopsis* (1.5 %). Janso and Carter (2010) isolated 123 endophytic actinobacteria from tropical native plants including ferns and club mosses collected from several locations in Papua New Guinea and Mborokua Island,

Solomon Islands. 16S rRNA gene sequence analysis revealed that 17 different genera were represented and rare genera such as *Sphaerisporangium* and *Planotetraspora*, which have never been previously reported to be endophytic, were prevalent.

There are approximately 12,000 species of moss distinguished by their multicellular rhizoids (Theissen et al. 2001). Mosses are abundant on the forest floor in a broad range of boreal forest types (Bach et al. 2009). A high diversity and complexity in phyllosphere bacterial communities was recently described for the sphagnum moss (Opelt et al. 2007). Park et al. (2013) studied the endophytic bacterial diversity of an Antarctic moss *Sanionia uncinata* through pyrosequencing of amplified 16S rRNA genes and showed that *Proteobacteria* was the most dominant phylum with 65.6 %, followed by *Bacteroidetes* (29.1 %) and *Actinobacteria* (11.7 %).

6 Diversity of Endophytic Actinobacteria in Aquatic Ecosystem

Aquatic ecosystems contribute to a large proportion of the planet's biotic productivity, and aquatic plants are largely an unexplored environment for endophytic actinobacterial diversity and their biotic potential.

Freshwater ecosystems cover 0.80 % of the Earth's surface and inhabit 0.009 % of its total water. They generate nearly 3 % of its net primary production (Alexander and Fairbridge 1999). Three basic types of freshwater ecosystems are lentic (include pools, ponds and lakes), lotic (streams and rivers) and wetlands. In the littoral zone of lakes, where rooted plants occur, ponds are typically small lakes of shallow water with abundant marsh and aquatic plants. Food webs are based both on free-floating algae and upon aquatic plants (Sculthorpe 1985; Chapman and Reiss 1998). However, the diversity of the microbial community, in particular endophytes, associated with planktons and aquatic plants in the freshwater ecosystems is poorly understood.

Wetlands are the most productive natural freshwater ecosystems in the world because of the proximity/availability of water and fertile (nutrient rich) soil. Hence, they support large numbers of plant and animal species. Wetlands are dominated by vascular plants that have adapted to saturated soil (Keddy 2010). Among the wetlands, the rice ecosystem microbial communities have been extensively studied due to its importance both for food production and also for its anaerobic methanogenesis causing global climate change (Bernstein et al. 2007).

Marine ecosystems cover approximately 71 % of the Earth's surface and contain approximately 97 % of the planet's water and an exceptional biological diversity, accounting for more than 95 % of the whole biosphere (Qasim 1999). Recent studies have identified a diverse community of actinobacteria associated with marine sponges and soft corals (Lee et al. 1998; Dharmaraj et al. 2010; Webster et al. 2001; EI-Bondkly et al. 2012; Nithyanand et al. 2011). However, as they are not considered to be plants, they are not included in this chapter.

Most of the research on seagrass root-associated microbiology includes communities present on the outside and inside of the root material; hence, the findings are not specific for endophytes only. Similar to results from terrestrial plants, actinobacteria were found to be one of the most abundant groups of bacteria in the roots of seagrass, such as *Zostera marina* (Jensen et al. 2007). Lee et al. (2008b) isolated *Phycicolagilvus* from living seaweed collected along the coast of Jeju, Republic of Korea, which represented a novel species of a new genus within the family *Microbacteriaceae*. From the seaweeds of the Gulf of Mannar, Saravanakumar et al. (2010) isolated 12 strains of actinobacteria, of which 9 represented the genus *Streptomyces* and 3 belonged to the genus *Micromonospora*, which showed strong antagonism against bacterial fish pathogens *Vibrio harveyi, V. fisheri, Aeromonas hydrophila* and *A. sobria*. Recently, Wu et al. (2012) reported that most of the 110 actinobacterial isolates from the seagrass, *Thalassia hemprichii*, harboured polyketide synthetase (PKS) and nonribosomal peptide synthetase (NRPS) gene sequences indicating their bioactive potential. Most of them belonged to ten genera of actinobacteria including *Streptomyces*, *Micromonospora*, *Saccharomonospora*, *Mycobacterium*, *Actinomycetospora*, *Nonomuraea*, *Verrucosispora*, *Nocardiopsis*, *Microbacterium and Glycomyces*.

As indicated before, the chemicals (e.g. NaCl and hypochlorite) used and the timing of treatment may vary depending upon the plant and organ type (Kaewkla and Franco 2013a), and a proper standardisation of sterilisation and isolation procedures appropriate for aquatic plants is essential for discovering the true diversity of their endophytes.

7 Methods for Diversity Analysis of Culturable Endophytic Actinobacteria

In the last 4 years alone, more than 50 new taxa have been identified from various terrestrial plants (Table 2.4). The identification of a pure actinobacterial culture is achieved with a polyphasic approach using techniques described in Fig. 2.2. However, not all of these techniques offer the discrimination required for the rapid characterisation of a large number of freshly isolated strains. In order to achieve this in an economical way, a combination of morphological, chemo-taxonomical and molecular fingerprinting methods are available for the characterisation and diversity analyses of actinobacteria (Embley and Stackebrandt 1994; Rademaker et al. 2000; Cook and Meyers 2003; Brusetti et al. 2008; Yuan et al. 2008).

Some of these methods can be employed to reduce the number of strains sent for sequencing and still be able to identify all the isolates. Culture morphology can be used to distinguish a number of genera such as *Micromonospora*, *Microbispora*, *Rhodococcus, Streptosporangium* and *Strepto-myces* spp., as well as a basis to form groupings of strains with similar morphological features. Representatives of each groups are subjected to molecular fingerprinting techniques such as RAPD (Mehling et al. 1995), AFLP, BOX or REP-PCR (Savelkoul et al. 1999; Rademaker et al. 2000) or the analysis of restriction patterns of PCR products of rRNA genes or ARDRA (Vaneechoutte et al. 1993) to identify strains that are similar to each other. Tian et al. (2007) used RFLP technique to characterise actinobacterial-specific 16S rRNA gene clone libraries constructed from the roots and stems of rice. RFLP analysis based on single digestion with restriction enzymes *SmaI* and *PstI* grouped clones with similar patterns together. Clones from each RFLP group were chosen for further identification by 16S rRNA gene sequencing. Amplified rDNA (Ribosomal DNA) Restriction Analysis (ARDRA) was originally developed by Vanee-choutte et al. (1993) to characterise *Mycobac-terium* species.

ARDRA has been used successfully in identifying several species of endophytic actinobacteria belonging to the genera *Actinomadura*, *Gordonia*, *Nocardia*, *Rhodococcus*, *Saccharomonospora*, *Sa ccharopolyspora*, *Streptomyces* and *Tsukamurella* (Steingrube et al. 1997; Wilson et al. 1998; Laurent et al. 1999; Harvey et al. 2001). Cook and Meyers (2003) identified four restriction endonucleases, *Sau*3AI, *Asn*I, *Kpn*I and *Sph*I, that significantly differentiated the genus *Streptomyces* from all other actinobacteria genera by using ARDRA. ARDRA can be useful in reducing ambiguity in isolate similarities based on morphological characterisations. Kaewkla and Franco (2013a) used ARDRA of partial 16S rRNA genes to distinguish both non-streptomycete- and streptomycete-like isolates obtained from Australian native trees. In this study, initial ARDRA with *HhaI*digestion yielded 13 ARDRA patterns for the total 579 isolates. However, second ARDRA patterns based on a second digestion with the enzymes *RsaI* and *PstI* more effectively differentiated the genera within the ARDRA patterns based on single enzyme digestion, indicating the necessity to use more than one restriction enzyme and judicious selection of isolates for identification by 16S rRNA gene sequencing.

Nimnoi et al. (2010a) employed random amplification of polymorphic DNA (RAPD) to determine the genetic relatedness up to the genus level for the endophytic actinobacterial isolates obtained from healthy shoots and roots of *Aquilaria crassna.* Though RAPD is a simple,

Table 2.4 New genera and species isolated as endophytic actinobacteria (from 2010 to till date)

Endophytic actinobacterial species	Name of the host plant	Plant types	Plant part	References
Actinoallomurus acacia	*Acacia auriculiformis*	Wattle tree	Leaves	Thamchaipenet et al. (2010)
Actinoallomurus oryzae	*Oryza sativa*	Rice	Roots	Indananda et al. (2011)
Actinomycetospora iriomotensis	–	Lichens	–	Yamamura et al. (2011a)
Actinomycetospora rishiriensis	–	Lichens	–	Yamamura et al. (2011b)
Actinophytocola oryzae	*Oryza sativa*	Rice	Roots	Indananda et al. (2010)
Actinoplanes rishiriensis	–	Lichens	–	Yamamura et al. (2012)
Actinopolymorpha pittospori	*Pittosporum phylliraedoies*	Australian apricot tree	Leaves	Kaewkla and Franco (2011b)
Allonocardiopsis opalescens	*Lonicera maackii*	Medicinal plant	Fruit	Du et al. (2013a)
Amycolatopsis endophytica	*Jatropha curcas*	Oil-seed	Seeds	Miao et al. (2011)
Amycolatopsis jiangsuensis	*Dendranthema indicum*	Coastal salt marsh plant	–	Xing et al. (2013)
Amycolatopsis samaneae	*Samanea saman*	Medicinal plant	Roots	Duangmal et al. (2011)
Brachybacterium saurashtrense	*Salicornia brachiata*	Extreme halophyte	Roots	Gontia et al. (2011)
Dietzia maris	*Viola mandshurica*	Manchurian violet	Roots	Kim et al. (2012)
Flindersiella endophytica	*Eucalyptus microcarpa*	Grey Box eucalyptus tree	Roots	Kaewkla and Franco (2011a)
Herbiconiux ginsengi	*Artemisia princeps var. orientalis*	Mugwort	Roots	Kim et al. (2012)
Jatrophihabitans endophyticus	*Jatropha curcas*	Oil-seed	Stem	Madhaiyan et al. (2013)
Jishengella endophytica	*Acanthus illicifolius*	Holy mangrove	Roots	Xie et al. (2010)
Kibdelosporangium phytohabitans	*Jatropha curcas*	Oil-seed	Roots	Xing et al. (2012a)
Kineococcus endophytica	*Limonium sinense*	Coastal halophyte	Roots	Bian et al. (2012b)
Kitasatospora viridis	*Lamium purpureum*	Purple henbit	Roots	Kim et al. (2012)
Kribbella endophytica	*Pittosporum phylliraedoies*	Australian apricot tree	Leaves	Kaewkla and Franco (2013b)
Micromonospora pisi	*Pisum sativum*	Pea	Root nodules	Garcia et al. (2010)
Micromonospora tulbaghiae	*Tulbaghia violacea*	Wild garlic	Leaves	Kirby and Meyers (2010)
Modestobacter roseus	*Salicornia europea*	Coastal halophyte	Roots	Qin et al. (2013a)
Nocardia callitridis	*Callitris preissii*	Pine tree	Roots	Kaewkla and Franco (2010c)
Nocardia endophytica	*Jatropha curcas*	Oil-seed	Roots	Xing et al. (2011)
Nocardioides caricicola	*Carex scabrifolia*	Halophyte	Roots	Song et al. (2011)
Nocardioides panzhihuaensis	*Jatropha curcas*	Oil-seed	Stem	Qin et al. (2012a)
Nocardioides perillae	*Perilla frutescens*	Medicinal plant	Roots	Du et al. (2013b)
Nonomuraea endophytica	*Artemisia annua*	Medicinal plant	Roots	Li et al. (2011b)
Phytohabitans flavus	–	Orchids	Roots	Inahashi et al. (2012)
Phytohabitan shouttuyneae	*Houttuynia cordata*	Orchids	Roots	Inahashi et al. (2012)
Phytohabitans rumicis	*Rumex acetosa*	Orchids	Roots	Inahashi et al. (2012)
Phytohabitans suffuscus	–	Orchids	Roots	Inahashi et al. (2010)
Phytomonospora endophytica	*Artemisia annua*	Medicinal plant	Roots	Li et al. (2011a)

(continued)

Table 2.4 (continued)

Endophytic actinobacterial species	Name of the host plant	Plant types	Plant part	References
Plantactinospora endophytica	*Camptotheca acuminata*	Happy tree	Leaves	Zhu et al. (2012)
Promicromonospora endophytica	*Eucalyptus microcarpa*	Grey Box eucalyptus tree	Roots	Kaewkla and Franco (2012)
Promicromonospora xylanilytica	*Maytenus austroyunnanensis*	Medicinal plant	Leaves	Qin et al. (2012b)
Pseudonocardia adelaidensis	*Eucalyptus microcarpa*	Grey Box eucalyptus tree	Stem	Kaewkla and Franco (2010a)
Pseudonocardia artemisiae	*Artemisia annua*	Medicinal plant	Roots	Zhao et al. (2011a)
Pseudonocardia bannensis	*Artemisia annua*	Medicinal plant	Roots	Zhao et al. (2011b)
Pseudonocardia eucalypti	*Eucalyptus camaldulensis*	Red gum tree	Roots	Kaewkla and Franco (2010b)
Pseudonocardia kunmingensis	*Artemisia annua*	Medicinal plant	Roots	Zhao et al. (2011d)
Pseudonocardia nantongensis	*Tamarix chinensis*	Coastal halophyte	Leaves	Xing et al. (2012b)
Pseudonocardia serianimatus	*Artemisia annua*	Medicinal plant	leaves	Zhao et al. (2011c)
Pseudonocardia sichuanensis	*Jatropha curcas*	Oil-seed	Roots	Qin et al. (2011)
Pseudonocardia tropica	*Maytenus austroyunnanensis*	Medicinal plant	Stem	Qin et al. (2010b)
Pseudonocardia xishanensis	*Artemisia annua*	Medicinal plant	Roots	Zhao et al. (2012a)
Rathayibacter festucae	*Conyza canadensis*	Horseweed	Roots	Kim et al. (2012)
Rhodococcus artemisiae	*Artemisia annua*	Medicinal plant	Roots	Zhao et al. (2012b)
Saccharopolyspora dendranthemae	*Dendranthema indicum*	Coastal salt marsh plant	–	Zhang et al. (2013)
Saccharopolyspora gloriosae	*Gloriosa superba*	Medicinal plant	Stem	Qin et al. (2010a)
Saccharothrix yanglingensis	*Cucumis sativus*	Cucumber	Roots	Yan et al. (2012)
Streptacidiphilus anmyonensis	*Chelidonium majus* var. *asiaticum*	Greater celandine	Roots	Kim et al. (2012)
Streptomyces artemisiae	*Artemisia annua*	Medicinal plant	Roots	Zhao et al. (2010a)
Streptomyces endophyticus	*Artemisia annua*	Medicinal plant	Roots	Li et al. (2013)
Streptomyces halophytocola	*Tamarix chinensis*	Coastal halophyte	Stem	Qin et al. (2013b)
Streptomyces phytohabitans	*Curcuma phaeocaulis*	Medicinal plant	Roots	Bian et al. (2012a)
Streptosporangium oxazolinicum	–	Orchids	Roots	Inahashi et al. (2011)
Tsukamurella suncheonensis	*Iris rossii* var. *rossii*	Caudate-bracted iris	Roots	Kim et al. (2012)

inexpensive and useful typing method for genetic studies of bacteria, it has low resolving power, limited applicability in species-specific comparisons and variable experimental reproducibility.

BOX-PCR is a version of the rep-PCR techniques that uses the BOX-A1R primer targeting the BOX dispersed-repeat motif, common in a number of actinobacterial groups (Van Belkum et al. 1998). The BOX-PCR genomic fingerprints generated from culturable isolates of endophytic actinobacteria permit identification, classification and differentiation to the species, subspecies and strain level. Yuan et al. (2008) characterised the endophytic actinobacteria isolated from medicinal plants through BOX-PCR fingerprinting and revealed more genetic diversity among the closely related strains belonging to the two genera, *Streptomyces* and *Micromonospora*. Endophytic actinobacterial isolates obtained from *Lupinus angustifolia*

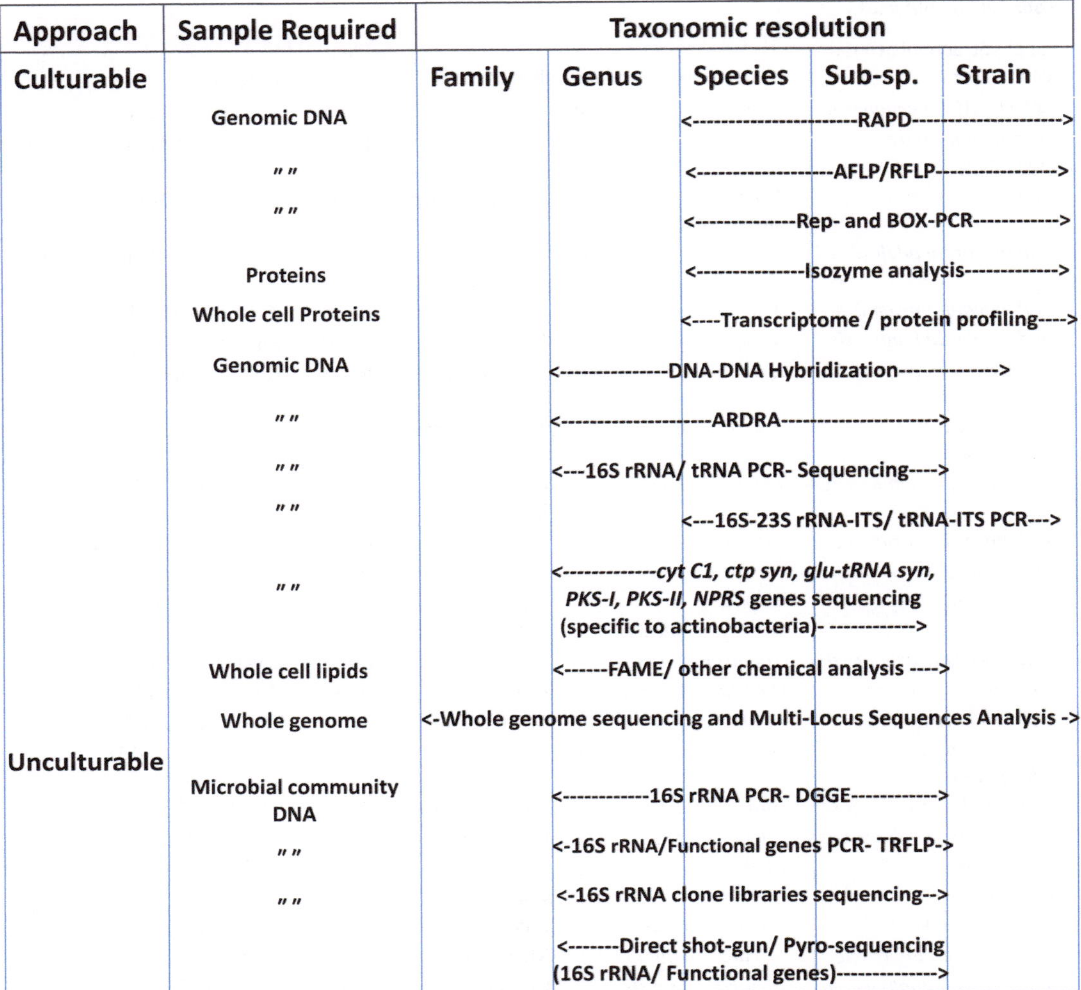

Approach	Sample Required	Family	Genus	Species	Sub-sp.	Strain
Culturable						
	Genomic DNA			<---RAPD--->		
	" "			<---AFLP/RFLP--->		
	" "			<---Rep- and BOX-PCR--->		
	Proteins			<---Isozyme analysis--->		
	Whole cell Proteins			<----Transcriptome / protein profiling---->		
	Genomic DNA		<---DNA-DNA Hybridization--->			
	" "		<---ARDRA--->			
	" "		<---16S rRNA/ tRNA PCR- Sequencing---->			
	" "			<---16S-23S rRNA-ITS/ tRNA-ITS PCR--->		
	" "		<---*cyt C1, ctp syn, glu-tRNA syn, PKS-I, PKS-II, NPRS* genes sequencing (specific to actinobacteria)--->			
	Whole cell lipids		<------FAME/ other chemical analysis ---->			
	Whole genome	<-Whole genome sequencing and Multi-Locus Sequences Analysis ->				
Unculturable						
	Microbial community DNA		<---16S rRNA PCR- DGGE--->			
	" "		<-16S rRNA/Functional genes PCR- TRFLP->			
	" "		<-16S rRNA clone libraries sequencing-->			
			<-------Direct shot-gun/ Pyro-sequencing (16S rRNA/ Functional genes)--->			

Fig. 2.2 Relative applicability of different molecular biological techniques used in the taxonomic identification and diversity analysis of endophytic actinobacteria (Modified from the Rademaker and De Bruijn 1997)

were analysed using BOX-PCR fingerprinting technique, and results revealed on unexpectedly high genetic diversity among the strains belonging to the genus *Micromonospora* (Trujillo et al. 2010). BOX-PCR patterns are not affected by the culture age of the strain to be analysed and have a similar or even better strain differentiation power than other molecular techniques (Kang and Dunne 2003). BOX-PCR is easier to perform and fingerprinting outputs can be easily analysed by computer-assisted methods. Recently, Brusetti et al. (2008) developed a fluorescent BOX-PCR, in which the amplified fluorescent-labelled products can be separated in an automated DNA sequencer which helps overcome limitations from poor band resolution on agarose gel electrophoresis.

Chemotaxonomical methods are more labour intensive, but the identification of the LL- or *meso*-form of the cell wall compound 2,6-diaminopimelic acid (DAP) can be effective in discriminating between *Streptomyces* and non-*Streptomyces* strains. The amino acid and sugar composition of cell walls provide information suitable for the classification of pure isolates of actinobacteria but are not diagnostic.

The fatty acid composition is another unique chemotaxonomic marker used for the identification and diversity characterisation of major genera of actinobacteria (Vestal and White 1989; Embley and Wait 1994). However, it is labour intensive and better suited to discriminating between species within a genus, although it can also be used to identify specific genera that are present in the Sherlock Microbial ID System (www.midi-inc.com), or when a small number of genera are present (González et al. 2005).

7.1 New Molecular Approaches for Strain Characterisation

In the last two decades, the whole genome sequence of number of bacteria has been decoded, and attempts are underway to test whether the data from whole genome comparison can be used for diversity characterisation and taxonomy of culturable bacteria. For example, pairwise comparison of complete whole genome sequences showed that the 'average nucleotide identity' (ANI) of all conserved genes between any two genomes correlated well with 16S rRNA sequence identity and DNA-DNA similarity values. It has also been shown that 70 % DNA-DNA similarity corresponds to 95 % ANI (Konstantinidis and Tiedje 2005). Moreover, all pairs of genomes showing 95 %, or higher, ANI also showed at least 98.5 % 16S rRNA gene identity (Goris et al. 2007). This approach of comparative genomics information has also been generated from the available whole genome sequences of well-known actinobacterial taxa including some of the endophytic actinobacterial genera like *Frankia, Leifsonia, Streptomyces* and *Nocardia* (Ventura et al. 2007).

Multilocus sequence analysis (MLSA), a phylogenetic characterisation based on sequence comparison of multiple housekeeping genes in bacterial genome, has been proposed as a replacement for DDH technique in the classification of prokaryotes (Gevers et al. 2006). In the recent *Bergey's Manual of Systematic Bacteriology*, the MLSA has been used in redefining phylogeny of actinobacterial genera like *Mycobacterium* and *Bifidobacterium* (Ventura et al. 2007). The concatenation of four gene fragments encompassing the 16S rRNA gene, *hsp65*, *rpoB* and *sod* has been used to create a supertree of the *Mycobacterium* genus, and species such as *Mycobacterium fortuitum* and *M. avium* are well separated by a super tree approach than using a single gene-based tree, i.e. 16S rRNA gene-based tree (Devulder et al. 2005). In the super tree of the genus *Bifidobacterium*, concatenation of seven conserved genes, i.e. *clpC*, *dnaB*, *dna*G, *dnaJ1*, *purF*, *rpoC* and *xfp*, has been used to infer its phylogeny (Ventura et al. 2006). Several recent MLSA studies showed that in addition to 16S rRNA gene, the concatenation of four genes such as *gyrB*, *rpoB*, *recA* and *atpD* genes has found useful in phylogeny of other actinobacterial genera like *Micromonospora* and *Steptomyces* (Rong et al. 2009; Rong and Huang 2010; Carro et al. 2012). More recently, Curtis and Meyers (2012) included the *relA* gene for the first time in MLSA of actinobacteria and generated the concatenated sequence super tree to examine the phylogenetic relationships of 17 type strains within the genus *Kribbella*, one of the known endophytic actinobacterial genus.

8 Culture-Independent Approaches for Diversity Analysis

Studies of diversity and functions of plant-associated microbes, especially prokaryotes, are impeded by difficulties in cultivating most of them, and endophytes inside host tissues are not easily amenable to biochemical or genetic analyses. Recent advances in methods for endophytic bacterial enrichment and direct applications of 16S rRNA gene-based culture-independent molecular techniques are helping to unravel the complex endophytic actinobacterial community (Table 2.5). Some of these methods include polymerase chain reaction (PCR)-based denaturing gradient gel electrophoresis (DGGE), terminal restriction fragment

Table 2.5 Endophytic actinobacteria from different plants identified using culture-independent methods

Plant species/ habitats	Method and source of microbial community DNA	Molecular techniques used	List of endophytic actinobacterial genera identified	References
Eaglewood tree (*Aquilaria crassna*)	Extraction of total DNA of root materials	PCR-DGGE	*Actinomadura*, *Nocardia*, *Nonomuraea*, *Pseudonocardia* and *Streptomyces*	Nimnoi et al. (2010b)
Grape vine (*Vitis vinifera*)	Endophyte enrichment from both leaves and roots and DNA extraction	PCR-DGGE	*Curobacterium* and *Streptomyces*	West et al. (2010)
Grape vine (*Vitis vinifera*)	Endophytes enrichment from whole plant and DNA extraction	16S rRNA gene clone libraries	*Curtobacterium*	Bulgari et al. (2009)
Medicinal tree (*Maytenus austroyunnanensis*)	Endophytes enrichment from root, stem and leaves and DNA extraction	16S rRNA gene clone libraries	*Actinokineospora*, *Marmoricola*, *Modestobacter*, *Pseudokineococcus*, *Pseudosporangium*, *Sanguibacter* and *Serinibacter*	Qin et al. (2012a, b, c)
Potato (*Solanum tuberosum*)	Bead beating of tubers and DNA extraction	PCR-DGGE (actinobacterial specific)	Mainly *Streptomyces*	Sessitsch et al. (2002)
Rice (*Oryza sativa*)	Extraction of total DNA of root and stem materials	16S rRNA gene clone libraries	*Actinoplanes*, *Amycolatopsis*, *Corynebacterium*, *Dactylosporangium*, *Frankia*, *Micromonospora*, *Mycobacterium*, *Nocardioides*, *Rhodococcus*, *Streptomyces* and other uncultured actinobacteria	Tian et al. (2007)
Soybean (*Glycine max*)	Enrichment through homogenisation roots, root nodules, stem and leaves, filtration and DNA extraction	16S rRNA gene clone libraries	Wide range of actinobacteria genera belonging to three suborders, namely, Frankineae, Propionibacterineae and Micrococcineae	Ikeda et al. (2009, 2010)
Wheat (*Triticum aestivum*)	Homogenisation of root samples with mini-bead beater and DNA extraction	PCR-TRFLP	*Arthrobacter*, *Kitasatospora*, *Micromonospora*, *Microbispora*, *Mycobacterium*, *Nocardia*, *Nocardioides*, *Streptomyces* and *Tsukamurella*	Conn and Franco (2004a)

length polymorphism (T-RFLP) analysis, construction and sequencing of 16S rRNA gene clone libraries and next-generation sequencing/ pyrosequencing. A combination of culturable and culture-independent approaches may be needed for in-depth understanding of the diversity and functional relevance of endophytic actinobacteria (Fig. 2.2).

8.1 Methods for Enrichment of Endophytes and Community DNA Isolation from Plants

Endophytic bacteria reside inside the plant tissues mainly in intercellular spaces, rarely in intracellular spaces and interior of vascular tissues (Thomas and Graham 1952). They are tightly attached to host cells and are difficult to extract and separate from plant tissues and prone to contamination from surface-associated bacteria. Mechanical removal of rhizoplane populations by vigorous shaking with glass beads can help overcome the contamination from surface bacteria (Reinhold et al. 1986). Initial studies on the unculturable endophytic diversity were carried out with the extraction of total DNA using general CTAB procedure with certain modifications (Xie et al. 1999; Sessitsch et al. 2002) and subsequent PCR amplification of 16S rRNA genes using prokaryotic universal primers (Dent et al. 2004; Sun et al. 2008). Since DNA obtained using such methods includes material from the plant nuclei, the plastids, the mitochondria and the plant-associated microbes, it is essential to design highly specific primers for endophytic bacteria alone. The high sequence homology between plant chloroplast 16S rRNA gene, mitochondrial 18S rRNA gene and bacterial 16S rRNA can cause interference with specific analysis of endophytic bacteria (Sun et al. 2008). Therefore, enrichment of endophytic bacteria prior to PCR amplification has been suggested to overcome the above-described problems and improve the sensitivity of analysis.

Jiao et al. (2006) enriched bacterial cells from plant tissues by enzymatic hydrolysis of the plant cell wall, followed by differential centrifugation. Subsequently, a variety of mild and specific enzymatic treatments have been successfully used to remove intact bacterial cells from the medicinal plant *Mallotus nudiflorus* (Wang et al. 2008) and grapevine leaf tissues (Bulgari et al. 2009). This method of endophyte enrichment has also helped in the culturing of rare/novel endophytic actinobacteria (Qin et al. 2009; Ikeda et al. 2009). Another technique suitable for enriching bacterial cells from fresh plant tissues was developed by using a bacterial cell extraction buffer containing Triton X-100 for tissue homogenisation with subsequent Nycodenz density gradient centrifugation. Here, the enrichment is based on the speculation that less green colour of the supernatant and interface is an indication of less contamination of plastids in the bacterial fraction obtained from homogenised plant samples (Ikeda et al. 2009). This enrichment technique has been successfully applied to clarify the diversity of endophytic actinobacterial communities in stems and leaves of soybean and rice (Ikeda et al. 2009, 2010). Recently, Nikolic et al. (2011) cut sterilised potato plant material into small pieces and then the endophytic bacteria were dislodged by overnight shaking at room temperature in 0.9 % NaCl. Bacteria were separated from the plant material by filtration and collected by centrifugation. The enrichment procedure allows the extraction of bacterial cells from large amounts of plant material thereby reducing variation associated with specific plant parts and collects rare members of the endophytic community. As a result next-generation sequencing operations which require large amounts of high-quality DNA can be conducted, e.g. for metagenomic analysis (Sessitsch et al. 2012).

8.2 Next-Generation Sequencing and Pyrosequencing

Recent developments in high-throughput sequencing (or next-generation sequencing) technologies enable rapid sequencing analysis of whole genomes and environmental DNA samples (Mardis 2008; Shendure and Ji 2008; Miller et al. 2009; Lauber et al. 2010; Robinson et al. 2010). Some of these methods include massively parallel signature sequencing or MPSS (Lynx Therapeutics), Polony sequencing (Agencourt Biosciences), 454 pyrosequencing (Life Sciences), Illumina (Solexa) sequencing (Illumina), SOLiD sequencing (Applied Biosystems), ion semiconductor sequencing (Ion Torrent Systems Inc.), DNA nanoball sequencing and HeliScope single molecule sequencing.

In 2010, pyrosequencing was used for the first time to examine the bacterial endophyte community in the roots of 12 different potato cultivars revealing an unprecedented level of diversity among the bacterial root endophytes. Interestingly, the presence of five of the ten most common eubacterial genera (*Rheinheimera*, *Dyadobacter*, *Devosia*, *Pedobacter* and *Pseudoxanthomonas*) revealed by pyrosequencing has not been previously reported as potato root endophytes (Manter et al. 2010). Analysis of endophytic bacterial diversity of an Antarctic moss, *Sanionia uncinata,* using 16S rRNA pyrosequencing technology, indicated that *Proteobacteria* was the most dominant phylum with 65.6 %, followed by *Bacteroidetes* (29.1 %) and *Actinobacteria* (11.7 %) (Park et al. 2013). Actinobacteria were found to be in higher abundance in the endophytic compartment (EC) of the *A. thaliana* rhizosphere microbiome, followed by *Proteobacteria, Firmicutes* and other minor bacterial taxa (Bulgarelli et al. 2012; Lundberg et al. 2012). Lower-order taxonomic analysis demonstrated that enrichment of a low-diversity actinobacteria community in the EC was driven by a subset of families, predominantly *Streptomy-cetaceae*, and the selective enrichment of actinobacteria in the roots community was suggested to depend on the colonisation cues from metabolically active host cells as well (Bulgarelli et al. 2012; Lundberg et al. 2012). These research advances in molecular biological techniques greatly improve our understanding of the complexity and ecological distributions of plant-associated actinobacteria. In spite of these advances, the true functional diversity and capabilities of actinobacteria in different endophytic habitats of various ecosystems remain to be fully discovered.

Acknowledgement Authors thank host institutions Department of Medical Biotechnology at Flinders University, CSIRO (VVSRG) and National Institute of Abiotic Stress Management (GV). First author also thanks DIISRTE, Australian Government, for providing 2012 Endeavour Research Fellowship.

References

Ahmadjian V (1993) The lichen symbiosis. Wiley, New York

Alexander DE, Fairbridge RW (1999) Encyclopedia of environmental science. Springer, Dordrecht, Germany. ISBN 0-412-74050-8

Alongi DM (2002) Present state and future of the world's mangrove forests. Environ Conserv 29:331–349

Ara I, Kudo T, Matsumoto A, Takahashi Y, Omura S (2007) *Nonomuraea maheshkhaliensis* sp. nov., a novel actinomycete isolated from mangrove rhizosphere mud. J Gen Appl Microbiol 53:159–166

Araujo WL, Marcon J, Maccheroni W Jr, Van Elsas JD, Van Vuurde JWL, Azevedo JL (2002) Diversity of endophytic bacterial populations and their interaction with *Xylella fastidiosa* in citrus plants. Appl Environ Microbiol 68:4906–4914

Bach LH, Frostegard A, Ohlson M (2009) Site identity and moss species as determinants of soil microbial community structures in Norway spruce forests across three vegetation zones. Plant Soil 318:81–91

Bascom-Slack CA, Ma C, Moore E, Babbs B, Fenn K, Greene JS, Hann BD, Keehner J, Kelley-Swift EG, Kembaiyan V, Lee SJ, Li P, Light DY, Lin EH, Schorn MA, Vekhter D, Boulanger LA, Hess WM, Vargas PN, Strobel GA, Strobel SA (2009) Multiple, novel biologically active endophytic actinomycetes isolated from upper Amazonian rainforests. Microb Ecol 58:374–383

Baskaran R, Mohan PM, Sivakumar K, Raghavan P, Sachithanandam V (2012) phyllosphere microbial populations of ten true mangrove species of the Andaman Island. Int J Microbiol Res 3:124–127

Benson DR, Silvester WB (1993) Biology of *Frankia* strain, actinomycetes symbionts of actinorrhizal plants. Microbiol Rev 57:293–319

Bentley SD, Chater KF, Cerdeno-Tarraga AM, Challis GL, Thomson NR, James KD et al (2002) Complete genome sequence of the model actinomycete *Streptomyces coelicolor* A3(2). Nature 417:141–147

Berg G, Krechel A, Ditz M, Faupel A, Ulrich A, Hallmann J (2005) Endophytic and ectophytic potato-associated bacterial communities differ in structure and antagonistic function against plant pathogenic fungi. FEMS Microbiol Ecol 51:215–229

Bernstein L, Bosch P, Canziani O, Chen Z, Christ R, Davidson O, Hare W, Huq S, Karoly D, Kattsov V (2007) IPCC, 2007: climate change 2007: synthesis report. Contribution of working groups I, II and III to the fourth assessment report of the Intergovernmental Panel on Climate Change. Intergovernmental Panel on Climate Change, Geneva. http://www.ipcc. ch/ipccreports/ar4-syr.htm

Bian GK, Feng ZZ, Qin S, Xing K, Wang Z, Cao CL, Liu CH, Dai CC, Jiang JH (2012a) *Kineococcus endophytica* sp. nov., a novel endophytic actinomycete isolated from a coastal halophyte in Jiangsu, China. Antonie Leeuwenhoek 102:621–628

Bian GK, Qin S, Yuan B, Zhang YJ, Xing K, Ju XY, Li WJ, Jiang JH (2012b) *Streptomyces phytohabitans* sp. nov., a novel endophytic actinomycete isolated from medicinal plant *Curcuma phaeocaulis*. Antonie Leeuwenhoek 102:289–296

Brusetti L, Malkhazova I, Gtari M, Tamagnini I, Borin S, Merabishvili M, Chanishvili N, Mora D, Cappitelli F, Daffonchio D (2008) Fluorescent-BOX-PCR for resolving bacterial genetic diversity, endemism and biogeography. BMC Microbiol 8:220–232

Bulgarelli D et al (2012) Revealing structure and assembly cues for *Arabidopsis* root-inhabiting bacterial microbiota. Nature 488:91–95

Bulgari D, Casati P, Brusetti L, Quaglino F, Brasca M, Daffonchio D, Bianco PA (2009) Endophytic bacterial diversity in grapevine (*Vitis vinifera* L.) leaves described by 16S rRNA gene sequence analysis and length heterogeneity-PCR. J Microbiol 47:393–401

Bunyoo C, Duangmal K, Nuntagij A, Thamchaipenet A (2009) Characterisation of endophytic actinomycetes isolated from wattle trees (*Acacia auriculiformis* A. Cunn. ex Benth.) in Thailand. Thai J Genet 2:155–163

Cao LX, Qiu ZQ, You JL, Tan HM, Zhou SN (2004) Isolation and characterization of endophytic *Streptomyces* strains from surface sterilized tomato (*Lycopersicon esculentum*) roots. Lett Appl Microbiol 39:425–430

Cao LX, Qiu ZQ, You JL, Tan HM, Zhou S (2005) Isolation and characterization of endophytic streptomycete antagonists of fusarium wilt pathogen from surface-sterilized banana roots. FEMS Microbiol Lett 247:147–152

Carro L, Spröer C, Alonso P, Trujillo ME (2012) Diversity of *Micromonospora* strains isolated from nitrogen fixing nodules and rhizosphere of *Pisum sativum* analyzed by multilocus sequence analysis. Syst Appl Microbiol 35:73–80

Castillo UF, Strobel GA, Mullenberg K, Condron MAM, Teplow DB, Folgiano V, Gallo M, Ferracane R, Mannina L, Viel S, Codde M, Robison R, Porter H, Jensen J (2006) Munumbicins E-4 and E-5: novel broad-spectrum antibiotics from *Streptomyces* NRRL 3052. FEMS Microbiol Lett 255:296–300

Chapman JL, Reiss MJ (1998) Ecology. Cambridge University Press, London

Chelius MK, Triplett EW (2001) The diversity of archaea and bacteria in association with the roots of *Zea mays* L. Microb Ecol 41:252–263

Chen M, Zhang L, Zhang X (2011) Isolation and inoculation of endophytic *Actinomycetes* in root nodules of *Elaeagnus angustifolia*. Mod Appl Sci 5:264–268

Cole ST, Brosch R, Parkhill J, Garnier T, Churcher C, Harris D et al (1998) Deciphering the biology of *Mycobacterium tuberculosis* from the complete genome sequence. Nature 393:537–544

Compant S, Duffy B, Nowak J, Clément C, Barka EA (2005) Use of plant growth-promoting bacteria for biocontrol of plant diseases: principles, mechanisms of action, and future prospects. Appl Environ Microbiol 71:4951–4959

Conn VM, Franco CMM (2004a) Analysis of the endophytic actinobacterial population in the roots of wheat (*Triticum aestivum* L.) by terminal restriction fragment length polymorphism and sequencing of 16S rRNA clones. Appl Environ Microbiol 70:1787–1794

Conn VM, Franco CMM (2004b) Effect of microbial inoculants on the indigenous actinobacterial endophyte population in the roots of wheat as determined by terminal restriction fragment length polymorphism. Appl Environ Microbiol 70:6407–6413

Conn VM, Walker AR, Franco CM (2008) Endophytic actinobacteria induce defense pathways in *Arabidopsis thaliana*. Mol Plant Microbe Interact 21:208–218

Cook AE, Meyers PR (2003) Rapid identification of filamentous actinomycetes to the genus level using genus-specific 16S rRNA gene restriction fragment patterns. Int J Syst Evol Microbiol 53:1907–1915

Coombs JT, Franco CMM (2003a) Isolation and identification of actinobacteria isolated from surface-sterilized wheat roots. Appl Environ Microbiol 69:5603–5608

Coombs JT, Franco CMM (2003b) Visualization of an endophytic *Streptomyces* species in wheat seed. Appl Environ Microbiol 69:4260–4262

Coombs JT, Michelsen PP, Franco CMM (2004) Evaluation of endophytic actinobacteria as antagonists of *Gaeumannomyces graminis* var. *tritici* in wheat. Biol Control 29:359–366

Curtis SM, Meyers PR (2012) Multilocus sequence analysis of the actinobacterial genus *Kribbella*. Syst Appl Microbiol 35:441–446

da Silva NMV, Pereira TM, Filho SA, Matsuura TV (2011) Taxonomic characterization and antimicrobial activity of actinomycetes associated with foliose lichens from the Amazonian Ecosystems. Aust J Basic Appl Sci 5:910–9181

de Araujo J, da Silva MAC, Azevedo JL (2000) Isolation of endophytic actinomycetes from roots and leaves of maize (*Zea mays* L.). Braz Arch Biol Technol 43:447–451

Dent KC, Stephen JR, Finch-Savage WE (2004) Molecular profiling of microbial communities associated with seeds of *Bera vulgaris* subsp. *vulgaris* (sugar beet). J Microbiol Methods 56:17–26

Devulder G, de Montclos DP, Flandrois JP (2005) A multigene approach to phylogenetic analysis using the genus *Mycobacterium* as a model. Int J Syst Evol Microbiol 55:293–302

Dharmaraj S, Ashokkumar B, Dhevendaran K (2010) Isolation of marine *Streptomyces* and the evaluation of its bioactive potential. Afr J Microbiol Res 4:240–248

Doumbou CL, Akimov V, Beaulieu C (1998) Selection and characterization of microorganisms utilizing Thaxtomin A, a phytotoxin produced by *Streptomyces scabies*. Appl Environ Microbiol 64:4313–4316

Downie JL (1994) Signalling strategies for nodulation of legumes by rhizobia. Trends Microbiol 2:318–324

Du HJ, Wei YZ, Su J, Liu HY, Ma BP, Guo BL, Zhang YQ, Yu LY (2013a) *Nocardioides perillae* sp. nov., isolated from surface-sterilized roots of *Perilla frutescens*. Int J Syst Evol Microbiol 63:1068–1072

Du HJ, Zhang YQ, Liu HY, Su J, Wei YZ, Ma BP, Guo BL, Yu LY (2013b) *Allonocardiopsis opalescens* gen. nov., sp. nov., a new member of the suborder *Streptosporangineae*, from the surface-sterilized fruit of a medicinal plant. Int J Syst Evol Microbiol 63:900–904

Duangmal K, Mingma R, Pathom-Aree W, Thamchaipenet A, Inahashi Y, Matsumoto A, Takahashi Y (2011) *Amycolatopsis samaneae* sp. nov., isolated from roots of *Samanea saman* (Jacq.) Merr. Int J Syst Evol Microbiol 61:951–955

Dyson P (2011) *Streptomyces*: molecular biology and biotechnology. Horizon Scientific Press, Norfolk, p 257

Eccleston GP, Brooks PR, Kurtböke DI (2008) The occurrence of bioactive micromonosporae in aquatic habitats of the Sunshine Coast in Australia. Mar Drugs 6:243–261

El-Bondkly AM, El-Gendy MMAA, Wiese J, Imhoff JF (2012) Phylogenetic diversity and antimicrobial activities of culturable endophytic actinobacteria isolated from different Egyptian marine sponges and soft corals. Aust J Basic Appl Sci 6:25–33

El-Tarabily KA, Nassar AH, Hardy GESJ, Sivasithamparam K (2009) Plant growth promotion and biological control of *Pythium aphanidermatum*, a pathogen of cucumber, by endophytic actinomycetes. J Appl Microbiol 106:13–26

Embley TM, Stackebrandt E (1994) The molecular phylogeny and systematics of the actinomycetes. Annu Rev Microbiol 48:257–289

Embley TM, Wait R (1994) Structural lipids of *Eubacteria*. Analysis of fatty acid methyl esters by capillary gas chromatography. In: Goodfellow M, Donnell O (eds) Chemical methods in prokaryotic systematics. Wiley, Chichester, pp 121–161

Ezra D, Castillo UF, Strobel GA, Hess WM, Porter H, Jensen JB, Condron MAM, Teplow DB, Sears J, Maranta M, Hunter M, Weber B, Yaver D (2004) Coronamycins, peptide antibiotics produced by a verticillate *Streptomyces* sp. (MSU-2110) endophytic on Monstera sp. Microbiology 150:785–793

Fialho de Oliveira M, Germano da Silva M, Van Der Sand ST (2010) Anti-phytopathogen potential of endophytic actinobacteria isolated from tomato plants (*Lycopersicon esculentum*) in southern Brazil, and characterization of *Streptomyces* sp. R18(6), a potential biocontrol agent. Res Microbiol 161:565–572

Fisher RF, Long SR (1992) Rhizobium-plant signal exchange. Nature 357:655–660

Franco C, Michelsen P, Percy N, Conn V, Listiana E, Moll S, Loria R, Coombs J (2007) Actinobacterial endophytes for improved crop performance. Aust Plant Pathol 36:524–531

Gao B, Gupta RS (2005) Conserved indels in protein sequences that are characteristic of the phylum *Actinobacteria*. Int J Syst Evol Microbiol 55:2401–2412

Gao B, Parmanathan R, Gupta RS (2006) Signature proteins that are distinctive characteristics of actinobacteria and their subgroups. Antonie Leeuwenhoek 90:69–91

Garcia LC, Martinez-Molina E, Trujillo ME (2010) *Micromonospora pisi* sp. nov., isolated from root nodules of *Pisum sativum*. Int J Syst Evol Microbiol 60:331–337

Garrity GM, Holt JG (2001) The road map to the manual. In: Boone DR, Castenholz RW, Garrity GM (eds) Bergey's manual of systematic bacteriology, vol 1, 2nd edn, The *Archaea* and the deeply branching and phototrophic bacteria. Springer, New York, pp 119–166

Gevers D, Dawyndt P, Vandamme P, Willems A, Vancanneyt M, Swings J, de Vos P (2006) Stepping stones towards a new prokaryotic taxonomy. Philos Trans R Soc Lond B 361:1911–1916

Gontia I, Kavita K, Schmid M, Hartmann A, Jha B (2011) *Brachybacterium saurashtrense* sp. nov., a halotolerant root-associated bacterium with plant growth-promoting potential. Int J Syst Evol Microbiol 61:2799–2804

González I, Ayuso-Sacido A, Anderson A, Genilloud O (2005) Actinomycetes isolated from lichens: evaluation of their diversity and detection of biosynthetic gene sequences. FEMS Microbiol Ecol 54:401–415

Goodfellow M, Williams ST (1983) Ecology of actinomycetes. Annu Rev Microbiol 37:189–216

Goris J, Konstantinidis KT, Klappenbach JA, Coenye T, Vandamme P, Tiedje JM (2007) DNA–DNA hybridization values and their relationship to whole-genome sequence similarities. Int J Syst Evol Microbiol 57:81–91

Gray EJ, Smith DL (2005) Intracellular and extracellular PGPR: commonalities and distinctions in the plant bacterium signalling process. Soil Biol Biochem 37:395–412

Guo B, Wang Y, Sun X, Tang K (2008) Bioactive natural products from endophytes: a review. Appl Biochem Microbiol 44:136–142

Gupta N, Mishra S, Basak UC (2009) Diversity of *Streptomyces* in mangrove ecosystem of Bhitarkanika. Iran J Microbiol 1:37–42

Hallman J, Quadt-Hallman A, Mahafee WF, Kloepper JW (1997) Bacterial endophytes in agricultural crops. Can J Microbiol 43:895–914

Hallmann J, Berg G, Schulz B (2006) Isolation procedures for endophytic microorganisms. In: Schulz BJE, Boyle CJC, Sieber TN (eds) Microbial root endophytes. Springer, New York, pp 299–314

Han L, Huang XS, Sattler I, Fu HZ, Grabley S, Lin WH (2007) Two new constituents from mangrove *Bruguiera gymnorrhiza*. J Asian Nat Prod Res 9:327–331

Harvey I, Cormier Y, Beaulieu C, Akimov VN, Mériaux A, Duchaine C (2001) Random amplified ribosomal DNA restriction analysis for rapid identification of thermophilic actinomycete-like bacteria involved in hypersensitivity pneumonitis. Syst Appl Microbiol 24:277–284

Hasegawa S, Meguro A, Shimizu M, Nishimura T, Kunoh H (2006) Endophytic actinomycetes and their interactions with host plants. Actinomycetologica 20:72–81

Hayakawa M (1990) Selective isolation methods and distribution of soil actinomycetes. Actinomycetologica 4:103–112

Hayward D, van Helden PD, Wiid IJF (2009) Glutamine synthetase sequence evolution in the mycobacteria and their use as molecular markers for *Actinobacteria* speciation. BMC Evol Biol 9:48–60

Higashide E, Asai M, Otsuka TS, Kozay Y, Hasegawa T, Kishi T et al (1977) Ansamitocins, a group of novel maytansinoid antibiotics with antitumour properties from *Nocardia*. Nature 270:721–722

Hiltner L (1904) Uber neuere Erfahrungen und Probleme auf dem Gebiete der Bodenbakteriologie unterbessonderer Berucksichtigung der Grundung und Brache. Arb Dtsch Landwirtsch Ges Berl 98:59–78

Hirsch P, Mevs U, Kroppenstedt RM, Schumann P, Stackebrandt E (2004) Crytpoendolithic actinomycetes from antarctic sandstone: *Micromonospora endolithica* sp. nov. and two isolates related to *Micromonospora coerulea* Jensen 1932. Syst Appl Microbiol 27:166–174

Hong K, Gao AH, Gao H, Zhang L, Lin HP, Li J, Yao XS, Goodfellow M, Ruan JS (2009) Actinomycetes for marine drug discovery isolated from mangrove soils and plants in China. Mar Drugs 7:24–44

Huang H, Li J, Hu Y, Fang Z, Zhang K, Bao S (2008) *Micromonospora rifamycinica* sp. nov., a novel actinomycete from mangrove sediment. Int J Syst Evol Microbiol 58:17–20

Huang XL, Zhuang L, Lin HP, Li J, Goodfellow M, Hong K (2012) Isolation and bioactivity of endophytic filamentous actinobacteria from tropical medicinal plants. Afr J Biotechnol 11:9855–9864

Igarashi Y, Iida T, Yoshida R, Furumai T (2002) Pteridic acids A and B, novel plant growth promoters with auxin-like activity from *Streptomyces hygroscopicus* TP-A0451. J Antibiot 55:764–767

Ikeda S, Kaneko T, Okubo T, Rallos LEE, Eda S, Mitsui H, Sato S, Nakamura Y, Tabata S, Minamisawa K (2009) Development of a bacterial cell enrichment method and its application to the community analysis in soybean stems. Microb Ecol 58:703–714

Ikeda S, Okubo T, Kaneko T, Inaba S, Maekawa T, Eda S, Sato S, Tabata S, Mitsui H, Minamisawa K (2010) Community shifts of soybean stem-associated bacteria responding to different nodulation phenotypes and N levels. ISME J 4:315–326

Inahashi Y, Matsumoto A, Danbara H, Omura S, Takahashi Y (2010) *Phytohabitans suffuscus* gen. nov., sp. nov., an actinomycete of the family *Micromonosporaceae* isolated from plant roots. Int J Syst Evol Microbiol 60:2652–2658

Inahashi Y, Matsumoto A, Omura S, Takahashi Y (2011) *Streptosporangium oxazolinicum* sp. nov., a novel endophytic actinomycete producing new antitrypanosomal antibiotics, spoxazomicins. J Antibiot (Tokyo) 64:297–302

Inahashi Y, Matsumoto A, Omura S, Takahashi Y (2012) *Phytohabitans flavus* sp. nov., *Phytohabitans rumicis* sp. nov. and *Phytohabitans houttuyneae* sp. nov., isolated from plant roots, and emended description of the genus *Phytohabitans*. Int J Syst Evol Microbiol 62:2717–2723

Indananda C, Matsumoto A, Inahashi Y, Takahashi Y, Duangmal K, Thamchaipenet A (2010) *Actinophytocola* gen. nov., a new genus of the family *Pseudonocardiaceae* and description of a new species, *Actinophytocola oryzae* sp. nov., isolated from root of Thai glutinous rice plant. Int J Syst Evol Microbiol 60:1141–1146

Indananda C, Thamchaipenet A, Matsumoto A, Duangmal K, Takahashi Y (2011) *Actinoallomurus oryzae* sp. nov., an endophytic actinomycete isolated from root of Thai jasmine rice plant. Int J Syst Evol Microbiol 61:737–741

Inderiati S, Franco CMM (2008) Isolation and identification of endophytic actinomycetes and their antifungal activity. J Biotechnol Res Trop Reg 1:1–6

Janso JE, Carter GT (2010) Biosynthetic potential of phylogenetically unique endophytic actinomycetes from tropical plants. Appl Environ Microbiol 76:4377–4386

Jensen SI, Kuhl M, Prieme A (2007) Different bacterial communities associated with the roots and bulk sediment of the seagrass *Zostera marina*. FEMS Microbiol Ecol 62:108–117

Jiao JY, Wang HX, Zeng Y, Shen YM (2006) Enrichment for microbes living in association with plant tissues. J Appl Microbiol 100:830–837

Kaewkla O, Franco CMM (2010a) *Pseudonocardia adelaidensis* sp. nov., an endophytic actinobacterium isolated from the surface sterilized stem of a grey box tree (*Eucalyptus microcarpa*). Int J Syst Evol Microbiol 60:2818–2822

Kaewkla O, Franco CMM (2010b) *Pseudonocardia eucalypti* sp. nov., an endophytic actinobacterium with a unique knobby spore surface, isolated from roots of a native Australian eucalyptus tree. Int J Syst Evol Microbiol 61:742–746

Kaewkla O, Franco CMM (2010c) *Nocardia callitridis* sp. nov., an endophytic actinobacterium isolated from a surface-sterilized root of an Australian native pine tree. Int J Syst Evol Microbiol 60:1532–1536

Kaewkla O, Franco CMM (2011a) *Flindersiella endophytica* gen. nov., sp. nov., an endophytic actinobacterium isolated from the root of Grey Box, an endemic eucalyptus tree. Int J Syst Evol Microbiol 61:2135–2140

Kaewkla O, Franco CMM (2011b) *Actinopolymorpha pittospori* sp. nov., an endophytic actinobacterium isolated from surface sterilized leaves of an Australian native apricot tree. Int J Syst Evol Microbiol 61:2616–2620

Kaewkla O, Franco CMM (2012) *Promicromonospora endophytica* sp. nov., an endophytic actinobacterium isolated from the root of an Australian native Grey Box tree. Int J Syst Evol Microbiol 62:1687–1691

Kaewkla O, Franco CMM (2013a) Rational approaches to improving the isolation of endophytic *Actinobacteria* from Australian native trees. Microb Ecol 65:384–393

Kaewkla O, Franco CMM (2013b) *Kribbella endophytica* sp. nov., an endophytic actinobacterium isolated from the surface-sterilized leaf of a native apricot tree. Int J Syst Evol Microbiol 63:1249–1253

Kafur A, Khan AB (2011) Isolation of endophytic actinomycetes from *Catharanthus roseus* (L.) G. Don leaves and their antimicrobial activity. Iran J Biotechnol 9:302–306

Kalkus J, Menne R, Reh M, Schlegel HG (1998) The terminal structures of linear plasmids from *Rhodococcus opacus*. Microbiology 144:1271–1279

Kang HP, Dunne WM (2003) Stability of repetitive-sequence PCR patterns with respect to culture age and subculture frequency. J Clin Microbiol 41:2694–2696

Keddy PA (2010) Wetland ecology, principles and conservation. Cambridge University Press, Cambridge, p 497. ISBN 978-0-521-51940-3

Khan Z, Doty SL (2009) Characterization of bacterial endophytes of sweet potato plants. Plant Soil 322:197–207

Kim T, Cho S, Han J, Shin YM, Lee HB, Kim SB (2012) Diversity and physiological properties of root endophytic actinobacteria in native herbaceous plants of Korea. J Microbiol 50:50–57

Kirby BM, Meyers PR (2010) *Micromonospora tulbaghiae* sp. nov., isolated from the leaves of wild garlic, *Tulbaghia violacea*. Int J Syst Evol Microbiol 60:1328–1333

Konstantinidis KT, Tiedje JM (2005) Genomic insights into the species definition for prokaryotes. Proc Natl Acad Sci USA 102:2567–2572

Kunisawa T (2007) Gene arrangements characteristic of the phylum Actinobacteria. Antonie Leeuwenhoek 92:359–365

Küster E, Williams ST (1964) Media for the isolation of streptomycetes: starch casein medium. Nature 202:928–929

Lauber CL, Zhou N, Gordon JI, Knight R, Fierer N (2010) Effect of storage conditions on the assessment of bacterial community structure in soil and human-associated samples. FEMS Microbiol Lett 307:80–86

Laurent FJ, Provost F, Boiron P (1999) Rapid identification of clinically relevant *Nocardia* species to genus level by 16S rRNA gene PCR. J Clin Microbiol 37:99–102

Lechevalier HA, Lechevalier MP (1967) Biology of actinomycetes. Annu Rev Microbiol 21:71–100

Lee HK, Lee DS, Lim J, Kim JS, Jung JH (1998) Topoisomerase I inhibitors from the *Streptomyces* sp. strain KM86-9B isolated from a marine sponge. Arch Pharm Res 21:729–733

Lee DW, Lee JM, Seo JP, Schumann P, Kim SJ, Lee SD (2008a) *Phycicola gilvus* gen. nov., sp. nov., an actinobacterium isolated from living seaweed. Int J Syst Evol Microbiol 58:1318–1323

Lee SO, Choi GJ, Choi YH, Jang KS, Park DJ, Kim CJ, Kim JC (2008b) Isolation and characterization of endophytic actinomycetes from Chinese cabbage roots as antagonists to *Plasmodiophora brassicae*. J Microbiol Biotechnol 18:1741–1746

Li J, Zhao GZ, Huang HY, Qin S, Zhu WY, Li WJ (2009) *Kineosporia mesophila* sp. nov., isolated from the surface sterilized stem of *Tripterygium wilfordii*. Int J Syst Evol Microbiol 59:3150–3154

Li J, Zhao GZ, Huang HY, Zhu WY, Lee JC, Xu LH, Kim CJ, Li WJ (2011a) *Nonomuraea endophytica* sp. nov., an endophytic actinomycete isolated from *Artemisia annua* L. Int J Syst Evol Microbiol 61:757–761

Li J, Zhao GZ, Zhu WY, Huang HY, Xu LH, Zhang S, Li WJ (2011b) *Phytomonospora endophytica* gen. nov., sp. nov., isolated from the roots of *Artemisia annua* L. Int J Syst Evol Microbiol 61:2967–2973

Li J, Zhao GZ, Huang HY, Qin S, Zhu WY, Zhao LX, Xu LH, Zhang S, Li WJ, Strobel G (2012) Isolation and characterization of culturable endophytic actinobacteria associated with *Artemisia annua* L. Antonie Leeuwenhoek 101:515–527

Li J, Zhao GZ, Zhu YW, Huang HY, Xu LH, Zhang S, Li WJ (2013) *Streptomyces endophyticus* sp. nov., an endophytic actinomycete isolated from *Artemisia annua* L. Int J Syst Evol Microbiol 63:224–229

Liba CM, Ferrara FIS, Manfio GP, Fantinatti-Garboggini F, Albuquerque RC, Pavan C, Ramos PL, Moreira CA, Barbosa HR (2006) Nitrogen-fixing chemo-organotrophic bacteria isolated from cyanobacteria-deprived lichens and their ability to solubilize phosphate and to release amino acids and phytohormones. J Appl Microbiol 101:1076–1086

Lin C, Lu C, Shen Y (2010) Three new 2-pyranone derivatives from mangrove endophytic *Actinomycete* strain *Nocardiopsis* sp. A00203. Rec Nat Prod 4:176–179

Lindow SE, Brandl MT (2003) Microbiology of the phyllosphere. Appl Environ Microbiol 69:1875–1883

Ludwig W, Klenk HP (2005) Overview: a phylogenetic backbone and taxonomic framework for procaryotic systematics. In: Brenner DJ, Krieg NR, Staley JT, Garrity GM (eds) Bergey's manual of systematic bacteriology, vol 2, 2nd edn, The *Proteobacteria*, Part A, introductory essays. Springer, New York, pp 49–65

Lundberg DS et al (2012) Defining the core *Arabidopsis thaliana* root microbiome. Nature 488:86–90

Madhaiyan M, Hu CJ, Kim SJ, Weon HY, Kwon SO, Ji L (2013) *Jatrophihabitans endophyticus* gen. nov., sp. nov., an endophytic actinobacterium isolated from a surface-sterilized stem of *Jatropha curcas* L. Int J Syst Evol Microbiol 63:1241–1248

Mangamuri UK, Muvva V, Poda S, Kamma S (2012) Isolation, identification and molecular characterization of rare actinomycetes from mangrove ecosystem of Nizampatnam. Malays J Microbiol 8:83–91

Manter DK, Delgado JA, Holm DG, Stong RA (2010) Pyrosequencing reveals a highly diverse and cultivar-specific bacterial endophyte community in potato roots. Microb Ecol 60:157–166

Mardis ER (2008) Next-generation DNA sequencing methods. Annu Rev Genomics Hum Genet 9:387–402

Mehling A, Wehmeier UF, Piepersberg W (1995) Application of random amplified polymorphic DNA (RAPD) assays in identifying conserved regions of actinomycete genomes. FEMS Microbiol Lett 128:119–125

Miao Q, Qin S, Bian GK, Yuan B, Xing K, Zhang YJ, Li Q, Tang SK, Li WJ, Jiang JHA (2011) *Amycolatopsis endophytica* sp. nov., a novel endophytic actinomycete isolated from oil-seed plant *Jatropha curcas* L. Antonie Leeuwenhoek 100:333–339

Miller SR, Strong AL, Jones KL, Ungerer MC (2009) Bar-coded pyrosequencing reveals shared bacterial community properties along the temperature gradients of two alkaline hot springs in Yellowstone National Park. Appl Environ Microbiol 75:4565–4572

Misk A, Franco CMM (2011) Biocontrol of chickpea root rot using endophytic *Actinobacteria*. BioControl 56:811–822

Monteiro-Vitorello CB, Camargo LE, Van Sluys MA, Kitajima JP et al (2004) The genome sequence of the gram-positive sugarcane pathogen *Leifsonia xyli subsp. xyli*. Mol Plant Microbe Interact 17:827–836

Montesinos E, Bonaterra A, Badosa E, Frances J, Alemany J, Llorente I, Moragrega C (2002) Plant-microbe interactions and the new biotechnological methods of plant disease control. Int Microbiol 5:169–175

Nikolic B, Schwab H, Sessitsch A (2011) Metagenomic analysis of the 1-aminocyclopropane-1-carboxylate deaminase gene (acdS) operon of an uncultured bacterial endophyte colonizing *Solanum tuberosum* L. Arch Microbiol 193:665–676

Nimnoi P, Pongsilp N, Lumyong S (2010a) Endophytic actinomycetes isolated from *Aquilaria crassna* Pierre ex Lec and screening of plant growth promoters production. World J Microbiol Biotechnol 26:193–203

Nimnoi P, Pongsilp N, Lumyong S (2010b) Genetic diversity and community of endophytic actinomycetes within the roots of *Aquilaria crassna* Pierre ex Lec assessed by *Actinomycetes*-specific PCR and PCR-DGGE of 16S rRNA gene. Biochem Syst Ecol 38:595–601

Nishimura T, Meguro A, Hasegawa S, Nakagawa Y, Shimizu M, Kunoh H (2002) An endophytic actinomycete, *Streptomyces* Sp. AOK-30, isolated from mountain laurel and its antifungal activity. J Gen Plant Pathol 68:390–397

Nithyanand P, Indhumathi T, Ravi AV, Pandian SK (2011) Culture independent characterization of bacteria associated with the mucus of the coral *Acropora digitifera* from the Gulf of Mannar. World J Microbiol Biotechnol 27:1399–1406

Normand P, Lapierre P, Tisa LS, Gogarten JP, Alloisio N, Bagnarol E et al (2007) Genome characteristics of facultatively symbiotic *Frankia* sp. strains reflect host range and host plant biogeography. Genome Res 17:7–15

Okazaki T (2003) Studies on actinomycetes isolated from plant leaves. In: Kurtböke DI (ed) Selective isolation of rare actinomycetes. Queensland Complete Printing Service, Nambour, pp 102–121

Okazaki T, Takahashi K, Kizuka M, Enokita R (1995) Studies on actinomycetes isolated from plant leaves. Annu Rev Sankyo Res Lab 47:97–106

Opelt K, Chobot V, Hadacek F, Schönmann S, Eberl L, Berg G (2007) Investigations of the structure and function of bacterial communities associated with Sphagnum mosses. Environ Microbiol 9:2795–2809

Park M, Leea H, Honga SG, Kima O (2013) Endophytic bacterial diversity of acn Antarctic moss, *Sanionia uncinata*. Antarct Sci 25:51–54

Qasim SZ (1999) The Indian Ocean: images and realities. Oxford and IBH, New Delhi, pp 57–90

Qin S, Li J, Zhao GZ, Chen HH, Xu LH, Li WJ (2008a) *Saccharopolyspora endophytica* sp. nov., an endophytic actinomycete isolated from the root of *Maytenus austroyunnanensis*. Syst Appl Microbiol 31:352–357

Qin S, Wang HB, Chen HH, Zhang YQ, Jiang CL, Xu LH, Li WJ (2008b) *Glycomyces endophyticus* sp. nov., an endophytic actinomycete isolated from the root of *Carex baccans* Nees. Int J Syst Evol Microbiol 58:2525–2528

Qin S, Li J, Chen HH, Zhao GZ, Zhu WY, Jiang CL, Xu LH, Li WJ (2009) Isolation, diversity, and antimicrobial activity of rare actinobacteria from medicinal plants of tropical rain forests in Xishuangbanna, China. Appl Environ Microbiol 75:6176–6186

Qin S, Chen HH, Lee JC, Kim CJ, Xu LH, Li WJ (2010a) *Saccharopolyspora gloriosa* sp. nov., a novel endophytic actinomycete isolated from the stem of *Gloriosa superba* L. Int J Syst Evol Microbiol 60:1147–1151

Qin S, Zhu WY, Jiang JH, Klenk HP, Li J, Zhao GZ, Xu LH, Li WJ (2010b) *Pseudonocardia tropica* sp. nov., an endophytic actinomycete isolated from the stem of *Maytenus austroyunnanensis*. Int J Syst Evol Microbiol 60:2524–2528

Qin S, Xing K, Fei SM, Lin Q, Chen XM, Cao CL, Sun Y, Wang Y, Li WJ, Jiang JH (2011) *Pseudonocardia sichuanensis* sp. nov., a novel endophytic actinomycete isolated from the root of *Jatropha curcas* L. Antonie Leeuwenhoek 99:395–401

Qin S, Chen H, Zhao G, Li J, Zhu W, Xu L, Jiang J, Li W (2012a) Abundant and diverse endophytic actinobacteria associated with medicinal plant *Maytenus austroyunnanensis* in Xishuangbanna tropical rainforest revealed by culture-dependent and culture-independent methods. Environ Microbiol Rep 4:522–531

Qin S, Yuan B, Zhang YJ, Bian GK, Tamura T, Sun BZ, Li WJ, Jiang JH (2012b) *Nocardioides panzhihuaensis* sp. nov., a novel endophytic actinomycete isolated from medicinal plant *Jatropha curcas* L. Antonie Leeuwenhoek 102:353–360

Qin S, Bian GK, Zhang YJ, Xing K, Cao CL, Liu CH, Dai CC, Li WJ, Jiang JH (2013a) *Modestobacter roseus* sp. nov., a novel endophytic actinomycete isolated from a coastal halophyte *Salicornia europaea* Linn. in Jiangsu, China and emended description of the genus. *Modestobacter*. Int J Syst Evol Microbiol 63:2197–2202

Qin S, Bian GK, Tamura T, Zhang YJ, Zhang WD, Cao CL, Jiang JH (2013b) *Streptomyces halophytocola* sp. nov., a novel endophytic actinomycete isolated from the surface-sterilized stems of a coastal halophyte

Tamarix chinensis Lour. in Jiangsu, China. Int J Syst Evol Microbiol 63:2770–2775

Rademaker JLW, de Bruijn FJ (1997) Characterization and classification of microbes by rep-PCR genomic fingerprinting and computer-assisted pattern analysis. In: Caetoan-Anollés G, Gresshoff PM (eds) DNA markers: protocols, applications and overviews. Wiley, New York, pp 151–171

Rademaker JLW, Hoste B, Louws FJ, Kersters K, Swings J, Vauterin L, Vauterin P, de Bruijn FJ (2000) Comparison of AFLP and rep-PCR genomic fingerprinting with DNA-DNA homology studies: *Xanthomonas* as a model system. Int J Syst Evol Microbiol 50:665–677

Rasche F, Trondl R, Naglreiter C, Reichenauer TG, Sessitsch A (2006a) Chilling and cultivar type affect the diversity of bacterial endophytes colonizing sweet pepper (*Capsicum anuum* L.). Can J Microbiol 52:1036–1045

Rasche F, Velvis H, Zachow C, Berg G, van Elsas JD, Sessitsch A (2006b) Impact of transgenic potatoes expressing antibacterial agents on bacterial endophytes is comparable to effects of wild type potatoes and changing environmental conditions. J Appl Ecol 43:555–566

Redenbach M, Scheel J, Schmidt U (2000) Chromosome topology and genome size of selected actinomycetes species. Antonie Leeuwenhoek 78:227–235

Reinhold B, Hurek T, Niemann EG, Fendrik I (1986) Close association of *Azospirillum* and diazotrophic rods with different root zones of Kallar grass. Appl Environ Microbiol 52:520–526

Reiter B, Sessitsch A (2006) The bacterial microflora in association with the wildflower *Crocus albiflorus*. Can J Microbiol 52:1–10

Reiter B, Pfeifer U, Schwab H, Sessitsch A (2002) Response of endophytic bacterial communities in potato plants to infection with *Erwinia carotovora* subsp. *atroseptica*. Appl Environ Microbiol 68:2261–2268

Richardson DHS, Cameron RP (2004) Cyanolichens: their response to pollution and possible management strategies for their conservation in northeastern North America. Northeast Nat 11:1–22

Robinson CJ, Bohannan BJ, Young VB (2010) From structure to function: the ecology of host-associated microbial communities. Microbiol Mol Biol Rev 74(3):453–476

Rong X, Huang Y (2010) Taxonomic evaluation of the *Streptomyces griseus* clade using multilocus sequence analysis and DNA–DNA hybridization, with proposal to combine 29 species and three subspecies as 11 genomic species. Int J Syst Evol Microbiol 60:696–703

Rong X, Gou Y, Huang Y (2009) Proposal to reclassify the *Streptomyces albidoflavus* clade on the basis of multilocus sequence analysis and DNA–DNA hybridization, and taxonomic elucidation of *Streptomyces griseus* subsp. *solvifaciens*. Syst Appl Microbiol 32:314–322

Rosenblueth M, Martínez-Romero E (2006) Bacterial endophytes and their interactions with hosts. Mol Plant Microbe Interact 19:827–837

Ryan RP, Germaine K, Franks A, Ryan DJ, Dowling DN (2008) Bacterial endophytes: recent development and applications. FEMS Microbiol Lett 278:1–9

Saravanakumar R, Moomeen HS, Ronald J, Kannan M (2010) Control of fish bacterial pathogens, by antagonistic marine Actinomycetes isolated from Gulf of Mannar Coast. World J Fish Mar Sci 2:275–279

Sardi P, Saracchi M, Quaroni S, Petrolini B, Borgonovi GE, Merli S (1992) Isolation of endophytic *Streptomyces* strains from surface-sterilized roots. Appl Environ Microbiol 58:2691–2693

Savelkoul PH, Aarts HJ, de Haas J, Dijkshoorn L, Duim B, Otsen M, Rademaker JL, Schouls L, Lenstra JA (1999) Amplified-fragment length polymorphism analysis: the state of an art. J Clin Microbiol 37:3083–3091

Schrempf H (2001) Recognition and degradation of chitin by streptomycetes. Antonie Leeuwenhoek 79:285–289

Sculthorpe CD (1985) The biology of aquatic vascular plants. Edward Arnold press, London

Sessitsch A, Reiter B, Pfeifer U, Wilhelm E (2002) Cultivation independent population analysis of bacterial endophytes in three potato varieties based on eubacterial and Actinomycetes-specific PCR of 16S rRNA genes. FEMS Microbiol Ecol 39:23–32

Sessitsch A, Reiter B, Berg G (2004) Endophytic bacterial communities of field-grown potato plants and their plant-growth-promoting and antagonistic abilities. Can J Microbiol 50:239–249

Sessitsch A, Hardoim P, Döring J, Weilharter A, Krause A, Woyke T, Mitter B, Hauberg-Lotte L, Friedrich F, Rahalkar M, Hurek T, Sarkar A, Bodrossy L, van Overbeek L, Brar D, van Elsas JD, Reinhold-Hurek B (2012) Functional characteristics of an endophyte community colonizing rice roots as revealed by metagenomic analysis. Mol Plant Microbe Interact 25:28–36

Shendure J, Ji H (2008) Next-generation DNA sequencing. Nat Biotechnol 26:1135–1145

Shimizu M, Nakagawa Y, Sato Y, Furumai T, Igarashi Y, Onaka H, Yoshida R, Kunoh H (2000) Studies on endophytic actinomycetes. (i). Streptomyces sp. isolated from rhododendron and its antifungal activity. J Gen Plant Pathol 66:360–366

Shirling EB, Gottlieb D (1966) Methods for characterization of *Streptomyces* species. Int J Syst Bacteriol 16:313–340

Song GC, Yasir M, Bibi F, Chung EJ, Jeon CO, Chung YR (2011) *Nocardioides caricicola* sp. nov., an endophytic bacterium isolated from a halophyte, *Carex scabrifolia* Steud. Int J Syst Evol Microbiol 61:105–109

Spalding M, Blasco F, Field C (1997) World mangrove atlas, Okinawa, Japan. In: The international society for Mangrove Ecosystem. 178. State Forest Report, Forest Survey of India, Dehra Dun, 1999

Stach JE, Bull AT (2005) Estimating and comparing the diversity of marine actinobacteria. Antonie Leeuwenhoek 87:3–9

Stackebrandt E, Schumann P (2000) Introduction to the taxonomy of actinobacteria. In: Dworkin M et al (eds)

The prokaryotes: an evolving electronic resource for the microbiological community. Springer, New York
Steingrube VA, Wilson RW, Brown BA, Jost KC Jr, Blacklock Z, Gibson JL, Wallace RJ Jr (1997) Rapid identification of the clinically significant species and taxa of aerobic actinomycetes, including *Actinomadura*, *Gordona*, *Nocardia*, *Rhodococcus*, *Streptomyces*, and *Tsukamurella* isolates, by DNA amplification and restriction endonuclease analysis. J Clin Microbiol 35:817–822
Strobel GA, Daisy B (2003) Bioprospecting for microbial endophytes and their natural products. Microbiol Mol Biol Rev 67:491–502
Sturz AV (1995) The role of endophytic bacteria during seed piece decay and potato tuberization. Plant Soil 175:257–263
Sturz AV, Christie BR, Matheson BG (1997) Associations of bacterial endophyte populations from red clover and potato crops with potential for beneficial allelopathy. Can J Microbiol 44:162–167
Sun L, Qiu F, Zhang X, Dai X, Dong X, Song W (2008) Endophytic bacterial diversity in rice (*Oryza sativa* L.) roots estimated by 16S rDNA sequence analysis. Microb Ecol 55:415–424
Surette MA, Sturz AV, Lada RR, Nowak J (2003) Bacterial endophytes in processing carrots (*Daucus carota* L. var. *sativus*): their localization, population density, biodiversity and their effects on plant growth. Plant Soil 253:381–390
Taechowisan T, Peberdy JF, Lumyong S (2003) Isolation of endophytic actinomycetes from selected plants and their antifungal activity. World J Microbiol Biotechnol 19:381–385
Tan HM, Cao LX, He ZF, Su GJ, Lin B, Zhou SN (2006) Isolation of endophytic actinomycetes from different cultivars of tomato and their activities against *Ralstonia solanacearum in vitro*. World J Microbiol Biotechnol 22:1275–1280
Tervet IW, Hollis JP (1948) Bacteria in the storage organs of healthy plants. Phytopathology 38:960–967
Thamchaipenet A, Indananda C, Bunyoo C, Duangmal K, Matsumoto A, Takahashi Y (2010) *Actinoallomurus acaciae* sp. nov., a novel endophytic actinomycete isolated from *Acacia auriculiformis* A. Cunn. ex Benth. in Thailand. Int J Syst Evol Microbiol 60:554–559
Theissen G, Munster T, Henschel K (2001) Why don't mosses flower? New Phytol 150:1–8
Thomas WD, Graham RW (1952) Bacteria in apparently healthy pinto beans. Phytopathology 42:214
Tian XL, Cao LX, Tan HM, Han WQ, Chen M, Liu YH, Zhou SN (2007) Diversity of cultivated and uncultivated actinobacterial endophytes in the stems and roots of rice. Microb Ecol 53:700–707
Tokala RK, Strap JL, Jung CM, Crawford DL, Salove MH, Deobald LA, Bailey JF, Morra MJ (2002) Novel plant–microbe rhizosphere interaction involving *Streptomyces lydicus* WYEC108 and the pea plant (*Pisum sativum*). Appl Environ Microbiol 68:2161–2171
Trujillo ME, Kroppenstedt RM, Fernández-Molinero C, Schumann P, Martínez-Molina E (2007) *Micromonospora lupini* sp. nov. and *Micromonospora saelicesensis* sp. nov., isolated from root nodules of *Lupinus angustifolius*. Int J Syst Evol Microbiol 57:2799–2804
Trujillo ME, Alonso-Vega P, Rodrıguez R, Carrol L, Cerda E, Alonso P, Martınez-Molina E (2010) The genus *Micromonospora* is widespread in legume root nodules: the example of *Lupinus angustifolius*. ISME J 4:1265–1281
Valdes M, Perez NO, Estrada-de Los Santos P, Caballero-Mellado J, Pena-Cabriales JJ, Normand P et al (2005) Non-Frankia actinomycetes isolated from surface sterilized roots of *Casuarina equisetifolia* fix nitrogen. Appl Environ Microbiol 71:460–466
Van Belkum A, Scherer S, Van Alphen L, Verbrugh H (1998) Short sequence DNA repeats in prokaryotic genomes. Microbiol Mol Biol Rev 62:275–293
Vaneechoutte M, de Beenhouwer H, Claeys G, Verschraegen G, de Rouck A, Paepe N, Elaichouni A, Portaels F (1993) Identification of *Mycobacterium* species by using amplified ribosomal DNA restriction analysis. J Clin Microbiol 31:2061–2065
Velazquez E, Rojas M, Lorite MJ, Rivas R, Zurdo-Pineiro JL, Heydrich M, Bedmar EJ (2008) Genetic diversity of endophytic bacteria which could be find in the apoplastic sap of medullary parenchym of the stem of healthy sugarcane plants. J Basic Microbiol 48:118–124
Ventura M, Canchaya C, Casale AD, Dellaglio F, Neviani E, Fitzgerald GF, van Sinderen D (2006) Analysis of bifidobacterial evolution using a multilocus approach. Int J Syst Evol Microbiol 56:2783–2792
Ventura M, Canchaya C, Tauch A, Chandra G, Fitzgerald GF, Chater KF, van Sinderen D (2007) Genomics of *Actinobacteria*: tracing the evolutionary history of an ancient phylum. Microbiol Mol Biol Rev 71:495–548
Verma VC, Gond SK, Kumar A, Mishra A, Kharwar RN, Gange AC (2009) Endophytic actinomycetes from *Azadirachta indica* A. *Juss*: isolation, diversity, and anti-microbial activity. Microb Ecol 57:749–756
Vestal JR, White DC (1989) Lipid analysis in microbial ecology. Bioscience 39:535–541
Vickers JC, Williams ST, Ross GW (1984) A taxonomic approach to selective isolation of streptomycetes from soil. In: Ortiz-Ortiz L, Bojalil LF, Yakoleff V (eds) Biological, biochemical and biomedical aspects of actinomycetes. Academic, Orlando, pp 553–561
Wang B, Qiu YL (2006) Phylogenetic distribution and evolution of mycorrhizas in land plants. Mycorrhiza 16:299–363
Wang HX, Geng ZL, Zeng Y, Shen YM (2008) Enrichment plant microbiota for a metagenomic library construction. Environ Microbiol 10:2684–2691
Webster NS, Wilson KJ, Blackall LL, Hill RT (2001) Phylogenetic diversity of bacteria associated with the marine sponge *Rhopaloeides odorabile*. Appl Environ Microbiol 67:434–444
West ER, Cother EJ, Steel CC, Ash GJ (2010) The characterization and diversity of bacterial endophytes of grapevine. Can J Microbiol 56:209–216

Whitman WB, Goodfellow M, Kämpfer P, Busse HJ, Trujillo ME, Ludwig W, Suzuki K, Parte A (2012) The *Actinobacteria*, Part A. In: Rainey FA, Kämpfer P, De Vos P, Chun J, Trujillo ME (eds) Bergey's manual of systematic bacteriology, 2nd edn. Springer, New York

Wilson D (1995) Endophyte–the evolution of a term, and clarification of its use and definition. Oikos 73: 274–276

Wilson RW, Steingrube VA, Brown BA, Wallace RJ Jr (1998) Clinical application of PCR-restriction enzyme pattern analysis for rapid identification of aerobic actinomycete isolates. J Clin Microbiol 36: 148–152

Wu H, Chen W, Wang G, Dai S, Zhou D, Zhao H, Guo Y, Ouyang Y, Li X (2012) Culture-dependent diversity of *Actinobacteria* associated with seagrass (*Thalassia hemprichii*). Afr J Microbiol Res 6:87–94

Xie Z-W, Ge S, Hong D-Y (1999) Preparation of DNA from silica gel dried mini-amount of leaves of *Oryza rufipogon* for RAPD study and total DNA bank construction. Acta Botanica Sinica 41:802–807

Xie QY, Wang C, Wang R, Qu Z, Lin HP, Goodfellow M, Hong K (2010) *Jishengella endophytica* gen. nov., sp. nov., a new member of the family *Micromonosporaceae*. Int J Syst Evol Microbiol 61:1153–1159

Xing K, Qin S, Fei SM, Lin Q, Bian GK, Miao Q, Wang Y, Cao CL, Tang SK, Jiang JH, Li WJ (2011) *Nocardia endophytica* sp. nov., an endophytic actinomycete isolated from the oil-seed plant *Jatropha curcas* L. Int J Syst Evol Microbiol 61:1854–1858

Xing K, Bian GK, Qin S, Klenk HP, Yuan B, Zhang YJ, Li WJ, Jiang JH (2012a) *Kibdelosporangium phytohabitans* sp. nov., a novel endophytic actinomycete isolated from oil-seed plant *Jatropha curcas* L. containing 1-aminocyclopropane-1-carboxylic acid deaminase. Antonie Leeuwenhoek 101:433–441

Xing K, Qin S, Bian GK, Zhang YJ, Zhang WD, Dai CC, Liu CH, Li WJ, Jiang JH (2012b) *Pseudonocardia nantongensis* sp. nov., a novel endophytic actinomycete isolated from the coastal halophyte *Tamarix chinensis* Lour. Antonie Leeuwenhoek 102:659–667

Xing K, Liu W, Zhang YJ, Bian GK, Zhang WD, Tamura T, Lee JS, Qin S, Jiang JH (2013) *Amycolatopsis jiangsuensis* sp. nov., a novel endophytic actinomycete isolated from a coastal plant in Jiangsu, China. Antonie Leeuwenhoek 103:433–439

Yamamura H, Ashizawa H, Nakagawa Y, Hamada M, Ishida Y, Otoguro M, Tamura T, Hayakawa M (2011a) *Actinomycetospora rishiriensis* sp. nov., isolated from a lichen. Int J Syst Evol Microbiol 61:2621–2625

Yamamura H, Ashizawa H, Nakagawa Y, Hamada M, Ishida Y, Otoguro M, Tamura T, Hayakawa M (2011b) *Actinomycetospora iriomotensis* sp. nov., a novel actinomycete isolated from a lichen sample. J Antibiot 64:289–292

Yamamura H, Shimizu A, Nakagawa Y, Hamada M, Otoguro M, Tamura T, Hayakawa M (2012) *Actinoplanes rishiriensis* sp. nov., a novel motile actinomycete isolated by rehydration and centrifugation method. J Antibiot (Tokyo) 65:249–253

Yan X, Huang LL, Tu X, Gao XN, Kang ZS (2012) *Saccharothrix yanglingensis* sp. nov., an antagonistic endophytic actinomycete isolated from cucumber plant. Antonie Leeuwenhoek 101:141–146

Yang CH, Crowley DE, Borneman J, Keen NT (2001) Microbial phyllosphere populations are more complex than previously realized. Proc Natl Acad Sci 98:3889–3894

Yuan HM, Zhang XP, Zhao K, Zhong K, Gu YF, Lindstrom K (2008) Genetic characterisation of endophytic actinobacteria isolated from the medicinal plants in Sichuan. Ann Microbiol 58:597–604

Zhang HW, Song YC, Tan RX (2006) Biology and chemistry of endophytes. Nat Prod Rep 23:753–771

Zhang YJ, Zhang WD, Qin S, Bian GK, Xing K, Li YF, Cao CL, Jiang JH (2013) *Saccharopolyspora dendranthemae* sp. nov. a halotolerant endophytic actinomycete isolated from a coastal salt marsh plant in Jiangsu, China. Antonie Leeuwenhoek 103:1369–1376

Zhao GZ, Li J, Qin S, Huang HY, Zhu WY, Xu LH, Li WJ (2010a) *Streptomyces artemisiae* sp. nov., a novel actinomycete isolated from surface-sterilized *Artemisia annua* L. tissue. Int J Syst Evol Microbiol 60:27–32

Zhao K, Penttinen P, Guan TW, Xiao J, Chen Q, Xu J, Lindström K, Zhang LL, Zhang XP, Strobel GA (2010b) The diversity and antimicrobial activity of endophytic actinomycetes isolated from medicinal plants in *Panxi plateau*. China Curr Microbiol 62:182–190

Zhao GZ, Li J, Huang HY, Zhu WY, Park DJ, Kim CJ, Xu LH, Li WJ (2011a) *Pseudonocardia kunmingensis* sp. nov., an actinobacterium isolated from surface-sterilized roots of *Artemisia annua* L. Int J Syst Evol Microbiol 61:2292–2297

Zhao GZ, Li J, Huang HY, Zhu WY, Zhao LX, Tang SK, Xu LH, Li WJ (2011b) *Pseudonocardia artemisiae* sp. nov., a novel actinobacterium isolated from surface-sterilized *Artemisia annua* L. Int J Syst Evol Microbiol 61:1061–1065

Zhao GZ, Li J, Zhu WY, Li XP, Tian SZ, Zhao LX, Xu LH, Li WJ (2011c) *Pseudonocardia bannaensis* sp. nov., a novel actinomycete isolated from the surface-sterilized roots of *Artemisia annua* L. Antonie Leeuwenhoek 100:35–42

Zhao GZ, Zhu WY, Li J, Xie Q, Xu LH, Li WJ (2011d) *Pseudonocardia serianimatus* sp. nov., a novel actinomycete isolated from the surface-sterilized leaves of *Artemisia annua* L. Antonie Leeuwenhoek 100: 521–528

Zhao GZ, Li J, Zhu WY, Wei DQ, Zhang JL, Xu LH, Li WJ (2012a) *Pseudonocardia xishanensis* sp. nov., an endophytic actinomycete isolated from the roots of *Artemisia annua* L. Int J Syst Evol Microbiol 62:2395–2399

Zhao GZ, Li J, Zhu YW, Tian SZ, Zhao LX, Yang LL, Xu LH, Li WJ (2012b) *Rhodococcus artemisiae* sp. nov., an endophytic actinobacterium isolated from the pharmaceutical plant *Artemisia annua* L. Int J Syst Evol Microbiol 62:900–905

Zhi XY, Li WJ, Stackebrandt E (2009) An update of the structure and 16S rRNA gene sequence-based definition of higher ranks of the class *Actinobacteria*, with the proposal of two new suborders and four new families and emended descriptions of the existing higher taxa. Int J Syst Evol Microbiol 59:589–608

Zhu WY, Zhao LX, Zhao GZ, Duan XW, Qin S, Li J, Xu LH, Li WJ (2012) *Plantactinospora endophytica* sp. nov., an actinomycete isolated from *Camptotheca acuminata* Decne., reclassification of *Actinaurispora siamensis* as *Plantactinospora siamensis* comb. nov. and emended descriptions of the genus *Plantactinospora* and *Plantactinospora mayteni*. Int J Syst Evol Microbiol 62:2435–2442

Zin NM, Loi CS, Sarmin NM, Rosli AN (2010) Cultivation-dependent characterization of endophytic actinomycetes. Res J Microbiol 5:717–724

3 Foliar Fungal Endophytes in Herbaceous Plants: A Marriage of Convenience?

Amanda F. Currie, James Wearn, Su Hodgson, Hilary Wendt, Sue Broughton, and Liang Jin

Abstract

Foliar fungal endophytes are widespread in herbaceous plants, although their interactions and ecological roles are little understood. They are phylogenetically and ecologically diverse, with the potential to be influential members of the biotic community. Compositionally, the endophyte community within a plant is determined by both the fungi (genotype, competitive ability, tissue specificity, infection location) and the host (genotype, variations in plant defences, geographical location). The plant–endophyte relationship is dynamic, as fungal composition varies temporally across months and seasons, with subsequent infections occurring after initial colonisation. Transmission generally occurs horizontally via air- or water-borne spores, with hyphae entering the host through stomata or through direct penetration. Contrasting to extensive mycorrhizal fungal colonisation in roots, infection by any one endophyte in aerial parts appears to be limited, due to plant defences, intra- or interspecific competition between endophytes and other factors governing niche occupancy. Fungal endophytes colonise host tissues for at least part of their life cycle, with no apparent outward pathology. Simultaneously, they can benefit their hosts through improved tolerance to biotic stress such as drought, enhanced photosynthesis and transpiration, protection against pathogens through induced plant systemic resistance and the deterrence of phytophagous invertebrates (depending on their feeding guild and degree of specialism). These benefits arise directly from endophyte metabolism or indirectly

A.F. Currie (✉) • S. Hodgson • H. Wendt
S. Broughton • L. Jin
School of Biological Sciences, Royal Holloway University of London, Egham, Surrey TW 20 EX, England, UK
e-mail: amanda.currie@rhul.ac.uk

J. Wearn
Herbarium, Library, Art and Archives, Royal Botanic Gardens, Kew, Richmond, Surrey TW9 3AE, UK

V.C. Verma and A.C. Gange (eds.), *Advances in Endophytic Research*,
DOI 10.1007/978-81-322-1575-2_3, © Springer India 2014

through the production of compounds that alter the host's physiology. Thus, the influence of fungal endophytes may pervade beyond their host plant, potentially affecting the nature of plant communities and that of higher tropic levels.

1 Introduction

Foliar fungal endophytes are widespread in herbaceous plants, although their interactions and ecological roles are little understood within these plants. They are phylogenetically and ecologically diverse, with the potential to be influential members of the biotic community. The role of microbes, especially those inhabiting the above-ground plant tissue, in shaping and altering plant communities has often been underestimated (Clay and Schardl 2002; van der Heijden 2004). Yet, only now, as techniques to sample, identify and subsequently monitor a plant's endophytic community have advanced, is there an increasing awareness of the complex dynamics taking place within ecosystems at all levels (e.g. Schulz and Boyle 2005; Hoffman and Arnold 2008; Gibert et al. 2013).

Foliar fungal endophytes in the majority of herbaceous plants belong to the non-clavicipitalean group (generally non-grass inhabiting). In a review undertaken by Rodriguez et al. (2009a), this group of endophytes was differentiated into Class II (infecting above- or belowground host tissues) and Class III (occurring primarily or exclusively in above-ground host tissues) (Rodriguez et al. 2009a). Class II endophytes are all members of the Dikarya (Ascomycota or Basidiomycota), most of which belong to the Ascomycota, as do the majority of endophytes in Class III, where they are common throughout the Pezizomycetes, Leotiomycetes, Eurotiomycetes, but especially the Sordariomycetes and Dothideomycetes.

Endophytes from each class dominate a particular host plant lineage or biome, e.g. endophytic Leotiomycetes are frequently isolated from conifers, whilst Sordariomycetes are found within woody tropical plants (Arnold et al. 2007; Higgins et al. 2007; Arnold and Lutzoni 2007; Rodriguez et al. 2009a). Some species are also Basidiomycotina, Deuteromycotina and Oomycetes, which are generally associated with woody plants (Petrini 1986; Chapela and Boddy 1988; Zheng and Jiang 1995; Sinclair and Cerkauskas 1996). Class III endophytes appear to be widespread in herbaceous plants, with the photosynthetic tissues of all plant species so far surveyed containing one or more endophytic species (Stone et al. 2000; Arnold 2007). Class III endophytes have been isolated from tropical leaves and plants (Lodge et al. 1996; Arnold et al. 2000) but also from nonvascular and seedless plants and trees and woody and herbaceous angiosperms in all biomes (Carroll and Carroll 1978; Stone 1988; Cabral et al. 1993; Fisher et al. 1995; Barnes and Shaw 2002; Saikkonen et al. 2003; Higgins et al. 2007; Gange et al. 2007). Class II endophytes appear to show less diversity within a host compared to Class III endophytes, but they can confer habitat-specific stress tolerance to a range of genetically different host plants as a result of specific local environmental pressures such as pH, temperature or salinity. Rodriguez et al. (2009a) suggests that Class II endophytes dramatically affect the ecophysiology of plants, allowing and enhancing rapid adaptation of host plants to otherwise unsuitable high-stress habitats. Plants in these habitats usually have very high infection frequencies (90–100 %) (Redman et al. 2001, 2002a; Rodriguez et al. 2008). Class III endophytes are known for their high diversity in host tissues, plants and populations. Examples of this are seen in tropical forests, where healthy leaves contain numerous infections (Lodge et al. 1996; Arnold and Herre 2003).

Class II endophytes have been found to enhance shoot and/or root biomass and improve growth and nutrient acquisition (Newsham 1994; Mucciarelli et al. 2003; Waller et al. 2005). This may be due to the production of plant hormones, stimulated by the interaction between fungi and host, or by the biosynthesis of these hormones by the endophytes themselves (Tudzynski and Sharon 2002). They provide tolerance to disease and pathogens (Narisawa et al. 2002; Campanile et al. 2007) and enhance resistance in some hosts, genotypes and environmental conditions, to drought, desiccation, heat and salinity (Redman et al. 2001; Kloepper 2002; Marquez et al. 2007; Rodriguez et al. 2008). Arnold et al. (2000) found that a range of Class III endophytes in a host plant had no influence on biomass, growth rate, root to shoot ratio or any other aspect to plant fitness. However, Webber (1981) showed that bark endophytes protected trees against Dutch elm disease, yet Schulz et al. (1998) demonstrated that some Class III endophytes impaired plant growth. Petrini (1991) suggests that unlike those in the clavicipitalean group, these endophytes may cause disease in a host plant after a period of latency, whilst others may only reproduce upon or after the onset of senescence or death of the host (Sinclair and Cerkauskas 1996). Host senescence often results in Class II and Class III endophytes rapidly emerging and sporulating, resulting in horizontal transmission of spores via wind and/or rain (Weber et al. 2004; Herre et al. 2005). Vertical transmission also occurs via the seed coat, seed or rhizomes (Redman et al. 2002a) in Class II endophytes and those in Class III (Posada and Vega 2005; Ganley and Newcombe 2006). Spores from Class III endophytes can also be dispersed in animal faeces or on their bodies (Monk and Samuels 1990; Arnold 2008; Fledman et al. 2008; Selosse et al. 2008).

How these foliar endophytes affect herbivorous insects and other invertebrates has received limited attention. Other fungal–plant–insect interactions have shown that feeding preferences by invertebrates can be influenced by the presence of certain endophytic fungi (Vicari et al. 2002; Roger et al. 2013) and that invertebrate fecundity (Gange et al. 1999) and survival are also affected (Currie et al. 2011; Nishida et al. 2010). These multitrophic interactions are an important aspect of community ecology that needs to be further investigated.

2 Relative Abundance and Sampling Efficiency

A review of the literature covering foliar fungal endophytes since 1979, combined with unpublished data from Hodgson (2010), showed that 69 species of herbaceous host plants from 47 genera across 24 plant families had been investigated for their endophytic communities (Fig. 3.1). This equates to only 5 % of the total number of currently accepted vascular plant families and is evident that further investigation of herbaceous host–foliar fungal endophyte interactions is required. Most of the genera surveyed are represented by one species only. Therefore, attempts at investigation of host specificity or patterns of endophyte distribution are limited by data currently available.

Host plants examined varied from native British plants such as common dock (*Rumex acetosa*) (Hodgson 2010) to ten species of tropical orchids (*Dendrobium* spp.) (Chen et al. 2011), as well as peanut (*Arachis hypogea*) (Suryanarayan and Murali 2006), and *Arabidopsis thaliana* (Junker et al. 2012).

Differences in endophyte isolation techniques may contribute to an unclear picture of endophyte abundance within herbaceous hosts. Cosmopolitan fungal species, which are isolated as endophytes from a wide variety of plants, are also characteristic of leaf epiphytic communities. Examples include *Alternaria* spp., *Cladosporium* spp. and *Epicoccum* spp. (e.g. Petrini 1991; Cabral et al. 1993; Gange et al. 2007; Wearn et al. 2012). They are isolated frequently and could be 'opportunistic' endophytes which have the ability to invade the plant more easily when it is under stress (Johnston 1998), possibly becoming pathogenic in these circumstances (Stone 1987; Carroll 1988; Schulz and Boyle 2005). Reviewing the literature showed that these opportunistic pathogens or saprobes were the most common endophytic genera to be isolated (Fisher and Petrini 1992), with species of *Alternaria* and *Fusarium* occurring in almost half of the hosts examined (Fig. 3.2). These generalist opportunistic species may at the onset appear to dominate the community composition, but endophyte

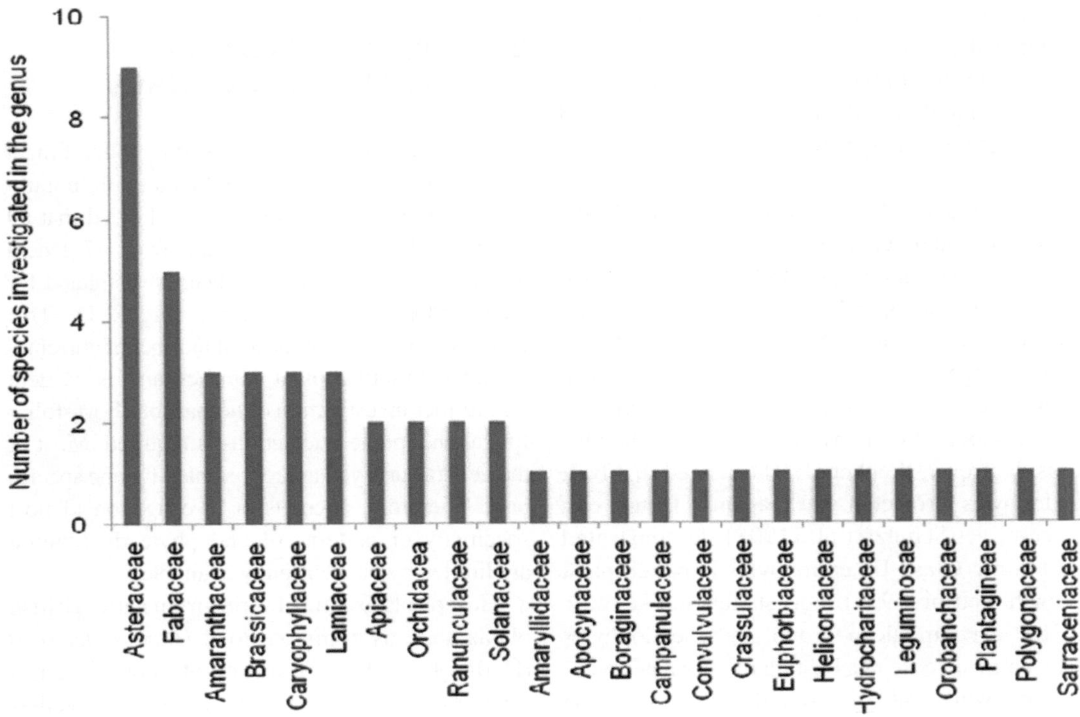

Fig. 3.1 Families of herbaceous plants appearing in the endophyte literature (World of Science)

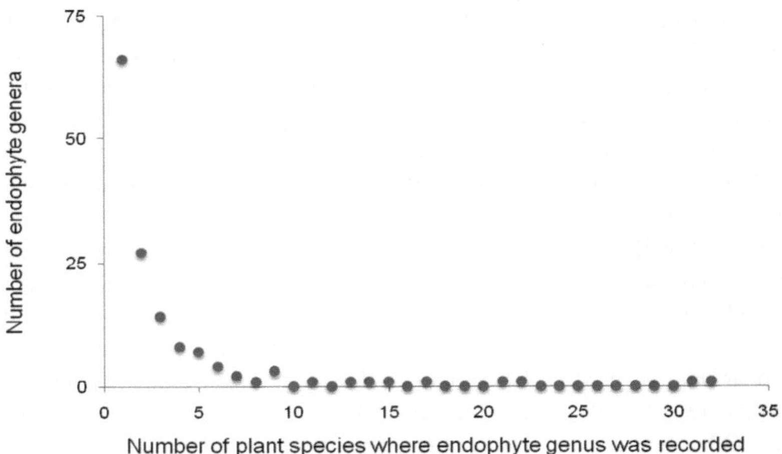

Fig. 3.2 Endophyte genus occurrence in plant hosts

community structure is actually skewed by the large proportion of apparently rare species. Around 70–90 % of endophytes are cultured only once or with very low abundances from any one plant. The host ranges and tissue preferences of these fungal species remain little known. This is very clear from studies undertaken in both the tropics, where leaves of tropical trees are considered to be hotspots of fungal diversity (e.g. Arnold and Lutzoni 2007), and temperate habitats (e.g. Gange et al. 2007; Wearn et al. 2012).

The literature survey revealed that approximately 320 endophyte species from 117 genera were associated with the plants investigated, 47 % of

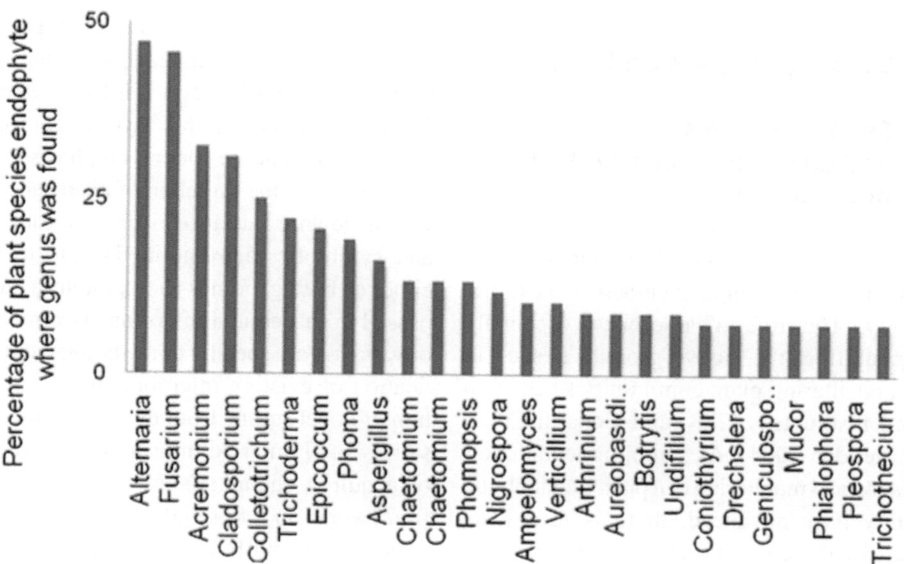

Fig. 3.3 Endophyte genera occurring in more than 5 % of plant species examined in papers published in Web of Science

which occur in one host genus only, with 65 % of endophytic species recorded as occurring in only one plant host (Fig. 3.3). Moreover, new host and national records, as well as species (even new classes of fungi) new to science, have been recovered as sampling effort has increased (e.g. Arnold and Lutzoni 2007; Wearn 2009; Gazis et al. 2012; Wearn et al. 2012) – but these fungi usually occur at very low levels of abundance in any given host plant or sampling regime.

Endophytes have been shown to exhibit traits similar to those found within bacteria, e.g. *Pseudoalteromonas tunicate*, where certain species are seemingly able to inhibit or facilitate the growth of others by the use of extracellular toxins and substances, with their presence ultimately changing the chemical composition of the substrate (Egan et al. 2000). Schulz et al. (2002) showed that 80 % of fungal endophytes tested exhibited at least one antibacterial, fungicidal, algicidal or herbicidal property. The metabolites produced by the endophytes depended upon the biotope in which they grew and to which they had adapted. Results from fungi growing in different biotopes showed they produced different metabolites with similar inhibitory functions. Gloer (1997) and Dreyfuss and Chapela (1994) found that the production of cyclosporin A, enchinocandin B, papulacandins and verrucarins varied with both habitat and substrate. Consequently, to avoid potential endophyte–endophyte competitive interactions, which could occur within a large piece of plant tissue during sampling and would result in the loss of one or more species when isolated on selective media, small segment sizes (ranging from 1 to 10 mm) of the selected plant tissue have been standard.

These very small segment sizes have been commonly used over the last 30+ years by a plethora of endophyte researchers around the globe. Moreover, this method was reinforced by a study completed by Carroll (1995) and Gamboa et al. (2002) who suggested that the optimal plant tissue segment size for endophyte isolation from tropical plants was less than 2.5 mm by 2.5 mm. Nevertheless, since these studies, many researchers have used segment sizes far larger than the recommended size. This can be justified given that temperate plants appear to harbour far fewer endophytes, which often exhibit a degree of organ and tissue specificity plus localised colonisation (Stone et al. 1994; Peláez et al. 1998; Carrol 1999; Deckert et al. 2001; Sieber 2002; Lu et al. 2004; Rodrigues et al. 2004). Therefore, a very small segment size may not be appropriate or optimal for all endophyte analyses, particularly those drawn from temperate hosts.

3 Existing in Space and Time

3.1 Symbiogenetics and Community Level Effects of Endophytes

As stated earlier, foliar fungal endophytes have the potential to be influential members of the biotic community. The role of microbes, especially those inhabiting the above-ground tissue, in shaping and altering plant communities has often been underestimated (Clay and Schardl 2002; Van der Heijden 2004). As sampling techniques, identification and monitoring of a plant's endophytic community have advanced, there is increasing awareness of the complex and dynamic interactions taking place within ecosystems at all levels (e.g. Schulz and Boyle 2005; Hoffman and Arnold 2008; Gibert et al. 2013). Schulz and Boyle (2005) published a review of the interactions between endophytes and their hosts, whilst more recently, Rodriguez et al. (2009b) provided a summary of endophyte diversity and function. The functional roles of endophytic fungi, when inhabiting plant organs, vary along a continuum from mutualistic to parasitic. Some live in apparent harmony with their hosts, whilst others are in constant combat with their host and/or cohabitants. Our awareness of the complexity of these diverse relationships has increased considerably, but much work is still required to enable a complete understanding of physiological pathways and evolutionary processes which have shaped this continuum and are continuing to do so.

Non-mycorrhizal fungal endophytes (both foliar and in the rhizosphere) engage in critical roles affecting above-ground to belowground (AG–BG) interactions within ecosystems, yet this taxonomically diverse group is absent from a recent review of multitrophic interactions (van Dam and Heil 2011) – evidence that recognition of the presence and importance of endophytes remains poor, even within the ecological community. Recent research has shown that herbaceous plant tissues infected with selected endophytic fungi can have positive or negative effects on invertebrate herbivore survival depending on the identities of all of the interacting organisms (Gange et al. 2012), whilst endophytic fungi can also host endohyphal bacteria (Hoffman and Arnold 2008), adding yet another layer of complexity.

The hologenome theory emphasises the role of microbes in the evolution of plants and animals, where genetic variation in the holobiont (host and symbiotic microbiotia) is able to occur in either or both genomes (hologenome) and can be passed to subsequent offspring. Two mechanisms of variation are specific to the hologenome, amplification of existing microorganisms and acquisition of novel strains from the environment, which satisfy the Lamarckian principle of 'inheritance of acquired characteristics' within a Darwinian framework and highlight the cooperation and competition within and between holobionts (Rosenberg et al. 2009). Accordingly, many microbes, including fungal endophytes, have the ability to change and modify the host plant at the genotypic, physiological and ecological level. Lucero et al. (2006) showed dramatic differences in the morphology and biomass of endophyte-infected *Bouteloua eripoda, Atriplex canescens* and *Sporobolus cryptandrus* plants (Poaceae). Additionally, Rodriguez et al. (2009b) noted that, traditionally, plant traits such as growth, stress tolerance and reproduction are 'treated as if they were genetic processes exclusive to the plant genome (intragenomic)'. However, as they point out, this view overlooks the ubiquity of the plant–endophyte association and ignores the fact that endophytes can have marked effects on the ability of plants to adapt to environmental stresses and can influence plasticity and phenology, as well as seedling establishment. Accordingly, Rodriguez et al. (2009b) assert that endophytes can affect population dynamics and reproductive success, stating: 'endophytes represent an intergenomic epigenetic form of plant gene regulation'. This conclusion stems in part from their work on several subspecies of sagebrush (*Artemesia tridenta*, Asteraceae) that occur in contiguous areas in North America, where the endophyte species associated with the sagebrush were habitat specific and transmitted on seed coats. These endophyte species interacted with the soils and different plant genotypes across the hybrid zones 'to confer enhanced plant reproduction in soil native to the endophyte and reduced reproduction in soil alien to the endophyte'. This 'intergenomic interaction provides a selective advantage, habitat specificity, and the means of restricting gene flow thus potentially leading to plant speciation'.

Endophytes have recently been linked to plant persistence strategies under environmental change. Gibert et al. (2012) found that endophytic interactions and water availability were linked to host plant survivorship (of *Lolium perenne*) and plant competition at population level. Gibert et al. (2013) investigated the effects of endophytic fungi on grass (*Festuca eskia*) persistence strategies in an alpine habitat, in order to predict responses of alpine plants to environmental change. Plants engaged in endophytic symbiosis (with the non-sexual form of *Epichloë festucae*) produced greater vegetative growth than non-endophytic plants and were able to shift their persistence strategy in response to increased levels of soil resources. The authors concluded that the presence of the fungal endophyte 'fine-tuned host persistence strategies according to soil resource level'. Rudgers et al. (2012) highlighted another level of complexity by modelling *Cinna arundinacea–Neotyphodium* interactions. They discovered that although an endophyte could negatively affect one aspect of grass plant fitness, this could be offset by a positive effect on another aspect, thereby producing a net benefit to the host. This can then be translated into population demographics. Thus, observation of single response parameters (in this case individual plant survival) could easily lead to incorrect conclusions from oversimplification. By considering regeneration, Rudgers et al. (2012) were able to elucidate the overall positive effect within their grass study system. Whether these outcomes occur throughout herbaceous host will need to be further investigated.

3.2 Influences of Host and Endophyte Genotypes

The endophyte community within a plant is determined by both the fungi (genotype, competitive ability, tissue specificity, infection location) and the host (genotype, variations in plant defences, geographical location). Numerous studies have shown that endophyte richness and diversity are influenced by a vast array of abiotic and biotic factors: the microclimate, microhabitat and geographic location (Carroll and Carroll 1978; Fisher et al. 1992; Rodrigues 1994; Higgins et al. 2007; and many others). More important, yet often overlooked, are the role of host and endophyte genotype and the complex interactions between the two, which shape plant and endophyte communities. Saikkonen (2007) likened a forest to the theory of island biogeography (MacArthur and Wilson 1967), where a forest was an archipelago, leaves were islands for endophyte infections arising from single-spore origin and trees could be monocormic, polycormic or clonal, which would influence the rate and pattern of infection and colonisation by endophytes. This analogy can easily be applied to any host plant species. A plant species may seem to be homogenous, but spatial and genetic differences can render some plants more, and others less, susceptible to endophyte infection and subsequent colonisation. For example, changing environmental conditions can influence the level of host susceptibility. Ahlholm et al. (2002a) observed this host–endophyte interaction in birch tree (*Betula* spp.), where the tree genotype directly influenced the diversity of the fungal endophyte *Venturia ditricha*. Host genotype-enhanced resistance or increased susceptibility has been studied extensively within the model organism *Arabidopsis thaliana*, especially in relation to pathogens. The enhanced disease resistance 1 (edr1) mutant has increased resistance to powdery mildew *Golovinomyces cichoracearum* (syn. *Erysiphe cichoracearum*) compared to wild type and other *Arabidopsis* genotypes (Wawrzynska et al. 2008, 2010; plus many others), whereas the enhanced disease susceptibility 1 (eds1) mutant is far more susceptible to *Hyaloperonospora parasitica* (syn. *Peronospora parasitica*), downy mildew (e.g. Parker et al. 1996, and others).

New genetic combinations are constantly being produced by sexual reproduction undertaken by both the host plant and endophyte species. This will inevitably lead to a degree of genetic incompatibility between host and endophyte, fuelling intensely complex interactions and an arms race between the two – the Red Queen hypothesis (Van Valen 1973). Furthermore as Saikkonen et al. (2004) suggested, these interactions may partially explain the varying levels of diversity and endophyte infection seen between pioneering and established, managed and natural, plus perennial and annual vegetation types. These genotypic differences among host and endophyte are probably the reason why endophyte transience is so prevalent and diversity is so varied within and between plants. Gundel et al. (2012) looked at the effects of gene flow in grass plants (*Lolium multiflorum*) on an endophytic symbiosis (with *Neotyphodium occultans*), testing the hypothesis that whilst gene flow produces hybrid vigour, it could reduce compatibility with endophytes and therefore the benefits derived from any symbiosis. Their experiments demonstrated an interaction between plant genotype and endophyte: the effects of the symbiosis at plant and population levels being controlled by the type and magnitude of environmental stress. Resistance to herbivory (by aphids) as a result of symbiosis in this case occurred independently of plant genotype, but resistance to herbicide differed between plant genotypes and was modified by levels of chemical application.

The benefits and roles provided by endophytes to the host have been well documented, especially in agronomic grasses and to a much lesser extent in trees and angiosperms (Malinowski et al. 1997; Waller et al. 2005; Porras-Alfaro et al. 2008; Rodriguez et al. 2008, and many others), with endophytes endowing the host with an extended phenotype (Yuan et al. 2010). Endophytes benefit their hosts through improved tolerance to biotic stress such as drought, enhanced photosynthesis and transpiration, protection against pathogens through induced plant systemic resistance and the deterrence of phytophagous invertebrates (depending on their feeding guild and degree of specialism). These benefits arise directly from endophyte metabolism or indirectly through the production of compounds that alter the host's physiology.

Studies focusing on natural grass populations suggest that the interaction ranges from antagonistic to mutualistic depending upon host and endophyte genotype and the prevailing environment (Agee and Hill 1994; Saikkonen et al. 1999, 2006; Clay and Holah 1999). However, the influence of the host and its internal environment on endophytic communities remains relatively unstudied. For many endophyte species, the host is likely to have a large impact, because the fungi are completely reliant on the plant for nutrition, protection and in some cases survival (Pan and Clay 2004). Genetic variation of the host was shown to enhance endophyte species richness and community composition within maize (Pan et al. 2008), and the percentage of condensed tannins within the bark of different *Populus* (Salicaceae) hybrids directly influenced the composition of the colonising endophyte community (Schweitzer et al. 2006). Likewise, host genotype has been shown to affect the richness and diversity of a wide range of other endophytic organisms, including mycorrhizal communities (Korkama et al. 2006), bacterial endophytes (Adams and Kloepper 2002), soil bacteria (Smith and Goodman 1999) and gut communities (Zoetendal et al. 2001; Vaahtovuo et al. 2003; Stewart et al. 2005).

Conversely, the plant does not always benefit from the presence of endophytes, and in some cases, plant fitness increased when the endophyte was absent (Faeth and Sullivan 2003; Ahlholm et al. 2002b; Redman et al. 2002b), and Christensen et al. (1997, 2002) found that host genotype strongly controlled the amount and distribution of fungal hyphae within leaf tissue. This suggests that the plant is often in control of the interaction and relationship. This genotypic control is not limited just to endophytes with studies by Rasche et al. (2006), Klerks et al. (2007) and Correa et al. (2007), all showing that phylloplane communities comprised of bacteria, yeasts and filamentous fungi were also influenced by different host cultivars and genotypes. Therefore, considering that the infection route of most endophytes colonising the aerial parts of herbaceous hosts is thought to start with spores landing on the plant's above-ground surfaces (see Sect. 3.4), infection, colonisation and diversity levels are being controlled and regulated by the host from the onset, often before the host tissue is even penetrated. If this is the case, it may be that the host maintains the infection, colonisation and persistence of certain endophyte species that confer some form of benefit.

As a result of the development of our understanding of the close linkages between multiple genomes across taxonomic boundaries, the discipline of 'symbiogenetics' has been established as 'a science studying the genetic control of interspecies interactions' (Tikhonovich and Provorov 2009). To view a plant in isolation is a gross oversimplification – it is a phytological superorganism. In order to unravel the complexity of the endophyte–plant system, we must 'think of individual plants as ecosystems of interacting microbes' (Wearn et al. 2012), which produce astounding chemical diversity and, in turn, 'a platform for integrative research training spanning the biological and physical sciences' (Bascom-Slack et al. 2012). Furthermore, interdisciplinary recognition of the 'phytological superorganism' has widespread implications for plant translocation, conservation and restoration activities and also food security.

3.3 Spatial Variability and Transience

Like most organisms, endophytic communities are affected by their geographic latitude and location (Carroll and Carroll 1978; Petrini et al. 1982; Frohlich and Hyde 1999; Gange et al. 2007; Higgins et al. 2007; Hoffman and Arnold 2008). However, endophytic fungal spore viability, dispersal and subsequent infection are influenced by factors such as the topography of the site, ambient climatic conditions and the microclimate created by the host (Saikkonen et al. 1998; Collado et al. 1999; Higgins et al. 2007; Göre and Bucak 2007; and many others). Thus, the surrounding vegetation, plant density and architecture, along with plant identity, genotype and condition, all play a part in influencing the nature of endophytic colonisation. Wearn et al. (2012) showed that very different endophyte communities can be found in plants of the same species at a single locality, across different seasons, and that endophytes show organ specificity within an individual host plant. Ganley and Newcombe (2006) found that several endophyte species appeared to be restricted to a specific area or tissue type within a plant, with highly localised endophyte infections being confined to the intercellular space between several plant cells, often increasing towards the basal part of the midrib of a broad leaf (Helander et al. 1993; Cannon and Simmons 2002) or the needle base in conifers (Bernstein and Carroll 1977). In some cases, these specialist endophytes may be restricted to a single host species, as seen with *Colletotrichum phyllachoroides*, which was confined to only the leafy tissue of *Suaeda fruiticosa* (Amaranthaceae) (Petrini 1986; Fisher and Petrini 1987a, b), or a single genus or family. These findings strongly indicate a fungal-host system that is in constant flux and where fungi come and go within space and time.

Endophyte composition of plants also differs spatially within and among sites. Arnold and Herre (2003) identified that higher endophyte infection levels occurred beneath a forest canopy when compared to hosts located in a clearing. Arnold et al. (2003) and Gange et al. (2007) both showed that endophyte assemblage differed with increasing distance between host plant localities. Morse et al. (2007) observed that the outcome of the endophyte–host interaction depended primarily upon the endophyte haplotype but was greatly affected by plant genotype and the local environment. Thus, the costs and benefits of harbouring endosymbionts shift when plants are subjected to varying environmental conditions (Faeth and Fagan 2002). Equally, Gange and Ayres (1999) showed that a mycorrhizal fungus could be mutualist, commensal or even parasitic depending upon soil chemistry or other prevailing environmental factors. Similarly, Faeth (2002) stated that the magnitude and direction of the host–endophyte interaction depended significantly upon the plant and endophyte genotype and the abiotic–biotic local environment, which will vary greatly in both time and space. For instance, the levels of disturbance and stability in the land use history, e.g. whether intensively farmed or left as a natural grassland, will have a massive impact upon the soil microbial community and the vegetation pool (White and Pickett 1985).

This will influence the quantity of spore inoculum available and, consequently, the amount of endophyte infection and subsequent colonisation. Gamboa and Bayman (2001) found higher levels of endophyte diversity in leaves of *Guarea guidonia* (Meliaceae) in a forest preserve compared to a disturbed forest area in Puerto Rico. Therefore, the cost of harbouring endophytes may be greater for plants in some localities, which would influence the level of endophyte diversity. As a result, it may be more beneficial for these plants to restrict endophyte colonisation, thus ensuring that limited resources are used to sustain their own fitness, because the true costs of accommodating endophytic fungi have yet to be elucidated (Clay 2004).

3.4 Transmission of Endophytes

Fungal endophytes colonise host tissues for at least part of their life cycle, with no apparent outward pathology. There are two main types of endophyte transmission route, vertical and horizontal. Carroll (1988) split endophytes into two groups based upon their mode of transmission: type 1 endophytes, constitutive mutualists, which are systemic and vertically transmitted, and type 2 endophytes, inducible mutualists which are horizontally transmitted. Vertical transmission, systemically from parental plant to progeny via the seed, appears to be the dominant mode and is thought to be restricted to only a few endophyte species within the family Clavicipitaceae, which inhabit grasses (Poaceae), sedges (Cyperaceae) and rushes (Juncaceae). *Neotyphodium* is an asexual, anamorphic endophyte, which is vertically transmitted within plants in these taxa. The related, sexually reproducing teleomoph, *Epichloë*, has the ability to be transmitted systemically and vertically via seeds or sexually and horizontally (airborne from plant to plant) via spores (Schardl et al. 1997). Vertically transmitted endophyte species always provide a greater degree of host protection than those species that are able to alternate between vertical and horizontal transmission (Bucheli and Leuchtmann 1996; Schardl and Clay 1997), probably because they are completely host dependent.

In contrast to the *Epichloë–Neotyphodium* interactions within grasses and their allies, other plant groups (liverworts, mosses, ferns, gymnosperms and angiosperms) are infected by a plethora of horizontally transmitted endophyte species (Stone 1987; Saikkonen et al. 1998; Davis et al. 2003; Davis and Shaw 2008). It should be noted that grasses too are hosts to nonsystemic, horizontally transmitted endophytes (Marquez et al. 2012).

Horizontal transmission generally occurs via air- or waterborne spores, with hyphae entering the host through stomata or through direct penetration. Since the host seems to have most of the control in the host–endophyte interaction, it is not unreasonable to suggest that horizontal transmission of sexually selected spores would inevitably benefit the endophyte species, allowing them to increase their genetic diversity and potentially stay one step ahead of the host in the ensuing arms race. This may explain why many endophyte species colonising most plants have undertaken this pathway. Nevertheless, vertical seed-borne transmission has also been identified in several non-graminoid plant species, e.g. pine (*Pinus* spp.), cowpea (*Vigna unguiculata*), cocoa (*Theobroma cacao*), chestnut (*Castanea* spp.) and African mopane seeds (*Colophospermum mopane*), and, unlike grass seeds, in these cases multiple endophyte species have been isolated (Washington et al. 1999; Posada and Vega 2005; Rodrigues and Menezes 2005; Ganley and Newcombe 2006; Jordaan et al. 2006). Several years earlier, Wilson and Carroll (1994) reported that endophytes had been found within acorns, suggestive of vertical transmission occurring alongside horizontal transmission. Similarly, Gallery et al. (2007) screened seeds from *Cecropia insignis* (Urticaceae) using culture-independent methods and recovered a diverse array of Ascomycota, which seemed to be consistent with the species isolated as foliar endophytes. It remains unknown how or why these endophytes have developed this systemic transmission route, but more importantly may be the question why does the host allow them to do so, unless they confer some benefit to the embryo and seedling?

Faeth and Bultman (2002) predicted that alkaloids produced by endophytes would be more beneficial and prevalent at the seed and seedling stage in grasses, when predation, pathogen attack and herbivory would have a direct effect on host fitness, a theory supported by Siegel et al. (1990), Bush et al. (1993), Welty et al. (1994) and Leuchtmann et al. (2000). This is consistent with the meta-analysis findings of Barton and Koricheva (2010), who showed that within woody plants, chemical defences increase at the seedling stage augmented by physical defences during the juvenile stage. However, within herbaceous plants, the level of secondary chemicals increased significantly throughout their entire ontogenetic trajectory.

3.5 Endophyte Succession

The plant–endophyte relationship is dynamic, as fungal composition varies temporally across months and seasons, with subsequent infections occurring after initial colonisation. Contrasting to extensive mycorrhizal fungal colonisation in roots, infection by any one endophyte in aerial parts appears to be limited, due to plant defences, intra- or interspecific competition between endophytes and other factors governing niche occupancy. Thus, endophytes undergo a form of succession, as species communities change. Ganley and Newcombe (2006) using regression analysis showed that the endophyte assemblage in *Pinus monticola* (Pinaceae) was influenced by the age of the host tree. Correspondingly, Hodgson (2010) showed marked differences in endophyte species richness in the largest/oldest leaves compared to that found in the smallest leaf and the relationship between plant height and leaf richness in *Rumex acetosa* (Polygonaceae). Here endophytes were undergoing succession, as fungal communities changed over time with increasing leaf age/size.

4 Foliar Endophytes and Abiotic Stress

Heat, drought and salt stress induce some similar plant responses including altered water relations, increased osmolyte production, production of signalling molecules such as abscisic acid (ABA) and the generation of reactive oxygen species (ROS) (Bohnert et al. 1995; Bray 1997; Wang et al. 2003; Apel and Hirt 2004).

Fungal endophytes may confer host protection against abiotic stresses, influence plant physiology and therefore enhance plant growth (Franken 2012). According to Rodriguez et al. (2009a), only Class II endophytes confer habitat-adapted stress tolerance to stresses such as pH, salinity or temperature and suggest that these endophytes are important for the survival of some plants in high-stress environments (Rodriguez et al. 2004) as mentioned earlier. Rodriguez et al. (2009a) state that Class III and Class IV endophytes have not been examined for their ability to confer stress tolerance. Much work has been carried out on the effects of these stresses to host plants infected with endophytes from the clavicipitalean group, probably as a result of the economic importance of the host plants.

4.1 Tolerance to Drought Stress

Plants respond to drought stress through a range of physiological and biochemical changes, and research has shown that fungal endophytes are able to increase a host plant's tolerance to drought stress, possibly through the enhancement of root development and leaf growth, regulating the opening and closing of stomata (Elbersen and West 2006; Swarthout et al. 2009), osmotic regulation (Bacon 1993) and improvement of the anti-oxidation protection system (Hamilton et al. 2012).

Studies have demonstrated that endophyte-infected plants are able to survive under conditions of drought stress whilst maintaining high yields. Arachevaleta et al. (1989) reported that under conditions of heavy drought, the biomass of plants infected with endophytes was higher, survival was improved and there was a significant increase in the ability to recover after drought, compared to endophyte-free plants, where 75 % of plants died. Conversely, Hesse et al. (2003) found that *Neotyphodium* spp. inoculating *L. perenne* could either promote or reduce plant growth when drought stress was applied, depending on the habitat that the host plant originated from. Malinowski and Belesky (2000) showed that

Neotyphodium spp. infecting tall fescue induced mechanisms of drought avoidance (morphological adaptations), drought tolerance (physiological and biochemical adaptations) and drought recovery in infected grasses.

Much work has been carried out on the clavicipitaceous and Class II endophytes and how they influence drought tolerance in their host plants. There appears to be a paucity of knowledge on whether Class III foliar endophytes will similarly affect a host plant's ability to withstand periods of drought.

4.2 Tolerance to Salinity Stress

For crop plants, soil salinity is one of the most significant abiotic stresses, as it reduces crop yield by more than 50 % (Boyer 1982; Bray et al. 2000). An example of this is the root-inhabiting endophyte *Piriformospora indica* which is able to increase plant growth under both normal and stress conditions (Waller et al. 2005; Schäfer et al. 2007). Within non-graminoid herbaceous plants, there appears to be no examination of the effects of foliar fungal endophytes on the stress caused by high soil salinity.

Similarly, as with other aspects of the ecological roles these non-graminoid foliar endophytic fungi have, there is a general paucity of research carried out on the effects – if any – that these fungi will have on the ability of their hosts to tolerate abiotic stress in various forms.

5 Invertebrates–Foliar Endophytes–Herbaceous Plants

5.1 Direct Interactions Between Endophytes and Invertebrates

Within the herbaceous plant–foliar endophyte system, herbivorous invertebrates have received little attention in the literature, compared to the interactions observed with graminoid host species (e.g. Lewis and Clements 1986; Clement et al. 2011; Crawford et al. 2010 and others). Endophytic fungi are important mediators of plant–herbivore interactions (Rajagopal and Suryanarayanan 2000; Omacini et al. 2001; Miller et al. 2002; Meister et al. 2006) and may enhance resistance to herbivorous insects (Breen 1994), by increasing the production of various alkaloid-based defence compounds within the host's tissues (Clay and Holah 1999; Faeth 2002) or through a change in the nutritional quality (e.g. phytosterols) of the plant (Bernays 1993). These changes to plant chemistry protect the host plant by deterring insect herbivores (Latch et al. 1985), reducing herbivory (Knoch et al. 1993), development rate (Valenzuela-Soto et al. 2010) and survival (Lacey and Neven 2006) and oviposition (Clay 1990). Ahmad et al. (1985) reported 100 % mortality rate of house crickets (*Acheta domesticus*) when they grazed perennial ryegrass (*Lolium perenne*) infected with the endophyte *Aremonium loliae*. In contrast, Jani et al. (2010) found that high alkaloid levels in the native grass *Achnatherum robustum* were associated with increased arthropod herbivore abundance and species richness. They suggest that these high alkaloid levels in native grasses may not protect the host from arthropod herbivores. The Argentine stem weevil (*Listronotus bonariensis*) had significantly higher feeding rates on perennial ryegrass infected with a *Gliocladium*-like endophyte compared to endophyte-free grass (Gaynor et al. 1983). Conversely, populations of this insect were negatively affected by the endophyte *Neotyphodium lolii* when it infected the same host grass (Prestidge et al. 1982). Gange et al. (2012) suggest that the degree of insect-feeding specialism and insect-feeding guild (Hartley and Gange 2009) are important factors when investigating the effect of foliar endophytes on insect herbivory. The majority of studies have focused on phloem feeders on grass hosts and their associated endophytes, where negative effects on insect performance were recorded (Wilkinson et al. 2000; Clement et al. 2001; Bultman et al. 2004; Hunt and Newman 2005; Krauss et al. 2007; Lehtonen et al. 2005; Meister et al. 2006; Züst et al. 2008). Contrary to this finding are the results recorded for chewing insects where they were either unaffected or positively influenced by the presence of foliar endophytes (Bultman and Bell 2003; Bultman et al. 2003; Davidson and Potter 1995; Williamson and Potter 1997). It is apparent that there is a continuum of endophyte effects on herbivorous invertebrates within the grass host system ranging from negative through null to positive outcomes. Table 3.1 shows the effects of various foliar endophytic fungi, infecting herbaceous plants, on specialist and generalist insects. With so few studies undertaken involving endophyte-infected herbaceous hosts and herbivorous invertebrates, it is at present impossible to draw any conclusions with regard to feeding specialism.

5.2 Higher Trophic Levels

Foliar fungal endophytes appear to be able to drive the shaping of multitrophic community structure. Evidence for this has again come from studies involving grass host species with no research being carried out using herbaceous hosts and their associated invertebrate herbivores. It appears that endophytes, and even individual isolates of endophytes being investigated, affect various aspects of parasitoid development and survival (Bultman et al. 1997, 2003). Predators attacking herbivorous invertebrates may also be affected by endophyte presence (Omacini et al. 2001; de Sassi et al. 2006). Hartley and Gange (2009) make the point that most of the studies so far undertaken on the multitrophic effects of endophytes have been controlled laboratory experiment, suggesting that investigation in the field may be difficult to carry out (Krauss et al. 2007). Gange et al. (2003) was the first to show that arbuscular mycorrhizal fungi can influence the performance of higher trophic levels, by increasing host plant size and as a result reducing the searching efficiency of the parasitoid for its host insect. This observation was also seen under field conditions. If there is an indirect effect of a root-inhabiting mycorrhizal fungus on a parasitoid, it seems likely that this should also happen to parasitoids, whose host insects are feeding on herbaceous foliage infected with endophytes.

These multitrophic interactions are an important aspect of community ecology that needs to be further studied, both in the laboratory and preferably in the field.

6 Conclusion

As we have demonstrated above, foliar endophytic fungi appear to be powerful, albeit underestimated, members of the biotic community. They are clearly influential in the performance of their host plants and should always be included in any plant performance investigation. As Wearn et al. (2012) suggest, plants themselves should be viewed as 'ecosystems of interacting microbes', so any research investigating plant parameters should always include an assessment of the endophyte community within the host plant. This reinforces Eriksen et al.'s (2002) study of traditionally managed boreal grasslands, which argues that 'plant interactions above and below ground' must be incorporated into management aimed at conserving species composition or restoration efforts aimed at the reintroduction of rare or vulnerable plant species. Therefore, it is hoped that future endophytic studies will advance our understanding of the role played by microbial communities in enabling, preserving or restoring diversity in the wide array of plant communities imperilled in the Anthropocene.

The three-way relationship between foliar fungal endophytes, herbaceous plants and herbivorous invertebrates requires further investigation to elucidate the ecological roles each member plays within this association, but also their impacts on higher trophic levels and the wider plant community, including host plant choice by insects (Gange et al. 2007). Included in this should be the assessment of any mycorrhizas present, as arbuscular mycorrhizal fungi are known to affect both herbivorous insects as well as the parasitoids that have so far been investigated (Gange et al. 2003).

Future research will need to take into account temporal variation of foliar endophytes and differences in geographical location even when

Table 3.1 The effects of foliar endophytic fungi, infecting herbaceous host plants, on generalist or specialist herbivorous invertebrates

Host plant	Endophyte	Invertebrate	Specialist/generalist	Outcome for invertebrate	References
Cucumis sativa	*Colletotrichum tropicale*	*Atta colombica*	Generalist	Leaves with low endophyte density were preferred	Estrada et al. (2013)
	Chaetomium cochlioides	*Mamestra brassicae*	Generalist	Reduced growth	Gange et al. (2012)
		Cassida rubiginosa	Specialist	Increased feeding	
Cirsium arvense	*Cladosporium cladosporioides*	*M. brassicae*	Generalist	No effect on feeding	
	C. cladoporioides+ *Trichoderma viride*	*C. rubiginosa*	Specialist	Increased feeding	
		C. rubiginosa	Specialist	Reduced feeding	
Merremia umbellata	*Glomerella cingulata*	*Chelymorpha altenans*	Generalist	No effect on feeding, but larvae fed diets augmented with high levels of this endophyte resulted in lower adult fecundity	van Bael et al. (2009a)
Merremia umbellata	*G. cingulata*	*Atta colombica*	Generalist	More time spent cutting leaves containing high levels of endophytes	van Bael et al. (2009b)
Centaurea stoebe	*Alternaria* sp. and *Epicoccum* sp.	*Larinus minutus*	Specialist	Inoculated plants were avoided	Newcombe et al. (2009)
Calotropis gigantea	*Colletotrichum gloeosporioides*	*Poecilocerus pictus*	Generalist	No effect on feeding	Devarajan and Suryanarayanan (2006)
Leucanthemum vulgare	*Chaetomium bostrychodes* and *C. cochlioides*	*Chromatomyia syngenesiae*	Generalist	Reduced fly attacks	Gange et al. (2007)
Cirsium Vulgare	*C. cochlioides*				
Gossypium hirsuta	'White coelomycete' and other 'morphospecies'	*Helicoverpa armigera*	Generalist	Reduced feeding and reduced larval growth	McGee (2002)

Where 'and' is used indicates that two endophyte species were investigated; '+' indicates the endophytes were used in a dual inoculation studying the same host plant, as well as endophyte specificity of host plants and their site of colonisation within host tissues.

The difference between non-herbaceous and herbaceous host plants is likely to have an impact on the composition and structure of the endophyte community infecting the host plant and, thus, their impact on herbivores and plant growth. However, as already stated, much more research is needed before any conclusions can be drawn in this area.

Endophytic research has the potential to offer insights into the complicated interactions of the natural world and may help to secure a better understanding of both the evolution and biogeography of a variety of plant taxa and that of other trophic levels, as well as helping to ensure appropriate conservation management. It is hoped that future work in this area will therefore assist our developing understanding of other members of the biotic community and of ecosystem function.

So is the relationship between foliar fungal endophytes and herbaceous plants a marriage of convenience? If the association is one of mutualism, then it appears to be a match made in heaven, with both partners seemingly gaining benefits. With a history of a very long-lived affair (of approx. 400 million years (Krings et al. 2007; Rodriguez and Redman 2008)), the relationship between foliar fungal endophytes and herbaceous hosts could perhaps be a marriage of true love. Other associations between plants and endophytes, however, may not be so harmonious, with partners remaining in the relationship whilst it is advantageous, to then defect to live with another. The fungus, through switching host, or the plant controlling its endophyte assembly, could drive this situation. At the extreme end of the relationship spectrum, one partner could turn on and destroy the other, such as a plant excluding costly endophytes, which are not paying their way, or an endophyte could be a latent pathogen or saprobe awaiting its opportunity to strike. Thus, 'true love', mutual support and dependence, infidelity and murder, plant host–endophyte relationships in herbaceous plants probably run the full gamut.

References

Adams PD, Kloepper JW (2002) Effect of host genotype on indigenous bacterial endophytes of cotton (*Gossypium hirsuta*). Plant Soil 240:181–189

Agee CS, Hill NS (1994) Ergovaline variability in *Acremonium*-infected tall fescue due to environmental and plant genotype. Crop Sci 34:221–226

Ahlholm JU, Helander ML, Henriksson J, Metzler M, Saikkonen K (2002a) Environmental conditions and host genotype direct genetic diversity of *Venturia ditricha*, a fungal endophyte of birch trees. Evolution 56:1566–1573

Ahlholm JU, Helander M, Lehtimäki S, Wäli P, Saikkonen K (2002b) Vertically transmitted fungal endophytes: different response of host-parasite systems to environmental conditions. Oikos 99:173–183

Ahmad S, Govindarajan S, Funk CR, Johnson-Cicalese JM (1985) Fatality of house crickets on perennial ryegrass infected with fungal endophyte. Entomol Exp Appl 39:183–190

Apel K, Hirt H (2004) Reactive oxygen species: metabolism, oxidative stress, and signal transduction. Annu Rev Plant Biol 55:373–399

Arachevaleta M, Bacon CW, Hoveland CS, Radcliffe E (1989) Effect of the tall fescue endophyte on plant response to environmental stress. Agron J 81:83–90

Arnold AE (2007) Understanding the diversity of foliar endophytic fungi: progress, challenges and frontiers. Fungal Biol Rev 21:51–66

Arnold AE (2008) Endophytic fungi: hidden components of tropical community ecology. In: Carson W, Schnitzer S (eds) Tropical forest community ecology. Blackwell Scientific Inc., Malden

Arnold AE, Herre EA (2003) Canopy cover and leaf age affect colonization by tropical fungal endophytes: ecological patterns and process in *Theobroma cacao* (Malvaceae). Mycologia 95:388–398

Arnold AE, Lutzoni F (2007) Diversity and host range of foliar fungal endophytes: are tropical leaves biodiversity hotspots? Ecology 88:541–549

Arnold AE, Maynard Z, Gilbert GS, Coley PD, Kursar TA (2000) Are tropical fungal endophytes hyperdiverse? Ecol Lett 3:267–274

Arnold AE, Mejia LC, Kyllo Dm Rojas E, Maynard Z, Robbins N, Herre EA (2003) Fungal endophytes limit pathogen damage in a tropical tree. Proc Natl Acad Sci USA 100(26):15649–15654

Arnold AE, Henk DA, Eells RL, Lutzoni F, Vilgalys R (2007) Diversity and phylogenetic affinities of foliar fungal endophytes in loblolly pine inferred by culturing and environmental PCR. Mycologia 99: 185–206

Bacon CW (1993) Abiotic stress tolerances (moisture, nutrients) and photosynthesis in endophyte-infected tall fescue. Agr Ecosyt Environ 44:123–141

Barnes SE, Shaw MW (2002) Factors affecting symptom production by latent *Botrytis cinerea* in *Primula* × *Polyantha*. Plant Pathol 5:746–754

Barton KE, Koricheva J (2010) The ontogeny of plant defense and herbivory: characterizing general patterns using meta-analysis. Am Nat 175:481–493

Bascom-Slack CA, Arnold AE, Strobel SA (2012) Student-directed discovery of the plant microbiome and its products. Science 338:485–486

Bernays EA (1993) Plant sterols and host-plant affiliations of herbivores. In: Bernays EA (ed) Insect–plant interactions, vol IV. CRC, Boca Raton, pp 45–57

Bernstein ME, Carroll GC (1977) Internal fungi in old-growth Douglas fir foliage. Can J Bot 55:644–653

Bohnert HJ, Nelson DE, Jensen RG (1995) Adaptations to environmental stresses. Plant Cell 7:1099–1111

Boyer JS (1982) Plant productivity and environment. Science 218:443–448

Bray EA (1997) Plant responses to water deficit. Trends Plant Sci 2:48–54

Bray EA, Bailey-Serres J, Weretilnyk E (2000) Responses to abiotic stresses. In: Gruissem W, Buchannan B, Jones R (eds) Biochemistry and molecular biology of plants. American Society of Plant Physiologists, Rockville, pp 1158–1249

Breen JP (1994) *Acremonium* endophyte interactions with enhanced plant resistance to insects. Annu Rev Entomol 39:401–423

Bucheli E, Leuchtmann A (1996) Evidence for genetic differentiation between choke-inducing and asymptomatic strains of *Epichloë* grass endophyte from *Brachypodium sylvaticum*. Evolution 50:1879–1887

Bultman TL, Bell GD (2003) Interaction between fungal endophytes and environmental stressors influences plant resistance to insects. Oikos 103:182–190

Bultman TL, Borowicz KL, Schneble RM, Coudron TA, Bush LP (1997) Effect of a fungal endophyte on the growth and survival of two *Euplectrus parasitoids*. Oikos 78(Fasc 1):170–176

Bultman TL, McNeill MR, Goldson SL (2003) Isolate-dependent impacts of fungal endophytes in a multi-trophic interaction. Oikos 102:491–496

Bultman TL, Bell GD, Martin WD (2004) A fungal endophyte mediates reversal of wound-induced resistance and constrains tolerance in a grass. Ecology 85:679–685

Bush LP, Gray S, Burham W (1993) Accumulation of alkaloids during growth of tall fescue. In: Proceedings XVII international grassland congression. Palmerston North, pp 1379–1381

Cabral D, Stone JK, Carroll GC (1993) The internal mycoflora of *Juncus* spp.: microscopic and cultural observations of infection patterns. Mycol Res 9:367–376

Campanile G, Ruscelli A, Luisi N (2007) Antagonistic activity of endophytic fungi towards *Diplodia corticola* assessed by *in vitro* and in planta test. Eur J Plant Pathol 117:237–246

Cannon PF, Simmons CM (2002) Diversity and host preference of leaf endophytic fungi in the Iwokrama Forest Reserve, Guyana. Mycologia 9:210–220

Carrol GC (1999) The foraging ascomycete. In: Abstracts, XVI international botanical congress, St Louis, p 309

Carroll GC (1988) Fungal endophytes in stems and leaves: from latent pathogen to mutualistic symbiont. Ecology 69:2–9

Carroll GC (1995) Forest endophytes: patterns and processes. Can J Bot 73(S1):1316–1324

Carroll GC, Carroll FE (1978) Studies on the incidences of coniferous needle endophytes in the Pacific Northwest. Can J Bot 56:3034–3043

Chapela IH, Boddy L (1988) Fungal colonization of attached beech branches. I. Early stages of development of fungal communities. New Phytol 110:39–45

Chen J, Hu K, Hou X, Guo S (2011) Endophytic fungal assemblages from 10 *Dendrobium* medicial plants (Orchidaceae). World J Microbiol Biot 27:1009–1016

Christensen MJ, Ball OJ-P, Bennett RJ, Schardl CL (1997) Fungal and host genotype effects on compatibility and vascular colonization by *Epichloë festucae*. Mycol Res 101:493–501

Christensen MJ, Bennette RJ, Schmid J (2002) Growth of *Epichloë/Neotyphodium* and endophytes in leaves of *Lolium* and *Festuca* grasses. Mycol Res 106:93–106

Clay K (1990) Fungal endophytes of grasses. Annu Rev Ecol Syst 21:275–279

Clay K (2004) Fungi and the food of the gods. Nature 427:401–402

Clay K, Holah J (1999) Fungal endophyte symbiosis and plant diversity in successional fields. Science 285:1742–1744

Clay K, Schardl CL (2002) Evolutionary origins and ecological consequences of endophyte symbiosis with grasses. Am Nat 160:99–127

Clement SL, Elberson LR, Youssef NN, Davitt CM (2001) Incidence and diversity of *Neotyphodium* fungal endophytes in tall fescue from Morocco, Tunisia and Sardinia. Crop Sci 41:570–576

Clement SL, Hu J, Steward AV, Wang B, Elberson LR (2011) Detrimental and neutral effects of a wild grass-fungal endophyte symbiotum on insect preference and performance. J Insect Sci 11:1–13

Collado J, Platas G, González I, Peláez F (1999) Geographical and seasonal influences on the distribution of fungal endophytes in *Quercus ilex*. New Phytol 144:525–532

Correa OS, Romero AM, Montecchia MS, Soria MA (2007) Tomato genotype and *Azospirillum* inoculation modulate the changes in bacterial communities associated with roots and leaves. J Appl Microbiol 102:781–786

Crawford KM, Land JM, Rudgers JA (2010) Fungal endophytes of native grasses decrease insect herbivore preference and performance. Oecologia 164:431–444

Currie AF, Murray PJ, Gange AC (2011) Is a specialist root-feeding insect affected by arbuscular mycorrhizal fungi? Appl Soil Ecol 47:77–83

Davidson AW, Potter DA (1995) Response of plant-feeding, predatory, and soil-inhabiting invertebrates to *Acremonium* endophyte and nitrogen fertilization in tall fescue turf. J Econ Entomol 88:367–379

Davis EC, Shaw AJ (2008) Biogeographic and phylogentic patterns in diversity of liverwort-associated endophytes. Am J Bot 95:914–924

Davis EC, Franklin JB, Shaw AJ, Vilgalys R (2003) Endophytic *Xylaria* (Xylariaceae) amongst liverworts and angiosperms: phylogenetics, distribution and symbiosis. Am J Bot 90:1661–1667 de Sassi C, Müller CB, Krauss J (2006) Fungal plant endosymbionts alter life history and reproductive success of aphid predators. Proc R Soc Edinb B 273:1301–1306

Deckert RJ, Melville L, Peterson RL (2001) Structural features of a *Lophodermium* endophyte during the cryptic life-cycle in the foliage of *Pinus strobus*. Mycol Res 105:991–997

Devarajan PT, Suryanarayanan TS (2006) Evidence for the role of phytophagous insects in dispersal of non-grass fungal endophytes. Fungal Divers 23:111–119

Dreyfuss MM, Chapela IH (1994) Potential of fungi in the discovery of novel, low molecular weight pharmaceuticals. In: Gullo VP (ed) The discovery of natural products with therapeutic potential. Butterworth-Heinemann, Stoneham, pp 49–80

Egan S, James S, Holmstrom C, Kjelleberg S (2000) Correlation between pigmentation and antifouling compounds produced by *Pseudoalteromonas tunicate*. Environ Microbiol 4:433–442

Elbersen HW, West CP (2006) Growth and water relations of field-grown tall fescue as influenced by drought and endophyte. Grass Forage Sci 51:333–342

Eriksen M, Bjureke KE, Dhillion SS (2002) Mycorrhizal plants of traditionally managed Boreal Grasslands in Norway. Mycorrhiza 12:117–123

Estrada C, Wcislo WT, van Bael S (2013) Symbiotic fungi alter plant chemistry that discourages leaf-cutting ants. New Phytol 198:241–251

Faeth SH (2002) Are endophytic fungi generally plant mutualists? Oikos 98:25–36

Faeth SH, Bultman TL (2002) Endophytic fungi and interactions among host plants, herbivores and natural enemies. In: Tscharnkte T, Hawkins BA (eds) Multitrophic level interactions. Cambridge University Press, Cambridge, pp 89–123

Faeth S, Fagan WF (2002) Fungal endophytes: common host plant symbionts but uncommon mutualists. Integr Comp Biol 42:360–368

Faeth SH, Sullivan TJ (2003) Mutualistic, asexual endophytes in a native grass are usually parasitic. Am Nat 161:310–325

Fisher P, Petrini O (1987a) Location of fungal endophytes in tissues of Suaeda fruticosa: a preliminary study. Trans Br Mycol Soc 89:246–249

Fisher PJ, Petrini O (1987b) Tissue specificity of fungi endophytic in Ulex europaeus. Sydowia 40:46–50

Fisher PJ, Petrini O (1992) Fungal saprobes and pathogens as endophytes of rice (*Oryza sativa* L.). New Phytol 120:137–143

Fisher PJ, Petrini O, Petrini LE, Descals E (1992) A preliminary study of fungi inhabiting xylem and whole stems of *Olea europea*. Sydowia 44:117–121

Fisher PJ, Graf F, Petrini LE, Sutton BC, Wookey PA (1995) Fungal endophytes of *Dryas octopetala* from the high arctic polar semidesert and from the Swiss Alps. Mycologia 87:319–323

Fledman TS, O'Brien H, Arnold AE (2008) Moth dispersal of mycoparasites and endophytes associated with *Claviceps paspali* and the grass *Paspalum* (Poaceae). Microb Ecol 56:742–750

Franken P (2012) The plant strengthening root endophyte *Piriformospora indica*: potential application and the biology behind. Appl Microbiol Bot 96: 1455–1464

Frohlich J, Hyde KD (1999) Biodiversity of palm fungi in the tropics: are global fungal diversity estimates realistic? Biodivers Conserv 8:977–1004

Gallery RE, Dalling JW, Arnold AE (2007) Diversity, host affinity, and distribution of seed-infecting fungi: a case study with *Ceropia*. Ecology 88:582–588

Gamboa MA, Bayman P (2001) Communities of endophytic fungi in leaves of a tropical timber tree (*Guarea guidonia*: Meliaceae). Biotropica 33:352–360

Gamboa MA, Laureano S, Bayman P (2002) Measuring diversity of endophytic fungi in leaf fragments: does size matter? Mycopathologia 156:41–45

Gange AC, Ayres RL (1999) On the relation between arbuscular mycorrhizal colonization and plant benefit. Oikos 87:615–621

Gange AC, Bower E, Brown VK (1999) Positive effects of an arbuscular mycorrhizal fungus on aphid life history traits. Oecologia 120:123–131

Gange AC, Brown VK, Aplin DM (2003) Multitrophic links between arbuscular mycorrhizal fungi and insect parasitoids. Ecol Lett 6:1051–1055

Gange AC, Dey S, Currie AF, Sutton BC (2007) Site- and Species-specific differences in endophyte occurrence in two herbaceous hosts. J Ecol 95:614–622

Gange AC, Eschen R, Wearn JA, Thawer A, Sutton BC (2012) Differential effects of foliar endophytic fungi on insect herbivores attacking a herbaceous plant. Oecologia 168:1023–1031

Ganley RJ, Newcombe G (2006) Fungal endophytes in seeds and needles of *Pinus monticola*. Mycol Res 110:318–327

Gaynor DL, Rowan DD, Latch GMC, Pilkington S (1983) Preliminary results on the biochemical relationship between adult Argentine stem weevil and two endophytes in ryegrass. Proc NZ Weed Pest Control Conf 36:220–224

Gazis R, Miadlikowska J, Lutzoni F, Arnold AE, Chaverri P (2012) Culture-based study of endophytes associated with rubber trees in Peru reveals a new class of Pezizomycotina: Xylonomycetes. Mol Phylogenet Evol 65:294–304

Gibert A, Volaire F, Barre P, Hazard L (2012) A fungal endophyte reinforces population adaptive differentiation in its host grass species. New Phytol 194:561–571

Gibert A, Magda D, Hazard L (2013) Endophytic fungus fine-tunes the persistence strategy of its alpine host grass in response to soil resource levels. Oikos 122:367–376

Gloer JB (1997) Applications of fungal ecology in the search for new bioactive natural products. In: Wicklow DT, Soderstrom BE (eds) The mycota, vol. IV. Environmental and microbial relationships. Springer, New York, pp 249–268

Göre ME, Bucak C (2007) Geographical and seasonal influences on the distribution of fungal endophytes in *Laurus nobilis*. For Pathol 37:281–288

Gundel PE, Martinez-Ghersa MA, Omacini M, Cuyeu R, Pagano E, Rios R, Ghersa CM (2012) Mutualism effectiveness and vertical transmission of symbiotic fungal endophytes in response to host genetic background. Evol Appl 5:838–849

Hamilton CE, Gundel PE, Helander M, Saikkonen K (2012) Endophytic mediation of reactive oxygen species and antioxidant activity in plants: a review. Fungal Divers 54:1–10

Hartley SE, Gange AC (2009) Impacts of plant symbiotic fungi on insect herbivores: mutualism in a multitrophic context. Annu Rev Entomol 54:323–342

Helander ML, Neuvonen S, Sieber T, Petrini O (1993) Simulated acid rain affects birch leaf endophyte populations. Microb Ecol 26:227–234

Herre EA, Van Bael SA, Maynard Z, Robbins N, Bischoff J, Arnold AE, Rojas E, Mejia LC, Cordero RA, Woodward C, Kyllo DA (2005) Tropical plants as chimera, some implications of foliar endophytic fungi for the study of host plant defense, physiology and genetics. In: Burslem DFRP, Pinard MA, Hartley SE (eds) Biotic interactions in the tropics. Cambridge University Press, Cambridge, pp 226–237

Hesse U, Schöberlein W, Wittenmayer L, Förster K, Warnstoff K, Diepenbrock W, Merbach W (2003) Effects of *Neotyphodium* endophytes on growth, reproduction and drought-stress tolerance of three *Lolium perenne* L. genotypes. Grass Forage Sci 58:407–415

Higgins KL, Arnold AE, Miadlikowska J, Sarvate SD, Lutzoni F (2007) Phylogenetic relationships, host affinity, and geographic structure of boreal and arctic endophytes from three major plant lineages. Mol Phylogenet Evol 42:543–555

Hodgson S (2010) Foliar fungal endophyte dynamics in herbaceous hosts. PhD thesis, University of London, London

Hoffman MT, Arnold AE (2008) Geographic locality and host identity shape fungal endophyte communities in cupressaceous trees. Mycol Res 112:331–344

Hunt MG, Newman JA (2005) Reduced herbivore resistance from a novel grass-endophyte association. J Appl Ecol 42:762–769

Jani AJ, Faeth SH, Gardner D (2010) Asexual endophytes and associated alkaloids alter arthropod community structure and increase herbivore abundances on a native grass. Ecol Lett 13:106–117

Johnston PR (1998) Leaf endophytes of manuka (*Leptospermum scoparium*). Mycol Res 102:1009–1016

Jordaan A, Taylor JE, Rossenkham R (2006) Occurrence and possible role of endophytic fungi associated with the seed pods of *Colophospermum mopane* (Fabaceae) in Botswana. S Afr J Bot 72:245–255

Junker C, Draeger S, Schultz B (2012) A fine line – endophytes or pathogens in *Arabidopsis* thaliana. Fungal Ecol 5:657–662

Klerks MN, Franz E, van Gent-Pelzer M, Zijlstra C, van Bruggen AHC (2007) Differential interaction of *Salmonella enterica* serovars with lettuce cultivars and plant-microbe factors influencing the colonization efficiency. ISME J 1:620–631

Kloepper JW (2002) Effect of host genotype on indigenous bacterial endophytes of cotton (*Gossypium hirsutum* L.). Plant Soil 240:181–189

Knoch TR, Faeth SH, Arnott DS (1993) Fungal endophytes: plant mutualists via seed predation and germination. Bull Ecol Soc Am 74(Abstr):313

Korkama T, Pakkanen A, Pennanen T (2006) Ectomycorrhizal community structure varies among Norway spruce (*Picea abies*) clones. New Phytol 171:815–824

Krauss J, Härri SA, Bush LP, Husi R, Bigler L, Power SA, Müller CB (2007) Effects of fertilizer, fungal endophytes and plant cultivar on the performance of insect herbivores and their natural enemies. Funct Ecol 21:107–116

Krings M, Taylor TN, Hass H, Kerp H, Dotzler N, Hermsen EJ (2007) Fungal endophytes in a 400-million-yr-old land plant: infection pathways, spatial distribution, and host responses. New Phytol 174:648–657

Lacey LA, Neven LG (2006) The potential of the fungus, *Muscodor albus*, as a microbial control agent of potato tuber moth (Lepidoptera: Gelechiidae) in stored potatoes. J Invertebr Pathol 9:195–198

Latch GCM, Christensen MJ, Gaynor DL (1985) Aphid detection of endophyte infection in tall fescue. NZ J Agric Res 28:129–132

Lehtonen P, Helander M, Saikkonen K (2005) Are endophyte-mediated effects on herbivores conditional on soil nutrients? Oecologia 142:38–45

Leuchtmann A, Schmidt D, Bush LP (2000) Different levels of protective alkaloids in grasses with stroma-forming and seed transmitted *Epichloë/Neotyphodium* endophytes. J Chem Ecol 26:1025–1036

Lewis GC, Clements RO (1986) A survey of ryegrass endophyte (*Acremonium loliae*) in the UK and its apparent ineffectuality on a seedling pest. J Agr Sci 107:633–663

Lodge DJ, Fisher PJ, Sutton BC (1996) Endophytic fungi of *Manikara bidentata* leaves in Puerto Rico. Mycologia 88:733–738

Lu G, Cannon PF, Reid A, Simmons CM (2004) Diversity and molecular relationships of endophytic *Colletotrichum* isolates from the Iwokrama Forest Reserve, Guyana. Mycol Res 108:53–63

Lucero ME, Barrow JR, Osuna P, Reyes I (2006) Plant-fungal interactions in arid and semi-arid ecosystems: large scale impacts from microscale process. J Arid Environ 65:276–284

MacArthur RH, Wilson EO (1967) The theory of island biogeography. Princeton University Press, Princeton

Malinowski DP, Belesky DP (2000) Adaptations of endophyte-infected cool-season grasses to environmental stresses: mechanisms of drought and mineral stress tolerance. Crop Sci 40:923–940

Malinowski D, Leuchtmann A, Schmidt D, Nösberger J (1997) Symbiosis with *Neotyphodium uncinatum* endophyte may increase the competitive ability of meadow fescue. Agron J 89:833–839

Marquez LM, Redman RS, Rodriguez RJ, Roossinck MJ (2007) A virus in a fungus in a plant – three way symbiosis required for thermal tolerance. Science 315:513–515

Marquez SS, Bills GF, Herrero N, Zabalgogeazcoa I (2012) Non-systemic fungal endophytes of grasses. Fungal Ecol 5:289–297

McGee PA (2002) Reduced growth and deterrence from feeding of the insect pest *Helicoverpa armigera* associated with fungal endophytes from cotton. Aus J Exp Agr 42:995–999

Meister B, Krauss J, Harri SA, Schneider MV, Müller CB (2006) Fungal endosymbionts affect aphid population size by reduction of adult life span and fecundity. Basic Appl Ecol 7:244–252

Miller JD, Mackenzie S, Foto M, Adams GW, Findlay JA (2002) Needles of white spruce inoculated with rugulosin-producing endophytes contain rugulosin reducing spruce budworm growth rate. Mycol Res 106:471–479

Monk KA, Samuels GJ (1990) Mycopaghy in grasshoppers (Orthoptera Acrididae) in Indo-Malayan rain forest. Biotropica 22:16–21

Morse LJ, Faeth SH, Day TA (2007) *Neotyphodium* interactions with a wild grass are driven mainly by endophyte haplotype. Funct Ecol 21:813–822

Mucciarelli M, Scannerini S, Bertea C, Maffei M (2003) *In vitro* and *in vivo* peppermint (*Mentha piperita*) growth promotion by nonmycorrhizal fungal colonisation. New Phytol 158:579–591

Narisawa LA, Kawamata H, Currah RS, Hashiba T (2002) Suppression of *Verticillium* wilt in eggplant by some fungal root endophytes. Eur J Plant Pathol 108: 103–109

Newcombe G, Shipunov A, Eigenbrode SD, Raghavendra AKH, Ding H, Anderson CL, Menjivar R, Crawford M, Schwarzlände M (2009) Endophytes influence protection and growth of an invasive plant. Commun Integr Biol 2:29–31

Newsham KK (1994) First record of intracellular sporulation by a coelomycete fungus. Mycol Res 98:1390–1392

Nishida T, Katayama N, Izumi N, Ohgushi T (2010) Arbuscular mycorrhizal fungi species-specificity affect induced plant responses to a spider mite. Popul Ecol 52:507–515

Omacini M, Chaneton EJ, Ghersa CM, Müller CB (2001) Symbiotic fungal endophytes control insect host-parasite interaction webs. Nature 409:78–81

Pan JJ, Clay K (2004) *Epichloë glyceriae* infection affects carbon translocation in the grass *Glyceria striata*. New Phytol 164:467–475

Pan JJ, Baumgarten AM, May G (2008) Effects of host plant environment and *Ustilago maydis* infection on the fungal endophyte community of maze (*Zea mays*). New Phytol 178:147–156

Parker JE, Holub EB, Forst LN, Falk A, Gunn ND, Daniels MJ (1996) Characterization of eds1, a mutation in *Arabidopsis* suppressing resistance to *Peronospora parasitica* specified by several different RPP genes. Plant Cell 8:2033–2046

Peláez F, Collado J, Arenal F, Basilio A, Cabello A, Diez Matas MT (1998) Endophytic fungi from plants living on gypsum soil as a source of secondary metabolites with antimicrobial activity. Mycol Res 102:755–761

Petrini O (1986) Taxonomy of endophytic fungi of aerial plant tissues. In: Fokenna NJ, van den Heuvel J (eds) Microbiology of the phylosphere. Cambridge University Press, Cambridge, pp 175–187

Petrini O (1991) Fungal endophytes of tree leaves. In: Andrews JH, Hirano SS (eds) Microbial Ecology of Leaves. Springer, New York, pp 179–197

Petrini O, Stone J, Carroll FE (1982) Endophytic fungi in evergreen shrubs in western Oregon: a preliminary study. Can J Bot 60:789–796

Porras-Alfaro A, Herrera J, Sinsabaugh RL, Odenbach KJ, Lowery T, Natvig DO (2008) Novel root fungal consortium associated with a dominant desert grass. Appl Environ Microbiol 74:2805–2813

Posada F, Vega FE (2005) Establishment of the fungal entomopathogen *Beauveria baasiana* (Ascomycota: Hyocreales) as an endophyte in cocoa seedlings (*Theobroma cocao*). Mycologia 97:1195–1200

Prestidge RA, Pottinger RP, Barker GM (1982) An association of *Lolium* endophyte with ryegrass resistance to Argentine stem weevil. In: Proceedings of the 35th New Zealand weed and pest control, Hamilton, pp 119–122

Rajagopal K, Suryanarayanan TS (2000) Isolation of endophytic fungi from leaves of neem (*Azadirachta indica* A. Juss). Curr Sci 78:1375–1377

Rasche F, Trondl R, Naglreiter C, Reichenauer TG, Sessitsch A (2006) Chilling and cultivar type affect the diversity of bacterial endophytes colonizing sweet pepper (*Capsicum anuum*). Can J Microbiol 52:1036–1045

Redman RS, Dunigan DD, Rodriguez RJ (2001) Fungal Symbiosis: from mutualism to parasitism, who controls the outcome, host or invader? New Phytol 151:705–716

Redman RS, Sheehan KB, Stout RG, Rodriguez RJ, Henson MJ (2002a) Thermotolerance conferred to plant host and fungal endophyte during mutualistic symbiosis. Science 298:1581

Redman RS, Roossinck MJ, Maher S, Andrews QC, Schneider WL, Rodriguez RJ (2002b) Field performance of cucurbit and tomato plants colonised by a non-pathogenic, mutualistic mutant (path-1) of *Colletotrichum magna* (Teleomorph: *Glomerella magna*; Jenkins and Winstead). Symbiosis 32:55–70

Rodrigues KF (1994) The foliar endophytes of the Amazonian palm *Euterpe oleracea*. Mycologia 86:376–385

Rodrigues AAC, Menezes M (2005) Identification and pathogenic characterization of endophytic *Fusarium* species from cowpea seeds. Mycopathologia 159:79–85

Rodrigues KF, Sieber TN, Gruenig CR, Holdenrieder O (2004) Characterization of *Guinnardia mangiferae* isolated from tropical plants based upon morphology, ISSR-PCR amplifications and ITSI-5.8S-ITS2 sequences. Mycol Res 108:45–52

Rodriguez R, Redman R (2008) More than 400 million years of evolution and some plants still can't make it on their own: plant stress tolerance via fungal symbiosis. J Exp Bot 59:1109–1114

Rodriguez RJ, Redman RS, Henson JM (2004) The role of fungal symbioses in the adaptation of plants to high stress environments. Mitig Adapt Strateg Glob Chang 9:261–272

Rodriguez RS, Henson J, Van Volkenburgh E, Hoy M, Wright L, Beckwith F, Kin YO, Redman RS (2008) Stress tolerance in plants via habitat-adapted symbiosis. ISME J 2:404–416

Rodriguez RJ, White JF Jr, Arnold AE, Redman RS (2009a) Fungal endophytes: diversity and functional roles. New Phytol 182:314–330

Rodriguez RJ, Freeman DC, McArthur ED, Ki YO, Redman RS (2009b) Symbiotic regulation of plant growth, development and reproduction. Commun Integr Biol 2:141–143

Roger A, Gétaz M, Rasmann S, Sanderd IR (2013) Identity and combinations of arbuscular mycorrhizal fungal isolates influence plant resistance and insect preference. Ecol Entomol. doi:10.1111/een.12022

Rosenberg E, Sharon G, Zilber-Rosenberg I (2009) The hologenome theory of evolution contains Lamarckian aspects within a Darwinian framework. Environ Microbiol 11:2959–2962

Rudgers JA, Miller TEX, Ziegler SM, Craven KD (2012) There are many ways to be a mutualist: endophytic fungus reduces plant survival but increases population growth. Ecology 93:565–574

Saikkonen K (2007) Forest structure and fungal endophytes. Fungal Biol Rev 21:67–74

Saikkonen K, Faeth SH, Helander M, Sullivan TJ (1998) Fungal endophytes: a continuum of interactions with host plants. Annu Rev Ecol Syst 29:319–343

Saikkonen K, Helander ML, Faeth SH, Schultthess F, Wilson D (1999) Endophyte-grass-herbivore interactions: the case of *Neotyphodium* endophytes in Arizona fescue populations. Oecologia 121:411–420

Saikkonen K, Helander ML, Rousi M (2003) Endophytic foliar in *Betula pendula*. For Pathol 31:321–329

Saikkonen K, Wali P, Helander M, Faeth SH (2004) Evolution of endophyte-plant symbioses. Trends Plant Sci 9:275–280

Saikkonen K, Lehtonen P, Helander M, Koricheva J, Faeth SH (2006) Model systems in ecology: dissecting the endophyte grass literature. Trends Plant Sci 11:428–433

Schäfer P, Khatabi B, Kogel KH (2007) Root cell death and systemic effects of *Piriformospora indica*: a study on mutualism. FEMS Microbiol Lett 275:1–7

Schardl CL, Clay K (1997) Evolution of mutalistic endophytes from plant pathogens. In: Carroll GC, Tudzynski P (eds) The mycota V. Part B, plant relationships. Springer, Berlin, pp 221–238

Schardl CL, Leuchtmann A, Chung KR, Penny D, Siegel MR (1997) Coevolution by common descent of fungal symbionts (*Epichloë* spp.) and grass hosts. Mol Biol Evol 14:133–143

Schulz C, Boyle C (2005) The endophytic continuum. Mycol Res 109:661–686

Schulz B, Guske S, Damman U, Boyle C (1998) Endophyte-host interactions II. Defining symbiosis of the endophyte-host interaction. Symbiosis 25:213–227

Schulz B, Boyle C, Draeger S, Römmert AK, Krohn K (2002) Endophytic fungi: a source of biologically active secondary metabolites. Mycol Res 106:996–1004

Schweitzer JA, Bailey JK, Bangert RK, Hart SC, Whitham TG (2006) The role of plant genetics in determining above- and below-ground microbial communities. In: Bailey MJ, Lilley AK, Timms-Wilson PTN, Spencer-Phillips PTN (eds) Microbial ecology of the aerial plant surface. CABI, Wallingford, pp 107–119

Selosse MA, Vohnik M, Chauvet E (2008) Out of the rivers: are some aquatic hyphomycetes plant endophytes? New Phytol 178:3–7

Sieber TN (2002) Fungal root endophytes. In: Waisel Y, Eshel A, Kafkafi U (eds) Plant roots: the hidden half. Marcel Dekker, New York, pp 887–917

Siegel MR, Latch GCM, Bush LP, Fannin FF, Rowen DD, Tapper BA, Bacon CW, Johnson MC (1990) Fungal endophyte-infected grasses: alkaloid accumulation and aphid response. J Chem Ecol 16:3301–3315

Sinclair JB, Cerkauskas RF (1996) Latent infection vs. endophyte colonization by fungi. In: Carris LM, Redlin CS (eds) Endophytic fungi in grasses and woody plants. American Phytopathological Society Press, St Paul

Smith KP, Goodman RM (1999) Host variation for interactions with beneficial plant-associated microbes. Annu Rev Phytopathol 37:473–491

Stewart JA, Chadwick VS, Murray A (2005) Investigations into the influence of host genetics on the predominant eubacteria in the faecal microflora of children. J Med Microbiol 54:1239–1242

Stone JK (1987) Initiation and development of latent infections by *Rhabdocline parkeri* on Douglas – fir. Can J Bot 65:2614–2621

Stone JK (1988) Fine structure of latent infections by *Rhabdocline parkeri* on Douglas fir, with observation of uninfected epidermal cells. Can J Bot 66:45–54

Stone JK, Viret O, Petrini O, Chapela I (1994) Histological studies of host penetration and colonization by endophytic fungi. In: Petrini O, Ouellette GB (eds) Host wall alterations by parasitic fungi. American Phytopathological Society Press, St Paul, pp 115–128

Stone JK, Bacon CW, White JF Jr (2000) An overview of endophytic microbes: endophytism defined. In: Bacon CW, White JF (eds) Microbial endophytes. Marcel Dekker, New York

Suryanarayan TS, Murali TS (2006) Incidence of *Leptosphaerulina crassiasca* in symptomless leaves of peanut in Southern India. J Basic Microbiol 46:305–309

Swarthout D, Harper E, Judd S, Gonthier D, Shyne R, Stowe T, Bultman T (2009) Measures of leaf-level water-use efficiency in drought stressed endophyte infected and non-infected tall fescue grasses. Environ Exp Bot 66:88–93

Tikhonovich IA, Provorov NA (2009) From plant-microbe interactions to symbiogenetics: a universal paradigm for the interspecies genetic integration. Ann Appl Biol 154:341–350

Tudzynski B, Sharon A (2002) Biosynthesis, biological role and application of fungal phyto-hormones. In: Osiewacz HD (ed) The mycota x. Industrial applications. Springer, Berlin

Vaahtovuo J, Toivanen P, Eerola E (2003) Bacterial composition of murine faecal microflora is indigenous and genetically guided. FEMS Microbiol Ecol 44:131–136

Valenzuela-Soto JH, Estrada-Hernandez MG, Ibarra-Laclette E, Delano-Frier JP (2010) Inoculation of tomato plants (*Solanum lycopersicum*) with growth-promoting *Bacillus* subtilis retards whitefly *Bemisia tabaci* development. Planta 231(2):397–410

van Bael SA, Valencia MC, Rojas EI, Gómez N, Windsor DM, Herre EA (2009a) Effects of foliar endophytic fungi on the preference and performance of the leaf beetle *Chelymorpha alternans* in Panama. Biotropica 4:221–225

van Bael SA, Fernández-Marín H, Valencia MC, Rojas EI, Wcislo WT, Herre EA (2009b) Two fungal symbioses collide: endophytic fungi are not welcome in leaf-cutting ant gardens. Proc R Soc Lond B276:2419–2426

van Dam NM, Heil M (2011) Multitrophic interactions below and above ground: en route to the next level. J Ecol 99:77–88

Van der Heijden MGA (2004) Arbuscular mycorrhizal fungi as support systems for seedling establishment in grassland. Ecol Lett 7:293–303

van Valen L (1973) A new evolutionary law. Evol Theory 1:1–30

Vicari M, Hatcher PE, Ayres PG (2002) Combined effect of foliar and mycorrhizal endophytes on an insect herbivore. Ecology 83:2452–2464

Waller F, Achatz B, Baltruschat H, Fodor J, Becker K, Fischer M, Hier T, Hückelhoven R, Neumann C, von Wettsein D, Franken P, Kogel KH (2005) The endophytic fungus *Piriformospora indica* reprograms barley to salt-stress tolerance, disease resistance and higher yield. Proc Natl Acad Sci USA 102:13386–13391

Wang W, Vincur B, Altman A (2003) Plant responses to drought, salinity and extreme temperatures: towards genetic engineering for stress tolerance. Planta 218:1–14

Washington WS, Stewart-Wade S, Hood V (1999) *Phomopsis castanea*, a seed-borne endophyte in chestnut trees. Aust J Bot 47:77–84

Wawrzynska A, Christiansen KM, Lan Y, Rodibaugh NL, Innes RW (2008) Powdery mildew resistance conferred by loss of the enhanced disease resistance1 protein kinase is suppressed by a missense mutation in keep on going, a regulator of abscisic acid signaling. Plant Physiol 148:1510–1522

Wawrzynska A, Rodibaugh NL, Innes RW (2010) Synergistic activation of defense responses in *Arabidopsis* by simultaneous loss of GSL5 callose synthase and the EDR1 protein kinase. Mol Plant Microbe Interact 23(5):578–584

Wearn J (2009) *Pithomyces chartarum* – a fungus on the up? Field Mycol 10:36–37

Wearn JA, Sutton BC, Morley NJ, Gange AC (2012) Species and organ specificity of fungal endophytes in herbaceous grassland plants. J Ecol 100: 1085–1092

Webber J (1981) A natural biological-control of Dutch elm disease. Nature 292:449–451

Weber RWS, Stenger E, Meffert A, Hahn M (2004) Brefeldin-A production by *Phoma medicaginis* in dead pre-colonized plant tissue: a strategy for habitat conquest. Mycol Res 108:662–671

Welty RE, Craig AM, Azevedo MD (1994) Variability of ergovaline in seeds and straw and endophyte infection in seeds among endophyte-infected genotypes of tall fescue. Plant Dis 78:845–849

White PS, Pickett STA (1985) Natural disturbance and patch dynamics: an introduction. In: Pickett STA, White PS (eds) The ecology of natural disturbance and patch dynamics. Academic Press, Orlando, pp 3–13

Wilkinson HH, Siegel MR, Blankenship JD, Mallory AC, Bush LP, Schardl CL (2000) Contribution of fungal loline alkaloids to protection from aphids in a grass-endophyte mutualism. Mol Plant Microbe Interact 1:1027–1033

Williamson RC, Potter DA (1997) Turfgrass species and endophyte effects on survival, development, and feeding preference of black cutworms (Lepidoptera: Noctuidae). J Econ Entomol 90:1290–1299

Wilson D, Carroll GC (1994) Infection studies of *Discula quercina* an endophyte of *Quercus garryana*. Mycologia 86:635–647

Yuan ZL, Zhang C-L, Lin F-C (2010) Role of diverse non-systemic fungal endophytes in plant performance and response to stress: progress and approaches. J Plant Growth Regul 29:116–126

Zheng RY, Jiang H (1995) *Rhizomucor endophyticus* sp. Nov., an endophytic zygomycetes from higher plants. Mycotaxon 56:455–466

Zoetendal EG, Akkermans ADL, Akkermans-van Vliet WM, de Visser JAGM, de Vos WM (2001) The host genotype affects the bacterial community in the human gastrointestinal tract. Microb Ecol Health Dis 13:129–134

Züst T, Härris SA, Müller CB (2008) Endophytic fungi decrease available resources for the aphid *Rhopalosiphum padi* and impair their ability to induce defences against predators. Ecol Entomol 33:80–85

Part II

Entomopathogenic Endophytes

4

Entomopathogenic and Nematophagous Fungal Endophytes

Enrique Quesada-Moraga, Noemí Herrero, and Íñigo Zabalgogeazcoa

Abstract

Biological control agents have received a considerable amount of attention as alternatives to chemicals for the development of new control methods but also due to the disparate ecological niches occupied by them. Entomopathogenic (EF) and nematophagous fungi (NF) enter their hosts directly via the cuticle or natural openings, what makes them attractive agents for biological pest control. These fungi have been traditionally viewed simply as animal predators, but recent studies show that a considerable number of fungal pathogens of invertebrates have an endophytic phase in their life cycles. Several taxa of EF and NF have been identified as naturally occurring endophytes and could be artificially inoculated in agricultural plant species. In addition, symbioses with some endophytic species positively affect plant growth and resistance against fungal pathogens. These additional ecological roles give a new perspective to the study of these organisms, because they are part of tritrophic interactions where plants, invertebrates, and fungi are closely involved. Understanding fungal-plant, fungal-pest, fungal-pathogen, and fungal-plant-pest interactions, plus the role of fungal viruses, that infect EF, could lead to the development of novel integrated crop production and protection tools.

E. Quesada-Moraga (✉)
Department of Agricultural and Forest Science and Resources, University of Córdoba (UCO), CeiA3, Campus de Rabanales, P.O. Box 3048, 14080 Córdoba, Spain
e-mail: equesada@uco.es

N. Herrero
Department of Plant Virology, Institute of Entomology, Biology Centre of the Academy of Sciences of the Czech Republic, 370 05 České Budějovice, Czech Republic

Í. Zabalgogeazcoa
Department of Abiotic Stress, Instituto de Recursos Naturales y Agrobiología de Salamanca (IRNASA-CSIC), P.O. Box 257, 37008 Salamanca, Spain

V.C. Verma and A.C. Gange (eds.), *Advances in Endophytic Research*,
DOI 10.1007/978-81-322-1575-2_4, © Springer India 2014

1 Introduction

The advent of chemical insecticides in the mid-twentieth century led to the idea that invertebrate pests could be completely eliminated from crops. Since then, a succession of compounds with insecticidal and nematicidal activity has been developed. Many early chemical pesticides were toxic and environmentally damaging; however, in recent years, due to an increase in public sensitivity toward environmental pollution and problems derived from the side effects of early materials, alternative benign insecticides were developed. Arising from the idea of sustainable agriculture and its main tool, integrated pest management, as well as the notion of organic farming, there now exists a demand for crop protection strategies more compatible with these principles, which are adopted by the Common Agricultural Policy of the European Community (EC) and its member states (i.e., Directive 2009/128/EC). As a prelude of this directive, EU Regulation (EC) 848/2008 led to a reduction of around 85 % of the registered pesticides (insecticides, fungicides, and herbicides). Thus, pest control has evolved to employ a variety of cultural, chemical, and biological tools for the management of pest invasions below an economic threshold.

2 Entomopathogenic and Nematophagous Fungi and Their Unusual Roles in the Ecosystem

Microbial control is considered to be the most viable alternative to synthetic chemical pesticides (Eilengberg and Hokkanen 2006). However, not all entomopathogenic microorganisms invade susceptible hosts in the same manner. For instance, viruses, bacteria, and protozoa must be ingested (Tanada and Kaya 1993), while entomopathogenic (EF) and nematophagous (NF) fungi may enter their hosts directly through the cuticle or by natural openings (Shah and Pell 2003; Goettel et al. 2005; Nordbring-Hertz et al. 2006; Charnley and Collins 2007). The adhesion of conidia or specialized fungal structures to the cuticle is only the beginning of the invasion. Germination, penetration into the host, modulation of cellular and humoral defenses, and fungal growth inside the hemocoel conclude with the death of the host, which is caused by nutrient depletion, the invasion of tissues and organs, and asphyxia due to the development of the fungus in the respiratory system and/or the production of toxic fungal metabolites. The life cycle of the fungus is completed with sporulation, when hyphae emerge from the cadaver and produce conidiophores and conidia, which allow for horizontal transmission (Goettel et al. 2005; Charnley and Collins 2007). Due to their mode of action and their natural presence in the soil (Quesada-Moraga et al. 2007) and in insect populations (Quesada-Moraga and Santiago-Álvarez 2008), EF are polyvalent biocontrol agents and are the only alternative for the biocontrol, among others, of sap-sucking insect and mite pests (i.e., Thysanoptera and Hemiptera), locusts and grasshoppers, and soil-dwelling insect pests (Charnley and Collins 2007; Santiago-Álvarez et al. 2008).

According to Faria and Wraight (2007), 171 mycoinsecticide products (fungus-based formulations targeting insects and mites) were commercially available in 2007. The fungi used in these products are primarily hypocrealean ascomycetes and include *Beauveria bassiana* (Bals.) Vuill., *B. brongniartii* (Sacc.) Petch, *Metarhizium anisopliae* (Metsch.) Sorokin., sensu lato, *M. acridum* (formerly *M. anisopliae* var. acridum) (Driver and Milner) J.F. Bischoff., Rehner, and Humber stat. nov., *Isaria fumosorosea* Wize (formerly *Paecilomyces fumosoroseus*), *Lecanicillium longisporum* and *muscarium* (Petch) R. Zare and W. Gams (formerly *Verticillium lecanii*), and *Hirsutella thompsonii* F.E. Fisher. Meanwhile, three main groups of nematophagous fungi are known: the nematode-trapping and the endoparasitic fungi that attack vermiform living nematodes by using specialized structures and the egg- and cyst-parasitic fungi that attack these stages with their hyphal tips. The performance of these biological control agents have varied.

While it has been hypothesized that many endophytic taxa might have an important role in ecosystems as decomposers, by switching their life strategy from an endophytic to a saprobic

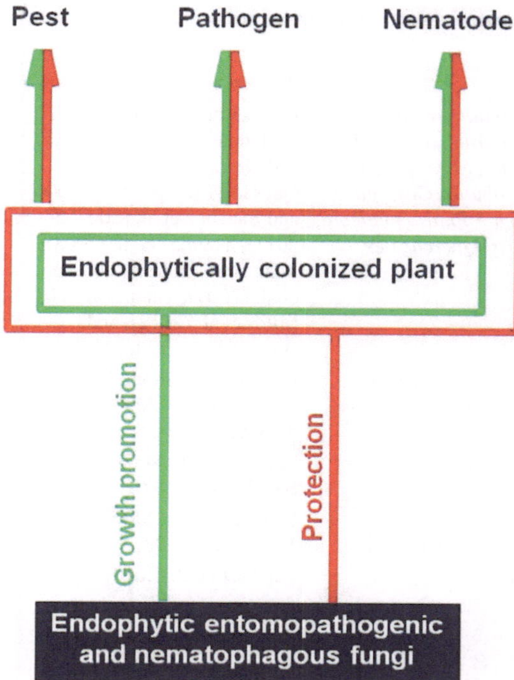

Fig. 4.1 Overview of the unusual ecological roles of entomopathogenic and nematophagous endophytic fungi. Growth promotion activities and their indirect effect on lower susceptibility to pest and diseases are represented in *green*, while direct protection against pest, pathogens, and nematodes are represented in *red*

mode (Promputtha et al. 2007; Parfitt et al. 2010; Purahong and Hyde 2011; Vázquez de Aldana et al. 2013), recent studies suggest that entomopathogenic and nematophagous endophytes might play additional unusual roles in the ecosystem (Vega et al. 2009), protecting plants against diseases, warding off insect pests, and increasing the fitness of plants by growth-promoting activities (Fig. 4.1).

Some fungal entomopathogens are naturally occurring endophytes, while others have been artificially inoculated into plants using various techniques. Nonetheless, in the majority of the aforementioned studies, fungal entomopathogens were introduced into the plant to act as biological control agents against specific pests, and the method of EF colonization was not determined. These fungi had no adverse effect on the growth of the plants (Tefera and Vidal 2009) probably because, as it occurs in endophyte-grass associations, hyphal growth is closely coordinated with host growth or, alternatively, because fungal growth is restrained by plant defenses, achieving the plant and the fungus a situation of "balanced antagonism" (Schulz and Boyle 2005; Rodriguez et al. 2009). However, little is known about (1) how the plants acquire in nature the endophytic EF, (2) the growth pattern that entomo- and nematopathogens display within the plant, and (3) the signaling pathways required for entry and growth in the plant and for the maintenance of a symbiotic interaction.

The implications for biological and agricultural science of the discovery of endophytic phases in many fungal species previously known as entomopathogenic and nematophagous are the main focus of this review.

3 Presence of Entomopathogenic and Nematophagous Fungal Endophytes in Nature

Evidence of the capability of some entomopathogens to have an endophytic phase in their life cycle comes from their isolation from surface-sterilized plant materials. In fact, surveys of fungal endophytes from several plant species indicate the presence of entomopathogenic and nematophagous species inside plant tissues (Table 4.1). In most of these surveys of natural plant hosts, nemato- and entomopathogens occur at a low frequency in populations of particular host plant species (i.e., Vázquez de Aldana et al. 2013). However, exceptions to this occur, and a predominance of some species of these fungi has been reported in the endophytic mycobiota of some plant hosts (Miles et al. 2012; Wearn et al. 2012).

Some entomopathogenic taxa have been identified as natural endophytes in more than one host plant species (i.e., *B. bassiana*, *Lecanicillium lecanii*) (Table 4.1), what indicates that they are multi-host endophytes capable of infecting several plant species. Therefore, although entomopathogenic endophytes might not be very abundant in particular plant species, some taxa like *B. bassiana* have a wide range of host plant species, becoming,

Table 4.1 Plant species where entomopathogens and nematophagous fungi have been identified as natural endophytes

Fungus	Plant host	References
Beauveria bassiana	*Abies beshanzuensis*,[a] *Ammophila arenaria*, *Carpinus caroliniana*, *Coffea arabica*, *Dactylis glomerata*, *Datura stramonium*, *Elymus farctus*, *Espeletia* spp., *Eucalyptus globulus*, *Gossypium hirsutum*, *Papaver somniferum*,[b] *Pinus monticola*, *Pinus radiata*, *Pinus sylvestris*, *Quercus ilex*, *Theobroma gileri*, *Zea mays*	Bills and Polishook (1991), Collado et al. (1999), Ganley and Newcombe (2006), Quesada-Moraga et al. (2006), and Sánchez et al. (2007, 2008, 2011), Thomas et al. (2008), Vega et al. (2008), Giordano et al. (2009), Yuan et al. (2011), Miles et al. (2012), and Brownbridge et al. (2012)
Lecanicillium lecanii	*Ammophila arenaria*, *Carpinus caroliniana*, *Dactylis glomerata*, *Elymus farctus*, *Gossypium hirsutum*	Bills and Polishook (1991), Sánchez et al. (2007, 2008), and de Souza Vieira et al. (2011)
Cordyceps sinensis	*Holcus lanatus*, *Theobroma gileri*	Thomas et al. (2008) and Sánchez et al. (2010)
Paecilomyces sp.	*Carpinus caroliniana*, *Dactylis glomerata*, *Holcus lanatus*, *Musa acuminata*, *Oryza sativa*	Sánchez et al. (2007, 2010) and Vega et al. (2008)
Tolypocladium cylindrosporum	*Festuca rubra*, *Holcus lanatus*	Sánchez et al. (2010)
Hirsutella aphidis	*Lolium perenne*	Authors, unpublished
Metarhizium anisopliae	*Cynodon dactylon*	Authors, unpublished
Cordyceps memorabilis	*Eucalyptus globulus*	Sánchez et al. (2011)
Clonostachys rosea	*Coffea arabica*, *Quercus myrsinifolia*	Vega et al. (2008) and Shirouzu et al. (2009)
Plectosphaerella cucumerina	*Phaseolus vulgaris*, *Cynodon dactylon*, *Ammophila arenaria*, *Elymus farctus*	Authors of *P. vulgaris*; Sánchez et al. (2010)
Isaria farinosa	*Pinus sylvestris*	Giordano et al. (2009)

[a]*Beauveria brongniartii*

[b]Fungus isolated from a mining insect found inside a plant

in general terms, a relatively common endophyte. The occurrence of entomopathogens and nematophagous fungi as natural endophytes also indicates that these fungi have complex life cycles, which can be completed in soil, invertebrates, and plants.

After some pioneering studies showed that the entomopathogen *B. bassiana* could be artificially inoculated into leaves of corn plants and behave as an endophyte (Bing and Lewis 1992), artificial inoculation of other EF and NF has been achieved in numerous plant hosts (Table 4.2). This has corroborated the multi-host nature as an endophyte of *B. bassiana* and other species and altered the rationale behind the use of entomopathogens as biocontrol agents in agriculture.

Recently, PCR quantification in parallel with confocal microscopy was used to monitor spatial and temporal patterns of leaf and stem colonization using a GFP-tagged transformant of the *B. bassiana* strain EABb 04/01-Tip. This work demonstrated that after leaf spray inoculation, *B. bassiana* effectively colonizes aerial tissues of opium poppy plants mainly through the intercellular space of the apoplast and is present even in the leaf trichomes, but fungal colonization was scarce and not uniform. A decline in endophytic colonization was also observed as time after the inoculation increased, although fungal structures still remained present in the leaf tissues (Landa et al. 2013). Hyphae were only observed at the leaf and stem intercellular spaces and circumscribed to determined zones of the parenchyma of the main leaf vein, but not reaching the vascular tissues, whereas we have previously observed fungal growth into the xylem vessels by scanning electron microscopy (Quesada-Moraga et al. 2006).

Table 4.2 Plant species where entomopathogenic and nematophagous endophytes have been artificially inoculated with success

Fungus	Host plant	References
Beauveria bassiana	*Coffea arabica, Corchorus olitorius, Musa* sp., *Papaver somniferum, Phoenix dactylifera, Pinus radiata, Sorghum bicolor, Theobroma cacao, Zea mays*	Akello et al. (2007), Bing and Lewis (1992), Biswas et al. (2012), Brownbridge et al. (2012), Gómez Vidal et al. (2006), Posada et al. (2007), Posada and Vega (2005), Quesada-Moraga et al. (2006), Tefera and Vidal (2009), and Wagner and Lewis (2000)
Lecanicillium lecanii	*Gossypium hirsutum*	Anderson et al. (2007)
Tolypocladium cylindrosporum	*Solanum lycopersicum, Phaseolus vulgaris*	Herrero et al. (2012b)
Lecanicillium dimorphum, Lecanicillium cf. *psalliotae*	*Phoenix dactylifera*	Gómez Vidal et al. (2006)
Pochonia chlamydosporia	*Hordeum vulgare, Solanum lycopersicum*	Bordallo et al. (2002)
Arthrobotrys oligospora	*Hordeum vulgare, Solanum lycopersicum*	Bordallo et al. (2002)
Arthrobotrys dactyloides	*Hordeum vulgare*	López Llorca et al. (2006)
Nematoctonus robustus	*Hordeum vulgare*	López Llorca et al. (2006)
Pleurotus djamor	*Hordeum vulgare*	López Llorca et al. (2006)
Metarhizium anisopliae	*Solanum lycopersicum*	García et al. (2011)
Metarhizium robertsii	*Panicum virgatum, Phaseolus vulgaris*	Sasan and Bidochka (2012)

The above study of colonization of opium poppy by *B. bassiana* shows that the leaf treatment with a fungal spore suspension is an effective technique to carry out this endophytic inoculation. Likewise, we have previously reported that conidial suspensions are also effective to systemically protect opium poppy against stem gall *I. luteipes* applied in seed dressings or soil sprays (Quesada-Moraga et al. 2009), and even the use of strain EABb 04/01-Tip in seed dressing to systemically protect opium poppy against the gall has been recently patented (Patent application number WO2010092223; Quesada-Moraga et al. 2010). However, it remains to be clearly elucidated which inoculation method guarantees the higher efficiency of establishment of the endophyte in opium poppy plants.

Although there are several studies dealing with fluorescent protein-tagged endophytic fungi-plant interactions in the literature, to the best of our knowledge, this is also the first application of a GFP-tagged *B. bassiana* strain for studying endophytic colonization. The penetration and colonization of barley roots by the nematophagous fungus *Pochonia chlamydosporia* has also been studied using GFP-transformed strains (Maciá Vicente et al. 2009). Also, Sasan and Bidochka (2012) have reported the competence in the rhizosphere of a *M. robertsii* GFP-tagged strain in switch grass (*Panicum virgatum*) and haricot beans (*Phaseolus vulgaris*), showing a long-term (60 days) association in which *M. robertsii* endophytically colonized cortical cells within bean roots, growing inter- and intracellularly.

The inocula most used for artificial inoculations are liquid or granular formulations of conidia obtained from liquid or solid mycelial cultures. Species like *B. bassiana* and *Tolypocladium cylindrosporum* sporulate profusely in artificial culture media, making easy the large-scale production of conidia. For instance, about 1.5×10^7 conidia of *T. cylindrosporum* can be obtained from 1 g of mycelium (Herrero et al. 2012b). The techniques that have been used for artificial inoculation are leaf spraying, injection in stems, soil drenching, or seed dressing with conidial suspensions (i.e., Bing and Lewis 1992; Tefera and Vidal 2009; Quesada-Moraga et al. 2006; Posada et al. 2007). Inoculation of plants has been achieved with all these methods, although with different rates of success, depending on several factors.

Evaluations of inoculation methods of endophytes indicate that several variables influence the efficiency of inoculation. The efficiency of endophytic colonization in sorghum after leaf spray, soil drench, or seed dressing inoculation was greatly affected by the substrate where plants were grown; colonization was very high for all three methods when the plant substrate was vermiculite, while only leaf spraying was satisfactory in non-sterile soil (Tefera and Vidal 2009). In coffee plants, injection was the most efficient method, better than drench or leaf spray; injection might help to avoid structural defenses of leaves and roots that cannot be overcome by the fungus (Posada et al. 2007). This is supported by the results of a study where slight wounding of leaf surfaces prior to inoculation improved the recovery of endophytic *Lecanicillium* in cotton (Anderson et al. 2007). From a commercial user point of view, methods like seed dressing or foliar sprays could be more efficient than injection or soil drenching. It is also possible that particular large-scale inoculation methods might be improved by means of inoculum formulations and other technical modifications of the process.

Differences among strains of *B. bassiana* in their capability to infect host plants (Posada and Vega 2005; Posada et al. 2007; Gómez Vidal et al. 2006) suggest that some fungal genotypes might be more specialized than others in carrying an endophytic lifestyle. Therefore, screening for competent endophytic strains might be as important as finding adequate inoculation methods in the development of endophytic entomopathogens as control agents. One current drawback to the use of endophytes as biocontrol agents comes from the variability observed in the endophytic persistence of the fungi after the inoculation. For instance, the presence of *T. cylindrosporum* decreased in bean and tomato leaves as time after inoculation increased (Herrero et al. 2012b). In the case of *B. bassiana*, its incidence in inoculated plants decreased in coffee, but not in corn, jute, or cocoa (Anderson and Lewis 1992; Biswas et al. 2012; Posada et al. 2007). Although the method of inoculation, as well as the fungal strain used, might be associated to the persistence of endophytes, the host plant genotype is likely to have a very important role in the compatibility of plant-endophyte associations. Most known fungal endophytes seem to colonize their host plants in a nonsystemic pattern (Rodriguez et al. 2009; Sánchez et al. 2012). It has been hypothesized that this limited growth occurs because of a situation of "balanced antagonism," where the host plant can restrain the growth of the fungus, and the fungus can modulate the effectiveness of plant defense mechanisms (Schulz and Boyle 2005). Research with EF and NF shows that in response to endophytic infection, plant defense responses like cell wall reinforcement occur (Bordallo et al. 2002; Maciá Vicente et al. 2009). Nevertheless, endophytes can overcome such defense responses, but perhaps only in particular plant genotypes. *B. bassiana* can modulate host defenses up to the point where its hosts are colonized systemically, like some grass endophytes of the genera *Neotyphodium* and *Epichloë* can do. In contrast with these grass endophytes, which are host specific, *B. bassiana* has a wide host range as an endophyte.

It has been reported that transmission of clavicipitaceous endophytes (C-endophytes) or class 1 endophytes is primarily vertical but also horizontal (Rodriguez et al. 2009). However, there are few studies, if any, on how entomopathogenic and nematophagous fungal endophytes are transmitted. Using the abovementioned species-specific two-step nested PCR for identifying and monitoring *B. bassiana* strain EABb 04/01-Tip, it has been recently found that this strain is transmitted vertically from maternal plants via seeds (Quesada-Moraga et al. 2013). Nonetheless, further studies elucidating the fate of the endophytic fungal inoculum in the ecosystem are needed for EF and NF.

4 Entomopathogenic Fungal Endophytes and Pest Control (Tritrophic Interaction Endophyte-Plant-Insect)

The idea that the establishment of EF *in planta* can confer systemic protection from herbivorous pests was proposed by Bing and Lewis (1992). In recent years, several studies reporting the use of EF as artificial endophytes (i.e., artificially inoculated into the plant rather than naturally infecting the plant) have appeared in the literature (Table 4.2). Such studies aimed at using these fungi as biological control agents against specific pests, whose life cycle (they feed internally producing extensive tunneling in stems, pseudostems, rhizomes, roots, seeds, etc.) seriously limits the effectiveness of chemical insecticides and other control methods. Compared to conventional biopesticides, the use of EF (i.e., *B. bassiana*) as artificial endophytes has the advantage of targeting the pest within the plant at reduced application costs because little inoculum is required in cases where colonization is systemic. Furthermore, the endophytic fungus is protected inside the plant from abiotic and biotic factors that would limit its use as an epiphyte. Most of these studies have only completed the first stage, that is, inoculation into the plant, although some of them have gone further, stating that the endophytic colonization of the plant by an entomopathogenic fungus affects the survivorship and development of the cryptic insects, while reducing plant damage (Akello et al. 2008; Bing and Lewis 1992; Quesada-Moraga et al. 2009).

The use of EF as plant endophytes is the only microbial control technique that targets the larvae that feed on the plant (Backman and Sikora 2008). One might expect that the biocontrolling abilities of endophytic EF are due to the infection of the insect upon feeding on the endophytically colonized plant. However, very few fungal-infected insects have been observed in the aforementioned studies; thus, apart from the antibiosis and feeding deterrence, it could be argued that endophytes could kill insects during the first stages of development by secreting toxic compounds *in planta*. Likewise, various species of endophytes are known to produce metabolites that deter insect feeding (Daisy et al. 2002), what suggests that the production of such compounds *in planta* might inhibit the insects from foraging on the plants (Vega et al. 2009). Regarding deterrence effects due to the presence of endophytic fungi, the following questions remain to be addressed: (1) Can EF induce pest control when colonizing the plant endophytically? (2) Can insects (including sucking ones) become infected when feeding within/on the plant? (3) How do endophytic entomopathogens affect insect behavior such as feeding and oviposition?

5 Entomopathogenic and Nematophagous Fungal Endophytes as Rhizosphere Colonizers (Tritrophic Interaction Endophyte-Plant-Invertebrate)

Entomopathogenic and nematophagous fungi of the order Hypocreales are ubiquitous members of the mycobiota of most terrestrial ecosystems and may play a key role in regulating soil-dwelling microherbivores (Humber 2008). Soil is a habitat for many potential insect and nematode hosts, some of which occur at high densities; thus, the continuity of the proximity of these fungi to potential hosts is a factor in the evolution of fungal predation of invertebrates (Humber 2008). Entomopathogenic species most frequently isolated from soils in temperate regions belong to the genera *Beauveria*, *Isaria*, and *Metarhizium* (Meyling and Eilenberg 2007; Quesada-Moraga et al. 2007); in the same habitats, nematophagous endophytes like *P. chlamydosporia* and *Clonostachys rosea* also occur (Bordallo et al. 2002; Wearn et al. 2012). As an environment, soil presents opportunities and challenges to EF. For instance, the soil protects fungi from damaging solar radiation and acts as a buffer against extremes in temperature and water availability (Rangel-Castro et al. 2005).

The potential of EF for the control of soil-dwelling insect pests in modern agriculture has attracted significant interest due to the following factors: (1) Approximately 80 % of insect pests spend part of their life cycle in the soil or come in contact with the soil during their lifetime (Tremblay 1994). (2) To control subterranean pests, more pesticide has to be applied to the soil than the canopy due to the buffering capacity of soils. (3) Few pesticides are available for the control of subterranean pests. Namely, many pesticides aimed toward the eradication of subterranean pests have been banned (e.g., methyl bromide, dazomet) or restricted (e.g., chlorpyrifos), and in addition, pests are likely to develop resistance through the extensive use of pesticides. (4) Soil pests are more difficult to control due to insufficient tools for their monitoring. As a result, insects and nematodes are likely to damage plants before control measures can be implemented. However, the prophylactic application of pesticides could be wasteful due to poor timing or targeting practices. (5) Moreover, feeding damage caused by subterranean pests allows for the entry of plant pathogens at the site of injury. Thus, the effective control of pests could reduce fungicide applications and lead to higher quality produce. (6) Most subterranean pests are polyphagous (attack a wide range of crops); therefore, control strategies developed on one crop could be extended to other crops.

In this context, plants and EF are involved in the dynamic process of coevolution. For example, the endophytic colonization of roots or saprotrophic growth on plant exudates has been observed. Thus, understanding whether EF may protect the plant against herbivores feeding on the roots and whether plants benefit from the presence of EF in the rhizosphere (i.e., by the parasitism of root-feeding pests) may allow to develop new pest control strategies. In sum, whether or not plants have evolved mechanisms that encourage the survival and development of EF in the rhizosphere remains a crucial question.

The ecology of these fungi in the rhizosphere is an understudied area of insect pathology. The rhizosphere is the soil region where the presence of root exudates influences the soil microbiota. It is in the rhizosphere that complex interactions between roots, root exudates, beneficial and pathogenic microorganisms, and invertebrates take place. There are three separate, but interacting, regions that make up the rhizosphere: the outer rhizosphere, the rhizoplane, and the roots (Bowen and Rovira 1999; Kennedy 1998). The outer rhizosphere contains the soil that is loosely adhered to the roots and it is the region where the root exudates influence the soil microbiota. The rhizoplane is the portion of the rhizosphere directly in contact with the root surface resulting in the soil being tightly adhered to the roots.

Tritrophic interactions may also involve entomopathogens, plant roots, and insects and have been found to operate below ground. Hence, understanding the dynamic interactions between insect pests, fungi, and host plants should be of key importance in the development and understanding of fungal entomopathogens as microbial control agents against root-feeding insects (Bruck 2010). In this tritrophic interaction, it is hypothesized that the behavior of the insect pest is modified in response to the presence of the rhizosphere-competent fungal entomopathogen. However, in this regard, there is contradictory evidence in the literature because it seems that the behavioral differences observed depend on both the fungal isolate and the insect species (Bruck 2010). St. Leger (2008) speculates that the presence of EF in the rhizosphere might provide a "repellent barrier" around plants roots. However, the opposite phenomenon, in which insects are attracted to plants when their rhizosphere is colonized, may also occur (Kepler and Bruck 2006). This last assumption makes sense from an evolutionary standpoint of the fungus, as the spores of EF in the soil are not able to actively seek out insect hosts, and if they are in fact utilizing the rhizosphere as a bridge between insect hosts, attracting the host would shorten the length of such bridge (Bruck 2010). To date, it is not known how the rhizosphere-competent EF attract the insects; this attraction may be mediated by the production of attractant compounds by either the fungus, the plant, or both, and whether such colonization has another effect beyond protecting the plant from root-feeding insects (i.e., white grubs and wireworms) as promoting plant growth by capturing essential micro- and macronutrients for plant development (i.e., iron, phosphorous, etc.).

The importance of NF or EF might be greater that what is known. Several reports indicate that predatory fungi are common constituents of plant roots in some species. For instance, *Clonostachys rosea* was the dominant endophytic species in roots of *Cirsium arvense* (Wearn et al. 2012). And one fourth of the 16 endophytic species identified in bean roots in Spain were nematophagous taxa (Arteaga et al. 2013). These results suggest that the association of fungi that predate invertebrate herbivores with plant roots might be a common event of mutualistic symbioses occurring in nature. This is a good example to support the bodyguard hypothesis, which states that plants use predators of their pathogens and pests for their own protection (Elliot et al. 2000).

6 Entomopathogenic Fungal Endophytes as Plant Growth Promoters (Interaction Endophyte-Plant)

The results of many studies suggest that the establishment of certain endophytic fungi *in planta* (entomopathogenic or not) has growth-promoting effects on a wide range of plants (Varma et al. 1999). However, the mechanism of these effects remains unknown, and several hypotheses have been suggested: (1) The mutualistic interaction between plant and fungi improves the ability of the plant to tolerate unfavorable conditions (Hesse et al. 2003; Rodríguez et al. 2008). (2) Endophytic fungi provide nutrients to the host plant through the transfer of nitrogen and the uptake of phosphorous and other minerals (Usuki and Narisawa 2007). (3) The association induces an improvement in the photosynthetic activity of the plant (Obledo et al. 2003). (4) Endophytic fungi are able to secrete compounds that affect plant development (Varma et al. 1999) and could be developed as bio-fertilizers.

In greenhouse experiments where invertebrates were absent, tomato plants treated with a soil drench of *M. anisopliae* had significantly greater shoot and root length and dry weight than controls, and the growth promotion observed was dependent on the fungal strains used (García et al. 2011). Growth promotion by *M. anisopliae* was also observed in soybean, where in addition the inoculated plants performed better than controls under salt stress (Khan et al. 2011). A significant promotion of root growth also occurred in the grass *Panicum virgatum* and in the legume *Phaseolus vulgaris* as a result of inoculation with *M. robertsii*, a common inhabitant of soils worldwide (Sasan and Bidochka 2012). Several mechanisms could be involved in plant growth promotion as a result of endophyte symbiosis. However, in the particular case of *Metarhizium*, it has been shown that in its endophytic phase, this entomopathogen transfers to the plant host nitrogen previously obtained from insect hosts (Behie et al. 2012).

Growth promotion activities have also been found by Sánchez-Rodríguez et al. (2012) in wheat *Triticum aestivum* due to colonization by an endophytic strain of *B. bassiana*. Three inoculation methods, soil treatment, seed dressing, and leaf spraying, were used. In two of the inoculation methods, soil treatment and seed dressing, the fungus was revealed as rhizosphere competent, with root re-isolation percentages ranging from 17 to 83 %. In contrast, the percentage of fungal re-isolation from leaf tissues was significantly higher in plants inoculated by leaf spraying, ranging from 8 to 75 %. Interestingly, the plant growth pattern in controls and inoculated plants was different, and at the end of the experiment, a general trend of higher plant height as the colonization of the plant was more intense was detected, reaching statistical significance in plants inoculated by leaf spraying. Likewise, manganese concentration, which also increased with the intensity of fungal colonization, was significantly higher in all inoculated plants. The possible origin of the plant height and manganese concentration increase in *B. bassiana*-inoculated plants, their implications for pest and disease control, and the promotion of plant growth are being investigated (Sánchez-Rodríguez et al. 2012). Also in wheat, an increase in root and shoot length of seedlings occurred as a result of the application of *L. lecanii* and *P. chlamydosporia*. This effect was attributed to the production of growth regulators in endophyte-infected plants (Monfort et al. 2005).

7 Entomopathogenic Fungal Endophytes as Plant Disease Antagonists (Tritrophic Interaction Endophyte-Plant-Pathogen)

Although the role of fungal endophytes as biological control agents of plant diseases is recognized, the role of endophytic EF in the biological control of plant pathogens is poorly studied. Nevertheless, as a result of endophytic colonization by entomopathogens, interference in the development of diseases caused by fungal and bacterial pathogens has been observed in plants (see Ownley et al. 2010). At the present time, reports of disease control *in planta* are scarce, but occur in different systems and deserve attention. For instance, seed treatment of tomato and cotton with *B. bassiana* resulted in reduced severity of damping-off caused by *Rhizoctonia solani* and *Pythium myriotylum* in seedlings (Ownley et al. 2000; Clark et al. 2006); and leaf disease caused by *Xanthomonas axonopodis* was reduced in cotton plants whose roots had been inoculated with *B. bassiana* (Griffin 2007). The development of powdery mildew was suppressed in cucumber and strawberry plants treated with *L. lecanii*, *L. longisporum*, and *Isaria fumosorosea* (Miller et al. 2004; Kavková and Curn 2005; Kim et al. 2008). The nematophagous endophyte *P. chlamydosporia* has been reported to compete with *Gaeumannomyces graminis var. tritici* for the colonization of wheat roots and to reduce the symptoms caused in roots by this pathogen (Monfort et al. 2005). In addition to these, there are several reports of in vitro studies where an inhibitory effect of *B. bassiana* and *Lecanicillium* spp. was observed on cultures of plant pathogens such as *Rhizoctonia solani*, *Fusarium oxysporum*, *Gaeumannomyces graminis var. tritici*, *Armillaria mellea, Rosellinia necatrix*, *Botrytis cinerea*, *Podosphaera fuliginea*, or *Pythium ultimum* (Ownley et al. 2010).

The above studies indicate that the mechanisms by which endophytic entomopathogens can interfere with the development plant diseases are varied, and hypothetically, the following could be involved: (1) production of antifungal compounds, (2) competition for space and nutrients, (3) mycoparasitism, or (4) induction of host defense responses (i.e., systemic acquired resistance) (Ownley et al. 2010). In addition, as a result of insect control, some vector-transmitted pathogens (i.e., viruses) could also be affected by the presence of entomopathogenic endophytes. In some cases, more than one of the above mechanisms could be involved in pathogen inhibition, for example, some *Lecanicillium* species are mycoparasites of plant pathogenic fungi but also produce antimicrobial compounds effective in vitro against other fungi (Vandermeer et al. 2009; Askary et al. 1997; Benhamou and Brodeur 2000); and as a result of their endophytic colonization of roots, induced resistance to powdery mildew has been observed in leaves (Hirano et al. 2008).

8 Mycoviruses Infecting Entomopathogenic Fungal Endophytes

Mycoviruses have been described in numerous fungal species having different lifestyles. In contrast to animal, plant, or bacterial viruses, an extracellular phase is unknown in the life cycle of mycoviruses, and no natural vectors are known for them. Nonetheless, mycoviruses have efficient means of transmission and dispersion, by means of anastomosis between compatible fungal strains or by vertical transmission to spores, what explains why viral infections are often persistent and prevalent in many fungal taxa (Ghabrial and Suzuki 2010). Mycovirus infections are commonly associated with the presence of double-stranded RNA (dsRNA) elements, because most known fungal viruses have either dsRNA genomes, or single-stranded RNA (ssRNA) genomes that produce dsRNA replicative intermediates (Morris and Dodds 1979).

A large variety of dsRNA elements has been observed in surveys of the presence of dsRNA in EF, what indicates that mycoviruses are common and diverse among these fungi (Table 4.3). In spite of this, very few virus genomes from entomopathogens have been completely sequenced up to date (Herrero and Zabalgogeazcoa 2011; Herrero et al. 2012a). The presence of mycoviruses is common among isolates of *B. bassiana*; different surveys found viral dsRNA in 15–67 % of the analyzed isolates (Table 4.1). *B. bassiana* victorivirus 1 (BbVV1), a mycovirus infecting this fungus, was detected in several fungal strains sampled in different geographical locations, what suggests that it is widespread (Herrero et al. 2012a). Viral dsRNA was also found in endophytic strains of

Table 4.3 Incidence of dsRNA elements in entomopathogenic fungi

Fungal species	Number of isolates analyzed	% isolates with ds RNA	ds RNA observed elements: Size range (kbp)	Number of elements	References
Beauveria bassiana	12	16.7	2.5	1	Melzer and Bidochka (1998)
Beauveria bassiana	34	20.6	1.0–5.0	2–5	Castrillo et al. (2004)
Beauveria bassiana	13	15.4	0.7–4.0	2	Dalzoto et al. (2006)
Beauveria bassiana	15	66.7	1.9–6.0	1–2	Herrero et al. (2009)
Beauveria bassiana	73	54.8	0.8–6.0	1–11	Herrero et al. (2012a)
Metarhizium anisopliae	73	38.4	0.5–5.2	1–8	Melzer and Bidochka (1998)
Metarhizium anisopliae	12	33.0	0.5–5.5	1–10	Giménez-Pecci et al. (2002)
Metarhizium anisopliae	7	42.9	0.78–4.1	1–8	Bogo et al. (1996)
Metarhizium anisopliae	41	4.9	0.75–3.5	9–13	Leal et al. (1994)
Metarhizium flavoviride	7	71.4	1.4–6.2	1–5	Martins et al. (1999)
Metarhizium flavoviride	6	83.3	0.5–5.2	1–8	Melzer and Bidochka (1998)
Paecilomyces spp.	19	52.6	0.5–5.6	2–7	Inglis and Valadares-Inglis (1997)
Paecilomyces fumosoroseus	12	25.0	0.5–4.5	3	Souza Azevedo et al. (2000)
Tolypocladium cylindrosporum	11	45.5	1.2–5.1	2–6	Herrero et al. (2011)
Lecanicillium lecanii	7	42.9	2.0–6.0	1–2	Herrero et al. (2009)
Lecanicillium lecanii	35	62.9	4.0–19	1–6	Sugimoto et al. (2003)

B. bassiana and of the entomopathogenic endophyte *T. cylindrosporum* (Herrero et al. 2009; Herrero and Zabalgogeazcoa 2011). A virus belonging to the same family as BbVV1, the victorivirus TcVV1, was completely sequenced in *T. cylindrosporum*. BbVV1 and TcVV1 have been found alone or together with other viruses in mixed infections, a common occurrence among mycoviruses (Ghabrial and Suzuki 2010). The variability of dsRNA patterns in *B. bassiana* strains obtained as endophytes was much lower than that observed in strains obtained from soil (Herrero et al. 2012a). This situation could be due to the existence of cryptic lineages of entomopathogens which could be more compatible with an endophytic lifestyle, and it could explain why particular sets of viruses are maintained in this group of entomopathogenic grass fungal endophytes.

There is not much knowledge regarding the function or effects of mycoviruses in their hosts. Most known mycoviruses are apparently symptomless; only a few are clearly detrimental to their hosts, producing malformations in mushrooms (Romaine and Goodin 2002) or reducing fertility or virulence (hypovirulence) in plant pathogens like *Cryphonectria parasitica* or *Rosellinia necatrix* (Chiba et al. 2009; Ghabrial and Suzuki 2009). In fact, some of the best-known mycoviruses are those causing hypovirulence in plant pathogenic fungi, because of their potential as biological control agents. Virus-induced hypovirulence against insects was suggested in the entomopathogens *M. anisopliae* and *B. bassiana*, but definite proof does not yet exist (Melzer and Bidochka 1998; Dalzoto et al. 2006). Nonetheless, the cases in which mycoviruses are detrimental to their hosts could be an exception to the rule. Actually, the high prevalence and persistence of mycoviruses among the major groups of fungi could indicate that their presence could be beneficial to their fungal hosts, as is known to occur with a number of non-fungal viruses that are able to establish mutualistic or neutral relationships with their hosts (Roossinck 2011). One example is the improvement of thermal tolerance of the plant host of the endophyte *Curvularia protuberata* when this fungus is infected by the mycovirus CThTV (Márquez et al. 2007).

Another study indicated that the mycovirus TcV1 might slightly affect the endophytic behavior of the entomopathogenic fungal endophyte *T. cylindrosporum* in different plant species (Herrero et al. 2012b).

Entomopathogens have developed complex interactions with arthropods and plants, but mycoviruses seem to play also a role, yet unknown, in the establishment of these relationships. In addition, viruses infecting arthropods and plants and having similar characteristics to those infecting fungi have been described (Isawa et al. 2011; Roossinck 2010); it would be interesting to know if they are able to move from kingdom to kingdom, influencing the intricate relationships formed among arthropods, plants, and fungi.

The new player that we present here in the form of mycovirus adds complexity to the knowledge of EF ecology, which is crucial for understanding their role in managed and natural ecosystems and for their successful development as biocontrol agents (Roy et al. 2010; Vega et al. 2009).

References

Akello J, Dubois T, Gold CS, Coyne D, Nakavuma J, Paparu P (2007) *Beauveria bassiana* (Balsamo) Vuillemin as an endophyte in tissue culture banana (*Musa* spp.). J Invertebr Pathol 96:34–42

Akello J, Dubois T, Coyne D, Kyamanywa S (2008) Effect of endophytic *Beauveria bassiana* on populations of the banana weevil, *Cosmopolites sordidus*, and their damage in tissue-cultured banana plants. Entomol Exp Appl 129:157–165

Anderson LA, Lewis LC (1992) Temporal relationships between *Zea mays*, *Ostrinia nubilalis* (*Lep. Pyralidae*) and endophytic *Beauveria bassiana*. Entomophaga 37:525–536

Anderson CMT, McGee PA, Nehl DB, Mensah RK (2007) The fungus *Lecanicillium lecanii* colonises the plant *Gossypium hirsutum* and the aphid *Aphis gossypii*. Aust Mycol 26:65–70

Arteaga N, Díaz Minguez JM, Gómez J, García I, Zabalgogeazcoa I (2013) Fungal endophytes of bean roots and their interactions with *Fusarium oxysporum* vascular wilts. In: Schneider C, Leifert C, Feldmann F (eds) Endophytes for plant protection: the state of the art. Deutsche Phytomedizinische Gesellschaft, Braunschweig, p 168

Askary H, Benhamou N, Brodeur J (1997) Ultrastructural and cytochemical investigations of the antagonistic effect of *Verticillium lecanii* on cucumber powdery mildew. Phytopathology 87:359–368

Backman PA, Sikora RA (2008) Endophytes: an emerging tool for biological control. Biol Control 46:1–3

Behie SW, Zelisko PM, Bidochka MJ (2012) Endophytic insect-parasitic fungi translocate nitrogen directly from insects to plants. Science 336:1576

Benhamou N, Brodeur J (2000) Evidence for antibiosis and induced host defense reactions in the interaction between *Verticillium lecanii* and *Penicillium digitatum*, the causal agent of green mold. Phytopathology 90:932–943

Bills GF, Polishook JD (1991) Microfungi from *Carpinus caroliniana*. Can J Bot 69:1477–1482

Bing LA, Lewis LC (1992) Temporal relationships between *Zea mays*, *Ostrinia nubilalis* (Lep.: *Pyralidae*) and endophytic *Beauveria bassiana*. Entomophaga 37:525–536

Biswas C, Dey P, Satpathy S, Satya P (2012) Establishment of the fungal entomopathogen *Beauveria bassiana* as a season long endophyte in jute (*Corchorus olitorius*) and its rapid detection using SCAR marker. BioControl 57:565–571

Bogo MR, Queiroz MV, Giménez DM, Azevedo JL, Schrank A (1996) Double-stranded RNA and isometric virus-like particles in the entomopathogenic fungus *Metarhizium anisopliae*. Micol Res 100:1468–1472

Bordallo JJ, Lopez-Llorca LV, Jansson HB, Salinas J, Persmark L, Asensio L (2002) Colonization of plant roots by egg-parasitic and nematode-trapping fungi. New Phytol 154:491–499

Bowen GD, Rovira AD (1999) The rhizosphere and its management to improve plant growth. Adv Agron 66:1–102

Brownbridge M, Reay SD, Nelson TL, Glare TR (2012) Persistence of *Beauveria bassiana* (ascomycota: Hypocreales) as an endophyte following inoculation of radiata pine seed and seedlings. Biol Control 61:194–200

Bruck DJ (2010) Fungal entomopathogens in the rhizosphere. BioControl 55:103–112

Castrillo LA, Griggs MH, Vandenberg JD (2004) Vegetative compatibility groups in indigenous and mass-released strains of the entomopathogenic fungus *Beauveria bassiana*: likelihood of recombination in the field. J Invertebr Pathol 86:26–37

Charnley AK, Collins SA (2007) Entomopathogenic fungi and their role in pest control. In: Kubicek CP, Druzhinina IS (eds) The mycota IV: environmental and microbial relationships. Springer, Berlin, pp 159–187

Chiba S, Salaipeth L, Lin YH, Sasaki A, Kanematsu S, Suzuki N (2009) A novel bipartite double-stranded RNA mycovirus from the white root rot fungus *Rosellinia necatrix*: molecular and biological characterization, taxonomic considerations, and potential for biological control. J Virol 83:12801–12812

Clark MM, Gwinn KD, Ownley BH (2006) Biological control of *Pythium myriotylum*. Phytopathology 96:S25

Collado J, Platas G, González I, Peláez F (1999) Geographical and seasonal influences on the distribution of fungal endophytes in *Quercus ilex*. New Phytol 14:525–532

Daisy BH, Strobel GA, Castillo U, Ezra D, Sears J, Weaver DK, Runyon JB (2002) Naphthalene, an insect repellent, is produced by *Muscodor vitigenus*, a novel endophytic fungus. Microbiology 148:3737–3741

Dalzoto PR, Glienke-Blanco C, Kava-Cordeiro V, Ribeiro JZ, Kitajima EW, Azevedo JL (2006) Horizontal transfer and hypovirulence associated with double-stranded RNA in *Beauveria bassiana*. Mycol Res 110:1475–1481

de Souza Vieira PD, de Souza Motta CM, Lima D, Torres JB, Maria Carolina Quecine MC, Azevedo JL, Tinti de Oliveira N (2011) Endophytic fungi associated with transgenic and non-transgenic cotton. Mycology 2:91–97

Eilengberg J, Hokkanen H (2006) An ecological and societal approach to biological control. Springer, Dordrecht

Elliot SL, Sabelis MW, Janssen A, van der Geest LPS, Beerling EAM, Fransen J (2000) Can plants use entomopathogens as bodyguards? Ecol Lett 3:228–235

Faria MR, Wraight SP (2007) Mycoinsecticides and mycoacaricides: a comprehensive list with worldwide coverage and international classification of formulation types. Biol Control 43:237–256

Ganley RJ, Newcombe G (2006) Fungal endophytes in seeds and needles of *Pinus monticola*. Mycol Res 110:318–327

García JE, Posadas JB, Perticari A, Lecuona RE (2011) *Metarhizium anisopliae* (Metschnikoff) Sorokin promotes growth and has endophytic activity in tomato plants. Adv Biol Res 5:22–27

Ghabrial SA, Suzuki N (2009) Viruses of plant pathogenic fungi. Annu Rev Phytopathol 47:353–384

Ghabrial SA, Suzuki N (2010) Fungal viruses. In: Mahy BWJ, Van Regenmortel MHV (eds) Desk encyclopedia of plant and fungal virology. Elsevier, Oxford, pp 517–524

Giménez-Pecci M, Bogo MR, Santi L, Moraes CK, Corrêa CT, Henning Vainstein M, Schrank A (2002) Charac-terization of mycoviruses and analyses of chitinase secretion in the biocontrol fungus *Metarhizium anisopliae*. Curr Microbiol 45:334–339

Giordano L, Gonthier P, Varese GC, Miserere L, Nicolotti G (2009) Mycobiota inhabiting sapwood of healthy and declining Scots pine (*Pinus sylvestris* L.) trees in the Alps. Fungal Divers 38:69–83

Goettel MS, Eilenberg J, Glare T (2005) Entomopathogenic fungi and their role in regulation of insect populations. In: Latrou K, Gilbert LB (eds) Comprehensive molecular insect science. Elsevier/Pergamon, Oxford

Gómez Vidal S, López llorca LV, Jansson HB, Salinas J (2006) Endophytic colonization of date palm (*Phoenix dactylifera L.*) leaves by entomopathogenic fungi. Micron 37:624–632

Griffin MR (2007) *Beauveria bassiana*, a cotton endophyte with biocontrol activity against seedling disease. PhD dissertation. University of Tennessee, Knoxville

Herrero N, Zabalgogeazcoa I (2011) Mycoviruses infecting the endophytic and entomopathogenic fungus *Tolypocladium cylindrosporum*. Virus Res 29:755–763

Herrero N, Sánchez Márquez S, Zabalgogeazcoa I (2009) Mycoviruses are common among different species of endophytic fungi of grasses. Arch Virol 154:327–330

Herrero N, Pérez-Sánchez R, Oleaga A, Zabalgogeazcoa I (2011) Tick pathogenicity, thermal tolerance and virus infection in Tolypocladium cylindrosporum. Ann Appl Biol 159:192–201

Herrero N, Dueñas E, Quesada-Moraga E, Zabalgogeazcoa I (2012a) Prevalence and diversity of viruses in the entomopathogenic fungus *Beauveria bassiana*. Appl Environ Microbiol 78:8523–8530

Herrero N, Sánchez Márquez S, Zabalgogeazcoa I (2012b) Mycovirus effect on the endophytic establishment of the entomopathogenic fungus *Tolypocladium cylindrosporum* in tomato and bean plants. BioControl 58:225–232

Hesse U, Schöberlein W, Wittenmayer L, Förster K, Warnstorff K, Diepenbrock W, Merbach U (2003) Effects of *Neotyphodium* endophytes on growth, reproduction and drought-stress tolerance of three *Lolium perenne L.* genotypes. Grass Forage Sci 58:407–415

Hirano E, Koike M, Aiuchi D, Tani M (2008) Pre-inoculation of cucumber roots with *Verticillium lecanii* (*Lecanicillium muscarium*) induces resistance to powdery mildew. Res Bull Obihiro Univ 29:82–94

Humber RA (2008) Evolution of entomopathogenicity in fungi. J Invertebr Pathol 98:262–266

Inglis PW, Valadares-Inglis MC (1997) Rapid isolation of double-stranded RNAs from entomopathogenic species of the fungus *Paecilomyces* using a commercial minicolumn system. J Virol Methods 67:113–116

Isawa H, Kuwata R, Hoshino K et al (2011) Identification and molecular characterization of a new nonsegmented double-stranded RNA virus isolated from *Culex* mosquitoes in Japan. Virus Res 155:147–155

Kavková M, Curn V (2005) *Paecilomyces fumosoroseus* (Deuteromycotina: Hyphomycetes) as a potential mycoparasite on *Sphaerotheca fuliginea* (Ascomycotina: *Erysiphales*). Mycopathology 159:53–63

Kennedy AC (1998) The rhizosphere and spermosphere. In: Sylvia DM, Fuhrmann JJ, Hartel PG, Zuberer DA (eds) Principles and applications of soil microbiology. Prentice Hall, Upper Saddle River, pp 389–407

Kepler RM, Bruck DJ (2006) Examination of the interaction between the black vine weevil (Coleoptera: *Curculionidae*) and an entomopathogenic fungus reveals a new tritrophic interaction. Environ Entomol 35:1021–1029

Khan AL, Hamayun M, Khan SA, Kang SM, Shinwari ZK, Kamran M, ur Rehman S, Kim JG, Lee IJ (2011) Pure culture of *Metarhizium anisopliae* LHL07 reprograms soybean to higher growth and mitigates salt stress. World J Microbiol Biotechnol 28:1483–1494

Kim JJ, Goettel MS, Gillespie DR (2008) Evaluation of *Lecanicillium longisporum*, Vertalec® for simultaneous suppression of cotton aphid, and *Sphaerotheca fuliginea*, on potted cucumbers. Biol Control 45:404–409

Landa BB, López-Díaz C, Jiménez-Fernández D, Montes-Borrego M, Muñoz-Ledesma FJ, Quesada-Moraga E (2013) *In-planta* detection and monitorization of colonization of an endophytic *Beauveria bassiana* strain by using a new-developed nested and quantitative PCR-based assay and confocal laser scanning microscopy. J Invertebr Pathol 114:128–138

Leal SCM, Bertioli DJ, Butt TM, Peberdy JF (1994) Characterization of isolates of the entomopathogenic fungus *Metarhizium anisopliae* by RAPD-PCR. Mycol Res 98:1077–1081

López Llorca LV, Jansson HB, Maciá Viocente JG, Salinas J (2006) Nematophagous fungi as root endophytes. In: Schulz B, Boyle C, Sieber T (eds) Microbial root endophytes, vol 9, Soil biology. Springer, Berlin, pp 191–206, Ch 11

Maciá Vicente JG, Jansson HB, Talbot NJ, López Llorca LV (2009) Real time PCR quantification and live-cell imaging of endophytic colonization of barley (*Hordeum vulgare*) roots by *Fusarium equiseti* and *Pochonia chlamydosporia*. New Phytol 182:213–228

Márquez LM, Redman RS, Rodriguez RJ, Roossinck MJ (2007) A virus in a fungus in a plant: three-way symbiosis required for thermal tolerance. Science 315:513–515

Martins MK, Furlaneto MC, Sosa-Gomez DR, Faria MR, Pelegrinelli Fungaro MH (1999) Double-stranded RNA in the entomopathogenic fungus *Metarhizium flavoviride*. Curr Genet 36:94–97

Melzer MJ, Bidochka MJ (1998) Diversity of double-stranded RNA viruses within populations of entomopathogenic fungi and potential implications for fungal growth and virulence. Mycologia 90:586–594

Meyling NV, Eilenberg J (2007) Ecology of the entomopathogenic fungi *Beauveria bassiana* and *Metarhizium anisopliae* in temperate agroecosystems: potential for conservation biological control. Biol Control 43:145–155

Miles LA, Lopera CA, González S, Cepero de García MC, Franco AE, Restrepo S (2012) Exploring the biocontrol potential of fungal endophytes from an Andean Colombian Paramo ecosystem. BioControl. doi:10.1007/s10526-012-9442-6

Miller TC, Gubler WD, Laemmlen FF, Geng S, Rizzo DM (2004) Potential for using *Lecanicillium lecanii* for suppression of strawberry powdery mildew. Biocontrol Sci Technol 14:215–220

Monfort E, López Llorca LV, Jansson HB, Salinas J, Park JO, Sivasithamparam K (2005) Colonisation of seminal roots of wheat and barley by egg-parasitic nematophagous fungi and their effects on *Gaeumannomyces graminis* var. *tritici* and development of root rot. Soil Biol Biochem 37:1229–1235

Morris TJ, Dodds JA (1979) Isolation and analysis of double-stranded RNA from virus-infected plant and fungal tissue. Phytopathology 69:854–858

Nordbring-Hertz B, Jansson HB, Tunlid A (2006) Nemato-phagous fungi. eLS. doi:10.1038/npg.els.0004293

Obledo EN, Barragán-Barragán LB, Gutiérrez-González P, Ramírez Hernández BC, Ramírez JJ, Rodríguez-Garay B (2003) Increased photosynthetic efficiency generated by fungal symbiosis in *Agave victoria-reginae*. Plant Cell Tissue Organ Cult 74:237–241

Ownley BH, Bishop DG, Pereira RM (2000) Biocontrol of *Rhizoctonia* damping-off of tomato with *Beauveria bassiana*. Phytopathology 90:S58

Ownley BH, Gwinn KD, Vega FE (2010) Endophytic fungal entomopathogens with activity against plant pathogens: ecology and evolution. BioControl 55:113–128

Parfitt D, Hunt J, Dockrell D, Rogers HJ, Boddy L (2010) Do all trees carry the seeds of their own destruction? PCR reveals numerous wood decay fungi latently present in sapwood of a wide range of angiosperm trees. Fungal Ecol 3:338–346

Posada F, Vega F (2005) Establishment of the fungal entomopathogen *Beauveria bassiana* (ascomycota: Hypocreales) as an endophyte in cocoa seedlings (*Theobroma cacao*). Mycologia 97:1195–1200

Posada FA, Aime MC, Peterson SW, Rehner SA, Vega FA (2007) Inoculation of coffee plants with the fungal entomopathogen *Beauveria bassiana* (Ascomycota: *Hypocreales*). Mycol Res 111:748–757

Promputtha I, Lumyong S, Dhanasekaran V, McKenzie EHC, Hyde KD, Jeewon R (2007) A phylogenetic evaluation of whether endophytes become saprotrophs at host senescence. Microb Ecol 53:579–590

Purahong W, Hyde KD (2011) Effects of fungal endophytes on grass and non-grass litter decomposition rates. Fungal Divers 47:1–7

Quesada-Moraga E, Santiago-Álvarez C (2008) Hongos Entomopatógenos. In: Urbaneja A, Jacas J (eds) Control biológico de plagas. Phytoma y Publicaciones de la Universidad Pública de Navarra, Navarra, pp 98–120

Quesada-Moraga E, Landa BB, Muñoz Ledesma J, Jiménez Díaz RM, Santiago Álvarez C (2006) Endophytic colonisation of opium poppy, *Papaver somniferum*, by an entomopathogenic *Beauveria bassiana* strain. Mycopathology 161:323–329

Quesada-Moraga E, Navas-Cortés JA, Maranhao EA, Ortiz-Urquiza A, Santiago-Álvarez C (2007) Factors affecting the occurrence and distribution of entomopathogenic fungi in natural and agricultural soils. Mycol Res 111:947–966

Quesada-Moraga E, Muñoz-Ledesma FJ, Santiago-Álvarez C (2009) Systemic protection of *Papaver somniferum L.* against *Iraella luteipes* (Hymenoptera: *Cynipidae*) by an endophytic strain of *Beauveria bassiana* (Ascomycota: Hypocreales). Environ Entomol 38:723–730

Quesada-Moraga E, Muñoz-Ledesma FJ, Santiago-Álvarez C (2010) Cepa de *Beauveria bassiana* y sus aplicaciones. Patent WO2010092223. Int. Cl: A01N 63/04, C12N 1/14, C12R 1/645

Quesada-Moraga E, Landa BB, López-Díaz C (2013) Vertical transmission of an endophytic strain of *Beauveria bassiana* (Ascomycota; Hypocreales) colonizing opium poppy *Papaver somniferum*. In: 14th meeting of the IOBV/wprs working group "Insect pathogens and entomoparasitic nematodes". Zagreb, June 2013

Rangel-Castro JI, Killham K, Ostle N, Nicol GW, Anderson IC, Scrimgeour CM, Ineson P, Meharg A, Prosser JI, Rangel-Castro JI (2005) Stable isotope probing analysis of the influence of liming on root exudate utilization by soil microorganisms. Environ Microbiol 7:828–838

Rodríguez RJ, Henson J, Van Volkenburgh E, Hoy M, Wright L, Beckwith F, Kim YO, Redman RS (2008)

Stress tolerance in plants via habitat-adapted symbiosis. ISME J 2:404–416

Rodriguez RJ, White JF Jr, Arnold AE, Redman RS (2009) Fungal endophytes: diversity and functional roles. New Phytol 182:314–330

Romaine CP, Goodin MM (2002) Unraveling the viral complex associated with *La France* disease of the cultivated mushroom, *Agaricus bisporus*. In: Tavantzis SM (ed) Concepts and applications in agriculture, forestry, and medicine dsRNA genetic elements. CRC Press, Boca Raton, pp 237–257

Roossinck MJ (2010) Lifestyles of plant viruses. Philos Trans R Soc B 365:1899–1905

Roossinck MJ (2011) The good viruses: viral mutualistic symbiosis. Nat Rev Microbiol 9:99–108

Roy HE, Brodie EL, Chandler D, Goettel MS, Pell JK, Wajnberg E, Vega FE (2010) Deep space and hidden depths: understanding the evolution and ecology of fungal entomopathogens. BioControl 55:1–6

Sánchez S, Bills GF, Zabalgogeazcoa I (2007) The endophytic mycobiota of *Dactylis glomerata*. Fungal Divers 27:171–195

Sánchez S, Bills GF, Zabalgogeazcoa I (2008) Diversity and structure of the fungal endophytic assemblages from two sympatric coastal grasses. Fungal Divers 33:87–100

Sánchez S, Bills GF, Domínguez Acuña L, Zabalgogeazcoa I (2010) Endophytic mycobiota of leaves and roots of the grass *Holcus lanatus*. Fungal Divers 41:115–123

Sánchez S, Bills GF, Zabalgogeazcoa I (2011) Fungal species diversity in juvenile and adult leaves of *Eucalyptus globulus* from plantations affected by *Mycosphaerella* leaf disease. Ann Appl Biol 158:177–187

Sánchez S, Bills GF, Herrero N, Zabalgogeazcoa I (2012) Non systemic fungal endophytes of grasses. Fungal Ecol 5:289–297

Sánchez-Rodríguez AR, López-Díaz C, del Campillo MC, Quesada-Moraga E (2012) Response of wheat Triticum aestivum to colonisation by an endophytic strain of the entomopathogenic ascomycete Beauveria bassiana (Ascomycota; Hypocreales). Environment workshops. UNIA. Plant-microbe-insect Interactions: from molecular mechanims to ecological implications, Baeza

Santiago-Álvarez C, Valverde-García P, Quesada-Moraga E (2008) Control biológico de langostas y saltamontes. In: Urbaneja A, Jacas J (eds) Control biológico de plagas. Phytoma y Publicaciones de la Universidad Pública de Navarra, Navarra, pp 165–178

Sasan RK, Bidochka MJ (2012) The insect-pathogenic fungus *Metarhizium robertsii* (*Clavicipitaceae*) is also an endophyte that stimulates plant root development. Am J Bot 99:101–107

Schulz B, Boyle C (2005) The endophytic continuum. Mycol Res 109:661–686

Shah PA, Pell JK (2003) Entomopathogenic fungi as biological control agents. Appl Microbiol Biotechnol 61:413–423

Shirouzu T, Hirose D, Fukasawa Y, Tokumasu S (2009) Fungal succession associated with the decay of leaves of an evergreen oak, *Quercus myrsinifolia*. Fungal Divers 34:87–109

Souza Azevedo AC, Sosa-Gomez DR, Rodrigues Faria M, Pelegrinelli Fungaro MH (2000) Effects of double-stranded RNA on virulence of *Paecilomyces fumosoroseus* (Deuteromycotina: Hyphomycetes) against the silverleaf whitefly, *Bemisia tabaci* strain B (Homoptera: *Aleyrodidae*). Genet Mol Biol 23:61–63

St. Leger RJ (2008) Studies on adaptations of *Metarhizium anisopliae* to life in the soil. J Invertebr Pathol 98:271–276

Sugimoto M, Koike M, Hiyama N, Nagao H (2003) Genetic, morphological, and virulence characterization of the entomopathogenic fungus *Verticillium lecanii*. J Invertebr Pathol 82:176–187

Tanada Y, Kaya HK (1993) Insect pathology. Academic, San Diego, p 665

Tefera T, Vidal S (2009) Effect of inoculation method and plant growth medium on endophytic colonization of sorghum by the entomopathogenic fungus *Beauveria bassiana*. Biocontrol 54:663–669

Thomas SE, Crozier J, Catherine Aime M, Evans HC, Holmes KA (2008) Molecular characterisation of fungal endophytic morphospecies associated with the indigenous forest tree, *Theobroma gileri*, in Ecuador. Mycol Res 112:852–860

Tremblay E (1994) Entomologia Applicata, vol 4. Liguori, Ercolano/Napoli

Usuki F, Narisawa H (2007) A mutualistic symbiosis between a dark septate endophytic fungus, *Heteroconium chaetospira*, and a nonmycorrhizal plant, Chinese cabbage. Mycologia 99:175–184

Vandermeer J, Perfecto I, Liere H (2009) Evidence for hyperparasitism of coffee rust (*Hemileia vastatrix*) by the entomogenous fungus, *Lecanicillium lecanii*, through a complex ecological web. Plant Pathol 58:636–664

Varma A, Verma S, Nirmal Sahay S, Bütehorn B, Franken P (1999) *Piriformospora indica*, a cultivable plant-growth-promoting root endophyte. Appl Environ Microbiol 65:2741–2744

Vázquez de Aldana BR, Bills G, Zabalgogeazcoa I (2013) Are endophytes an important link between airborne spores and allergen exposure? Fungal Divers 60:33–42

Vega FE, Posada F, Aima MC, Pava-Ripoll M, Infate F, Rehner SA (2008) Entomopathogenic fungal endophytes. Biol Control 46:72–82

Vega FE, Goettel MS, Blackwell M, Chandler D, Jackson MA, Keller S, Koike M, Maniania NK, Monzon A, Ownley BH, Pell JK, Rangel DEN, Roy HE (2009) Fungal entomopathogens: new insights on their ecology. Fungal Ecol 2:149–159

Wagner BL, Lewis LC (2000) Colonization of corn, *Zea mays*, by the entomopathogenic fungus *Beauveria bassiana*. Appl Environ Microbiol 66:3468–3473

Wearn JA, Sutton BC, Morley NJ, Gange AC (2012) Species and organ specificity of fungal endophytes in herbaceous grassland plants. J Ecol 100:1085–1092

Yuan ZL, Rao LB, Chen YC, Zhang CL, Wu YG (2011) From pattern to process: species and functional diversity in fungal endophytes of *Abies beshanzuensis*. Fungal Biol 115:197–213

Part III

Host-Endophyte Interactions

5 Interactions of Meristem-Associated Endophytic Bacteria

Johanna Pohjanen, Janne J. Koskimäki, and Anna Maria Pirttilä

Abstract

Generally, all endophytes should be considered as a community that interacts with other symbiotic organisms, such as mycorrhiza. Even though an endophyte may colonize the plant systematically, communities colonizing the plant shoots normally differ to a degree from the root-associated endophytes. Meristem-associated shoot endophytic bacteria are often found as contaminants in plant tissue cultures started from shoot tips (buds) or embryos. Whereas root endophytic bacteria are reasonably well studied with respect to location and interactions with the host, not much is known about endophytes associated with shoot meristems. Endophytic bacteria have been localized in the meristematic tissues of buds and flowers by in situ hybridization and transmission electron microscopy. Meristem-associated endophytes may share some growth-promoting traits with the root endophytes, but likely additional mechanisms of actions exist. For example, such endophytes can produce adenine derivatives that induce growth of the host tissue. These endophytes may also affect the plant development by various ways. Some of them can co-synthesize secondary metabolites together with the plant host. Many more mechanisms remain to be determined by methods such as genomics and metabolomics, which are valuable tools for characterizing the interactions between the plant and endophytic bacteria.

1 Introduction

The studies on endophytic bacteria are often done on the plant root tissues (Rosenblueth and Martinez-Romero 2006). However, the root-associated communities typically differ from the shoot-associated ones on their diversity and function (Moore et al. 2006; Mano et al. 2006, 2007;

J. Pohjanen • J.J. Koskimäki • A.M. Pirttilä (✉)
Department of Biology, University of Oulu,
PO Box 3000, 90014 Oulu, Finland
e-mail: am.pirttila@oulu.fi

V.C. Verma and A.C. Gange (eds.), *Advances in Endophytic Research*,
DOI 10.1007/978-81-322-1575-2_5,

Izumi et al. 2008; Yrjälä et al. 2010; Compant et al. 2011). The study by Yrjälä et al. (2010) on hybrid aspen seedlings showed that the most frequently cultured leaf endophyte was *Methylobacterium fujisawaense*, whereas the roots mainly contained bacterial species of *Burkholderia fungorum, Pseudomonas koreensis*, and *Rahnella aquatilis*. Izumi et al. (2008) compared the endophytic populations of pine, birch, and rowan in the below- and aboveground tissues using cultivation-dependent and cultivation-independent analyses. They found a clear difference between the bacterial communities and also showed that a higher number of strains are found in the roots than in the stem and leaf tissues, whereas there was no difference between stem and leaf communities. Cultivation-dependent analyses of grape vine (Compant et al. 2011) and rice (Mano et al. 2006, 2007) have given similar results. In this chapter, the shoot tissues, especially the meristematic tissues in shoot tips (buds), flowers, seeds, and seedlings, are discussed with respect to endophytic bacteria and their interactions with the plant host, possibly affecting plant growth and development. A number of growth-promoting traits are shared between epiphytes and endophytes, as some of the species do occupy both niches. However, most endophytes inhabit only the specific niche of the plant interior (Izumi et al. 2008; Yrjälä et al. 2010), and more than likely, they have specific traits and roles within the plant tissue. In this chapter, we discuss the role of bacterial endophytes in the plant shoot tissues in the light of the most recent discoveries.

2 Plant Shoot-Associated Endophytes

The endophytic bacteria of shoot tissues are often isolated from plant tissue cultures, which are started from the meristems of the shoot tips, or seed embryos. For example, endophytic bacteria have been detected in the tissue cultures of papaya (Thomas et al. 2007), banana (Thomas et al. 2008), hazelnut (Reed et al. 1998), sour cherry (Kamoun et al. 1998), various species of poplar, larch, black locust and Norway spruce (Van Aken et al. 2004; Ulrich et al. 2008), and Scots pine (Laukkanen et al. 2000; Pirttilä et al. 2000). The range of bacterial species isolated from plant tissue cultures is wide, *Paenibacillus, Bacillus, Pseudomonas*, and *Methylobacterium* probably being the most commonly reported genera (Pirttilä et al. 2000; Ulrich et al. 2008).

2.1 Shoot Tissues as a Niche for Endophytic Bacteria

Compared to roots, plant shoot tissues are exposed to UV radiation, rapidly fluctuating temperatures and alternations in relative humidity. Shoot tissues contain more methanol, as methanol is mostly produced by the shoot tissues, contributing to methanol emissions to the atmosphere (Nemecek-Marshall et al. 1995). When exogenously applied to shoots, methanol induces plant growth, whereas root application results in toxic effects for the plant (Ramírez et al. 2006). Another significant difference between the shoot and root tissues as a niche for endophytes is photosynthesis, which exclusively occurs in the shoots. The few studies performed suggest that photosynthetic products are not consumed by endophytic bacteria, neither is photosynthetic efficiency affected by them. For example, the poplar endophyte *Enterobacter* sp. 638 has no effect on photosynthesis, stomatal conductance, photosynthetic water use efficiency or the maximum and operating efficiency of photosystem II (Rogers et al. 2012). Another example is the endophyte *Methylobacterium extorquens* DSM13060, isolated from shoot tips of Scots pine, which cannot utilize glucose or fructose as the energy source (Pirttilä et al. 2000). It is not well understood how the endophytes of shoot tissues enter the plant. Likely, some strains enter from the leaf surface through the epiderm or stomatal cells. In this case, their origin would be the water or air (wind). A number of shoot endophytes can be vertically transmitted, that is, through the seeds, although this has not exclusively been proved. Endophytes have been isolated from the seeds and even pollen (Cankar et al. 2005;

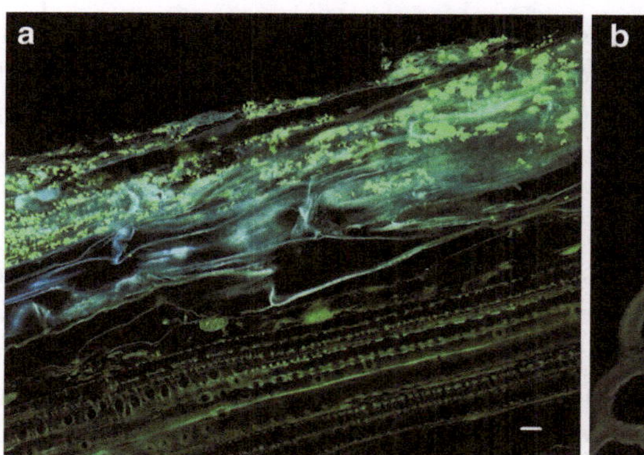

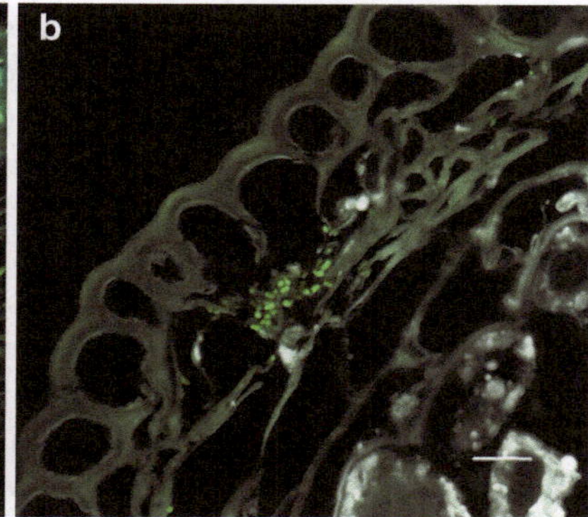

Fig. 5.1 Colonization of Scots pine seedling by GFP-tagged *Methylobacterium extorquens* DSM13060. (**a**) A longitudinal section of the pine root epiderm and cortex highly colonized by the bacteria 12 days after inoculation. (**b**) A cross section of the shoot, showing bacteria inside the cortex tissue 7 months after inoculation (scale bar = 10 μm)

Madmony et al. 2005; Pirttilä 2011), and a seed-inoculated endophyte has been shown to colonize the seedling tissues in *Eucalyptus* (Ferreira et al. 2008). A third likely source is the soil. When endophytes colonize plant shoots through the roots, they need to find a way to transfer further to the shoot tissues, and xylem has been proposed in several studies as the means of transportation after the first discovery of bacteria inhabiting xylem vessels (Bell et al. 1995). Our recent studies on colonization of Scots pine seedlings by the GFP-tagged *M. extorquens* DSM13060 indicate that all three routes can occur (Fig. 5.1; Koskimäki et al. unpublished).

2.2 Detection and Localization of Endophytic Bacteria in Shoot Meristematic Tissues

The traditional methods developed for the detection of endophytes relied on techniques dependent on plating of the bacteria. For example, surface-sterilized plant tissue was plated and the colonies growing on the medium after a specific incubation time were studied further. The endophytic bacteria associated with meristematic tissues were often isolated from plant tissue cultures, which had been started from surface-sterilized plant material. As a result, only cultivable strains were typically studied further, and the methods were also selective for species that preferred the growth conditions used. However, most endophytes are likely not cultivable (Koskimäki et al. 2010; Tejesvi et al. 2010) and a higher number of endophytes have been found by culture-independent methods than by culture-dependent ones (Yang et al. 2001; Podolich et al. 2007; Tejesvi et al. 2010; Yashiro et al. 2011). Therefore, culture-independent techniques, such as in situ hybridization (Pirttilä et al. 2000), and PCR-based methods, for example, denaturing gradient gel electrophoresis (DGGE) (Yang et al. 2001; Izumi et al. 2008), restriction fragment length polymorphism (RFLP) (Ardanov et al. 2012), and direct sequencing (Koskimäki et al. 2010), have been developed and applied for the study of single endophytic bacterial strains or whole communities. However, the methods based on amplification of bacterial 16S rDNA are often hampered by the similarity between bacterial, plant mitochondrial, and chloroplast sequences and need careful designing of primers specific for the bacteria (Sessitsch et al. 2002; Ardanov et al. 2012). Endophytes can be localized in the plant tissue by various microscopic methods. Transmission electron microscopy (TEM) enables very high magnification of the plant tissue and study of the location of bacteria in the cellular compartments, although distinguishing the bacterial cells in the sample requires specific expertise. Another weakness of the method is that TEM gives no information on the species of the endophytic organism. By TEM, endophytic bacteria have been detected in ultrathin sections of buds of linden (*Tilia cordata* L.) and needles of blue spruce (Doronina et al. 2004; Pirttilä et al. 2008).

In situ hybridization can be used for localization of endophytic bacteria by species, genus, class, or phylum. Pirttilä et al. (2000, 2003) developed oligonucleotide probes to detect endophytes in pine tissues. Using probes specific for eubacteria, *Methylobacterium* spp., a *Pseudomonas fluorescens* subgroup, and *Mycobacterium* spp., the corresponding endophytes were identified in the cells of scale primordia, the meristems, and around the resin ducts of Scots pine buds (Pirttilä et al. 2000, 2003, 2005) and in the cells of growing callus culture (Pirttilä et al. 2002). The advantage of using the in situ hybridization method is that besides localizing the microbes, it reflects the changes in the metabolic activity of the microbes when the probes are hybridized to transcripts such as ribosomal RNA (DeLong et al. 1989). Therefore, the location and metabolic activity of endophytes in the Scots pine shoot tips were dependent on the growth season when studied by in situ hybridization throughout the year. Endophytes were not detected at all during pine dormancy and rarely found in the elongating shoot tips during growth season. The highest endophytic metabolic rates were detected in tissues of spring and autumn, prior to growth or differentiation of the bud (Pirttilä et al. 2005). In addition to buds, endophytic bacteria are detected in reproductive organs. Madmony et al. (2005) isolated *Enterobacter cloacae* from pollen and fertilized ovules of different *Pinus* sp., and Pirttilä (2011) detected endophytes in inflorescences and seed embryos of *Pinus sylvestris*. Bacteria in the genera *Pseudomonas* and *Rahnella* were found in seeds of Norway spruce (Cankar et al. 2005). Furthermore, several endophytic bacterial species in the taxa *Gammaproteobacteria* (relatives of *Pseudomonas* sp.) and *Firmicutes* (relatives of *Bacillus pumilus* and *B. cereus* group members) have been isolated from flowers, fruits, and seeds of grapevine (Compant et al. 2011). In another recent study, species of *Kocuria*, *Acinetobacter*, *Enterobacter*, and *Staphylococcus* were isolated from seeds, endocarp, and mesocarp of different *Carica papaya* variety fruits (Krishnan et al. 2012). Johnston-Monje and Raizada (2011) studied recently the endophytic microbes in the seeds of various *Zea* sp. and by culture-independent methods identified *Clostridium* and *Paenibacillus* spp., and by culturing, bacteria in the genera *Enterobacter*, *Methylobacterium*, *Pantoea*, and *Pseudomonas*. Molecular methods provide additional tools for studying bacterial colonization and localization. These methods have commonly been used for studying microbes in the rhizosphere. Genetic tagging of endophytic bacteria with genes encoding for fluorescent reporter proteins allows detailed monitoring of the colonization process inside the plant tissues by using laser scanning confocal microscopy (LSCM) (Poonguzhali et al. 2008; Prieto et al. 2011). Broad host-range plasmid vectors and transposon systems with stable site-directed insertions to bacterial chromosome provide several alternatives suitable for transformation of most bacterial species (Koch et al. 2001; Ramos et al. 2011). Advances in the development of novel reporter protein derivates, which are brighter and more photostable than the conventional ones, have supplied new means to overcome the extensive autofluorescence of plant tissues, which often hinders the colonization studies by LSCM (Shaner et al. 2007; Lagendijk et al. 2010). Combination of LSCM with advanced genetic tagging methods presents a valid, noninvasive alternative for complex endophyte-host interaction studies to be performed with live or fixed plant tissues. In our recent interaction study, a dual labeling strategy was used to monitor simultaneously the endophytic colonization and gene expression of *Methylobacterium extorquens* DSM13060 in Scots pine (*Pinus sylvestris* L.) seedlings. *M. extorquens* DSM13060 was tagged chromosomally with green fluorescent protein (eGFP) under constantly active promoter by using Tn5 transposon. To assess the bacterial gene activity during the endophytic lifestyle, another reporter protein "mCherry" regulated by a selected promoter region was subsequently transformed to the same bacterial strain. Activation of the mCherry reporter verified that the selected promoter and the gene regulated by it were functioning in the endophytic conditions. At the same time, the dual reporter experiment provided detailed information about methylobacterial colonization and localization in the pine tissues (Fig. 5.1, Koskimäki et al. unpublished).

3 Interactions of Shoot Meristem-Associated Endophytes with Plant Host

Methanol present in the shoot tissues creates a good carbon source specifically for methylotrophs, which can utilize methanol and methane as the energy source (Fall 1996; Fall and Benson 1996). Because methanol is toxic for the plant, the removal by methylotrophs may already have significant benefits for the plant. Methanol applied to the plant surface increases plant shoot growth (Nonomura and Benson 1991; Ramírez et al. 2006), which suggests that methylotrophic bacteria transform methanol to products beneficial for the plant. For example, *Methylobacterium* spp. can participate in the biosynthesis of compounds commonly known as plant products (Zabetakis 1997; Koutsompogeras et al. 2007). Endophytic bacteria were recently detected in the receptacle vascular tissue and in the cells of achenes of raw strawberry. This study indicated that the biosynthesis of the strawberry flavor compounds DHMF and mesifuran is aided by the bacterial methanol dehydrogenase, as the bacterial methanol dehydrogenase and plant DMHF biosynthesis genes were localized by in situ hybridization in the same tissues or cells of the strawberry receptacle (Nasopoulou 2012). Independent of methylotrophy, many studies have reported the positive effect of shoot endophytic bacteria on tissue organogenesis and embryogenesis (Visser et al. 1994; Murthy et al. 1999; Pirttilä et al. 2004; Pohjanen et al. 2013). However, rarely specific, individual compounds are identified responsible for such effects. Phytohormones produced by endophytes are the most popular compounds suggested responsible for the morphological effects on plant host.

3.1 Endophytic Products

Production of plant growth hormones is typical for all plant-associated microbes. However, even though a microbe can produce plant growth hormones, it cannot be generalized to promote growth on all plant hosts, but the result depends on mutual interactions, as was discovered on *Solanum nigrum* endophytic bacteria (Long et al. 2008). Whereas gibberellin production can be considered a typical trait for root-associated bacteria, epiphytic and root endophytic bacteria most typically synthesize and secrete auxins (Brandl and Lindow 1996; Bastián et al. 1998; Costacurta et al. 1998; Doronina et al. 2002; Gamalero et al. 2003; Ivanova et al. 2001, 2008; Merzaeva and Shirokikh 2010). However, IAA has been identified as a product of a few endophyte species isolated from shoots. For example, the shoot endophytic *Pseudomonas stutzeri* strain producing IAA has been isolated from *Echinacea* tissue culture (Lata et al. 2006). The endophyte of poplar, *M. populi*, and the endophyte of pollen grains of *Pinus* spp., *Enterobacter cloacae*, are reported to produce IAA (Madmony et al. 2005; Taghavi et al. 2009). A number of pathogenic and beneficial plant-associated bacteria synthesize cytokinins (Akiyoshi et al. 1987; Timmusk et al. 1999; Garcia de Salamone et al. 2001). Methylotrophic epiphytic bacteria such as *Methylovorus mays* and *Methylobacterium mesophilicum* JCM 2829 also synthesize cytokinins (Ivanova et al. 2000, 2008). These results would indicate a significant role for plant growth hormones such as cytokinins in the plant growth promotion by plant-associated microbes. However, when cytokinin production and plant growth promotion were studied in the type strain *Methylobacterium extorquens* AM1, results indicated that cytokinin production might not be the factor contributing to plant growth (Koenig et al. 2002). *M. extorquens* was reported to produce tRNA-derived trans-zeatin, but when cytokinin-null (*miaA*) mutants incapable of cytokinin synthesis were generated, they stimulated germination of the heat-treated soybean seeds at the same level as the wild-type bacteria (Koenig et al. 2002).

Plant growth hormone production is not common to all endophytes, especially those associated with meristematic tissues. Even in the strains producing plant growth hormones, the levels vary greatly (Ivanova et al. 2008). These results indicate that other possibly more prominent methods of growth promotion by endophytes exist. The endophytes isolated from Scots pine shoot tips, *Methylobacterium extorquens* DSM13060 and *Pseudomonas synxantha* DSM13080, produce compounds that extend the viability and affect the morphology of callus tissues in vitro (Pirttilä et al. 2004). The most common plant growth hormones were not identified responsible for these effects, but adenine and adenine ribosides were produced by *M. extorquens* DSM13060 (Pirttilä et al. 2004). Adenine induces plant growth in tissue culture, but the mode of action is unknown (George and Sherrington 1984). Adenine riboside is the metabolite of adenine (Baumann et al. 1994) and found abundant in the vascular cambial region of *Pinus sylvestris* (Moritz and Sundberg 1996). Therefore, adenine and adenine riboside are potential plant-growth-promoting products of shoot endophytes. A trait often associated with endophytic bacteria is production of the enzyme aminocyclopropane-1-carboxylate (ACC) deaminase. This enzyme transforms the ethylene precursor ACC to ammonia and 2-oxobutanoate, preventing ethylene signaling. Ethylene is a plant hormone acting in seed germination and various stresses, such as bacterial colonization. It has been suggested that ACC deaminase increases plant growth and development in stressful conditions by decreasing plant ethylene levels (Glick 2005). For example, the root endophyte *Burkholderia phytofirmans* PsJN carries a gene encoding ACC deaminase, and inactivation of this gene results in loss of the ability to promote root elongation in canola seedlings (Sun et al. 2009). Whereas the ACC deaminase-carrying endophytes are often isolated and studied in the rhizosphere or roots, a recent study performed on cut flowers indicates that bacteria were able to colonize the shoot where ACC deaminase prolonged flowering (Ali et al. 2012). However, an analysis of sequenced endophyte genomes suggests that ACC deaminase is less important than anticipated (Frank 2011). The *Methylobacterium extorquens* DSM13060 isolated from Scots pine buds carries the gene for ACC deaminase. When activation of this gene was studied by promoter fusion with a fluorescent protein, it was rarely active during endophyte colonization of pine seedlings (Koskimäki et al. unpublished). This might indicate a smaller role of ACC deaminase in the plant shoot-colonizing endophytes. Epiphytic methylotrophs can synthesize vitamin B_{12} (Nishio et al. 1977; Ivanova et al. 2006, 2008), which has been suggested a plant-growth-promoting product of endophytes, as well (Ivanova et al. 2008). Vitamin B_{12} comprises a group of compounds that have trivalent cobalt as the cofactor. Generally, vitamin B_{12} is the coenzyme for isomerization and transmethylation reactions in the biosynthesis of compounds containing methyl groups. Enzymes requiring vitamin B_{12} as the coenzyme are found in many flowering plants that cannot synthesize vitamin B_{12} (Holland and Polacco 1994). In mosses, epiphytic methylotrophs increase the biomass, amount, length, and the degree of branching of gametophytes (Koopman and Kutschera 2005), which are also obtained by exogenously applied vitamin B_{12} (Basile et al. 1985). However, our recent reporter gene studies on the shoot endophyte *M. extorquens* DSM13060 suggest a smaller role for bacterial vitamin B_{12} production in the plant-endophyte interaction, than previously suggested (Koskimäki et al. unpublished).

3.2 Interaction Web in the Full Plant Microbiome

The interactions between various plant-associated microbes are often studied in isolated in vitro conditions using single strains. These studies are usually concentrated on the roots because of the well-known benefits of root fungal and bacterial symbionts, mycorrhiza, and rhizobia, respectively. Mutualistic interactions can be found between mycorrhizal fungi and a group of bacteria, called mycorrhizal helper bacteria (MHB; Garbaye 1994). Furthermore, interactions between different plant-growth-promoting rhizobacteria (PGPR, Bashan and de-Bashan 2005) have been shown beneficial for the host plant (Madhaiyan et al. 2010). These microbes usually improve the growth and nutrition of the plant and, in the case of MHB, also the growth and sporulation of the fungal partner. Similarly, the mycorrhizal fungus can promote growth of the bacterial partner. For example, in *Pinus halepensis* roots, the ectomycorrhizal fungus *Suillus granulatus* improved the survival of *Pseudomonas fluorescens* in areas where the fungal colonization was the highest (Rincón et al. 2005). The interaction between microbes is often specific for the species or the strain. Studies combining epiphytic *Methylobacterium oryzae* strains with different rhizobacteria (Madhaiyan et al. 2010) or with arbuscular mycorrhiza (Kim et al. 2010) showed that the positive growth effect was dependent on the combination of microbes. Similarly, the root endophytic bacteria *Pseudomonas aeruginosa* and *Burkholderia cepacia* of oil palm were shown to act as mycorrhizal helper bacteria on two arbuscular mycorrhizal fungi, *Glomus clarum* and *Glomus intraradices*, but to exhibit antagonism on the pathogen *Ganoderma boninense* (Sundram et al. 2011). Although the microbial communities differ in the aerial parts from those of the roots (Izumi et al. 2008; Yrjälä et al. 2010) and there is a very low number of published examples of microorganisms interacting in the plant shoot tissues, a similar interaction between various members likely exists. For example, parallel to bacteria found in the hyphae of mycorrhizal fungi in the rhizosphere, Hoffman and Arnold (2010) revealed bacteria inhabiting the living hyphae of foliar endophytic fungi. Furthermore, Araújo et al. (2001) isolated several endophytic species from leaf tissues of citrus rootstocks and found that *Guignardia citricarpa*, one of the most abundant fungi among the isolates, stimulated growth of the endophytic *P. agglomerans* but had an inhibitory effect on growth of some endophytic *Bacillus* species.

Microbes can prevent or inhibit the growth of other strains by several ways. Direct growth inhibition can occur through secreted compounds, but antagonism includes also the competition for colonization sites, nutrients, and minerals (reviewed by Berg 2009). Endophytic *Bacillus subtilis* strain from the stem of the giant hogweed (*Heracleum sosnowskyi*, Manden) produces antifungal lipopeptide antibiotics and is able to protect tomato against the fungal pathogen causing tomato foot and root rot (Malfanova et al. 2011, 2012). *Bacillus mojavensis* isolated from kernels of maize is able to inhibit growth of the pathogenic fungus *Fusarium verticillioides* and reduce mycotoxin production (Bacon et al. 2001; Bacon and Hinton 1999), and a number of *B. mojavensis* strains were shown to produce a mixture of surfactins, which are toxic to several pathogens (Bacon and Hinton 2011). Another example comes from our study on shoot endophytic *Methylobacterium* sp. IMBG290, which induced resistance against the pathogen *Pectobacterium atrosepticum* in potato. The resistance was not due to produced toxins but dependent on the inoculum density of *Methylobacterium* sp., which was associated with changes in the structure of the existing, innate endophyte community. The changes correlated with resistance or susceptibility, suggesting that the whole endophytic community acted on the plant responses (Ardanov et al. 2012). Interaction between symbiotic microorganisms can also occur across various plant compartments (Novas et al. 2009; Liu et al. 2011), such as roots and shoot tips. These examples demonstrate that an endophyte strain isolated from the host plant should never be considered as an organism interacting with the plant host alone, but as a member of the full plant microbiome.

4 Conclusions

The plant shoot-colonizing bacterial endophytes are considerably less studied than bacteria living in the roots or in the rhizosphere. Due to easy access to culturable isolates in the root tissues, the great majority of studies worldwide are concentrated on root-colonizing endophytes (Rosenblueth and Martinez-Romero 2006). However, the shoot meristems can be considered one of the most important tissues of the plant, responsible for growth and development of new leaves and stems. The finding of bacterial endophytes in these tissues suggests that a balanced interaction is essential for their proper function. How is the plant regulating the endophytes colonizing these tissues, and which role are the microbes playing in plant development? It is known that symbiotic microbes affect the development of animals (Troll et al. 2009). As endophytes have been occupying the plant interior for more than 400 million years (Krings et al. 2007), mutual evolution must have driven ways to subsist, adapt, and eventually refine the interaction to a balanced state. Development of genomic tools is effectively opening the doors to the secret world of bacterial endophytes and allowing further studies on their life inside the plant, as we have described in this chapter. Metabolomics is another tool that can provide a systemic view of the plant-microbe interaction at the level where genomics has no access (Scherling et al. 2009; Fester et al. 2011). Knowledge gained with these powerful methods will be helpful in defining the details of the plant-endophyte interaction in the plant shoot meristems.

Acknowledgments *Societas pro Fauna et Flora Fennica*, The Finnish Cultural Foundation, North Ostrobothnia Regional Fund, Tauno Tönning Foundation, and Niemi Foundation are thanked for financial support to J. Pohjanen and J. J. Koskimäki. We would also like to thank Dr. Ellen L. Lagendijk and Dr. Ole Nybroe, M.Sc. Emmi-Leena Ihantola, and M.Sc. Pavlo Ardanov.

References

Akiyoshi DE, Regier DA, Gordon MP (1987) Cytokinin production by *Agrobacterium* and *Pseudomonas* spp. J Bacteriol 169:4242–4248

Ali S, Charles TC, Glick BR (2012) Delay of flower senescence by bacterial endophytes expressing 1-aminocyclopropane-1-carboxylate deaminase. J Appl Microbiol 113:1139–1144

Araújo WL, Maccheroni W Jr, Aguilar-Vildoso CI et al (2001) Variability and interactions between endophytic bacteria and fungi isolated from leaf tissues of citrus rootstocks. Can J Microbiol 47:229–236

Ardanov P, Sessitsch A, Häggman H et al (2012) Methylobacterium-induced endophyte community changes correspond with protection of plants against pathogen attack. PLoS One 7(10):e46802

Bacon CW, Hinton DM (1999) Use of *Bacillus subtilis* as an endophyte for the control of diseases caused by fungi. US Patent and Trademark Office November 30, 1999

Bacon CW, Hinton DM (2011) Bacillus mojavensis: its endophytic nature, the surfactins, and their role in the plant response to infection by Fusarium verticillioides. In: Maheshwari DK (ed) Bacteria in agrobiology: plant growth responses. Springer, Berlin/Heidelberg, pp 21–39. doi:10.1007/978-3-642-20332-9_2

Bacon CW, Yates IE, Hinton DM et al (2001) Biological control of *Fusarium moniliforme* in maize. Environ Health Perspect 109:325

Bashan Y, de-Bashan L (2005) Bacteria/plant growth-promotion. Encycl Soil Environ 1:103–115

Basile DV, Basile MR, Li QY et al (1985) Vitamin B_{12}-stimulated growth and development of *Jungermannia leiantha* Grolle and *Gymnocolea inflata* (Huds.) Dum. (Hepaticae). Bryologist 88:77–81

Bastián F, Cohen A, Piccoli P et al (1998) Production of indole-3-acetic acid and gibberellins A1 and A3 by *Acetobacter diazotrophicus* and *Herbaspirillum seropedicae* in chemically-defined cultures. Plant Growth Regul 24:7–11

Baumann TW, Schulthess BH, Linden A et al (1994) Structure and metabolism of 7-β-D-glucopyranosyladenine – the product of a cytokinin-sparing reaction? Phytochemistry 36:537–542

Bell CR, Dickie GA, Harvey WLG et al (1995) Endophytic bacteria in grapevine. Can J Microbiol 41:46–53

Berg G (2009) Plant–microbe interactions promoting plant growth and health: perspectives for controlled use of microorganisms in agriculture. Appl Microbiol Biotechnol 84:11–18

Brandl MT, Lindow SE (1996) Cloning and characterization of a locus encoding an indolepyruvate decarboxylase involved in indole-3-acetic acid synthesis in *Erwinia herbicola*. Appl Environ Microbiol 62:4121–4128

Cankar K, Kraigher H, Ravnikar M et al (2005) Bacterial endophytes from seed of Norway spruce (*Picea abies* L. Karst). FEMS Microbiol Lett 244:341–345

Compant S, Mitter B, Colli-Mull JG et al (2011) Endophytes of grapevine flowers, berries, and seeds: identification of cultivable bacteria, comparison with other plant parts, and visualization of niches of colonization. Microb Ecol 62:188–197

Costacurta A, Mazzafera P, Rosato Y (1998) Indole-3-acetic acid biosynthesis by *Xanthomonas axonopodis* pv. *citri* is increased in the presence of plant leaf extracts. FEMS Microbiol Lett 159:215–220

DeLong EF, Wickham GS, Pace NR (1989) Phylogenetic stains: ribosomal RNA-based probes for the identification of single cells. Science 243:1360–1363

Doronina NV, Ivanova EG, Trotsenko YA (2002) New evidence for the ability of methylobacteria and methanotrophs to synthesize auxins. Mikrobiologiya 71:130–132

Doronina NV, Ivanova EG, Suzina NE et al (2004) Methanotrophs and methylobacteria are found in woody plant tissues within the winter period. Mikrobiologiya 73:702–709

Fall R (1996) Cycling of methanol between plants, methylotrophs and the atmosphere. In: Lidstrom ME, Tabita FR (eds) Microbial growth on C1 compounds. Kluwer Academic Publishers, Dordrecht, pp 343–350

Fall R, Benson AA (1996) Leaf methanol – the simplest natural product from plants. Trends Plant Sci 1:296–301

Ferreira A, Quecine MC, Lacava PT et al (2008) Diversity of endophytic bacteria from *Eucalyptus* species seed and colonization of seedlings by *Pantoea agglomerans*. FEMS Microbiol Lett 287:8–14

Fester T, Fetzer I, Buchert S et al (2011) Towards a systemic metabolic signature of the arbuscular mycorrhizal interaction. Oecologia 167:913–924

Frank AC (2011) The genomes of endophytic bacteria. In: Pirttilä AM, Frank AC (eds) Endophytes of forest trees: biology and applications, vol 80, 1st edn, Forestry sciences. Springer, New York, pp 107–136

Gamalero E, Fracchia L, Cavaletto M et al (2003) Characterization of functional traits of two fluorescent pseudomonads isolated from basidiomes of ectomycorrhizal fungi. Soil Biol Biochem 35:55–65

Garbaye J (1994) Tansley review No. 76. Helper bacteria: a new dimension to the mycorrhizal symbiosis. New Phytol 128(2):197–210

Garcia de Salamone IE, Hynes RK, Nelson LM (2001) Cytokinin production by plant growth promoting rhizobacteria and selected mutants. Can J Microbiol 47:404–411

George EF, Sherrington PD (1984) Plant propagation by tissue culture methods. Handbook and directory of commercial laboratories. Eastern Press, Reading/Berks

Glick BR (2005) Modulation of plant ethylene levels by the bacterial enzyme ACC deaminase. FEMS Microbiol Lett 25:1–7

Hoffman MT, Arnold AE (2010) Diverse bacteria inhabit living hyphae of phylogenetically diverse fungal endophytes. Appl Environ Microbiol 76:4063–4075

Holland MA, Polacco JC (1994) PPFMs and other covert contamination: is there more to plant physiology than just plant? Annu Rev Plant Phys Plant Mol Biol 45:197–209

Ivanova EG, Doronina NV, Shepelyakovskaya AO et al (2000) Facultative and obligate aerobic methylobacteria synthesize cytokinins. Mikrobiologiya 69:764–769

Ivanova EG, Doronina NV, Trotsenko YA (2001) Aerobic methylobacteria are capable of synthesizing auxins. Mikrobiologiya 70:452–458

Ivanova EG, Fedorov DN, Doronina NV et al (2006) Production of vitamin B_{12} in aerobic methylotrophic bacteria. Mikrobiologiya 75:494–496

Ivanova EG, Pirttilä AM, Fedorov DNF et al (2008) Association of methylotrophic bacteria with plants: metabolic aspects. In: Sorvari S, Pirttilä AM (eds) Prospects and applications for plant associated microbes. A laboratory manual, part A: bacteria. Biobien Innovations, Turku, pp 225–231

Izumi HIH, Anderson ICAIC, Killham KKK et al (2008) Diversity of predominant endophytic bacteria in European deciduous and coniferous trees. Can J Microbiol 54:173–179

Johnston-Monje D, Raizada MN (2011) Conservation and diversity of seed associated endophytes in *Zea* across boundaries of evolution, ethnography and ecology. PLoS One 6(6):e20396

Kamoun R, Lepoivre P, Boxus P (1998) Evidence for the occurrence of endophytic prokaryotic contaminants in micropropagated plantlets of *Prunus cerasus* cv. 'Montgomery'. Plant Cell Tissue Organ Cult 52:57–59

Kim K, Yim W, Trivedi P et al (2010) Synergistic effects of inoculating arbuscular mycorrhizal fungi and *Methylobacterium oryzae* strains on growth and nutrient uptake of red pepper (*Capsicum annuum* L.). Plant Soil 327:429–440

Koch B, Jensen LE, Nybroe O (2001) A panel of Tn7-based vectors for insertion of the gfp marker gene or for delivery of cloned DNA into Gram-negative bacteria at a neutral chromosomal site. J Microbiol Method 45:187–195

Koenig RL, Morris RO, Polacco JC (2002) tRNA is the source of low-level trans-zeatin production in *Methylobacterium* spp. J Bacteriol 184:1832–1842

Koopman V, Kutschera U (2005) *In vitro* regeneration of sunflower plants: effects of a *Methylobacterium* strain on organ development. J Appl Bot Food Qual 79:59–62

Koskimäki JJ, Nylund S, Suorsa M et al (2010) Mycobacterial endophytes are enriched during micropropagation of *Pogonatherum paniceum*. Environ Microbiol Rep 2:619–624

Koutsompogeras P, Kyriacou A, Zabetakis I (2007) The formation of 2, 5-dimethyl-4-hydroxy-2*H*-furan-3-one by cell free extracts of *Methylobacterium extorquens* and strawberry (*Fragaria* × *ananassa* cv. Elsanta). Food Chem 104:1654–1661

Krings M, Taylor TN, Hass H et al (2007) Fungal endophytes in a 400–million-yr-old land plant: infection pathways, spatial distribution, and host responses. New Phytol 174:648–657

Krishnan P, Bhat R, Kush A et al (2012) Isolation and functional characterization of bacterial endophytes from *Carica papaya* fruits. J Appl Microbiol 113: 308–317

Lagendijk EL, Validov S, Lamers GEM et al (2010) Genetic tools for tagging Gram-negative bacteria with mCherry for visualization *in-vitro* and in natural habitat, biofilms and pathogenicity studies. FEMS Microbiol Lett 305:81–90

Lata H, Li XC, Silva B et al (2006) Identification of IAA-producing endophytic bacteria from micropropagated *Echinacea* plants using 16S rRNA sequencing. Plant Cell Tissue Organ Cult 85:353–359

Laukkanen H, Soini H, Kontunen-Soppela S et al (2000) A mycobacterium isolated from tissue cultures of mature *Pinus sylvestris* interferes with growth of Scots pine seedlings. Tree Physiol 20:915–920

Liu Q, Parsons AJ, Xue H et al (2011) Competition between foliar *Neotyphodium lolii* endophytes and mycorrhizal *Glomus* spp. fungi in *Lolium perenne* depends on resource supply and host carbohydrate content. Funct Ecol 25:910–920

Long HH, Schmidt DD, Baldwin IT (2008) Native bacterial endophytes promote host growth in a species-specific manner; phytohormone manipulations do not result in common growth responses. PLoS One 3:e2702

Madhaiyan M, Poonguzhali S, Kang BG et al (2010) Effect of co-inoculation of methylotrophic *Methylo-bacterium oryzae* with *Azospirillum brasilense* and *Burkholderia pyrrocinia* on the growth and nutrient uptake of tomato, red pepper and rice. Plant Soil 328:71–82

Madmony A, Chernin L, Pleban S et al (2005) *Enterobacter cloacae*, an obligatory endophyte of pollen grains of Mediterranean pines. Folia Microbiol 50:209–216

Malfanova N, Kamilova F, Validov S et al (2011) Characterization of *Bacillus subtilis* HC8, a novel plant-beneficial endophytic strain from giant hogweed. Microb Biotechnol 4:523–532

Malfanova N, Franzil L, Lugtenberg B et al (2012) Cyclic lipopeptide profile of the plant-beneficial endophytic bacterium *Bacillus subtilis* HC8. Arch Microbiol 194:893–899

Mano H, Tanaka F, Watanabe A et al (2006) Culturable surface and endophytic bacterial flora of the maturing seeds of rice plants (*Oryza sativa*) cultivated in a paddy field. Microb Environ 21:86–100

Mano H, Tanaka F, Nakamura C et al (2007) Culturable endophytic bacterial flora of the maturing leaves and roots of rice plants (*Oryza sativa*) cultivated in a paddy field. Microb Environ 22:175–185

Merzaeva OV, Shirokikh IG (2010) The production of auxins by the endophytic bacteria of winter rye. Appl Biochem Microbiol 46:51–57

Moore FP, Barac T, Borremans B et al (2006) Endophytic bacterial diversity in poplar trees growing on a BTEX-contaminated site: the characterization of isolates with potential to enhance phytoremediation. Syst Appl Microbiol 29:539–556

Moritz T, Sundberg B (1996) Endogenous cytokinins in the vascular cambial region of *Pinus sylvestris* during activity and dormancy. Physiol Plant 98:693–698

Murthy BNS, Vettakkorumakankav NN, KrishnaRaj S et al (1999) Characterization of somatic embryogenesis in *Pelargonium* × *hortorum* mediated by a bacterium. Plant Cell Rep 18:607–613

Nasopoulou C (2012) Study of strawberry (*F. ananassa*) and *M. extorquens* cells for the biosynthesis of strawberry flavor. Scientific Report of COST STSM Reference Number: COST-STSMFA1103-10547 and COST Action: FA1103. 1

Nemecek-Marshall M, MacDonald RC, Franzen JJ, Wojciechowski CL, Fall R (1995) Methanol emission from leaves (enzymatic detection of gas-phase methanol and relation of methanol fluxes to stomatal conductance and leaf development). Plant Physiol 108:1359–1368

Nishio N, Tanaka M, Matsuno R et al (1977) Production of vitamin B_{12} by methanol-utilizing bacteria, *Pseudomonas* AM-1 and *Microcyclus eburneus*. Ferment Technol 55:200–203

Nonomura AM, Benson AA (1991) The path of carbon in photosynthesis: improved crop yields with methanol. Proc Natl Acad Sci USA 89:9794–9798

Novas MV, Iannone LJ, Godeas AM, Cabral D (2009) Positive association between mycorrhiza and foliar endophytes in Poa bonariensis, a native grass. Mycol Prog 8:75–81

Pirttilä AM (2011) Colonization of tree shoots by endophytic fungi. In: Pirttilä AM, Sorvari S (eds) Prospects and applications for plant-associated microbes. A laboratory manual, Part B: fungi. Biobien Innovations, Turku, pp 93–95

Pirttilä AM, Laukkanen H, Pospiech H et al (2000) Detection of intracellular bacteria in the buds of Scotch pine (*Pinus sylvestris* L.) by *in situ* hybridization. Appl Environ Microbiol 66:3073–3077

Pirttilä AM, Laukkanen H, Hohtola A (2002) Chitinase production in pine callus (*Pinus sylvestris* L.): a defense reaction against endophytes? Planta 214:848–852

Pirttilä AM, Pospiech H, Laukkanen H et al (2003) Two endophytic fungi in different tissues of Scots pine buds (*Pinus sylvestris* L.). Microb Ecol 45:53–62

Pirttilä AM, Joensuu P, Pospiech P et al (2004) Endophytic products affect morphology and mitigate browning of callus cultures of Scots pine (*Pinus sylvestris* L.). Physiol Plant 121:305–312

Pirttilä AM, Pospiech H, Laukkanen H et al (2005) Seasonal variation in location and population structure of endophytes in buds of Scots pine. Tree Physiol 25:289–297

Pirttilä AM, Hohtola A, Ivanova EG et al (2008) Identification and localization of methylotrophic plant-associated bacteria. In: Sorvari S, Pirttilä AM (eds) Prospects and applications for plant associated microbes. A laboratory manual, part A: bacteria. Biobien Innovations, Turku, pp 218–224

Podolich O, Ardanov P, Voznyuk T et al (2007) Endophytic bacteria from potato activated by exogenic non-pathogenic bacteria. Biopolym Cell 23:21–27

Pohjanen J, Koskimäki JJ, Sutela S et al. (2013) Interaction between ectomycorrhizal fungi and endophytic *Methylobacterium* affects nutrition and growth of Scots pine seedlings *in vitro* (Manuscript, submitted)

Poonguzhali S, Madhaiyan M, Yim WJ et al (2008) Colonization pattern of plant root and leaf surfaces visualized by use of green-fluorescent-marked strain of *Methylobacterium suomiense* and its persistence in rhizosphere. Appl Microbiol Biotechnol 78:1033–1043

Prieto P, Schilirò E, Maldonado-González MM et al (2011) Root hairs play a key role in the endophytic colonization of olive roots by *Pseudomonas* spp. with biocontrol activity. Microb Ecol 62:435–445

Ramírez I, Dorta F, Espinoza V et al (2006) Effects of foliar and root applications of methanol on the growth of *Arabidopsis*, tobacco and tomato plants. J Plant Growth Regul 25:30–44

Ramos HJ, Yates MG, Pedrosa FO et al (2011) Strategies for bacterial tagging and gene expression in plant-host colonization studies. Soil Biol Biochem 43:1626–1638

Reed BM, Mentzer J, Tanprasert P et al (1998) Internal bacterial contamination of micropropagated hazelnut: identification and antibiotic treatment. Plant Cell Tissue Organ Cult 52:67–70

Rincón A, Ruiz-Díez B, García-Fraile S et al (2005) Colonisation of *Pinus halepensis* roots by *Pseudomonas fluorescens* and interaction with the ectomycorrhizal fungus *Suillus granulatus*. FEMS Microbiol Ecol 51:303–311

Rogers A, McDonald K, Muehlbauer MF et al (2012) Inoculation of hybrid poplar with the endophytic bacterium *Enterobacter* sp. 638 increases biomass but does not impact leaf level physiology. GCB Bioenergy 4:364–370

Rosenblueth M, Martinez-Romero E (2006) Bacterial endophytes and their interactions with hosts. Mol Plant Microbe Interact 19:827–837

Scherling C, Ulrich K, Ewald D et al (2009) Metabolic signature of the beneficial interaction of the endophyte *Paenibacillus* sp. isolate and *in vitro*–grown poplar plants revealed by metabolomics. Mol Plant Microbe Interact 22:1032–1037

Sessitsch A, Reiter B, Pfeifer U, Wilhelm E (2002) Cultivation independent population analysis of bacterial endophytes in three potato varieties based on eubacterial and Actinomycetes-specific PCR of 16S rRNA genes. FEMS Microbiol Ecol 39:23–32

Shaner NC, Patterson GH, Davidson MW (2007) Advances in fluorescent protein technology. J Cell Sci 120:4247–4260

Sun Y, Cheng Z, Glick BR (2009) The presence of a 1-aminocyclopropane-1-carboxylate (ACC) deaminase deletion mutation alters the physiology of the endophytic plant growth-promoting bacterium Burkholderia phytofirmans PsJN. FEMS Microbiol Lett 296:131–136

Sundram S, Meon S, Seman IA, Othman R (2011) Symbiotic interaction of endophytic bacteria with arbuscular mycorrhizal fungi and its antagonistic effect on *Ganoderma boninense*. J Microbiol 49:551–557

Taghavi A, Garafola C, Monchy S et al (2009) Genome survey and characterization of endophytic bacteria exhibiting a beneficial effect on growth and development of poplar trees. Appl Environ Microbiol 75:748–757

Tejesvi MV, Ruotsalainen AL, Markkola AM et al (2010) Root endophytes along a primary succession gradient in northern Finland. Fungal Diver 41:125–134

Thomas P, Kumari S, Swarna GK et al (2007) Ubiquitous presence of fastidious endophytic bacteria in field shoots and index-negative apparently clean shoot-tip cultures of papaya. Plant Cell Rep 26:1491–1499

Thomas P, Swarna GK, Roy PK et al (2008) Identification of culturable and originally non-culturable endophytic bacteria isolated from shoot tip cultures of banana cv Grand Naine. Plant Cell Tissue Organ Cult 93:55–63

Timmusk S, Nicander B, Granhall U et al (1999) Cytokinin production by *Paenibacillus polymyxa*. Soil Biol Biochem 31:1847–1852

Troll JV, Adin DM, Wier AM et al (2009) Peptidoglycan induces loss of a nuclear peptidoglycan recognition protein during host tissue development in a beneficial animal-bacterial symbiosis. Cell Microbiol 11:1114–1127

Ulrich K, Ulrich A, Ewald D (2008) *Paenibacillus*- a predominant endophytic bacterium colonizing tissue cultures of woody plants. Plant Cell Tissue Organ Cult 93:347–351

Van Aken B, Peres CM, Doty SL et al (2004) *Methylobacterium populi* sp. nov., a novel aerobic, pink-pigmented, facultatively methylotrophic, methane-utilizing bacterium isolated from poplar trees (*Populus deltoides* × *nigra* DN34). Int J Syst Evol Microbiol 54:1191–1196

Visser C, Murthy BNS, Odumeru J et al (1994) Modulation of somatic embryogenesis in hypocotyl cultures of geranium (*Pelargonium* × *hortorum* Bailey) cv. Ringo Rose by a bacterium. In Vitro Cell Dev Biol 30P:140–143

Yang CH, Crowley DE, Borneman J et al (2001) Microbial phyllosphere populations are more complex than previously realized. Proc Natl Acad Sci USA 98: 3889–3894

Yashiro E, Spear RN, McManus PS (2011) Culture-dependent and culture-independent assessment of bacteria in the apple phyllosphere. J Appl Microbiol 110(5):1284–1296

Yrjälä K, Mancano G, Fortelius C et al (2010) The incidence of *Burkholderia* in epiphytic and endophytic bacterial cenoses in hybrid aspen grown on sandy peat. Boreal Environ Res 15:81–96

Zabetakis I (1997) Enhancement of flavour biosynthesis from strawberry (*Fragaria* × *ananassa*) callus cultures by *Methylobacterium* species. Plant Cell Tissue Organ Cult 50:179–183

Unraveling the Dark Septate Endophyte Functions: Insights from the *Arabidopsis* Model

6

Keerthi Mandyam and Ari Jumpponen

Abstract

The global occurrence of plant root-associated fungal endophytes and their great abundance in many habitats necessitate studies to decipher their potential functions. Improved understanding of the basic endophyte ecology including host range, host preference, and host responses to endophyte colonization has been made possible through populations of endophytes (e.g., *Periconia macrospinosa* and *Microdochium* sp.) isolated from North American native tallgrass prairie. The recent demonstration of the endophyte symbiosis of the model plant *Arabidopsis thaliana* has provided additional tools to further elucidate the ecology of these endophytes. The availability of a large number of *Arabidopsis* ecotypes and mutants, microarrays, and databases allows the molecular dissection of endophyte symbiosis to better understand the importance of fungal endophytes in host nutrient uptake, defenses, and/or responses to pathogens and stress. In this chapter, we discuss the ecology and functions of endophytic fungi through experiments utilizing the *Arabidopsis* model system. We draw parallels with another deeply dissected *Piriformospora indica* root endophyte symbiosis, which has been demonstrated to promote growth of model and non-model plants.

1 Introduction

Plants typically host a variety of microbial endophytes. Plant roots in particular maintain broad assortments of microbial communities exemplified by actinobacteria, plant growth-promoting rhizobacteria (PGPR), nitrogen-fixing rhizobia, as well as fungi including a variety of mycorrhizal and endophytic fungi. Some fungal endophytes (e.g., clavicipitaceous foliar endophytes) have been the subject of rigorous investigations, and much is understood about their role as plant symbionts largely owing to their potential applications with crop or forage species. However, dark septate endophyte (DSE) fungi are an exception, and despite their similarities with mycorrhizal fungi – global occurrence, high colonization

K. Mandyam • A. Jumpponen (✉)
Division of Biology, Kansas State University,
Ackert Hall, Manhattan, KS 66506, USA
e-mail: ari@ksu.edu

V.C. Verma and A.C. Gange (eds.), *Advances in Endophytic Research*,
DOI 10.1007/978-81-322-1575-2_6, © Springer India 2014

rates, and broad host range (Jumpponen and Trappe 1998a; Jumpponen 2001) – there is a conspicuous lack of substantial data on their ecological significance (Rodriguez et al. 2009).

Several factors may have contributed to the paucity of information on the DSE fungi. *First*, the DSE fungi have been notoriously difficult to identify. Most do not produce sexual stages, and many fail to produce asexual spores even after induced conidiogenesis. This inevitably leads to lack of morphological traits to assist in taxonomic assignment of these fungi. *Phialocephala fortinii* is perhaps the most well-known DSE fungus (Addy et al. 2005). Although it has been morphologically defined, the taxon likely consists of a number of cryptic taxa. This has necessitated the use of various molecular markers to provide species rank (Grünig et al. 2008b). Similarly to *P. fortinii*, close relatives that also form DSE symbioses (e.g., *Acephala applanata*) fail to sporulate (Grünig et al. 2008a). As a result, these are assigned to *Phialocephala fortinii* s.l. – *Acephala applanata* species complex (PAC) that constitute the bulk of DSE communities colonizing conifers and Ericaceae in the Northern Hemisphere including Europe, North America, and Asia (Jumpponen and Trappe 1998a; Grünig et al. 2008b; Queloz et al. 2011; Walker et al. 2011). *Second*, thus far no evidence has been presented for an interface that would facilitate nutrient exchange between the host and its DSE symbionts (Petersen et al. 2008). The few ultrastructural studies that have focused on DSE fungi have failed to pinpoint host-derived perifungal membrane, the hallmark of biotrophic interactions (Petersen et al. 2008). This is in contrast to mycorrhizal fungi that are defined by their structural attributes when colonizing the host as well as by their presumed function in nutrient transfer to host plants via identifiable fungal interfaces (Bonfante 1984, 2001; Parniske 2000; Genre et al. 2008; Smith and Read 2008). A further complication is that the DSE colonization may vary depending on the host and environment making the identification of such interfaces even more challenging. Some host cell wall responses are characteristic of interactions with necrotrophic or hemibiotrophic fungi. This combined with the necrotic cytoplasm (Petersen et al. 2008) suggest a lifestyle not indicative of a mutualistic symbiont. However, in some cases, DSE colonization has shown similarities with mycorrhizal colonization like narrowing of hyphae when traversing plant cell wall, accumulation of polyphosphate (Yu et al. 2001), or development of mantle and Hartig net as seen in ectomycorrhiza (Petersen et al. 2008). Until major structural interfaces or novel colonization morphologies that unequivocally signify transfer of nutrients to the plant hosts, DSE symbiosis is unlikely to elicit major interest in research that seeks to find putative mutualists that can be used in applications that clearly benefit commercially important plants. *Third*, host responses to DSE colonization are highly variable, and growth promotion is rarely consistent by PAC fungi (Alberton et al. 2010; Newsham 2011; Mayerhofer et al. 2013). Most early studies on DSE effects on host growth and/or nutrient uptake have focused on PAC strains (see Kageyama et al. 2008; Petersen et al. 2008; Newsham 2011), and the variable responses were confounded by different experimental designs or experimental systems (Jumpponen and Trappe 1998b; Grünig et al. 2008a; Newsham 2011). Although the PAC and other DSE fungi elicit host responses along the mutualism-parasitism continuum (Jumpponen 2001; Grünig et al. 2008a; Mandyam et al. 2013), positive growth responses in empirical studies can be rare or absent. For example, Norway spruce inoculated with various PAC strains responded mainly negatively or neutrally (Tellenbach et al. 2011). Interestingly, in that study, the host responses varied with the source of the inoculant fungus: PAC native to the host tended to be more virulent than those originating from outside their natural range (Tellenbach et al. 2011).

The isolation and identification of cryptic PAC species as well as non-PAC DSE from grasslands in North America and Europe (see Sect. 4) provide new opportunities to study the function of DSE fungi. Including many fungal strains and host plants, manipulation of experimental conditions (e.g., temperature, salinity), and combining many DSE fungi or DSE and mycorrhizal fungi have provided new insights into the DSE symbioses. For example, Reininger et al. (2012) concluded that many PAC strains may not negatively affect host growth and that when nonparasitic PAC strains are co-inoculated with parasitic ones, the virulence of latter is reduced. These results highlight a further complication in making clear statements about the DSE function: the experimental outcomes are often synergistic and context dependent. In essence, the combinations of DSE and host genotypes control the outcome of the symbiotic interaction, which may be further modulated by environmental conditions (see Newsham 2011). Some recent studies underline this context dependency from the perspective of host species selection. When four PAC strains were inoculated into one of the three tree species, they had negligible effects on birch or Douglas fir growth but clearly inhibited spruce growth (Reininger et al. 2012; Reininger and Sieber 2012). Interestingly, fungal biomass was negatively correlated with plant biomass and dependent on strain-host combination so that spruce maintained a greater PAC biomass than birch (Reininger et al. 2012). These studies suggest differing compatibilities among DSE fungi and their hosts (see also Mandyam et al. 2012) but also the importance of considering the cost-benefit tradeoffs for DSE symbioses.

While hosts may differ in their responses to DSE fungi, these differences may also result from interactions within the fungal communities. It is important to bear in mind that simple deductive studies, while important as a first step, fail to account for the complexities of the hyperdiverse soil environments. Combinations of different root-associated fungal guilds provide further insight. When the ectomycorrhizal fungus, *Laccaria bicolor*, was inoculated on spruce in combination with PAC fungi, PAC colonization declined and the ectomycorrhizal fungus likely compensated the adverse PAC effects on spruce biomass (Reininger and Sieber 2012). This was also true for some combinations of PAC fungi but tended to depend on host and the experimental condition (Reininger and Sieber 2012). While empirical studies address such issues, meta-analyses provide the power to seek generalities across many independent studies. Recent meta-analyses (Newsham 2011; Mayerhofer et al. 2013) primarily concluded that DSE fungi rarely negatively affect host growth and tend to enhance total shoot and root biomass or shoot nitrogen and phosphorous contents. It is notable that while shoot nitrogen tended to increase as a result of DSE inoculation, the increase was far greater if nitrogen was supplied in organic forms. While the meta-analyses are exciting, their conclusions depend on selection of studies and/or analytical tools chosen for inference (Newsham 2011; Mayerhofer et al. 2013) and some debate on their conclusions remains (Alberton et al. 2010; Newsham 2011).

The root-colonizing fungal communities tend to be diverse and are comprised of taxa that may or may not form DSE associations (Herrera et al. 2010; Mandyam et al. 2010; Knapp et al. 2012). Among the root-colonizing fungi, *Periconia macrospinosa* and *Microdochium* sp. are DSE isolated from native grasslands in North America (Mandyam et al. 2010) and Europe (Knapp et al. 2012). Inoculation studies with multiple strains of these fungi and 12 native grasses and forbs not only confirmed the broad host range of these DSE fungi but also suggested that grasses in general may be more heavily colonized and respond to colonization more positively than forbs (Mandyam et al. 2012). The broad host range further motivated experiments that took advantage of well-established model plant *Arabidopsis thaliana*. The use of 38 DSE strains and three different *Arabidopsis* ecotypes confirmed the broad host range of DSE fungi (Mandyam et al. 2013). More importantly, these studies indicated that the host responses were mainly neutral – few host-fungus interactions were negative, fewer yet were positive (Mandyam et al. 2013). However, even the highly controlled laboratory reinoculation studies can lead to incongruent conclusions: further studies that included six *Arabidopsis* accessions and three *Periconia* strains indicated that positive responses were far more common than neutral ones and that no inoculation treatment inhibited growth compared to the non-inoculated controls (Fig. 6.1). Perhaps one important conclusion emerging from those *Arabidopsis* experiments is that evaluating multiple combinations of

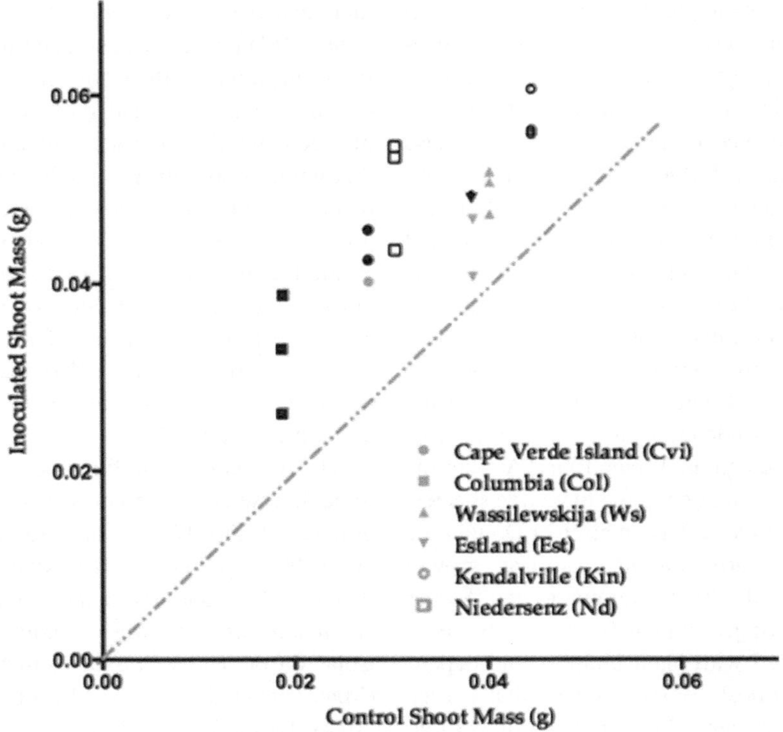

Fig. 6.1 Responses of six *Arabidopsis thaliana* accessions to inoculation with three *Periconia macrospinosa* strains. *X*-axis indicates the shoot dry weight on non-inoculated controls after 6-week incubation, *Y*-axis that of the inoculated plants. The *dashed line* is the isocline indicating equal masses in control and inoculated treatments; values above line suggest positive responses to inoculation; values below negative responses. *Black symbols* indicate significant differences between the control and inoculated treatments (Dunnett's test; $\alpha = 0.05$). Methods for these studies are like those described in Mandyam et al. (2013)

hosts and DSE fungi is imperative for drawing meaningful inferences about DSE symbiosis.

In their earlier studies, Mandyam and Jumpponen (2005) hypothesized that DSE are multifunctional and have the potential to promote host growth and facilitate nutrient uptake, improve host resistance to biotic and abiotic stressors, or partly control plant community dynamics via differential host responses. Like mycorrhizal fungi, DSE may induce plant defenses or otherwise increase host tolerance of the pathogens, while not necessarily enhancing host growth. Such interactions are exemplified in studies wherein inoculation with PAC fungi has a negative impact on Norway spruce growth, but increases pathogen tolerance (Tellenbach et al. 2013). In those studies, *Phialocephala europa* was found to produce a variety of plant growth-promoting compounds as well as antibiotics that controlled oomycete pathogen *Phytophthora citricola*. Similarly, multiple *Phialocephala subalpina* strains reduced spruce mortality and severity of oomycete *P. citricola* and *Pythium undulatum* diseases (Tellenbach and Sieber 2012). Complex interactions among soil- and root-inhabiting organisms are likely common. For example, exudates from *Drechslera* sp., a DSE colonizing the grass *Lolium multiflorum*, stimulated hyphal branching and extramatrical hyphae of arbuscular mycorrhizal fungus *Gigaspora rosea* thus modulating this symbiosis (Scervino et al. 2008).

The symbiosis between DSE and *Arabidopsis* as well as the ease with which *Arabidopsis* forms symbioses with DSE fungi isolated from a tallgrass prairie (Mandyam et al. 2013) permit testing hypotheses put forth by Mandyam and Jumpponen (2005). We propose that this model plant serves to further elucidate the DSE functions. The *Arabidopsis* model has already provided a glimpse of the molecular mechanisms underlying beneficial plant symbioses with PGPR, actinobacteria, and basidiomycetes (Sect. 2). Our recent data suggest that *Arabidopsis* (Sect. 4.1) serves as a model for analyzing DSE symbiosis. The natural microbial communities associated with *Arabidopsis* roots are also discussed to emphasize the validity of the *Arabidopsis* system in studying DSE fungi and their function in symbiotic associations.

2 *Arabidopsis* Model to Study Host-Endophyte Symbiosis

Arabidopsis model has been used extensively to study plant-pathogen interactions – particularly to elucidate defense responses (Nishimura and Dangl 2010). *The Arabidopsis Book* regularly publishes reviews on a range of topics including plant-pathogen or plant-microbe interactions (e.g., Betsuyaku et al. 2011; Day and Knepper 2010; Laluk and Mengiste 2010; Micali et al. 2008; Thilmony et al. 2002). In addition to pathogen interactions, *Arabidopsis* has provided insight into beneficial plant-microbe symbioses. In this section, we provide a backdrop for the use of *Arabidopsis* in dissecting DSE symbiosis.

Arabidopsis and its numerous signaling mutants have improved our understanding of growth promotion and induced systemic resistance (ISR) by PGPR. Not only rhizobacterial colonization but also their volatile organic compounds (VOC) can promote growth (Ryu et al. 2003, 2005). In vivo and in vitro studies with various *Arabidopsis* mutants in symbiosis with eight PGPR strains indicated that brassinosteroid, indole acetic acid (IAA), salicylic acid (SA), and gibberellin signaling were involved in growth promotion in vivo and ethylene signaling in growth promotion in vitro (Ryu et al. 2005). The in vitro growth promotion depended on the physical distance of bacterial inoculum from the roots suggesting the diffusion of growth-promoting compounds. Ryu et al. (2003) found that PGPR produce strain-specific VOCs and the strains that induced the greatest growth promotion (*Bacillus subtilis* GB03 and *Bacillus amyloliquefaciens* IN937a) produced 2,3-butanediol and acetoin with butanediol that enhance leaf surface area. The use of three *Arabidopsis* ecotypes plus various signaling mutants pinpointed that cytokinin signaling and not brassinosteroid, gibberellic acid, or ethylene signaling was responsible for the growth promotion by VOC. Rhizobacterial VOCs did not contain auxin or other known growth-regulating compounds and promoted *Arabidopsis* growth by auxin homeostasis as well as by cell expansion (Zhang et al. 2007). The analysis of *Arabidopsis* proteome when exposed to VOCs from *B. subtilis* GB03 showed upregulation of ethylene biosynthesis, antioxidant proteins, and proteins in jasmonic acid (JA) and SA signaling (Kwon et al. 2010). Rhizobacterial VOCs not only promote plant growth but also play a role in ISR: 2,3-butanediol from *B. subtilis* GB03 and *B. amyloliquefaciens* IN937a activated ISR 4 days after inoculation (dai), significantly reducing *Erwinia carotovora* disease severity, and this ISR activation was dependent on ethylene signaling independent of SA or JA signaling (Ryu et al. 2004a). In the case of viral disease caused by cucumber mosaic virus (CMV), PGPR *Serratia marcescens* and *Bacillus pumilus* reduced symptom severity and CMV accumulation in *Arabidopsis* leaves (Ryu et al. 2004b). *Arabidopsis* signaling and PGPR mutants showed that ISR by *S. marcescens* against CMV was dependent on JA signaling independently of SA or NPR1 signaling.

In addition to the inoculation studies with single PGPR, their mixtures have been evaluated using *Arabidopsis*. The commercial biopreparation Bioyield, containing a mixture of growth-promoting *B. subtilis* GB03 and ISR agent *B. amyloliquefaciens* IN937a, confirmed the growth promotion independent of any known hormone signaling (Ryu et al. 2007). These studies also showed that the protection from CMV was independent of SA signaling, even though protection from the bacterial pathogen *Pseudomonas syringae* did depend on SA signaling (Ryu et al. 2007).

These studies demonstrated differences in *Arabidopsis* responses to single bacterium and mixtures: growth promotion was independent of ethylene signaling when *Arabidopsis* was inoculated with a mixture of two PGPR, but not so when inoculated with a single PGPR (Ryu et al. 2005). Interestingly, the inoculation treatments did not affect CMV titer but reduced CMV symptom severity, suggesting that PGPR improved host tolerance rather than resistance. To study *Arabidopsis* genes specifically expressed during root colonization by ISR-inducing rhizobacteria, fluorescent *Pseudomonas* spp. induced a thaumatin-like gene AtTLP1 in the root vascular bundle along with a protein homologous with pathogenesis-related protein 5 (PR-5) known to have antimicrobial properties but not involved in ISR (Léon-Kloosterziel et al. 2005). Further experiments with defense signaling mutants suggested SA signaling in *Pseudomonas chlororaphis* VOC-butanediol-induced drought tolerance combined with *Arabidopsis* stomatal closure (Cho et al. 2008).

Arabidopsis mutants have also improved our understanding of the PGPR effects on root architecture and the roles of auxin and ethylene in PGPR symbiosis. *Arabidopsis* growth-promoting *Bacillus megaterium* inhibited primary root growth by reducing cell elongation and proliferation in the root meristem (López-Bucio et al. 2007; Shi et al. 2010). These responses were combined with increases in lateral root number, lateral root growth, and root hair length. The responses were similar when *Arabidopsis* was inoculated with *S. marcescens*, except that plants maintained normal cell elongation (López-Bucio et al. 2007; Shi et al. 2010). Although changes in root architecture were similar to those mediated by auxin, auxin and ethylene signaling mutants showed that the changes were independent of these signaling pathways (López-Bucio et al. 2007). Further studies by Shi et al. (2010) provided a more complex picture: the changes in root architecture were controlled by both auxin-dependent and auxin-independent signaling as well as ethylene; JA and SA signaling were involved in induction of second order lateral roots. The PGPR also have more global effects on *Arabidopsis* gene expression. When inoculated with growth-promoting *Pseudomonas fluorescens*, at least 95 *Arabidopsis* genes were significantly upregulated and 105 downregulated (Wang et al. 2005). Various genes involved in metabolism, signal transduction, stress response, putative auxin signaling, and nodulin-like genes were among those that were upregulated, whereas ethylene genes were downregulated. Additionally, genes involved in ISR were differentially expressed as a result of the inoculation treatments. Although the PGPR have been isolated from rhizospheres of different plant species and may find valuable applications for a number of economically important plants, *Arabidopsis* ultimately has provided the mechanistic understanding how various pathways and mechanisms are involved in PGPR effects. Similarly to PGPR, *Arabidopsis* model has also permitted a deeper understanding of ISR by actinobacteria against plant pathogens (Conn et al. 2008; Lin et al. 2012).

In addition to bacterial endophytes, *Arabidopsis* has also been a valuable tool in studying fungal symbioses. For example, *Paraphaeosphaeria quadriseptata* – an endophyte of desert cacti and also of maize – can improve thermotolerance in *Arabidopsis* via production of monocilin 1 (MON-1) that inhibits plant heat shock protein 90 (HSP90) (McLellan et al. 2007). Even the study of ectomycorrhizal systems can benefit from the use of powerful *Arabidopsis* system. For example, hypaphorine, an exudate from ectomycorrhizal fungus *Pisolithus tinctorius*, is involved in symbiosis-related root differentiation and elicits similar responses in ectomycorrhizal *Eucalyptus globulus* and *Arabidopsis* that do not form mycorrhizas of any kind (Reboutier et al. 2002). In addition to these symbioses, *Arabidopsis* has been used to dissect *Piriformospora indica* root symbiosis. *Piriformospora* is a sebacinaceaous fungus from the Thar Desert in India (Varma et al. 1999) colonizing a range of plants often improving growth, fitness, and tolerance to biotic and abiotic stressors (see Franken 2012; Qiang et al. 2012a). We highlight the key findings that have emerged from these studies in the following section.

2.1 Case Study: *Arabidopsis-Piriformospora* Symbiosis

In the last decade, a mutualistic root endophyte *P. indica* has been the subject of keen investigation. This is due to its taxonomic position and interesting attributes: *Piriformospora* is a Basidiomycete in Sebacinales, a globally distributed order with mycorrhizal fungi as well as non-mycorrhizal root endophytes (Selosse et al. 2009; Weiss et al. 2011). In line with the lack of host specificity in Sebacinales, *P. indica* colonizes bryophytes, pteridophytes, gymnosperms, angiosperms, monocots and dicots, and even non-mycorrhizal members of Brassicaceae (Franken 2012; Oelmüller et al. 2009; Qiang et al. 2012a). However, unlike many mycorrhizal fungi, *P. indica* grows axenically, making it particularly attractive for applications for many crop species that form arbuscular mycorrhizal symbioses. Most importantly, *P. indica* is a mutualist that can promote host growth and fitness plus improve host tolerance to biotic and abiotic stress (Franken 2012; Qiang et al. 2012a).

The colonization of non-mycorrhizal *Arabidopsis* in the wild by a sebacinalean fungus (Weiss et al. 2011; see Sect. 2.1) and the successful in vitro resynthesis of *Arabidopsis* colonization by *P. indica* (Peškan-Berghöfer et al. 2004; Shahollari et al. 2007) have provided a model for fungal mutualisms in *Arabidopsis*. Although a glimpse into the molecular mechanisms underlying *P. indica* growth promotion and improved host stress tolerance has been achieved using barley, wheat, tomato, tobacco, or Chinese cabbage (Baltruschat et al. 2008; Barazani et al. 2005; Deshmukh et al. 2006; Felle et al. 2009; Hilbert et al. 2012; Schäfer et al. 2009; Sherameti et al. 2005; Sun et al. 2010; Waller et al. 2005, 2008), the in-depth analyses of the *P. indica* symbiosis, defense signaling, role of phytohormones, ISR, and differential gene expression has been only achieved through the *Arabidopsis* model.

Peškan-Berghöfer et al. (2004) first reported *Arabidopsis* root colonization by *P. indica* and the resultant growth enhancement. Colonized plants, when transferred to soil, grew faster, had more leaves, flowered earlier, set seed earlier, and had higher seed yield than the control plants. The shoot dry weight significantly increased after only 8 dai coinciding with distinct changes in root architecture. The inoculated plants had longer and thinner roots, and many root proteins were upregulated. Additionally, the colonized plants seemed to be more stress tolerant: on Murashige-Skoog medium, *P. indica* inoculated plants tolerated 200 μM concentrations of cadmium.

The colonization and life history attributes of *P. indica* are unique. Genome and transcriptome analyses suggest a combination of biotrophic and saprotrophic lifestyles. Biotrophy is supported by accumulation of small, secreted proteins such as lectin-like proteins, absence of nitrogen metabolizing genes, as well as absence of genes coding for secondary metabolism, for example, polyketide synthase and non-ribosomal peptide synthase. Saprotrophy is supported by the presence of cell wall degrading enzymes and metallopeptidases (Zuccaro et al. 2011). Consistent with the genomic analyses, studies with barley revealed that the initial establishment of a biotrophic interaction was facilitated by overcoming or suppressing host defenses followed by host cell death to permit the establishment of symbiosis (Deshmukh et al. 2006; Hilbert et al. 2012; Lahrmann and Zuccaro 2012; Oelmüller et al. 2009; Qiang et al. 2012a; Schäfer et al. 2007, 2009). Establishment of *P. indica* symbiosis in *Arabidopsis* knockout line for a gene homologue of "does not make infections" (DMI1) protein that channels ion fluxes across plasma membrane critical in legume mycorrhizal and rhizobial symbioses indicated a colonization mechanism unlike any known previously (Shahollari et al. 2007). Recently, Jacobs et al. (2011) confirmed a unique biphasic colonization in *Arabidopsis* and proposed a model involving four stages: (1) extracellular colonization of root surface at 1 dai by chlamydospore germination, (2) biotrophic colonization phase of rhizodermal, cortical cells and root hairs at <3 dai without any ultrastructural changes wherein fungal hyphae is covered by host plasma membrane, (3) cell death at >3 dai and fungal reproduction by external sporulation at 7 dai, and (4) internal sporulation by 14 dai in root maturation zone. Further, the use of at least 12 mutants and a GFP-tagged *Arabidopsis* line (GFP-tmKKXX) permitted detailed analyses of the novel microbially mediated cell death. During colonization, *P. indica* induces endoplasmic reticulum (ER) swelling and vacuolar collapse, but suppresses ER stress signaling, that is, unfolded protein response (UPR), leading to vacuolar processing enzyme (VPE)/caspase 1-like-mediated cell death during the establishment of symbiosis without impairing root function or development (Qiang et al. 2012b). After cell death, the fungus continues to colonize adjoining cells. Typical localized immune response cell death attributes including hypersensitive response (HR) (whole cell auto fluorescence or browning of colonized cells due to accumulation of phenolics, mitochondrial swelling, increased vesicle formation, and protoplast shrinkage) was lacking, suggesting a novel "compatibility-based cell death" rather than immunity-related cell death (Jacobs et al. 2011).

The unique *Arabidopsis* mutants, microarrays, and other tools have permitted the elucidation of molecular mechanisms behind *P. indica-mediated* growth promotion.

(1) The growth responses of *Arabidopsis* mutants that differ in root shoot ratios, lengths, or architectures (e.g., *ahk2 ahk3*, 35S::CKX1, 35::CKX2, *sur1-1*, *tfl2*) are similar suggesting that *P. indica* promotes growth independently of *Arabidopsis* root architecture (Vadassery et al. 2008).
(2) Under laboratory resynthesis, *P. indica-colonized Arabidopsis* fresh weight increased by 21 %. When transplanted to soil, the inoculated plant seed production was 22 % greater compared to uninoculated plants (Shahollari et al. 2007). Development of a *P. indica* insensitive *Arabidopsis* mutant (*pii-2*) lead to the discovery that a leucine-rich repeat (LRR) protein containing ER retention signal and another atypical receptor protein At5g1650 present in the ER/plasma membrane continuum in roots are required for growth promotion and enhancement of seed production. Also, sphingolipids involved in plasma membrane signaling are required for *P. indica* symbiosis as demonstrated by growth reduction in sphingosine kinase knockout lines (Shahollari et al. 2007).
(3) Although the exact role of PYK10 (β-glucosidase located in the ER) is unknown, it is required for the *P. indica-enhanced* growth and fitness of *Arabidopsis* (Sherameti et al. 2008a). Based on the upregulation of plant defensin protein (PDF1.2), a defense response and the downregulation of LRR1 protein (a marker for beneficial interaction under reduced PYK10 levels), and the fact that PYK10 is similar to PEN2 (a glycosyl hydrolase known to restrict ascomycetous pathogens in *Arabidopsis*), PYK10 may control fungal colonization to maintain mutualisms. Since PYK10 released from endosomal system and forms a multimeric complex with PBP1 (a cytoplasm protein when *Arabidopsis* tissue is damaged), the substrate for PYK10 could only come in contact with the enzyme when cell and cellular components are damaged during the contact of host and fungal symbionts. This is highly likely considering the unique *P. indica* colonization pattern as outlined in Qiang et al. (2012b).
(4) At least two ROP (RHO-related GTPases) proteins ROP1 and ROP6 and ROP1-interacting protein RIC4 were stimulated in *P. indica*-colonized lateral roots compared to uncolonized plants. Experiments with *Arabidopsis* mutants showed that these proteins were involved in hormone-mediated seed germination, root growth and root hair development, and F-actin bundle formation in the roots (Venus and Oelmüler 2012). The unimpaired fungal root colonization coupled with lack of growth promotion in ROP knockouts or overexpressors indicated the role of ROP1, ROP6, and RIC4 in maintaining beneficial interaction via stimulation of F-actin bundle formation. Since Ca^{2+} cellular elevation is one early signaling event of *P. indica* mutualism (Vadassery et al. 2009b) and ROP6 controls root hair growth by affecting Ca^{2+} gradient, the upregulation of Ca^{2+} inducible/Ca^{2+}/calmodulin-binding

protein (CBP60g) that is induced by both pathogens and microbe-associated molecular patterns (MAMP) in *P. indica*-colonized wild-type *Arabidopsis* suggests that CBP60g maintains host defense responses until it recognizes *P. indica* as a mutualist. This is further evidenced by lesser upregulation in *rop* mutants.

(5) Indole-3-acetaldoxime (IAOx)-derived compounds partly control mutualistic fungal root colonization (Nongbri et al. 2012). Phytoalexins are plant antimicrobial substances produced at pathogen infection sites or when exposed to oxidative stress by abiotic factors. Camalexin is the main phytoalexin in *Arabidopsis*, and IAOx is an intermediate in the camalexin-producing metabolic pathway. The camalexin production requires transcription factor WRKY33 to activate camalexin biosynthesis and is catalyzed by both cytochrome P450 and PAD3. Nongbri et al. (2012) observed much lower *Arabidopsis* camalexin levels during *P. indica* colonization compared to pathogens. This coincided with expression of cytochrome P450 (CYP79B2, CYP79B3, CYP71A13), PAD3, and WRKY33. Camalexin and IAOx-deficient mutants were more heavily colonized by *P. indica*, lacked any growth promotion, and had an upregulation of a suite of defense responses (pathogenesis-related protein (PR1, PR3), plant defensin PDF1.2, phenylalanine ammonia lyase, and germin), implying a role of IAOx-derived compounds in the early stages of colonization as well as in maintaining beneficial symbiosis. Additionally, the cellular Ca^{2+} elevation during early colonization and the concomitant mitogen-activated protein kinase (MAPK) (Vadassery et al. 2009b) activation seem essential for production of IAOx-derived metabolites.

(6) In *Arabidopsis* roots, OXI1 (oxidative signal inducible 1) is a protein kinase involved in oxidative-burst-mediated pathogen resistance and induced by H_2O_2 and PDK1 (3-phosphoinositide-dependent kinase). PDK1 is activated by phosphatidic acid (PA) produced by phospholipase D (PLD). *Piriformospora indica* colonization triggers PA synthesis in *Arabidopsis*, upregulates OXI1 and PDK1 genes, and suppresses H_2O_2 production and defense genes. Expected growth enhancement by *P. indica* does not occur in *Arabidopsis oxi1*, *pdk1* (*pdk1.1 pdk1.2*), or *pldα1* and *pldδ* mutants, suggesting that PLD-PDK1-OXI1 cascade is essential for mutualistic interaction (Camehl et al. 2011).

(7) Ascorbate-glutathione cycle offers protection from reactive oxygen species like H_2O_2 by maintaining high ascorbate levels in the cells. Ascorbate peroxidase uses ascorbate to reduce H_2O_2 by producing monodehydroascorbate which is reduced back to ascorbate by monodehydroascorbate reductase (MDAR). If dehydroascorbate is produced, it is converted to ascorbate by dehydroascorbate reductase (DHAR). Essentially, MDAR and DHAR maintain reduced state of ascorbate. Vadassery et al. (2009a) observed MDAR2 and DHAR5 upregulation in *Arabidopsis* as a result of *P. indica* or its cell extract, coinciding with a 1.5-fold increase in root ascorbate level, lack of growth promotion, retarded flower development, and seed yield. Experiment with *mdar2* and *dhar5* mutants identified that MDAR2 and DHAR5 enzymes are also involved in maintaining the mutualistic symbiosis.

(8) Although *P. indica* produces auxin (Sirrenburg et al. 2007; Vadassery et al. 2008), it is not essential for the *Arabidopsis* growth promotion (Lee et al. 2011). Further, the endogenous levels of free and conjugated IAA in *Arabidopsis* roots with or without *P. indica* do not differ as suggested by lack of responses in the DR5-*GUS* reporter system and most of auxin-responsive genes to *P. indica* colonization (Vadassery et al. 2008). Overall, auxin production is an unlikely candidate for the observed positive growth responses in *Arabidopsis*: *tfl2*, *ilr1-1*, *cyp79b2b3*, and *nit1-3* mutants with reduced auxin levels respond to *P. indica* inoculation like the wild type (Vadassery et al. 2008). While auxin may not be the

primary mechanism of growth promotion in *Arabidopsis*, its role may vary among the plant species as exemplified by its importance for *Brassica campestris* growth response to *P. indica* (Lee et al. 2011). However, fungal modulation of auxin levels and metabolism has been observed in auxin overproducing *Arabidopsis* mutant *sur1-1* whose dwarf phenotype was restored by *P. indica*. In sum, it remains unclear whether the auxin-induced growth promotion is controlled directly by fungus or indirectly as a plant response to the fungus.

(9) In addition to the potential and differing roles of auxin in the symbiosis between *P. indica* and its hosts, other plant growth hormones may also be important: cytokinin-responsive gene ARR5 was expressed at 54 % greater levels in *P. indica*-inoculated *Arabidopsis* roots compared to fungus-free controls (Vadassery et al. 2008). Cytokinin biosynthesis-defective *Arabidopsis* mutants provided further evidence that trans-Zeatin may be critical for symbiotic growth promotion and that cytokinin receptor combination CRE1/AKH2 is crucial for symbiotic growth promotion.

(10) Finally, ethylene signaling and ethylene-targeted transcription factor (ETF) are essential for maintaining mutualistic symbiosis (Camehl et al. 2010; Khatabi et al. 2012), highlighting the overall complexity of the hormonal plant responses to fungal colonization.

Not only the living fungus but its extracts enhance plant growth. *Arabidopsis* responds to autoclaved *P. indica* cell wall extract (CWE) like it does to fungal colonization (Vadassery et al. 2009b). *Piriformospora indica* inoculation increased *Arabidopsis* growth by 36 %, whereas CWE increased shoot growth by 15 % and root growth by 21 % after 10 dai. Similarly to inoculation with the fungus, CWE induced upregulation of LRR1 (Shahollari et al. 2007), 2-nitro-propane-dioxygenease (Sherameti et al. 2005), monodehydroascorbate reductase (MDAR2), and dehydroascorbate reductase (DHAR5) genes in roots and shoots (Vadassery et al. 2009a). Ascorbate biosynthesis gene MIOX, Ca^{2+} sensor CIPK13, and Ca^{2+} signaling calmodulin-like genes (CML) were all upregulated, resulting in elevated cytosolic and nuclear Ca^{2+} levels and upregulation of Ca^{2+}-MAPK (Vadassery et al. 2009b). Further, similarly to the fungus, CWE elicited no immune responses such as H_2O_2 production or activation of defense-related genes (Vadassery et al. 2009b). Liquid cultures of *P. indica* produce IAA and the fungus, CWE, and its ethyl acetate extract change root architecture (root stunting and extensive root branching) similarly to a treatment with 18 nmol IAA (Sirrenburg et al. 2007). Although its importance may be uncertain, *P. indica* IAA (325 pmol/g dry weight) and cytokinin (403 pmol/g of dry weight) production is considerable (Vadassery et al. 2008). The hormones produced by *P. indica* can significantly alter resource plant allocation: although shoot biomass remained unaltered after 14 dai, the inoculation significantly reduced (86.5 %) the main root length compared to controls (Stein et al. 2008). Despite *P. indica* hormone production, like with the living fungus, CWE-induced growth is not attributable to fungal auxin or sugars (Lee et al. 2011). Chemical composition of CWE is not clear although it is heat stable, is partially inhibited by trypsin, and remains unaltered by chitinase and glucanase (Vadassery et al. 2009b).

Interestingly, *P. indica* harbors an endobacterium, α-proteobacterial *Rhizobium radiobacter* (*Agrobacterium tumefaciens* or *Agrobacterium radiobacter*) (Sharma et al. 2008). This endobacterium was present in the original *P. indica* isolate and is vertically transmitted via spores. Although present in low frequency within the fungus (up to 0.035 ng bacterial DNA per 100 ng of fungal DNA), it cannot be eliminated by antibiotics, via single spore or hyphal tip isolations or exposure of fungal protoplasts to antibiotics. However, the bacterium can be repeatedly isolated and grown axenically. In pure culture, it is capable of producing up to 40 μg/ml of IAA after 24 h at 25 °C in the presence of tryptophan. When inoculated on barley, the bacterium can enhance growth by 17 % and reduce biotrophic pathogen

Blumeria graminis pustules by 54 %. *Piriformospora indica*-colonized barley has 27 % growth enhancement and 64 % reduction in *B. graminis* pustules. Curiously, the bacterium hosts *virD2* gene coding Ti plasmid without the isopentenyltransferase (*ipt*) gene associated with cytokinin biosynthesis. This likely explains the lack of pathogenicity of the bacterium. Since it has not been possible to obtain bacteria-free fungus, it is not possible to ascertain if the beneficial effects are due to the fungus or its endosymbiont (Sharma et al. 2008).

In addition to enhancing plant growth, *P. indica* can improve host tolerance to biotic and abiotic stress. At least 13 known diseases in five different crop species and *Arabidopsis* are suppressed by *P. indica* inoculation (see review by Qiang et al. 2012a). *Piriformospora indica* induces ISR in *Arabidopsis.* ISR is usually elicited by nonpathogenic microbes, for example, rhizobacteria. The application of nonpathogenic microbe to one part of the plant elicits JA or ethylene defense signaling to reduce disease severity from many different pathogens in distant plant parts. In contrast, systemic acquired resistance (SAR) is elicited by necrotizing pathogens or chemical elicitors by inducing the SA pathway along with expression of PR proteins (Pieterse et al. 1998). By ISR, *P. indica* root colonization can reduce the severity of powdery mildew of *Arabidopsis* leaves caused by *Golovinomyces orontii* (Stein et al. 2008). *Arabidopsis* JA mutants (jasmonate-resistant 1, *jar1-1;* or jasmonate-insensitive mutant 1, *jin1*) maintained higher pathogen loads, whereas mutants unable to accumulate SA (*NahG*) or the mutant incapable of expressing PR genes (*npr1-3*) maintained lesser pathogen loads, suggesting the importance of JA defense in ISR (Stein et al. 2008). After 14 dai, *P. indica* colonization of *Arabidopsis* roots did not alter the expression of SA-responsive *PR1* and *PR5* or ET-responsive ethylene response factor (*ERF1*), but marginally lowered the levels of JA-responsive vegetative storage protein (*VSP*), *PDF1.2,* and lipoxygenase 2 (*LOX2*) in the leaves. After 3 days of pathogen challenge, *VSP* increases eightfold, confirming a JA-mediated systemic response that was absent in the *jin1* mutant defective in *VSP* expression. Similarly, in barley, *P. indica* promoted resistance to powdery mildew caused by *Blumeria graminis* by ISR (Waller et al. 2008). It is likely that the *P. indica* primes defense-related genes *PR1*, *PR2*, and *PR5* or a heat shock protein (HSP70), which have antifungal properties (Molitor et al. 2011).

In addition to biotic stressors, *P. indica-colonized* plants tolerate abiotic stress (e.g., drought) better than fungus-free controls (Sherameti et al. 2008b). The improved drought stress also leads to much improved fitness – after 84 h drought stress, 46.5 % of *P. indica-inoculated Arabidopsis* produced seed when the plants were transplanted into soil, whereas none of the control plants even survived the stress. *Piriformospora indica* primes *Arabidopsis* aerial parts by upregulation of at least nine genes involved in drought stress. Transfer of colonized plants to soil also showed that at least three of those genes (PLD, calcineurin B-like protein CBL1, and histone acetyl transferase HAT) remain upregulated longer than in the fungus-free controls. The *P. indica*-colonized plants also accumulate MDAR2 in the ascorbate cycle faster than the uninoculated controls (Vadassery et al. 2009b). Interestingly, *mdar2* and *dhar5* mutants were more heavily colonized and had very high levels of antifungal PDF1.2 in leaves. In contrast, wild-type plants, whether colonized or not, did not express PDF1.2 under stress, suggesting that MDAR2 and DHAR5 enzymes are not only required for maintaining a beneficial symbiosis but are also essential to repress defense gene expression to prevent the shift from mutualism to parasitism (Vadassery et al. 2009b).

In summary, the use of at least 30 different *Arabidopsis* mutants has permitted the detailed dissection of the importance of auxin, ethylene, abscisic acid, cytokinin, defense compounds, and other proteins in maintaining beneficial symbiosis and/or growth enhancement or ISR by *P. indica.* These studies and the described examples underline the power afforded by the use of model plants and *Arabidopsis* in particular for more detailed and improved understanding of the root-symbiotic endophytes.

3 Native Endophytes of *Arabidopsis*

Arabidopsis has served as a model plant for more than 25 years. Although recent developments have made it feasible to use economically relevant plant species as tractable models, *Arabidopsis* continues to provide breakthroughs (Jones et al. 2008; Koornneef and Meinke 2010). *Arabidopsis* model has provided a strong foundation for understanding basic mechanisms of plant-microbe interactions. Yet, the microbial endophytes and epiphytes native to *Arabidopsis* have been largely ignored. Thus far, only a handful of studies have investigated bacterial and fungal endophytes of *Arabidopsis* (Table 6.1). Like any other plant, different *Arabidopsis* organs maintain an assortment of microbial endophytes that may include pathogens, nonpathogenic saprobes, and even mutualistic endophytes, such as PGPR or nitrogen-fixing bacteria. In fact, in many Brassicaceae – including *Arabidopsis* – genes involved in mycorrhizal and rhizobial symbioses (e.g., nodulation signaling pathway 1 and 2, NSP1 and NSP2) are conserved and may play a role in plant-microbe interactions (Hayward et al. 2012).

Table 6.1 Microbial endophytes native to *Arabidopsis thaliana*

Microbial type/ plant part	Identity of endophytes	Notes	References
Bacteria			
Rhizosphere	Alphaproteobacteria (Rhizobiales 27 %), Acidobacteria (17 %), Bacteroidetes (14 %), Gammaproteobacteria (12 %; Xanthomonadaceae, Pseudomonadaceae), Betaproteobacteria (10 %; Burkholderiales), Verrucomicrobia (7 %), Actinobacteria (5 %), Gemmatimonadetes (5 %), Deltaproteobacteria (3 %)		Micallef et al. (2009a)
Rhizosphere	Actinobacteria, Proteobacteria, Cyanobacteria, Acidobacteria, Chloroflexi, Firmicutes, Bacteroidetes, Gemmatimonadetes		Sugiyama et al. (2013), Bulgarelli et al. (2012), Lundberg et al. (2012), Bressan et al. (2009), and Badri et al. (2013)
Root	Alpha-, Beta-, and Gammaproteobacteria, Actinobacteria, Bacteroidetes, Firmicutes, and Cyanobacteria	DNA- and RNA-based metagenomic sequencing	Bulgarelli et al. (2012) and Lundberg et al. (2012)
Root	Gammaproteobacteria (*Pseudomonas* sp. G62)	Natural PGPR	Schwachtje et al. (2011)
Root	Alphaproteobacteria (*Agrobacterium tumefaciens*, *Agrobacterium* sp., *Rhizobium* sp.), Betaproteobacteria (*Bordetella holmesii*, *Achromobacter* sp., *Pelomonas puraquae*, *Duganella* sp., *Herbaspirillum* sp., *Zoogloea* sp.), Gammaproteobacteria (*Dokdonella* sp.), Acidobacteria, Archaea	Partial 16S and 18S rRNA sequences from dominant DGGE bands	Bressan et al. (2009)

(continued)

Table 6.1 (continued)

Microbial type/ plant part	Identity of endophytes	Notes	References
Leaves	Alphaproteobacteria (*Agrobacterium tumefaciens*, *Sphingomonas* sp.) Gammaproteobacteria (*Pseudomonas* sp., *Xanthomonas* sp.,), Bacteroidetes (*Flavobacterium* sp.), Firmicutes (*Bacillus* sp.), Actinobacteria (*Nocardia corynebacterioides*, *Arthrobacter* sp., *Curtobacterium flaccumfaciens*, *Streptomyces chartreusis*, *Rhodococcus erythropolis*, *Frigoribacterium* sp., *Agromyces salentinus*)	Culturing; natural bacterial pathogens and putative nonpathogens	Traw et al. (2007) and Kniskern et al. (2007)
Leaf	*Pseudomonas viridiflava* ps. *syringae*	Culturing and identification; natural leaf pathogen	Jakob et al. (2002)
Seed	Alphaproteobacteria: *Rhizobium* sp. (40 %), *Sphingomonas* sp. (33 %), *Sinorhizobium* sp. (8 %), *Methylobacterium* sp. (14 %), Betaproteobacteria: *Acidovorax* sp. (1 %), *Variovorax* sp. (0.76 %), Actinobacteria: *Micrococcus* sp. (0.01 %), Firmicutes: *Bacillus* sp. (2 %), *Staphylococcus* sp. (0.52 %)	Endophytes cultivated and identified by 16S rDNA-based method	Truyens et al. (2012)
Fungi			
Rhizosphere	Basidiomycota (*Marchandiobasidium aurantiacum*), Ascomycota (*Stilbocrea macrostoma*), Chytridiomycota	Partial 16S and 18S rRNA sequences from dominant DGGE bands	Bressan et al. (2009)
Root	Unidentified DSE fungi	Microscopy	Mandyam et al. (2013)
Root	Basidiomycota (group B Sebacinales fungus)	rDNA-based detection	Weiss et al. (2011)
Root	Basidiomycota (*Marchandiobasidium aurantiacum*), Ascomycota (*Nectria* sp.), Chytridiomycota (*Olpidium brassicae* and other unculturable chytrids)	Partial 16S and 18S rRNA sequences from dominant DGGE bands	Bressan et al. (2009)
Leaves, stem, roots	Ascomycota (total 38 genera, 13 from roots remaining from leaves and stems)	Culturing and morphological identification	Junker et al. (2012)
Leaves, siliques	Ascomycota (26 genera), Basidiomycota (*Leucosporidium* sp.), Zygomycota (*Mortierella* sp.)	Isolation and rRNA sequencing	Garcia et al. (2012)

3.1 Bacterial Endophytes

Culture-dependent and culture-independent studies have identified bacterial endophytes naturally present in *Arabidopsis*. *Arabidopsis* grown in non-native potting soils supported culturable eubacterial population densities of 2×10^7 to 1×10^9 cfu/g in rhizosphere and *Pseudomonas* spp. populations of 5×10^5 to 5×10^7 cfu/root (Doornbos et al. 2011). *Arabidopsis* seed endophytes were comprised of bacterial populations as high as 10^7 cfu/g and dominated by α-proteobacteria and β-proteobacteria with fewer Actinobacteria and Firmicutes (Truyens et al. 2012; Table 6.1). Multi-generation exposure of *Arabidopsis* seeds to cadmium altered the bacterial community composition and relative abundance of endophytes. However, many of the dominant α-proteobacteria persisted, suggesting the presence of a "core microbiome" in *Arabidopsis* seeds. These persistent bacteria included (1) at least two *Rhizobium* spp. and three *Sphingomonas* spp. capable of fixing nitrogen, (2) at least three *Rhizobium* spp. that increased root length and root growth rate, and (3) many isolates capable of producing siderophores, IAA, organic acids, and mineralization of phytates (Truyens et al. 2012).

Many culturable bacteria from *Arabidopsis* leaves represented groups present in seeds (Kniskern et al. 2007; Traw et al. 2007; Table 6.1). In contrast to the seeds, leaves were dominated by γ-proteobacteria (*Pseudomonas* and *Xanthomonas*) and α-proteobacteria (*Agrobacterium*). *Pseudomonas viridiflava*, *Pseudomonas syringae*, and *Xanthomonas campestris*, which were previously identified as natural pathogens in *Arabidopsis* populations of Midwestern USA (Jakob et al. 2002; Tsuji and Somerville 1992), along with *Agrobacterium tumefaciens*, also a plant pathogen with broad host range, accounted for 80 % of the leaf endophytes. Similar bacterial groups were among epiphytic bacteria dominated by *X. campestris*, *A. tumefaciens*, *Flavobacterium* spp., and *Nocardia corynebacterioides* (Kniskern et al. 2007); the latter two are common epiphytic saprobes (Beattie and Lindow 1995). Although there was considerable overlap, the endophytic and epiphytic communities were distinct. Perhaps somewhat unsurprisingly, the epiphytes were also more species rich and diverse than the endophytes, but largely unaffected by the host genotype. Plant pathogens and epiphytic saprobes often comprise a large proportion in the endophytic and epiphytic communities (Beattie and Lindow 1995; Ercolani 1978) as indicated by the strong presence of *X. campestris* and *A. tumefaciens* as epiphytes and endophytes (Kniskern et al. 2007).

In contrast to the limited attention endophytes in aboveground plant parts have received, at least half a dozen studies have recorded the abundance and diversity of root bacterial communities by employing culture-independent, next-generation sequencing. In addition to bacteria commonly isolated from seeds and leaves (e.g., α-proteobacteria), the root communities included Acidobacteria and Archaea (Bressan et al. 2009; Micallef et al. 2009a; Table 6.1). The root communities were dominated by *Rhizobium* sp. and *Agrobacterium* sp. irrespective of host genotype (Bressan et al. 2009; Micallef et al. 2009a).

The most exhaustive studies of plant root microbiomes have been carried out in *Arabidopsis* using metagenomics (Bulgarelli et al. 2012; Lundberg et al. 2012). Despite different *Arabidopsis* accessions, soils from different continents, primers targeting different 16S regions, or different portions of roots, these studies arrived at similar conclusions about the rhizobiome (Hirsch and Mauchline 2012): (1) different plant-free soils supported similar bacterial communities dominated by Proteobacteria and included large proportions of Acidobacteria, Actinobacteria, Bacteroidetes, Firmicutes, and Gemmatimonadetes; (2) soil types qualitatively and quantitatively influence rhizosphere bacteria and endophytic bacteria are most likely not *Arabidopsis* specific but a subset of soil bacteria; (3) rhizoplane-attached bacteria are derived from soil bacteria; (4) host genotype exerts only quantitative controls on a small subset of root endophytic bacteria, and both host genotype and plant age are less important than soil type in defining root bacterial communities (Lundberg et al. 2012); and (5) ~40 % of *Arabidopsis* root endophytic bacteria respond to plant cell wall cues and ~60 % respond to root exudates (Bulgarelli et al. 2012). Although the host genotype may have little effect in defining the rhizosphere communities, the microbial responsiveness to host roots reflects host control in attracting microbes. This selection of plant tissue-associated microorganisms (Lundberg et al. 2012) may have functional significance (Sugiyama et al. 2013), as *Streptomycetaceae*, for example, produce antimicrobial metabolites and many Proteobacteria including plant growth-promoting taxa were enriched in roots in response to root exudates. The functional significance of the bacterial rhizobiome in plant performance was supported by the variable bacterial effects on the growth of three *Arabidopsis* ecotypes (Sugiyama et al. 2013).

Examples also exist for *Arabidopsis-associated* PGPR. Schwachtje et al. (2011) reported that *Pseudomonas* sp. G62 from natural populations of *Arabidopsis* in Golm, Germany, were capable of auxin production and ACC deaminase activity, phosphate solubilization, and production of siderophores, but were unable to fix nitrogen. This strain enhanced growth of different *Arabidopsis* ecotypes by sucrose accumulation in roots independently of disease resistance induction (Schwachtje et al. 2011). *Arabidopsis* root exudates stimulated the growth of this PGPR strain, and, as a result, it is localized in root hairs and rhizoplane with little colonization of the root cortex. Further transcription studies found that this rhizobacterium induced host responses similar to those observed during carbohydrate starvation even when sugar levels were not depleted (Schwachtje et al. 2011). The loss of carbon in root exudates was compensated by increased photosynthesis correlated with increased leaf area.

Root exudates can influence rhizosphere microbes. Recent community and metabolome data support rhizobiome modulation by *Arabidopsis* root exudates. (1) *Arabidopsis* or its root exudates alone recruit similar soil fungal communities in native soils (Broeckling et al. 2008). (2) Qualitative and quantitative changes in root exudates alter root endophyte and rhizosphere microbial assemblage (Badri et al. 2009; Bressan et al. 2009). Root exudates are a cocktail of various phenolics, amino acids, sugars, and sugar alcohols among other compounds. At least 107 compounds were reported *Arabidopsis* Col-0 ecotype, and they may vary over time (Chaparro et al. 2013). Minor shifts in the host genetic makeup (e.g., single mutation of the ATP-binding cassette (ABC) transporter) can alter the ratios of phytochemicals. For example, increased phenolics and decreased sugars can lead to abundance of beneficial bacteria like PGPR, nitrogen-fixing bacteria, or bacteria involved in heavy metal remediation become abundant in rhizosphere (Badri et al. 2009). Similarly, alterations in chemical composition of the root exudates can impact the composition of Rhizobiaceae communities (Bressan et al. 2009). (3) Natural variation among *Arabidopsis* ecotypes (Micallef et al. 2009a) or developmental stages (Chaparro et al. 2013; Micallef et al. 2009b) can change the root exudates adequately to shift rhizosphere bacterial communities. Metatranscriptomic analyses revealed that microbial functional genes in the rhizosphere that were involved in the secondary metabolite and amino acid metabolism were correlated with plant root exudates (Chaparro et al. 2013). The metabolites and the rhizobiome transcriptional profiles corresponded with the host developmental stage (Chaparro et al. 2013) suggesting a coupling between the host metabolism and microbial activity in the rhizosphere. (4) Finally, root exudates alone can alter the rhizobiome microbes (Badri et al. 2013). For example, phenolic compounds can attract specific microbes, whereas sugars, sugar alcohols, and amino acids are general attractants to a variety of microbes (Badri et al. 2013). However, it seems that the core microbiome of Proteobacteria and Bacteroidetes (Bulgarelli et al. 2012; Lundberg et al. 2012) is persistent and minimally influenced by the differences in the suites of root phytochemicals (Badri et al. 2013).

In sum, a broad suite of bacteria inhabits *Arabidopsis* tissues. These microbial communities are comprised of potential pathogens, growth promoters, saprobes, and nitrogen fixers. The microbial communities likely include a stable core. It is notable that endophyte and epiphyte as well as rhizosphere and root communities overlap considerably. Surprisingly and interestingly, nitrogen-fixing bacteria are present in seeds and roots. However, nitrogen fixation in situ remains to be demonstrated. Finally, studies using a number of accessions in native soils that have reported accession-specific endophytes or rhizosphere bacterial assemblages may permit teasing apart the influence of plant genotype, edaphic factors, plant defense signaling, root exudates, and root architecture on microbial communities.

3.2 Fungal Endophytes

The interest in *Arabidopsis* microbiome is a recent development, and bacterial communities have received more attention than fungi. Very few studies have examined the fungal endophyte presence in *Arabidopsis* (Table 6.1). Using DGGE fingerprinting, Bressan et al. (2009) identified various ascomycete (e.g., *Nectria*), basidiomycete (e.g., *Marchandobasidium*), and chytridiomycete (e.g., *Olpidium*) genera in *Arabidopsis* roots. Similarly to bacteria, some taxa occurred in the roots and rhizosphere, and stable isotope labeling studies suggested that fungal communities were influenced by *Arabidopsis* root exudates. Culture-dependent surveys in Germany (Junker et al. 2012) and Spain (Garcia et al. 2012) identified common fungal endophytes. These studies pointed out that both leaf and root tissues hosted diverse fungal communities (Table 6.1) and indicated some temporal variability (Junker et al. 2012). While pathogenic *Leptosphaeria maculans* was the most commonly isolated fungus in the first year of the 2-year study, other genera (*Phoma* and *Phomopsis*) dominated the samples in the second year (Junker et al. 2012). Subsequent laboratory resyntheses confirmed that many of these fungi were *Arabidopsis* pathogens. In addition to temporal variability, the fungal communities that colonize *Arabidopsis* are spatially variable (Garcia et al. 2012). Fungal isolates from leaves and siliques of five Spanish *Arabidopsis* populations were largely comprised of *Alternaria*, *Embellisia*, and *Cladosporium*. Majority of the fungal isolates from *Arabidopsis* are ascomycetes, and few represent basidiomycetes or basal fungal lineages (Table 6.1; Garcia et al. 2012). Despite the low frequency in tissues, basidiomycetes in the order Sebacinales are interesting: They form mycorrhizas and endophytic associations with many hosts and have been recently reported in *Arabidopsis* under natural conditions (Weiss et al. 2011). It is notable that *P. indica* (also Sebacinales) forms a stable symbiosis with *Arabidopsis*, opening thus the toolbox afforded by the well-characterized model plant (see Sect. 2.1). Interestingly, Mandyam et al. (2013) observed DSE colonization of *Arabidopsis* in native European soils. Finally, the composition of fungal communities is likely driven by a suite of factors including plant intrinsic (e.g., exudates – Bressan et al. 2009) and extrinsic controls (precipitation, seasonality, plant phenology, plant age – Junker et al. 2012; Garcia et al. 2012).

Although limited, the available data unequivocally show that *Arabidopsis* roots host diverse microbial communities, including some potentially beneficial symbionts. Even though the data on *Arabidopsis* microbiome is only starting to accumulate, many meaningful extrapolations can be made using fungi from native plants if they are also able to colonize *Arabidopsis*.

4 Dark Septate Root Endophytes (DSE) in Grasslands

Konza Prairie Long Term Ecological Research (LTER) site in the Flint Hills region of eastern Kansas represents the native tallgrass prairie of North America. C_4 photosynthetic grasses (*Andropogon gerardii*, *Sorghastrum nutans*, *Schizachyrium scoparium*) typically dominate the vegetation. Although the native tallgrass plants predominantly form arbuscular mycorrhizas (Hartnett et al. 1993, 1994; Hetrick et al. 1988, 1992), DSE fungi are likely as abundant as the mycorrhizal fungi (Mandyam and Jumpponen 2008). North American tallgrass prairie ecosystem seems to support DSE communities distinct from those routinely isolated from temperate and boreal forests which are dominated by fungi with affinities within the PAC in the Northern Hemisphere (Queloz et al. 2011). *Periconia* and *Microdochium* repeatedly and commonly isolated from the grassland ecosystem formed characteristic DSE structures (Mandyam et al. 2010). Grasslands may in general host fungal endophyte communities that differ from those observed in forested ecosystems (see Khidir et al. 2010; Kageyama et al. 2008). However, woodland steppes and savannas may have an endophyte composition that includes taxa typical to both biomes (Knapp et al. 2012).

Functionally, the DSE fungi from the grasslands align well with the present understanding of the DSE fungi: they colonize native grasses and forbs suggesting a broad host range (Mandyam et al. 2012) resulting in host growth responses as predicted along the mutualism-parasitism continuum (Mandyam et al. 2012, 2013). Grasses also tended to respond more positively than the forbs and also host a greater DSE colonization in field-collected roots. The greater grass affinity to DSE leaves much to speculation, and its ecological significance is uncertain. We use recent study by Knapp et al. (2012) as an example. They isolated similar DSE from both native and invasive plants supporting broad host range (see also Mandyam et al. 2012). Further, based on the similarity of DSE fungi in the Hungarian Plain compared to those in North American grasslands, Knapp et al. (2012) hypothesized that semiarid grasslands share common dominant DSE across continents. On one hand, this may indicate fungal adaptation to these biomes and their plant communities. However, on the other hand, one is tempted to argue the opposite: fungi play a role in structuring the grassland biomes and their plant communities. In either case, it is highly likely that biomes globally may share fungal communities. This argument is supported by the PAC fungi reported from 44 undisturbed or naturally regenerated forest sites across Europe, America, and Asia and by the lack of biogeographic patterns in these populations in the Northern Hemisphere (Queloz et al. 2011). Combinations of culture-dependent and independent tools are critical for identifying DSE fungi. However, they may not provide comprehensive views of the fungal communities as indicated by poor representation of known DSE fungi in roots that host DSE colonization (see Jumpponen 2011).

Based on the limited available data from grasslands, DSE fungi in grassland biomes may be distinct. Lack of reports on PAC in the grasslands prohibits statements about their absence or dominance in these systems (but see Knapp et al. 2012). However, in the interest of broader and improved understanding of the DSE fungi, it is mandatory to include non-PAC DSE fungi from grassland ecosystems.

4.1 DSE Symbiosis of *Arabidopsis*

Arabidopsis does form DSE symbioses as indicated by melanized inter- and intracellular hyphae and microsclerotia in native European soils (Mandyam et al. 2013). Further greenhouse studies with native tallgrass prairie soil confirmed low but persistent colonization in three *Arabidopsis* ecotypes. These observations and the ease of inoculating DSE onto *Arabidopsis* in laboratory resynthesis system justify the use of *Arabidopsis* as a model for DSE symbiosis. Depending on the fungal taxon and strain, the *Arabidopsis* roots were extensively colonized by chlamydospores (*Microdochium* sp.) or melanized microsclerotia (*Periconia macrospinosa*). The disparity in the colonization levels in the greenhouse and laboratory resynthesis studies is likely due to the use of pure cultures of DSE fungi and artificial growth conditions (see Junker et al. 2012). Notably, the *Arabidopsis* growth responses (Mandyam et al. 2013) were very similar to those observed for forbs to the native tallgrass prairie (Mandyam et al. 2012): responses were variable and represented the mutualism-parasitism continuum. Further, three DSE isolates significantly increased shoot dry weight in three accessions (Columbia, Kendallville, Niederlenz), two of the three isolates increased shoot dry weight in one accession (Cape Verde Island) and had nonsignificant or minimal effects on the remaining two ecotypes (Estland and Wassilewskija) (Fig. 6.1). *Arabidopsis* may also permit more general conclusions about DSE symbiosis. Root endophytes isolated from natural *Arabidopsis* populations include fungi with no adverse effects on the host (Junker et al. 2012). The responses that varied from negative to neutral in that study may have been partly attributable to the use of 2 % sugar in the growth medium exacerbating the negative interactions.

The expedience of the *Arabidopsis* model also permits expansion of inference to populations of fungi. For example, Mandyam et al. (2013) used 34 strains of *P. macrospinosa* in symbiosis with three *A. thaliana* ecotypes (Col, Cvi, and Kin) and concluded that (1) at the fungal population level, the DSE symbiosis was either neutral (Kin) or negative (Col and Cvi); (2) at the fungal strain level, growth responses varied from positive, neutral, or negative within each ecotype; and (3) conspecific isolates elicit a range of growth responses, namely, any fungal strain could elicit a range of responses depending on host genotype (see Fig. 2 in Mandyam et al. 2013). Only one of the used strains was pathogenic when inoculated on one of the three *Arabidopsis* accessions (Cvi), whereas it reduced (Col) or increased (Kin) shoot biomass in the remaining two accessions. The variable responses with mostly neutral responses, few negative, and even fewer positive responses in *Arabidopsis* (Mandyam et al. 2013) are similar to the view emerging from studies that paired fungi and plants native to tallgrass prairie (Mandyam et al. 2012). Conversely, the use of DSE fungi that were not isolated from *Arabidopsis* do not result in disproportionately large proportion of parasitic, pathogenic, or mutualistic responses further supporting *Arabidopsis* as an appropriate model for this system.

Many studies that use laboratory resyntheses record simple growth responses (shoot biomass, root biomass, root/shoot length, proportion of healthy leaves, and inflorescences). However, similarly to PAC fungi, DSE from grasslands may have minor effects on growth, yet providing other benefits (e.g., improved stress tolerance; Mandyam and Jumpponen 2005; Kageyama et al. 2008). For example, when *Arabidopsis* plants first inoculated with *P. macrospinosa* were exposed 5 weeks later to either a necrotrophic fungal pathogen (*Botrytis cinerea*) or bacterial hemibiotrophic pathogen (*Pseudomonas syringae* ES 4326), DSE priming significantly reduced the progression of *B. cinerea* leaf necrosis only (see figure 3 in Kageyama et al. 2008). Based largely on the robust evidence from the *Arabidopsis* model, pathogen lifestyle is considered a predictor of defense responses invoked in the host (McDowell and Dangl 2000): SA defense signaling is mainly effective against biotrophic pathogens, whereas JA defense pathways provide resistance against necrotrophs and generalist chewing insects (Glazebrook 2005; McDowell and Dangl 2000; Thomma et al. 2001). For *Arabidopsis*, *Pseudomonas syringae* is largely considered a biotroph although it may initially be biotrophic and later necrotrophic (Alfano and Collmer 1996). The DSE-induced resistance to a necrotrophic *B. cinerea* and the lack of suppression of a hemibiotrophic *P. syringae* can be explained using the framework of mycorrhiza-induced resistance (MIR; Pozo and Azcón-Aguilar 2007). Similarly to the suppression of SA in the *Rhizobium*-legume symbiosis (Stacey et al. 2006), obligately biotrophic arbuscular mycorrhizal fungi require partial SA defense suppression for colonization. This increases host susceptibility to biotrophs, whereas the fully established mycorrhizal symbiosis enhances JA levels (Hause et al. 2007) resulting in an increased resistance to necrotrophs and wounding insects. Similarly, rhizobacteria-induced resistance decreases susceptibility to necrotrophic pathogens by regulating JA signaling, although there may be considerable cross talk between SA and JA defense signaling (Kloepper and Ryu 2006). The significant suppression of *Botrytis* necrosis and the slight increase in hemibiotrophic *Pseudomonas* titer though not statistically significant follow the MIR model. The DSE induction of defense responses – especially upregulation of JA and ET signaling (Fig. 6.2) – is further supported by the preliminary microarray studies using *Arabidopsis* Kin-1 and *Microdochium* sp. that neither stimulate nor inhibit this host's growth.

In addition to mycorrhizas and the rhizobial mutualists, well-known biocontrol agents including *Trichoderma* induce ISR against a variety of fungal and bacterial pathogens (Alfano et al. 2007; Harman et al. 2004a; Woo et al. 2006). This suggests that ISR induction may be a general response to colonization and therefore detectable in DSE-colonized *Arabidopsis*. To exemplify, cucumber root colonization by *Trichoderma asperellum* modulated genes in JA/ET signaling pathways and reduced *P. syringae* pv. *lachrymans* leaf necrosis (Shoresh et al. 2005). Similarly, *Trichoderma virens* and *Trichoderma harzianum* increased maize resistance against the leaf pathogen *Colletotrichum graminicola* (Djonovic et al. 2007; Harman et al. 2004b). *Trichoderma asperellum* symbiosis activates mitogen-activated protein kinase (MAPK) that is involved in plant defense signal transduction and upstream of JA/ET signaling molecules (Shoresh et al. 2006). Both SA and JA can reduce host susceptibility to *P. syringae*, but SA is more effective than JA (Shoresh et al. 2005;

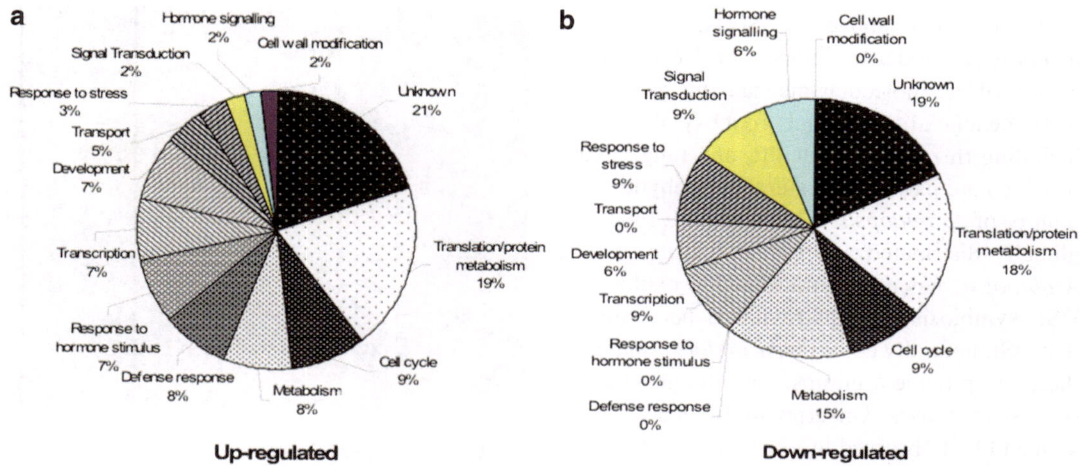

Fig. 6.2 Differential gene expression of *Arabidopsis* Kin-1 ecotype inoculated with the root colonizing fungus *Microdochium* that elicits neither positive nor negative growth response. Upregulated (**a**) and downregulated (**b**) genes in Arabidopsis inoculated with Microdochium relative to a fungus-free mock inoculated controls after 2-week incubation. The experimental treatment included four replicates and each replicate was a composite of nine plants

Thaler et al. 1999, 2004; Zhao et al. 2003). The lack of resistance in our DSE-colonized *Arabidopsis* with enhanced JA defense signaling is consistent with the lack of *P. syringae* susceptibility to JA-mediated defenses in *Arabidopsis* (Glazebrook 2005; McDowell and Dangl 2000; Thomma et al. 2001). However, the JA-mediated reduction of *P. syringae* necrosis in tomato (Thaler et al. 2004) and the reduced cucumber *P. syringae* necrosis due to *Trichoderma*-mediated JA defenses contrast our studies and underscore the importance of controlling for the host genotypic background in predicting defense responses.

Our unpublished microarray data from the shoots of 2-week-old *Arabidopsis-Microdochium* symbiosis identified 168 upregulated genes and 33 downregulated genes (Fig. 6.2). Protein metabolism, cell cycle, defense response, transcription, transport, stress response, hormone signaling (especially JA and ET), signal transduction, and cell wall modification genes were among those that were upregulated. In contrast, metabolism, protein metabolism, transcription, stress response, hormone signaling (especially auxin signaling), and signal transduction were downregulated (Fig. 6.2). Metabolomic analyses of similar tissues identified 222 upregulated polar and nonpolar metabolites and 269 downregulated metabolites, which at the time of writing remain unannotated. The differential gene expression in *Arabidopsis*-DSE symbiosis is similar to that in plants colonized by *Trichoderma. Trichoderma harzianum* Rifai strain T22 colonization upregulated 114 and downregulated 50 genes in maize shoots (Shoresh and Harman 2008). Genes involved in carbohydrate metabolism, photosynthesis, stress, amino acid, cell wall metabolism, transcription, and JA and ET signaling were upregulated (Shoresh and Harman 2008). Similarly, *T. hamatum* upregulated cell wall, defense, stress, and RNA metabolism genes in tomato, even in the absence of any growth response (Alfano et al. 2007). Overall, whether the root endophytes are rhizobacteria, mycorrhiza, *Trichoderma*, or DSE fungi, the commonly observed genes that are upregulated include defense responses, metabolism (carbon, protein, or nitrogen), stress, and hormone regulation (Alfano et al. 2007; Cartieaux et al. 2003; Duplessis et al. 2005; Gallou et al. 2012; Johansson et al. 2004; Le Quere et al. 2005; Shoresh and Harman 2008; Wang et al. 2005). The analyses of root endophytes suggest a considerable overlap in their effects on the host.

As outlined above in Sects. 2 and 3, *Arabidopsis* model has allowed a deeper appreciation of the molecular mechanisms associated with beneficial microbial root symbioses – including rhizobacteria, PGPR, and fungi, many of which are part of the natural endophytic community of *Arabidopsis*. Although DSE fungi are globally distributed and colonize many plants, *Arabidopsis* has not been utilized to explore the DSE symbiosis. Studies reviewed here suggest that DSE naturally colonize *Arabidopsis* and may therefore provide a glimpse into the generalities of host responses. Unsurprisingly, gene expression and ISR observed in *Arabidopsis*-DSE symbiosis are similar to those observed with other root endophytes. *Arabidopsis* model provides means to dissect host genomic, proteomic, and metabolomic responses to DSE fungi originating from *Arabidopsis* or from any other host plant to demystify this obscure symbiosis.

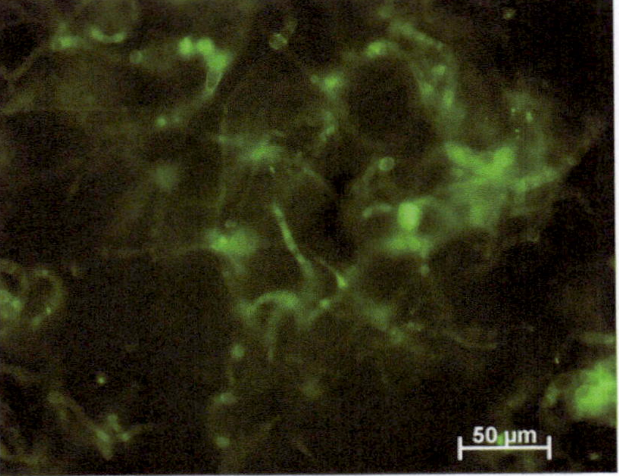

Fig. 6.3 GFP-tagged Periconia macrospinosa expressing the green fluorescent protein

4.2 Evaluation of DSE Function Using *Arabidopsis* Host

Nitrogen-fixing bacteria, PGPR, actinobacteria, and mycorrhizal fungi are usually considered mutualistic. Clearly, the jury is still out to determine where DSE fungi belong in the mutualism-parasitism continuum. Mandyam and Jumpponen (2005) and Kageyama et al. (2008) reviewed the potential functions of DSE fungi and considered that DSE fungi may be multifunctional beyond improving host growth or facilitating host nutrient uptake.

In previous sections, we have leaned on the *P. indica-Arabidopsis* model to outline the potential of the model system to expand our understanding of novel and poorly understood symbioses. If one were to recruit the *Arabidopsis* for DSE symbiosis, what would the primary questions be? We propose three crucial starting points. *First*, can we designate life histories for DSE fungi? Microscopic data largely suggested necrotrophism or biotrophism in many hosts as the absence of interfaces that would permit nutrient exchange casted doubts about mutualism (Petersen et al. 2008). Experiments similar to those with *P. indica* and the green fluorescent protein (GFP)-tagged *Arabidopsis* used by Jacobs et al. (2011) and Qiang et al. (2012b) will likely prove invaluable in following the DSE colonization. Alternatively, GFP-tagged DSE fungi can provide a mycocentric view of the colonization, for example (Gorfer et al. 2007), GFP-tagged *Cadophora finlandica* and *P. fortinii* that were stable in repeated subculturing. Their studies visualized *P. fortinii* appressorium or spore development on pine root surface. Our preliminary attempts to GFP-tag *P. macrospinosa* resulted in fluorescent strains that unfortunately remained unstable in subculturing (Fig. 6.3). However, comparisons of the colonization process among different DSE are of utmost importance; they represent phylogenetically diverse organisms (Jumpponen and Trappe 1998a). Studies that compare strains that either promote or inhibit host growth would likely provide important cues on why such differences exist. Use of one host species (*Arabidopsis*) and even genetically homogeneous inbred lines would permit direct comparisons among the root-colonizing fungi. Additional genome and transcriptome studies would also prove valuable in determining the fungal lifestyle strategies. Such studies are exemplified by *P. indica*, whose unique traits combine biotrophy with saprotrophy-

hemibiotrophy usually associated with phytopathogens (Zuccaro et al. 2011). *Second*, what are the causes of the observed host growth responses – fungal or plant-mediated hormonal control, facilitation of nutrient uptake, or changes in root or shoot anatomy? Only a small proportion of DSE isolates enhanced *Arabidopsis* growth in laboratory resynthesis studies (Mandyam et al. 2013). Comparisons among hormonal signaling in *Arabidopsis*, root and shoot architecture, and nutrient transporter mutants can provide preliminary cues to elucidate these questions as with *P. indica* (see Sect. 2.1). Similarly, the well-annotated *Arabidopsis* microarrays can shed light on the host transcriptomic responses to the DSE fungi (see above). *Finally*, how do the DSE fungi improve host tolerance to biotic and abiotic stressors? Our preliminary data (Sect. 4.1) suggest that DSE can improve *Arabidopsis* tolerance to necrotrophic pathogens, but not to biotrophic ones. *Arabidopsis* defense signaling or defense hormone mutants provide tools to understand the defense signaling operating at different stages of fungal colonization and/or the cross talk between different signaling pathways. Similarly, *Arabidopsis* mutants provide tools to evaluate DSE modulation of the host tolerance to stress caused by heat, drought, or environmental contaminants.

Arabidopsis model can provide a multitude of tools for in-depth analyses of DSE symbiosis. Such studies can initially be modeled after numerous examples that interrogate pathogenic or mutualistic interactions (see Sect. 2). Although the *Arabidopsis* model is perhaps imperfect to evaluate PAC from boreal forest conifers or DSE from tallgrass prairie, many fundamental molecular mechanisms are conserved in plants. An elegant example of this is the presence of microbial symbiosis gene NSP in *Arabidopsis* implied important in plant-microbe interactions (Hayward et al. 2012). Furthermore, the plentitude of advantages afforded by a well-developed model should not be ignored. Once mechanisms have been identified in a model system, testing ecologically more accurate hypotheses with natural hosts may become easier. The similarities in the mechanisms underlying *P. indica* symbiosis with *Arabidopsis* and barley highlight this.

5 Conclusions

DSE and AM fungi share global distribution, broad host ranges, great abundance, and colonization of root tissues simultaneously with other endophytes. Yet, the DSE symbioses remain obscure and our understanding of them cursory. Mycorrhizal fungi are vital for ecosystem functioning because of the benefits they provide to the plants. This is despite the carbon cost incurred for maintenance of the symbiosis and the variable host responses along the mutualism-parasitism continuum. In comparison, DSE fungi have been mostly ignored despite their potential for providing similar services to the host plants. Experiments with native prairie plants and DSE fungi clearly indicate that these fungi are rarely detrimental to hosts. Rather, they tend to have small but variable effects on host growth ranging from inhibition to stimulation. Studies utilizing *Arabidopsis* and native DSE strains provided results congruent with those with native hosts, suggesting that *Arabidopsis* likely provides an appropriate model to further interrogate the DSE symbiosis. Additional *Arabidopsis* transcriptome studies suggest that DSE are similar to other root colonizing organisms including rhizobacteria, mycorrhizas, and other potentially beneficial fungi. As an example, the ISR in DSE symbiosis is similar to that MIR in mycorrhizal symbiosis. Overall, many observations suggest that DSE may be important in improving host tolerance to biotic stress. The studies reviewed here highlight potential afforded by the exploitation of *Arabidopsis* model with its available "omics" resources. Uses of these tools will likely permit major strides in unraveling the role of DSE in alleviating abiotic and biotic stress and mechanisms of growth promotion or depression.

Acknowledgments This work was supported in part by the National Science Foundation Grants No. 0344838 and 0221489 (to AJ).

References

Addy HD, Piercey MM, Currah RS (2005) Microfungal endophytes in roots. Can J Bot 83:1–13

Alberton O, Kuyper TW, Summerbell RC (2010) Dark septate root endophytic fungi increase growth of Scots pine seedlings under elevated CO_2 through enhanced nitrogen use efficiency. Plant Soil 328:459–470

Alfano JR, Collmer A (1996) Bacterial pathogens in plants: life up against the wall. Plant Cell 8:1683–1698

Alfano G, Lewis Ivey ML, Cakir C, Bos JIB, Miller SA, Madden LV, Kamoun S, Hitink HAJ (2007) Systemic modulation of gene expression in tomato by *Trichoderma hamatum* 382. Phytopathology 97:429–437

Badri DV, Quintana N, El Kassis EG, Kim HK, Choi YH, Sugiyama A, Verpoorte R, Martinoia E, Manter DK, Vivanco JM (2009) An ABC transporter mutation alters root exudation of phytochemicals that provoke an overhaul of natural soil microbiota. Plant Physiol 151:2006–2017

Badri DV, Chaparro JM, Zhang R, Shen Q, Vivanco JM (2013) Application of natural blends of phytochemicals derived from the root exudates of *Arabidopsis* to the soil reveal that phenolic related compounds predominantly modulate the soil microbiome. J Biol Chem. doi:10.1074/jbc.M112.433300

Baltruschat H, Fodor J, Harrach BD, Niemczyk E, Barna B, Gullner G, Janeczko A, Kogel KH, Schafer P, Schwarczinger I, Zuccaro A, Skoczowski A (2008) Salt tolerance of barely induced by the root endophyte *Piriformospora indica* is associated with a strong increase in antioxidants. New Phytol 180:501–510

Barazani O, Benderoth M, Groten K, Kuhlemeier C, Baldwin IT (2005) *Piriformospora indica* and *Sebacina vermifera* increase growth performance at the expense of herbivore resistance in *Nicotiana attenuata*. Oecologia 146:234–243

Beattie GA, Lindow SE (1995) The secret life of foliar bacterial pathogens on leaves. Annu Rev Phytopathol 33:145–172

Betsuyaku S, Sawa S, Yamada M (2011) The function of the CLE peptide in plant development and plant-microbe interactions. Arab Book 9:e0149

Bonfante P (1984) Anatomy and morphology of VA Mycorrhizae. In: Powell CL, Joseph Bagyaraj D (eds) VA mycorrhiza. CRC Press, Conway, pp 5–33

Bonfante P (2001) At the interface between mycorrhizal fungi and plants: the structural organization of cell wall, plasma membrane and cytoskeleton. In: Hock B (ed) Mycota, IX fungal associations. Springer, Berlin, pp 45–91

Bressan M, Roncato MA, Bellvert F, Comte G, Haichar FZ, Achouak W, Berge O (2009) Exogenous glucosinolate produced by *Arabidopsis thaliana* has an impact on microbes in the rhizosphere and plant roots. ISME J 3:1243–1257

Broeckling CD, Broz AK, Bergelson J, Manter DK, Vivanco JM (2008) Root exudates regulate soil fungal community composition and diversity. Appl Environ Microbiol 74:738–744

Bulgarelli A, Rott M, Schlaeppi K, Loren van Themaat E, Ahmadinejad N, Assenza F, Rauf P, Huettel B, Reinhardt R, Schmelzer E, Peplies J, Gloeckner FO, Amann R, Eickhorst T, Schulze-Lefert P (2012) Revealing structure and assembly cues for *Arabidopsis* root-inhabiting bacterial microbiota. Nature 488:91–95

Camehl I, Sherameti I, Venus Y, Bethke A, Varma A, Lee J, Oelmüller R (2010) Ethylene signaling and ethylene-targeted transcription factors are required to balance beneficial and nonbeneficial traits in the symbiosis between the endophytic fungus *Piriformospora indica* and *Arabidopsis thaliana*. New Phytol 185: 1062–1073

Camehl I, Drzewiecki C, Vadassery J, Shahollari B, Sherameti I et al (2011) The OXI1 kinase pathway mediates *Piriformospora indica*-induced growth promotion in *Arabidopsis*. PLoS Pathog 7:e1002051

Cartieaux F, Thibaud MC, Zimmerli L, Lessard P, Sarrobert C, David P, Gerbaud A, Robaglia C, Somerville S, Nussaume L (2003) Transcriptome analysis of *Arabidopsis* colonized by a plant growth promoting rhizobacterium reveals a general effect on disease resistance. Plant J 36:177–188

Chaparro JM, Badri DV, Bakker MG, Sugiyama A, Manter DK, Vivanco JM (2013) Root exudation of phytochemicals in *Arabidopsis* follows specific patterns that are developmentally programmed and correlate with soil microbial functions. PLoS One 8(2):e55731

Cho SM, Kang BR, Han SH, Anderson AJ, Park JY, Lee YH, Cho BH, Yang KY, Ryu CM, Kim YC (2008) 2R,3R-butanediol, a bacterial volatile produced by *Pseudomonas chlororaphis* O6, is involved in induction of systemic tolerance to drought in *Arabidopsis thaliana*. Mol Plant Microbe Interact 21:1067–1075

Conn VM, Walkr AR, Franco CMM (2008) Endophytic actinobacteria induce defense pathways in *Arabidopsis thaliana*. Mol Plant Microbe Interact 21:208–218

Day B, Knepper C (2010) From perception to activation: the molecular-genetic and biochemical landscape of disease resistance signaling in plants. Arab Book 8:e012

Deshmukh S, Hückelhoven R, Schäfer P, Imani J, Sharma M, Weiss M, Waller F, Kogel KH (2006) The root endophytic fungus *Piriformospora indica* requires host cell death for proliferation during mutualistic symbiosis with barley. Proc Natl Acad Sci USA 103:18450–18457

Djonovic S, Vargas WA, Kolomiets MV, Horndeski M, Wiest A, Kenerley CM (2007) A proteinaceous elicitor Sm1 from the beneficial fungus *Trichoderma virens* is required for induced systemic resistance in maize. Plant Physiol 145:875–889

Doornbos RF, Geraats BPJ, Kuamae EE, Van Loon LC, Bakker PAHM (2011) Effects of jasmonic acid, ethylene and salicylic acid signaling on the rhizosphere bacterial community of *Arabidopsis thaliana*. Mol Plant Microbe Interact 24:395–407

Duplessis S, Courty P, Tagu D, Martin F (2005) Transcript patterns associated with ectomycorrhiza development in *Eucalyptus globulus* and *Pisolithus microcarpus*. New Phytol 165:599–611

Ercolani GL (1978) *Pseudomonas savastanoi* and other bacteria colonizing the surface of olive leaves in the field. J Gen Appl Microbiol 109:245–257

Felle HH, Waller F, Molitor A, Kogel KH (2009) The mycorrhiza fungus *Piriformospora indica* induces fast root-surface pH signaling and primes systemic alkanization of the leaf apoplast upon powdery mildew infection. Mol Plant Microbe Interact 22:1179–1185

Franken P (2012) The plant strengthening root endophyte *Piriformospora indica*: potential application and the biology behind. Appl Microbiol Biotechnol 96:1455–1464

Gallou A, Declereck S, Cranenbrouck S (2012) Transcriptional regulation of defence genes and involvement of the WRKY transcription factor in arbuscular mycorrhizal potato root colonization. Funct Integr Genomics 12:183–198

Garcia E, Alonso A, Platas G, Sacristan S (2012) The endophytic mycobiota of *Arabidopsis thaliana*. Fungal Divers. doi:10.1007/s13225-012-0219-0

Genre A, Chabaud M, Faccio A, Barker DG, Bonfante P (2008) Prepenetration apparatus assembly precedes and predicts the colonization patterns of arbuscular mycorrhizal fungi and the root cortex of both *Medicago truncatula* and *Daucus carota*. Plant Cell 20:1407–1420

Glazebrook J (2005) Contrasting mechanisms of defense against biotrophic and necrotrophic pathogens. Annu Rev Phytopathol 43:205–227

Gorfer M, Klaubauf S, Bandian D, Strauss J (2007) *Cadophora finlandia* and *Phialocephala fortinii*: *Agrobacterium*-mediated transformation and functional GFP expression. Mycol Res 111:850–855

Grünig CR, Queloz V, Sieber TN, Holdenrieder O (2008a) *Phialocephala fortinii* s.l.-*Acephala applanata* species complex in tree roots: classification, population biology and ecology. Botany 86:1355–1369

Grünig CR, Duò A, Sieber TN, Holdenrieder O (2008b) Assignment of species rank to six reproductively isolated cryptic species of the *Phialocephala fortinii* s.l.–*Acephala applanata* species complex. Mycologia 100:47–67

Harman GE, Howell CR, Viterbo A, Chet I, Lorito M (2004a) *Trichoderma* species – opportunistic, avirulent plant symbionts. Nat Rev Microbiol 2:43–56

Harman GE, Petzoldt R, Comis A, Chen J (2004b) Interactions between *Trichoderma harzianum* strain T22 and maize inbred line Mo17 and effects of these interactions on diseases caused by *Pythium ultimum* and *Colletotrichum graminicola*. Phytopathology 94:147–153

Hartnett DC, Hetrick BAD, Wilson GWT, Gibson DJ (1993) Mycorrhizal influence on intra- and interspecific neighbour interactions among co-occurring prairie grasses. J Ecol 81:787–795

Hartnett DC, Samanus RJ, Fischer LE, Hetrick BAD (1994) Plant demographic responses to mycorrhizal symbiosis in tallgrass prairie. Oecologia 99:21–26

Hause B, Mrosk C, Isayenkov S, Stracck D (2007) Jasmonates in arbuscular mycorrhizal interactions. Phytochemistry 58:101–110

Hayward A, Vighnesh G, Delay C, Samian MR, Manoli S, Stiller J, McKenzie M, Edwards D, Batley J (2012) Second-generation sequencing for gene discovery in the Brassicaceae. Plant Biotechnol 10:750–759

Herrera J, Khidir HH, Eudy DM, Porras-Alfaro A, Natvig DO, Sinsabaugh RL (2010) Shifting fungal endophyte communities colonize *Bouteloua gracilis*: effect of host tissue and geographical distribution. Mycologia 102:1012–1026

Hetrick BAD, Kitt DG, Wilson G (1988) Mycorrhizal dependence and growth of warm-season and cool-season tallgrass prairie plants. Can J Bot 66:1376–1380

Hetrick BAD, Wilson GT, Todd TC (1992) Relationships of mycorrhizal symbiosis, rooting strategy and phenology among tallgrass prairie forbs. Can J Bot 70:1521–1528

Hilbert M, Voll LM, Ding Y, Hofmann J, Sharma M, Zuccaro A (2012) Indole derivative production by the root endophyte *Piriformospora indica* is not required for growth promotion but for biotrophic colonization of barley roots. New Phytol 196:520–534

Hirsch PR, Mauchline TH (2012) Who's who in the plant root microbiome. Nat Biotechnol 30:961–962

Jacobs S, Zechmann B, Molitor A, Trujillo M, Petutsching E, Lipka V, Kogel KH, Schafer P (2011) Broad spectrum suppression of innate immunity is required for colonization of *Arabidopsis* roots by the fungus *Piriformospora indica*. Plant Physiol 156:726–740

Jakob K, Goss EM, Araki H, Van T, Kreitman M, Bergelson J (2002) *Pseudomonas viridiflava* and *P. syringae*-natural pathogens of *Arabidopsis thaliana*. Mol Plant Microbe Interact 15:1195–1203

Johansson T, Le Quéré A, Ahren D, Söderström B, Erlandsson R, Lundberg J, Uhlén M, Tunlid A (2004) Transcriptional responses to *Paxillus involutus* and *Betula pendula* during formation of ectomycorrhizal root tissue. Mol Plant Microbe Interact 17:202–215

Jones AM, Chory J, Dangl JL, Esteelle M, Jacobson SE, Meyerowitz EM, Nordborg M, Weigel D (2008) The impact of *Arabidopsis* on human health: diversifying our portfolio. Cell 133:939–943

Jumpponen A (2001) Dark septate endophytes – are they mycorrhizal? Mycorrhiza 11:207–211

Jumpponen A (2011) Analysis of ribosomal RNA indicates seasonal fungal community dynamics in *Andropogon gerardii* roots. Mycorrhiza 21:453–464

Jumpponen A, Trappe JM (1998a) Dark septate endophytes: a review of facultative biotrophic root colonizing fungi. New Phytol 140:295–310

Jumpponen A, Trappe JM (1998b) Performance of *Pinus contorta* inoculated with two strains of root endophytic fungus *Phialocephala fortinii*: effects of resynthesis system and glucose concentration. Can J Bot 76:1205–1213

Junker C, Draeger S, Schulz B (2012) A fine line-endophytes or pathogens in *Arabidopsis thaliana*. Fungal Ecol 5:657–662

Kageyama SA, Mandyam K, Jumpponen A (2008) Diversity, function and potential functions of root associated endophytes. In: Varma A (ed) Mycorrhiza, 3rd edn. Springer, Berlin, pp 29–58

Khatabi B, Molitor A, Lindermayr C, Pfiffi S, Durner J, von Wettstein D, Kogel KH, Schäfer P (2012) Ethylene supports colonization of plant roots by the mutualistic fungus *Piriformospora indica*. PLoS One 7:e35502

Khidir HH, Eudy DM, Porras-Alfaro A, Herrera J, Natvig DO, Sinsabaugh RL (2010) A general suite of fungal endophytes dominate the roots of two dominant grasses in a semiarid grassland. J Arid Environ 74:35–42

Kloepper JW, Ryu CM (2006) Bacterial endophytes as elicitors of induced systemic resistance. In: Schulz B, Boyle C, Sieber TN (eds) Microbial root endophytes, vol 9, Soil biology. Springer, Berlin/Heidelberg, pp 33–52

Knapp DG, Pintye A, Kovács GM (2012) The dark side is not fastidious – dark septate endophytic fungi of native and invasive plants of semiarid sandy areas. PLoS One 7:e32570

Kniskern JM, Traw MB, Bergelson J (2007) Salicylic acid and jasmonic acid signaling defense pathways reduce natural bacterial diversity on *Arabidopsis thaliana*. Mol Plant Microbe Interact 20:1512–1522

Koornneef M, Meinke D (2010) The development of *Arabidopsis* as a model plant. Plant J 61:909–921

Kwon YS, Ryu C-M, Lee S, Han KS, Lee JH, Lee K, Chung WS, Jeong M-J, Kim HK, Bae D-W (2010) Proteome analysis of *Arabidopsis* seedlings exposed to bacterial volatiles. Planta 232:1355–1370

Lahrmann U, Zuccaro A (2012) Opprimo ergo sum-evasion and suppression in the root endophytic fungus *Piriformospora indica*. Mol Plant Microbe Interact 25:727–737

Laluk K, Mengiste T (2010) Necrotroph attacks on plants: wanton destruction or covert extortion? Arab Book 8:e0136

Le Quere A, Wright DP, Soderstrom B, Tunlid A, Johansson T (2005) Global patterns of gene regulation associated with the development of ectomycorrhiza between birch (*Betula pendula* Roth.) and *Paxillus involutus* (Batsch) Fr. Mol Plant Microbe Interact 18:659–673

Lee YC, Johnson JM, Chien CT, Sun C, Cai D, Lou B, Oelmuller R, Yeh KW (2011) Growth promotion of the Chinese cabbage and *Arabidopsis* by *Piriformospora indica* is not stimulated by mycelium-synthesized auxin. Mol Plant Microbe Interact 24:421–431

Léon-Kloosterziel KM, Verhagen BWM, Keurentjes JJB, VanPelt JA, Rep M, VanLoon LC, Pieterse CMJ (2005) Colonization of the *Arabidopsis* rhizosphere by fluorescent pseudomonas spp. activates a root-specific, ethylene-responsive PR-5 gene in the vascular bundle. Plant Mol Biol 57:731–748

Lin L, Ge HM, Yan T, Qin YH, Tan RX (2012) Thaxtomin A-deficient endophytic *Streptomyces* sp. enhances plant disease resistance to pathogenic Streptomyces scabies. Plants 236:1849–1861

López-Bucio J, Campos-Cuevas JC, Hernández-Calderon E, Velásquez-Becerra C, Farías-Rodríguez R, Macías-Rodríguez LI, Valencia-Cantero E (2007) *Bacillus megaterium* rhizobacteria promote growth and alter root-system architecture through an auxin- and ethylene-independent signaling mechanism in *Arabidopsis thaliana*. Mol Plant Microbe Interact 20:207–217

Lundberg DS, Lebeis SL, Paredes SH, Yourstone S, Gehring J, Malfatti S, Tremblay J, Engelbrektson A, Kunin V, Glavina del Rio T, Edgar RC, Eickhorst T, Ley RE, Hugenholtz P, Tringe SG, Dangl JL (2012) Defining the core *Arabidopsis thaliana* root microbiome. Nature 488:86–90

Mandyam K, Jumpponen A (2005) Abundance and possible functions of the root- colonizing dark septate endophytic fungi. In: Summerbell R, Currah RS, Sigler L (eds) The missing lineages: phylogeny and ecology of endophytic and other enigmatic root- associated fungi, vol 53, Studies in mycology., pp 173–189

Mandyam K, Jumpponen A (2008) Seasonal and temporal dynamics of arbuscular mycorrhizal and dark septate endophytic fungi in a tallgrass prairie ecosystem are minimally affected by nitrogen enrichment. Mycorrhiza 18:145–155

Mandyam K, Loughin T, Jumpponen A (2010) Isolation and morphological and metabolic characterization of common endophytes in annually burned tallgrass prairie. Mycologia 102:813–821

Mandyam K, Fox C, Jumpponen A (2012) Septate endophyte colonization and host responses of grasses and forbs native to a tallgrass prairie. Mycorrhiza 22:109–119

Mandyam K, Roe J, Jumpponen A (2013) *Arabidopsis thaliana* model system reveals a continuum of responses to root endophyte colonization. Fungal Biol 117:25–260

Mayerhofer MS, Kernaghan G, Harper KA (2013) The effects of fungal root endophytes in plant growth: a meta-analysis. Mycorrhiza 23:119–128

McDowell JM, Dangl JL (2000) Signal transduction in the plant immune response. Trends Biochem Sci 25:79–82

McLellan CA, Turbyville TJ, Wijeratne EMK, Kerschen A, Vierling E, Queitsch C, Whitesell L, Gunatilaka AAL (2007) A rhizosphere fungus enhances *Arabidopsis* thermotolerance through production of an HSP90 inhibitor. Plant Physiol 145:174–182

Micali C, Göllner K, Humphry M, Consonni C, Panstruga R (2008) Powdery mildew disease of Arabidopsis: a paradigm for the interaction between plants and biotrophic fungi. Arab Book 6:e0115

Micallef SA, Channer S, Shiiaris MP, Colon-Carmona A (2009a) Plant age and genotype impact the progression of bacterial community succession in the *Arabidopsis* rhizosphere. Plant Signal Behav 4:777–780

Micallef SA, Shiaris MP, Colon-Carmona A (2009b) Influence of *Arabidopsis thaliana* accessions on rhizobacterial communities and natural variation in root exudates. J Exp Bot 60:1729–1742

Molitor A, Zajic D, Voll LM, Kühnemann JP, Samans B, Kogel KH, Waller F (2011) Barley leaf transcriptome and metabolite analysis reveals new aspects of compatibility and *Piriformospora indica*-mediated systemic induced resistance to powdery mildew. Mol Plant Microbe Interact 24:1427–1439

Newsham KK (2011) A meta-analysis of plant responses to dark septate root endophytes. New Phytol 190:783–793

Nishimura MT, Dangl JL (2010) *Arabidopsis* and the plant immune system. Plant J 61:1053–1066

Nongbri P, Johnson JM, Sherameti I, Glawischnig E, Halkier BA, Oelmüller R (2012) Indole-3-acetaldoxime-derived compounds restrict root colonization in the beneficial interaction between *Arabidopsis* roots and the endophyte *Piriformospora indica*. Mol Plant Microbe Interact 25:1186–1197

Oelmüller R, Sherameti I, Tripathi S, Varma A (2009) *Piriformospora indica*, a cultivable root endophyte with multiple biotechnological applications. Symbiosis 49:1–17

Parniske M (2000) Intracellular accommodation of microbes by plants: a common developmental program for symbiosis and disease? Curr Opin Plant Biol 3:320–328

Peškan-Berghöfer T, Shahollari B, Giong PH, Hehl S, Markert C, Blanke V, Kost G, Varma A, Oelmüller R (2004) Association of *Piriformospora indica* with *Arabidopsis thaliana* roots represents a novel system to study beneficial plant-microbe interactions and involves early plant protein modifications in the endoplasmic reticulum and at the plasma membrane. Physiol Planta 122:465–477

Petersen RL, Wagg C, Paulter M (2008) Associations between microfungal endophytes and roots: do structural features indicate function? Botany 86:445–456

Pieterse CMJ, vanWees SCM, van Pelt JA, Knoester M, Laan R, Gerrits H, Weisbeek PJ, van Loon LC (1998) A novel signaling pathway controlling induced systemic resistance in *Arabidopsis*. Plant Cell 10:1571–1580

Pozo MJ, Azcón-Aguilar C (2007) Unraveling mycorrhiza-induced resistance. Curr Opin Plant Biol 10:393–398

Qiang X, Weiss M, Kogel KH, Schafer P (2012a) *Piriformospora indica*-a mutualistic basidiomycete with an exceptionally large plant host range. Mol Plant Pathol 13:508–518

Qiang X, Zechmann B, Reitz MU, Kogel KH, Schäfer P (2012b) The mutualistic fungus *Piriformospora indica* colonizes *Arabidopsis* roots by inducing an endoplasmic reticulum stress-triggered caspase-dependent cell death. Plant Cell 24:794–809

Queloz V, Sieber TN, Holdenrieder O, McDonald BA, Grunig CR (2011) No biogeographical pattern for a root-associated fungal species complex. Glob Ecol Biogeogr 20:160–169

Reboutier D, Bianchi M, Brault M, Roux C, Dauphin A, Rona JP, Legué V, Lapeyrie F, Bouteau F (2002) The indolic compound hypaphorine produced by ectomycorrhizal fungus interferes with auxin action and evokes early responses in non-host *Arabidopsis thaliana*. Mol Plant Microbe Interact 15:932–938

Reininger V, Sieber TN (2012) Mycorrhiza reduces adverse effects of dark septate endophytes (DSE) on growth of conifers. PLoS One 7:e42865

Reininger V, Grunig CR, Sieber TN (2012) Host species and strain combination determine growth reduction of spruce and birch seedlings colonized by root-associated dark septate endophytes. Environ Microbiol 14:1064–1076

Rodriguez RJ, White JF Jr, Arnold AE, Redman RS (2009) Fungal endophytes: diversity and functional roles. New Phytol 182:314–330

Ryu CM, Farag M, Hu CH, Reddy MS, Wei H-S, Pare P, Kloepper JW (2003) Bacterial volatiles promote growth in *Arabidopsis*. Proc Natl Acad Sci USA 100:4927–4932

Ryu CM, Farag M, Hu CH, Reddy MS, Pare P, Kloepper JW (2004a) Bacterial volatiles induced systemic resistance in *Arabidopsis*. Plant Physiol 134:1017–1026

Ryu C-M, Murphy JF, Mysore KS, Kloepper JW (2004b) Plant Growth-Promoting Rhizobacteria Protect Systemically *Arabidopsis thaliana* against Cucumber mosaic virus by a salicylic acid and NPR1-independent and jasmonic acid-dependent signaling pathway. Plant J 39:381–392

Ryu CM, Hu CH, Locy RD, Kloepper JW (2005) Study of mechanisms for plant growth promotion elicited by rhizobacteria in *Arabidopsis thaliana*. Plant Soil 286:285–292

Ryu C-M, Murphy JF, Reddy MS, Kloepper JW (2007) A two strain mixture of rhizobacteria elicits induction of systemic resistance against *Pseudomonas syringae* and Cucumber mosaic virus coupled to promote plant growth on *Arabidopsis thaliana*. J Microbiol Biotechnol 17:280–286

Scervino JM, Gottlieb A, Silvani VA, Pérgola M, Fernández L, Godeas AM (2008) Exudates of dark septate endophyte (DSE) modulate the development of the arbuscular mycorrhizal fungus (AMF) *Gigaspora rosea*. Soil Biol Biochem 41:1753–1756

Schäfer P, Khatabi B, Kogl KH (2007) Root cell death and systemic effects of *Piriformospora indica*: a study on mutualism. FEMS Microbiol Lett 275:1–7

Schäfer P, Pfiffi S, Voll LM, Zajic D, Chandler PM, Waller F, Scholz U, Pons-Kühnemann J, Sonnewald S, Sonnewald U, Kogel KH (2009) Manipulation of plant innate immunity and gibberellin as factor of compatibility in the mutualistic association of barley roots with *Piriformospora indica*. Plant J 59:461–474

Schwachtje J, Karojet S, Thormahlen I, Bernholz C, Kunz S, Brouwer S, Shwochow M, Kohl K, van Dongen JT (2011) A naturally associated rhizobacterium of *Arabidopsis thaliana* induces a starvation-like transcriptional response while promoting growth. PLoS One 6(12):e29382

Selosse MA, Dubois MP, Alavrez A (2009) Do sebacinales commonly associate with plant roots as endophytes? Mycol Res 113:1062–1069

Shahollari B, Vadassery J, Varma A, Oelmüller R (2007) A leucine-rich repeat protein is required for growth promotion and enhanced seed production mediated by the endophytic fungus *Piriformospora indica* in *Arabidopsis thaliana*. Plant J 50:1–13

Sharma M, Schmid M, Rothballer M, Hause G, Zuccaro A, Imani J, Kampfer P, Domann E, Schafer P, Hartmann A, Kogel KH (2008) Detection and identification of the bacteria intimately associated with fungi of the order Sebacinales. Cell Microbiol 10:2235–2246

Sherameti I, Shahollari B, Venus Y, Altschmied L, Varma A, Oelmüller R (2005) The endophytic fungus *Piriformospora indica* stimulates the expression of nitrate reductase and the starch-degrading enzyme glucan-water dikinase in tobacco and *Arabidopsis* roots through a homeodomain transcription factor which binds to a conserved motif in their promoters. J Biol Chem 280:2641–2647

Sherameti I, Venus Y, Drzewiecki C, Tripathi S, Dan VM, Nitz I, Varma A, Grundler FM, Oelmuller R (2008a) PYK10, a b-glucosidase located in the endoplasmatic reticulum, is crucial for the beneficial interaction between *Arabidopsis thaliana* and the endophytic fungus *Piriformospora indica*. Plant J 54:428–439

Sherameti I, Tripathi S, Varma A, Oelmuller R (2008b) the root-colonizing endophyte *Piriformospora indica* confers drought tolerance in *Arabidopsis* by stimulating the expression of drought stress-related genes in leaves. Mol Plant Microbe Interact 21:799–807

Shi CL, Park HB, Lee JB, Ryu S, Ryu CM (2010) Inhibition of primary roots and stimulation of lateral root development in Arabidopsis thaliana by the rhizobacterium *Serratia marcescens* 90-166 is through both auxin-dependent and –independent signaling pathways. Mol Cells 29:251–258

Shoresh M, Harman GE (2008) The molecular basis of shoot responses of maize seedlings to *Trichoderma harzianum* T22 inoculation of the root: a proteomic approach. Plant Physiol 147:2147–2163

Shoresh M, Yedidia I, Chet I (2005) Involvement of jasmonic acid/ethylene signaling pathway in the systemic resistance induced in cucumber by *Trichoderma asperellum* T203. Phytopathology 95:76–84

Shoresh M, Gal-On A, Leibman D, Chet I (2006) Characterization of a mitogen-activated protein kinase gene from cucumber required for *Trichoderma-conferred* plant resistance. Plant Physiol 142:1169–1179

Sirrenburg A, Göbel C, Grond S, Czempinski N, Ratzinger A, Karlovsky P, Santos P, Feussner I, Pawlowski K (2007) *Piriformospora indica* affects plant growth by auxin production. Physiol Planta 131:581–589

Smith SE, Read DJ (2008) Mycorrhizal symbiosis, 3rd edn. Academic, London

Stacey G, McAlwin CB, Kim SY, Olivares J, Soto MJ (2006) Effects of endogenous salicylic acid on nodulation in the model legumes *Lotus japonicas* and *Medicago truncatula*. Plant Physiol 141:1473–1481

Stein E, Molitor A, Kogel KH, Wallr F (2008) Systemic resistance in Arabidopsis by the mycorrhizal fungus *Piriformospora indica* requires jasmonic acid signaling and the cytoplasmic function of NPR1. Plant Cell Physiol 49:1747–1751

Sugiyama A, Bakker MG, Badri DV, Manter DK, Vivanco JM (2013) Relationships between *Arabidopsis* genotype-specific biomass accumulation and associated soil microbial communities. Botany 91:123–126

Sun C, Johnson JM, Cai D, Sherameti I, Oelmüller R, Lou B (2010) *Piriformospora indica* confers drought resistance in Chinese cabbage leaves by stimulating antioxidant enzymes, the expression of drought-related genes and the plastid-localized CAS protein. J Plant Physiol 167:1009–1017

Tellenbach C, Sieber TN (2012) Do colonization by dark septate endophytes and elevated temperature affect pathogenicity of oomycetes? FEMS Microbiol Ecol 82:157–168

Tellenbach C, Grunig CR, Sieber TN (2011) Negative effects on survival and performance of Norway spruce seedlings colonized by dark septate root endophytes are primarily isolate-dependent. Environ Microbiol 1:2508–2517

Tellenbach C, Sumarah MW, Grünig CR, Miller JD (2013) Inhibition of *Phytophthora* species by secondary metabolites produced by the dark septate endophyte *Phialocephala europaea*. Fungal Ecol 6:12–18

Thaler JS, Fidantsef AL, Duffey SS, Bostock RM (1999) Trade-offs in plant defense against pathogens and herbivores: a field demonstration of chemical elicitors of induced resistance. J Chem Ecol 25:1597–1609

Thaler JS, Owen B, Higgins VJ (2004) The role of the jasmonate response in plant susceptibility to diverse pathogens with a range of lifestyles. Plant Physiol 135:530–538

Thilmony R, Fumiaki Katagiri F, He SY (2002) The *Arabidopsis thaliana-Pseudomonas syringae* interaction. Arab Book 1:e0039

Thomma B, Penninckx I, Broekaert WF, Cammue BPA (2001) The complexity of disease signaling in Arabidopsis. Curr Opin Immunol 13:63–68

Traw BM, Knsikern JM, Bergelson J (2007) SAR increases fitness of *Arabidopsis thaliana* in the presence of natural bacterial pathogens. Evolution 61:2444–2449

Truyens S, Weyens N, Cuypers A, Vangronsveld J (2012) Changes in the population of seed bacteria of transgenerationally Cd-exposed *Arabidopsis thaliana*. Plant Biol. doi:10.1111/j.1438-8677.2012.00711.x

Tsuji J, Somerville SC (1992) First report of the natural infection of *Arabidopsis thaliana* by *Xanthomonas campestris* pv. *campestris*. Plant Dis 761:539

Vadassery J, Ritter C, Venus Y, Camehl I, Varma A, Shahollari B, Novák O, Strnad M, Ludwig-Müller J, Oelmüller R (2008) The role of auxins and cytokinins in the mutualistic interaction between *Arabidopsis* and *Piriformospora indica*. Mol Plant Microbe Interact 21:1371–1383

Vadassery J, Ranf S, Drzewiecki C, Mithöfer A, Mazarsa C, Scheel D, Lee J, Oelmüller R (2009a) A cell wall extract from the endophytic fungus *Piriformospora indica* promotes growth of *Arabidopsis* seedlings and induces intracellular calcium elevation in roots. Plant J 59:193–206

Vadassery J, Tripathi S, Prasad R, Varma A, Oelmüller R (2009b) Monodehydroascorbate reductase 2 and dehydroascorbate reductase 5 are crucial for a mutualistic interaction between *Piriformospora indica* and *Arabidopsis*. J Plant Physiol 166:1263–1274

Varma A, Sudha S, Franken P (1999) *Piriformospora indica*-a cultivable plant growth promoting root endophyte with similarities to arbuscular mycorrhizal fungi. Appl Environ Microbiol 65:2741–2744

Venus Y, Oelmüler R (2012) *Arabidopsis* ROP1 and ROP5 influence germination time, root morphology, the formation of F-actin bundles and symbiotic fungal interactions. Mol Plant. doi:10.1093/mp/sss101

Walker JF, Aldrich-Wolfe L, Riffel A, Barbare H, Simpson NB, Trowbridge J, Jumpponen A (2011) Diverse helotiales associated with the roots of three species of arctic ericaceae provide no evidence for host specificity. New Phytol 191:515–527

Waller F, Achatz B, Baltruschat H, Fodor J, Becker K, Fischer M, Heier T, Hückelhoven R, Neumann C, Wettstein DV, Franken P, Kogel KH (2005) The endophytic fungus Piriformospora indica reprograms barley to salt-stress tolerance, disease resistance and higher yield. Proc Natl Acad Sci USA 102:13386–13391

Waller F, Mukherjee K, Deshmukh SD, Achatz B, Sharma M, Schäfer P, Kogel KH (2008) Systemic and local modulation of plant responses by *Piriformospora indica* and related Sebacinales species. J Plant Physiol 165:60–70

Wang Y, Ohara Y, Nakayashiki H, Tosa Y, Mayama S (2005) Microarray analysis of the gene expression profile induced by the endophytic plant growth-promoting rhizobacteria, *Pseudomonas fluorescens* FPT9601-T5 in *Arabidopsis*. Mol Plant Microbe Interact 18:385–396

Weiss M, Sykorova Z, Garnica S, Riess K, Martos F, Krause C, Oberwinkler F, Bauer R, Redecker D (2011) Sebacinales everywhere: previously overlooked ubiquitous fungal endophytes. PLoS One 6:e16793

Woo SL, Scala F, Ruocco M, Lorito M (2006) The molecular biology of the interactions between *Trichoderma* spp., pathogenic fungi and plants. Phytopathology 96:181–185

Yu T, Nassuth A, Peterson RL (2001) Characterization of the interaction between the dark septate fungus *Phialocephala fortinii* and *Asparagus officinalis* roots. Can J Bot 47:741–753

Zhang H, Kim MS, Krishnamachari V, Payton P, Sun Y, Grimson M, Farag MA, Ryu CM, Allen R, Melo IS, Pare PW (2007) Rhizobacterial volatile emissions regulate auxin homeostasis and cell expansion in *Arabidopsis*. Planta 226:839–851

Zhao YF, Thilmony R, Bender CL, Schaller A, He SY, Howe GA (2003) Virulence systems of *Pseudomonas syringae* pv. in tomato promote bacterial speck disease in tomato by targeting the jasmonate signaling pathway. Plant J 36:485–499

Zuccaro A, Lahrmann U, Guldener U, Langen G, Pfiffi S, Biedenkopf D, Wong P, Samans B, Grimm C, Basiewicz M, Murat C, Martin F, Kogel KH (2011) Endophytic life strategies decoded by genome and transcriptome analyses of the mutualistic root symbiont *Piriformospora indica*. PLoS Pathog 7:e1002290

7 Asexual Endophytes of Grasses: Invisible Symbionts, Visible Imprints in the Host Neighborhood

Marina Omacini

Abstract

Asexual fungi from the genus *Neotyphodium*, relatives of the sexual *epichloë* species (Clavicipitaceae, Ascomycota), are symbionts of several cool-season grasses inhabiting virtually all terrestrial ecosystems. The host plants incur carbon costs to sustain this symbiosis, but, in return, they obtain multiple benefits from the fungal partners, above all, protection from herbivores. These endophytes are often considered to be defensive mutualists or private protectors because they produce a considerable range of secondary metabolites which prove to be toxic to livestock or deterrent to insects. Over the past decade, ecologists have begun to recognize the critical role played by this grass–endophyte symbiosis in the structure and functioning of natural and human-made communities. In this chapter, I will identify different pathways through which the presence of endophytic plants or their dead tissues (litter) can alter the fitness of nonsymbiotic plants. Those pathways lead to show how these symbionts impact on the establishment and productivity of nonsymbiotic neighbors and the interaction of the latter with multiple above- and belowground ecosystem components. A set of recent studies performed with plants of *Lolium multiflorum* associated with *Neotyphodium occultans* will provide experimental evidence to those effects. Finally, I will discuss the relevance of placing these pathways under the spotlight in order to understand the processes that determine the frequency of symbiotic plants within a population. Estimating endophyte impacts on host fitness must consider advantages or disadvantages transferred to conspecific plants in the neighborhood, coexisting as a consequence of inefficiencies during the transmission from plants to seeds.

M. Omacini (✉)
IFEVA, Facultad de Agronomía, Universidad de Buenos Aires, CONICET Av. San Martín 4453, Buenos Aires, CABA (1417), Argentina
e-mail: omacini@agro.uba.ar

V.C. Verma and A.C. Gange (eds.), *Advances in Endophytic Research*,
DOI 10.1007/978-81-322-1575-2_7, © Springer India 2014

1 Introduction

Symbionts play key roles in terrestrial ecosystems and influence a large number of important components and processes at different spatial and temporal scales (Clay and Holah 1999; Omacini et al. 2005; van der Heijden et al. 2006, 2008; Hartley and Gange 2009; Kothamasi et al. 2010). In many symbioses, participating organisms gain multiple benefits through novel traits or complex metabolic capabilities that either displays on its own (Clay and Schardl 2002; Rodriguez and Redman 2008).

For instance, microbes that form intimate symbiotic associations with plants can stimulate plant productivity by supplying them with limiting nutrients and/or by protecting them from natural enemies such as pathogens or herbivores (Thrall et al. 2007; Douglas 2010). It is expected that phenotypic alterations from symbioses increase fitness of the plants and the microbes involved, with consequences on diversity, structure, and composition of plant community and both partners evolution (Sachs et al. 2004; van der Heijden et al. 2008; Douglas 2010). The consequences of symbiont effects will vary if the microbe confers private benefits to one partner or if the benefits are available to many community members.

Plant symbionts are a taxonomical diverse group of bacteria, fungi, and virus with contrasting life history traits (Douglas 2010). Many of these microorganisms rely completely on host plant for carbon and asymptomatically colonize in or on living tissues. It is important to point out that new symbioses are reported every year for the same host, and recent studies show that symbiont effects depend not only on the presence of other host symbionts but also on the symbiont symbionts (e.g., Márquez et al. 2007). Most studies focus on provider symbionts, above all mycorrhizal fungi and nitrogen-fixing bacteria (Omacini et al. 2012).

There is a growing interest on a small group of species of the family "Clavicipitaceae" that reside entirely within aerial tissues of cool-season grasses (Poaceae) (Schardl et al. 2004; Rodriguez et al. 2009; Schardl 2010). These fungi are often considered to be defensive mutualists or protective symbionts because they produce a considerable range of secondary metabolites which prove to be toxic to livestock or deterrent to insects (Clay 1988, 2009; Clay and Schardl 2002; Schardl et al. 2004; Schardl 2010). While potential host species of these fungal endophytes are widely recognized as an important component of natural and human-made communities around the world, the impact of the grass–endophyte symbiosis on community structure and function is still poorly understood (Omacini et al. 2005; Cheplick and Faeth 2009; Saikkonen et al. 2010).

In this chapter, I explore different pathways through which the presence of endophytic plants or their dead tissues (litter) can alter the fitness of nonsymbiotic plants and I identify research gaps and propose new avenues of research. First, I briefly introduce the symbiosis between grasses and asexual endophytes and discuss the impact of endophyte on the host grass interaction with its biotic and abiotic environment. Second, I describe how these symbionts may impact on the performance of nonsymbiotic neighbors and the interaction of the latter with multiple above- and belowground ecosystem components. Third, I resort to results from a set of recent studies performed with plants of Italian ryegrass (*Lolium multiflorum*) associated with the endophyte *Neotyphodium occultans* to provide experimental evidences of these potential effects. My ultimate aim is to highlight the significance of endophyte–grass symbiosis at the neighborhood level for understanding the processes determining symbiosis frequency and structuring plant communities (van der Putten et al. 2001; Stanton 2003; Palmer et al. 2010). I end with conclusions and identify future research priorities.

2 Overview of Grass–Endophyte Symbioses

Many definitions of endophyte have been proposed since the term was first introduced by de Bary, referring to any organism occurring within plant tissues (de Bary 1879). The most commonly used is the one by Petrini (1991): "all organisms inhabiting plant organs that, at some time in their life, can colonize internal plant tissues without causing apparent harm to the host." Great controversy has surrounded the definition of the boundaries of such a broad concept considering that any organism isolated from all plants studied to date would conformed an endophyte (Wilson 1995). Thus, some authors, whose opinion I endorse, restricted the term endophyte to any fungi that reside entirely within plant tissues (i.e., roots, stems, and/or leaves) although they can emerge to sporulate at plant or host-tissue senescence, excluding mycorrhizal fungi (Carroll 1988; Cabral et al. 1993; Rodriguez et al. 2009). Under this definition, two major groups can be recognized differing in their taxonomy, plant hosts, and ecological functions: the clavicipitaceous endophytes which infect cool-season grasses and the non-clavicipitaceous endophytes which can be recovered from asymptomatic tissues of nonvascular plants, ferns, conifers, and angiosperms (Rodriguez et al. 2009).

Clavicipitaceous fungi (phylum Ascomycota), in particular the *Epichloë*, include many species that are exclusively endophytic symbionts of cool-season grasses (Schardl 2010). These fungi live within aerial plant tissues, among cells, and colonize host ovaries. They were first isolated from seeds of *Lolium temulentum* (Vogl 1980). Up to the present, more than 200 species have been documented as hosts (Leuchtmann 2006; Saikkonen et al. 1998; Rudgers and Orr 2009; Schardl 2010), although there is an inaccurate estimation of the number or percentage of Poaceae that form symbiotic associations with this type of endophytes. Symbiotic plants are common in many natural and seminatural ecosystems in different biomes around the world, from tropical rain forests to the high artic, and their frequencies within host populations range from 0 to 100 % (Iannone et al. 2011). There are many life-forms among the host species, including annual and perennial, palatable or unpalatable, and native or exotic, introduced accidentally or intentionally.

Among endophytic fungi of grasses, *Neotyphodium* species (formerly *Acremonium*, Glenn et al. 1996) are asexual forms that are exclusively vertically transmitted from symbiotic plants to their seeds; the said fungi do not sporulate at plant or host-tissue senescence (Fig. 7.1). According to phylogenetic studies, they have arisen independently from hybridization of sexual ancestors of *Epichloë* genus (Moon et al. 2000; Schardl 2010). Sexual endophytes may produce collars of mycelium (stromata) on host stems under certain contexts, spreading contagiously and reducing host seed set. Instead, asexual endophytes belong to a group previously recognized as Type III or class I (White 1987; Rodriguez et al. 2009; but see also Tadych et al. 2007), whose fitness is determined by its effects on host fitness (i.e., seed production) and its efficiency to grow in elongating grass leaves (Christensen et al. 2008) and to be transmitted across host life history stages (e.g., the number of symbiotic seeds produced by the host plant) (Fig. 7.1) (Ravel et al. 1997; Gundel et al. 2008, 2009).

Recent studies show that failures in endophyte transmission depend on the host stage, the species considered, and the environmental condition (e.g., Afkhami and Rudgers 2008; García Parisi et al. 2012). As the association between *Neotyphodium* and grasses appears to be essential for the fungus while facultative for host plants, evolutionary theory predicts that this type of symbiotic interaction should result in mutualism, where both participating organisms benefit (Clay and Schardl 2002; Douglas 2010). The host plant provides nutrition and means of propagation to the endophytic fungus while the symbiont confers the plant with resistance to herbivory and tolerance to diverse causes of abiotic stress probably mediated through bioactive alkaloids and antioxidants (summarized by Malinowski and Belesky 2000; Clay and Schardl 2002; Kuldau and Bacon 2008; White and Torres 2009). Interest for grass–endophyte symbioses greatly increased around the 1970s when toxicosis in cattle grazing on tall fescue and perennial ryegrass was associated with the presence of these asexual endophytes (Bacon and Hill 1997). But not all

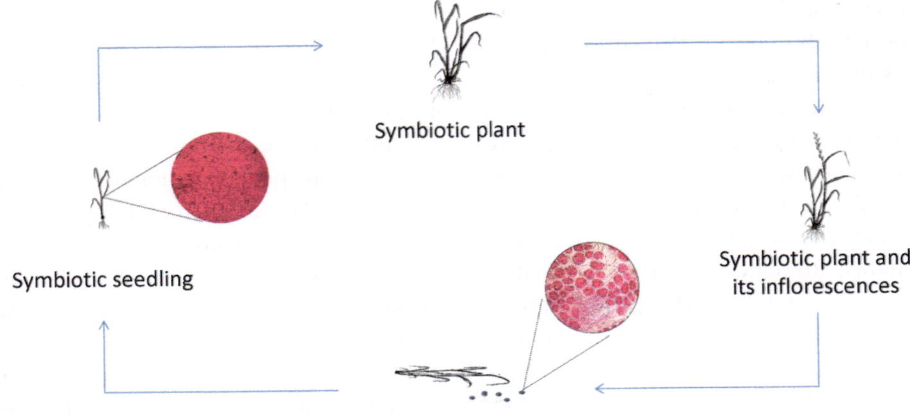

Fig. 7.1 Life cycle of asexual endophytes (Clavicipitaceae, Ascomycota) in relation to the life cycle of their host grasses. They transmit only vertically without suppressing seed production by the symbiotic host. The right angle turn in each *arrow* indicates that the symbiont may be lost by imperfect transmission in the passage from one plant stage to another. The photos show hyphae of the endophyte *Neotyphodium occultans* in seed or seedling tissues of the annual grass *Lolium multiflorum* (i.e., hyphae present in the aleurone layer or intercellular spaces respectively). Both photographs were taken at 200× on a light microscope (Photo credit: M. Rabadán)

grass–*Neotyphodium* symbioses are toxic for cattle (e.g., *Lolium multiflorum*–*Neotyphodium occultans*), and a number of researchers are searching for nontoxic associations, which are more productive and tolerant to stresses than the nonsymbiotic counterpart.

Despite the profound effects that asexual endophytes have on the communities in which they occur, they are still often ignored in community theory. Bearing this in mind, this chapter examines the numerous interactions that *Neotyphodium* endophytes mediate between host and nonhost plant and other organisms. Besides, I discuss their impacts on the abiotic environment to highlight the far-reaching consequences of these interactions for community structure and function. Considering this review is primarily concerned with neighborhood level interactions, I only briefly refer to some endophyte impacts on individual host plants.

More details of the multiple direct impacts, their physiological basis, and the genetic and environmental contingency of the grass–endophyte relationship can be found in, for example, Malinowski and Belesky (2000), Cheplick and Faeth (2009), Rasmussen et al. (2009), and Brosi et al. (2011).

3 From Host Plant to Neighborhood

In order to understand the linkages between endophyte–grass symbiosis and ecosystem function, a hierarchy of ecological domains needs to be recognized, ranging from individual host plant to the whole biotic community (Fig. 7.2). Between both extremes, the neighborhood level can be defined as an entity comprising a diverse assemblage of species that belong to different trophic levels interacting continuously with one another and with their biotic and abiotic surroundings. Within this framework, the whole community is made up by the result of the information issued by these assemblages dynamically connected by horizontal movement of organism and abiotic resources (Wiens 1989). The said hierarchy allows us to trace the many consequences of endophyte presence at different spatial and temporal scales (Omacini et al. 2005). Previous studies show that grass–endophyte symbiosis may have strong impacts on the structure and functioning of plant and arthropod communities (e.g., Clay and Holah 1999; Omacini et al. 2001; Clay et al. 2005; Finkes et al. 2006;

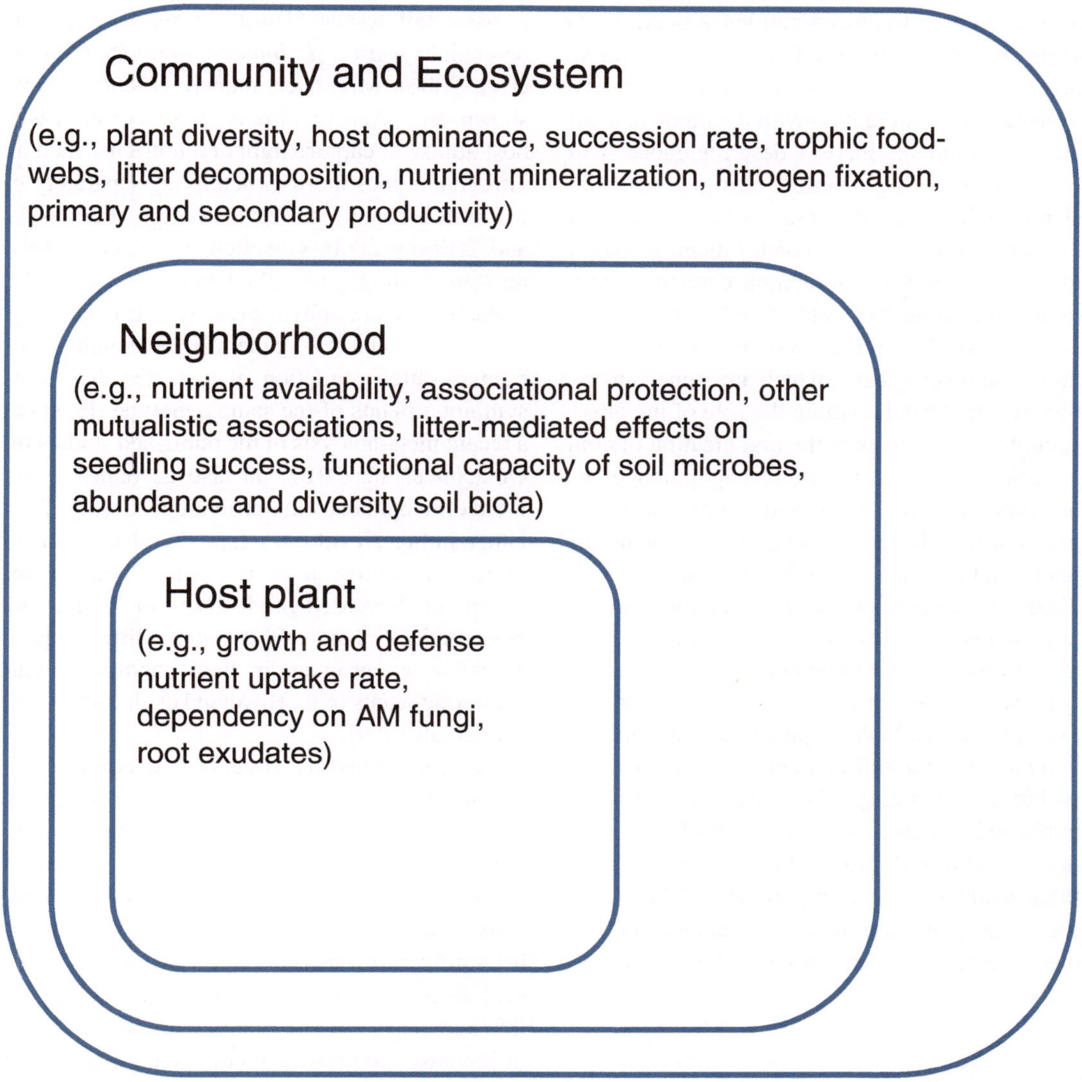

Fig. 7.2 Three hierarchical levels in which the influence of endophytic fungi can be detected at different spatial and temporal scales and examples of potential effects on attributes and processes for each ecological domain. The neighborhood is defined as an entity comprising a diverse assemblage of species concurring with the host grass within a relative small area and sharing a common resource pool (Modified from Johnson et al. 1997)

Rudgers and Clay 2007, Rudgers et al. 2010, Fig. 7.2). Short-term endophytes impact on plant composition and diversity has been mainly associated to an increase on host performance in the presence of herbivores, which leads to changes in competitive interactions between symbiotic and nonsymbiotic plants with cascading effects. But there are multiple above- and belowground species that can also respond to endophyte presence and impact on community organization (Rudgers et al. 2010).

Here, I analyze, at least, three different pathways through which endophyte impact on host level can modify the interaction between host and nonhost plants and of nonhost plants with invertebrate herbivores, root symbionts, and pathogens at neighborhood level. I focus on direct and indirect interactions between plants in close proximity or between generations of plants in the same site, in which the influence of a symbiotic plant on another nonsymbiotic plant can involve a change in its interaction with the biotic and abiotic environment. First, I describe those pathways considering those changes in host characteristics and attributes that are relevant to the study at this level. Second, I mention experiments that simplify the complex assemblage to reduce the amount of data to be handled. In this step, the whole complex system is broken down into several subsystems which are then examined one by one. Understanding the role of the grass–endophyte symbiosis in the organization of communities is limited in art. Extrapolation of the outcomes of small-scale and simplified experiments with only few associations is difficult and interpretations are unreliable (Saikkonen et al. 2006). In addition, considering that the effects of any symbiosis are relative to the environmental conditions and developmental stage of all the species involved, these experiments offer only examples of endophyte potential influences on complex and multidirectional biotic interactions within a community. Currently, the number of published studies is insufficient; further research is required in order to establish the generality of these insights and the importance of these pathways on plant community structure and the mechanisms underpinning the described responses.

3.1 Pathways

The first pathway considers changes in resource availability generated by symbiotic plants that can reduce directly the fitness of nonsymbiotic plants in the neighborhood. Despite a small number of experiments, empirical work using symbiotic and nonsymbiotic plants living together and separately supports that endophyte presence positively affects host competitive ability (Omacini et al. 2005; Cheplick and Faeth 2009; but see Saikkonen et al. 2006). The majority of these studies have focused on two agronomical and economically important forage grasses: tall fescue (*Festuca arundinacea*) and perennial ryegrass (*Lolium perenne*) (Saikkonen et al. 2006). Moreover, numerous studies have shown that *Neotyphodium* presence increases host ability to capture light and nutrients such as nitrogen or phosphorus when living separately in controlled conditions (reviewed by Malinowski and Belesky 2000; Cheplick and Faeth 2009), thus suggesting plant–plant interactions can be impacted by endophyte presence. The fact that symbiotic plants (E+) generally have higher root biomass and production of exudates than nonsymbiotic plants of the same genotype (E−) (see a recent meta-analysis of the published studies by Omacini et al. 2012) can also account for the more efficient host soil nutrient uptake (Fig. 7.3). Other indirect evidences that E+ plants can be more competitive than E− plants come from reports of decreased productivity or biomass of nonsymbiotic plants of the same or different genotypes when growing in field conditions with symbiotic plants (e.g., Hoveland et al. 1999; Clay and Holah 1999).

Authors generally suggest that competition for resources was the predominant mechanism of symbiosis negative impacts on nonsymbiotic plants mainly when those effects were detected in the absence of aboveground herbivores (e.g., Clay et al. 2005; Omacini et al. 2006). The studies rarely rule out other possible mechanisms such as the production of allelopathic or diffusible substances by E+ plants that may decrease or increase the growth of concurring plants (but see Sutherland et al. 1999; Koulman et al. 2007; Vázquez-de-Aldana et al. 2011; Mersch and Cahoon 2012). Those processes can also help to explain the results of other studies where endophyte effects on host growth and reproductive capacity were negative or varied with environmental conditions (Saikkonen et al. 2006; Cheplick and Faeth 2009).

The second pathway deals with the potential indirect effects through the impact of symbiotic plants on the activity of above- and belowground ecosystem components that interact with the hosts and their neighboring plants (Fig. 7.3). Those effects can determine a short-

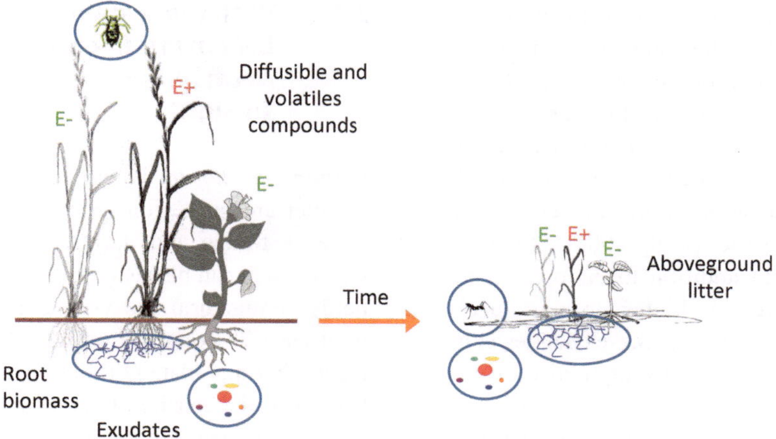

Fig. 7.3 Conceptual diagram showing how endophytic plants can modify the performance of neighboring and subsequent plants. Aphids, leaf-cutting ants, arbuscular mycorrhizal fungi, and soil biota represent organisms that may interact negatively and positively with symbiotic plants and nonsymbiotic plants of the same or different species at the neighborhood level

term shift in competitive ability favor E− plants. The grass–endophyte symbiosis has been considered primarily protective (i.e., a defensive mutualism since Clay 1988), enabling the host plant to reduce herbivory through the production of a considerable diversity of toxic alkaloids. Among them, there are several classes that deter or kill invertebrates exclusively and others that are also toxic to mammalian grazers (summarized by Schardl 2010).

Protective roles of loline and peramine alkaloids against insects are well established by choice or preference experiments (e.g., Bultman et al. 2009; Jensen et al. 2009) and genetic tests (Wilkinson et al. 2000; Tanaka et al. 2005). In general, although invertebrate herbivores show a significant preference for E− plants, the degree of response depended on the identity of grass, endophyte, and herbivore species; developmental stage of the participating organisms; and environmental conditions. Considerable evidence exists showing that endophyte negative impact on grass–herbivore interactions, may, in turn, affect invertebrate species of higher trophic levels within the grassland community (e.g., Omacini et al. 2001; de Sassi et al. 2006; Bultman et al. 2009; Hartley and Gange 2009). Furthermore, Popay et al. (2003) proposed that E+ plants may increase the vulnerability of E− to belowground herbivores, based on observation of grass grub (*Costelytra zealandica*) choice between roots of maize plants growing with E+ or E− meadow fescue (*Festuca pratensis*) plants.

Recent studies show that the protection provided by these fungal endosymbionts to the host plant can be extended to other plants in the neighborhood. For instance, Koulman et al. (2007) were able to detect alkaloids in E+ plant guttation fluid and the fluids flowing out of tissue damages. Additionally, Lehtonen and collaborators (2005) detected the transfer of endophyte-produced defensive alkaloids from a grass to a hemiparasitic plant which reduced herbivory susceptibility of those plants attached to E+ plants. In spite of the fact they have not been studied yet, these results also suggest that endophyte-mediated changes in the degree of herbivory and/or abundance of herbivores may protect E+ and E− plants from infections of virus such as the barley yellow dwarf virus transferred by aphids (Lehtonen et al. 2006).

Furthermore, there are many other endophyte-mediated alterations on host biochemistry and physiology that may have consequences on plants resistance or tolerance to biotic stresses and may modify the interaction of concurring plants with other organisms (e.g., Rasmussen et al. 2008, 2009). For instance, although such impacts are yet to be directly measured, previous studies have shown variation of odor or volatiles organic compounds by the presence of endophytic fungi (Yue et al. 2001; Steinebrunner et al. 2008). In addition to endophyte indirect effects observed at higher trophic levels via deterrence to herbivores, these changes in microenvironmental conditions might also explain the observed differential response of natural enemies of herbivores (i.e., aphid parasitoids) and their predators to endophyte presence (Omacini et al. 2001; Bultman et al. 2009).

A further way, the third pathway, in which symbiotic plants may impact on both E+ and E– plants, is through inputs of their dead tissues (litter). Litter accumulation and decomposition have long been considered as complex and important factors in controlling vegetation structure and ecosystem function (Grime 1979; Facelli and Pickett 1991a, b; Wardle et al. 1997). Endophyte presence is known to impact on the quantity and quality of litter deposited by E+ plants and its rate of decomposition (e.g., Omacini et al. 2004; Lemons et al. 2005, but see Omacini et al. 2012). Apparently, alkaloids may not be responsible for endophyte-associated reductions in decomposition rates (Siegrist et al. 2010). Previous studies have shown significant effects of the litter deposited by symbiotic plants or their aqueous extracts on the activity of soil organisms (Lemons et al. 2005; Antunes et al. 2008; Omacini et al. 2004) and on seedling establishment (Omacini et al. 2009). For instance, Antunes et al. (2008) detected that mycorrhizal colonization of E– seedlings of *Bromus inermis* was significantly reduced when emerged through litter produced by E+ tall fescue plants or when watered with water-soluble substances leached from that litter. These interactions including seedlings from one generation and dead tissues of individual from the previous generations suggest a previously unrecognized effect of the grass–endophyte symbiosis on host population dynamics and community structure.

3.2 With the Symbiosis Between *Lolium multiflorum* and *Neotyphodium occultans* in Sight

Lolium multiflorum is an annual grass from the Mediterranean region commercialized worldwide as forage and introduced accidentally as weed which can be easily found in roadsides, old fields, seminatural ecosystems, and wheat fields in at least 78 countries (accessed through GBIF Data Portal, data.gbif.org, January 2013). Currently, the density of this grass is promoted to improve winter forage production in natural grasslands of Argentina (de Battista 2005; Rodriguez and Jacobo 2010). Together with tall fescue, it is the only endophyte symbiotic species in the current flora of the flooding Pampa region, since resident native grasses are not associated with clavicipitaceous endophytes (Gundel et al. 2009; Iannone et al. 2011). Particularly, Italian ryegrass populations have shown high levels of *N. occultans* incidence (>90 %) in grazed and ungrazed humid mesophytic meadows (Gundel et al. 2009). Unlike tall fescue, no record exists of cattle intoxication. However, it is supposed that this species can produce dramatic and potentially permanent alterations in community structure, function, and composition because it is of a life-form, habit, and phenology not previously abundant in the native community (Chaneton et al. 2002).

During the last 12 years, we used different experimental approaches to study endophyte impacts on the traits of this annual grass and their relation with multiple above- and belowground ecosystem members. As it was predicted by studies with other associations, we detected an endophyte positive effect on Italian ryegrass vegetative and reproductive biomass, and seedling establishment under diverse microenvironmental conditions created by litter or water availability (Omacini et al. 2006, 2009) and under field conditions (Uchitel et al. 2011). We also observed an endophyte-mediated reduction on host colonization by fungal pathogens and by certain species of invertebrate herbivores with significant consequences on energy flow through

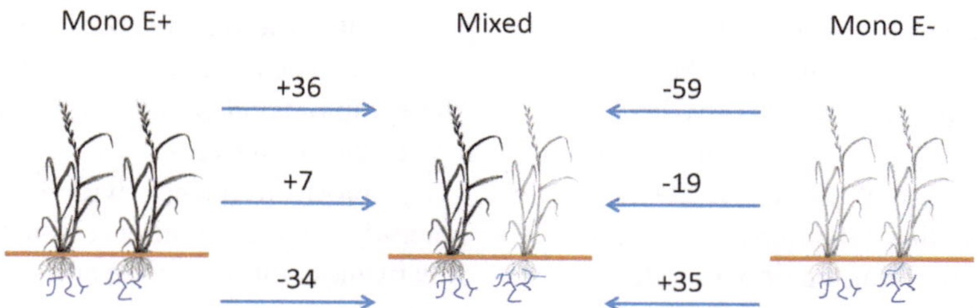

Fig. 7.4 Endophyte impact on the aboveground vegetative and reproductive biomass and colonization by arbuscular mycorrhizal fungi of nonsymbiotic *Lolium multiflorum* plants, comparing their response in monocultures of plants with the same symbiotic level and in mixtures of symbiotic and nonsymbiotic plants (E+ and E–, respectively). *Numbers* above the *arrows* indicate the magnitude (in percentage) and direction of change (increase +; decrease –) in Omacini et al. (2009) study

insect food webs (Omacini et al. 2001). The presence of this endophyte also exerted significant effects on host seed predation by rodents (Uchitel et al. 2011) and the activity and function of soil microbial communities (Omacini et al. 2006; Casas et al. 2011). Furthermore, through outdoor microcosm experiments, we detected substantial consequences on ecosystem processes such as litter decomposition (Omacini et al. 2001, 2004).

Prior experiences with this grass–endophyte symbiosis allow us to exemplify how the presence of symbiotic *L. multiflorum* plants significantly reduced growth and seed production of conspecific plants (Fig. 7.4). In Omacini et al. (2009) experiment, symbiotic and nonsymbiotic plants were growing together and separated in soils with and without arbuscular mycorrhizal fungi (AMF) inoculation in order to evaluate plant fitness and ability to be colonized by another symbiont when comparing mixed scenarios with monocultures. A reduction on E+ host colonization by AMF was detected that modify neither host growth nor competitive ability (Omacini et al. 2006). In both scenarios, E+ plants had lower levels of mycorrhizal colonization. However, in the mixtures, there was a significant reduction in mycorrhizal colonization in E+ plants and an equivalent increment in E− conspecific neighbors (Fig. 7.4). Competitive relationships between foliar endophyte and mycorrhizal fungi have been suggested by several studies with different grass–endophyte associations (Omacini et al. 2012, but see Novas et al. 2009; Larimer et al. 2012). Further on, Rasmussen et al. (2007) showed that this tripartite interaction will affect alkaloid accumulation which may have an impact on host–herbivore interaction (see Barker 1987; Vicari et al. 2002; Mack and Rudgers 2008). Future efforts quantifying changes in the rhizosphere are needed to elucidate the mechanisms that condition interaction between concurring nonsymbiotic plants and their symbionts (i.e., AM fungi or rhyzobia bacteria) and to understand how these processes contribute to both fitness of plants sharing the same space and resources and ecosystem productivity.

Recent experiments showed that the protection provided by these symbionts to Italian ryegrass can be extended to other attractive plants in the neighborhood. Garcia Parisi and collaborators (unpublished data) measured the natural infestation of aphids in white clover (*Trifolium repens*) plants surrounded by Italian ryegrass plants with contrasting endophyte symbiotic levels (i.e., E+ and E− plants) and found that the presence of E+ plants can reduce three times the proportion of both species leaves with aphids. Finally, we detected that endophyte can confer associational protection to the seedlings emerged through the litter previously produced by symbiotic plants of this annual grass (Omacini et al. 2009). Although the litter produced by E+ plants may inhibit the emergence of either E+ or E− plants, it may create an environment that decreases the seedlings damaged (Fig. 7.3). Litter produced by symbiotic plants reduced by 50 % the proportion of E+ and E− seedlings attacked by leaf-cutting ants compared to litter produced by nonsymbiotic plants (see also White et al. 2001). These results suggest that alkaloids, the purported mechanism underlying herbivore deterrence, were examined in a rudimentary fashion and did not match to expectation that only E+ plants are protected because of high alkaloids. Along with resources competition, chemical cues may play a role in the protection against natural enemies conferred by endophytes in multiple generations of symbiotic and nonsymbiotic plants.

4 The Challenge of Scaling Up

The point under discussion in this section will be using the information covered so far for predicting the changes in symbiosis frequency within host population and its consequences at community level. Many symbioses are considered multifunctional, with potential benefits to the host plant, derived from multiple mechanisms (Newsham et al. 1995; Schardl et al. 2004). Although multifunctionality is frequently cited for grass–endophyte symbioses, it is typical to quantify either a single function provided by the symbiont exclusively to the partner or endophyte effects on a certain stage of host life cycle (Gundel et al. 2009; Rudgers et al. 2012). As discussed above, endophyte impacts on host plants can simultaneously affect a large number of different concurring species as well as the physical environment around them (i.e., soil water, nutrients, pH). Such effects may have further consequences in altering the resource supply to and behavior of multiple organisms, including subsequent generations of plants. Examples of endophyte effects are documented for species that can potentially reduce or increase growth and reproduction (antagonistic or beneficial species, respectively) of symbiotic and nonsymbiotic plants of the same or different genotype (see Fig. 7.3). Thus, endophyte can modify signals the community receives from all the lower hierarchical levels which are not proportional when compared to its location and size. Thus, such impacts can occur when endophyte is, apparently, a private symbiont and, additionally, a minor component of the ecosystems.

Symbiosis frequency within a population is generally explained after estimating symbiont effectiveness through comparing the response of symbiotic and nonsymbiotic plants living separately under controlled and simplified conditions. It is appealing to use these data to account for the symbiosis persistence in the ecosystem or the difference in the percentage of symbiotic plants in diverse environments. However, this procedure may result in over simplifications leading to misunderstandings. In this respect, Gundel and collaborators (2008, 2011) highlighted endophyte-transmission efficiency as an additional mechanism neglected that could also influence infection frequencies in local populations of grasses. On the one hand, subtle differences in seed production between symbiotic and nonsymbiotic plants can be enough for the persistence of an endophyte perfectly transmitted through generations. On the other hand, where endophyte fails to colonize all host seeds, the host plant contributes to soil bank with nonsymbiotic seeds (e.g., García Parisi et al. 2012, see Fig. 7.1). Thus, an increment in host seed production can determine a reduction in symbiosis frequency. However, we should additionally consider the impact of a symbiotic plant on the difference in fitness between symbiotic and nonsymbiotic plants in order to estimate endophyte frequency more accurately. Thus, the biological importance of symbioses to neighboring and subsequent plants is central to currently understanding processes determining symbiotic plants frequency.

5 Conclusions and Perspectives

Asexual endophytes of grasses as well as other plant symbionts should be considered as important determinants of plant community structure and ecosystem functioning (Stanton 2003; Kothamasi et al. 2010; van der Heijden et al. 2008). The host grass gains benefits through novel traits or complex metabolic capabilities, which are not displayed by nonsymbiotic plants. Research on grass–endophyte symbioses has emphasized fitness benefits to host plants; however, those benefits can impact positively or negatively on other plants in close proximity through alternative and complementary pathways. In this chapter, I have demonstrated that these endophytic fungi can have considerable impact on multiple above- and belowground ecosystem components and processes. I review and discuss mechanisms traditionally and consider and propose novel mechanisms yet to be evaluated. Evaluation of potential effects should also take into account as many aspects of host life history as possible. Despite not being in contact with soil system, endophytes may also greatly impact on soil biota and soil resources: this can have further consequences for concurring organisms, litter decomposition, and subsequent generations of plants. Although the mechanisms underpinning these potential responses remain elusive, prior experience with the association *Lolium multiflorum–Neotyphodium occultans* makes it clear that both direct and indirect effects of symbiotic plants can be powerful factors in determining floristic, structural, and dynamical community properties. Effects on communities of host plant species may range from simple competitive replacement of one or more species, to loss or reduction of whole guilds, to major conversion of communities structure and organization.

Considering the multiple potential effects of grass–endophyte symbioses beyond the host level and its lifetime, a practical dimension emerges. The outcome of grass–endophyte interaction is unpredictable with our present perception, which makes it inaccurate to estimate symbiosis effectiveness according to an inventory of effects only on the host plant. To better comprehend how endophyte influences host dynamics, symbiosis persistence, and ecosystem functioning, several key questions still need to be answered. First, understanding how the presence of the grass–endophyte symbiosis modifies the performance of concurring nonhost plants is a major challenge for the future. Answering this question, however, is complicated and will require the development of experimental systems, allowing the manipulation of symbiotic plants density and their products (e.g., volatile organic compounds, exudates) without influencing other factors and contamination from the outside. Up to the present, the process that has been mainly considered is endophyte impact on competitive balance between host and nonhost plants and their ultimate effect at community level. Second, recent studies show that symbiosis may induce changes in litter quality and quantity which may have some impact on seedlings establishment and their interaction with, for example, arbuscular mycorrhizal fungi. It is important to understand whether and how such changes in microenvironmental conditions might feedback on plant composition and productivity. Further on, experimental studies are necessary to contrast the hypothesis that endophyte chemical weapons can function as allelopathic agents and also as mediators of multitrophic interactions and can be a major force driving the reduction of plant diversity in productive communities. Third, the resistance or susceptibility to other interacting organisms that a nonsymbiotic plant can receive by sharing a microsite with an endophytic plant provide much-needed insight into how genetic, biotic, and abiotic interactions affect the outcome of grass–endophyte symbioses and how these interactions, in turn, can influence management strategies within an agronomic context. Grass–endophyte symbiosis or its residues may play an important but hitherto unknown role in the associational protection against herbivores or pathogens in nonhost forage species or crops.

In conclusion, fungal endophytes are clearly major and key components of many ecosystems as a natural and invisible endosymbionts of grasses of widespread interest to ecological and agricultural research. They should not be ignored in community study or theory given the considerable extent of their impacts, even when they are minor components of ecosystems. Up the present, few predictions about extended endophyte benefits to nonsymbiotic plants can be strongly supported given available studies and data. The paucity of generalization regarding community effects still presents a substantial dilemma for community managers but helps to identify research priorities and innovations for more sustainable practices than the current agrochemicals inputs (Tikhonovich and Provorov 2009; Andrews et al. 2011; Thrall et al. 2011).

Acknowledgments I am grateful to the lab group; most ideas presented here were developed through our helpful discussions on the issues dealt with this essay. I wish to thank Claudio M. Ghersa, Luis I. Perez, Pablo Garcia Parisi, and Beatriz Santos for thought-provoking conversations and helpful comments on earlier versions, and to Pablo Roset for many details that improved the manuscript. Preparation of this chapter was facilitated by grants from the University of Buenos Aires (UBA), the National Research Council (CONICET), and the National Scientific and Technological Promotion (FONCYT). Mirta Rabadán provided photos.

References

Afkhami M, Rudgers J (2008) Symbiosis lost: imperfect vertical transmission of fungal endophytes in grasses. Am Nat 175:405–416

Andrews MA, Edwards GRA, Ridgway HJA, Cameron KCA, Di HJA, Raven JA (2011) Positive plant microbial interactions in perennial ryegrass dairy pasture systems. Ann Appl Biol 159:79–92

Antunes P, Miller J, Carvalho L, Klironomos JN, Newman J (2008) Even after death the endophytic fungus of *Schedonorus phoenix* reduces the arbuscular mycorrhizas of other plants. Funct Ecol 22:912–918

Bacon CW, Hill NS (1997) *Neotyphodium*/grass interactions. Plenum, New York

Barker GM (1987) Mycorrhizal infection influences *Acremonium*-induced resistance to Argentine stem weevil. Proc N Z Weed Pest Control Conf 40:199–203

Brosi GB, McCulley RL, Bush LP, Nelson JA, Classen AT, Norby RJ (2011) Effects of multiple climate change factors on the tall fescue-fungal endophyte symbiosis: infection frequency and tissue chemistry. New Phytol 189:797–805

Bultman TLAB, Rodstrom JLB, Radabaugh KRB, VanDop JDB, Librizzi JMB, Longwell LLBC, Pulas CBD, Grant LBE, Sullivan TJB (2009) Influence of genetic variation in the fungal endophyte of a grass on an herbivore and its parasitoid. Entomol Exp Appl 130:173–180

Cabral D, Stone J, Carroll G (1993) The internal mycobiota of *Juncus* spp: microscopic and cultural observations of infections patterns. Mycol Res 97:367–376

Carroll G (1988) Fungal endophytes in stems and leaves: from latent pathogen to mutualistic symbiont. Ecology 69:10–16

Casas MC, Omacini M, Montecchia MS, Correa OS (2011) Soil microbial community responses to the fungal endophyte Neotyphodium in Italian ryegrass. Plant Soil 340:347–355

Chaneton E, Perelman S, Omacini M, Leon RJC (2002) Grazing, environmental heterogeneity and alien plant invasions in temperate Pampa grasslands. Biol Invasions 4:7–24

Cheplick G, Faeth S (2009) Ecology and evolution of the grass-endophyte symbiosis. Oxford University Press, New York

Christensen MJ, Zhang X, Scott B (2008) Regulation switching of *Epichloe typhina* within elongating perennial ryegrass leaves. Mycol Res 112:1056–1062

Clay K (1988) Fungal endophytes of grasses: a defensive mutualism between plants and fungi. Ecology 69:10–16

Clay K (2009) Defensive mutualism and grass endophytes: still valid after all these years. In: White JF, Torres MR (eds) Defensive mutualism in microbial symbiosis. CRC Press, Boca Raton, pp 9–20

Clay K, Holah J (1999) Fungal endophyte symbiosis and plant diversity in successional fields. Science 285: 1742–1744

Clay K, Schardl CL (2002) Evolutionary origins and ecological consequences of endophyte symbiosis with grasses. Am Midl Nat 160:99–127

Clay K, Holah J, Rudgers JA (2005) Herbivores cause a rapid increases in hereditary symbiosis and alter plant community composition. Proc Natl Acad Sci U S A 102:12465–12470

de Bary A (1879) Die Erscheinung der Symbiose. Verlag von Karl J. Trubner, Strassburg

de Battista J (2005) *Neotyphodium* research and application: current trends in South America. In: Roberts C, West CP, Spiers D (eds) *Neotyphodium* in cool-season grasses. Blackwell publishing, Iowa, pp 65–71

de Sassi C, Muller C, Krauss J (2006) Fungal plant endosymbionnts alter life history and reproductive success of aphid predators. Proc R Soc B 273:1301–1306

Douglas A (2010) The symbiotic habit. Princeton University Press, Princeton

Facelli JM, Pickett STA (1991a) Plant litter: light interception and effects on an old field plant community. Ecology 72:1024–1031

Facelli JM, Pickett STA (1991b) Plant litter: its dynamics and effects on plant community structure. Bot Rev 57:1–31

Finkes L, Cady A, Mulroy J, Clay K, Rudgers J (2006) Plant-fungus mutualism affects spider composition in successional fields. Ecol Lett 9:347–356

García Parisi PA, Casas C, Gundel PE, Omacini M (2012) Consequences of grazing on the vertical transmission of a fungal *Neotyphodium* symbiont in an annual grass population. Aust Ecol 37:620–628

Glenn A, Bacon C, Price R, Hanlin R (1996) Molecular phylogeny of *Acremonium* and its taxonomic implications. Mycologia 88:369–383

Grime JP (1979) Plant strategies and vegetation processes. Wiley, Chichester

Gundel P, Batista WH, Texeira M, Martinez-Ghersa MA, Omacini M, Ghersa CM (2008) *Neotyphodium* endophyte infection frequency in annual grass populations: relative importance of mutualism and transmission efficiency. Proc R Soc Lond B Biol Sci 275:897–905

Gundel P, Garibald LA, Tognetti PM, Aragón R, Ghersa CM, Omacini M (2009) Imperfect vertical transmission of the endophyte *Neotyphodium* in exotic grasses in grasslands of the flooding Pampa. Microb Ecol 57:740–748

Gundel PE, Rudgers JA, Ghersa CM (2011) Incorporating the process of vertical transmission into understanding of host-symbiont dynamics. Oikos 120:1121–1128

Hartley S, Gange AC (2009) Impacts of plant symbiotic fungi on insect herbivores: mutualism in a multitrophic context. Annu Rev Entomol 54:323–342

Hoveland C, Bouton J, Durham R (1999) Fungal endophyte effects on production of legumes in association with tall fescue. Agron J 91:897–902

Iannone L, White J Jr, Giussani L, Cabral D, Novas MV (2011) Diversity and distribution of *Neotyphodium-infected* grasses in Argentina. Mycol Prog 10:9–19

Jensen JGA, Popay AJA, Tapper BAB (2009) Argentine stem weevil adults are affected by meadow fescue endophyte and its loline alkaloids. N Z Plant Prot 62:12–18

Johnson N, Graham J, Smith F (1997) Functioning of mycorrhizal associations along the mutualism-parasitism continuum. New Phytol 135:575–585

Kothamasi D, Kiers ET, van der Heijden MGA (2010) Mutualisms and community organization. In: Verhoef HA, Morin PJ (eds) Community ecology, processes, models, and applications. Oxford University Press, Oxford, pp 179–192

Koulman A, Lane GA, Christensen MJ, Fraser K, Tapper BA (2007) Peramine and other fungal alkaloids are exuded in the guttation fluid of endophyte-infected grasses. Phytochemistry 68:355–360

Kuldau G, Bacon C (2008) Clavicipitaceous endophytes: their ability to enhance resistance of grasses to multiple stresses. Biol Control 46:57–71

Larimer AL, Bever JD, Clay K (2012) Consequences of simultaneous interactions of fungal endophytes and arbuscular mycorrhizal fungi with a shared host grass. Oikos 121:2090–2096

Lehtonen P, Helander M, Siddiqui SA, Lehton K, Saikkonen K (2005) Are endophyte-mediated effects on herbivores conditional on soil nutrients? Oecologia 142:38–45

Lehtonen P, Helander M, Siddiqui S, Lehton K, Saikkonen K (2006) Endophytic fungus decreases plant virus infections in meadow ryegrass (*Lolium pratense*). Biol Lett 2:620–623

Lemons A, Clay K, Rudgers JA (2005) Connecting plant-microbial interactions above and belowground: a fungal endophyte affects decomposition. Oecologia 145:595–604

Leuchtmann A (2006) Systematics, distribution and host specificity of grass endophytes. Nat Toxins 1:150–162

Mack K, Rudgers J (2008) Balancing multiple mutualists: asymmetric interactions among plants, arbuscular mycorrhizal fungi and fungal endophytes. Oikos 117:310–320

Malinowski DP, Belesky DP (2000) Adaptations of endophyte-infected cool-season grasses to environmental stresses: mechanisms of drought and mineral stress tolerance. Crop Sci 40:923–940

Márquez SS, Bills GF, Zabalgogeazcoa I (2007) The endophytic mycobiota of the grass *Dactylis glomerata*. Fungal Divers 27:171–195

Mersch SM, Cahoon AB (2012) Biomass and tiller growth responses to competition between Ky31 and MaxQ *Festuca arundinacea* cultivars and response of Ky31 to exogenously applied liquid preparation of *Neotyphodium coenophialum* under glasshouse conditions. Grass Forage Sci 67:299–304

Moon CD, Scott B, Schardl CL, Christensen MJ (2000) Evolutionary origins of *Epichloë* endophytes from annual ryegrasses. Mycologia 92:1103–1118

Newsham KK, Fitter A, Watkinson A (1995) Multi-functionality and biodiversity in arbuscular mycorrhizas. Trends Ecol Evol 10:407–411

Novas MV, Iannone L, Godeas A, Cabral D (2009) Positive association between mycorrhiza and foliar endophytes in *Poa bonariensis*, a native grass. Mycol Prog 8:75–81

Omacini M, Chaneton EJ, Ghersa CM, Muller CB (2001) Symbiotic fungal endophytes control insect host-parasite interaction webs. Nature 409:78–81

Omacini M, Chaneton E, Ghersa CM, Otero P (2004) Do foliar endophytes affect grass litter decomposition? A microcosm approach using *Lolium multiflorum*. Oikos 104:581–590

Omacini M, Chaneton EJ, Ghersa CM (2005) A framework for understanding the ecosystem consequences of endophyte grass symbioses. In: Roberts C, West CP, Spiers D (eds) Neotyphodium in cool-season grasses current research & applications. Ames, Blackwell publishing, Iowa, pp 141–161

Omacini M, Eggers T, Bonkowsky M, Gange A, Jones TH (2006) Leaf endophytes affect mycorrhizal status and growth of co-infected and neighbouring plants. Funct Ecol 20:226–232

Omacini M, Chaneton EJ, Bush L, Ghersa CM (2009) A fungal endosymbiont affects host ant recruitment through seed and litter mediated mechanisms. Funct Ecol 23:1148–1156

Omacini M, Semmartin M, Pérez LI, Gundel PE (2012) Grass-endophyte symbiosis: a neglected aboveground interaction with multiple belowground consequences. Iowa, Appl Soil Ecol Iowa 61:273–279

Palmer TM, Foak DF, Stanton ML, Bronstein JL, Kiers ET, Young TP, Goheen JR, Pringle RM (2010) Synergy of multiple partners, including freeloaders, increases host fitness in a multispecies mutualism. Proc Natl Acad Sci 107:17234–17239

Petrini O (1991) Fungal endophytes of tree leaves. In: Andrews JH, Hirano SS (eds) Microbial ecology of leaves. Springer, New York, pp 179–197

Popay AI, Townsend RJ, Fletcher L (2003) The effect of endophyte (*Neotyphodium uncinatum*) in meadow fescue on grass grub larvae. N Z Plant Prot 56: 123–128

Rasmussen S, Parsons AJ, Bassett S, Christensen MJ, Hume DE, Johnson LJ, Johnson RD, Simpson WR, Stacke C, Voisey CR, Xue H, Newman JA (2007) High nitrogen supply and carbohydrate content reduce fungal endophyte and alkaloid concentration in *Lolium perenne*. New Phytol 173:787–797

Rasmussen S, Parsons A, Poppy A, Xue H, Newman J (2008) Plant-endophyte-herbivore interactions: more than just alkaloids? Plant Signal Behav 3:974–977

Rasmussen S, Parsons AJ, Newman JA (2009) Metabolomics analysis of the *Lolium perenne-Neotyphodium lolii* symbiosis: more than just alkaloids? Phytochem Rev 8:535–550

Ravel C, Michalakis Y, Charmet G (1997) The effect of imperfect transmission on the frequency of mutualistic seed-borne endophytes in natural populations of grasses. Oikos 80:18–24

Rodriguez A, Jacobo E (2010) Glyphosate application change plant functional groups proportion and reduces floristic richness and diversity in flooding Pampa rangeland (Argentina). Agric Ecosyst Environ 138:222–231

Rodriguez R, Redman R (2008) More than 400 million years of evolution and some plants still can't make it on their own: plant stress tolerance via fungal symbiosis. J Exp Bot 59:1109–1114

Rodriguez R, White J, Arnold A, Redman R (2009) Fungal endophytes: diversity and functional roles. New Phytol 182:314–330

Rudgers J, Clay K (2007) Endophyte symbiosis with tall fescue: how strong are the impacts on communities and ecosystems? Fungal Biol Rev 21:107–124

Rudgers J, Orr S (2009) Non-native grass alters growth of native tree species via leaf and soil microbes. J Ecol 97:247–255

Rudgers J, Fischer S, Clay K (2010) Managing plant symbioses: fungal endophyte genotype alters plant community composition. J Appl Ecol 47:468–477

Rudgers J, Miller T, Ziegler S, Craven K (2012) There are many ways to be a mutualist: endophytic fungus reduces plant survival but increases population growth. Ecology 93:565–574

Sachs JL, Mueller UC, Wilcox TP, Bull JJ (2004) The evolution of cooperation. Q Rev Biol 79:135–160

Saikkonen K, Faeth S, Helander M, Sullivan T (1998) Fungal endophytes: a continuum of interactions with host plants. Ann Rev Ecol Syst 29:319–343

Saikkonen K, Lehtonen P, Helander M, Koricheva J, Faeth S (2006) Model systems in ecology: dissecting the endophyte-grass literature. Trends Plant Sci 11:428–433

Saikkonen K, Saari S, Helander M (2010) Defensive mutualism between plants and endophytic fungi? Fungal Divers 41:101–113

Schardl C (2010) The *Epichloae*, Symbionts of the grass subfamily Pooideae. Ann Mo Bot Gard 97:646–665

Schardl CL, Leuchtmann A, Spiering M (2004) Symbioses of grasses with seedborne fungal endophytes. Annu Rev Plant Biol 55:315–340

Siegrist J, McCulley R, Bush L, Phillips T (2010) Alkaloids may not be responsible for endophyte-associated reductions in tall fescue decomposition rates. Funct Ecol 24:460–468

Stanton M (2003) Interacting guilds: moving beyond the pairwise perspective on mutualisms. Am Nat 162:10–23

Steinebrunner F, Schiestl FP, Leuchtmann A (2008) Ecological role of volatiles produced by *Epichloë*: differences in antifungal toxicity. FEMS Microbiol Ecol 64:307–316

Sutherland BL, Hume DE, Tapper BA (1999) Allelopathic effects of endophyte-infected perennial ryegrass extracts on white clover seedlings. N Z J Agric Res 42:19–26

Tadych M, Bergen M, Dugan FM, White JF (2007) Evaluation of the potential role of water in spread of conidia of the Neotyphodium endophyte of Poa ampla. Mycol Res 11:466–472

Tanaka A, Tapper B, Popay A, Parker EJ, Scott B (2005) A symbiosis expressed non-ribosomal peptide synthetase from a mutualistic fungal endophyte of perennial ryegrass confers protection to the symbiotum from insect herbivory. Mol Microbiol 57:1036–1050

Thrall P, Hochberg ME, Burdon J, Bever JD (2007) Co-evolution of symbiotic mutualists and parasites in a community context. Trends Ecol Evol 22:120–126

Thrall PH, Oakeshott JG, Fitt G, Southerton S, Burdon JJ, Sheppard A, Russell RJ, Zalucki M, Heino M, Denison RF (2011) Evolution in agriculture – the application of evolutionary approaches to the management of biotic interactions in agro-ecosystems. Evol Appl 4:200–215

Tikhonovich I, Provorov N (2009) From plant-microbe interactions to symbiogenetics: a universal paradigm for the interspecies genetic integration. Ann Appl Biol 154:341–350

Uchitel A, Omacini M, Chaneton E (2011) Inherited fungal symbionts enhance establishment of an invasive annual grass across successional habitats. Oecologia 165:465–475

van Der Heijden MGA, Streitwolf-Engel R, Riedl R, Siegrist S, Neudecker A, Ineichen K, Boller T, Wiemken A, Sanders AR (2006) The mycorrhizal contribution to plant productivity, plant nutrition and soil structure in experimental grassland. New Phytol 172:739–752

van Der Heijden MGA, Bardgett RD, van Straalen NM (2008) The unseen majority: soil microbes as drivers of plant diversity and productivity in terrestrial ecosystems. Ecol Lett 11:296–310

van der Putten WH, Vet LEM, Harvey JH, Wackers FL (2001) Linking above and below ground multitrophic interactions of plants, herbivores, pathogens, and their antagonists. Trends Ecol Evol 16:547–554

Vázquez-de-Aldana BR, Romo M, García-Ciudad A, Petisco C, García-Criado B (2011) Infection with the fungal endophyte *Epichloë festucae* may alter the allelopathic potential of red fescue. Ann Appl Biol 159:281–290

Vicari M, Hatcher P, Ayres A (2002) Combined effect of foliar and mycorrhizal endophytes on an insect herbivore. Ecology 83:2452–2464

Vogl RJ (1980) The ecological factors that produce perturbation dependent ecosystems. In: Cairns J Jr (ed) The recovery process in damaged ecosystems. Ann Arbor Science, Arbor Ann, pp 63–94

Wardle DA, Bonner KI, Nicholson KS (1997) Biodiversity and plant litter: experimental evidence which does not support the view that enhanced species richness improves ecosystem function. Oikos 79:247–258

White JF (1987) Wide spread distribution of endophytes in the Poaceae. Plant Dis 71:340–342

White JF Jr, Torres M (2009) Defensive mutualism in microbial symbiosis. CRC Press, Boca Raton

White BS, Sullivan R, Balady G, Gianfagna T, Yue Q, Meyer W, Cabral D (2001) A fungal endosymbiont of the grass *Bromus setifolius*: distribution in some Andean populations, identification and examination of beneficial properties. Symbiosis 31:241–257

Wiens JA (1989) Spatial scaling in ecology. Funct Ecol 3:385–397

Wilkinson HH, Siegel DI, Blankenship JP, Mallory AC, Bush JP, Schardl CL (2000) Contribution of fungal loline alkaloids to protection form aphids in a grass-endophyte mutualism. Mol Plant Microbe Interact 13:1027–1033

Wilson D (1995) Endophyte – the evolution of a term, and clarification of its use and definition. Oikos 73:274–276

Yue Q, Wang C, Gianfagna T, Meyer W (2001) Volatile compounds of endophyte-free and infected tall fescue (*Festuca arundinaceae* Schreb.). Phytochemistry 58:935–941

Part IV

Bioactive Compounds from Endophytes

8 Microbial Endophytes: Their Resilience for Innovative Treatment Solution to Neglected Tropical Diseases

Ajay Kumar Verma, Saurabh Kumar Patel, Chandra Bhan Pratap, Mayank Gangwar, and Gopal Nath

Abstract

Fungal endophytes are relatively overlooked as a platform for discovering bioactive molecules against some major neglected tropical diseases, until some recent reports. Looking their potential as prolific producer of bioactive compounds against array of diseases and ailments makes them a suitable platform for such explorations. A major part of third world countries are facing growing problems of neglected tropical diseases (NTDs). More than two billion people of tropical and subtropical countries are facing serious health problems caused by lymphatic filariasis, onchocerciasis, echinococcosis, and other helminthic and zoonotic infections. Increasing side effects and appearance of resistance to the synthetic anthelmintics stimulates researchers for exploration of novel natural alternatives from medicinal plants and their associated endophytic microbes as a useful alternative. In this chapter, some aspects with respect to novel chemistry of endophytes and their structure activity relationship (SAR) toward tropical diseases like antiparasitic, antimalarial, and other neglected tropical diseases have been discussed.

1 Introduction

A wide diversity of endophytic fungi is isolated from the internal healthy tissues of almost every terrestrial and aquatic plants studied so far and even also recovered from red and brown algae (Raghukumar et al. 1992). Endophytic fungi are present in almost every plant parts, such as leaves, root, stem, and rhizome. Mostly the asymptomatic tissues are considered to have endophytes either fungal, bacterial, or actinobacteria, and that is why they are often called as latent pathogens. It is however very unclear about this specific lifestyle

A.K. Verma • S.K. Patel • C.B. Pratap • G. Nath (✉)
Department of Microbiology, Institute of Medical Sciences, Banaras Hindu University, Varanasi 221 005, UP, India
e-mail: gopalnath@gmail.com

M. Gangwar
Department of Microbiology, Institute of Medical Sciences, Banaras Hindu University, Varanasi 221 005, UP, India

Department of Pharmacology, Institute of Medical Sciences, Banaras Hindu University, Varanasi 221 005, UP, India

V.C. Verma and A.C. Gange (eds.), *Advances in Endophytic Research*,
DOI 10.1007/978-81-322-1575-2_8, © Springer India 2014

of microbial endophytes, as it was evidenced that under certain conditions these microbes can switch from endophyte to pathogen or vice versa. There are two microbiomes residing just few millimeter distance the phylloplane and endophytic microbes have significant differences in their microbial community, diversity and spatial distribution, some time it is very hard to have a clear separation between these two microbiome (Porras-Alfaro and Bayman 2011).

Endophytes colonize plant tissue and remain within the tissue, except that fruiting bodies that may emerge through the surface of the plant tissues. Leaves may be colonized by a variety of fungi just within a few weeks of its emergence. The colonies remain asymptomatic, and in some perennial plant, it may have a very long life. Fungal endophytes represent nearly every taxonomic representative from all divisions of fungi, that is, ascomycetes, hyphomycetes, and relatively less reported basidiomycetes; however, the dominant microfungi were from hyphomycetes and sac forming ascomycetes. Many of the listed endophytes so far are representative genera from common soil fungi that cause disease in plants and animals; this phylogenetic evidence is used to suggest that endophytes have evolved from pathogens or vice versa. The mechanisms of host recognition and development of colonization may also be common among closely related endophytic and pathogenic fungi (Redline and Carris 1985). Endophytic fungi can be biotrophic mutualists, benign commensals, decomposers, or latent pathogens (Promputtha et al. 2007). All plants in the natural environment can shelter endophytic fungi, including algae, mosses, ferns, conifers, and angiosperms. This fungal group appears to significantly influence the lifestyle of its host (Rodriguez et al. 2009).

Taxonomically, most of the endophytic fungi belong to the phylum Ascomycota and its associated anamorphs, while some species belong to the phyla Basidiomycota and Zygomycota (Huang et al. 2001). Endophytic fungi are an important source of bioactive natural molecules. These bioactive metabolites have an array of biological activities and could be the starting materials or lead structures for the development of pharmaceutical or agrochemical products (Baker et al. 2000). There have been many studies on the diversity, ecology, and biotechnological applications of endophytic fungi in grasses and wood plants in temperate environments. However, there is limited information about the diversity of endophytic fungal communities in tropical forests, which are endowed with a rich biodiversity of flora. Dreyfuss and Chapela (1994) have estimated that approximately 1.3 million species of endophytic fungi remain to be discovered. The substances produced by endophytic fungi originate from different biosynthetic pathways, including isoprenoid, polyketide, and amino acid, and belong to diverse structural groups, such as terpenoids, steroids, xanthones, quinones, phenols, isocoumarins, benzopyranones, tetralones, cytochalasins, and enniatins (Schulz et al. 2002; Verma et al. 2009). Indeed, these bioactive molecules represent a chemical reservoir for discovering new compounds, such as antibiotic, antioxidant, immunomodulating, anticancer, and antiparasitic compounds, for use in the pharmaceutical and agrochemical industries.

2 Endophytic Fungi in Treatment of Tropical Diseases

Tropical diseases (TDs) cause over 500,000 deaths annually and are estimated to result in a greater number of lost disability-adjusted life span than malaria and tuberculosis (Hotez et al. 2006, 2009). The unconcerned or neglected tropical diseases (NTDs) are a group of chronic, debilitating, and poverty-promoting parasitic, bacterial, viral, and fungal infections, sporadic in the poorest people living in third world countries (Hotez and Yamey 2009). The major causes of these diseases are lack of sanitation, unhygienic water supply, malnutrition, and above all illiteracy and low economic status of the prevalence area. In fact, the link with poverty is so strong that the prevalence of these diseases serves as an indicator of the level of a country's socioeconomic development (WHO 2006). Tropical diseases occur in impoverished settings and are chronic conditions; victims can harbor chronic

TDs for years or decades, frequently resulting in disability, disfigurement, and stigmatization (Hotez and Yamey 2009).

Drug resistance of pathogens causing fatal diseases has increased in recent years, which is a prime aspect to be addressed by researchers. Evidently, scientists have provided the public health cause with many effective drugs and vaccines, but the battle against these notorious microbes is still far. Diseases caused by microbes (bacteria, viruses, fungi, protozoans, and prokaryotes) such as respiratory infections, HIV/AIDS, tuberculosis, and malaria and diseases such as cancer account for many infections leading to death. These microbes obtained resistance to many of the first-line drugs used for the treatment. Resistance to these first-line drugs has forced to change the treatment to more expensive second- or third-line drug alternatives. If resistances to these drugs also emerge, we have no more options for treatment. An intensive search for newer and more effective lead agents to deal with these problems is now seriously underway. One such renewable source apart from the medicinal plants is the endophytic microbes. Some of the interesting compounds produced by endophytic microbes are taxol, cryptocin, cryptocandin, jesterone, oocydin, isopestacin, pseudomycins and ambuic acid, and many more. The reason endophytes mimic the chemistry of their respective hosts and make the same bioactive natural products or derivatives has been attributed to the possible intergeneric genetic exchange between higher plants and the endophytic microbe. Usually the host-endophyte associations are symptomless, as the latter do not interfere with the host biological or physiological dealings. In some cases, especially in grasses, it is observed that secondary metabolites produced by the inhabiting endophytes show beneficial effects on the growth of the host. Nowadays endophytic metabolites are looking as new source for bioactive molecule against major parasitic diseases and NTDs (Hotez et al. 2006). Interruption and default of therapies against TDs are still important obstacles to disease control in many endemic countries, with consequences for both patients and control programs; low adherence results in potential remaining sources of infection, incomplete curing, and irreversible complications and may lead to multidrug resistance (Heukelbach et al. 2011).

The most important viral unconcerned TDs are dengue and yellow fevers (Hotez et al. 2008). Tropical climates have experienced a great resurgence in dengue fever in recent years, and it appears to be spreading to new areas (Carroll et al. 2007). The WHO reports that two-fifths of the world's population is at risk of dengue infection, with an increase in the annual number of cases (Murrell et al. 2011). There is no specific treatment available for dengue fever so far. Dengue is an increasing concern because of the lack of a vaccine that protects against all dengue serotypes (WHO 2006). The increase in dengue infections and the prevalence of all four circulating dengue serotypes has contributed to a rise in the incidence of dengue hemorrhagic fever (Murrell et al. 2011). Paracoccidioidomycosis (PCM), a kind of mycoses, is also responsible for major public health and economic burden in Latin America (Hotez et al. 2008). The available drugs most commonly used for treatment of PCM are sulfonamides, ketoconazole, itraconazole, and amphotericin B. A long extended period of treatment are required; apart from increasing concerns about drug toxicity, the cost of treatment and unacceptable rates of noncompliance with these therapies further complicate the situation (Travassos et al. 2008).

Some helminth parasites are also among most common agents of human infection in developing countries, like schistosomiasis, cysticercosis, hydatidosis, and onchocerciasis. There are two major phyla of helminths, which include the major intestinal worms, filarial worms (*Wuchereria bancrofti*) that cause lymphatic filariasis and onchocerciasis and platyhelminthes such as the schistosomes and the agent of cysticercosis (Hotez et al. 2008). The drugs albendazole, oxamniquine, praziquantel, and ivermectin are the only available drugs to treat helminthiasis so far (Hotez et al. 2008). Increasing development in molecular techniques has led to the identification of new targets for the discovery and development of anthelmintic drugs. Yellow fever originated in Africa and was imported to Europe and the Americas as a consequence of the slave trade between these continents (Gardner and Ryman 2010). Interest in developing new inactivated vaccines has been spurred by the recognition of rare but serious and sometimes fatal adverse events following live-virus vaccination (Hayes 2010).

Leishmania (*Trypanosomatidae*) are protozoan parasites that cause high morbidity and mortality levels and are recognized by the WHO as a major tropical public health problem (Asford 1997). Currently no vaccines for leishmaniasis are present, and the drugs available for leishmaniasis treatment are toxic, expensive, and sometimes ineffective (Croft and Coombs 2003). Chagas disease (American *Trypanosomiasis*) is caused by the hemoflagellate protozoan *Trypanosoma cruzi* and transmitted to humans either by blood-sucking triatomine vectors, blood transfusion, or congenital transmission. The geographical distribution of human *T. cruzi* infection extends from the southern United States and Mexico to southern Argentina (WHO 1991). There is evidence that trypanocidal drug treatment with nitrofuran and imidazole compounds can treat acute *T. cruzi* infection, but further studies are needed to develop new trypanocidal drugs (Reyes and Vallejo 2005).

3 Tropical Endophytic Fungal Diversity

There are 1.3 million species of endophytic fungi alone, the majority of which are likely found in tropical ecosystems. This estimate is supported by various studies that have sought to characterize the fungal communities associated with tropical plants. Fungal endophytic communities are divided into two basic groups: generalists that are found in high abundance and singletons that are found in low abundance. Tropical plants are expected to shelter a highly diverse population of endophytic fungi, but few tropical plants have been screened for their presence. Studies have shown that tropical plants shelter a great diversity of singleton species (Dreyfuss and Chapela 1994). The greatest fungal diversity probably occurs in tropical forests, where highly diverse populations of angiosperms are present (Arnold et al. 2000). The magnitude of fungal diversity in tropical forests is still unclear, and new species remain to be described (Hawksworth 2004). In support of this proposal, a large number of fungal endophytic species have been described in association with plants in Asia, Australia, Africa, Central and South America, Mexico, and some Pacific and Atlantic Islands. However, the diversity of endophytic fungi can vary across different biomes of a tropical forest. Suryanarayanan et al. (2002) showed that the endophytic fungal assemblage of a dry tropical forest had much less endophyte diversity than a wet tropical forest. Arnold et al. (2000) suggested that endophytic fungi are hyperdiverse and about 1.5 million species may be an underestimate of their magnitude. In addition, the taxonomic placement of tropical fungi has been confounded by misidentifications made in comparison with temperate fungal communities, including the endophytic fungal community present in the leaves of tropical plants (Arnold et al. 2001). Endophytic fungi can be passive residents or act as an assemblage of latent pathogens in their host (Ganley et al. 2004).

Endophytic fungi have been categorized into two main groups based on differences in evolution, taxonomy, plant hosts, and ecological functions: clavicipitaceous, which are able to infect only some species of grasses, and nonclavicipitaceous, which are found in the asymptomatic tissues of bryophytes, ferns, gymnosperms, and angiosperms (Rodriguez et al. 2001). Clavicipitaceous endophytes belong to the family *Clavicipitaceae* (*Hypocreales*; *Ascomycota*), many species of which are known to produce bioactive molecules (mainly of the genera *Cordyceps*, *Balansia*, *Epichloë*/*Neotyphodium*, *Claviceps*, and *Myriogenospora*). In contrast, nonclavicipitaceous endophytes are a large group that have not been well defined taxonomically, but the majority of the species belong to the phyla *Ascomycota* and *Basidiomycota*, represented by the genera *Alternaria*, *Arthrobotrys*, *Aspergillus*, *Cladosporium*, *Colletotrichum*, *Coprinellus*, *Curvularia*, *Fusarium*, *Paecilomyces*, *Penicillium*, *Phanerochaete*, *Phoma*, among others. Species of these two endophytic groups have been investigated for their ability to produce various molecules, and species living in association with tropical plants have been shown to be effective producers of bioactive metabolites.

3.1 Isolation and Identification Method of Endophytic Fungi

The methods used to isolate endophytic fungi vary in the technique used for surface disinfection of the host plant tissue (leaves, stems, roots, bark, flowers, fruits, and seeds) and the choice of culture media. The disinfection process can influence the detection of endophytic fungi; in general, the plant surface is disinfected with a strong oxidant or surfactant agent for a specific period of time. The most commonly used agents include 1–4 % detergent, 3 % H_2O_2, 2–10 % NaOCl, or 70–95 % ethanol. The culture medium is another important limiting factor. Commonly used media include potato dextrose agar (PDA), malt extract agar (MEA), yeast malt agar (YMA), and Sabouraud agar (SA), supplemented with antibacterial agents (chloramphenicol, penicillin, ampicillin, tetracycline, streptomycin, among others) to suppress contaminating bacteria. After isolation, the endophytic fungi, including the bioactive species, must be identified correctly. Macro- and micromorphological cultural characteristics, molecular analysis, and metabolite profiles are the main criteria that are used to identify endophyte fungal taxonomy. The identification of endophytic fungi relies significantly on the taxonomic expertise of the mycologist and frequently requires polyphasic taxonomy. In tropical regions, multiple endophytic fungal species are recovered and are commonly grouped based on similar culture characteristics into morphospecies, which represent a functional taxonomic unit for endophytic fungal species (Arnold et al. 2000). After characterization as a morphospecies, endophytic fungi are submitted to molecular grouping using microsatellite markers that are detected with $(GTG)_5$, M13, or EI primers based on PCR-fingerprinting methods that amplify genomic segments different from the repeat region itself (Lieckfeldt and Seifert 2000). Most endophytic fungi (about 50 %) do not produce conidia or spores when cultured on common mycological media. In these cases, endophytic fungi can frequently be identified based on the sequence of the internal transcribed spacer (ITS) region of the large subunit of the rRNA gene. Molecular techniques are a powerful tool for identifying the endophytic genera and species of non-sporulating fungi. After sequencing the ITS1–5.8S–ITS2 region, the sequence of the endophytic fungus is compared with the sequences of other taxa deposited in public databases. The GenBank database is a major source of nucleotide sequences.

Endophytic fungi produce a large number of metabolites, and certain molecules are very consistently found in species of a few genera when cultured under standard conditions. According to Larsen et al. (2005), fungal isolates of different species have different chemotypes, which can be differentiated or grouped by modern methods for dereplication analysis. The chemical analysis includes techniques such as thin layer chromatography (TLC), gas chromatography (GC), high performance liquid chromatography (HPLC), mass spectrometry (MS), and nuclear magnetic resonance (NMR), alone or in combination with bioinformatics tools.

3.2 Different Fermentation Technique and Crude Extract Production

Filamentous fungi have long been known for their versatility to produce an array of secondary metabolites, which have potentially useful attributes. Recent focus on fungal genomics coupled with advances in detection and molecular manipulation has revolutionized this field. Secondary metabolites are compounds with varied chemical structures that are usually produced only during the stationary phase of growth (Robinson et al. 2001). These compounds do not have a physiological role during exponential phase, and their production starts when a key nutrient source, such as carbon, nitrogen, or phosphate, is exhausted (Barrios-González and Mejia 1996). In the last two decades, there has been a period of rapid discovery of new biological activities of these compounds and appropriate modern strategies for their identification (Petrini et al. 1992).

The culture medium is chosen based on the purpose and the species under investigation. Liquid media are preferably used for physiological studies, but agar media are more convenient and practical for the rapid screening of many isolates (Hölker et al. 2004). To isolate endophytic fungal secondary metabolites, fermentation techniques such as submerged fermentation (SmF) (or Liquid fermentation) and solid-state fermentation (SSF) have become widely used (Table 8.1) (Barrios-González and Mejia 1996; Pandey 2003; Hölker et al. 2005).

Table 8.1 Comparison of the main characteristics of solid-state fermentation and submerged fermentation

Microorganism and substrates	Need to agitated continuously
1. Water usage	Unlimited use
2. Oxygen supply	Aeration
3. Volume to fermentation mash	Larger
4. Liquid waste produced	Larger
5. Physical energy requirement	High
6. Human energy requirement	Low
7. Capital investment	High

The SmF and SSF techniques differ, both can be used to identify secondary metabolites produced by endophytic fungi (Fig. 8.1). However, the appropriateness of a given technique should be evaluated based on the aim of the study and the available resources. In addition, optimal parameters for both techniques, such as incubation conditions, medium composition, agitation, temperature, and pH, must be standardized to improve process efficiency and maintain reproducibility.

3.3 Host Specificity of Tropical Endophytic Fungi

Endophytic relationships may have begun from the time that higher plants first appeared hundreds of millions of years ago. Evidence of

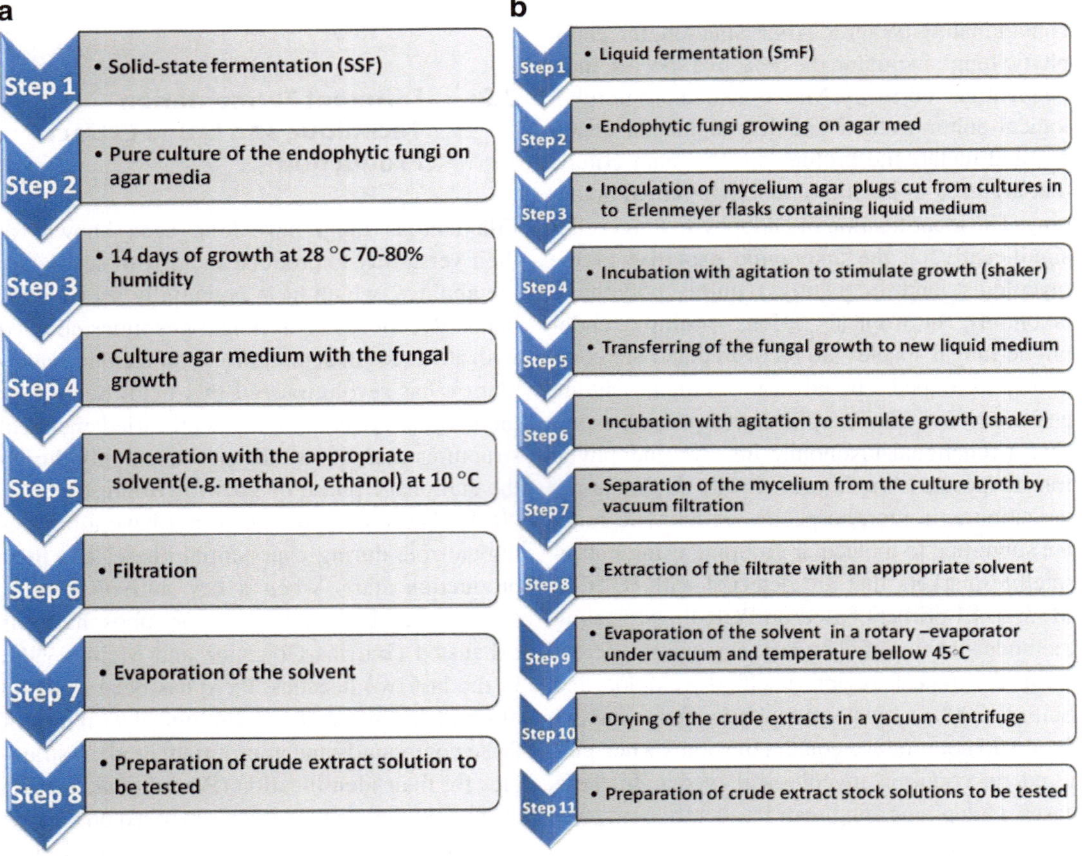

Fig. 8.1 (**a**) Solid-state fermentation and (**b**) liquid fermentation processes for obtaining endophytic fungal secondary metabolites

plant-associated fungi has been discovered in fossilized tissues of stems and leaves (Taylor et al. 1999). As a consequence of these long-term associations, some of these microorganisms may have developed a mechanism of genetic cross systems that allow the exchange of information between themselves and the higher plant. This exchange would allow the fungi to more efficiently cope with the environmental conditions and perhaps increase compatibility with the plant host. The dependent evolution of endophytic fungi may have allowed them to better adapt to the plant such that the fungi could contribute to the relationship by performing protective functions against pathogens and insects (Petrini et al. 1992; Strobel and Daisy 2003; Gunatilaka 2005). Tropical and temperate forests are considered to be the most diverse terrestrial ecosystems, with the greatest number and diversity of endophytic fungi (Strobel 2002). The constant innovation present in ecosystems where the evolutionary race to survive is the most active may result in the production of a plethora of chemical molecules (Strobel 2006). Tropical rainforests are an important example of this type of environment: there is great competition, resources are limited, and selection pressure is at its peak. Consequently, there is a high probability that fungi associated with tropical hosts may be a source of novel molecular structures and compounds that are active against neglected diseases. Each of the approximately 300,000 known plant species may host at least one endophytic fungus. As tropical and subtropical regions harbor most of the world's plant diversity, endophytic fungal diversity in this climatic zone is also higher, and all vascular plant species examined to date possess an endophytic fungus. Reasonable guidelines should govern the plant selection strategy for the discovery of bioactive endophytic fungi, which would include plants that are found in unique environmental settings, have ethnobotanical histories, or are endemic or growing in regions of high diversity according to Strobel (2003). Plant endophytic fungi are defined as the fungi which spend the whole or part of their lifecycle colonizing inter- and/or intracellularly inside the healthy tissues of the host plants, typically causing no apparent symptoms of disease. They are important components of plant micro-ecosystems (Tan et al. 2001; Zhang et al. 2006; Rodriguez et al. 2009). Plant endophytic fungi have been found in each plant species examined, and it is estimated that there are over one million fungal endophytes existed in the nature (Petrini 1991).

Plant endophytic fungi have been recognized as an important and novel resource of natural bioactive products with potential application in agriculture, medicine, and food industry (Strobel et al. 2004; Gunatilaka 2006; Verma et al. 2009). Since the "gold" bioactive compound paclitaxel (taxol) discovered from the endophytic fungus *Taxomyces andreanae* in 1993 (Stierle et al. 1993), many scientists have been increasing their interests in studying fungal endophytes as potential producers of novel and biologically active compounds. In the past two decades, many valuable bioactive compounds with antimicrobial, insecticidal, cytotoxic, and anticancer activities have been successfully discovered from the endophytic fungi. These bioactive compounds could be classified as alkaloids, terpenoids, steroids, quinones, lignans, phenols, and lactones (Xu et al. 2008). During the long period of coevolution, a friendly relationship was gradually set up between each endophytic fungus and its host plant. The host plant can supply plenteous nutriment and easeful habitation for the survival of its endophytes. On the other hand, the endophytes would produce a number of bioactive compounds for helping the host plants to resist external biotic and abiotic stresses, and benefiting for the host growth in return (Silvia et al. 2007). Some endophytic fungi have developed the ability to produce the same or similar bioactive substances as those originated from the host plants. This is beneficial for us to study the relations between the endophytes and their host plants and to develop a substitutable approach for efficiently producing these scarce and valuable bioactive compounds (Zhao et al. 2011; Gunatilaka 2006). For example, an endophytic fungus from *Juglans mandshurica* isolated named FSN006 from inner bark of the plant and was identified for their antitumor activity against liver cancer cell HepG2. In addition, there was preeminent selective inhibiting effect against the normal liver cell strain HL-7702 and its caner counter strain HepG2. The inhibiting effect against strain HL-7702 was only one quarter of that against HepG2 at the concentration of IC_{50}, showing its higher efficacy and lower toxicity (Li et al. 2009). Another important Brazilian medicinal plant *Stryphnodendron adstringens* (Mart) displayed high richness, diversity, and low dominance indices against cancer cell line and amastigote forms of *Leishmania amazonensis* and proved their fungal endophytes to be a new source for novel anticancer drug (Carvalho et al. 2012). Sclerotiorin is an important bioactive compound isolated through high throughput screening (HTP) from an endophytic fungus *Cephalotheca faveolata.* Sclerotiorin was found to be potent antiproliferative and found to induce apoptosis in colon cancer (HCT-116) cells through BAX activation and downregulation of BCL-2 which cleaved caspase-3 causing apoptosis of cancer cells (Giridharan et al. 2012). Some researchers from National Park, Pahang, isolated endophytic fungal extract which induces apoptosis against HCT116, MCF-7, and K562 cell lines with IC50 values less than 17 μg/mL. Molecular analysis, based on ITS1 and ITS4 sequencing, revealed that these fungus belongs to ascomycetes (Hazalin et al. 2012). In vitro anticancer activities of some endophytic fungi have been reported from Panama against MCF-7 cells; some of their strains show promising activity and lack of toxicity in the assays (Martínez-Luis et al. 2011). Endophytic fungi, that is, *Aspergillus flavus*, *A. niger*, *Fusarium oxysporum*, and *F. solani,* were isolated from different parts of *Crotalaria pallida* which were extracted in methanol by Soxhlet and Microwave-Assisted Extraction (MAE) to detect the amount of coumarin. Isolated compounds were analyzed by HPLC methods and column chromatography which was used for in vitro anti-HIV and anticancer activity assessment. In vitro anti-HIV activity was done against glycohydrolase enzyme (α-glucosidase, β-glucorinidase and lysozyme). The o-coumaric acid showed strong inhibition of these viral replicating enzymes and displayed potent anti-HIV activity (Umashankar et al. 2012). Plant endophytic fungi, as a novel and important microbial resource for producing bioactive compounds originally from their hosts, have attracted many researchers' attentions on their theoretical study as well as potential applications. After more than two decades of research, much progress has been achieved, though there are still many issues (i.e., increasing compound yield in fermentation culture, elucidating biosynthetic pathway of the compounds in the endophytic fungi) needed to be further clarified and resolved (Fig. 8.2).

4 Bioactive Compounds Against NTDs

Fungal metabolites have primarily served as lead structures for the development of anticancer, antifungal, and antibacterial agents, but recently a few reports of anti-parasiticidal activity especially against hydatid cyst *Echinococcus granulosus* have also been recently reported (Verma et al. 2013). Although new drugs are needed to treat all aspects of leishmaniasis, the scientific literature on the bioprospecting of endophytic fungi of tropical rainforests is limited. Indian ecosystems are potentially more diverse source of endophytic fungi that are able to produce bioactive prototype molecules for developing prodrugs to combat NTDs. Chemical investigation of a new endophytic fungus *Mycosphaerella sp.* nov. strain F2140 associated with the foliage of the plant *Psychotria horizontalis* (Rubiaceae) in Panama produces cercosporin and its analogue. These compounds were tested in vitro to determine their antiparasitic activity against the causal agents of malaria (*Plasmodium falciparum*), leishmaniasis (*Leishmania donovani*), and Chagas disease (*Trypanosoma cruzi*). Also, the cytotoxicity and potential anticancer activity of these compounds were evaluated using mammalian Vero cells and MCF7 cancer cell lines, respectively. Some of these derivatives displayed high potency and are active with no toxicity at

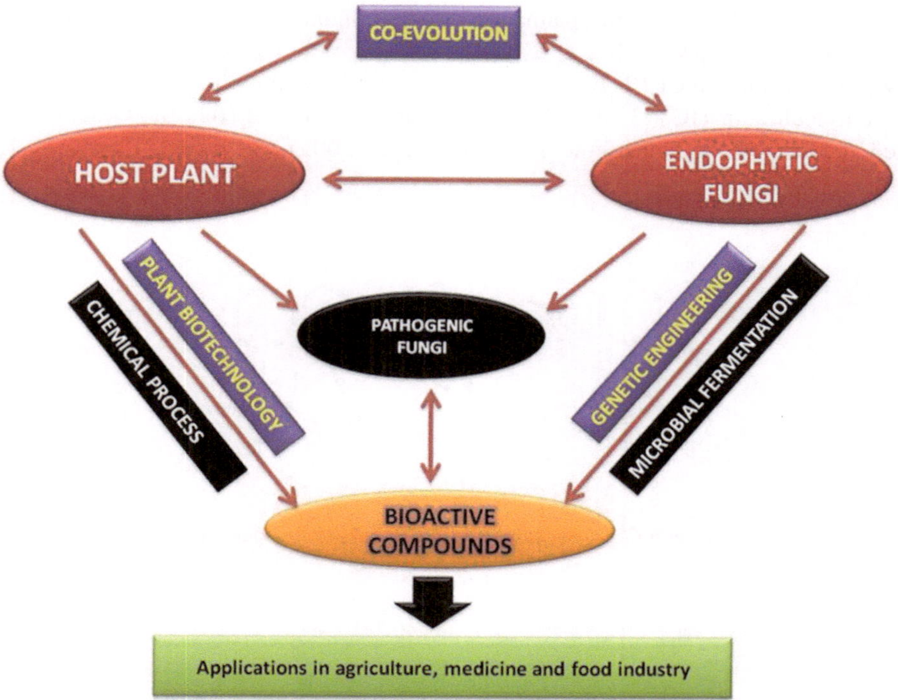

Fig. 8.2 Outline of the bioactive compounds from both endophytic fungi and their host plants along with their potential applications

tested concentrations (Moreno et al. 2011). Some fungal endophytes from Panama has been reported for their antiparasitic activity against *Leishmania donovani*, *Plasmodium falciparum*, and *Trypanosoma cruzi* and found to have significant antimalarial activity with relatively low toxicity (Martínez-Luis et al. 2011). Endophytic fungus *Aspergillus* sp. strain F1544 reported to have anti-leishmanial activity from its isolated compounds pseurotin A, 14-norpseurotin A, FD-838, pseurotin D, and fumoquinone B (Martínez-Luis and Cherigo 2012). Citrinin, a polyketide isolated from *Penicillium janthinellum* from the fruit of *Melia azedarach* (*Meliaceae*) in Brazil, was previously found in *Penicillium citrinum* and several other *Aspergillus* species (Vrabcheva et al. 2000) and significantly inhibited *Leishmania mexicana* at a concentration of 40 μg/mL (Marinho et al. 2005).

Citrinin

The endophytic fungus *Edenia* sp. was isolated from mature leaves of *Petrea volubilis* (Verbenaceae), which was collected from the Coiba National Park in Panama. Bioassay-guided fractionation of organic extracts of *Edenia* sp. led to the isolation of the anti-leishmanial compounds preussomerin EG1 (IC_{50} 0.12 μM), palmarumycin CP2 (IC_{50} 3.93 μM), palmarumycin CP17 (IC_{50} 1.34 μM), palmarumycin CP18 (IC_{50} 0.62 μM), CJ-12, 37 (IC_{50} 8.40 μM), palmarumycin CP19 (IC_{50} 11.6 μM), and 5-methylochracin (IC_{50} 33.4 μM), which inhibited the growth of amastigote forms of *Leishmania donovani*. Preussomerin EG1 was the most active substance and inhibited growth of *L. donovani* with a potency similar to that of amphotericin B (IC_{50} 0.09 μM) (Martínez-Luis et al. 2009).

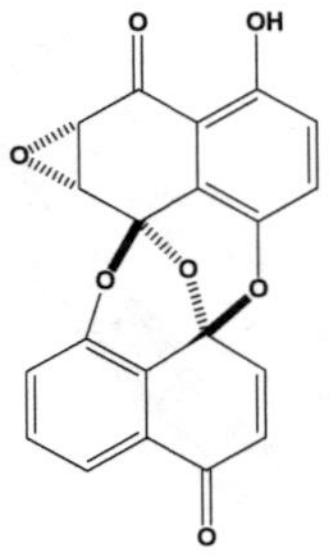

Preussomerin G 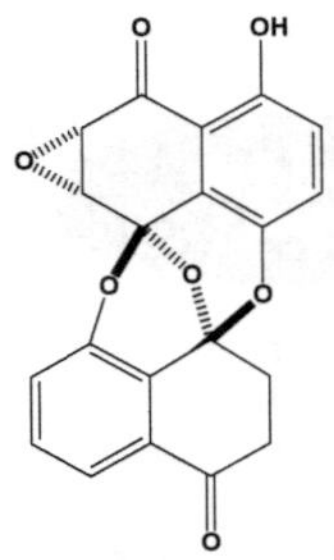 Preussomerin H 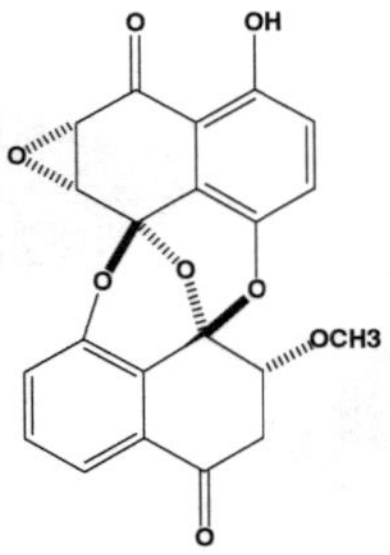 Preussomerin I

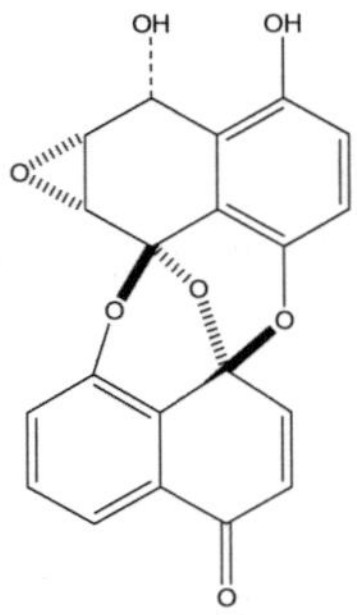

Preussomerin D 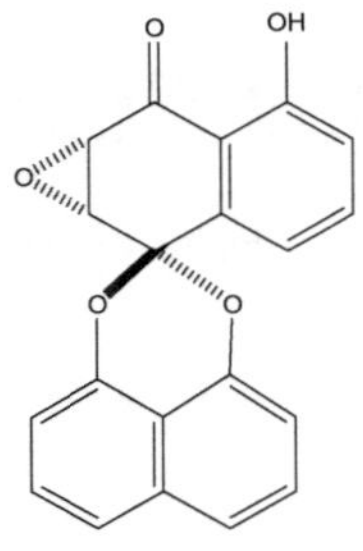 deoxypreussomerin A (Palmarumycin C_2) deoxypreussomerin B (PalmarumycinCP$_2$)

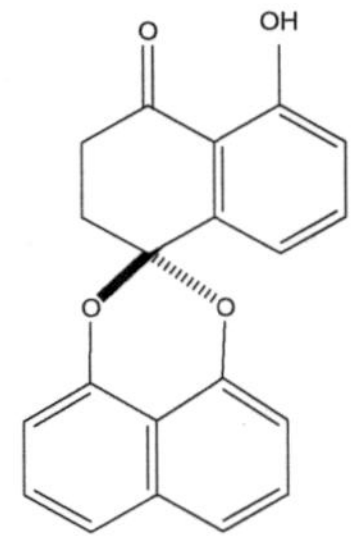

Endophytic *Cochliobolus* sp. obtained from the plant *Piptadenia adiantoides* produces cochlioquinone A and isocochlioquinone A. Both compounds were active in an assay against *L. amazonensis*, with EC_{50} values of 1.7 and 4.1 μM, respectively (Campos et al. 2008).

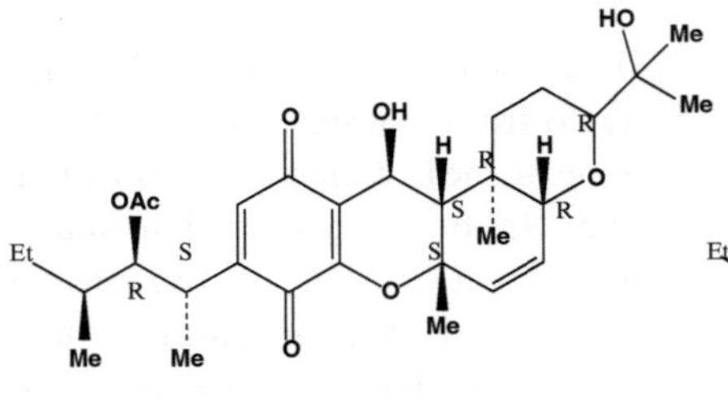

Cochlioquinone A 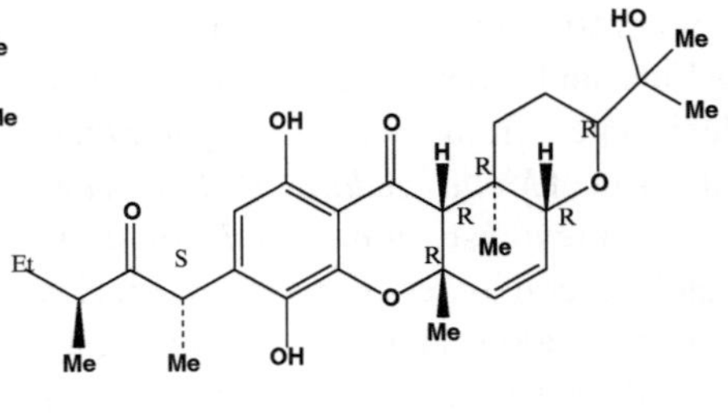 Isocochlioquinone A

Grandisin, a tetrahydrofuran lignan isolated from *Piper solmsianum* (*Piperaceae*) (Martins et al. 2003) and *Virola surinamensis* (*Myristicaceae*), has potent trypanocidal activity against the trypomastigote form of *T. cruzi* at 5 μg/mL (Lopes et al. 1998). Biotransformation of this compound by the endophytic fungus *Phomopsis* sp. obtained from *Viguiera arenaria* yielded the compound 3,4-dimethyl-2-(4′-hydroxy-3′,5′-dimethoxyphenyl)-5-methoxy-tetrahydrofuran. It had trypanocidal activity (IC_{50} 9.8 μmol/mL) similar to its natural precursor (IC_{50} 3.7 μmol/mL) (Verza et al. 2009).

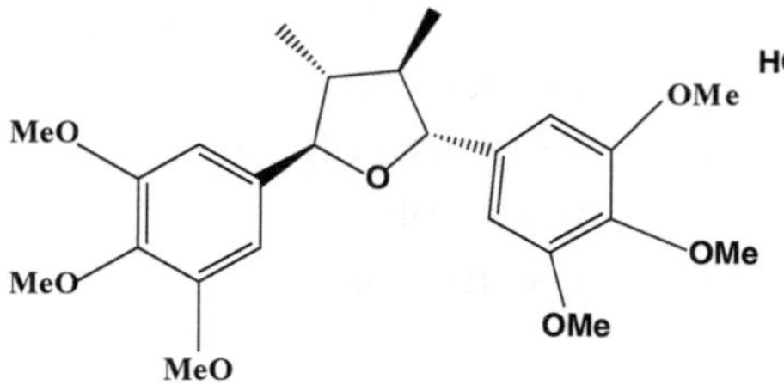

Grandicin

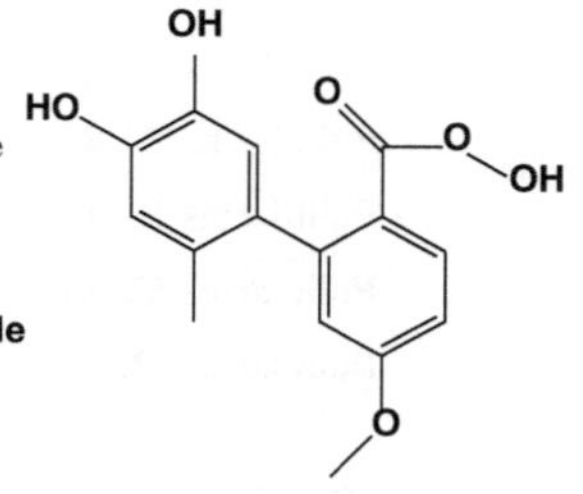

Altenusin

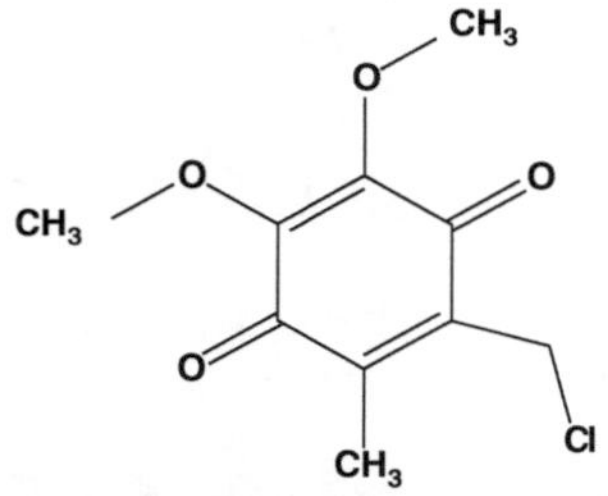

(2-chloro-5-methoxy-3-methylcyclohexa-2,5-diene-1,4-dione)

Altenusin is a metabolite obtained from the organic extract of a broth culture of the endophytic fungus *Alternaria* sp. UFMGCB 55, which was isolated from a plant known to contain trypanocidal compounds, *Trixis vauthieri*. This fungus inhibited TryR enzymatic activity with an $IC_{50\ \text{value}}$ of 4.3 mM (Cota et al. 2008). The endophytic fungus *Diaporthe phaseolorum*, recovered from *Viguiera arenaria*, displayed promising results by inhibiting the parasitic enzyme gGAPDH (95 %) at 100 μg/mL (Guimarães et al. 2008). An organohalogen natural product 2-chloro-5-methoxy-3-methylcyclohexa-2,5-diene-1,4-dione and a quinine derivative 7-hydroxy-8-methoxy-3,6-dimethyldibenzofuran-1,4-dione were obtained from the organic extract of *Xylaria* sp. PBR-30. This endophytic fungus was isolated from healthy leaves of *Sandoricum koetjape* (*Meliaceae*). These natural products had in vitro activity against *P. falciparum* (K1, multidrug-resistant strain), with IC_{50} values of 1.84 and 6.68 μM (Tansuwan et al. 2007).

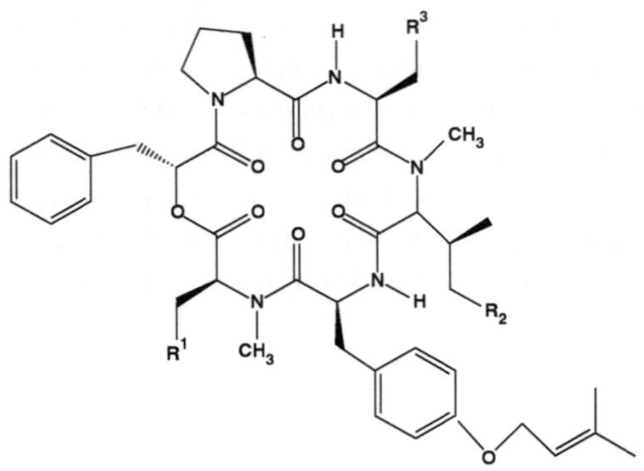

Pullularins A, R1 = H, R2 = CH_3, R3 = OH

Pullularins B, R1 = CH_3, R2 = CH_3, R3 = OH

Pullularins C, R1 = H, R2 = H, R3 = OH

Pullularins D, R1 = H, R2 = CH_3, R3 = OH

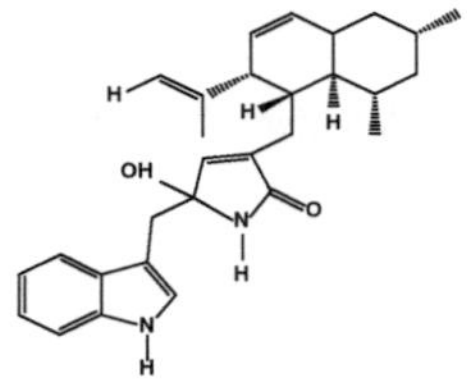

Codinaeopsin 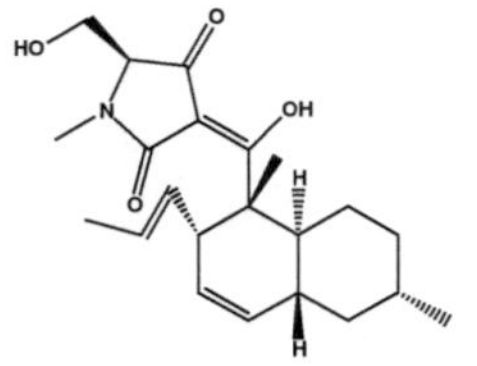 Equisetin 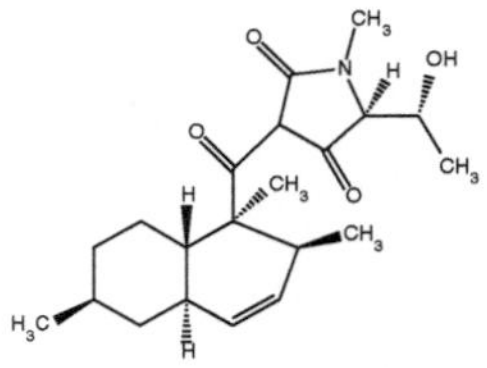 Cryptocin

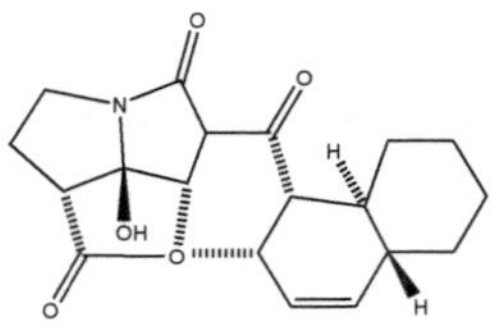

UCS1025A 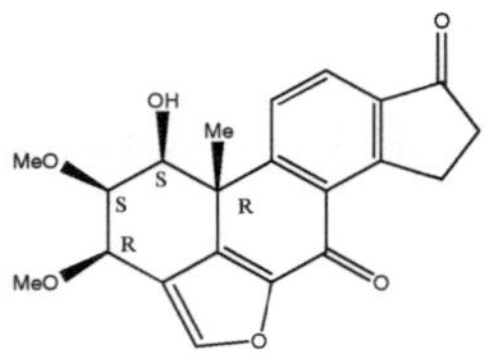 Viridiol

The yeast-like fungus *Aureobasidium pullulans*, which was isolated from leaves of *Calophyllum* sp. collected in Narathiwat Province, Thailand, produces the cyclohexadepsipeptides pullularins A–D. Pullularin A exhibited antimalarial activity (IC_{50} 3.6 μg/mL) and moderate antituberculosis activity (MIC 25 μg/mL). Pullularin B exhibited considerable antimalarial activity (IC_{50} 3.3 μg/mL), but this substance and pullularin C exhibited weaker activities in other assays when compared with pullularin A. The low lipophilicity of a deprenyl analogue of pullularin A may explain the inactivity of this substance in all of the assays (Isaka et al. 2007). *Codinaeopsis gonytrichoides* was isolated from *Vochysia guatemalensis* (*Vochysiaceae*), a white yemeri tree collected in Costa Rica. A new tryptophan-polyketide hybrid named codinaeopsin, which contains an unusual heterocyclic unit linking indole and decalin fragments, was isolated from the crude extract of this endophytic fungus. Codinaeopsin is active against the 3D7 strain of *P. falciparum* with an IC_{50} value of 2.3 μg/mL (4.7 μM). Codinaeopsin has the same scaffold as the HIV-integrase inhibitor equisetin, the antifungal agent cryptocin, and the telomerase inhibitor UCS1025A. These compounds have a linear fragment joined to amino acids or N-methyl amino acids (Kontnik and Clardy 2008). Stems of *Melaleuca quinquenervia* (*Myrtaceae*), collected from Toohey Forest, Australia, were examined for fungal content. Chemical investigations of a fermentation culture from the endophytic fungus *Pestalotiopsis* sp. yielded three caprolactams, which were named pestalactams A–C. Pestalactams A and B displayed modest in vitro selectivity against chloroquine-resistant (IC_{50} 41.3 and 36.3 μM, respectively) and chloroquine-sensitive (IC_{50} 16.2 and 20.7 μM, respectively) cell lines of the malaria causing parasite *P. falciparum* versus neonatal foreskin fibroblasts (NFF, IC_{50} 20.2 and 12.8 μM, respectively) (Davis et al. 2010). *Chalara alabamensis*, an anamorphic fungus, was isolated from the host plant *Asterogyne martiana* (Arecaceae), which was collected in Costa Rica. The dichloromethane extract of this fungus inhibited PfHsp86, an essential protein-folding chaperone from *P. falciparum*, with an EC_{50} value of 24 μg/mL. The only active compound isolated from the extract was viridiol, a steroidal furan with an EC_{50} value of 1.2 μg/mL (Cao et al. 2010).

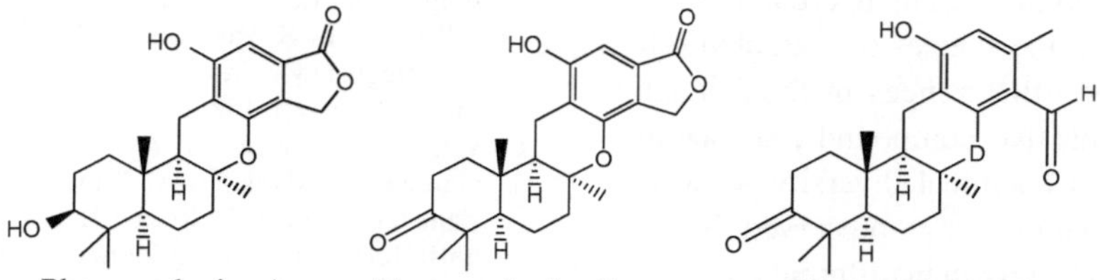

Phomoarcherins A Phomoarcherins B Phomoarcherins C

Phomoarcherins A–C were isolated from the endophytic fungus *Phomopsis archeri*. These structures were established on the basis of spectroscopic evidence. Compound phomoarcherins B having antimalarial activity against *Plasmodium falciparum* with an IC50 value of 0.79 μg/mL (Hemtasin et al. 2011).

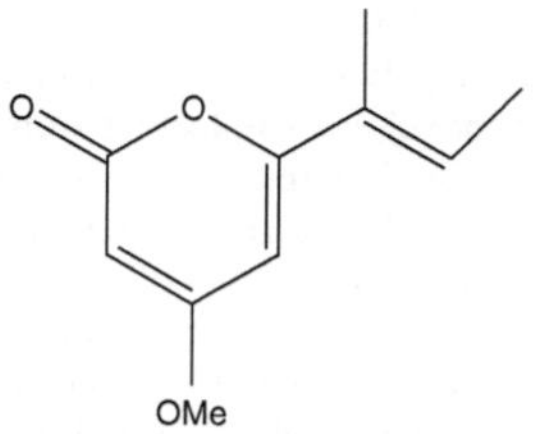

Pestalopyrone

Pestalopyrone, 6-(1′-methylprop-1′-enyl)-4-methoxy-2-pyrone, which was isolated from a Costa Rican endophytic fungus, *Phomatospora bellaminuta*, had activity against *P. falciparum* in an assay with an IC_{50} value of 37 μM and is a promising candidate for a prototype molecule for antimalarial drugs (Cao and Clardy 2011).

5 Conclusions

Plant endophytic fungi, as a novel and abundant microorganism resource, owning the special ability to produce the same compounds as originated from their host plants, as well as other bioactive compounds, have increased many investigators interesting in both basic research and applied fields. In the past two decades, scientists mainly focused on the investigation of fungal endophytes for diversity, relationships with their host plants. Recently interest has been generated in seeking for natural bioactive compounds originated from the endophytic fungi and improving the productivity of some potential candidates by taking advantage of genetic engineering, microbial fermentation projects, and other measures.

For the development of new drugs, few bioactive molecules discovered from tropical endophytic fungi were included in studies. The metabolites described in this review were used against concerned tropical diseases, because they are able to act against eukaryotic cells such as cancer cells, immune system cells, cells infected with viruses, and some human pathogenic fungi. Indeed, the use of the tropical endophytic fungi as novel scaffolds for the development of new drugs against neglected diseases represents a challenge to researchers of several scientific areas. The study of this fungal group offers some unique advantages such as: (1) endophytic fungi have a complex relationship with their host plant and produce bioactive metabolites; (2) there are a lot of different isolates of same species of endophytic fungi, which also have differences in their capability to produce bioactive compounds, so taxonomic diversity brings chemical diversity as well; (3) tropical endophytic fungi preserved in culture collections can be grown in different conditions of nutrients, temperature, pH, agitation, and aeration to optimize and recover the high amount of crude extracts, as well as bioactive pure compounds; and (4) if the crude extract and fractions produced by endophytic fungi do not display toxic activities, they can be used as therapeutic agents.

Unexplored natural environment is an excellent source of bioactive compounds that can act as the scaffold for commercial drugs. By taking advantage of new genomic, proteomic, and drug design techniques, endophytic fungal communities associated with tropical forest plants, with their high diversity of species and their diverse genetic and metabolic pathways, may be resources for intelligent screening for discovering new drugs to treat unconcerned tropical diseases.

References

Arnold A, Maynard Z, Gilbert G, Coley P, Kursar T (2000) Are tropical endophytes hyperdiverse? Ecol Lett 3:267–274

Arnold A, Maynard Z, Gilbert G (2001) Fungal endophytes in dicotyledonous neotropical trees: patterns of abundance and diversity. Mycol Res 105:1502–1507

Asford R (1997) The leishmaniases as model zoonoses. Ann Trop Med Parasitol 91:693–702

Baker D, Mocek U, Garr C (2000) Natural products vs. combinatorials: a case study. In: Wrigley SK, Hayes MA, Thomas R, Chrystal EJT, Nicholson N (eds) Biodiversity: new leads for pharmaceutical and agrochemical industries. The Royal Society of Chemistry, Cambridge, pp 66–72

Barrios-González J, Mejia A (1996) Production of secondary metabolites by solid state fermentation. Biotechnol Ann Rev 2:83–121

Campos F, Rosa L, Cota B, Caligiorne R, Rabello A, Alves T, Rosa C, Zani C (2008) Leishmanicidal metabolites from *Cochliobolus* sp., an endophytic fungus isolated from Piptadenia adiantoides (Fabaceae). PLoS Negl Trop Dis 2:348

Cao S, Clardy J (2011) New naphthoquinones and a new d-lactone produced by endophytic fungi from Costa Rica. Tetrahedron Lett 52:2206–2208

Cao S, Ross L, Tamayo G, Clardy J (2010) Asterogynins: secondary metabolites from a Costa Rican endophytic fungus. Org Lett 12:4661–4663

Carroll I, Toovey S, Van Gompel A (2007) Dengue fever and pregnancy-A review and comment. Travel Med Infect Dis 5:183–188

Carvalho CR, Gonçalves VN, Pereira CB, Johann S, Galliza IV, Alves TMA, Rabello A, Sobral MEG, Zani CL, Rosa CA, Rosa LH (2012) The diversity, antimicrobial and anticancer activity of endophytic fungi associated with the medicinal plant *Stryphnodendron adstringens* (Mart.) Coville (Fabaceae) from the Brazilian savannah. Symbiosis 57:95–107

Cota B, Rosa L, Caligiorne R, Rabello T, Rosa C, Zani C (2008) Altenusin, a biphenyl isolated from the endophytic fungus *Alternaria* sp. inhibits trypanothione reductase from *Trypanosoma cruzi*. FEMS Microbiol Lett 285:177–182

Croft S, Coombs G (2003) Leishmaniasis-current chemotherapy and recent advances in the search for novel drugs. Trends Parasitol 19:502–508

Davis R, Carroll A, Andrews K, Boyle G, Tran T, Healy P, Kalaitzis J, Shivas R (2010) Pestalactams A–C: novel caprolactams from the endophytic fungus. Org Biomol Chem 8:1785–1790

Dreyfuss MM, Chapela IH (1994) Potential of fungi in the discovery of novel, low-molecular weight pharmaceuticals. In: Gullo VP (ed) The discovery of natural products with therapeutic potential. Butterworth -Heinemann, London, pp 49–80

Ganley R, Brunsfeld S, Newcombe G (2004) A community of unknown, endophytic fungi in western white pine. Proc Natl Acad Sci U S A 101:10107–10112

Gardner D, Ryman F (2010) Yellow fever: a reemerging threat. Clin Lab Med 30:237–260

Giridharan P, Verekar SA, Khanna A, Mishra PD, Deshmukh SK (2012) Anti-cancer activity of sclerotiorin, isolated from an endophytic fungus *Cephalotheca faveolata* Yaguchi, Nishim & Udagawa. Indian J Exp Biol 50:464–468

Guimarães DO, Borges WS, Kawano CY, Ribeiro PH, Goldman GH, Nomizo A, Thiemann OH, Oliva G, Lopes NP, Pupo MT (2008) Biological activities from extracts of endophytic fungi isolated from *Viguiera arenaria* and *Tithonia diversifolia*. FEMS Immunol Med Microbiol 52:134–144

Gunatilaka A (2005) Natural products from plants-associated microorganisms: distribution, structural diversity, bioactivity, and applications of their occurrence. J Nat Prod 69:509–526

Gunatilaka AAL (2006) Natural products from plant-associated microorganisms: distribution, structural diversity, bioactivity, and implications of their occurrence. J Nat Prod 69:509–526

Hawksworth D (2004) Fungal diversity and its implications for genetic resource collections. Stud Mycol 50:9–18

Hayes E (2010) Is it time for a new yellow fever vaccine? Vaccine 28:8073–8076

Hazalin NA, Ramasamy K, Lim SM, Cole AL, Majeed AB (2012) Induction of apoptosis against cancer cell lines by four ascomycetes (endophytes) from Malaysian rainforest. Phytomedicine 19:609–617

Hemtasin C, Kanokmedhakul S, Kanokmedhakul K, Hahnvajanawong C, Soytong K, Prabpai S, Kongsaeree P (2011) Cytotoxic pentacyclic and tetracyclic Aromatic sesquiterpenes from *Phomopsis archeri*. J Nat Prod 74:609–613

Heukelbach J, Chichava O, Oliveira A, Häfner K, Walther F (2011) Interruption and defaulting of multidrug therapy against Leprosy: population-based study in Brazil's savannah region. PLoS Negl Trop Dis 5:e1031

Hölker U, Hofer M, Lenz J (2004) Biotechnological advantages of laboratory-scale solid-state fermentation with fungi. Appl Microbiol Biotechnol 64:175–186

Hölker U, Hofer M, Lenz J (2005) Solid-state fermentation – are there any biotechnological advantages? Curr Opin Microbiol 8:301–306

Hotez P, Yamey G (2009) The evolving scope of neglected tropical diseases. PLoS Negl Trop Dis 3:e379

Hotez P, Stoever K, Fenwick A, Molyneux D, Savioli L (2006) The Lancet's chronic diseases series. Lancet 367:563–564

Hotez P, Bottazzi M, Franco-Paredes C, Ault S, Periago M (2008) The neglected tropical diseases of Latin America and the Caribbean: a review of disease burden and distribution and a roadmap for control and elimination. PLoS Negl Trop Dis 2:e300

Hotez P, Fenwick A, Savioli L, Molyneux D (2009) Rescuing the bottom billion through control of neglected tropical diseases. Lancet 373:570–575

Huang Y, Wang J, Li G, Zheng Z, Su W (2001) Antitumor and antifungal activities in endophytic fungi isolated from pharmaceutical plants. FEMS Immunol Med Microbiol 31:163–167

Isaka M, Berkaew P, Intereya K, Komwijit S, Sathitkunanon T (2007) Antiplasmodial and antiviral cyclohexadepsipeptides from the endophytic fungus *Pullularia* sp. BCC 8613. Tetrahedron 63:6855–6860

Kontnik R, Clardy J (2008) Codinaeopsin, an antimalarial fungal polyketide. Org Lett 10:4149–4151

Larsen T, Smedgaard J, Nielsen K, Hansen M, Frisvad J (2005) Phenotypic taxonomy and metabolite profiling in microbial drug discovery. Nat Prod Rep 22:672–695

Li M, Wu Y, Jiang F, Yu X, Tang K, Miao Z (2009) Isolation, identification and anticancer activity of an endophytic fungi from *Juglans mandshurica*. Zhongguo Zhong Yao Za Zhi 34:1623–1627

Lieckfeldt E, Seifert KA (2000) An evaluation of the use of ITS sequences in the taxonomy of the hypocreales. Stud Mycol 45:35–44

Lopes N, Albuquerque S, Kato M, Yoshida M (1998) Flavonoids and lignans from *Virola surinamensis* twigs and their *in-vitro* activity against *Trypanosoma cruzi*. Planta 64:667–669

Marinho A, Rodrigues-Filho E, Moitinho M, Santos L (2005) Biologically active polyketides produced by *Penicillium janthinellum* isolated as an endophytic fungus from fruits of *Melia azedarach*. J Braz Chem Soc 16:280–283

Martínez-Luis S, Cherigo L (2012) Antiparasitic and anticancer constituents of the endophytic fungus *Aspergillus* sp. strain F1544. Nat Prod Commun 7:165–168

Martínez-Luis S, Cherigo L, Spadafora C, Gerwick W, Cubilla-Rios L (2009) Additional anti-leishmanial constituents of the Panamanian endophytic fungus *Edenia* sp. Revista Latinoamericana de Química 37:104–114

Martínez-Luis S, Cherigo L, Higginbotham S, Arnold E, Spadafora C, Ibañez A, Gerwick WH, Cubilla-Rios L (2011) Screening and evaluation of antiparasitic and *in-vitro* anticancer activities of Panamanian endophytic fungi. Int Microbiol 14:95–102

Martins R, Latone L, Sartorelli P, Kato M (2003) Trypanocidal tetrahydrofuran lignans from inflorescences of *Piper solmsianum*. Phytochemistry 64:667–670

Moreno E, Varughese T, Spadafora C, Arnold AE, Coley PD, Kursar TA, Gerwick WH, Cubilla-Rios L (2011) Chemical constituents of the new endophytic fungus *Mycosphaerella* sp. nov. and their anti-parasitic activity. Nat Prod Commun 6:835–840

Murrell S, Wu S, Butler M (2011) Review of dengue virus and the development of a vaccine. Biotechnol Adv 29:239–247

Pandey A (2003) Solid-state fermentation. Biochem Eng J 13:81–84

Petrini O (1991) Fungal endophytes of tree leaves. In: Andrews NJ, Hirano SS (eds) Microbial ecology of leaves. Springer, New York, pp 179–197

Petrini O, Sieber T, Toti L, Viret O (1992) Ecology metabolite production and substrate utilization in endophytic fungi. Nat Toxins 1:185–196

Porras-Alfaro A, Bayman P (2011) Hidden fungi, emergent properties: endophytes and microbiomes. Annu Rev Phytopathol 49:291–315

Promputtha I, Lumyong S, Dhanasekaran V, Huge E, Mackenzie C, Hyde K, Jeewon R (2007) Phylogenetic evaluation of whether endophytes become saprotrophs at host senescence. Microb Ecol 53:579–590

Raghukumar C, Nagarkar S, Raghukumar S (1992) Association of thraustochytrids and fungi with living marine algae. Mycol Res 96:542–546

Redline RC, Carris LM (1985) Endophytic fungi of grasses and woody plants. APS Press, St Paul

Reyes P, Vallejo M (2005) Trypanocidal drugs for late stage, symptomatic Chagas disease (*Trypanosoma cruzi infection*). Cochrane Database Syst Rev 1:1–16

Robinson T, Singh D, Nigan P (2001) Solid-state fermentation: a promising microbial technology for secondary metabolites production. Appl Microbiol Biotechnol 55:284–289

Rodriguez S, Querol X, Alastues A, Kallos G, Kakaliagou O (2001) Saharan dust contribution to PM10 and TSP levels in southern and eastern Spain. Atmos Environ 35:2433–2447

Rodriguez RJ, White JF Jr, Arnold AE, Redman RS (2009) Fungal endophytes: diversity and functional roles. New Phytol 182:314–330

Schulz B, Boyle C, Draeger S, Römmert A, Krohn K (2002) Endophytic fungi: a source of novel biologically active secondary metabolites. Mycol Res 106: 996–1004

Silvia F, Sturdikova M, Muckova M (2007) Bioactive secondary metabolites produced by microorganisms associated with plants. Biologia 62:251–257

Stierle A, Strobel G, Stierle D (1993) Taxol and taxane production by *Taxomyces andreanae*, an endophytic fungus of Pacific yew. Science 260:214–216

Strobel G (2002) Rainforest endophytes and bioactive products. Crit Rev Biotechnol 22:315–333

Strobel G (2003) Endophytes as sources of bioactive products. Microbes Infect 6:535–544

Strobel G (2006) Harnessing endophytes for industrial microbiology. Curr Opin Microbiol 9:240–244

Strobel G, Daisy B (2003) Bioprospecting for microbial endophytes and their natural products. Microbiol Mol Biol Rev 67:491–502

Strobel GB, Daisy U, Castillo U, Harper J (2004) Natural products from endophytic microorganisms. J Nat Prod 67:257–268

Suryanarayanan S, Murali T, Venkatesan G (2002) Occurrence and distribution of fungal endophytes in tropical across a rainfall gradient. Can J Bot 80: 818–826

Tan S, Chen X, Zhang A, Li D (2001) Isolation and characterization of DNA microsatellites from cotton bollworm (*Helicoverpa armigera* Hübner). Mol Ecol Notes 1:243–244

Tansuwan S, Pornpakakul S, Roengsumran S, Petsom A, Muangsin N, Sihanonta P, Chaichit N (2007) Antimalarial benzoquinones from an endophytic fungus *Xylaria* sp. J Nat Prod 70:1620–1623

Taylor J, Hyde K, Jones E (1999) Endophytic fungi associated with the temperate palm, *Trachycarpus fortunei*, within and outside its natural geographic range. New Phytol 142:335–346

Travassos L, Taborda C, Colombo A (2008) Treatment options for paracoccidioidomycosis and new strategies investigated. Exp Rev Anti Infect Ther 6:251–262

Umashankar R, Govindappa M, Ramachandra YL (2012) *In-vitro* antioxidant and anti-HIV activity of endophytic Coumarin from *Crotalaria pallida* Aiton. Planta Med 78:102

Verma VC, Kharwar RN, Strobel GA (2009) Chemical and functional diversity of natural products from plant associated fungal endophytes. Nat Prod Commun 4:1511–1532

Verma VC, Gangwar M, Nath G (2013) Osmoregulatory and tegumental ultrastructural damages to protoscoleces of hydatid cysts *Echinococcus granulosus* induced by fungal endophytes. J Parasit Dis. doi:10.1007/s12639-013-0271-z

Verza M, Arakawa N, Lopes N, Kato M, Pupo M, Saida S, Carvalho I (2009) Biotransformation of a tetrahydrofuran lignan by the endophytic fungus *Phomopsis* sp. J Braz Chem Soc 20:195–200

Vrabcheva T, Usleber E, Dietrich R, Maertlbauer E (2000) Co-occurrence of ochratoxin A and citrinin in cereals from Bulgarian villages with a history of Balkan endemic nephropathy. J Agric Food Chem 48: 2483–2488

World Health Organization (1991) Control of Chagas' disease. WHO Technical report series No. 811. World Health Organization, Geneva

World Health Organization (2006) Manuscript of neglected tropical diseases. Annual report. World Health Organization, Geneva

Xu L, Zhou L, Zhao J, Li J, Li X, Wang J (2008) Fungal endophytes from *Dioscorea zingiberensis* rhizomes and their antibacterial activity. Lett Appl Microbiol 46:68–72

Zhang HW, Song YC, Tan RX (2006) Biology and chemistry of endophytes. Nat Prod Rep 23:753–771

Zhao J, Shan T, Mou Y, Zhao L (2011) Plant derived bioactive compounds produced by endophytic fungi. Mini Rev Med Chem 11:159–168

Endophytes and Plant Secondary Metabolite Synthesis: Molecular and Evolutionary Perspective

9

P. Mohana Kumara, S. Shweta, M.M. Vasanthakumari, N. Sachin, B.L. Manjunatha, S. Sagar Jadhav, G. Ravikanth, K.N. Ganeshaiah, and R. Uma Shaanker

Abstract

The distribution of endophytes (fungal and bacterial) is ubiquitous and almost without exception; the endophytes have been reported from all tissues, including leaves, stems, roots, flowers and fruits. As typical symptomless organisms, in contrast to their pathogenic counterparts, they pose a serious challenge in explaining their continued maintenance in plants. How do plants tolerate them? And how do the endophytes contain the plant defences? But a more intriguing and enigmatic issue with many endophytes is the fact that they mimic the production of specific plant-associated secondary metabolites (e.g. taxol, camptothecin and rohitukine) in culture, independent of the host tissue. Several theories including the possibility of horizontal gene transfer from the respective hosts have

P. Mohana Kumara • S. Shweta
M.M. Vasanthakumari • N. Sachin
B.L. Manjunatha
Department of Crop Physiology, University of Agricultural Sciences, GKVK Campus, Bangalore 560065, India

School of Ecology and Conservation, University of Agricultural Sciences, GKVK Campus, Bangalore 560065, India

S. Sagar Jadhav
Department of Crop Physiology, University of Agricultural Sciences, GKVK Campus, Bangalore 560065, India

G. Ravikanth
School of Ecology and Conservation, University of Agricultural Sciences, GKVK Campus, Bangalore 560065, India

Ashoka Trust for Research in Ecology and the Environment, Royal Enclave, Srirampura, Jakkur Post, Bangalore 560064, India

K.N. Ganeshaiah
School of Ecology and Conservation, University of Agricultural Sciences, GKVK Campus, Bangalore 560065, India

Ashoka Trust for Research in Ecology and the Environment, Royal Enclave, Srirampura, Jakkur Post, Bangalore 560064, India

Department of Forestry and Environmental Sciences, University of Agricultural Sciences, GKVK Campus, Bangalore 560065, India

R. Uma Shaanker (✉)
Department of Crop Physiology, University of Agricultural Sciences, GKVK Campus, Bangalore 560065, India

School of Ecology and Conservation, University of Agricultural Sciences, GKVK Campus, Bangalore 560065, India

Ashoka Trust for Research in Ecology and the Environment, Royal Enclave, Srirampura, Jakkur Post, Bangalore 560064, India
e-mail: umashaanker@gmail.com

V.C. Verma and A.C. Gange (eds.), *Advances in Endophytic Research*,
DOI 10.1007/978-81-322-1575-2_9, © Springer India 2014

been proposed, but none has so far been supported. In this paper, we critically review studies on endophytes producing plant secondary metabolites and explore the possible molecular mechanisms. By analysing the pathway genes for a few major metabolites, including taxol and camptothecin, we show that a far more intricate molecular mechanism might be involved in the production of the secondary metabolites by the endophytes. We show that these molecular mechanisms could have arisen through the evolutionary interactions of the endophytes with their respective host plants. We discuss these findings in the context of the current interest in harnessing endophytes as alternative sources of plant secondary metabolites.

1 Introduction

Endophytic fungi, often referred to as "symptomless fungi", occur ubiquitously in plants (Arnold and Lutzoni 2007). They reside in intercellular spaces of stems, petioles, roots and leaves of plants without causing any obvious negative effects (Bacon and White 2000). Almost without exception, endophytes have been reported from all plants and parts of plants investigated. Woody plants or trees often contain a greater diversity of endophytes compared to herbaceous plants (Petrini et al. 1992; Gaylord et al. 1996; Faeth and Hammon 1997; Saikkonen et al. 1998; Arnold et al. 2000). The endophytes may be transferred either horizontally through airborne spores or vertically through seeds (Hartley and Gange 2009). Like other fungi, endophytic fungi are heterotrophic and obtain their metabolic carbon from their host plants. In fact, in the absence of overt pathogenesis, endophytic fungi are believed to be engaged in some mutualistic and/or symbiotic relationship with the host, with the latter benefiting in such an engagement.

Many plant processes have been attributed to be shaped by endophytic fungal association. For example, endophytic fungi are suggested to have played a major role in structuring plant communities and in shaping processes such as colonisation, competition, coexistence and soil nutrient dynamics (Saikkonen et al. 1998; Schulz et al. 2002). Several studies have demonstrated the role of endophytic fungi in imparting tolerance to plants against abiotic and biotic stresses (Bae et al. 2009; Arnold et al. 2003). Unlike in mutualism or symbiosis, the association between plants and their endophytic fungi is not strong. Endophytic fungal diversity is shaped by environmental or habitat conditions in which the plants take residence (Vega et al. 2010). For instance, incidence, diversity and host breadth of endophytic fungi were shown to increase with latitude (Arnold 2007; Arnold and Lutzoni 2007). Plants at higher latitudes were mainly comprised by members of Ascomycota, while those in lower tropical latitudes with a number of different species (Arnold and Lutzoni 2007).

Few studies have suggested an active symbiotic association between plants and their endophytic fungal associates. The close synchronisation of fungal and host reproduction in certain grasses, for example, is often cited as an example of possible co-evolution between the fungus and its host (Schardl et al. 1991; Moricca and Ragazzi 2008). Lack of plant defence reactions against endophytic fungi (Christensen et al. 2002) as well as the ability of endophytes to produce bioactive metabolites mimicking those produced by their respective host plants (Stierle et al. 1993; Amna et al. 2006; Eyberger et al. 2006; Kusari et al. 2009a, b; Shweta et al. 2010; 2013a, b; Mohana Kumara et al. 2012) also indicate the possibility of a more sustained or evolved relationship between the endophytic fungi and their host plants.

2 Plant Secondary Metabolite Production by Endophytic Fungi

Besides their role in aiding several plant growth processes, endophytic fungi are known to produce a larger number of metabolites as demonstrated by a number of fungal culture studies (Tan and Zou 2001). The metabolites including alkaloids, steroids, terpenoids, isocoumarins, quinones, flavonoids, phenylpropanoids, lignans, peptides, phenolics, aliphatics and volatile organic compounds have raised tremendous interest especially from the possibility of exploiting the fungi as source of pharmaceutically important compounds (Tan and Zou 2001; Gunatilaka 2006; Zhang et al. 2006). Many of these compounds intriguingly are the same as produced by the respective host plants, suggesting a possible genetic cross talk between the host and the endophytes (Fig. 9.1). Stierle et al. (1993) for the first time showed that an endophytic fungus, *Taxomyces andreanae*, isolated from the yew plant, *Taxus brevifolia*, also produced paclitaxel, the multibillion dollar anticancer compound, just as is produced by the yew plant. In a way this demonstration stoked an unprecedented interest in endophytic fungi, especially from the possibility that these could serve as alternative sources of important plant-based metabolites (Rubini et al. 2005; Tan and Zou 2001; Strobel and Daisy 2003).

Following Stierle et al. (1993), a number of studies have demonstrated the production of host-mimicking secondary metabolite production by their endophytic fungal associates. These metabolites include besides taxol, camptothecin, podophyllotoxin, vinblastine, hypericin, diosgenin, azadirachtin and rohitukine. Nineteen genera of endophytic fungi (*Alternaria, Aspergillus, Botryodiplodia, Botrytis, Cladosporium, Ectostroma, Fusarium, Metarhizium, Monochaetia, Mucor, Ozonium, Papulaspora, Periconia, Pestalotia, Pestalotiopsis, Phyllosticta, Pithomyces, Taxomyces* and *Tubercularia*) isolated from taxol-producing plants have been shown to produce paclitaxel and its analogues (i.e. baccatin III, 10-deacetylbaccatin III) (Zhao et al. 2010).

An endophytic fungus isolated from *Catharanthus roseus* produced in culture the antileukaemic compound, vincristine (Yang et al. 2004). Similarly, podophyllotoxin, a natural product precursor of useful anticancer agents, was found to be produced by *Trametes hirsuta,* an endophyte from *Podophyllum hexandrum* (Puri et al. 2006) and also by the endophytic fungus *Phialocephala fortinii* associated with *Podophyllum peltatum* (Eyberger et al. 2006).

An endophytic fungus *Shiraia* sp. Slf14 isolated from *Huperzia serrata* produced huperzine (Zhu et al. 2010). Endophytic fungus isolated from the stems of *Hypericum perforatum* produced hypericin and emodin (Kusari et al. 2009a, b). Other bioactive molecules isolated from endophytic fungus include huperzine A, α-irone, β-irone, diosgenin, hypericin and toosendanin (Zhao et al. 2010). Several endophytic fungi producing camptothecin (CPT), an anticancer alkaloid, have been isolated from CPT-producing plants (Puri et al. 2005; Amna et al. 2006; Rehman et al. 2008; Kusari et al. 2009a, b; Shweta et al. 2010). For example, *Entrophospora infrequens* and *Neurospora* sp. isolated from *Nothapodytes foetida* were found to produce camptothecin in culture (Puri et al. 2005; Rehman et al. 2008). Kusari et al. (2009a, b) reported the production of CPT, 9-methoxy camptothecin and 10-hydroxycamptothecin, by the endophytic fungus *Fusarium solani* isolated from *Camptotheca acuminata.* Two strains of *Fusarium solani* from *Apodytes dimidiata* were found to produce CPT, 9-methoxy camptothecin and 10-hydroxycamptothecin (Shweta et al. 2010). More recently, Shwetha et al. (2013a, b) demonstrated the production of camptothecin by other endophytes isolated from *Miquelia dentata*, a plant recently found to produce the highest amount of camptothecin in its fruits (Ramesha et al. 2013).

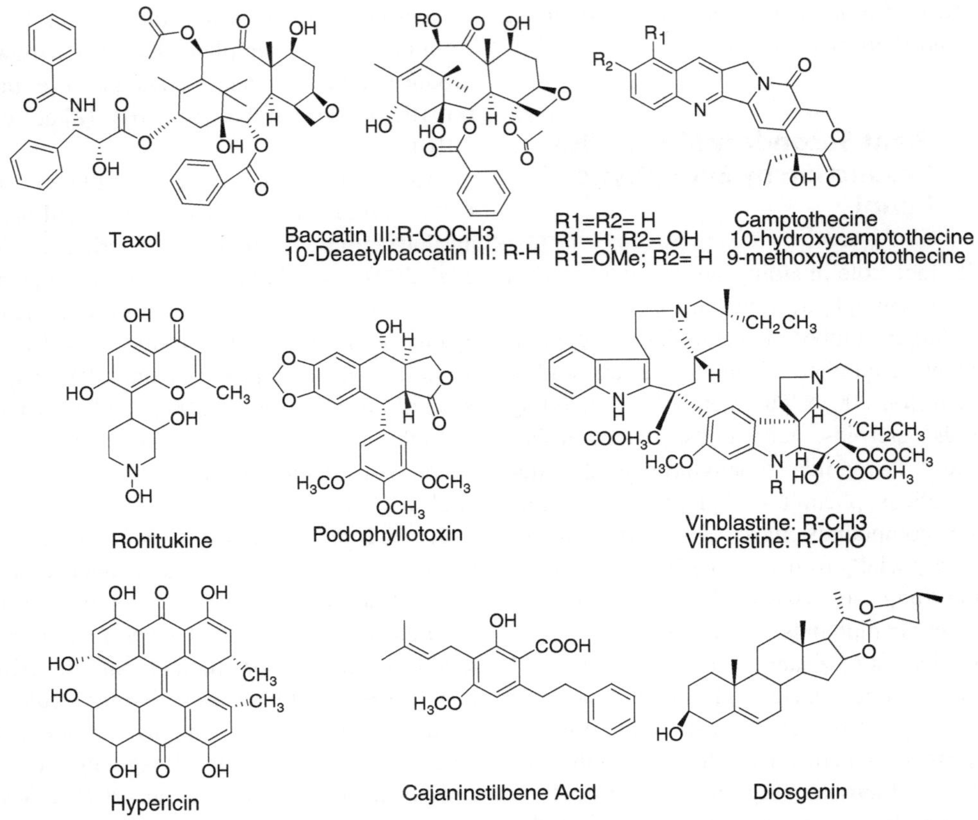

Fig. 9.1 Some major plant secondary metabolites produced by endophytic fungi

3 Attenuation of Endophytic Fungi

Despite the avalanche of studies reporting the production of host-mimicking secondary metabolites by endophytic fungal associates, none have led to realising endophytes as a viable source of the plant-based secondary metabolites (Priti et al. 2009). On subculturing the fungi in axenic medium, the endophytes tend to lose their ability to produce the secondary metabolites (Table 9.1). This process referred to as attenuation is a common phenomenon observed in many fungi, bacteria and viruses. With particular reference to endophytes, successive cultures of the endophytic fungi *Periconia* sp. isolated from *Torreya grandifolia* resulted in the attenuation of taxol production (Li et al. 1998). Successive cultures of several endophytic fungi isolated from *Nothapodytes nimmoniana*, a plant producing camptothecin, resulted in the attenuation of camptothecin production (Gurudatt et al. 2010). Similarly, Kusari et al. (2009a, b) reported attenuation of camptothecin production by the endophytic fungi, *Fusarium solani*, isolated from *Camptotheca acuminata*. The attenuation of endophytic fungi has become a serious impediment to the use of endophytic fungi as alternative sources of plant secondary metabolites. Among the various reasons, it is hypothesised that the attenuation could be due to a lack of host-specific stimuli when the fungi are cultured in axenic medium and/or due to silencing of genes in axenic cultures (Priti et al. 2009). Several efforts to reverse the attenuation by supplementing the axenic cultures with their respective host extracts have not been very promising.

Table 9.1 Attenuation of production of plant secondary metabolites by endophytic fungi

Compound	Fungus/bacterial strains	Host plant	Metabolite production over subculture generation 1st	2nd	3rd	4th	5th	Reference
Camptothecin	*Fusarium solani*	*Camptotheca acuminata*	600 μg/100 g	570 μg/100 g	60 μg/100 g	100 μg/100 g	50 μg/100 g	Kusari and Spiteller (2010)
	UAS023	*Nothapodytes nimmoniana*	–	3 μg/100 g	1.7 μg/100 g	1 μg/100 g	–	Gurudatt et al. (2010)
	UAS015	*Nothapodytes nimmoniana*	–	2 μg/100 g	1.4 μg/100 g	0.8 μg/100 g	–	
	UAS001	*Nothapodytes nimmoniana*	–	1.5 μg/100 g	0.8 μg/100 g	0.55 μg/100 g	–	
	UAS013	*Nothapodytes nimmoniana*	–	0.5 μg/100 g	0.40 μg/100 g	0.50 μg/100 g	–	
Taxol	*Periconia* sp.	*Torreya grandifolia*	350 ng/L	330 ng/L	280 ng/L	200 ng/L	118 ng/L	Li et al. (1998)
Rohitukine	*Fusarium proliferatum*	*Dysoxylum binectariferum*	186 μg/100 g	120 μg/100 g	50 μg/100 g	–	–	Mohana Kumara et al. (2012)

A serious handicap in addressing the problem of attenuation lies in the lack of clarity underlying the synthesis of plant secondary metabolite by the fungi. It has been suggested that the production of the plant secondary metabolite by the endophytic fungi could be due to a genetic recombination between the fungi and its host plant in evolutionary time. However, there is as yet no evidence for such horizontal transfer of genes coding for secondary metabolites between plant and fungi (Heinig et al. 2013).

4 Do Endophytic Fungi Possess Genetic Machinery to Synthesise Plant Secondary Metabolites?

In the recent past several attempts have been made to unravel the biosynthetic pathway of plant secondary metabolites with specific reference to terpenoid indole alkaloid and to ask if these are in fact recovered in the endophytic fungi. Here we briefly recapitulate certain recent studies that have a bearing on the production of plant secondary metabolites by endophytes.

Plant secondary metabolites are basically derived from three pathways, namely, polyketide, shikimate and mevalonate pathways. While the polyketide pathway contributes to the synthesis of phenols, quinine and prostaglandins, the shikimate pathway is responsible for the synthesis of aromatic amino acid. All the three pathways are present in plants, fungi and bacteria but not in animals. Together these pathways easily explain most of the plant secondary metabolite diversity. For example, the decarboxylated form of tryptophan, tryptamine, on condensation with secologanin, a monoterpenoid glucoside, gives rise to nitrogenous glucoside, strictosidine. Over 1,000 indole alkaloids, including quinine, strychnine and the anticancer compounds vinblastine, vincristine and camptothecin are derived from strictosidine (Cordell 1974). While a number of pathway genes upstream of strictosidine have been unravelled in plants, for most metabolites, the downstream pathway genes are unexplored.

In the recent past, few studies have attempted to unravel plant secondary metabolite pathway genes in endophytic fungi. Here we briefly review the pathway genes for three specific metabolites, namely, camptothecin, taxol and gibberellic acid, in plants and then relate it to the current understanding of the pathways in endophytes.

4.1 Camptothecin (CPT)

Camptothecin (CPT) is a quinoline alkaloid first isolated from *Camptotheca acuminata*, a deciduous tree native to China and Tibet. The bark of the tree is extensively used in traditional Chinese medicine (Wall et al. 1966). Later, camptothecin was discovered in several other species belonging to the families Icacinaceae, Rubiaceae, Apocynaceae and Loganiaceae, with the highest concentration reported from *Nothapodytes nimmoniana* (Graham) Mabb. (Icacinaceae) [0.3 % by dry weight (DW) in its bark] (Govindachari and Viswanathan 1972; Uma Shaanker et al. 2008).

The biosynthetic pathway of CPT in plants is only partially characterised (Yamazaki et al. 2003, 2004). Strictosidine, a precursor of terpenoid indole alkaloids, is considered as the precursor for CPT. The synthesis of strictosidine is catalysed by strictosidine synthase (*STR*), an enzyme committed for CPT biosynthesis (Fig. 9.2). This gene was isolated and identified from *Ophiorrhiza pumila* (Yamazaki et al. 2003), *Catharanthus roseus* (McKnight et al. 1990) and *Rauvolfia serpentina* (Kutchan et al. 1988). More recently, Sun et al. (2011) cloned and characterised three putative genes involved in CPT biosynthesis, namely, geraniol-10-hydroxylase, secologanin synthase and strictosidine synthase from *C. acuminata*.

Attempts to unravel the secondary metabolite genes and their clusters in fungi using whole genome sequences (SMURF; secondary metabolite unique regions finder, KEGG: Kyoto Encyclopedia of Genes and Genomes and FUNGI path v3.0) failed to detect the presence of *STR* as also other downstream genes of terpenoid indole alkaloids including those, for example, known in the biosynthesis of vindoline, reserpine and ajmalicine (Kanehisa et al. 2004; Khaldi et al. 2010; Sandrine et al. 2010; Sachin et al. 2013). These results were also confirmed by the FUNGI3 pathway database where orthologs of *STR* and other downstream genes could not be located in any of the fungal genomes analysed. The databases also failed to detect any of the genes responsible for taxol biosynthesis (Sachin et al. 2013).

Recently, an attempt was made to unravel the CPT biosynthetic gene from a CPT-producing endophytic fungus *Fusarium solani* isolated from *C. acuminata* (Kusari et al. 2011). Significantly while they could locate the presence of *TDC* (shikimate pathway) and *G10H* and *SLS* (mevalonate pathway) genes, they could not locate the presence of the crucial gene, *STR* in the endophyte. Studies on the evolution of STR and SSLs (strictosidine synthase like proteins) show that these proteins have been recovered from algae, cyanobacteria and insects but never from fungi (Sachin et al. 2013). Yet the endophyte was shown to produce CPT. Kusari et al. (2011) suggested that the endophyte might be using the host STR to synthesise CPT. However, as Sachin et al. (2013) argued, this suggestion is inconceivable, considering the fact that the endophyte was able to synthesise CPT in axenic cultures for several generations in absence of the host tissue where obviously the fungus cannot access the host *STR*.

4.2 Taxol

Paclitaxel, a highly functionalised diterpenoid, occurs naturally in *Taxus* (Yew) plants (Suffness 1995). Paclitaxel and some of its derivatives represent the first major group of anticancer agents that was reported to be produced by endophytes. Paclitaxel precludes tubulin molecules from depolymerising during the process of cell division (Schiff and Horowitz 1980).

In plants, taxol is produced by a series of enzymatic conversion of diterpenoid precursor geranylgeranyl diphosphate (GGPP) by the plastidial methylerythritol phosphate pathway (Fig. 9.2); among them three genes, namely, *ts* (involved in formation of the taxane skeleton), *dbat* (involved in baccatin III formation) and *bapt* (involved in phenylpropanoyl side chain formation at C13) are regarded as key to taxol biosynthesis (Xiong et al. 2013).

Several attempts have been made to unravel the taxol biosynthetic pathway in the fungal endophytes. For example, Zhang et al. (2009) showed the presence of the gene 10-deacetylbaccatin-III-10-O-acetyl transferase responsible for taxol biosynthesis in the endophyte *Cladosporium cladosporioides* MD213 isolated from *Taxus media* (yew species). Furthermore, Staniek et al. (2009) reported the presence of the *Taxadiene synthase* (*txs*), a gene unique to the formation of the primary taxane skeleton, as well as *phenylpropanoyltransferase* (*bapt*) gene encoding of the final acylation of the core structure of taxol present in the endophytic fungus *Taxomyces andreanae* isolated from *Taxus*. More recently, Xiong et al. (2013) showed that in three taxol-producing endophytes isolated from Anglojap Yew, *T. media*, the fungus gave positive hits for the three key genes, *ts*, *dbat* and *bapt*. However, the homology of these genes with those of *T. media* was low (between 40 and 44 % for *ts* and *bapt*), indicating that the genes in the endophytes may have independently evolved in the endophytes (Xiong et al. 2013).

However, in another recent study, Heinig et al. (2013) using three probes, taxadiene synthase, taxane-5α-hydroxylase and taxane-13α-hydroxylase, all involved in taxol biosynthesis, found none of the genes in the endophytic genome sequence. This result was further confirmed by sequencing the endophytes; none of the contigs had any significant homology to the genes responsible for taxane biosynthesis in the yew plant. Analysis of FUNGI3 pathway database also failed to detect any of the genes responsible for taxol biosynthesis (Sachin et al. 2013). Based on these studies, Heinig et al. (2013) concluded that the endophytes recovered from *Taxus* species do not have the ability to produce taxol endogenously and that the reported presence of taxol in the endophytes is probably due to the residual taxanes absorbed by the endophyte cell wall structures.

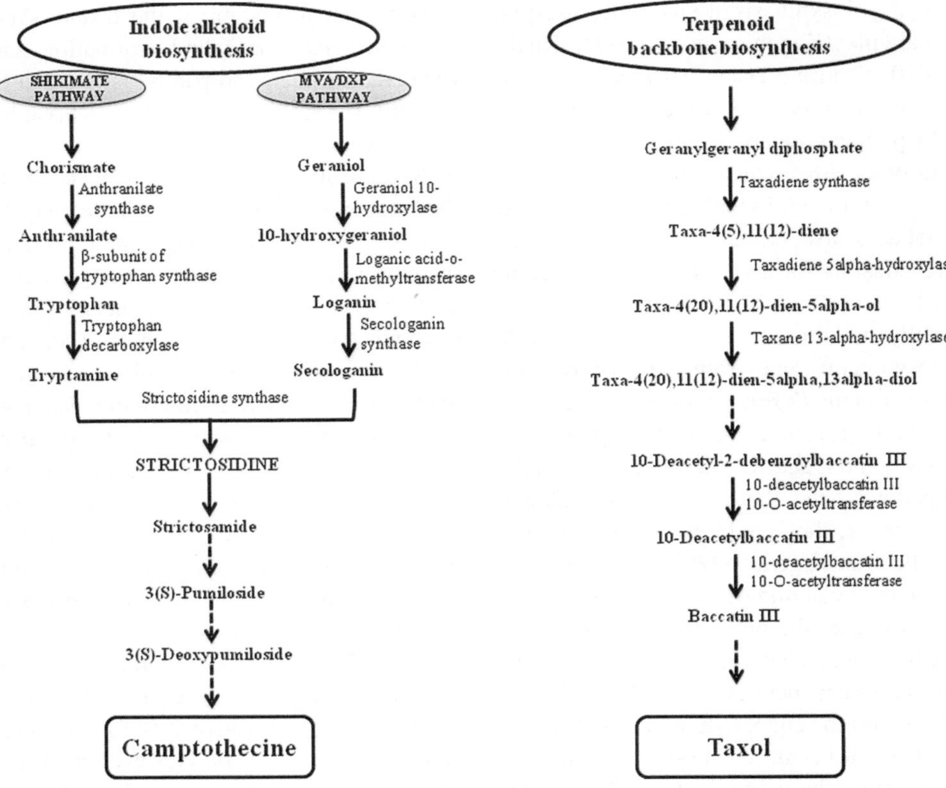

Fig. 9.2 Schematic representation of the biosynthetic pathways of indole alkaloid, camptothecin and diterpenoid taxol in plants

4.3 Gibberellic Acid

Unlike many other plant secondary metabolites, the biosynthetic pathway of gibberellic acid (GA) is well known. GA has been shown to be produced by plant, fungi and even bacteria. Intriguingly while the GA produced by all of them is structurally identical, they are all synthesised from diverse pathways. For example, while the basic pathway of GA biosynthesis in both plants and fungi are the same until GA12-aldehyde, the two differ later on in the process of making active GAs (GA1 or GA3). In plants, conversion of GGDP to active GAs requires the presence of 3 terpene synthases, two 450s and a soluble 20 DDS. In contrast, in the fungus, the synthesis is made by only 1 bifunctional terpene cyclase (CPSD/KS) and by P450s.

These results suggest that the biosynthetic pathways in plants and fungi may have evolved independently (Bömke and Tudzynski 2009). GA production has also been reported from endophytic fungi, *F. proliferatum,* isolated from orchid roots. The fungus was found to contain the GA biosynthetic gene P450-4. Strains of *F. proliferatum* that did not produce GA were found to have mutations at some of the genes involved in GA biosynthesis. Complementation of these fungi with those producing GA restored the GA production. These studies therefore clearly negate a long-held view that the fungi acquired the genes for GA biosynthesis from higher plants through horizontal gene transfer (Bömke and Tudzynski 2009).

In summary, despite the overwhelmingly large number of reports of endophytic fungi producing plant secondary metabolites, there seems to be little evidence of them having the pathway genes associated with the synthesis of these metabolites.

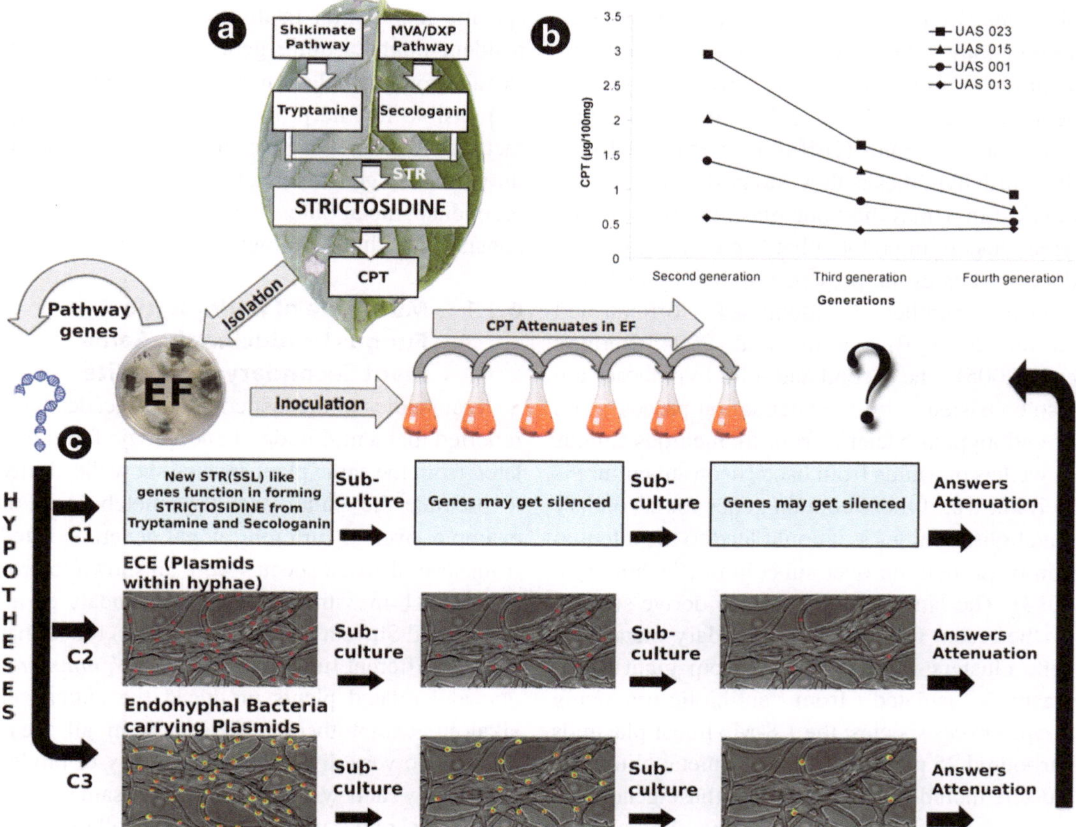

Fig. 9.3 Schematic representation of the hypotheses describing the mechanism of production of plant secondary metabolites by endophytic fungi. (**a**) Formation of strictosidine by the enzyme strictosidine synthase (STR) in the plant. (**b**) Typical attenuation of camptothecin (CPT) by CPT-producing endophytic fungi from *Nothapodytes nimmoniana* 5. (**c**) Flow of events on isolation of endophytic fungi from the plant. (C1), Hypothesis 1; (C2), Hypothesis 2 (*red circles* represent the extra-chromosomal elements (ECEs) carried within the fungal mycelia); and (C3), Hypothesis 3. *Yellow* dots are endohyphal bacteria presumed to carry plasmids (*red*) bearing secondary metabolite genes (Adapted from Sachin et al. 2013)

5 Alternative Hypotheses

Recently, Sachin et al. (2013) proposed three alternative hypotheses that could explain the observed production of plant secondary metabolites by endophytic fungi with specific reference to CPT-producing endophytes (Fig. 9.3). In its generic sense, these hypotheses can explain the production of other plant secondary metabolites, such as taxol, as well as by endophytes.

All the hypotheses can explain the observed attenuation of production of secondary metabolite upon subculturing the endophytes. According to the first hypothesis, the key gene strictosidine synthase (*STR*) involved in camptothecin biosynthesis in plants may be replaced by a new strictosidine synthase-like (SSL) gene or an entirely different protein. Under this condition, conventional methods to clone the gene (using degenerate primers) from the endophytes or searching for homologous segments in the endophyte genome sequence may be futile. Thus, in the absence of the known *STR* gene, endophytes can still synthesise camptothecin in culture. Silencing of the gene by differential methylation during subculture generations can easily explain the observed attenuation. In the second hypothesis, Sachin et al. (2013) proposed that the endophytic fungi may carry the critical gene clusters for secondary metabolite synthesis in extra-chromosomal elements (ECEs) or plasmids. The genes themselves may have evolved independently or acquired from the host tissue through horizontal gene transfer. Evidence exists to indicate that fungi carry plasmids (Griffithis 1995). Finally, in the third hypothesis they suggested that endophytic fungi may harbour plasmids bearing the gene clusters in endohyphal bacteria. Again evidence indicates the presence of endohyphal bacteria in a number of endophytes (Hoffman and Arnold 2010; Bianciotto et al. 2004; Bertaux et al. 2005). The second and third hypotheses are also consistent with the attenuation process; loss of endohyphal bacteria from filamentous fungus as well as plasmids from bacteria on subculture is well known; if these contain genes for secondary metabolite synthesis, it could lead to the attenuation of production over subculture (Sachin et al. 2013). The latter two hypotheses derive support by the recent findings that secondary metabolite gene clusters were recovered from giant linear plasmids isolated from antibiotic-producing *Streptomyces* species; the 1.8-Mb linear plasmids harboured 25 putative secondary metabolite gene clusters including polyketide synthase gene clusters and those coding for terpene synthases or cyclases (Medema et al. 2010).

6 Spin-Off or an Evolutionary Adaptation?

Just why do endophytic fungi produce plant secondary metabolites mimicking those produced by their host plants? Surprisingly despite more than hundred published studies on endophytic fungi producing plant secondary metabolites, no attempt has been made to address this issue. Here we propose two alternative explanations.

6.1 Spin-Off Hypothesis

In this hypothesis we argue that the production of plant secondary metabolites by endophytic fungi mimicking those produced by their host plant may merely be a spin-off by virtue of the fungus being located in the host tissue producing the specific metabolite. In this scenario, the fungal resident might acquire gene clusters located in extra-chromosomal elements (ECEs) or plasmids responsible for the production of the secondary metabolite from the host plant and integrate it into its own genetic machinery to produce the secondary metabolites. The spin-off theory is consistent with the following observations.

6.1.1 Multitude of Endophytic Fungus Producing the Same Plant Secondary Metabolite

Several studies in the last couple of decades have reported that a multitude of endophytic fungi isolated from the same plant species show the ability to produce the same secondary metabolite. For example, over 19 different fungal genera isolated from several *Taxus* species were all shown to produce taxol in vitro. Similarly, Gurudatt et al. (2010) and Shweta et al. (2010, 2013a) found that over 28 different fungi isolated from *N. nimmoniana* and related plants produced the anticancer alkaloid, camptothecin. The fungi in all these cases were very diverse taxonomically or phylogenetically and yet produced the same compound. It is rather unlikely that such collaboration between the host and the diverse fungi could have arisen through mutualistic/symbiotic relationship. Normally for the latter to happen, the mutualistic interactions span over evolutionary time and often engage one or few taxonomically defined groups such as those evident from well-documented relationships between ant-pollinator mutualism, ant-fungal gardens and fig and fig-wasp mutualism.

6.1.2 Attenuation of Production over Subculture Generations Through Loss of Genes or Silencing of Genes

The spin-off theory is also consistent with the phenomenon of attenuation of fungi. Assuming that the endophytes have acquired the metabolic machinery for the production of the secondary metabolite from the host plant through extra-chromosomal elements, it is easy to visualise their gradual loss over subculture generation leading to the attenuation of production of the secondary metabolite. There is now ample evidence to indicate the loss of extra-chromosomal elements or plasmids through subculture of bacteria (Alex and Michael 1998). Alternatively, the attenuation could also be due to silencing of the acquired genes in the endophytes in absence of the host plant.

6.2 Evolutionary Adaptation

In contrast to the spin-off theory, the ability of the fungus to produce secondary metabolites similar to that produced by the host plants could be a result of an evolutionary adaptation that imparts fitness to either or both the partners, the fungus and the host plant. Accordingly in this scenario, we argue that once in the host plant, the endophyte acquires critical gene clusters in extra-chromosomal elements or plasmids responsible for the synthesis of the secondary metabolites and incorporates them to produce the metabolite in its own tissues, in addition to them being synthesised in the host plant tissue as well. Thus, a fungus resident in *Taxus baccata* could use the gene clusters borne on a plasmid and incorporate it in its own biosynthesis for adaptive significance – of deterring other fungal invasions. This is clearly of evolutionary significance to the fungus concerned. However, unlike the spin-off hypothesis, the evolutionary adaptation-based argument would be expected to favour at most a few endophytic associations (not a multitude) with a host plant considering that the host plant-endophyte relationship may have evolved over an evolutionary timescale. Available evidence however indicates that by far the relationship between host and endophyte is one (host) to many (endophyte) relationships. Clearly more research is required to generate critical evidence for these processes.

7 Conclusions

In conclusion, it appears that the mechanism by which endophytes produce secondary metabolites that mimic those produced by their host plants is far from clear. Efforts to unravel the pathway genes in the endophytes, for instance, involved in taxol and camptothecin synthesis, have failed to detect critical genes corresponding to those that exist in plants. Yet, the endophytes isolated from the respective host plants produce the metabolites in culture, identical to those produced by their host plants. Reviewing some recent studies, we argue that the mechanisms leading to the synthesis of the host-specific metabolites by endophytes could be enabled by gene clusters carried in extra-chromosomal elements or plasmids. On invasion into host tissues, these plasmids could be incorporated into the endophyte leading to the production of the metabolites in axenic cultures. However, on subculturing the endophytes, the plasmids are cured or shed off, leading to the attenuation of production of the metabolite by the endophyte. But why do endophytes in the first place produce host-specific metabolites? We propose two possibilities, one based on a simple spin-off theory and the other based on an intrinsic evolutionary adaptation to the endophyte and to the host. Clearly more research is required to unravel the mechanism by which endophytes produce secondary metabolites mimicking those produced by their host plants. Until then, there is a little hope to use endophytes as alternative sources of plant metabolites, a promise that has been in the air for over two decades now.

Acknowledgements The work reported in the paper has been supported by grants from the Department of Biotechnology, Government of India. Thanks are due to the members of the School of Ecology and Conservation for brainstorming some of the ideas presented in the chapter.

References

Alex MS, Michael JB (1998) Bacterial fitness and plasmid loss: the importance of culture conditions and plasmid size. Can J Microbiol 44:351–355

Amna T, Puri SC, Verma V, Sharma JP, Khajuria RK, Musarrat J, Spiteller M, Qazi GN (2006) Bioreactor studies on the endophytic fungus *Entrophospora infrequens* for the production of an anticancer alkaloid camptothecin. Can J Microbiol 52:189–196

Arnold AE (2007) Understanding the diversity of foliar endophytic fungi: progress, challenges, and frontiers. Fungal Biol Rev 21:51–66

Arnold AE, Lutzoni F (2007) Diversity and host range of foliar fungal endophytes: are tropical leaves biodiversity hotspots? Ecology 88:541–549

Arnold AE, Maynard Z, Gilbert GS, Coley PD, Kursat TA (2000) Are tropical fungal endophytes hyperdiverse? Ecol Lett 3:267–274

Arnold AE, Mejia LC, Kyllo D, Rojas EI, Maynard Z, Robbins N, Herre EA (2003) Fungal endophyte limit pathogen damage in tropical trees. Proc Natl Acad Sci U S A 100:15649–15654

Bacon CW, White JF (2000) Microbial endophytes. Marcel Deker Inc., New York, p 487

Bae H, Sicher RC, Kim MS, Kim SH, Strem MD, Melnick RL (2009) The beneficial endophyte *Trichoderma hamatum* isolate DIS 219b promotes growth and delays the onset of the drought response in *Theobroma cacao*. J Exp Bot 60:3279–3295

Bertaux J, Schmid M, Hutzler P, Hartmann A, Garbaye J, Frey-Klett P (2005) Occurrence and distribution of endobacteria in the plant-associated mycelium of the ectomycorrhizal fungus *Laccaria bicolor* S238N. Environ Microbiol 7:1786–1795

Bianciotto V, Genre A, Jargeat P, Cumini E, Becard G, Bonfante P (2004) An obligately endosymbiotic mycorrhizal fungus itself harbors obligately intracellular bacteria. Appl Environ Microbiol 70:3600–3608

Bömke C, Tudzynski B (2009) Diversity, regulation and evolution of the gibberellin biosynthetic pathway in fungi compared to plants and bacteria. Phytochemistry 70:1876–1893

Christensen MJ, Bennett RJ, Schmid J (2002) Growth of *Epichloë/Neotyphodium* and p-endophytes in leaves of *Lolium* and *Festuca* grasses. Mycol Res 106:93–106

Cordell GA (1974) The biosynthesis of indole alkaloids. Lloydia 37:219–298

Eyberger AL, Dondapati R, Porter JR (2006) Endophyte fungal isolates from *Podophyllum peltatum* produce podophyllotoxin. J Nat Prod 69:1121–1124

Faeth SH, Hammon KE (1997) Fungal endophytes in oak trees. I. Long-term patterns of abundance and associations with leafminers. Ecology 78:810–819

Gaylord ES, Preszler RW, Boecklen WJ (1996) Interactions between host plants, endophytic fungi, and a phytophagous insect in an oak (*Quercus Grisea* × *Q. Gambelii*) hybrid zone. Oecologia 105:336–342

Govindachari TR, Viswanathan N (1972) 9-methoxy camptothecin A new alkaloid from *Mappia foetida*. Indian J Chem 10:453–454

Griffithis AJF (1995) Natural plasmids of filamentous fungi. Microbiol Rev 59:673–685

Gunatilaka AAL (2006) Natural products from plant-associated microorganisms: distribution, structural diversity, bioactivity and implication of their occurrence. J Nat Prod 69:509–526

Gurudatt PS, Priti V, Shweta S, Ramesha BT, Ravikanth G, Vasudeva R, Amna T, Deepika S, Ganeshaiah KN, Uma Shaanker R, Puri S, Qazi GN (2010) Attenuation of camptothecin production and negative relation between hyphal biomass and camptothecin content in endophytic fungal strains isolated from *Nothapodytes nimmoniana* Grahm (Icacinaceae). Curr Sci 98: 1006–1009

Hartley SE, Gange AC (2009) Impacts of plant symbiotic fungi on insect herbivores: mutualism in a multitrophic context. Annu Rev Entomol 54:323–342

Heinig U, Scholz S, Jennewein S (2013) Getting to the bottom of Taxol biosynthesis by fungi. Fungal Divers 60:161–170

Hoffman MT, Arnold AE (2010) Diverse bacteria inhabit living hyphae of phylogenetically diverse fungal endophytes. Appl Environ Microbiol 76:4063–4075

Kanehisa M, Goto S, Kawashima S, Okuno Y, Hattori M (2004) The KEGG resource for deciphering the genome. Nucleic Acids Res 32(database issue):D277–D280

Khaldi N, Seifuddin FT, Turner G, Haft D, Nierman WC, Wolfe KH, Fedorova ND (2010) SMURF: genomic mapping of fungal secondary metabolite clusters. Fungal Genet Biol 47:736–741

Kusari S, Zühlke S, Spiteller M (2009a) An endophytic fungus from Camptotheca acuminata that produces camptothecin and analogues. J Nat Prod 72:2–7

Kusari S, Zuhlke S, Kosuth J, Cellarova E, Spiteller M (2009b) Light independent metabolomics of endophytic *Thielavia subthermophila* provides insight into microbial hypericin biosynthesis. J Nat Prod 72: 1825–1835

Kusari S, Spiteller M (2010) Do endophyte metabolomics reflect the phytochemical diversity of the host plant? A test of the hypothesis with Hypericum species. In: Patro LR (ed) Biodiversity conservation and management. Discovery Publishing House Pvt. Ltd., New Delhi, pp 1–28

Kusari S, Zuhlke S, Spiteller M (2011) Effect of artificial reconstitution of the interaction between the plant *Camptotheca acuminata* and the fungal endophyte *Fusarium solani* on Camptothecin biosynthesis. J Nat Prod 74:764–775

Kusari S, Verma VC, Lamshoeft M, Spiteller M (2012) An endophytic fungus from *Azadirachta indica* A. Juss. that produces azadirachtin. World J Microbiol Biotechnol 28:1287–1294

Kutchan TM, Hampp N, Lottspeich F, Beyreuther K, Zenk MH (1988) The cDNA clone for strictosidine synthase from *Rauvolfia serpentina* DNA sequence determination and expression in *Escherichia coli*. FEBS Lett 237:40–44

Li JY, Sidhu RS, Ford E, Hess WM, Strobel GA (1998) The induction of taxol production in the endophytic fungus *Periconia* sp. from *Torreya grandifolia*. J Ind Microbiol 20:259–264

McKnight TD, Roessner CA, Devagupta R, Scott AI, Nessler CL (1990) Nucleotide sequence of a cDNA encoding the vacuolar protein strictosidine synthase from *Catharanthus roseus*. Nucleic Acids Res 18:4939

Medema MH, Trefzer A, Kovalchuk A, van den Berg M, Müller U, Heijne W, Wu L, Alam MT, Ronning CM, Nierman WC, Bovenberg RA, Breitling R, Takano E (2010) The Sequence of a 1.8-Mb bacterial linear plasmid reveals a rich evolutionary reservoir of secondary metabolic pathways. Genome Biol Evol 2:212–224

Mohana Kumara P, Zuehlke S, Priti V, Ramesha BT, Shweta S, Ravikanth G, Vasudeva R, Santhoshkumar TR, Spiteller M, Uma Shaanker R (2012) *Fusarium proliferatum, an endophytic fungus* from *Dysoxylum binectariferum* Hook.f, produces rohitukine, a chromane alkaloid possessing anti-cancer activity. Antonie Van Leeuwenhoek 101:323–329

Moricca S, Ragazzi A (2008) Fungal endophytes in Mediterranean oak forests: a lesson from *Discula quercina*. Phytopathology 98:380–386

Petrini O, Sieber TH, Toti L, Viret O (1992) Ecology, metabolite production, and substrate utilization in endophytic fungi. Nat Toxins 1:185–196

Priti V, Ramesha BT, Shweta S, Ravikanth G, Ganeshaiah KN, Suryanarayanan TS, Uma Shaanker R (2009) How promising are endophytic fungi as alternative sources of plant secondary metabolites. Curr Sci 97:4–6

Puri SC, Verma V, Amna T, Qazi GN, Spiteller M (2005) An endophytic fungus from *Nothapodytes foetida* that produces camptothecin. J Nat Prod 68:1717–1719

Puri SC, Nazir A, Chawla R, Arora R, Riyaz-Ul-Hasan S, Amna T, Ahmed B, Verma V, Singh S, Sagar R, Sharma A, Kumar R, Sharma RK, Qazi GN (2006) The endophytic fungus *Trametes hirsuta* as a novel alternative source of podophyllotoxin and related aryl tetralin lignans. J Biotechnol 122:494–510

Ramesha BT, Suma HK, Senthilkumar U, Priti V, Ravikanth G, Vasudeva R, Santhosh Kumar TR, Ganeshaiah KN, Uma Shaanker R (2013) New plant sources of the anticancer alkaloid, camptothecine from the Icacinaceae taxa, India. Phytomedicine 20:521–527

Rehman S, Shawl AS, Kour A, Andrabi R, Sudan P, Sultan P, Verma V, Qazi GN (2008) An endophytic *Neurospora* sp. from *Nothapodytes foetida* producing camptothecin. Appl Biochem Microbiol 44:203–209

Rubini MR, Silva-Ribeiro RT, Pomella AWV, Maki CS, Araujowl WS (2005) Diversity of endophytic fungal community of cacao (*Theobroma cacao* L.) and biological control of *Crinipellis perniciosa*, causal agent of witches' broom disease. Int J Biol Sci 1:24–33

Sachin N, Manjunatha BL, Mohana Kumara P, Ravikanth G, Shweta S, Suryanarayanan TS, Ganeshaiah KN, Uma Shaanker R (2013) Do endophytic fungi possess pathway genes for plant secondary metabolites? Curr Sci 104:178–182

Saikkonen K, Faeth SH, Helander M, Sullivan TJ (1998) Fungal endophytes: a continuum of interactions with host plants. Annu Rev Ecol Syst 29:319–343

Sandrine G, Bernard L, Olivier L (2010) FUNGI path: a tool to assess fungal metabolic pathways predicted by orthology. BMC Genomics 11:81

Schardl CL, Liu JS, White JF, Finkel RA, An Z, Siegel MR (1991) Molecular phylogenetic relationships of nonpathogenic grass mycosymbionts and clavicipitaceous plant pathogens. Plant Syst Evol 178:27–41

Schiff PB, Horowitz SB (1980) Taxol stabilizes microtubules in mouse fibroblast cells. Proc Natl Acad Sci U S A 77:1561–1565

Schulz B, Boyle C, Draeger S, Rommert AK, Krohn K (2002) Endophytic fungi: a source of novel biologically active secondary metabolites. Mycol Res 106:996–1004

Shweta S, Zuehlke S, Ramesha BT, Priti V, Mohana Kumar P, Ravikanth G, Spiteller M, Vasudeva R, Uma Shaanker R (2010) Endophytic fungal strains of *Fusarium solani*, from *Apodytes dimidiata* E. Mey. ex Arn (Icacinaceae) produce camptothecin, 10-hydroxycamptothecin and 9-methoxycamptothecin. Phytochemistry 71:117–122

Shweta S, Gurumurthy BR, Ravikanth G, Ramanan US, Shivanna MB (2013a) Endophytic fungi from *Miquelia dentata* Bedd produce the anti-cancer alkaloid, camptothecine. Phytomedicine 20:337–342

Shweta S, HimaBindu J, Raghu J, Suma HK, Manjunatha BL, Mohana Kumara P, Ravikanth G, Nataraja KN, Ganeshaiah KN, Uma Shaanker R (2013b) Isolation of endophytic bacteria producing the anti-cancer alkaloid camptothecine from *Miquelia dentata* Bedd (Icacinaceae). Phytomedicine. doi:org/10.1016/j.phymed.2013.04.004

Staniek A, Woerdenbag HJ, Kayser O (2009) *Taxomyces andreanae*: a presumed paclitaxel producer demystified? Planta Med 75:1561–1566

Stierle A, Strobel GA, Stierle D (1993) Taxol and taxane production by *Taxomyces andreanae*, an endophytic fungus of Pacific yew. Science 260:214–216

Strobel G, Daisy B (2003) Bioprospecting for microbial endophytes and their natural products. Microbiol Mol Biol Rev 67:491–502

Suffness M (1995) Taxol, science and applications. CRC Press, Boca Raton

Sun Y, Lu H, Li Y, Sun C, Song J, Niu Y, Zhu Y, Dong L, Lv A, Tramontano E, Chen S (2011) Pyrosequencing of the Camptotheca acuminata transcriptome reveals putative genes involved in camptothecin biosynthesis and transport. BMC Genomics C7–533(12):1–11

Tan RX, Zou WX (2001) Endophytes: a rich source of functional metabolites. Nat Prod Rep 18:448–459

Uma Shaanker R, Ramesha BT, Ravikanth G, Gunaga R, Vasudeva R, Ganeshaiah KN (2008) Chemical profiling of *Nothapodytes nimmoniana* for Camptothecin, an important anticancer alkaloid: toward the development of a sustainable production system. Bioactive Mol Med Plants 2008:197–213

Vega FE, Simpkins A, Aime MC, Posada F, Peterson SW, Rehner SA, Infante F, Castillo A, Arnold AE (2010) Fungal endophyte diversity in coffee plants from Colombia, Hawaii, Mexico and Puerto Rico. Fungal Ecol 3:122–138

Wall ME, Wani MC, Cook CE (1966) Plant antitumor agents-I. The isolation and structure of camptothecin, a novel alkaloidal leukemia and tumor inhibitor from *Camptotheca acuminata*. J Am Chem Soc 88: 3888–3890

Xiong Z-Q, Yang Y-Y, Zhao N, Wang Y (2013) Diversity of endophytic fungi and screening of fungal paclitaxel producer from Anglojap yew, Taxus x media. BMC Microbiol 13:71

Yamazaki Y, Sudo H, Yamazaki M, Aimi N, Saito K (2003) Camptothecin biosynthetic genes in hairy roots of *Ophiorrhiza pumila*: cloning, characterization and differential expression in tissues and by stress compounds. Plant Cell Physiol 44:395–403

Yamazaki Y, Kitajima M, Arita M, Takayama H, Sudo H, Yamazaki M, Aimi N, Saito K (2004) Biosynthesis of camptothecin. *In silico* and *in vivo* tracer study from [$^{1-13}$C] glucose. Plant Physiol 134:161–170

Yang X, Zhang L, Guo B, And Guo S (2004) Preliminary study of vincristine-endophytic fungus isolated from leaves of *Catharanthus roseus*. Zhongcaoyao 35:79–81

Zhang HW, Song YC, Tan RX (2006) Biology and chemistry of endophytes. Nat Prod Rep 23:753–771

Zhang P, Zhou P, Yu LJ (2009) An endophytic Taxol producing fungus from *Taxus media*, *Cladosporium cladosporioides* MD2. Curr Microbiol 59:227–232

Zhao J, Zhou L, Wang J, Shan T, Zhong L, Liu X, Gao X (2010) Endophytic fungi for producing bioactive compounds originally from their host plants. In: Mendez-Vilas A (ed) Current research, technology and education in tropics in applied microbiology and microbial biotechnology, Microbiology book series – Number 2

Zhu D, Wang J, Zeng Q, Zhang Z, Yan R (2010) A novel endophytic Huperzine A–producing fungus, *Shiraia* sp. Slf14, isolated from *Huperzia serrata*. J Appl Microbiol 109:1469–1478

Endophytes as a Novel Source of Bioactive New Structures

10

Mahmoud Fahmi Elsebai, Mysore V. Tejesvi, and Anna Maria Pirttilä

Abstract

Fighting the existing and emerging diseases is one of the big challenges of this age, as the appearance of drug-resistant pathogens is an alarming phenomenon, globally. To address this matter of urgency, researchers and pharmaceutical companies have to revive efforts to develop completely new classes of pharmaceuticals. Natural products have proved a fascinating resource in the continued search for new drug candidates. Among various natural sources, microorganisms represent a sustainable and reproductive source of bioactive compounds, where endophytes are considered a hidden component. Endophytes have fascinating potential for a source of new drug leads as they have capacity to synthesize organic compound of diverse structural features. Most of the promising natural products are available only in extremely small quantities, which necessitate substantial efforts to produce required amounts for pharmacological testing. In addition, many natural products have highly complex structures, complicating commercial production through chemical synthesis. The majority of such drug candidates remains pharmacologically undeveloped due to the perceived supply problem and anticipated higher production costs. Therefore, new methods and techniques such as metagenomics and metatranscriptomics are needed to facilitate production of such compounds for pharmaceutical industry.

1 Introduction

Despite the discovery of many effective drugs, a number of health problems still remain uncured. These problems include various types of cancer, viral infections such as HIV and HCV, severe fungal and bacterial infections, Parkinson's and Alzheimer's diseases, depression, obesity, cardiovascular diseases, inflammatory disorders, and many others. Therefore, the search for novel

M.F. Elsebai
Department of Biology, University of Oulu, PO Box 3000, FIN-90014 Oulu, Finland

Faculty of Pharmacy, Department of Pharmacognosy, Mansoura University, Mansoura, Egypt

M.V. Tejesvi • A.M. Pirttilä (✉)
Department of Biology, University of Oulu, PO Box 3000, FIN-90014 Oulu, Finland
e-mail: am.pirttila@oulu.fi

V.C. Verma and A.C. Gange (eds.), *Advances in Endophytic Research*, DOI 10.1007/978-81-322-1575-2_10, © Springer India 2014

therapeutic agents continues, and there is a need to discover new structural leads for drug development. Natural products offer a good opportunity both for a direct therapeutic effect and for the discovery of lead compounds that provide the basis and inspiration for the semisynthesis or total synthesis of effective drugs. Investigation of endophytic compounds, especially fungal ones, resulted in the discovery of structurally novel natural products with interesting biological activities, evidenced by the increased number of published reviews and the release of compounds in the clinical market (Mayer et al. 2011; Rateb and Ebel 2011). Unfortunately, these fascinating machineries synthesizing diverse structures still remain under-explored for new drug discovery.

2 Importance of Microbes

Throughout the history, microorganisms have been of considerable environmental and economic importance to mankind. Microbes, being single-cell entities, possess unique traits since all functions of life such as reproduction, assimilation, digestion, and growth take place in a highly condensed form within a single cell, evidenced by their ability to grow fast. This may explain why these extremely small, single-cell organisms are highly productive and their enzyme systems can catalyze a wide variety of chemical reactions, some being so complicated, that they cannot be reproduced in the synthetic chemical laboratories.

Microorganisms are proven sources of potential drug candidates; besides, they are identified as significant agents for biotransformation and fermentation. For example, *Candida utilis* is used for food and fodder yeast production, production of secondary metabolites, genetic engineering, and microbial pesticides against entomopathogenic fungi, bacteria, and viruses. Vinegar production is perhaps the oldest and best-known example of microbial oxidation. Similarly, food manufacturing processes such as beer, wine, cheese, saccharifying grains, and leavening of bread all involve the use of beneficial microbes. The further exploitation, such as the production of certain alcohols, vitamins, alkaloids, organic and amino acids, antibiotics, cortisone, and nucleotides, is much more recent. The most promising natural products are available only in extremely small quantities, and, therefore, substantial efforts are needed to provide sufficient amounts for pharmacological testing. For example, one ton of a *Lissodendoryx* sp. sponge was collected to obtain 310 mg of the anticancer compound halichondrin B (Piel 2006). In addition, many natural products have highly complex structures, complicating commercial production through chemical synthesis. Therefore, the majority of such drug candidates remain pharmacologically undeveloped. Pharmaceutical companies hesitate to pursue such bioactive natural products due to the perceived supply problem (Paterson and Anderson 2005) leading to high production costs. Therefore, microorganisms, such as bacteria, cyanobacteria, and fungi, have mainly attracted attention as potential lead compound producers (Lam 2007). However, culturable microorganisms can be manipulated and processed due to their small size and huge reproduction capabilities. Scaling up and mass production are relatively easy in microorganisms that can be grown in large volume. Many microorganisms can be stored for an indefinite time, ensuring availability of the targeted source organism. Microorganisms can be manipulated both physicochemically and genetically to increase yields of desired natural products (Kharwar et al. 2011).

Microorganisms produce secondary metabolites for many reasons, such as predation and defense against invading pathogens. A fascinating example is the isolation of isatin from the shrimp *Palaemon macrodactylus*. The surface of the shrimp embryos is consistently covered by a bacterium of the genus *Alteromonas* which is the real producer of isatin. Treatment of the embryos with antibacterials inhibited bacterial growth and leads to death of the embryos from infection by the fungus *Lagenidium callinectes*, indicating that the bacterial metabolite isatin protected the shrimp embryos against fungal infection (Kelecom 2002).

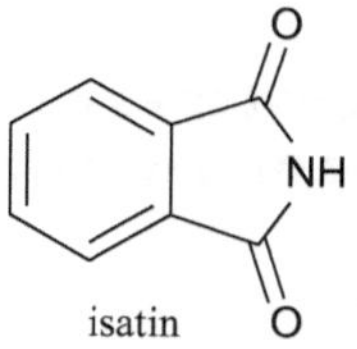

Many drugs currently in the pharmaceutical market, especially antibiotics, have been reported from microorganisms. The most famous antibiotic is the penicillin produced by the fungus *Penicillium chrysogenum* (previously known as *P. notatum*), which was discovered by the Nobel laureate Alexander Fleming in 1928. The clinical use of penicillin in the 1950s opened up a new era in drug discovery, followed by the isolation of a huge number of antibiotics from soil microbes, for example, cephalosporins from *Cephalosporium* species. Later the chemical derivatization of antibiotics that were discovered until the early 1970s established new generations of clinically useful antibiotics (Overbye and Barrett 2005). In 1949, Harold Raistrick initiated the first systematic study of fungal metabolites and recognized fungi as a prolific source of natural products (Saleem et al. 2007).

Kelecom (2002) predicted a relationship between the type of secondary metabolite and the source of microbe, rather than the microorganisms themselves. The latter was exemplified by the fungi in the genus *Aspergillus* that produce fumiquinazoline derivatives if they are obtained from fish, sesquiterpene nitrobenzoate derivatives if they originate from algae, and indole diketopiperazine derivatives if they are isolated from sponges. Kelecom (2002) also reported that bacteria produce almost equally antitumour and antibacterial compounds, but fungi are richer sources of anticancer metabolites than antibacterial compounds. Therefore, when cytotoxic compounds are desired, for example, sediment bacteria, algal fungi or spongeal fungi should be preferred in a marine environment. If antibacterial compounds are searched for, one should prefer bacteria over fungi.

Genetic studies have shown that fungi are more closely related to animals than to plants. The main difference is that the fungal cell walls contain mainly chitin. However, Jones et al. (2011) reported a new species of fungi without chitin in their cell walls, and hence we are in front of a novel intermediate form, which redefines the fungal tree of life. This form does not produce a chitin-rich cell wall during any of the life cycle stages observed and therefore does not conform to the standard fungal body plan and is named *Cryptomycota*. This new issue desires more attention from the endophytic researchers for drug discovery and also the possibility of finding these new fungi in the marine environment.

3 Marine Fungi

Most reviews on bioactive compounds from endophytes have overlooked marine sources, and, therefore, we take this source in further consideration. Marine fungi are a form of ecological, and not a taxonomic, group of fungi that is divided into two groups, that is, obligate and facultative, a classical definition that is still universally accepted. Obligate marine fungi are those that grows and sporulate exclusively in marine water, while facultative marine fungi are those from freshwater or terrestrial milieus able to grow and possibly also to sporulate in the marine environment after some physiological adaptations (Raghukumar 2008). All marine habitats can host fungal strains, for example, marine plants (algae, sea grasses, driftwood, and mangrove plants), marine invertebrates (sponges, corals, bivalves, and crustaceans), vertebrates (fishes), and inorganic matter (soil, sediments). It is estimated that in general 74,000 fungal species have been described so far, and the overall expected global fungal diversity amounts to 1.5 million species. The fungal diversity in individual habitats or regions is considerably underestimated, for example, marine fungi from sediments are not observed easily by microscope due to their tendency to form aggregates (Rateb and Ebel 2011). Since algae and sponges are the most prevalent sources of marine fungi for chemical studies, they are subjected to meticulous studies on their fungal communities. The organisms that live and thrive in spite of pronounced pressures can, to a high degree, be expected to produce metabolites, which might be of interest for the development of drugs to cure various diseases. Therefore, many researchers are interested in bio-prospecting such unusual environments in a quest to find exotic and unique metabolite-producing organisms. The extreme conditions in terrestrial and marine environments are encountered in the form of high temperatures at the tropical areas, elevated hydrostatic pressure and low temperatures in the deep sea, low temperatures in sea ice, high temperature and elevated hydrostatic pressure with high concentrations of metals in hydrothermal vents, and hypersaline water bodies and hypoxic conditions in coastal as well as offshore waters, not to mention deep-sea sediments, oil-contaminated sites, and sites grazed by both terrestrial and marine animals (Raghukumar 2008). Examples of the extremophilic microbes (extremophiles) include acidophiles (acidic sulfur hot springs), alkaliphiles (alkaline lakes), halophiles (salt lakes), hypo- and hyperthermophiles (deep-sea vents), and psychrophiles (alpine lakes, arctic and antarctic waters) (Cragg et al. 2009).

4 Endophytes as a Potential Source of Natural Products

Among the natural sources, the potential of endophytes in drug discovery has been identified within the past decade (Pirttilä and Frank 2011). Isolation and identification of metabolites from microorganisms, especially endophytic fungi and bacteria, are rapidly growing, as can be observed from the increased number of reviews, patents, and original research articles published every year in this modern field of drug discovery (Tejesvi and Pirttilä 2011). The presence of many fungal metabolites in the pharmaceutical market indicates the potential of microorganisms as a valuable source of lead drugs, for example, the antibacterial terpenoid fusidic acid (Fucidin®), the antibiotic polyketide griseofulvin (Likuden M®), semisynthetic or synthetic penicillins and cephalosporins, chloramphenicol, macrolides, statins, as well as the ergot alkaloids such as ergotamine (Ergo-Kranit®) (Hamilton-Miller 2008; Butler 2008; Parry et al. 2011).

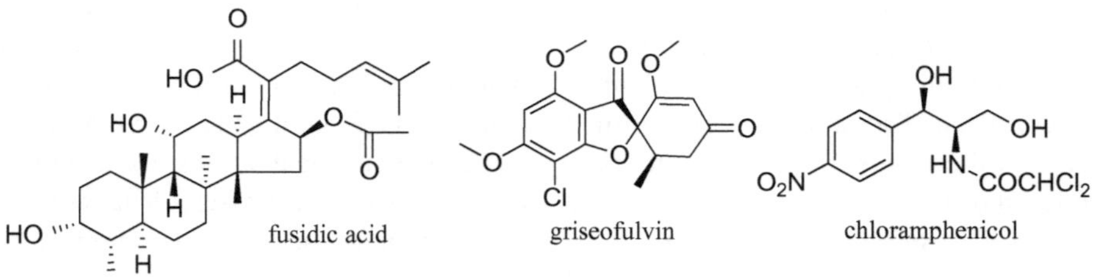

fusidic acid griseofulvin chloramphenicol

Furthermore, endophytes have recently obtained attention in bio-inoculation to increase the plant growth (biomass) and production of key plant secondary metabolites. This is exemplified by bio-inoculation of bacterial endophytes to the plant *Catharanthus roseus*. *C. roseus* is particularly well known for its therapeutically useful terpenoid indole alkaloids, including the anticancer bisindole alkaloids vinblastine and vincristine, as well as other alkaloids, such as ajmalicine and serpentine (Tiwari et al. 2013).

5 Chemistry and Pharmacology of Endophytes

Endophytic fungi are considered as the hidden members of the microbial world and represent an underutilized resource for new compounds. They produce diverse structural metabolites such as polyketides, alkaloids, peptides, proteins, lipids, shikimates, glycosides, isoprenoids, and hybrids of these metabolites (Mayer et al. 2011; Rateb and Ebel 2011). These metabolites exhibit diverse pharmacological activities. In this sense, appreciation of endophytic fungi is much greater than that of other endophytes. Global biodiversity of endophytic fungi is enormous, and more than 100 fungal strains have been isolated from some plant taxa (Tejesvi et al. 2011) and have received less attention than soil microbes or plant pathogens, because they exist asymptomatically in the plant tissue (Kharwar et al. 2011).

5.1 Cytotoxic and Antimicrobial Endophytic Metabolites

Cancer is a major cause of death worldwide; likewise, microbial infections have become a serious health threat. Specifically, development of resistance toward current antibiotics is a significant problem in the treatment of infectious diseases. Therefore, the discovery and development of new antibiotics is becoming a high priority in biomedical research (Saleem et al. 2010; Zhang et al. 2009). Since the discovery and application of penicillin, antibiotics have saved billions of lives and played an important role in human history. Many pathogenic microorganisms, for example, methicillin-resistant *Staphylococcus aureus* (MRSA) and vancomycin-resistant *Enterococcus faecium* (VREF), have developed resistance toward current antibiotics, and the spread of resistance has become exceedingly serious. Meanwhile, some new emerging infectious diseases, for example, cryptococcal meningitis and toxoplasmosis, have emerged and become prevalent. All these new problems demand an increasing amount of novel antibiotics to be discovered (Zhang et al. 2009; Alekshun and Levy 2007). The nonnitrogenous methyl phenalenones produced by the endophytic fungus *Coniothyrium cereale* have shown valuable cytotoxic and antimicrobial activities. In antimicrobial assays, conioscleroderolide, coniosclerodione, (–)-cereolactone, and (–)-scleroderolide showed activity against *Staphylococcus aureus* SG 511 with MIC values of 23.8, 65.7, 52.0, and 23.8 μM, respectively. In agar diffusion assays, *Z*-coniosclerodinol, (*S*,*S*)-sclerodinol, and coniolactone inhibited (>15 mm) the growth of *Mycobacterium phlei*. (–)-Trypethelone strongly inhibited the growth of *M. phlei*, *S. aureus,* and *E. coli* with inhibition zones of 18, 14, and 12 mm, respectively (Elsebai et al. 2011a, b). In cytotoxic assays, using an MTT assay with mouse fibroblast cells, the compounds (–)-sclerodione and (–)-trypethelone exhibited significant activity with an IC_{50} value of 6.4 and 7.5 μM, respectively. Cytotoxicity was also determined using an epithelial bladder carcinoma cell line, in which the compounds conioscleroderolide and (–)-scleroderolide exhibited very weak in vitro cytotoxicity with IC_{50} values of 27 and 41 μM, respectively (Elsebai et al. 2011a, b).

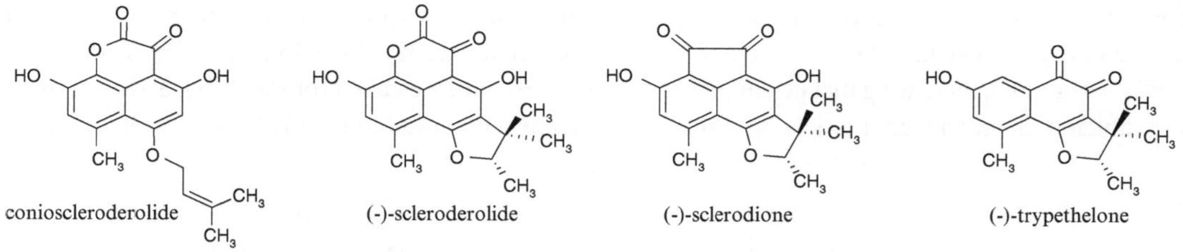

conioscleroderolide (-)-scleroderolide (-)-sclerodione (-)-trypethelone

Metabolites of the endophytic fungus *Coniothyrium cereale* with cytotoxic and antimicrobial activities

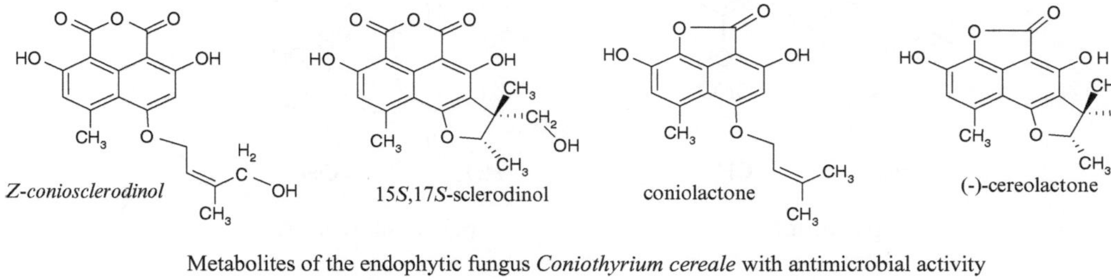

Z-coniosclerodinol 15*S*,17*S*-sclerodinol coniolactone (-)-cereolactone

Metabolites of the endophytic fungus *Coniothyrium cereale* with antimicrobial activity

Studies on endophytic fungi indicate that they are prolific producers of bioactive natural products. After the isolation of taxol (potent microtubule stabilizer) from the endophyte of northwestern Pacific yew (Stierle et al. 1993), researchers have reported several other important anticancer agents from fungal endophytes, such as camptothecin and its analogues, vincristine, and podophyllotoxin (Kharwar et al. 2011). Taxol was originally isolated from the host plant *Taxus brevifolia* and then reported as a product of the endophytic fungus *Taxomyces andreanae* (Stierle et al. 1993). Also podophyllotoxin was originally isolated from the rhizomes of the host plant *Podophyllum peltatum* and later identified as products of the endophytes *Phialocephala fortinii* (Eyberger et al. 2006) and *Fusarium oxysporum* of *Juniperus recurva* (Tejesvi et al. 2011). Podophyllotoxin is a valuable natural product as the lead for several therapeutic agents, including the clinically used anticancer drugs teniposide and etoposide (Canel et al. 2000).

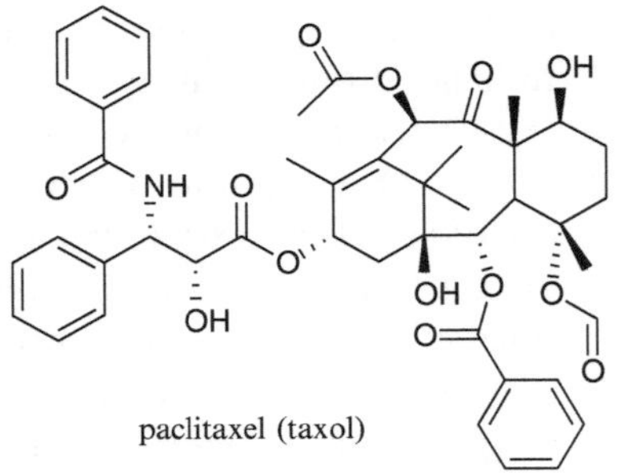

paclitaxel (taxol)

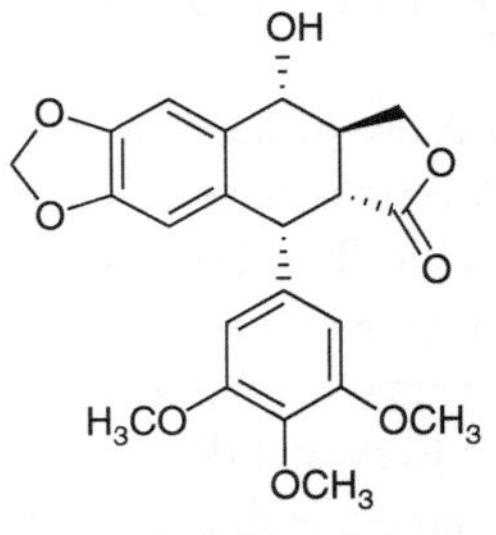

podophyllotoxin

Another example demonstrating the potential of endophytes for natural products discovery is pestalone. Pestalone is a chlorinated benzophenone antibiotic that was produced by a co-cultured endophytic algal marine fungus/unicellular marine bacterium strain CNJ-328. Pestalone exhibits moderate in vitro cytotoxicity and shows potent antibiotic activity against methicillin-resistant *Staphylococcus aureus* (MIC=37 ng/mL) and vancomycin-resistant *Enterococcus faecium* (MIC=78 ng/mL), indicating that pestalone should be evaluated in advanced models of infectious diseases (Cueto et al. 2001). Pestalachlorides A–C, three new chlorinated benzophenone derivatives, have been isolated from cultures of the endophytic fungus *Pestalotiopsis adusta*. Pestalachloride A was obtained as a mixture of two inseparable isomers, whereas pestalachloride C was found to be a racemic mixture. Pestalachloride A and B displayed significant antifungal activities against some plant pathogens (Li et al. 2008). Pestalone can be readily converted into pestalachloride A by a simple treatment with ammonia at pH 8 (Slavov et al. 2010).

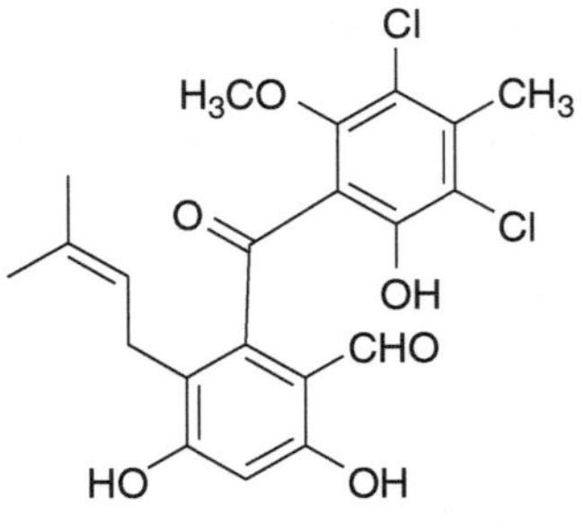

pestalone

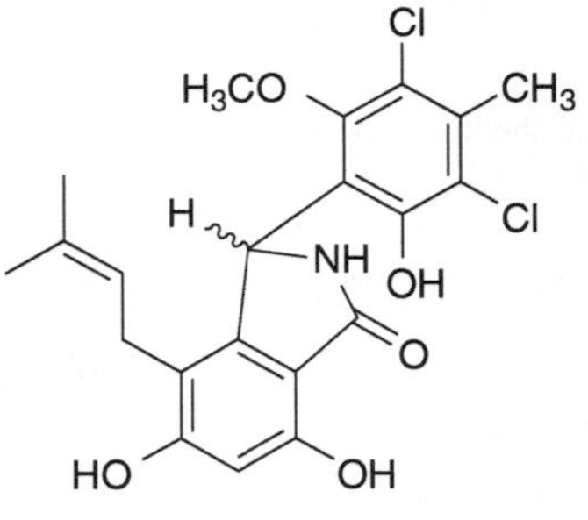

pestalachlorideA

The novel compound KL-4 was isolated from the fungal endophytic extract of the medicinal plant *Gloriosa superba* and was subjected to antimicrobial and anticancer activities. It showed broad spectrum as antifungal and significantly inhibited leukemic cancer cell line THP-1 and breast cancer cell line with IC_{50} 30 and 50 μg/mL, respectively, and was found to possess potency comparable to standard anticancer agents mitomycin-c and 5-FU. Compound KL-4 also inhibited lung cancer cell lines A-549 and CV-1 (Budhiraja et al. 2013).

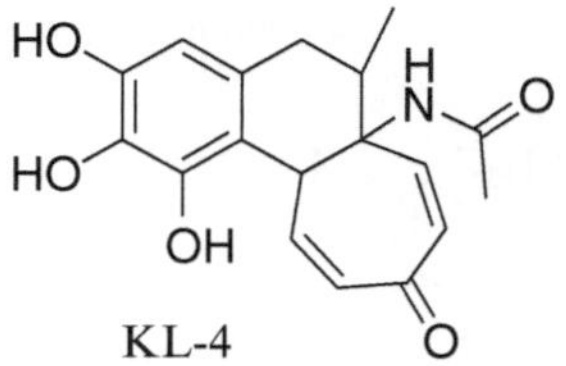

KL-4

Another recent example of endophytic products with cytotoxic activity is the endophytic fungus *Cephalotheca faveolata* isolated from leaves of *Eugenia jambolana* Lam (Lamiaceae) from India. Sclerotiorin exhibited antiproliferative activity against different cancer cell lines and induced apoptosis in colon cancer (HCT-116) (Giridharan et al. 2012).

The endophytic fungus *Pichia guilliermondii* Ppf9 derived from the medicinal plant *Paris polyphylla* var. *yunnanensis* produces interesting antimicrobial steroids and one nordammarane triterpenoids. Helvolic acid exhibited the strongest antibacterial activity against all tested bacteria with MIC values ranging from 1.56 to 50 μg/mL and IC_{50} values from 0.98 to 33.19 μg/mL. It also exhibited a strong inhibitory activity on the spore germination of *Magnaporthe oryzae* with an IC_{50} value of 7.20 μg/mL (Zhao et al. 2010).

5.2 Endophytic Metabolites Acting on Human Leukocyte Elastase Enzyme (HLE)

The excessive and uncontrolled human leukocyte elastase (HLE) activity may result in several pathological states such as chronic obstructive pulmonary disease (COPD), pulmonary emphysema, rheumatoid arthritis, and cystic fibrosis (Korkmaz et al. 2008). The detailed analysis of the marine endophytic fungus *Phaeosphaeria spartinae* resulted in discovery of active compounds, named spartinoxide and prenyl-hydroxyl benzoic acid, toward HLE (Elsebai et al. 2010). Analysis of the products of another marine

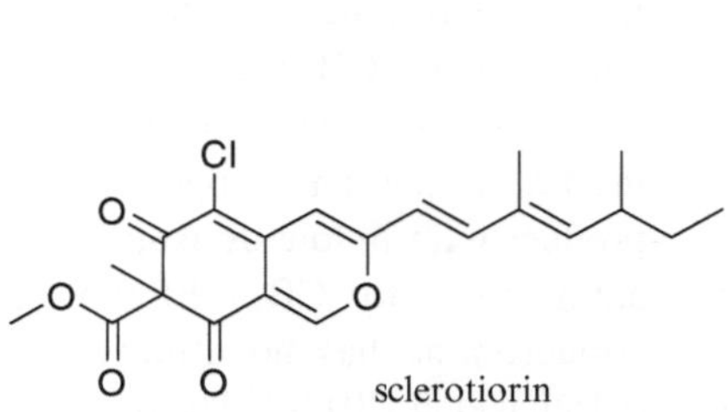

sclerotiorin

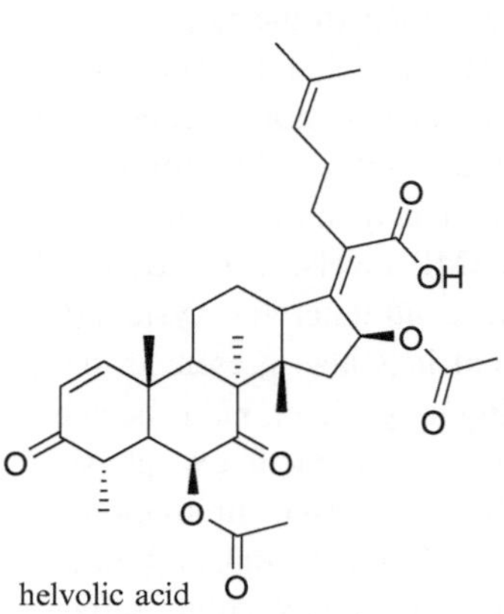

helvolic acid

endophytic fungus *Coniothyrium cereale* resulted in discovery of a series of compounds with a polyketidic methyl phenalenones. The alkaloidal nitrogenous derivatives, named (–)-cereolactam, (–)-cereoaldomine, and conioimide, have valuable activities toward HLE (Elsebai et al. 2011b, 2012).

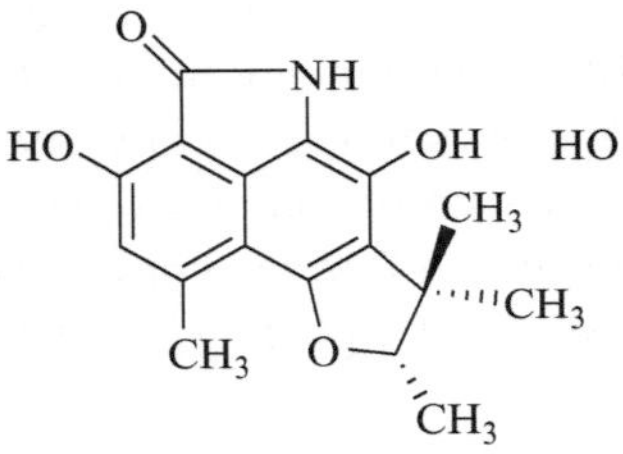

(-)-cereolactam 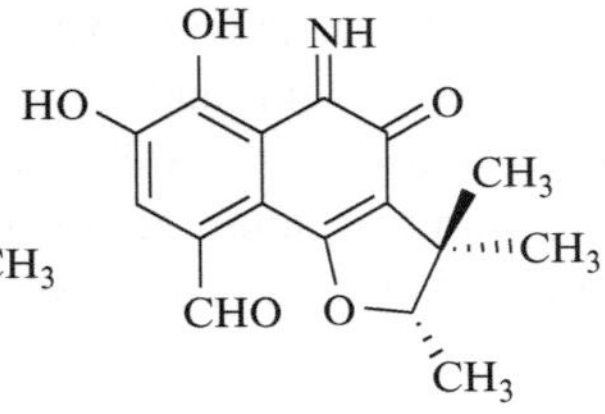 (-)-cereoaldomine 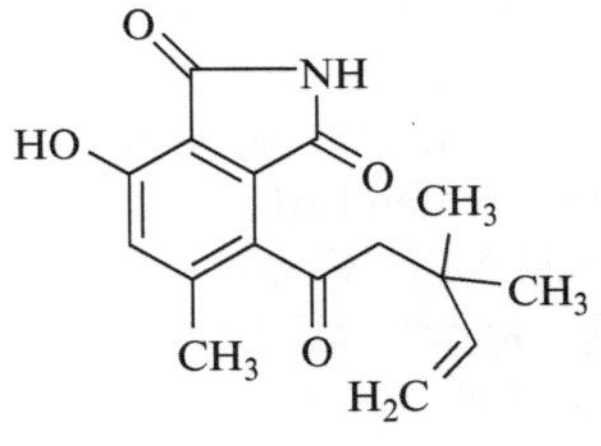 conioimide

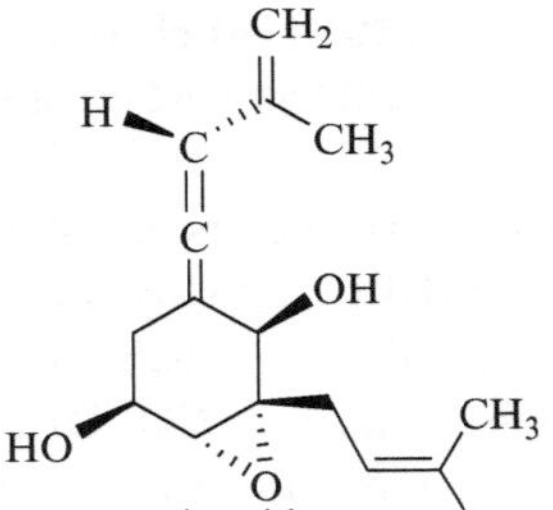

spartinoxide 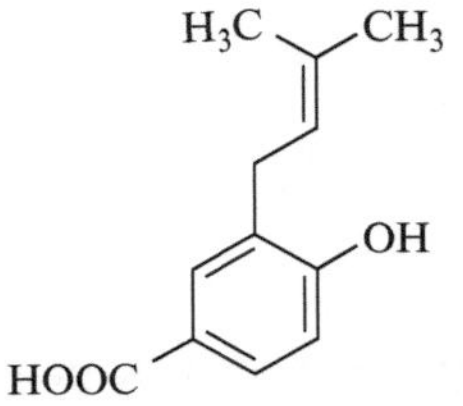 prenyl-OH-benzoicacid

5.3 Endophytic Metabolites Acting on Cannabinoid Receptors

Cannabinoid receptors are located in the cell membrane and belong to the G protein-coupled receptor (GPCR) super family. They are divided into two distinct cannabinoid receptor subtypes designated CB_1 and CB_2 and inhibition of adenylate cyclase upon activation results in reduced intracellular cAMP levels. The CB_1 receptor is expressed in the central nervous system (CNS) in high density, but it is also present in peripheral tissues including lungs, liver, kidneys, and adipocytes. CB_1 activation mediates physiological responses such as analgesia, stimulation of appetite, and euphoria. CB_1 antagonists show appetite-suppressing and antischizopathic effects. The CB_2 receptor is mainly present in organs and cells of the immune system including spleen, tonsils, and thymus, and its activation results in analgesic and anti-inflammatory effects. Here, we report a highly potent and selective cannabinoid receptor ligand, a CB_1 antagonist, from a novel source, namely, a fungal source (Elsebai et al. 2011c). Peripherally acting CB_1 receptor antagonists without CNS penetration would be promising drugs for the treatment of metabolic disorders associated with abdominal obesity, as they would avoid side effects caused by central CB_1 receptor activation, for example, depression, anxiety, and stress disorders (Elsebai et al. 2011c). The endophytic *Auxarthron reticulatum* produces an alkaloid amauromine, which is a potent selective antagonist for CB_1 receptors with a K_i value 178 nM. To the best of our knowledge, amauromine is the first fungal and exogenous dipeptide natural product with indole derivative that has selective antagonism to CB_1 (Elsebai et al. 2011c). Amauromine has no affinity to CB_2 receptors (Elsebai et al. 2011c). The origin of the ligands of cannabinoid receptors can be classified into three groups: (1) endocannabinoids, such as N-arachidonoylethanolamine; (2) phytocannabinoids, such as Δ^9-tetrahydrocannabinol (Δ^9-THC); and (3) synthetic cannabinoids such as the agonist nabilone, a synthetic analogue of Δ^9-THC, and the antagonist rimonabant, which is synthetically produced. Amauromine, hence, represents the fourth source of cannabinoid ligands, namely, fungal origin.

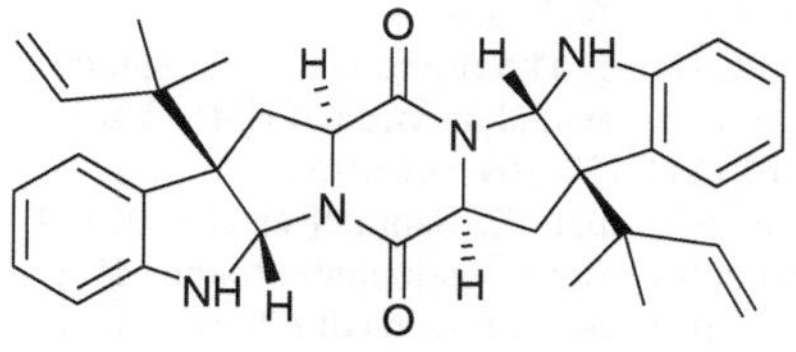

amauromine

In functional assays that measured forskolin-induced cAMP accumulation in CHO cells expressing the human CB_1 receptor, amauromine had no agonistic effect. However, amauromine (300 nM) led to a significant rightward shift of the concentration-response curve for the potent CB receptor agonist CP55, 940 in inhibiting forskolin-induced cAMP accumulation at the G_i protein-coupled CB_1 receptor. A K_b value of 66.6 nM was determined for amauromine. Many synthetic indoles are known to have affinity for CB receptors and exhibit high cannabimimetic effects on CB_2, but only weak or no affinity to CB_1 receptors. In contrast to such previous results, the compound amauromine functions as a selective antagonist to CB_1 receptor (Elsebai et al. 2011c).

6 Culture-Independent Methods for Searching New Endophytic Metabolites

In general, microorganisms are ubiquitous and widely distributed in the nature, playing an important role in the regulation and maintenance of ecological processes. However, it is estimated that <1 % of microorganisms can be cultivated using standard laboratory techniques (Amann et al. 1995). As a result, the culture-dependent methods bias our view on microbial diversity and majority of the prokaryotic phyla are unculturable (Connon and Giovannoni 2002). During the past two decades, the application of molecular methods by polymerase chain reaction (PCR) amplification of ribosomal (rRNA) and conserved protein-encoding genes, such as beta tubulins, histone (Tejesvi and Prakash 2009), ribulose-1,5-bisphosphate carboxylase/oxygenase (RuBisCO) (Chen et al. 2009), sulfate thioesterase/thiohydrolase (soxB) (Chen et al. 2009), and methyl-coenzyme reductase (mcrA) (Vianna et al. 2009), has revolutionized the identification of microbial communities in the environments. Majority of the studies have relied on the restriction fragment length polymorphisms (RFLP) (Laguerre et al. 1994), single-strand-conformation polymorphism (SSCP) (Lee et al. 1996), denaturing/temperature gradient gel electrophoresis (DGGE/TGGE) (Muyzer 1999), terminal restriction fragment length polymorphisms (T-RFLP), (Dunbar et al. 2000), and quantitative PCR (qPCR) (Takai and Horikoshi 2000). These traditional methods have been widely applied over the past two decades, but they can only be employed for detection or identification of microbes and are not suitable for functional screening and identification of gene-encoded peptides or metabolites. Meanwhile, the emergence of next-generation sequencing methods (pyrosequencing) has propelled many exciting fields such as single-cell genomics (Blainey 2013), metagenomics (Felczykowska et al. 2012), transcriptomics (McGettigan 2013), metatranscriptomics (Jang et al. 2012), and metaproteomics (Muth et al. 2012). These techniques will enable the functional screening and identification of candidate gene-encoded proteins and peptides from endophytes for application in agriculture, food, and pharmaceuticals. Fungi are known to have an excellent potential for the production of diverse secondary metabolites. For instance, the genome sequences of the *Aspergillus fumigatus*, *A. nidulans,* and *A. oryzae* have revealed the presence of 28 (*A. fumigatus*) to 48 (*A. oryzae*) gene clusters with polyketide synthase and nonribosomal peptide synthetase genes (Keller et al. 2005). Several metabolites and peptides have been discovered recently from the environment using metagenomic tools. Notably, antibacterials such as violacein, indigo, nocardamine, and turbomycins were all discovered from soil libraries using metagenomics (Banik and Brady 2010). With respect to endophytes, care should be taken to differentiate endophytic products from those of the plant host. For example, in our study on screening for defensins from *Picea glauca* EST libraries, a peptide with high similarity to plectasin (Mygind et al. 2005) was identified. Further studies indicated that this defensin, named endopiceasin, likely originated from an endophytic fungus of *P. glauca* (Picart et al. 2012). We recently developed a method to separate endophytic DNA from that of the host for metagenomic purposes. This way, we discovered a novel gene *En-MAP1* of fungal origin from the plant *Empetrum nigrum* L., having no significant similarity to other known sequences (Tejesvi et al., unpublished). The folded, expressed protein itself had no antibacterial activity, but its tryptic digests exhibited antimicrobial activity against *S. aureus* and *E. coli*.

7 Conclusion

Endophytes have already shown to be a potent source for discovery of bioactive compounds, but still new and innovative approaches are needed for natural product-based drug discovery to become successful again. The current efficiency in identifying new drugs from endophytes is poor and there should be systematic approaches for isolation and development of bioactive compounds. Increasing number of novel methods combined with tools of metagenomics, metatranscriptomics, and metaproteomics should be employed to mine microbial genomes from the environment for returning to the golden age of natural product discovery.

Acknowledgments This chapter is dedicated to Profs. Drs. Magda Nasr and Hassan-Elrady A. Saad, Faculty of Pharmacy, Mansoura University. The Egyptian Government is thanked for financial support to Dr. M. Elsebai.

References

Alekshun MN, Levy SB (2007) Molecular mechanisms of antibacterial multidrug resistance. Cell 128: 1037–1050

Amann RI, Ludwig W, Schleifer KH (1995) Phylogenetic identification and in situ detection of individual microbial cells without cultivation. Microbiol Rev 59:143–169

Banik JJ, Brady SF (2010) Recent application of metagenomic approaches toward the discovery of antimicrobials and other bioactive small molecules. Curr Opin Microbiol 13:603–609

Blainey PC (2013) The future is now: single-cell genomics of bacteria and archaea. FEMS Microbiol Rev. doi:10.1111/1574-6976.12015

Budhiraja A, Nepali K, Sapra S et al (2013) Bioactive metabolites from an endophytic fungus of *Aspergillus* species isolated from seeds of *Gloriosa superba* Linn. Med Chem Res 22:323–329

Butler MS (2008) Natural products to drugs: natural product-derived compounds in clinical trials. Nat Prod Rep 25:475–516

Canel C, Moraes RM, Dayan FE et al (2000) Molecules of interest: podophyllotoxin. Phytochemistry 54: 115–120

Chen Y, Wu L, Boden R et al (2009) Life without light: microbial diversity and evidence of sulfur- and ammonium-based chemolithotrophy in Movile Cave. ISME J 3:1093–1104

Connon SA, Giovannoni SJ (2002) High-throughput methods for culturing microorganisms in very-low-nutrient media yield diverse new marine isolates. Appl Environ Microbiol 68:3878–3885

Cragg GM, Grothaus PG, Newman DJ (2009) Impact of natural products on developing new anti-cancer agents. Chem Rev 109:3012–3043

Cueto M, Jensen PR, Kauffman C et al (2001) Pestalone, a new antibiotic produced by a marine fungus in response to bacterial challenge. J Nat Prod 64: 1444–1446

Dunbar J, Ticknor LO, Kuske CR (2000) Assessment of microbial diversity in four southwestern United States soils by 16S rRNA gene terminal restriction fragment analysis. Appl Environ Microbiol 66:2943–2950

Elsebai MF, Kehraus S, Gütschow M et al (2010) Spartinoxide, a new enantiomer of A82775C with inhibitory activity toward HLE from the marine-derived fungus *Phaeosphaeria spartinae*. Nat Prod Commun 5:1071–1076

Elsebai MF, Lindequist U, Gütschow M et al (2011a) Antimicrobial phenalenone derivatives from the marine-derived fungus *Coniothyrium cereale*. Org Biomol Chem 9:802–808

Elsebai MF, Natzeem L, Mohamed IE et al (2011b) HLE-inhibitory alkaloids with a polyketide skeleton from the marine-derived fungus *Coniothyrium cereale*. J Nat Prod 74:2282–2285

Elsebai MF, Rempel V, Schnakenburg G et al (2011c) Metabolites from the marine-derived fungus *Auxarthron reticulatum*: identification of a potent and selective cannabinoid CB_1 receptor antagonist. Med Chem Lett 2:866–869

Elsebai MF, Nazir M, Marcourt L et al (2012) Novel polyketide skeletons from the marine alga-derived fungus *Coniothyrium cereale*. Eur J Org Chem 31:6197–6203

Eyberger AL, Dondapati R, Porter JR (2006) Endophyte fungal isolates from *Podophyllum peltatum* produce podophyllotoxin. J Nat Prod 69:1121–1124

Felczykowska A, Bloch SK, Nejman-Falenczyk B et al (2012) Metagenomic approach in the investigation of new bioactive compounds in the marine environment. Acta Biochim Pol 59:501–505

Giridharan P, Verekar SA, Khanna A et al (2012) Anticancer activity of sclerotiorin, isolated from an endophytic fungus *Cephalotheca faveolata* Yaguchi, Nishim. & Udagawa. Indian J Exp Biol 50:464–468

Hamilton-Miller JMT (2008) Development of the semi-synthetic penicillins and cephalosporins. Int J Antimicrob Agents 31:189–192

Jang HB, Kim YK, Del Castillo CS et al (2012) RNA-Seq-based metatranscriptomic and microscopic investigation reveals novel metalloproteases of *Neobodo* sp. as potential virulence factors for soft tunic syndrome in *Halocynthia roretzi*. PloS One 7(12):e52379

Jones MDM, Forn I, Gadelha C et al (2011) Discovery of novel intermediate forms redefines the fungal tree of life. Nature 474:200–206

Kelecom A (2002) Secondary metabolites from marine microorganisms. Acad Bras Cienc 74:151–170

Keller NP, Turner G, Bennett JW (2005) Fungal secondary metabolism – from biochemistry to genomics. Nat Rev Microbiol 3:937–947

Kharwar RN, Mishra A, Gond SK et al (2011) Anti-cancer compounds derived from fungal endophytes: their importance and future challenges. Nat Prod Rep 28:1208–1228

Korkmaz B, Moreau T, Gauthier F (2008) Neutrophil elastase, proteinase 3 and cathepsin G: physicochemical properties, activity and physiopathological functions. Biochimie 90:227–242

Laguerre G, Allard MR, Revoy F et al (1994) Rapid identification of rhizobia by restriction fragment length polymorphism analysis of PCR-amplified 16S rRNA genes. Appl Environ Microbiol 60:56–63

Lam KS (2007) New aspects of natural products in drug discovery. Trends Microbiol 15:279–289

Lee DH, Zo YG, Kim SJ (1996) Nonradioactive method to study genetic profiles of natural bacterial communities by PCR-single-strand-conformation polymorphism. Appl Environ Microbiol 62:3112–3120

Li E, Jiang L, Guo L, Zhang H et al (2008) Pestalachlorides A-C, antifungal metabolites from the plant endophytic fungus *Pestalotiopsis adusta*. Bioorg Med Chem 1:7894–7899

Mayer AMS, Rodriguez AD, Berlinck RGS et al (2011) Marine pharmacology in 2007-8: marine compounds with antibacterial, anticoagulant, antifungal, anti-inflammatory, antimalarial, antiprotozoal, antituberculosis, and antiviral activities; affecting the immune and nervous system, and other miscellaneous mechanisms of action. Comp Biochem Physiol C 153:191–222

McGettigan PA (2013) Transcriptomics in the RNA-seq era. Curr Opin Chem Biol. doi:10.1016/j.cbpa.2012.12.008

Muth T, Benndorf D, Reichl U et al (2012) Searching for a needle in a stack of needles: challenges in metaproteomics data analysis. Mol Biosyst. doi:10.1039/C2MB25415H

Muyzer G (1999) DGGE/TGGE a method for identifying genes from natural ecosystems. Curr Opin Chem Biol 2:317–322

Mygind PH, Fischer RL, Schnorr KM et al (2005) Plectasin is a peptide antibiotic with therapeutic potential from a saprophytic fungus. Nature 437: 975–980

Overbye KM, Barrett JF (2005) Antibiotics: where did we go wrong. Drug Discov Today 10:45–52

Parry R, Nishino S, Spain J (2011) Naturally-occurring nitro compounds. Nat Prod Rep 28:152–167

Paterson I, Anderson EA (2005) The renaissance of natural products as drug candidates. Science 310: 451–453

Picart P, Pirttilä AM, Raventos D et al (2012) Identification of novel defensin-encoding genes of *Picea glauca*: isolation and characterization of Pgd2, a conserved spruce defensin with strong antifungal activity. BMC Plant Biol 12:1–14

Piel J (2006) Combinatorial biosynthesis in symbiotic bacteria. Nat Chem Biol 2:661–662

Pirttilä AM, Frank AC (2011) Endophytes of forest trees, biology and applications. Springer, Berlin

Raghukumar C (2008) Marine fungal biotechnology: an ecological perspective. Fungal Divers 31:19–35

Rateb ME, Ebel R (2011) Secondary metabolites of fungi from marine habitats. Nat Prod Rep 28:290–344

Saleem M, Ali MS, Hussain S et al (2007) Marine natural products of fungal origin. Nat Prod Rep 24: 1142–1152

Saleem M, Nazir M, Ali MS et al (2010) Antimicrobial natural products: an update on future antibiotic drug candidates. Nat Prod Rep 27:238–254

Slavov N, Cvengroš J, Neudörfl JM et al (2010) Total synthesis of the marine antibiotic pestalone and its surprisingly facile conversion into pestalalactone and pestalachloride A. Angew Chem Int Ed 49:7588–7591

Stierle A, Strobel G, Stierle D (1993) Taxol and taxane production by *Taxomyces andreanae*, an endophytic fungus of Pacific yew. Science 260:214–216

Takai K, Horikoshi K (2000) Rapid detection and quantification of members of the archaeal community by quantitative PCR using fluorogenic probes. Appl Environ Microbiol 66:5066–5072

Tejesvi MV, Pirttilä AM (2011) Potential of tree endophytes as sources for new drug compounds. In: Pirttilä AM, Frank AC (eds) Endophytes of forest trees: biology and applications. Springer, New York, pp 295–312

Tejesvi MV, Prakash HS (2009) Phylogenetic tools for the identification of fungi. In: Sridhar KR (ed) Frontiers in fungal ecology, diversity and metabolites, 1st edn. I. K. International Pvt Ltd., New Delhi, pp 285–299

Tejesvi MV, Kajula M, Mattila S et al (2011) Bioactivity and genetic diversity of endophytic fungi in

Rhododendron tomentosum Harmaja. Fungal Divers 47:97–107

Tiwari R, Awasthi A, Mall M et al (2013) Bacterial endophyte-mediated enhancement of in planta content of key terpenoid indole alkaloids and growth parameters of *Catharanthus roseus*. Ind Crop Prod 43:306–310

Vianna ME, Conrads G, Gomes BP et al (2009) T-RFLP-based mcrA gene analysis of methanogenic archaea in association with oral infections and evidence of a novel *Methanobrevibacter* phylotype. Oral Microbiol Immun 24:417–422

Zhang Y, Mu J, Feng Y et al (2009) Broad-spectrum antimicrobial epiphytic and endophytic fungi from marine organisms: isolation, bioassay and taxonomy. Mar Drugs 7:97–112

Zhao J, Mou Y, Shan T et al (2010) Antimicrobial metabolites from the endophytic fungus *Pichia guilliermondii* isolated from *Paris polyphylla* var. *yunnanensis*. Molecules 15:7961–7970

Host-Mimetic Metabolomics of Endophytes: Looking Back into the Future

11

Vijay Chandra Verma, Satya Prakash, Rana Gopal Singh, and Alan Christopher Gange

Abstract

Endophytic research is now gaining pace together with the technological advancement and refinements. The phenomenal potential of endophytes as prolific producer of a wide range of bioactive compounds occupies a complimentary domain of natural product research. The discovery of paclitaxel (Taxol) as bioactive natural product of endophytic origin seems to draw indisputable attention not only for their antitumor activity but as potential microbial alternative for this high in-demand drug. Plenty of opinion is given by the enthusiasts on microbial production of paclitaxel as phylogenetic process and driving paradigm of evolution; however, skeptics described it as phylogenetic anomalies. But despite being highly controversial, the horizontal gene transfer (HGT) theory still seems quite justifiable. Let's have another example: "maytansinoid," a potent cytotoxic agent, was isolated and characterized from microbial endophyte of the same plant; however in both cases, further investigations recorded their occurrence not only in same host but also from deferent distant hosts and even from different endophytes. So the report of taxane and related taxoids from a taxonomically distant host raises several questions. One may assume that this might be due to evolutionary invention; however, it is very unlikely to accept that all modules of gene responsible for biosynthesis of these molecules invented in microbial systems during long evolutionary

V.C. Verma (✉)
DST-FAST Track Young Scientist, Centre of Experimental Medicine and Surgery (CEMS), Institute of Medical Sciences, Banaras Hindu University, Varanasi 221 005, UP, India
e-mail: chandravcv@gmail.com

S. Prakash
National Facility of Tribal and Herbal Medicine, Institute of Medical Sciences, Banaras Hindu University, Varanasi 221 005, UP, India

R.G. Singh
Department of Nephrology, Institute of Medical Sciences, Banaras Hindu University, Varanasi 221 005, UP, India

A.C. Gange
School of Biological Sciences, Royal Holloway University of London, Egham, Surrey, TW20 0EX, UK

V.C. Verma and A.C. Gange (eds.), *Advances in Endophytic Research*,
DOI 10.1007/978-81-322-1575-2_11,

symbiosis. With this chapter we are trying to get into the mechanistic aspects of host-specific chemicals synthesized by endophytic microbes together with our experience with isolation and characterization of host-specific compounds like piperine and azadirachtin. Nevertheless, the significance of this potential of endophytes cannot be ignored, as it provides not only alternative source to existing pharmaceuticals but also on the other hand save the valuable biodiversity of highly medicinal plants.

1 Introduction

Traditionally humans have always relied on natural remedies for treating and healing of their ailments. Nature has been the ultimate source of substances that reduce human sufferings, and many of the world's tribal peoples have a better understanding of native medicinal flora for this purpose. Some of the most erudite peoples in this regard are those in the Amazon basin, the highlands of Papua New Guinea, and the Aborigines of Australia, each of which has its own pharmacopeia of medicinal knowledge (Isaacs 1994). They learn through nature and used medicinal plants as possible cure, which drive pharmaceutical companies for their chemical bioprospection and formulation as drugs. Many medicinal plants on earth are the only source for some well-known pharmaceutically important metabolites and thus are overexploited and facing challenge of their existence. Additionally, a plant source provides a very trace-level production of desired metabolite that too is obtained after rigorous purification procedure, starting with huge raw material. Endophytic microbes emerge as an alternative source of host-specific molecules and thus provide new platform for harvesting desired molecule of plant origin from microbial sources (Zhao et al. 2011). Every plant examined to date harbors at least one species of endophytic fungus and many plants, especially woody plants, and may contain literally hundreds or thousands of species (Petrini et al. 1992; Geylord et al. 1996; Faeth and Hammon 1997; Saikkonen et al. 1998; Arnold et al. 2000). With the discovery of the "Taxol" from an endophytic fungi *Taxomyces andreanae* of pacific yew plant (*Taxus brevifolia*) by Stierle and colleagues, a new era of research in endophytic biology opens (Stierle et al. 1993), since before that pacific yew plant is the only source of "Taxol" known. So it is gradually established that certain endophytes during their long mutualistic symbiosis somehow acquires potential to produce the phytochemicals mimetic to those as their host have. This is one of the strong reasons to the selection of medicinal plants for endophytic study that have therapeutic as well as ethnobotanical history backed by some very strong phytomolecule. Some well-known systems are like *Catharanthus roseus-vincristine, Azadirachta indica-azadirachtin, Camptotheca foetida-camptothecin, Curcuma longa-curcumin,* etc. After the discovery of taxol from endophyte, plenty of other reports have been made to identify endophytic microbes as sources of host-mimetic natural products. Endophytes producing camptothecin (CPT) and its structural analogs (Puri et al. 2005; Shweta et al. 2010), anticancer pro-drugs podophyllotoxin (Eyberger et al. 2006; Puri et al. 2006), antimycobacterial piperine (Verma et al. 2011), and natural insecticides azadirachtin (Kusari et al. 2012) are some recent reports that followed the same hypothesis. However, the ability of endophytes to produce these chemicals raises several intriguing questions that have yet to be answered, including (1) whether the compound is first synthesized by the plant or by the fungus and whether there is a transfer of genetic information between the two. (2) Furthermore, we need to determine if microbes are capable of communicating with each other within the plant, and whether chemical production occurs as a result of cross talk between microbes. (3) It remains a question of further research as to how and why certain endophytes produce an metabolite identical to that derived from its host while not others. (4) As suggested by Tan and Zou, there may be the possibility of genetic recombination between endophyte and the host, in the course of evolutionary symbiosis, and if this happens, then the exact mechanism should be resolved (Tan and Zou 2001).

The potential of endophytic microbes in synthesizing bioactive compounds within their host plants is a very significant aspect of the host-endophyte interaction (Verma et al. 2009; Gunatilaka 2006; Guo et al. 2008; Strobel 2002; Strobel et al. 2004). The discovery of fungal endophytes with potential of producing host-specific metabolites may have significant scientific and industrial applications. However, there are several factor associated in its industrial implications; one major factor is the metabolic regulation of fungus in axenic culture. It is observed that after second or third successive generations, this potential was substantially attenuated, and the production of interested molecule dropped below the limit of detection. This might be assumed due to lack of host stimuli in axenic cultures or rapid inactivation or transformation of the interested molecule. Meanwhile, if endophytes can produce the same rare and important bioactive compounds as their host plants, this would not only reduce the need to harvest slow-growing and possibly rare plants but also help to preserve the world's ever-diminishing biodiversity. But it requires further insight research about the biosynthetic pathway dissection of these two molecules in host as well as in microbes. This however provides an exciting platform for further scientific exploration within both the ecological and biochemical contexts.

2 Host-Mimetic Metabolism

First we would like to discuss our own experience with this hypothesis as we too find some endophyte host system that validates this hypothesis. Piperine and azadirachtins were obtained from endophytic microbes *Periconia* and *Eupenicillium* sp. isolated from their respective host, that is, *Piper longum* L. and *Azadirachta indica* A. Juss.

2.1 Azadirachtin and Its Analog from Endophytes of *Azadirachta indica* A. Juss

Azadirachtin (**1**) is a well-known insecticide found in three species of the neem tree, *Azadirachta indica* A. Juss., *A. excelsa* (Jack) Jacobs, and *A. siamensis* Valeton (Meliaceae). It is chemically interesting because of its complex structure and the challenges of its chemical synthesis, while biologically interesting because it has feeding-deterrent and growth-disrupting activity for a wide range of insects causing huge loss of standing crops. Azadirachtin is a highly oxygenated tetranortriterpenoid that has eight condensed rings, of which three are carbocyclic and five are heterocyclic. It has plethora of oxygen functionality, comprising enol ether, acetal, hemiacetal, one acid–base-sensitive hemiketal, one strained and sterically hindered epoxide, and tetrasubstituted oxirane, together with variety of carboxylic esters. It contains 16 stereogenic centers, 7 of which are fully substituted (Ley 1994; Ley et al. 1993). It takes about 16 years for its first structural elucidation and refinements (Butterworth and Morgan 1968 and Butterworth et al. 1972) and 25 years for its chemical synthesis (Veitch et al. 2007a, b). The azadirachtin has been synthesized chemically from a common intermediate "epoxide-2"; this molecule alone has the potential as an intermediate to synthesize compounds from all three groups of limonoids: azadirachtin, azadirachtol, and meliacarpin from the neem tree (Veitch et al. 2007a, b, 2008; Devkumar and Kumar 2008). Within cellular metabolism, azadirachtin is formed via the "iso-prenoid pathway" (IPP) and follows a relay route through mevalonate, squalene, and apotirucallol to a series of oxidation, ring cleavage, and degradation reactions (Kraus et al. 1985).

Azadirachtin affects insects as antifeedant, insect growth regulator, and sterilant (Mordue et al. 1998). The dihydrofuran acetal moiety of the azadirachtin molecule is supposed to be responsible for the antifeedant activity, while the decalin fragment is responsible for insect growth regulation (Aldhous 1992) (Fig. 11.1). Azadirachtin functions at cellular level by disrupting protein synthesis, more precisely at molecular level by

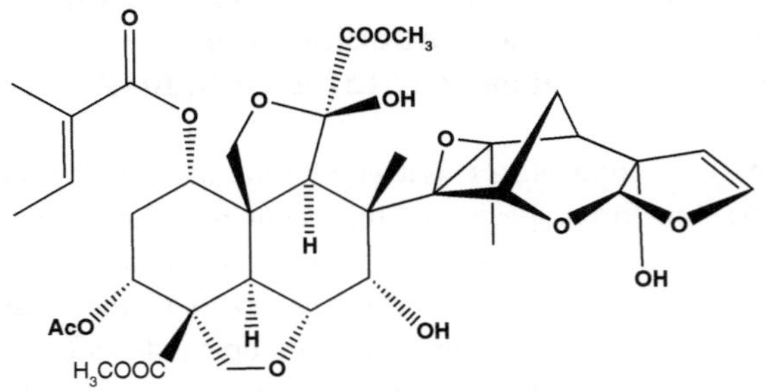

Azadirachtin(1)

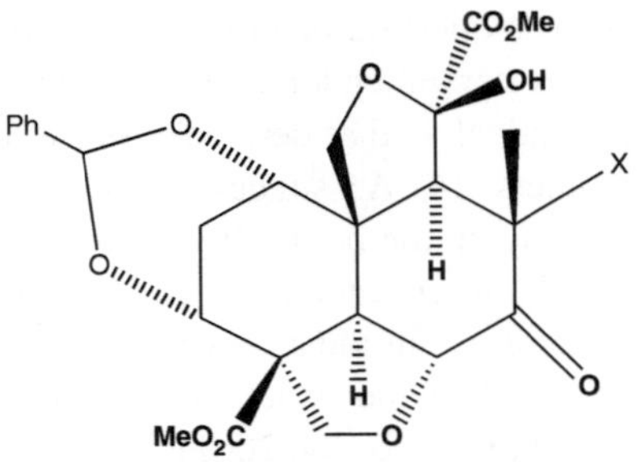

Decalin portion

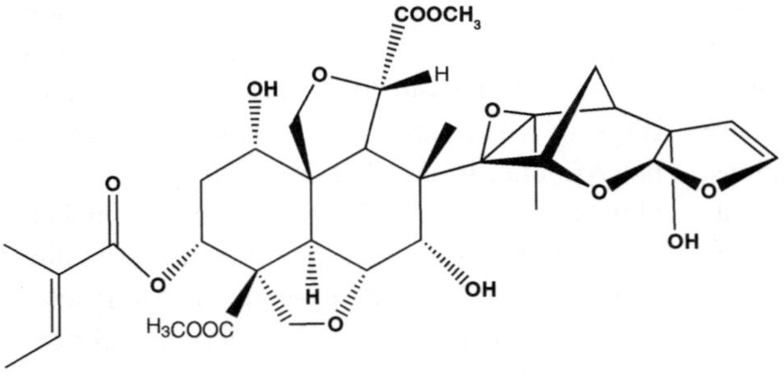

3-tigloylazadirachtin (2)

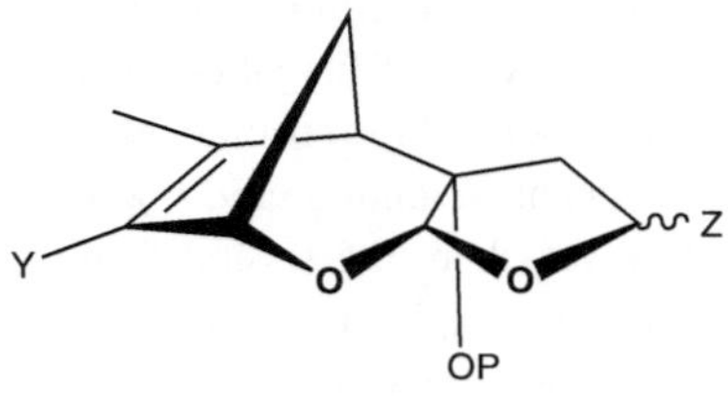

Hydroxy-furan acetal fragment

Fig. 11.1 Azadirachtin (**1**) and 3-tigloylazadirachtol (**2**) molecules and their decalin and hydroxyfuran acetal fragments

Table 11.1 Different isomeric forms of azadirachtins; these are closely related stereo isomers

Azadirachtin B–G	Rambold (1988), Rembold et al. (1987), Klenk et al. (1986), Kraus et al. (1987), and Govindachari et al. (1997)
Azadirachtin H and I	Ramaji et al. (1996) and Govindachari et al. (1992a)
Azadirachtin J and K	Govindachari et al. (1992b, 1996)
Azadirachtin L	Kalinowski et al. (1993)
Azadirachtin M and N	Luo et al. (1999)
Azadirachtin O–Q	Kanokmedhakul et al. (2005)

altering the transcription and translation of protein expressed during rapid protein synthesis (Mordue and Nisbet 2000). Azadirachtin has several structurally related isomers (Table 11.1). Azadirachtin A and its several congeners are having significant biological activity specifically insecticidal and nematicidal (Morgan 2009; Klenk et al. 1986; Butterworth and Morgan 1968).

Thus looking at commercial and industrial relevance of azadirachtin, extensive efforts have been made in recent years to facilitate the production of this highly desired molecule by adopting some novel biotechnological approaches such as callus culture (Prakash et al. 2002; Rafiq and Dahot 2010), cell culture (Jarvis et al. 1997), and hairy root culture of neem plant (Satdive et al. 2007). As a new alternative approach, we report for the first time the production of azadirachtin (**1**) and its analog 3-tigloylazadirachtol (**2**) by an endophytic fungus *Eupenicillium pervum* from neem plant (*A. indica*), using high-resolution mass spectrometry.

The endophytic fungus Eupenicillium pervum was isolated from surface-sterilized stem tissues of neem plant. The identification of the compounds in the fungal biomass was achieved using unique ion fragmentation patterns with LC-HRMS[3] and by comparison with the authentic reference standards. This fungus was screened based on its potential of secreting very prominent

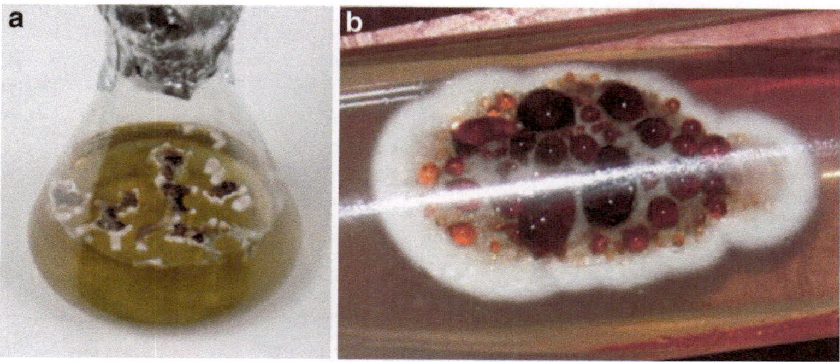

Fig. 11.2 The colony morphology of the endophytic fungi *Eupenicillium pervum*, (**a**) under liquid and (**b**) plate culture (Image courtesy Vijay Verma)

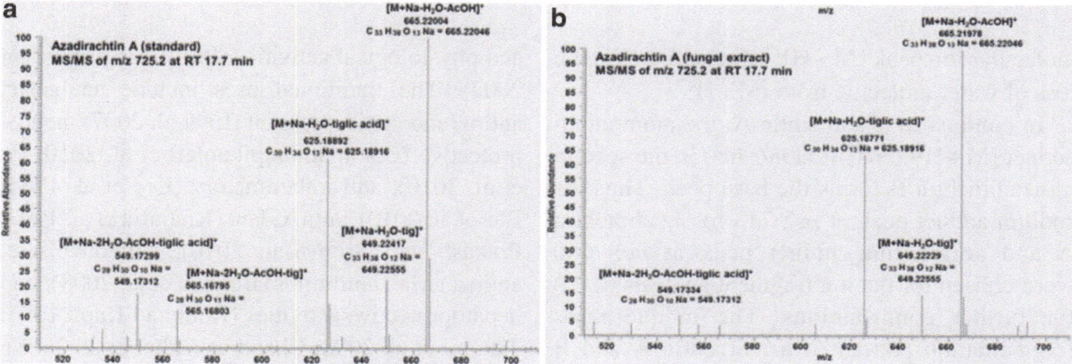

Fig. 11.3 High-resolution MS^2 product ions of azadirachtin A (**a**) authentic reference standard, (**b**) azadirachtin A produced by the endophytic fungus (After Kusari et al. 2012)

dark-pink exudates on culture plates (Fig. 11.2). The colonies were very slow growing, initially cottony white which later turned into light pink due to secretion of exudates as it becomes mature. Additionally, looking into its prominent exudates' secretion potential, a "micro-extraction method" was adapted for the exudates. The exudate droplets were picked directly from the culture plate by using gel-loading tips of micropipette; about 10 ml of exudates were collected from 7-day-old 5 culture plates. These exudates were extracted in ethyl acetate (10 ml) twice and condensed to dryness; after redissolving the extract in 2 ml of methanol, high-resolution mass spectrometry was performed. In comparison to the cell-free extract, these exudates extract show very prominent signals for both azadirachtin (Fig. 11.3) and 3-tigloylazadirachtol (Fig. 11.4). This indicates that the azadirachtin production is extracellular in nature. The LC-MS chromatogram for the fungal extract and the standard azadirachtin A shows strong signals at retention time 17.73 min, which correspond to the molecular ion $[M+H]^+$ peak at *m/z* 721, while *m/z* 703 represents the removal of one water molecule from $[M+H]^+$, and *m/z* 743 is the sodium adduct peak.

The mass spectrum of the azadirachtin A standard showed only a weak protonated molecule $[M+H]^+$ at 721 but relatively intense sodium adduct ions $[M+Na]^+$ at *m/z* 743 caused by the traces of sodium in the solvent. The base peak *m/z* 703 was formed by the elimination of the water molecule $[M+H\text{-}H_2O]^+$. Similarly, Fig. 11.4 represents mass spectroscopy of fungal extract, and for standard 3-tigloylazadirachol, strong peak at retention time 18.73 min was observed which corresponds in fungal extract at 18.68 min. A very intense peak at *m/z* 680 was observed both in standard and in fungal extract that corresponds to the ammonium adduct. The peak at *m/z* 662 represents the

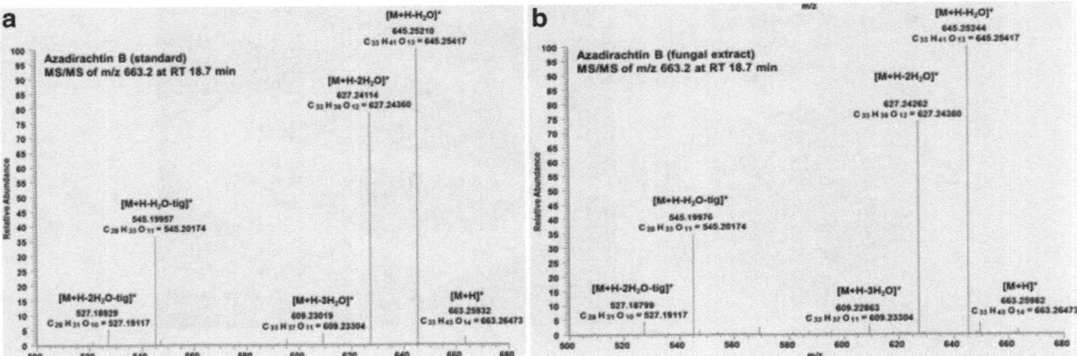

Fig. 11.4 High-resolution MS^2 product ions of 3-tigloyalazadirachtol (**a**) authentic reference standard (**b**) 3-tigloyalazadirachtol produced by the endophytic fungus (After Kusari et al. 2012)

molecular ion peak $[M+H]^+$ while *m/z* 645 is the loss of water molecule from $[M+H]^+$.

In contrast to azadirachtin A, the ammonium adduct $[M+H+(NH_4)]^+$ at *m/z* 680 in the spectra of azadirachtin B forms the base peak. Thus the sodium adduct peak at *m/z* 743 for azadirachtin A and ammonium adduct peak at *m/z* 680 were chosen for the ion fragment analysis (CID) for further confirmations. The unique mass fragmentation pattern of azadirachtin A and B recorded with the electrospray probe in positive ion mode (Figs. 11.3 and 11.4) provided additional evidences to the presence of these structures in the fungal extract. Also for more confirmation, the plant extracts were also run with the same parameters. The unique fragmentation patterns for these molecules both in plant as well as in fungal extracts in accordance with the reference standards confirm the presence of these molecules in fungal extracts. The high-resolution measurements confirmed the molecular formulas of the compounds **1** $[M+H]^+$ 720.28714 ($C_{35}H_{44}O_{16}$) and **2** $[M+H]^+$ 663.28497 ($C_{33}H_{44}O_{14}$) and the characteristic fragments (Kusari et al. 2012).

2.2 Piperine from Endophytic *Periconia* sp. of *Piper longum* L.

In our effort we have isolated piperine from *Piper longum* plant, and this is also a first report from this host (Verma et al. 2011). Piperine is a piperidine derivative with multiple pharmacological and physiological activities (Pie 1983; Srinivasan 2007). The traditional uses include analgesic, antipyretic, antidepressant (Li et al. 2007), neuroprotective (Chonpathompikunlert et al. 2010; Fu et al. 2010), anti-inflammatory (Lee et al. 1984; Bae et al. 2010), antioxidant (Khajuria et al. 1997; Prakash and Srinivasan 2010), anticonvulsant, antibacterial, antitumor (Bezerra et al. 2008), and hepatoprotective activities (Koul and Kapil 1993; Takumi et al. 2008; Chandrashekhar et al. 2008). Piperine provides protection against seizures in epilepsy (Timmers 1994) and has been shown to enhance the bioavailability of several drugs, such as sulfadiazine, tetracycline, streptomycin (Hu and Davies 1997), rifampicin, pyrazinamide, isoniazid (Karan et al. 1988), ethambutol (Zutshi et al. 1985), and phenytoin (Bano et al. 1987). The pepper plant has formed the basis of many traditional formulations that have been in existence for thousands of years in the Indian system of medicine called "Ayurveda." The pepper and its phytoconstituents play an essential role in healthcare in several other traditional systems of medicine in many other countries (Pie 1983). Thus, given the importance of piperine and other related alkaloids, we attempted to isolate endophytic fungi from *P. longum* plants, with the objective to screen and isolate strains that have the potential to produce piperine and related alkaloids, as an alternative source other than their host. The only report other than our own is the isolation and characterization of piperine from an *Ulocladium* sp. (Dahiya et al. 1988); however it does provide details about biology and ecology of the fungus.

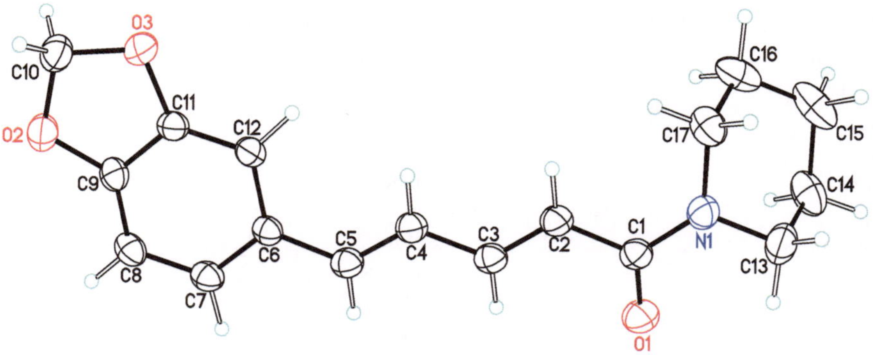

Fig. 11.5 Crystal structure of piperine, $C_{17}H_{19}NO_3$, based on single crystal X-ray crystallography (Image courtesy Prof. Emil Lobkovsky)

We screen an endophytic *Periconia* sp. from the leaf tissue of *Piper longum* L. which has the potential to produce alkaloid "Piperine" in trace amount (Verma et al. 2011). Interestingly this fungal piperine has significant antimycobacterial activity against *Mycobacterium tuberculosis* (1.74 μg/ml) and *M. smegmatis* (2.62 μg/ml). We attempt further purification and obtained crystals of fungal piperine and obtained crystallographic analysis. The single crystals of size $0.60 \times 0.50 \times 0.40$ mm^3 belongs to the space group P2(1)/n of monoclinic family with unit cell dimensions: $a = 8.6712(6)$, $b = 13.4428(8)$, and $c = 12.9744(9)$A°. A total of 15,940 reflections were collected, 4,430 of which were symmetry independent $R_{(int)} = 0.0275$, with 608 "strong" reflections (Fig. 11.5). These crystal parameters are in very close proximity to the natural/parent piperine from the host plant. The molecular structure was obtained with the details of parameters as obtained from crystallography (Table 11.2).

This fungal piperine shows very promising potential as antimycobacterial agent, as it showed prominent growth inhibitory activity against two strains of *Mycobacterium*. In *M. tuberculosis,* all treatments differed from each other, and there was a clear trend of increasing inhibition with increased fungal piperine concentration ($F_{3,8} = 248.3$, $P < 0.001$). A very similar pattern was seen with *M. smegmatis* ($F_{3,8} = 452.6$, $P < 0.001$). In general, stronger inhibition was observed with *M. tuberculosis,* than with *M. smegmatis*. The mycobacterial bioassay established by us and the MICs by the alamar blue assay provided a very clear indication of antimycobacterial activity with purified fungal piperine. These bioactivities confirm that the fungal piperine is as active and functional as piperine from the host plant.

3 Dissecting Host-Mimetic Metabolism in Endophytic Microbes

Hereby, two intriguing examples from the endophytic world are discussed. Both instances deal with the occurrence of identical natural products in unrelated taxa, namely, the host and the invader.

3.1 Taxol: The Dissection of Biosynthetic Pathway in Endophytes

Taxol is the best ever example of phytomimetic metabolite which has been extensively investigated (Wildung and Croteau 1996). The biosynthetic machinery for taxol synthesis *in planta* is considered to require 19 enzymatic steps initiated by universal terpenoids precursor geranylgeranyl diphosphate (GGPP) which itself is derived from isopentenyl pyrophosphate (IPP). The first step is the conversion of GGPP to taxa-4(5), 11(12)-diene by taxadiene synthase (TS). Another enzyme phenylpropanoyl transferase (BAPT) provides for the final attachment of phenolic side chain (Fig. 11.6);

Table 11.2 Crystal data and structure refinement parameters for fungal piperine

Empirical formula	$C_{17} H_{19} N O_3$	
Formula weight	285.33	
Temperature	173(2) K	
Wavelength	0.71073 Å	
Crystal system	Monoclinic	
Space group	P2 (1)/*n*	
Unit cell dimensions	a=8.6712(6)Å	α=90°
	b=13.4428(8) Å	β=107.998(3)°
	c=12.9744(9) Å	γ=90°
Volume	1,438.36(16) Å^3	
Z	4	
Density (calculated)	1.318 Mg/m^3	
Absorption coefficient	0.090 mm^{-1}	
F (000)	608	
Crystal size	0.60×0.50×0.40 mm^3	
Theta range for data collection	2.24–30.65°	
Index ranges	$-11 \leq h \leq 12, -17 \leq k \leq 19, -18 \leq l \leq 15$	
Reflections collected	15,940	
Independent reflections	4,430 [$R_{(int)}$=0.0275]	
Completeness to theta=30.65°	99.7 %	
Absorption correction	Semi-empirical from equivalents	
Max. and min. transmission	0.9648 and 0.9479	
Refinement method	Full-matrix least squares on F^2	
Data/restraints/parameters	4,430/0/266	
Goodness-of-fit on F^2	1.005	
Final R indices [I >2sigma(I)]	R1=0.0413, wR2=0.1121	
R indices (all data)	R1=0.0596, wR2=0.1278	
Largest diff. peak and hole	0.345 and –0.173 e. Å^{-3}	

it is this side chain which is responsible for the anticancer efficacy (Walker et al. 2002). These two enzymes were encoded by *txs* and *bapt* genes, which have been extensively studied and reported in the plant *Taxus brevifolia*; however the *txs* gene has later been also reported to be present in *Taxomyces andreanae* (Staniek et al. 2009; Zhou et al. 2007; Zhang et al. 2008). Recent reports have claimed the isolation of a fungal gene-encoding taxadiene synthase, the sequence similarity (96 %) being surprisingly high when compared to plants. Again it was also confirmed by a protein of the expected molecular weight (110 kDa) of TS in a paclitaxel-producing fungal strain using a plant anti-TS antibody (Soliman et al. 2011). Two pathways can generate IPP and GGPP precursors for the taxane ring of paclitaxel: the classical mevalonate (MVA) pathway, which is cytosolic and active in all organisms including fungi, and 2-C-methyl-D-erythritol-4-phosphate (MEP or DXP) pathway, which is exclusively chloroplastic (plastidic) and bacterial. The MVA pathway generates the precursors for sesquiterpenes, including steroids and triterpenes, and it also generates the precursor GGPP for diterpenoids in fungi and yeast. To distinguish whether the taxane ring system of plant paclitaxel is derived primarily from the cytosolic MVA or plastidic MEP pathways or both, Soliman et al. (2011) studies chemical inhibitors targeting enzymes specific to each of these pathways (Fig. 11.6). For example, lovastatin and fosmidomycin were used to block 3-hydroxy-3-methylglutaryl-coenzyme A reductase (HMGR) and 1-deoxy-D-xylulose-5-phosphate reductoisomerase (DXR), key enzymes in the MVA (cyto-

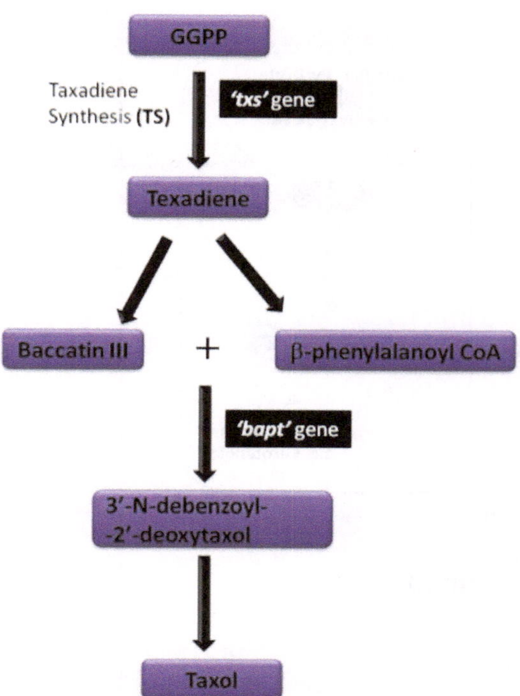

Fig. 11.6 A simplified outline of paclitaxel biosynthesis indicating crucial steps, respective genes, and their product enzymes

solic) and MEP (plastid) branches, respectively, of the terpenoid biosynthetic pathway (Fig. 11.7). Paclitaxel production was significantly lowered by lovastatin and fosmidomycin treatment, suggesting that both MVA and MEP pathways contribute to taxane ring of paclitaxel biosynthesis. Surprisingly, inhibition by fosmidomycin suggested that fungal paclitaxel production absolutely requires DXR, an enzyme in the MEP pathway normally found in plants and bacteria. Inhibition by compactin identified HMGR, a rate-limiting enzyme in the fungal mevalonate pathway, as being required for fungal paclitaxel biosynthesis, similar to plants, where it contributes to the terpenoid ring system. Similarly, identification and differential expression of a gene-encoding HMGS confirmed the importance of the mevalonate pathway in fungal paclitaxel biosynthesis. Surprisingly, inhibition by fosmidomycin suggested that fungal paclitaxel production absolutely requires DXR, an enzyme in the MEP pathway normally found in plants and bacteria. Three additional types of evidence support the unexpected conclusion that a fungus might possess enzymes in the MEP pathway. First, a plant anti-DXR antibody cross-reacted with a fungal peptide of the correct molecular weight. Second, apparent fungal DXR expression correlated to changes in paclitaxel production based on elicitor treatment and fungal age. Finally, a gene-encoding DXS, the enzyme that immediately precedes DXR in the MEP pathway, was identified in another fungus, *Aspergillus* (Hans et al. 2004). The presence of many common steps in biosynthesis of paclitaxel suggests that a lateral transfer of genetic information shaped the evolutionary trajectory of taxonomically unrelated, yet coexisting, species; however it remains inconclusive. Moreover, advocating HGT to be a driving force in the evolution of fungal gene clusters – a phenomenon now considered a hallmark characteristic of secondary metabolic biosynthetic pathways (Lawrence and Roth 1996; Rosewich and Kistler 2000) – raises several questions as to whether the genes responsible for paclitaxel formation in *Taxomyces andreanae* are in a contiguous cluster. Identifying the regulatory mechanisms will be of considerable future interest and will provide further insight into the true nature of the fine-tuned equilibrium of plant-microbe interactions (Staniek et al. 2008; Hines and Zahn 2009).

3.2 Maytansine: From *Maytenus serrata*, a Story Different to Taxol

Maytansine is a potent cytotoxic metabolite recovered from an Ethiopian shrub *Maytenus serrata* (Kupchan et al. 1972) and other higher plants also (Wani et al. 1973; Ahmed et al. 1981; Powel et al. 1982). This is later on also obtained from Gram-positive actinomycetes *Actinosynnema pretiosum* (Higashide et al. 1977; Asai et al. 1978).

One could assume that the biosynthesis of these unique natural products has been repeatedly invented during evolution. However, the fact that approximately 48 genes are involved in the microbial synthesis of maytansinoids (Yu et al. 2002) makes it highly unlikely. Nevertheless, before invoking HGT, alternative and often equally plausible

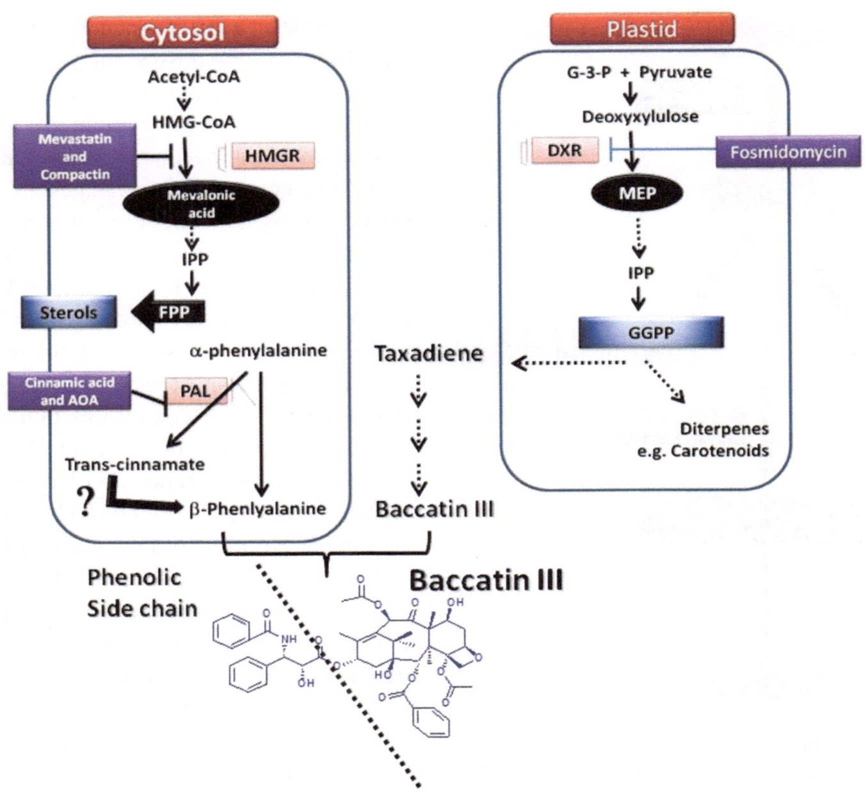

Fig. 11.7 The cytosolic MVA pathway and plastid MEP pathway contribute terpenoid precursors for paclitaxel biosynthesis (Redrawn from Soliman et al. 2011)

Fig. 11.8 Structure of (**a**) maytansine and (**b**) structurally similar ansamitocin P-3

explanations ought to be thoroughly considered first. In case of maytansinoids, all evidence seems to point to them being ultimately produced by plant-associated microorganisms. Maytansine, the unique parent compound (Fig. 11.8a), was found neither in cell suspension cultures from *Maytenus buchananii* (Kutney et al. 1981) nor in callus cultures raised from *Maytenus wallichiana* (Dymowski and Furmanowa 1990) and *Putterlickia verrucosa* (Pullen et al. 2003), unlikely to the taxol case where taxol is also present in the suspension culture. It is noteworthy, however, that these 19-membered macrocyclic lactams are closely related to ansamycin antibiotics of microbial origin, such as Rifamycin B and geldanamycin (Rinehart and Shield 1976). In fact, the aforementioned similarity stimulated a search for maytansinoid-producing microorganisms, ultimately leading to the isolation of ansamitocins (Fig. 11.8b) from the Actinomycetes, *Actinosynnema pretiosum* ssp. *pretiosum* and a mutant strain *Actinosynnema pretiosum* ssp. *auranticum* (Higashide et al. 1977; Asai et al. 1978). This is in line with the result of an in-depth search for the unique gene involved in maytansinoid biosynthesis, encoding for 3-amino-5-hydroxy-benzoic acid (AHBA) synthase, in *Putterlickia verrucosa cell* cultures. An extensive PCR-based homology screen gave negative results only (Pullen et al. 2003). These observations point to the conclusion that plants do not produce maytansinoids ab initio. However, an active role of the plant in an overall biosynthesis cannot be excluded, as it seems likely that the host converts a bacterially synthesized precursor into the final, biologically active compound. Secondly, it is possible that maytansine is only produced as a consequence of a pathogen attack on the plant. The plants may contain a biologically inactive bacterially produced precursor, which is only converted into the potent final product in response to a signal resulting from the attack. Alternatively, and more plausibly, the bacterial production of the maytansinoid precursor could be triggered by a plant signal in response to the pathogen aggression (Cassady et al. 2004). On the contrary, the bio-formation of paclitaxel seems to be a genuine feature of the yew host, as ample evidence supporting the production of the diterpenoid by sterile cell suspension cultures of *Taxus* has been provided (e.g., Ketchum and Gibson 1996; Ketchum and Croteau 1998; Yukimune et al. 2000; Wu and Lin 2003; Naill and Roberts 2005; Khosroushahi et al. 2006; Vongpaseuth and Roberts 2007). This conclusion is further supported by the aforementioned work of Croteau and his associates who succeeded in the isolation of paclitaxel biosynthetic genes of plant origin. Interestingly, the taxadiene synthase gene has a long N-terminal targeting sequence for localization to and processing in the plastids, indicating that this gene is plant derived rather than a fungal product (Koepp et al. 1995; Walker and Croteau 2001). Accordingly, an extensive PCR-based screen for taxadiene synthase gene in *Taxomyces andreanae*, the very first presumed endophytic taxane producer (Stierle et al. 1993), failed to provide any positive results (Staniek, unpublished data).

4 Horizontal Gene Transfer (HGT)

Transmission of genetic material in between two distinct evolutionary lineages that lead to the genomic innovation between several microbial lineages is a phenomena well known as horizontal gene transfer, HGT (Andersson 2005; Keeling and Palmer 2008; Jain et al. 2003). A genomic analysis of many microbes as protozoans suggests that universal eukaryotic features, such as the possession of linear chromatin-based chromosomes, intron-exon gene structures, and the nuclear envelope are not barriers to HGT. Equipped with this essential genetic information and modern tools to manipulate the biosynthetic machinery, the research on microbial paclitaxel synthesizers could enter a novel combinatorial stage. While several heterologous systems including *Escherichia coli* (Huang et al. 2001), *Saccharomyces cerevisiae* (DeJong et al. 2006; Engels et al. 2008), and *Pichia pastoris* (Schmeer and Jennewein 2009) were already exploited for expression of plant-derived genes encoding early paclitaxel biosynthetic enzymes, to engineer and co-mobilize a functional gene cluster in a parent producer microorganism affords the advantage of all the regulatory elements being present and functional. This suggests that specific plant environment may be required for the induction of paclitaxel biosynthetic genes in the fungal symbiont. Identifying these triggering mechanisms will be of considerable future interest, not only providing further insight into the true nature of the fine-tuned equilibrium of plant-microbe interactions, but also revealing their tremendous potential as possible new therapeutics. A recent report communicates a unique endeavor to reestablish the intriguing co-habitat by proposing a promising co-culture system for *Taxus chinensis* var. *mairei* and its endophyte *Fusarium mairei* (Li et al. 2009). The next challenge lies in the further integration of these approaches to develop a comprehensive view of how life history traits of both "players" interact with the environment to shape evolutionary road map (Burdon and Thrall 2009).

5 Current Progress and Future Challenges

Biodiversity: a precious source of novelty, not only in terms of unraveling numerous mysteries of nature – discovering a plethora of yet undescribed species, their evolutionary backgrounds, genetics, and ecology, as well as the richness of thus implied new, potentially valuable molecules – but also a revolution of thought, an expanded view promising to transform glimpses of reductionist research of the past years into snapshots of a dynamic world of systems biology, where cells grow, divide, and produce, or organisms develop, differentiate, and begin to deviate from the norm (Kate and Laird 2000; Stephanopoulos et al. 2004; Kayser and Quax 2007). Endophytic microbes seem to fit perfectly into this natural "warehouse," only a small part of which we have been able to tap into so far. The production of bioactive compounds by endophytes, especially those exclusively mimetic to their host plants, is highly significant for molecular and biochemical perspective, apart from their ecological importance. This potential of microbial endophytes makes them high in demand as alternative and sustainable source for valuable phytochemical. These microbes might be used as platform for investigation, application, and implication of desired compounds from specific host plants (Kusari and Spiteller 2011).

The recent genomics revolution to which a better term could be the "Revolomics" has given momentum to considerable progress in the development of new technologies addressing specifically the concerns in natural product research: whole genome sequence mining (Lautru et al. 2005) and genome scanning as an alternative approach, providing an efficient way to discover natural product biosynthetic gene clusters without having the complete genome sequence (Zazopoulos et al. 2003); advances in microbial cell fermentation technology (Zengler et al. 2005; Weuster-Botzl et al. 2007); and metagenomics as a valuable alternative offering cultivation-independent approaches (Schloss and Handelsman 2005). In recent years ample successes in heterologous expression and metabolic engineering (Alper et al. 2005; Schmidt et al. 2005; Wenzel et al. 2005; DeJong et al. 2006; Julsing et al. 2006; Li and Unsöld 2006; Lindahl et al. 2006; Nims et al. 2006; Ro et al. 2006) have been observed, the latter being in fact perceived as a progenitor of functional genomics and systems biology (Stephanopoulos et al. 2004; Tyo et al. 2007). With advancement in technologies, we believe that in the future we could be able to elucidate the basic machinery of this host-mimetic synthesis of important pharmaceuticals.

Acknowledgments VCV gratefully acknowledges the financial support from University Grant Commission (wide letter No. F. 4-2/2006/13-552/2011/BSR) and Council of Scientific and Industrial Research (CSIR), New Delhi. VCV is also thankful to the Department of Science and Technology (DST), for the recognition as "Fast track young scientist" (wide letter No.: SERC/LS-515/2011).

References

Ahmed MS, Fong HH, Soejarto DD, Dobberstein RH, Waller DP, Moreno-Azorero RJ (1981) High-performance liquid chromatographic separation and quantitation of maytansinoids in *Maytenus ilicifolia*. J Chromatogr 213:340–344

Aldhous P (1992) Neem chemicals: the pieces fall in place. Science 258:893

Alper H, Jin YS, Moxley JF, Stephanopoulos G (2005) Identifying gene targets for the metabolic engineering of lycopene biosynthesis in *Escherichia coli*. Metab Eng 7:155–164

Andersson JO (2005) Lateral gene transfer in eukaryotes. Cell Mol Life Sci 62:1182–1197

Arnold AE, Maynord Z, Gilbert G, Coley PD, Kursar TA (2000) Are tropical fungal endophytes hyperdiverse? Ecol Lett 3:167–274

Asai M, Mizuta E, Izawa M, Haibara K, Kishi T (1978) Isolation, chemical characterization and structure of ansamitocin, a new antitumour antibiotic. Tetrahedron 35:1079–1085

Bae GS et al (2010) Inhibition of lipopolysaccharide-induced inflammatory responses by piperine. Eur J Pharmacol 642:154–162

Bano G, Amla V, Raina RK, Zutshi U, Chopra CL (1987) The effect of piperine on pharmacokinetics of phenytoin in healthy volunteers. Planta Med 53:568–569

Bezerra DP et al (2008) *In vitro* and *in vivo* antitumor effect of 5-FU combined with piplartine and piperine. J Appl Toxicol 28:156–163

Burdon JJ, Thrall PH (2009) Co-evolution of plants and their pathogens in natural habitats. Science 324:755–756

Butterworth JH, Morgan ED (1968) Isolation of a substance that suppresses feeding in locusts. J Chem Soc Chem Commun 1968:23–24

Butterworth JH, Morgan ED, Percy GR (1972) The structure of azadirachtin; the functional groups. J Chem Soc Perkin Trans 1:2445–2450

Cassady JM, Chan KK, Floss HG, Leistner E (2004) Recent developments in the maytansinoids antitumour agents. Chem Pharm Bull 52:1–26

Chandrashekhar RB, Jithan A, Narsimha RY, Malla RV (2008) Fabrication and investigations on hepatoprotective activity of sustained release biodegradable piperine microspheres. Int J Pharm Sci NanoTechnol 1:87–96

Chonpathompikunlert P, Wattanathorn J, Muchimapura S (2010) Piperine, the main alkaloid of Thai black pepper, protects against neurodegeneration and cognitive impairment in animal model of cognitive deficit like condition of Alzheimer's disease. Food Chem Toxicol 48:798–802

Dahiya JS, Woods DL, Tiwari JP (1988) Piperine from an *Ulocladium* sp. Phytochemistry 27:2366

DeJong JM, Liu Y, Bollon AP, Jennewein S, Williams D, Croteau R (2006) Genetic engineering of taxol biosynthetic genes in *Saccharomyces cerevisiae*. Biotechnol Bioeng 93:212–224

Devkumar C, Kumar R (2008) Total synthesis of azadirachtin: a chemical odyssey. Curr Sci 95:573–575

Dymowski W, Furmanowa M (1990) Investigating cytostatic substances in tissue of plants *Maytenus molina in in vitro* cultures. Chromatographic test of extracts from callus of *Maytenus wallichiana*. Acta Pol Pharm 47:51–54

Engels B, Dahm P, Jennewein S (2008) Metabolic engineering of taxadiene biosynthesis in yeast as a first step towards Taxol (paclitaxel) production. Metab Eng 10:201–206

Eyberger AL, Dondapati R, Porter JR (2006) Endophyte fungal isolates from *Podophyllum peltatum* produce podophyllotoxin. J Nat Prod 69:1121–1124

Faeth SH, Hammon KE (1997) Fungal endophytes in oak tree; long term pattern of abundance and association with leaf miners. Ecology 78:810–819

Fu M, Sun ZH, Zuo HC (2010) Neuroprotective effect of piperine on primarily cultured hippocampal neurons. Biol Pharm Bull 33:598–603

Geylord ES, Preszler RW, Boecklen WJ (1996) Interactions between host plants, endophytic fungi, and a phytophagous insect in an Oak (*Quercus grisea* × *Q. gambelii*) hybrid zone. Oecologia 105:336–342

Govindachari TR, Sandhya G, Raj SPG (1992a) Azadirachtin H and I: two new tetranortriterpenoid from *Azadirachta indica*. J Nat Prod 55:596–601

Govindachari TR, Sandhya G, Raj SPG (1992b) Structure of azadirachtin K, a new tetranortriterpenoid from *Azadirachta indica*. Indian J Chem Sect B 31:295–298

Govindachari TR, Gopalakrishnan G, Rajan SS, Kabaleeswaran V, Lessinger L (1996) Molecular and crystal structure of azadirachtin-H. Acta Crystallogr Sect B Struct Sci B52:145–150

Govindachari TR, Gopalakrishnan G, Suresh G (1997) Purification of azadirachtin-B (3- tigloylazadirachtol) by preparative high performance liquid chromatography, using the recycling mode. J Liq Chromatogr Relat Technol 20:1633–1636

Gunatilaka AAL (2006) Natural products from plant-associated microorganisms: distribution, structural diversity, bioactivity and implications of their occurrence. J Nat Prod 69:509–526

Guo B, Wang Y, Sun X, Tang K (2008) Bioactive natural products from endophytes: a review. Appl Biochem Microbiol 44:136–142

Hans J, Hause B, Strack D, Walter MH (2004) Cloning characterization and immunolocalization of a mycorrhizal inducible 1-Deoxy-D-Xylulose 5-phosphate reductoisomerase in arbuscule containing cells of Maize. Plant Physiol 134:614–624

Higashide E, Asai M, Ootsu K, Tanida S, Kozay Y, Hasegawa T, Kishi T, Sugino Y, Yoneda M (1977) Ansamitocins, a group of novel maytansinoid antibiotics with anti-tumour properties from *Nocardia*. Nature 270:721–722

Hines PJ, Zahn LM (2009) What's bugging plants? Introduction to special issue. Science 324:741

Hu RQ, Davies JA (1997) Effects of *Piper nigrum* L. on epileptiform activity in cortical wedges prepared from DBA/2 mice. Phytother Res 11:222–225

Huang Q, Roessner CA, Croteau R, Scott AI (2001) Engineering *Escherichia coli* for the synthesis of taxadiene, a key intermediate in the biosynthesis of Taxol. Bioorg Med Chem 9:2237–2242

Isaacs J (1994) Bush food aboriginal food and herbal medicine. Lansdowne Publishing Pty. Ltd., Sydney

Jain R, Rivera MC, Moore JE, Lake JA (2003) Horizontal gene transfer accelerates genome innovation and evolution. Mol Biol Evol 20:1598–1602

Jarvis AP, Morgan ED, van der Esch SA, Vitali F, Lay SV, Pape A (1997) Identification of azadirachtin in tissue-cultured cells of neem (*A. indica*). Nat Prod Lett 10:95–98

Julsing MK, Koulman A, Woerdenbag HJ, Quax WJ, Kayser O (2006) Com-binatorial biosynthesis of medicinal plant secondary metabolites. Biomol Eng 23:265–279

Kalinowski HO, Ermel K, Schmutterer H (1993) Strukturaufklärung eines azadirachtin derivates aus dem Marrangobaum *Azadirachta excelsa* durch NMR-spektroskopie. Liebigs Ann Chem 1993:1033–1035

Kanokmedhakul S, Kanokmedhakul K, Prajuabsuk T, Panichajakul S, Panyamee P, Prabpai S, Kongsaeree P (2005) Azadirachtin derivatives from seed kernels of *Azadirachta excelsa*. J Nat Prod 68:1047–1050

Karan RS, Bhargava VK, Garg SK (1988) Effect of piperine on the pharmacokinetic profile of isoniazid in rabbits. Indian J Pharmacol 30:254–256

Kate KT, Laird SA (eds) (2000) The commercial use of biodiversity: access to genetic resources and benefit sharing. Earthscan Publications Ltd., London

Kayser O, Quax WJ (eds) (2007) Medicinal plant biotechnology: from basic research to industrial applications. Wiley-VCH Verlag Gmbh & Co. KGaA, Weinheim

Keeling PJ, Palmer JD (2008) Horizontal gene transfer in eukaryotic evolution. Nat Rev Genet 9:605–618

Ketchum RE, Croteau R (1998) Recent progress toward an understanding of taxol biosynthesis in plant cell cultures. In: Ageta H, Aimi N, Ebizuka Y, Fujita T, Honda G (eds) Towards natural medicine research in the 21st century. Proceeding of the international symposium on natural medicines. Elsevier, Amsterdam, pp 339–348

Ketchum RE, Gibson DM (1996) Paclitaxel production in suspension cell cultures of Taxus. Plant Cell Tissue Organ Cult 46:9–16

Khajuria A, Thusus N, Zutshi U, Bedi KL (1997) Antioxidant potential of piperine on oxidant induced alterations in rat intestinal lumen. Indian Drugs 34:557–563

Khosroushahi AY, Valizadeh M, Ghasempour A, Khosrowshahli M, Naghdibadi H, Dadpour MR, Omidi Y (2006) Improved taxol production by combination of inducing factors in suspension cell culture of *Taxus baccata*. Cell Biol Int 30:262–269

Klenk A, Bokel M, Kraus W (1986) 3-tigloylazadirachtol (tigloyl = 2-methyl crotonoyl), an insect growth regulating constituent of *Azadirachta indica*. J Chem Soc Chem Commun 7:523–524

Koepp AE, Hezari M, Zajicek J, Vogel BS, LaFever RE, Lewis NG, Croteau R (1995) Cyclization of geranylgeranyl diphosphate to taxa-4(5),11(12)-diene is the committed step of taxol biosynthesis in Pacific yew. J Biol Chem 270:8686–8690

Koul IB, Kapil A (1993) Evaluation of the liver protective potential of piperine, an active principle of black and long peppers. Planta Med 59:413–417

Kraus H, Bokel M, Klank A, Pohnl H (1985) The structure of azadirachtin and 22, 23-dihydro-23β-methoxyazadirachtin. Tetrahedron Lett 26:6435–6438

Kraus W, Bokel M, Bruhn A, Cramer R, Klaiber I, Klenk A, Nagl G, Pohnl H, Sadlo H, Vogler B (1987) Structure determination by NMR of azadirachtin and related compounds from *Azadirachta indica* A. Juss. (Meliaceae). Tetrahedron 43:2817–2830

Kupchan SM, Komoda Y, Court WA, Thomas GJ, Smith RM, Karim A, Gilmore CJ, Haltiwanger RC, Bryan RF (1972) Maytansine, a novel antileukaemic ansa macrolide from *Maytenus ovatus*. J Am Chem Soc 94:1355–1356

Kusari S, Spiteller M (2011) Are we ready for industrial production of bioactive plant secondary metabolites utilizing endophytes? Nat Prod Rep 28:1203–1207

Kusari S, Verma VC, Lamshöft M, Spiteller M (2012) An endophytic fungus from Azadirachta indica A. Juss. that produces azadirachtin. World J Microbiol Biotechnol 28:1287–1294

Kutney JP, Beale MH, Salisbury PJ, Stuart KL, Worth BR, Townsley PM, Chalmers WT, Nilson K, Jacoli GG (1981) Isolation and characterization of natural products from plant tissue cultures of *Maytenus buchananii*. Phytochemistry 4:653–657

Lautru S, Deeth RJ, Bailey LM, Challis GL (2005) Discovery of a new pep-tide natural product by *Streptomyces coelicolor* genome mining. Nat Chem Biol 1:265–269

Lawrence JG, Roth JR (1996) Selfish operons: horizontal transfer may drive the evolution of gene clusters. Genetics 143:1843–1860

Lee EB, Shin KH, Woo WS (1984) Pharmacological study on piperine. Arch Pharmacal Res 7:127–132

Ley SV (1994) Synthesis and the chemistry of the insect antifeedant azadirachtin. Pure Appl Chem 66:2099–2102

Ley SV, Denholm AA, Wood A (1993) The chemistry of azadirachtin. Nat Prod Rep 10:109–157

Li SM, Unsöld IA (2006) Post-genome research on the biosynthesis of ergot alkaloids. Planta Med 72:1117–1120

Li S, Wang C, Wang M, Li W, Matsumoto K, Tang Y (2007) Antidepressant like effects of piperine in chronic mild stress treated mice and its possible mechanisms. Life Sci 80:1373–1381

Li YC, Tao WY, Cheng L (2009) Paclitaxel production using co-culture of *Taxus* suspension cells and paclitaxel-producing endophytic fungi in a co-bioreactor. Appl Microbiol Biotechnol 83:233–239

Lindahl AL, Olsson ME, Mercke P, Tollbom O, Schelin J, Brodelius M, Brodelius PE (2006) Production of the artemisin in precursor amorpha-4,11-diene by engineered *Saccharomyces cerevisiae*. Biotechnol Lett 28:571–580

Luo X, Ma Y, Wu S, Wu D (1999) Two novel azadirachtin derivatives from *Azadirachta indica*. J Nat Prod 62:1022–1024

Mordue AJ, Nisbet AJ (2000) Azadirachtin from the neem tree *Azadirachta indica*: its action against insects. An Soc Entomol Bras 29:615–632

Mordue AJ, Simmonds MSJ, Ley SV, Blaney WM, Nasiruddin M, Nisbet AJ (1998) Actions of azadirachtin, a plant allelochemical against insects. Pestic Sci 54:277–284

Morgan ED (2009) Azadirachtin: a scientific goldmine. Bioorg Med Chem 17:4096–4105

Naill MC, Roberts SC (2005) Cell cycle analysis of *Taxus* suspension cultures at the single cell level as an indicator of culture heterogeneity. Biotechnol Bioeng 90: 491–500

Nims E, Dubois CP, Roberts SC, Walker EL (2006) Expression profiling of genes involved in paclitaxel biosynthesis for targeted metabolic engineering. Metab Eng 8:385–394

Petrini O, Sieber TN, Toti L, Viret O (1992) Ecology, metabolite production, and substrate utilization in endophytic fungi. Nat Toxins 1:185–196

Pie YQ (1983) A review of pharmacology and clinical use of piperine and its derivatives. Epilepsia 24:177–183

Powel RG, Weisleder D, Smith CR, Kozlowski J, Rohwedder WK (1982) Treflorine, trenudine, and N-methyltrenudone: novel maytansinoids tumour inhibitors containing two fused macrocyclic rings. J Am Chem Soc 104:4929–4934

Prakash UN, Srinivasan K (2010) Gastrointestinal protective effect of dietary spices during ethanol-induced oxidant stress in experimental rats. Appl Physiol Nutr Metab 35:134–141

Prakash G, Bhojwani SS, Shrivastava AK (2002) Production of azadirachtin from tissue culture: state of the art and future prospects. Biotechnol Bioprocess Eng 7:185–193

Pullen CB, Schmitz P, Hoffmann D, Meurer K, Boettcher T, von Bamberg D, Pereira AM, de Castro França S, Hauser M, Geertsema H, van Wyk A, Mahmud T, Floss HG, Leistner E (2003) Occurrence and non-detectability of maytansinoids in individual plants of the genera *Maytenus and Putterlickia*. Phytochemistry 62:377–387

Puri SC, Verma V, Amna T, Qazi GN, Spiteller M (2005) An endophytic fungus from *Nothapodytes foetida* that produces camptothecin. J Nat Prod 68: 1717–1719

Puri SC, Nazir A, Chawla R, Arora R, Riyaz-ulHasan S, Amna T, Ahmed B, Verma V, Singh S, Sagar R, Sharma A, Kumar R, Sharma RK, Qazi GN (2006) The endophytic fungus *Trametes hirsuta as* a novel alternative source of podophyllotoxin and related aryl tetralin lignans. J Biotechnol 122:494–510

Rafiq M, Dahot M (2010) Callus and azadirachtin related limnoids production through *in vitro* culture of neem (*Azadirachta indica* A. Juss). Afr J Biotechnol 9:449–453

Ramaji N, Venkatakrishnan K, Madyastha KM (1996) 11-Epi- azadirachtin H from *Azadirachta indica*. Phytochemistry 42:561–562

Rambold H (1988) Isomeric azadirachtin and their mode of action. In: Jacobson J (ed) Focus on phytochemical pesticides, vol 1, The neem tree. CRC Press, London

Rembold H, Forster H, Sonnenbichler (1987) Z.Z. Naturforsch C, 42: 4–6

Rinehart KL, Shield LS (1976) Chemistry of the ansamycin antibiotics. In: Herz W, Grisebach H, Kirby GW (eds) Progress in the chemistry of organic natural products. Springer, New York, pp 232–300

Ro DK, Paradise EM, Ouellet M, Fisher KJ, Newman KL, Ndungu JM, Ho KA, Eachus RA, Ham TS, Kirby J, Chang MC, Withers ST, Shiba Y, Sarpong R, Keasling JD (2006) Production of the antimalarial drug precursor artemisinic acid in engineered yeast. Nature 440:940–943

Rosewich UL, Kistler HC (2000) Role of horizontal gene transfer in the evolution of fungi. Ann Rev Phytopathol 38:325–363

Saikkonen K, Faeth SH, Helander ML, Sullivan TJ (1998) Fungal endophytes: a continuum of interactions with host plants. Ann Rev Ecol Syst 29:319–343

Satdive RK, Fulzele DP, Eapen S (2007) Enhanced production of azadirachtin by hairy root cultures of *Azadirachta indica* A Juss. by elicitation and media optimization. J Biotechnol 128:281–289

Schloss PD, Handelsman J (2005) Metagenomics for studying unculturable microorganisms: cutting the Gordian knot. Genome Biol 6:229–233

Schmeer H, Jennewein S (2009) Bioorganic synthesis of the key taxoid pre-cursor taxa-4(5),11(12)-diene using a one-pot, two enzyme catalyzed re-actions. Enzyme Engineering XX, Groningen, the Netherlands

Schmidt EW, Nelson JT, Rasko DA, Sudek S, Eisen JA, Haygood MG, Ravel J (2005) Patellamide A and C biosynthesis by a microcin-like pathway in *Prochloron didemni*, the cyanobacterial symbiont of *Lissoclinum patella*. Proc Nat Acad Sci USA 102:7315–7320

Shweta S, Zühlke S, Ramesha BT, Priti V, Kumar PM, Ravikanth G, Spiteller M, Vasudeva R, Shaanker RU (2010) Endophytic fungal strains of *Fusarium solani*, from *Apodytes dimidiate* E. Mey. ex Arn (Icacinaceae) produce camptothecin, 10-hydroxycamptothecin and 9-methoxycamptothecin. Phytochemistry 71:117–122

Soliman SSM, Tsao R, Raizada MN (2011) Chemical inhibitors suggests endophytic fungal paclitaxel is derived from both mevalonate and non-mevalonate-like pathways. J Nat Prod 74:2497–2504

Srinivasan K (2007) Black pepper and its pungent principle-piperine: a review of diverse physiological effects. Crit Rev Food Sci Nutr 47:735–748

Staniek A, Woerdenbag HJ, Kayser O (2008) Endophytes: exploiting biodiversity for the improvement of natural product-based drug discovery. J Plant Interact 3:75–93

Staniek A, Woerdenbag HJ, Kayser O (2009) *Taxomyces andreanae*: a presumed paclitaxel producer demystified? Planta Med 75:1561–1566

Stephanopoulos G, Alper H, Moxley J (2004) Exploiting biological complexity for strain improvement through systems biology. Nat Biotechnol 22:1261–1267

Stierle A, Strobel GA, Stierle D (1993) Taxol and taxane production by *Taxomyces andreanae*, an endophytic fungus of Pacific yew. Science 260:214–216

Strobel GA (2002) Microbial gifts from the rain forest. Can J Plant Pathol 24:14–20

Strobel GA, Daisy B, Castillo U, Harper J (2004) Natural products from endophytic fungi. J Nat Prod 67:257–268

Takumi I, Mizuho I, Siho T, Yuzi I, Hideyuki Y (2008) Piperine, a pepper ingredient, improves the hepatic increase in free fatty acids caused by 2, 3, 7, 8-tetrachlorodibenzo-p-dioxin. J Health Sci 54:551–558

Tan RX, Zou WX (2001) Endophytes: a rich source of functional metabolites. Nat Prod Rep 18:448–459

Timmers L (1994) Herbal medicines used against epilepsy in developing countries; Publication Number PUG/94-4. Publicaties Wetenschapswinkel Geneesmiddelen, Vrouwen

Tyo KE, Alper HS, Stephanopoulos GN (2007) Expanding the metabolic engineering toolbox: more options to engineer cells. Trends Biotechnol 25:132–137

Veitch GE, Beckmann E, Burke BJ, Boyer A, Ayats C, Ley SV (2007a) A relay route for the synthesis of azadirachtin. Angew Chem Int Ed 46:7633–7635

Veitch GE, Beckmann E, Burke BJ, Boyer A, Ayats C, Ley SV (2007b) Synthesis of azadirachtin: a long but successful journey. Angew Chem Int Ed 46:7629–7632

Veitch GE, Boyer A, Ley SV (2008) The azadirachtin story. Angew Chem Int Ed 47:9402–9429

Verma VC, Kharwar RN, Strobel GA (2009) Chemical and functional diversity of natural products from plant associated endophytic fungi. Nat Prod Commun 4:1511–1532

Verma VC, Lobkovsky E, Gange AC, Singh SK, Prakash S (2011) Piperine production by endophytic *Periconia* sp. isolated from *Piper longum* L. J Antibiot 64: 427–431

Vongpaseuth K, Roberts SC (2007) Advancements in the understanding of paclitaxel metabolism in tissue culture. Curr Pharm Biotechnol 8:219–236

Walker K, Croteau R (2001) Taxol biosynthetic genes. Phytochemistry 58:1–7

Walker K, Long R, Croteau R (2002) The final acylation step in taxol biosynthesis: cloning of the taxoid C13-side chain N-benzoyltransferase from Taxus. Proc Nat Acad Sci USA 99:9166–9171

Wani MC, Taylor HL, Wall ME (1973) Plant antitumour agents: colubrinol acetate and colubrinol, antileukaemic ansa macrolides from *Colubrina texensis*. J Chem Soc Chem Commun 1973:390

Wenzel SC, Gross F, Zhang Y, Fu J, Stewart F, Müller R (2005) Heterologous expression of a myxobacterial natural products assembly line in *Pseudomonads* via Red/ET recombineering. Chem Biol 12:349–356

Weuster-Botzl D, Hekmat D, Puskeiler R, Franco-Lara E (2007) Enabling technologies: fermentation and downstream processing. Adv Biochem Eng Biotechnol 105:205–247

Wildung MR, Croteau R (1996) A cDNA clone for taxadiene synthase, the diterpene cyclase that catalyzes the committed step of taxol biosynthesis. J Biol Chem 271:9201–9204

Wu J, Lin L (2003) Enhancement of taxol production and release in *Taxus chinensis* cell cultures by ultrasound, methyl jasmonate and in situsol-vent extraction. Appl Microbiol Biotechnol 62:151–155

Yu TW, Bai L, Clade D, Hoffmann D, Toelzer S, Trinh KQ, Xu J, Moss SJ, Leistner E, Floss HG (2002) The biosynthetic gene cluster of the maytansinoid antitumour agent ansamitocin from *Actinosynnema pretiosum*. Proc Nat Acad Sci USA 99:7968–7973

Yukimune Y, Hara Y, Nomura E, Seto H, Yoshida S (2000) The configuration of methyl jasmonate affects paclitaxel and baccatin III production in *Taxus* cells. Phytochemistry 54:13–17

Zazopoulos E, Huang K, Staffa A, Liu W, Bachmann BO, Nonaka K, Ahlert J, Thorson JS, Shen B, Farnet CM (2003) A genomics-guided approach for discovering and expressing cryptic metabolic pathways. Nat Biotechnol 21:187–190

Zengler K, Walcher M, Clark G, Haller I, Toledo G, Holland T, Mathur EJ, Woodnutt G, Short J, Keller M (2005) High-throughput cultivation of microorganisms using microcapsules. Method Enzymol 397:124–130

Zhang P, Zhou P, Jiang C, Yu H, Yu LJ (2008) Screening of taxol-producing fungi based on PCR amplification from *Taxus*. Biotechnol Lett 30:2119–2123

Zhao J, Shan T, Mou Y, Zhao L (2011) Plant derived bioactive compounds produced by endophytic fungi. Mini Rev Med Chem 11:159–168

Zhou X, Wang Z, Jiang K, Wei Y, Lin J, Sun X, Tang K (2007) Screening of taxol-producing endophytic fungi from *Taxus chinensis* var. *mairei*. Prikl Biokhim Mikrobiol 43:490–494

Zutshi RK, Singh R, Zutshi U, Johri RK, Atal CK (1985) Influence of piperine on rifampicin blood levels in patients of pulmonary tuberculosis. J Assoc Phys India 33:223–224

Myconanosynthesis: Redefining the Role of Microbial Endophytes 12

Santosh Kumar Singh, Amit Ranjan, Shriti Singh, Swechha Anand, and Vijay Chandra Verma

Abstract

In recent years, a surge of interest was observed in synthesizing nanoparticles and other highly structured nanomaterials using microbes. Plenty of reports in cited domain claims synthesis of nanomaterials with desired shape, size and architecture through fungi, bacteria and actinomycetes. More precisely, fungi are frequently reported for their pivotal potential in bioreduction of the aqueous metal ions into their respective nanomaterials. The sporadic reports of nanomaterial synthesis from fungi led to the development of 'myconanotechnology' as a new domain of nanotechnology. This newly emerging domain of nanotechnology attracts not only the microbiologist but also material chemists and technologists, because of safe, sustainable and non-toxic 'green chemistry' associated with it. There is possibility of getting a total control over shape and size in a microbial system more easily than chemical methods. So far, a number of fungal strains have been reported for this potential among which some most common are *Aspergillus*, *Fusarium*, *Colletotrichum*, *Penicillium*, *Verticillium*, etc. However, the exact mechanism of this mycoreduction is not known so far, but it is speculated that fungal enzymes and/or metabolites are usually responsible for reduction of metal ions into their respective nanoparticles. Although many soil and pathogenic fungi have been reported as nano-factories of desired metals, relatively few reports are available about the synthesis of nanomaterials using fungal endophytes. It's surprising since fungal endophytes occupying the unusual habitat have potential to survive under stress conditions and thus must have set of enzymes and metabolites not found in

S.K. Singh (✉) • A. Ranjan • S. Singh • S. Anand
Centre of Experimental Medicine and Surgery, Institute of Medical Sciences, Banaras Hindu University, Varanasi 221 005, UP, India
e-mail: singhsk71@yahoo.com

V.C. Verma
DST-FAST Track Young Scientist, Centre of Experimental Medicine and Surgery (CEMS), Institute of Medical Sciences, Banaras Hindu University, Varanasi 221 005, UP, India
e-mail: chandravcv@gmail.com

V.C. Verma and A.C. Gange (eds.), *Advances in Endophytic Research*,
DOI 10.1007/978-81-322-1575-2_12, © Springer India 2014

their wild-type counterparts. For this reason, fungal endophytes could be a better candidate for synthesizing nanomaterials. We, in this review, provide a brief review of recent account about endophyte-mediated synthesis of nanomaterials.

1 Introduction

Nanomaterials are at the leading edge of the rapidly developing field of nanotechnology. The development of reliable experimental protocols for the synthesis of nanomaterials over a range of compositions, sizes, with high monodispersity is one of the most challenging issues in current nanotechnology researches. In this context, the current drive to develop green technologies in material synthesis is of considerable importance. Recent studies on the use of microorganisms in the synthesis of nanoparticles are a relatively new exciting and fascinating area of research having potential for array of advantages over the conventional chemical synthesis. In recent years, the use of fungi and actinomycetes in the synthesis of nanoparticles has become more sporadic as 'green synthesis' of nanomaterials (Mandal et al. 2006; Verma et al. 2009; Gade et al. 2010). After 2009, significant interest has been paid by scientists to develop new protocols for synthesizing nanomaterials by fungi; this was observed by increasing the number of publications and citations published in this duration (Fig. 12.1). In fact, this interdisciplinary field of the so-called nanobiotechnology now emerges as a pivotal technology that intersects the different domains of sciences at a single platform.

Several reports are now available about the fungus-mediated biosynthesis of metal nanoparticles especially noble metals like gold and silver. Advantages of these green protocols include tight controlled, highly reproducible, non-toxic by-products, highly stable on room temperature and biocompatible. The biosynthetic method employing microbes has received some attention as a simple and viable alternative to chemical procedures and physical methods synthesizing metal nanoparticles only in recent years. Several fungi were evaluated for their potential to bioreduce the aqueous ionic solutions of metals into

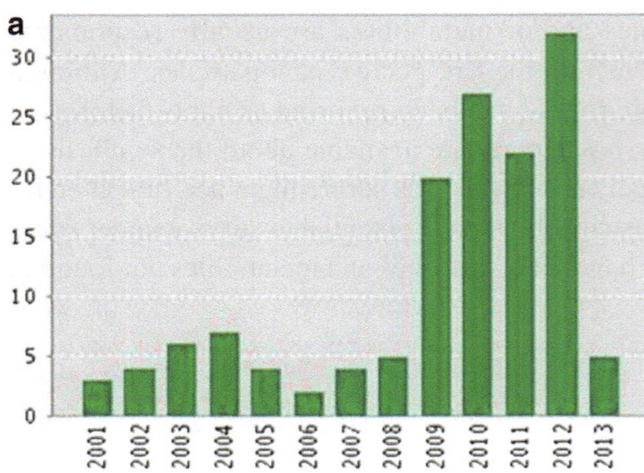

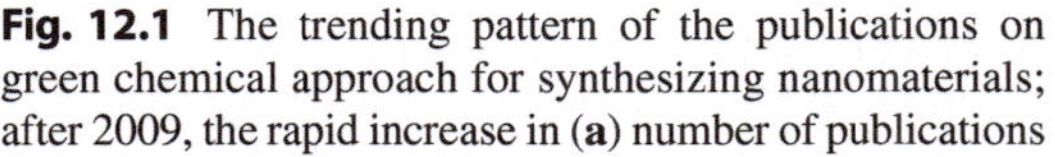

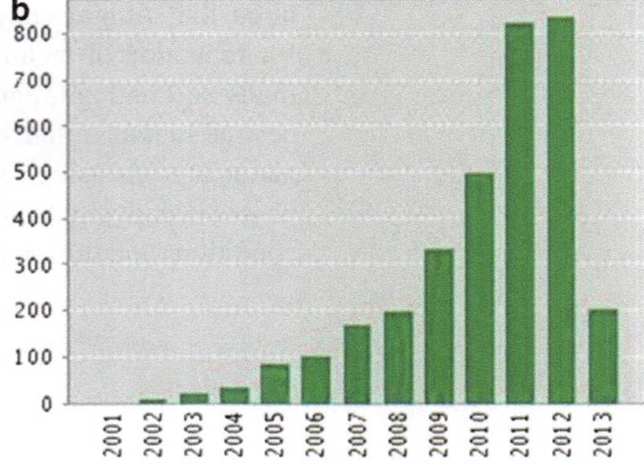

Fig. 12.1 The trending pattern of the publications on green chemical approach for synthesizing nanomaterials; after 2009, the rapid increase in (**a**) number of publications and (**b**) number of citations indicates the importance of this approach (Data sourced from web of science accessed on April 3rd 2013)

Table 12.1 A list of most recent fungal strains used for synthesis of nanoparticles (contains report from 2010 onwards)

Fungi	Nanoparticle	Size	Morphology	References
A. alternata	Se	30±5	Spherical	Sarkar et al. (2011b)
A. alternata	Ag	12±5	Spherical	Sarkar et al. (2011a)
A. alternata	Ag	35–90	ND	Acharya et al. (2011)
A. clavatus	Ag	10–25	Spherical, hexagonal	Verma et al. (2010)
A. clavatus	Au	20–35	Triangular	Verma et al. (2011)
A. flavus	Ag	17±5.9	Spherical	Jain et al. (2010)
A. flavus	Ag	7	Spherical	Moharrer et al. (2012)
A. fumigatus	Ag	7–19	Variable shapes	Navazi et al. (2010)
A. oryzae var. viridis	Au	10–60	Triangle, hexagon	Binupriya et al. (2010)
A. terreus	Ag	1–20	Spherical	Li et al. (2012)
Bipolaris nodulosa	Ag	10–60	Semi-pentagonal	Saha et al. (2010)
Cochliobolus lunatus	Ag	3–21	Spherical	Salunkhe et al. (2011)
C. versicolor	CdS	100	Spherical	Chen et al. (2011)
F. oxysporum	Au	20–40	Multishaped	Anitha and Palanivelu (2011)
F. oxysporum	Ag	20–70	Multishaped	Pandiarajan et al. (2010)
Neurospora crassa	Ag/Au	11	Spherical	Castro-Longoria et al. (2011)
Fusarium solani	Ag	3–8	Spherical	El-Rafie et al. (2010)
L. lecanii	Ag	45–100	Spherical	Namasivayam and Avimanyu (2011)
N. oryzae	Ag	30–90	Spherical	Saha et al. (2011)
Penicillium sp.	Ag	52–104	Multishaped	Hemath et al. (2010)
Penicillium sp.	Au	30–50	Spherical	Du et al. (2011)
Pestalotia sp.	Ag	10–40	Spherical	Raheman et al. (2011)
Phanerochaete chrysosporium	Au	10–100	Spherical	Sanghi et al. (2011)
Phoma sorghina	Ag	30–40	Rods	Gade et al. (2011)
R. stolonifer	Ag	5–50	Spherical	Afreen and Ranganath (2011)
S. rolfsii	Au	25	Triangle	
Trichoderma harzianum	Ag	30–50	Spherical	Singh and Balaji (2011)
Trichoderma reesei	Ag	5–50	Multishaped	Vahabi et al. (2011)
T. viride	Ag	5–40	Spherical	Fayaz et al. (2010)
Tricholoma crassum	Ag	5–50	Spherical	Ray et al. (2011)

their respective nanoparticles such as *Fusarium* spp. (Ahmad et al. 2003a; Ingle et al. 2008; Sawle et al. 2008), *Aspergillus* spp. (Verma et al. 2010, 2011; Vigneshwaran et al. 2007; Gade et al. 2008) and *Verticillium* sp. (Mukherjee et al. 2001). Similarly, many actinomycetes were also reported to synthesize nanomaterials like *Thermomonospora* sp. (Sastry et al. 2003) and *Rhodococcus* sp. (Ahmad et al. 2003b). However, all these strains mentioned above were not endophytic in nature (Table 12.1). As this fact was established that the fungal system can efficiently produce nanomaterials which are safe and eco-friendly, many scientists started to think of the efficacy of selected strains that have precise control over the shape and size of nanomaterials, and endophytic microbes are then evaluated for this purpose.

Table 12.2 List of most recent microbial endophytes that have been reported for synthesis of nanoparticles

SI	Host plant	Endophytic microbe	Nanoparticle	References
1	*Bauhinia variegata*	*Penicillium citrinum*	Au	Alappat et al. (2012)
		Colletotrichum gloeosporioides		
		Colletotrichum lindemuthianum		
		Phyllosticta sp.		
2	*Garcinia xanthochymus*	*Bacillus cereus*	Ag	Sunkar and Nachiyar (2012a, b)
3	*Phellodendron amurense*	*Epicoccum nigrum*	Ag	Qian et al. (2012)
4	*Avicennia marina*	*Aspergillus conicus*	Ag	Bharathidasan and Panneerselvam (2012)
	Suaeda monoica	*Penicillium janthinellum*		
	Rhizophora mucronata	*Phomopsis* sp.		
5	*Syzygium cumini*	*Pestalotia* sp.	Ag	Raheman et al. (2011)
6	*Pelargonium graveolens*	*Colletotrichum* sp.	Au	Shankar et al. (2003)
7	*Centella asiatica*	*Penicillium* sp.	Ag	Devi et al. (2012)
8	*Azadirachta indica*	*Aspergillus clavatus*	Ag	Verma et al. (2010)
9	*A. indica*	*Aspergillus clavatus*	Au	Verma et al. (2011)
10	*A. indica*	*Saccharomonospora* sp.	Au	Verma et al. (2013)
11	*Piper nigrum*	*Bordetella* sp.	Ag	Thomas et al. (2012)

2 Microbial Endophytes in Nanoparticles Synthesis

With increasing interest in the mycosynthesis of nanomaterials, apart from several soil and other fungi from other sources, the microbial endophytes were also been recently used for their potential in myconanosynthesis (Table 12.2). The first report of endophytic fungi used for synthesis of nanoparticles was from endophytic *Colletotrichum* sp. from *Pelargonium graveolens* (geranium leaf) to bioreduction of chloroaurate ions into gold nanoparticles. According to their report, they obtained different shape of gold nanoparticles instead of a specific shape; however, they didn't claim to get a control over the shape. The majority of nano-gold obtained were decahedral and icosahedral in shape ranging in size from 20 to 40 nm; interestingly, they have obtained multiply twinned particles (MTPs). As part of mechanism, they examined the extract of *Pelargonium graveolens* by FTIR and obtained strong bands at 1,658, 1,543 and 1,240 cm^{-1}. They hypothesized that these strong bands correlate with presence of polypeptide/proteins that are earlier reported in capping and stabilizing agents (Shankar et al. 2003). It is well known that certain proteins can bind to gold nanoparticles either through free amine groups or cysteine residues in the proteins. Endophytic fungi *Aspergillus clavatus* isolated from *Azadirachta indica* plant have also been reported to synthesize silver nanoparticles which have significant antibacterial and antifungal activity (Verma et al. 2010). Mostly spherical and hexagonal nanoparticles as well as clusters of nanoparticles were observed. Simple analysis of the topography suggests that the nanoparticles were spherically shaped with a height range of 2–6 nm and width range of 30–60 nm. This report shows that AgNP-embedded composite film exhibited clear inhibition zones when targeted against *Candida albicans* seeded plates, while no inhibition zone was observed in the controls (composite films of AgNP solution as positive control and composite films without nanoparticles as negative control). Maximum of 16 mm inhibition zone was observed for *C. albicans*. The minimum inhibitory concentration (MIC) was in the range of 5.8 μg ml^{-1} for *C. albicans*, while the minimum bactericidal concentration (MBC) was about 9.7 μg ml^{-1}. While other reports claim MIC about 4.8 while MBC 6.2 μg ml^{-1}, but that involved *Streptococcus mutans* as the test organism, thus, it is likely that *S. mutans* is more sensitive to the silver nanoparticles than *C. albicans*. The AgNPs are also found effective against *Pseudomonas fluorescens* with inhibition zone of 14 mm, while against *Escherichia coli,* it has 10 mm inhibition spectrum.

In an effort, endophytic *Pestalotia* sp. isolated from leaf tissues of *Syzygium cumini* was evaluated for its potential in synthesizing silver nanoparticles. The extracellular synthesis of AgNPs was observed with average size of 10–40 nm and mostly spherical and polydisperse (Raheman et al. 2011). A detailed in vitro antibacterial activity was also performed in this study, and interestingly, a combination of AgNPs along with antibiotics showed a manyfold increase in their antibacterial activity. AgNPs produced from endophytic fungus *Pestalotia* sp. were carried out without antibiotics and in combination with commercially available antibiotics gentamicin and sulphamethizole against *S. aureus* and *S. typhi*. AgNPs without antibiotics showed antibacterial activity but the efficiency was found to be increased significantly in combination with antibiotics. AgNPs in combination with gentamicin showed maximum activity (30 mm) (increase in fold area, 0.23) against *S. aureus* followed by sulphamethizole (25 mm) (increase in fold area, 0.18). Similar results were reported in case of *S. typhi* where silver nanoparticles in combination with gentamicin (28 mm) (increase in fold area, 0.15) showed more activity than combination of silver nanoparticles and sulphamethizole (24 mm) (increase in fold area, 0.08). To confirm the role of extract in nanosynthesis, an FTIR study of extract of *Pestalotia* sp. was performed, which confirms that amino acid residues and peptides of proteins have the stronger ability to bind with metal, so that the proteins could most possibly form a coat covering the metal nanoparticles, that is, capping of silver nanoparticles to prevent agglomeration of the particles and stabilizing in the medium (Raheman et al. 2011).

Besides many highly medicinal plants, some reports are available in which endophytes of mangrove plants were investigated for bioreduction of metal ion solutions into corresponding nanoparticles. Foliar endophytes of three selected mangrove plants *Avicennia marina*, *Suaeda monoica* and *Rhizophora mucronata* were isolated from Karankadu, Ramanathapuram district in south India. The authors have randomly selected three endophytes from each plant *Aspergillus conicus*, *Penicillium janthinellum* and *Phomopsis,* respectively, for the synthesis of silver nanoparticles. The lyophilized nanoparticle samples were analyzed in FTIR to identify the possible biomolecules responsible for the reduction of the Ag+ ions by the cell filtrate. The representative spectra of nanoparticles obtained manifest absorption peak located at about 3,843.68 cm^{-1} (–NH group of amines), 3,597.73 cm^{-1} (–OH group of phenols), 2,080.65 cm^{-1} (aromatic-CH stretching), 1,631.66 cm^{-1} (–NHCO of amide) and 767.16 cm^{-1} (C–CI) (Table 12.3). Silver nanoparticle synthesized by *Aspergillus conicus*, *Penicillium janthinellum* and *Phomopsis* shows better zone of inhibition when tested against several bacterial pathogens such as *Salmonella typhi*, *Staphylococcus aureus*, *Bacillus subtilis* and *Vibrio cholerae*. It was observed that in *n-butanol*, there are 10 nm inhibition zones for *Vibrio cholerae* and *Staphylococcus aureus* followed by about 15 mm for *Salmonella typhi;* however, the methanol extracts show moderate activity against tested pathogens with 10 and 5 mm of inhibition zones (Bharathidasan and Panneerselvam 2012).

Many other recent reports indicate several new endophytic strains for rapid synthesis of nanoparticles. Here, we briefly named a few more. An endophytic *Penicillium* sp. isolated from *Centella asiatica* has also been reported to produce silver nanoparticle and also has spectrum of antibacterial activity against an array of pathogenic strains such as *Proteus mirabilis*, *Shigella dysenteriae*, *Klebsiella pneumoniae*, *Staphylococcus aureus* and *Candida albicans* (Devi et al. 2012). Four endophytic fungi *Penicillium citrinum*, *Colletotrichum gloeo-*

Table 12.3 The FTIR analysis of the extracts from endophytic *Aspergillus conicus*, *Penicillium janthinellum* and *Phomopsis* sp. and assignments of various functional groups

S.N.	Group frequency cm^{-1} of the sample	Functional group assignment
1	3,406.98	N–H stretch, primary two bands, amine N–H stretching
2	2,925.89	Chelating compound Co–H stretching vibration-free OH
3	2,861.34	C–H alkalines, C–H stretching vibrations two band (aldehyde)
4	2,356.28	Hydrocarbon chromophone, C–H stretching (alkane)
5	2,145.24	–N=C=N– stretching vibrations, diamides
6	1,731.49	Cyclic, β-lactams, dilute solution
7	1,647.67	C–C alkene/ketone stretching β dilution, –N=N– stretching
8	1,559.32	N–H, amine salt, β-diketone, primary amide –N–H, Coo-aromatic
9	1,449.90	Aromatic
10	1,379.47	Coo-anion, OH phenol (sulphonyl chlorides)
11	1,073.88	(C–F) halogen compound C–X stretching vibrations

sporioides, *Colletotrichum lindemuthianum*, and *Phyllosticta* sp. isolated from *Bauhinia variegata* were investigated for biosynthesis of gold nanoparticle (Alappat et al. 2012). Out of the three strains tested, only *Penicillium citrinum* was found to synthesize the gold nanoparticles.

Apart from fungal endophytes, there are also reported bacterial endophytes used for the synthesis of nanoparticles. Endophytic bacterium *Bacillus cereus* isolated from leaf of *Garcinia xanthochymus* was used to synthesize silver nanoparticle and also evaluate antibacterial activity against many pathogenic strains like *Escherichia coli*, *Pseudomonas aeruginosa*, *Staphylococcus aureus*, *Salmonella typhi* and *Klebsiella pneumoniae*. The silver nanoparticles range from 20 to 40 nm in size; however, the shape is not mentioned in the report. The authors have convincingly state that with EDX analysis, the signals for C, N and O indicate the presence of proteins as a capping material on the surface of silver nanoparticles (Sunkar and Nachiyar 2012a, b). Also in another report, endophytic bacterium (unidentified) from *Coffee arabica* has been also reported to synthesize silver nanoparticles with significant antibacterial activity (Baker and Shreedharmurthy 2012a). Endophytic *Bordetella* sp. isolated from *Piper nigrum* was also found to have the ability to biofabricate extracellular silver nanoparticles at room temperature. The antibacterial potential of silver nanoparticles synthesized was also tested against pathogens like *Salmonella paratyphi*, *Vibrio cholera* and *Staphylococcus aureus* (Thomas et al. 2012).

However, the above-mentioned reports are mainly restricted to the synthesis of nanoparticles, and except a few, none have tried to investigate possibilities to get a control over the shape and size of biogenic nanoparticles by endophytes. Recently, a report come on the shape-controlled synthesis of gold nanoparticle from endophytic fungi *Aspergillus clavatus* (Verma et al. 2011). This endophytic strain was isolated from *Azadirachta indica* plant; earlier, the same authors have reported the same fungus for the synthesis of silver nanoparticles (Verma et al. 2010). The most interesting outcome of this study was to get fairly monodisperse nanotriangles of gold. The authors have shown that this particular strain can be used to modulate for getting a shape-controlled synthesis of gold nanoparticles (Table 12.2).

They have measured a single gold nanotriangle using atomic force microscopy (AFM); the purified gold nanotriangles showed a particle size distribution ranging from 20 to 35 nm with an average particle size of 30 ± 2 nm. Among the triangles also, they obtained a set of different types of triangles like sharp angle triangle, snipped angle triangle, and truncated triangles. This is the first report that thoroughly investigates the possibility of getting shape-controlled synthesis of nanoparticles from endophytes (Fig. 12.2). In effort to elucidate the mechanism of the biogenic synthesis of nanoparticles, the same authors have also investigated another endophytic actinomycetes *Saccharomonospora* sp. (Verma et al. 2013). In this study, the authors have convincingly shown

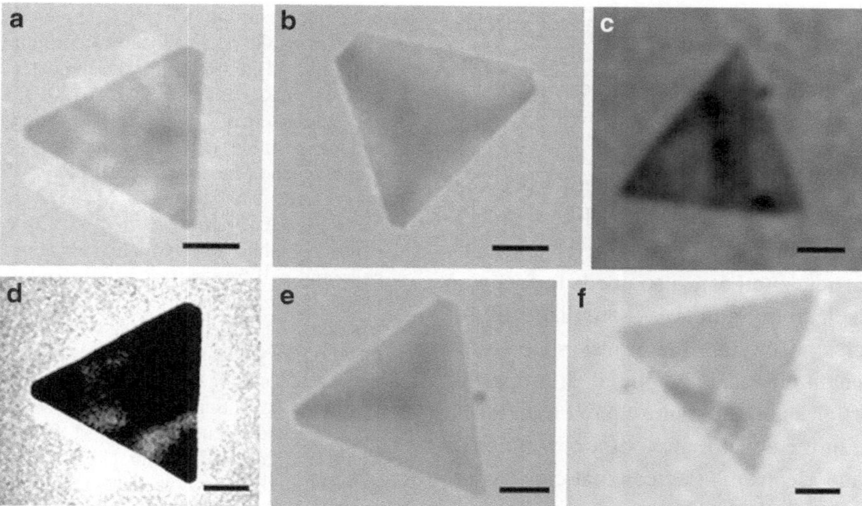

Fig. 12.2 TEM images of gold nanotriangles synthesized endophytic fungi *Aspergillus clavatus*. The different types of edge/tip margins of gold nanotriangles were obtained such as *sharp-edged triangles* (**a**, **c**, **f**), *truncated triangles* (**b**, **d**, **e**) and *snipped triangles* (**a**–**b**, **d**–**e**) (Adapted from Verma et al. 2011)

the role of proteins in bioreduction and stabilization of gold nanoparticles. They perform SDS protein profiling to investigate the role of protein. They identify two proteins corresponding to 42 and 50 kD that gradually decline from the reduction mixture.

3 Concluding Remarks

Size-dependent properties exhibited by the nanoparticles make them to have attractive medical and technological applications. Nanoparticles differ significantly from the bulk materials in its physical, chemical and biological properties due to their large fraction of surface atoms, large surface energy and spatial confinement. In addition to its advantages as non-toxic, and environmentally benign synthetic procedures, biological methods of nanoparticle synthesis provide particles with good control over the size distribution. The variation in physico-chemical properties of the nanoparticles synthesized through diverse microbial sources can affect its applications. This indicates the importance of exploration of novel untapped microbial sources for the identification of potential strains with nanoparticle-synthesizing property. Endophytes, by occupying the unique habitat, are greatly unexplored in terms of its nanoparticle-synthesizing properties. Although endophytic bacteria and fungi were greatly studied for its bioactive compounds and also for its use as biocontrol agent, its studies on nanoparticle synthesis are very limited. The interface between endophytes and nanomaterials is a relatively new and unexplored area (Baker and Shreedharmurthy 2012b). Improved scientific knowledge and implementation of new technologies unfold interaction of nano-revolution with biological entities, and the role of microbes in bio- and green synthesis of nanoparticles seems to have drawn unequivocal attention with a view of reformulating the novel strategies as alternatives for conventional methods for the synthesis of nanoparticles which are bound with various implications such as expensive costs and toxicity risks on health from environmental contaminants.

The biosynthetic route for the synthesis of metal nanoparticles using fungi is a simple process involving the reaction of microbial endophyte culture with aqueous solutions of metal ions. But there are a number of questions, which need to be addressed. The synthesis process points out that there are a number of reducing agents involved in the reduction of metal ions and corresponding formation of nanoparticles. These reducing agents also affect the size and shape of nanoparticles; hence, there is a need to investigate the exact mechanism involved in the biosynthesis of nanoparticles. Studies on the synthesis of nanoparticles of specific size and shape depend on different factors like temperature and light intensity. Biosynthetic approach for nanoparticle synthesis also needs to focus on the shape selectivity and size monodispersity of nanoparticles. Studying the novel shape- and size-dependent physical and chemical properties of nanoparticles and their subsequent interaction could help in the development of a new range of photonic and electronic devices that can control and manipulate light at nanoscale.

Acknowledgements Authors gratefully acknowledge the financial support from Department of Science and Technology (DST), New Delhi, India. Thanks are also due to Professor-in-Charge, Centre of Experimental Medicine and Surgery, for his support.

References

Acharya K, Chattopadhyay D, Patra S, Sarkar J, Saha S (2011) Mycosynthesis of silver nanoparticles and investigation of their antimicrobial activity. J NanoSci NanoEng Appl 1:17–26

Afreen RV, Ranganath E (2011) Synthesis of monodispersed silver nanoparticles by *Rhizopus stolonifer* and its antibacterial activity against MDR strains of *Pseudomonas aeruginosa* from burnt patients. Int J Environ Sci 1:1582–1592

Ahmad A, Mukherjee P, Senapati S, Mandal D, Khan MI, Kumar R, Sastry M (2003a) Extracellular biosynthesis of silver nanoparticles using the fungus *Fusarium oxysporum*. Colloids Surf B 28:313–318

Ahmad A, Senapati S, Khan MI, Kumar R, Sastry M (2003b) Extracellular biosynthesis of monodisperse gold nanoparticles by a novel extremophilic actinomycete *Thermomonospora* sp. Langmuir 19:3550–3553

Alappat CF, Kannan KP, Vasanthi NS (2012) Biosynthesis of Au nanoparticles using the endophytic fungi isolated from *Bauhinia variegata* L. Eng Sci Tech Int J 2:377–380

Anitha TS, Palanivelu P (2011) Synthesis and structural characterization of polydisperse silver and multishaped gold nanoparticles using *Fusarium oxysporum* MTCC 284. Dig J Nanomater Biostruct 6:1587–1595

Baker S, Shreedharmurthy S (2012a) Antimicrobial activity and biosynthesis of nanoparticles by endophytic bacterium inhabiting *Coffee arabica* L. Sci J Biol Sci 1:107–113

Baker S, Shreedharmurthy S (2012b) Endophytes: toward a vision in synthesis of nanoparticle for future therapeutic agents. Int J Bio Inorg Hybrid Nanomater 1:67–77

Bharathidasan R, Panneerselvam A (2012) Biosynthesis and characterization of silver nanoparticles using endophytic fungi *Aspergillus conicus*, *Penicillium janthinellum* and *Phomopsis* sp. IJPSR 3:3163–3169

Binupriya AR, Sathishkumarb M, Vijayaraghavanb K, Yuna S-I (2010) Bioreduction of trivalent aurum to nano-crystalline gold particles by active and inactive cells and cell-free extract of *Aspergillus oryzae* var. *viridis*. J Hazard Mater 177:539–545

Castro-Longoria E, Vilchis-Nestor AR, Avalos-Borja M (2011) Biosynthesis of silver, gold and bimetallic nanoparticles using the filamentous fungus *Neurospora crassa*. Colloids Surf B 83:42–48

Chen GQ, Zou ZJ, Zeng GM, Yan M, Fan JQ, Chen AW, Yang F, Zhang WJ, Wang L (2011) Coarsening of extracellularly biosynthesized cadmium crystal particles induced by thioacetamide in solution. Chemosphere 83:1201–1207

Devi NN, Shankar DP, Sutha S (2012) Biomimetic synthesis of silver nanoparticles from an endophytic fungus and their antimicrobial efficacy. Int J Biomed Adv Res 3:409–415

Du L, Xian L, Feng JX (2011) Rapid extra-/intracellular biosynthesis of gold nanoparticles by the fungus *Penicillium* sp. J Nanopart Res 13:921–930

El-Rafie MH, Mohamed AA, Shaheen THI, Hebeish A (2010) Antimicrobial effect of silver nanoparticles produced by fungal process on cotton fabrics. Carbohydr Polym 80:779–782

Fayaz M, Balaji K, Girilal M, Yadav R, Kalaichelvan PT, Venketesan R (2010) Biogenic synthesis of silver nanoparticles and their synergistic effect with antibiotics: a study against gram-positive and gram-negative bacteria. Nanomed Nanotechnol Biol Med 6:103–109

Gade A, Bonde PP, Ingle AP, Marcato P, Duran N, Rai MK (2008) Exploitation of *Aspergillus niger* for synthesis of silver nanoparticles. J Biobased Mater Bioenergy 2:1–5

Gade A, Ingle A, Whiteley C, Rai M (2010) Mycogenic metal nanoparticles: progress and applications. Biotechnol Lett 32:593–600

Gade A, Rai M, Kulkarni S (2011) *Phoma sorghina*, a phytopathogen mediated synthesis of unique silver rods. Int J Green Nanotechnol 3:153–159

Hemath NKS, Kumar G, Karthik L, Bhaskara Rao KV (2010) Extracellular biosynthesis of silver nanoparticles using the filamentous fungus *Penicillium* sp. Arch Appl Sci Res 2:161–167

Ingle A, Gade A, Pierrat S, Sönnichsen C, Rai M (2008) Mycosynthesis of silver nanoparticles using the fungus *Fusarium acuminatum* and its activity against some human pathogenic bacteria. Curr Nanosci 4:141–144

Jain N, Bhargava A, Majumdar S, Tarafdar JC, Panwar J (2010) Extracellular biosynthesis and characterization of silver nanoparticles using *Aspergillus flavus* NJP08: a mechanism perspective. Nanoscale 3:635–641

Li G, He D, Qian Y, Guan B, Gao S, Cui Y, Yokoyama K, Wang L (2012) Fungus-mediated green synthesis of silver nanoparticles using *Aspergillus terreus*. Int J Mol Sci 13:466–476

Mandal D, Bolander ME, Mukhopadhyay D, Sarkar G, Mukherjee P (2006) The use of microorganisms for the formation of metal nanoparticles and their application. Appl Environ Biotechnol 69:485–492

Moharrer S, Mohammad B, Gharamohammad RA, Yargol M (2012) Biological synthesis of silver nanoparticles by *Aspergillus flavus*, isolated from soil of Ahar copper mine. Indian J Sci Technol 5:2443–24447

Mukherjee P, Ahmad A, Mandal D, Senapati S, Sainkar SR, Khan MI, Ramani R, Parischa R, Ajayakumar PV, Alam M, Sastry M, Kumar R (2001) Bioreduction of $AuCl_4$ ions by the fungus *Verticillium* sp. and surface trapping of the gold nanoparticles formed. Angew Chem Int Ed Engl 40:3585–3588

Namasivayam SKR, Avimanyu (2011) Silver nanoparticle synthesis from *Lecanicillium lecanii* and evolutionary treatment on cotton fabrics by measuring their improved antibacterial activity with antibiotics against *Staphylococcus aureus* (ATCC 29213) and E. coli (ATCC 25922) strains. Int J Pharm Pharm Sci 3:190–195

Navazi ZR, Pazouki M, Halek FS (2010) Investigation of culture conditions for biosynthesis of silver nanoparticles using *Aspergillus fumigates*. Iran J Biotechnol 8:61

Pandiarajan G, Govindaraj R, Kumar M, Ganesan V (2010) Biosynthesis of silver nanoparticles using silver nitrate through biotransformation. J Ecobiotechnol 2:13–18

Qian Y, Yu H, He D, Yang H, Wang W, Wan X, Wang L (2012) Biosynthesis of silver nanoparticles by the endophytic fungus *Epicoccum nigrum* and their activity against pathogenic fungi. Bioprocess Biosyst Eng. doi:10.1007/s00449-013-0937-z

Raheman F, Deshmukh S, Ingle A, Gade A, Rai M (2011) Silver nanoparticles: Novel antimicrobial agent synthesized from an endophytic fungus *Pestalotia* sp. isolated from leaves of *Syzygium cumini* (L). Nano Biomed Eng 3:174–178

Ray S, Sarkar S, Kundu S (2011) Extracellular biosynthesis of silver nanoparticles using the mycorrhizal mushroom *Tricholoma crassum* BERK, its antimicrobial activity against pathogenic bacteria and fungus, including multidrug resistant plant and human bacteria. Dig J Nanomater Biostruct 6:1299

Saha S, Sarkar J, Chattopadhyay D, Patra S, Chakraborty A, Acharya K (2010) Production of silver nanoparticles by a phytopathogenic fungus *Bipolaris nodulosa* and its antimicrobial activity. Dig J Nanomater Biostruct 5:887–895

Saha S, Chattopadhyay D, Acharya K (2011) Preparation of silver nanoparticles by bio-reduction using Nigrospora oryzae culture filtrate and its antimicrobial activity. Dig J Nanomater Biostruct 6:1519–1528

Salunkhe RB, Patil SV, Salunke BK, Patil CD, Sonawane AM (2011) Studies on silver accumulation and nanoparticle synthesis by *Cochliobolus lunatus*. Appl Biochem Biotechnol 165:221–234

Sanghi R, Verma P, Puri S (2011) Enzymatic formation of gold nanoparticles using *Phanerochaete chrysosporium*. Adv Chem Eng Sci 1:154–162

Sarkar J, Chattopadhyay D, Patra S, Deo SS, Sinha S, Ghosh M, Mukherjee A, Acharya K (2011a) *Alternaria alternata* mediated synthesis of protein capped silver nanoparticles and their genotoxic activity. Dig J Nanomater Biostruct 6:563–573

Sarkar J, Dey P, Saha S, Acharya K (2011b) Mycosynthesis of selenium nanoparticles. IET Micro Nano Lett 6:599–602

Sastry M, Ahmad A, Khan MI, Kumar R (2003) Biosynthesis of metal nanoparticles using fungi and actinomycete. Curr Sci 85:162–170

Sawle BD, Salimath B, Deshpande R, Bedre MD, Prabhakar KB, Venkataraman A (2008) Biosynthesis and stabilization of Au and Au–Ag alloy nanoparticles by fungus *Fusarium semitectum*. Sci Technol Adv Mater 9:9035012

Shankar SS, Ahmad A, Pasricha R, Sastry M (2003) Bioreduction of chloroaurate ions by *Geranium* leaves and its endophytic fungus yields gold nanoparticles of different shapes. J Mater Chem 13:1822–1826

Singh P, Balaji R (2011) Biological synthesis and characterization of silver nanoparticles using the fungus *Trichoderma harzianum*. Asian J Exp Biol Sci 2:600–605

Sunkar S, Nachiyar CV (2012a) Microbial synthesis and characterization of silver nanoparticles using the endophytic bacterium *Bacillus cereus*: a novel source in the benign synthesis. GJMR 12:43–49

Sunkar S, Nachiyar CV (2012b) Biogenesis of antibacterial silver nanoparticles using the endophytic bacterium *Bacillus cereus* isolated from *Garcinia xanthochymus*. Asian Pac J Trop Biomed 2:930–934

Thomas R, Jasim B, Mathew J, Radhakrishnan EK (2012) Extracellular synthesis of sliver nanoparticles by endophytic *Bordetella* sp. isolated from *Piper nigrum* and its antibacterial activity analysis. Nano Biomed Eng 4:183–187

Vahabi K, Ali MG, Karimi S (2011) Biosynthesis of silver nanoparticles by fungus *Trichoderma reesei* a route for large-scale production of AgNPs. Insci J 1:65–79

Verma VC, Kharwar RN, Strobel GA (2009) Chemical and functional diversity of natural products from plant associated endophytes. Nat Prod Commun 4:1511–1532

Verma VC, Kharwar RN, Gange AC (2010) Biosynthesis of antimicrobial silver nanoparticles by the endophytic fungus *Aspergillus clavatus*. Nanomedicine (UK) 5:33–40

Verma VC, Singh SK, Solanki R, Prakash S (2011) Biofabrication of anisotropic gold nanotriangles using extract of endophytic *Aspergillus clavatus* as a dual functional reductant and stabilizer. Nanoscale Res Lett 6:16

Verma VC, Anand S, Ulrichs C, Singh SK (2013) Biogenic gold nanotriangles from *Saccharomonospora* sp., an endophytic actinomycetes of *Azadirachta indica* A. Juss. Int Nano Lett 3:21

Vigneshwaran N, Ashtaputre NM, Varadarajan PV, Nachane RP, Paralikar KM, Balasubramanya RH (2007) Biological synthesis of silver nanoparticles using the fungus *Aspergillus flavus*. Mater Lett 61:1413–1418

Part V

Bio-control and Bioremediation

Biological Control of Insect-Pest and Diseases by Endophytes

13

Paulo Teixeira Lacava and João Lúcio Azevedo

Abstract

The natural and biological control of insect-pests and diseases affecting cultivated plants has gained much attention in the past decades as a way of reducing the use of pesticides in agriculture. Biocontrol has been frequently used in tropical countries, such as Brazil, and it is supported by the development of local basic and applied research. In this context, tropical endophytes have attracted special attention to develop their roles to control of pest insect and plant diseases. Endophytic symbiotic microorganisms are defined in different ways and a recent definition includes all of the culturable microorganisms that inhabit inner parts of plant tissues causing no harm to their hosts. They can be divided in two groups: those that do not generate external structures from the host and those able to develop external structures such as nodules of N_2 fixing bacteria and mycorrhizal fungi. Endophytes have important roles in the plant host protection, acting against predators and pathogens. They protect host plants against herbivores such as cattle and pest insect. They also may increase plant resistance to pathogens that produce antimicrobial agents and plant-growth hormones and have other effects countering biotic and abiotic stresses. Endophytic microorganisms were first studied in plants in temperate regions but more recently have been also studied in plants from tropical regions. In this chapter, we focus on examples of endophytic bacteria and fungi, especially those that may control pest insects and plant diseases by antagonistic effects, production of enzymes, or introduction of heterologous genes by recombinant DNA technology.

P.T. Lacava (✉)
Centre of Biological Sciences and Health, Federal University of São Carlos, Rodovia Washington Luís, km 235, PO BOX 676, 13565-905 São Carlos, SP, Brazil
e-mail: ptlacava@ufscar.br

J.L. Azevedo
Department of Genetics, Escola Superior de Agricultura "Luiz de Queiroz", University of São Paulo, Av. Pádua Dias 11, PO BOX 83, 13400-970 Piracicaba, SP, Brazil

V.C. Verma and A.C. Gange (eds.), *Advances in Endophytic Research*,
DOI 10.1007/978-81-322-1575-2_13, © Springer India 2014

1 Introduction

The term endophyte is applied to microorganisms that live within plant tissues for all or part of their life cycles and cause no apparent infections or symptoms of disease (Wilson 1995; Azevedo et al. 2000; Bacon and White 2000; Saikkonen et al. 2004). Hallmann et al. (1997) describe endophytes as those organisms that can be isolated from surface-sterilized plant parts or extracted from inner tissues and that cause no damage to the host plant. In addition, Azevedo and Araújo (2007) suggested that endophytes are all microorganisms, culturable or not, that inhabit the interior of plant tissues, cause no harm to the host, and do not develop external structures. More recently, Mendes and Azevedo (2007) defined endophytic microorganisms in the same way as other authors (Hallmann et al. 1997; Azevedo et al. 2000; Azevedo and Araújo 2007) but suggested a division of endophytes in two types: Type I, or endophytes that do not develop external structures, and Type II, or endophytes that develop external structures. Endophytic bacteria have been isolated from many different plant species (Lodewyckx et al. 2002; Idris et al. 2004; Rosenblueth and Martinez-Romero 2006; Barzanti et al. 2007; Sheng et al. 2008; Mastretta et al. 2009). Also fungal endophytes have been isolated from lichens, moss, ferns, gymnosperms, monocotyledonous, and dicotyledonous plants, growing in different environments (Petrini 1986; Petrini et al. 1990). More recently they have been frequently isolated from plants growing in tropical and subtropical regions. According to Azevedo and Araújo (2007), more than 50 plant species studied from these regions and hundreds of different species of fungi were isolated and these numbers are constantly increasing (Bernardi-Wenzel et al. 2010; Gazis and Chavern 2010; Suryanarayanan et al. 2011; Radji et al. 2011; Orlandelli et al. 2012; Rhoden et al. 2012; Garcia et al. 2012). This category of microorganisms may stimulate host growth through several mechanisms, including biological control; induction of systemic resistance to pathogens; nitrogen fixation; production of growth regulators, antimicrobial products, and enzymes; and enhancement of mineral nutrients or water uptake (Ryan et al. 2008). Additionally, the endophytic microorganisms isolated from plants that hyperaccumulate metals exhibit tolerance to high metal concentrations (Idris et al. 2004; Rajkumar et al. 2009). There is a great deal of interest in understanding endophyte diversity and the role of endophytic microorganisms in plant and microbial ecology, evolutionary biology, and applied research, ranging from biological control to bioprospecting for genes (Azevedo et al. 2000; Araújo et al. 2008). In the past two decades, a lot of information on the role of endophytic microorganisms in nature has been collected. The ability to colonize internal host tissues has made endophytes valuable as a tool to improve crop performance. In this review, we address the major topics concerning the biocontrol potential of endophytes in agrobiology systems.

2 Endophytic Bacteria from Different Host Plants

Reported endophytes include both Gram-positive and Gram-negative bacteria and the classes *Alpha-*, *Beta-*, and *Gammaproteobacteria*, *Actinobacteria*, *Firmicutes*, and *Bacteroidetes* (Lodewyckx et al. 2002; Bacon and Hinton 2006). Approximately 300,000 plant species growing in unexplored areas of the earth are host to one or more endophytes (Araújo et al. 2001), and the presence of biodiverse endophytes in huge numbers plays an important role in the ecosystems with the greatest biodiversity, such as tropical and temperate rainforests (Arachevaleta et al. 1989), which are found extensively in Brazil and possess almost 20 % of its biotechnological source materials (Araújo et al. 2002). Endophytic bacteria have been isolated from a variety of plants, as reviewed by Sturz et al. (2000) and Hallmann et al. (1997). Plants harboring endophytes were reported in a review by Rosenblueth and Martinez-Romero (2006) of bacterial endophytes and their interactions with hosts but, most likely, there is not a single plant species devoid of endophytes. The few examples of apparent absence of endophytes suggest that some microorganisms are not easily isolated or cultured. The diversity of endophytic bacterial species has been largely based on culture techniques. Culture-independent analysis of bacterial populations inside citrus plants also suggests that bacterial endophytic populations are much more diverse than previously realized (Araújo et al. 2002; Lacava et al. 2006). Various reports concerning endophytic bacteria in agricultural plants have demonstrated that the use of fingerprinting techniques and clone analysis can provide additional information for analyzing the community composition of endophytic bacteria (Chelius and Triplett 2001; Garbeva et al. 2001; Seghers et al. 2004; Sessitsch et al. 2004). Culture-independent molecular approaches based on 16S rRNA gene analysis, such as PCR amplification of 16S rDNAs, amplified ribosomal DNA restriction analysis (ARDRA), denaturing gradient gel electrophoresis (DGGE), and terminal restriction fragment length polymorphism (T-RFLP), have been successfully used for bacterial community analysis in a great variety of environments, including soil ecosystems (Dunbar et al. 1999), marine environments (Cottrell and Kirchman 2000), rhizospheres (Smalla et al. 2001), foods (Cocolin et al. 2002), and human intestines (Kibe et al. 2005), to overcome the limitations of culture-dependent approaches. However, these culture-independent approaches used on endophytic bacteria have met with limited success due to disturbances from chloroplast 16S rDNA and mitochondrial 18S rDNA. Recently, Sessitsch et al. (2012) suggested a new approach to study the functional characteristics of endophytic bacteria. The authors presented the first metagenomic approach to analyze an endophytic bacterial community inside roots of rice. They asserted that assessing microbial functions is impeded by difficulties in cultivating most prokaryotes, and endophytes inside host tissues are not always amenable to biochemical or genetic analyses (Mano and Morisaki 2008; Weyens et al. 2009). From the results of Sessitsch et al. (2012), metagenome sequences were obtained from endophytic cells extracted from the roots of field-grown plants (rice). Putative functions were deduced from protein domains or similarity analyses of protein-encoding gene fragments, and this allowed insight into the capacities of endophytic cells. Prominent features included flagella, plant-polymer-degrading enzymes, protein secretion systems, iron acquisition and storage, quorum sensing, and detoxification of reactive oxygen species. In this metagenome analysis, endophytes might be involved in the entire nitrogen cycle as protein domains involved in N_2-fixation, denitrification, and nitrification because genes involved in these cases were detected and expressed. Finally, the authors concluded that a deeper understanding of endophytic functions and mechanisms for their establishment in the endosphere could be exploited to improve agricultural management practices with respect to biocontrol, bioremediation, and plant nutrition. They suggested the metagenome approach as a method alternative to cultivation for the study of the role of bacterial endophytes that reside inside host plants.

3 Localization Inside of Host Plants

Endophytic bacteria appear to originate from seeds (Pleban et al. 1995; Adams and Kloepper 1996), vegetative planting material (Dong et al. 1994), rhizosphere soil (Sturz 1995; Hallmann et al. 1997; Mahaffee and Kloepper 1997), and the phylloplane (Beattie and Lindow 1995). With the exception of seed-transmitted bacteria, which are already present in the plant, potential endophytes must first colonize the root surface prior to entering the plant. The initial processes of colonization of plant tissue by endophytic bacteria can be via stoma, lenticels, areas of emergence of lateral roots, and germinating radicles (Huang 1986). Several authors have reported colonization of the secondary root emergence zone by bacterial endophytes (Reinhold and Hurek 1988; Wiehe et al. 1994; Mahaffe et al. 1997). Various bacterial endophytes have been reported to live within cells, in intercellular spaces, or in the vascular systems of plants (Hallmann et al. 1997; James and Olivares 1998; Reinhold-Hurek and

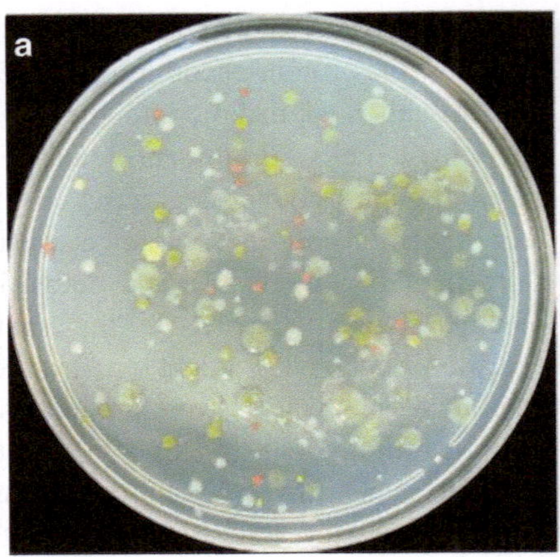

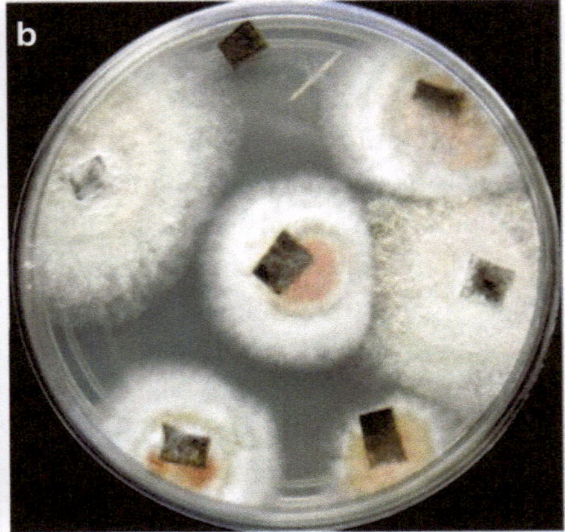

Fig. 13.1 Diversity of culturable endophytic microorganisms isolated from leaf tissue of mangrove plants (*Rhizophora mangle*). (**a**) Primary isolation of endophytic bacteria from *R. mangle*. (**b**) Primary isolation of endophytic fungi from *R. mangle*

Hurek 1998; Sturz et al. 2000; Rosenblueth and Martinez-Romero 2006; Gai et al. 2009). Although endophyte populations vary in different plants according to many factors, bacterial populations are generally larger in roots and smaller in stems and leaves (Lamb et al. 1996). Additionally, the population density of endophytic bacteria found in plants depends on the plant species, genotype, and tissue, the growth stage and specialization of the bacteria, differences in colonization pathway, and mutual exclusion of different bacterial populations (Sturz et al. 1997). According to Strobel and Daisy (2003), many factors change endophytic biology, including the season, the age of the host plant, the environment, and the location. The processes of colonization of plant tissue by endophytic bacteria are complex and include host recognition, spore germination, penetration, and colonization, and the sources of endophytic colonization are diverse, ranging from transmission via seeds (Ferreira et al. 2008) and vegetative planting material to entrance from the surrounding environment, such as the rhizosphere and phyllosphere. However, there is interest in finding bacterial strains with biological control or plant-growth-promoting capabilities. If these bacteria can be found in internal plant tissues, as they can in the rhizosphere, these bacteria may have the unique capacity to elicit beneficial effects from within the plants. As new beneficial bacterial strains are identified, delivery of these strains to specific plant tissues will be needed. To use endophytic bacteria in practical agronomic production, reliable and practical methods of inoculation must be developed. Several delivery systems have been reported for endophytic bacteria (van Der Peer et al. 1990; Kumar and Dube 1992; Musson 1994). In our studies, we have used culture-dependent approaches based on media culture (Fig. 13.1) and fluorescent microscopy (Fig. 13.2) to determinate the localization of endophytic bacteria in host plants. The endophytic bacterium *Methylobacterium mesophylicum* (strain SR1.6/6) in *Catharanthus roseus* and *Nicotiana clevelandii* plants was made visible by scanning electron microscopy (SEM). The highest densities were observed in the roots and hypocotyl, suggesting that these sites may be the most important points of entry for strain SR1.6/6 in both plants. Remarkably, cells adhering to the plants were immersed in a mucilaginous layer, suggesting that strain SR1.6/6 is able to form a biofilm on the root and hypocotyl surfaces of both plants (Andreote et al. 2006). Lacava et al. (2007b), using fluorescence microscopy, revealed that *Klebsiella pneumoniae* strain Kp342 colonized the xylem vessels of *Citrus sinensis* roots and

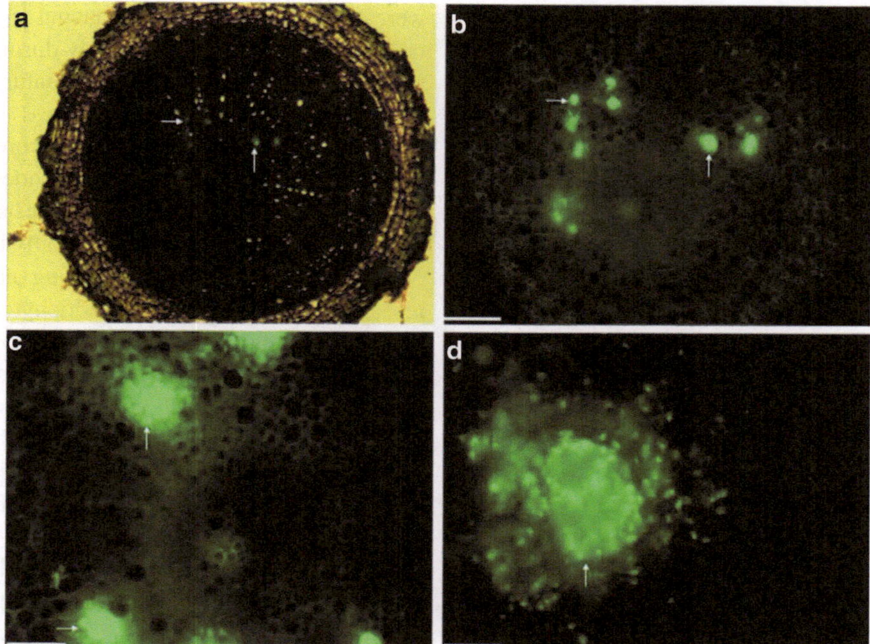

Fig. 13.2 Transverse section of *Citrus sinensis* roots. Series of images demonstrating colonization by GFP-labeled *Klebsiella pneumoniae* 342 strain (**a**, **b**, **c**, **d**). *Arrows* point to GFP-tagged bacterial cells. Bars, 50 μm (Modified Lacava et al. 2007a)

branches, and it was able to colonize the xylem vessels of *C. roseus* branches and roots. Previous reports have described the ability of *K. pneumoniae* to colonize the roots and vascular tissue of plants (Dong et al. 2003). Based on isolation and fluorescence microscopy, Lacava et al. (2007a) suggested that *C. roseus* could be used as a model plant to study the interaction between endophytic bacteria and host plants. Ferreira et al. (2008) reported an endophytic bacterial community residing in *Eucalyptus* seeds and the transmission of these bacteria from seeds to seedlings. The authors suggested that endophytic bacteria can be transmitted vertically from seeds to seedlings, assuring the support of the bacterial community in the host plant. The authors evaluated the characteristics of colonization of endophytic bacteria by isolation and fluorescence microscopy. Gai et al. (2009) reported the localization of the endophytic bacterium *M. mesophilicum* in *C. roseus* and the transmission of this endophyte by *Bucephalogonia xanthophis* using isolation and fluorescence microscopy. *C. roseus* is a model plant for the study of interactions between endophytic bacteria and *Xylella fastidiosa*, the causal agent of citrus variegated chlorosis, and *B. xanthophis* is an insect vector that transmits *X. fastidiosa* to citrus plants (Hartung et al. 1994).

4 Endophytic Bacteria: Biotechnological Potential

A better understanding of endophytic bacteria may help to elucidate their function and potential role in developing sustainable systems of crop production (Sun et al. 2008). Bacteria interact with plants in four ways: as pathogens, symbionts, epiphytes, or endophytes. Of these four types of bacteria–plant interactions, endophytic interactions are the least studied and least understood (Iniguez et al. 2005). Endophytic bacteria are of biotechnological and agronomic interest because they can enhance plant growth and improve the nutrition of plants, and they can also control pests and plant diseases (Boddey et al. 2003; Sevilla et al. 2001; Azevedo et al. 2000). Endophytes may increase crop yields, remove contaminants, inhibit pathogens, and produce fixed nitrogen or novel substances (Rosenblueth and Martinez-Romero 2006). The repertoire of their effects and functions in plants has not been comprehensively defined. The challenge and goal is to be able to manage microbial communities that favor plant colonization by beneficial bacteria. This will be possible when better knowledge of endophyte ecology and plant–endophyte molecular interactions is attained. The endophyte–host relationship is believed to be complex and most likely varies from host to host and microorganism to microorganism (Boursnell 1950). Many experiments have been conducted to compare how endophyte-infected plants and noninfected plants behave in response to environmental stress and attack by insect and animal predators (Owen and Hundley 2004). Furthermore, endophyte-infected plants often grow faster than noninfected ones (Cheplick et al. 1989). This effect is at least in part due to the endophytes' production of phytohormones, such as indole-3-acetic acid (IAA), cytokines, and other plant-growth-promoting substances (Tan and Zou 2001), and the fact that endophytes enhance the hosts' uptake of nutritional elements such as nitrogen (Reis et al. 2000) and phosphorus (Malinowski and Belesky 1999).

The search for interesting natural biological activities has been the basis for the development of various applications in biotechnology and agriculture. The microbial world, and endophytes in particular, reflects a genetic and metabolic biodiversity, which has not yet been thoroughly explored.

4.1 Endophytic Fungi: Isolation, Localization, and Biotechnological Potential

Fungi were the first microorganisms described as endophytes (de Bary 1866) but at that time they were considered neutral, not causing any benefits or harm to their plant hosts. Only during the last two decades of the twentieth century it was shown that endophytic fungi have important roles, protecting plants against herbivores including cattle and insects. They also provide nutrients to the host and increase plant resistance to drought, cold, and pathogens. So, only in the last 30 years, there were an increasing number of research studies dealing with endophytes. As previously mentioned they were found to occur in every plant till now studied. It is estimated that there are about 1.5 million of fungal species in our planet (Hawksworth 2001) and only a small percentage of them have been described. As the majority of fungal species are valuable from environmental and biotechnological point of views and as endophytic fungi were isolated only from few, among the 300,000 existing plant species, endophytes are a potential source as producers of new antibiotics, enzymes, dyes, and many other useful compounds. They also can be valuable as biological controllers of pests and diseases and increase plant-growth vigor by producing hormones or providing nutrients to the host. Several reviews cover different aspects of fungal endophytes (Azevedo et al. 2000; Azevedo and Araújo 2007; Vega et al. 2008; Suryanarayanan 2011; Suryanarayanan et al. 2012). As already mentioned for bacteria, with few differences, fungi are found in seed, stems, leaves, and other plant organs and tissues. Besides vertical transmission as from seeds, colonization began with penetration of the fungus from natural or artificial openings as root emission zone, stomata, or injuries caused by root growth, agricultural practices, or insects. After penetration endophytes can be found all over the plant. Isolation of endophytes from plants is easily made by using appropriated fungal culture media with plant fragments previously surface treated to eliminate epiphytic microorganisms and, after incubation, fungi are transferred to new media and purified. Details of different methods of isolation and purification can be found in a practical guide organized by Araújo et al. (2010). Molecular approaches to recover culture-independent data from fungi are now used, opening new ways to detect valuable characteristics of endophytic fungi. The processes described using bacteria may be applied with appropriated modifications, when fungi are considered (Araújo et al. 2010). Considering the classic and modern molecular approaches, the biotechnological potential of endophytic microorganisms for the production of pharmaceutical products, biological control, plant-growth promotion, enzymes, and other products is continuously growing. Some examples of biotechnology and agronomic uses of endophytic microorganisms were already mentioned for endophytic bacteria and most of them may also be applied to endophytic fungi. A more detailed aspect, that is, use of endophytic fungi for biological control of pests and diseases, will be further discussed.

4.2 Biological Control of Insect-Pests and Plant Diseases by Endophytic Microorganisms

The control of insect-pests and diseases by means of biological processes, such as the use of entomopathogenic microorganisms or those that inhibit/antagonize microorganisms pathogenic to plants, is an alternative that may help to reduce or eliminate the use of chemical products in agriculture (Azevedo et al. 2000). Agriculture by its own nature is anti-ecological, and, with the use of chemical fertilizers, insecticides, fungicides, herbicides, and antibiotics on a large scale, profound biological modifications have been occurring. Products such as insecticides and fungicides aim to control pests and phytopathogenic microorganisms. However, they are responsible for eliminating important species of insects that control other pests and microorganisms that are performing a crucial role in the environment, inhibiting the growth and the multiplication of other microorganisms. One group of microorganisms that is affected by these anthropogenic modifications is the endophytes. The natural and biological control of pests and diseases affecting cultivated plants has gained much attention in the past decades as a way of reducing the use of chemical products in agriculture. Biological control has been frequently used in Brazil, and it is supported by the development of basic and applied research on this field not only in our country but also in South America, as shown by several reviews (Lecuona 1996; Alves 1998; Melo and Azevedo 1998). The use of agrochemicals, although decreasing the impact of insects and phytopathogenic microorganisms, still represents a high risk for field workers and consumers. In this review we will first focus on examples of endophytic bacteria, especially those that may control insect-pests and plant diseases by antagonistic effects, production of enzymes, or introduction of heterologous genes by recombinant DNA technology followed by examples of endophytic fungi control of plant pests and diseases.

4.2.1 Biocontrol of Plant Diseases by Antagonistic Endophytic Bacteria

Recent studies have indicated that biological control of bacterial wilt disease could be achieved using antagonistic bacteria (Fig. 13.3). Different bacterial species, namely, *Alcaligenes* spp. and *Kluyvera* spp. (Assis et al. 1998), *Pseudomonas fluorescens*, *P. alcaligenes*, *P. putida*, *Flavobacterium* spp. and *Bacillus megaterium* (Reiter et al. 2002), *B. pumilus* (Benhamou et al. 1998) and *Microbacterium* spp., *Clavibacter michiganensis*, *Curtobacterium* spp., and *B. subtilis* (Zinniel et al. 2002), have been reported as endophytes and were inhibitory to plant pathogens. Toyota and Kimura (2000) have reported the suppressive effect of some antagonistic bacteria on *R. solanacearum*. Moreover, Ciampi-Panno et al. (1989) have demonstrated the use of antagonistic microbes in the control of *R. solanacearum* under field conditions. Ramesh et al. (2009) have suggested that Pseudomonads are the major antagonistic endophytic bacteria that suppress the bacterial wilt pathogen, *Ralstonia solanacearum*, in eggplant (*Solanum melongena* L.). Twenty-eight bacterial isolates that effectively inhibited *R. solanacearum* were characterized and identified in vitro (Ramesh et al. 2009). More than 50 % of these isolates were *Pseudomonas*

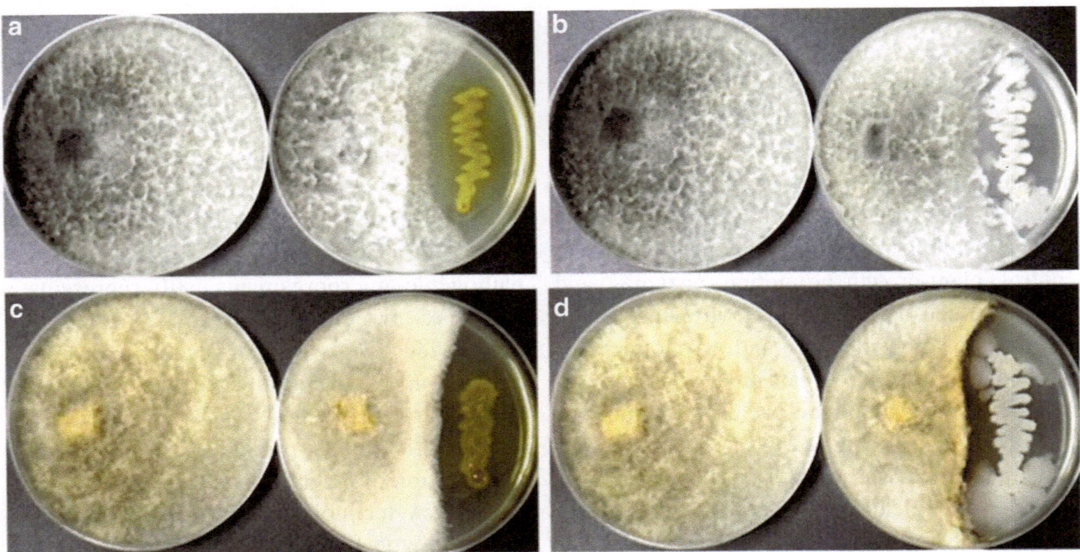

Fig. 13.3 In vitro antagonistic activities of endophytic bacteria isolated from *Vitis labrusca* against phytopathogenic fungi. (**a**–**b**) Antagonist activity against *Ceratocystis paradoxa* and (**c**–**d**) against *Rhizoctonia solani*

fluorescens. In greenhouse experiments, the plants treated with *Pseudomonas* isolates (EB9, EB67), *Enterobacter* isolates (EB44, EB89), and *Bacillus* isolates (EC4, EC13) reduced the incidence of wilt by more than 70 %. All the selected isolates reduced damping by more than 50 % and improved the growth of seedlings in the nursery stage. Large-scale field evaluations and detailed knowledge of antagonistic mechanisms could provide an effective biocontrol solution for bacterial wilt of solanaceous crops. In our study, we suggested that the endophytic bacteria *Curtobacterium flaccumfaciens*, isolated from citrus plants (Araújo et al. 2001), can inhibit *X. fastidiosa*, a phytopathogenic bacterium that is the causal agent of citrus variegated chlorosis (CVC) (Schaad et al. 2004), both in vitro (Lacava et al. 2004) and in vivo (Lacava et al. 2007b), when inoculated in the model plant *C. roseus* (Monteiro et al. 2001). *C. roseus* has been used to study the interaction between endophytic bacteria and *X. fastidiosa* in greenhouse environments (Lacava et al. 2006; Andreote et al. 2006). To characterize the interactions of *X. fastidiosa* and the endophytic bacteria *C. flaccumfaciens* in vivo, *C. roseus* plants were inoculated separately with *C. flaccumfaciens*, *X. fastidiosa*, and both bacteria together (Lacava et al. 2007b). The number of flowers produced by the plants, the heights of the plants, and the exhibited disease symptoms were evaluated. *X. fastidiosa* induced stunting and reduced the number of flowers produced by *C. roseus*. When *C. flaccumfaciens* was inoculated together with *X. fastidiosa*, no stunting was observed. The number of flowers produced by our doubly inoculated plants was an intermediate between the number produced by the plants inoculated with either of the bacteria separately. These data indicate that *C. flaccumfaciens*, an endophytic bacterium, interacted with *X. fastidiosa* in *C. roseus* and reduced the severity of the disease symptoms induced by *X. fastidiosa* (Fig. 13.4). The identification of biological sources for the control of plant pathogenic fungi remains an important objective for sustainable agricultural practices. In a recent project with financial support from several Brazilian agencies (Foundation of Support the Research of the State of Amazonas [FAPEAM] and the State of São Paulo Research Foundation [FAPESP – Grant/Process no. 09/53376-2]), we screened the antagonistic activity in vitro of endophytic bacteria

Fig. 13.4 (**a**) Disease symptoms induced in *Catharanthus roseus* plants 2 months after inoculation with *Xylella fastidiosa* (*right*). A symptom-free plant doubly inoculated *X. fastidiosa* and *C. flaccumfaciens* (*left*). Leaf stunting and chlorosis induced in *C. roseus* leaves 2 months after inoculation with (**b**) *X. fastidiosa* (*left*). (**c**) Symptom-free leaves from a plant doubly inoculated with *X. fastidiosa* and *C. flaccumfaciens* (*right*) (Modified Lacava et al. 2007b)

versus *Colletotrichum* sp., the causal agent of anthracnose disease (Silva et al. 2004) of guarana (*Paullinia cupana* var. *sorbilis* Mart. Ducke). Fruits from guarana are of both economic and social importance in Brazil. Sodas, syrups, juices, and several pharmaceutical products are made from guarana toasted grains (Ângelo et al. 2008). A significant decrease in the area of guarana production, particularly in the Brazilian Amazon region, can be attributed to anthracnose disease. In this study, the endophytic bacteria used in the antagonism test were isolated from guarana plants. We found some endophytic isolates from guarana with antagonism activity against *Colletotrichum* sp. in our preliminary results.

4.2.2 Endophytic Actinobacteria in the Control of Phytopathogens

Endophytic actinobacteria have been isolated from a wide variety of plants, and the most frequently isolated species belong to the genera *Microbispora*, *Nocardia*, *Micromonospora*, and *Streptomyces*, the last of which is by far the most abundantly observed (Sardi et al. 1992; Taechowisan et al. 2003). Actually, the best studied genus of actinobacteria is *Streptomyces* (Seipke et al. 2012), which has a complex developmental life cycle (Flärdh and Buttner 2009) and produces numerous secondary metabolites (Challis and Hopwood 2003). Endophytic *Streptomyces* bacteria are not simply plant

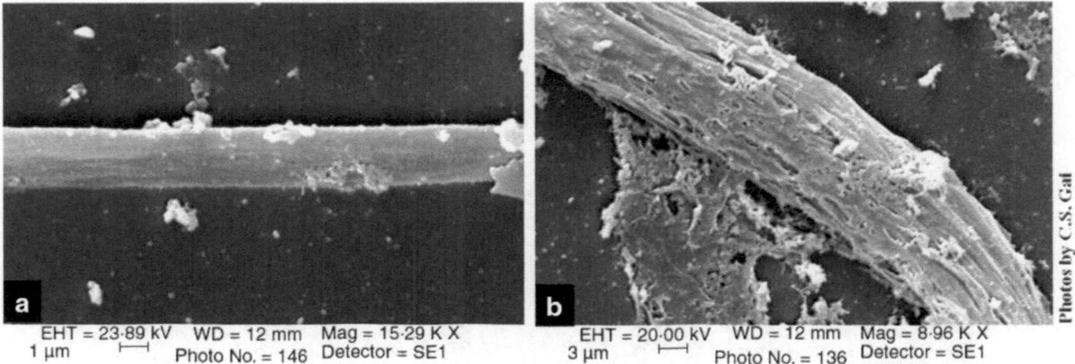

Fig. 13.5 Scanning electronic microscopic analysis of *Colletotrichum sublineolum*. (**a**) Control: fungi hyphae on saline solution. (**b**) Chitinase action: hyphae fungi, after incubation at 28 °C for 3 h in crude extract chitinolytic of A8 strain. Bars indicate 10 μm (Modified Quecine et al. 2008). Photos authorized by authors

commensals but confer beneficial traits to their hosts that primarily fall into two categories: growth promotion and protection from phytopathogens. Members of the genus *Streptomyces* are prolific producers of antimicrobial compounds, and endophytic *Streptomycetes* are no exception (Seipke et al. 2012). Numerous endophytic *Streptomyces* isolates inhibit the growth of fungal phytopathogens both in vitro and *in planta*, and this antibiosis has been proposed as one of the mechanisms by which endophytes suppress plant diseases (Sardi et al. 1992; Coombs and Franco 2003; Taechowisan et al. 2003; Franco et al. 2007). Endophytic actinobacteria (Sardi et al. 1992; Coombs and Franco 2003; El-Tarabily 2003; Rosenblueth and Martinez-Romero 2006) have been isolated from within the living tissues of various plant species. These endophytes have been shown to protect plants against different plant pathogens including *Rhizoctonia solani* and *Verticillium dahliae* (Krechel et al. 2002), *Plectosporium tabacinum* (El-Tarabily 2003), *Gaeumannomyces graminis* var. *tritici* and *R. solani* (Coombs et al. 2004), *Fusarium oxysporum* (Cao et al. 2005), *Pythium aphanidermatum* (El-Tarabily et al. 2009), and *Botrytis cinerea* and *Curvularia lunata* (Kafur and Khan 2011).

Quecine et al. (2008) evaluated chitinase production by endophytic actinobacteria and the potential of this for the control of phytopathogenic fungi. Actinobacteria are used extensively in the pharmaceutical industry and agriculture owing to their great diversity of enzyme production. In this study, endophytic *Streptomyces* strains were grown on minimal medium supplemented with chitin, and chitinase production was quantified. The strains were screened for any activity towards phytopathogenic fungi with a dual-culture assay in vitro. The correlation between chitinase production and pathogen inhibition was calculated and further confirmed on *Colletotrichum sublineolum* cell walls by scanning electron microscopy. Quecine et al. (2008) report a genetic correlation between chitinase production and the biocontrol potential of endophytic actinobacteria in an antagonistic interaction with different phytopathogens, suggesting that this control could occur inside the host plant (Fig. 13.5). Additionally, a genetic correlation between chitinase production and pathogen inhibition was demonstrated. Finally, these results provide an enhanced understanding of endophytic *Streptomyces* and its potential as a biocontrol agent.

4.2.3 Endophytic Actinobacteria in the Control of Insect-Pests

The actinomycetes are a widely exploited group of microorganisms that can produce enzymes and antibiotics for agricultural applications such as eco-friendly crop protection. Among the actinomycetes, *Streptomyces* spp. are particularly efficient in the breakdown of chitin via chitinolytic enzymes (Bhattacharya et al. 2007; Quecine et al. 2008). During the past decade, several reports described this chitinolytic activity, and the corresponding genes responsible have been isolated and characterized (Robbins et al. 1998; Tsujibo et al. 1993; Christodoulou et al. 2001; Barboza-Corona et al. 2003; Kim et al. 2003). There is a wide variety of chitinases and a correspondingly large range of optimal temperatures and pH values for chitinase activity to determinate how well suited the chitinase is for pest control applications (Kramer and Muthukrishnan 1997). Our research group reported the partial characterization of the chitinolytic extract produced by an endophytic *Streptomyces* sp. strain (A8) (Quecine et al. 2011). The extract produced by the A8 strain was also tested against *Anthonomus grandis* Boheman (Coleoptera: Curculionidae), the cotton boll weevil (Quecine et al. 2011). The chitinase crude extract from the A8 strain was cultured for 5 days in a minimal liquid medium supplemented with chitin. The extract was partially characterized by standard methods. The chitinolytic extract had an optimum temperature of 66 °C and an optimum pH between 4 and 9 (approximately 80 % of relative activity). We also characterized the temperature and pH stability and measured the effects of enzyme inhibitors. The filtered chitinolytic extract was added to an artificial boll weevil diet. Boll weevil development from the egg stage to the adult stage was prolonged, and the percentage of adults that emerged was approximately 66 % less than on the control diet. This study showed that the larval development of *A. grandis* was inhibited by the presence of characterized chitinolytic extract in the artificial diet. This work provides an experimental basis for using the chitinase from an endophytic *Streptomyces* sp. as an alternative to controlling the plant pest *A. grandis*. In this context, the cotton boll weevil, *A. grandis*, is major pest that affects cotton production in the Americas (Martins et al. 2007, 2008). It is typically controlled with chemical agents, but these chemicals are expensive and may disrupt predator and parasitoid populations due to their broad-spectrum activities (Burton 2006; Wolkers et al. 2006). Consequently, it is necessary to search for safer alternatives for boll weevil control. Biological and other control strategies to decrease the damage to cotton crops by the boll weevil are encouraged in integrated pest management strategies, which utilize insecticides that are more selective (Pimenta et al. 1997).

4.2.4 Biological Control of Pests by Endophytic Fungi

The first report showing that endophytic microorganisms play an important role to control insects was reported by Webber (1981) which showed that *Phomopsis oblonga,* an endophytic fungus, protected elm trees against the beetle *Physocnemum brevillineum* which is a vector of the Elm Dutch disease caused by the pathogenic fungus *Ceratocystis ulmi.* Other early reports were published; the Azevedo et al. (2000) review presents several examples of fungal endophytes controlling insect-pests. Besides insects, endophytic fungi are able to produce toxins which protect plants against herbivorous domestic mammals. This was first demonstrated by Bacon et al. (1977) showing a correlation between the endophyte *Epichloe typhina* producing a toxin in the host plant *Festuca arundinacea*. Inoculation of the entomopathogenic fungus *Beauveria bassiana* was carried out in *Zea mays* (maize) to control *Ostrinia nubilalis*, the European corn borer (Lewis and Cossentine 1986; Bing and Lewis 1991), using aqueous and granular formulations. Also the fungus *B. bassiana* was later found as endophyte in several plant species and probably plays an important role to avoid attack of insect-pests against plants. However, *O. nubilalis* feeding on maize with *B. bassiana* as endophyte showed a low percentage of insects with mycoses (Bing and Lewis 1993) and it was proposed, as no conidia was found inside the host plant, that the mode of action involves fungal metabolites, which cause insect feeding deterrence or antibiosis (Wagner and Lewis 2000; Cherry et al. 2004). Several papers and reviews reported the presence of entomopathogenic microorganisms as endophytes, occurring in host plants, some of them with great agricultural importance. The reviews of Vega et al. (2008, 2009) present some examples of entomopathogenic endophytic fungi isolated from several host plants.

Entomopathogenic endophytic microorganisms were also isolated from our group in Brazil, and the results obtained with some of them will be reported. One or more known as insect and nematode controllers fungi as *Beauveria, Cladosporium, Cordyceps, Paecilomyces, Verticillium* (*Lecanicillium*), among others were quite frequently isolated from several studied plant hosts. Among plants of agricultural importance, these fungi were found in *Citrus* spp. (Glienke-Blanco et al. 2002), *Glycine max* (Pimentel 2001), *Theobroma cacao* (Rubini et al. 2005), *Saccharum* (Stuart et al. 2010), *Vitis labrusca* (Brum et al. 2012), *Coffea arabica* (Ciraulo 2011), and *Zea mays* (Pimentel 2001; Pamphile and Azevedo 2002). *B. bassiana* strains B95 and B157 isolated from maize were further studied. Morphological characterization and molecular characterization showed that both strains resembled *B. bassiana* but could not be exactly classified as the *B. bassiana* used as controls. Distinctions between B95 and *B. bassiana* could be explained by the fact that, as it is known there are small differences between endophytic and direct insect isolated fungi, it is an endophytic. However, strain B 157 showed to be distinct from others and was classified as *Beauveria amorpha*. These strains (Campos et al. 2005; Sia 2006) were used against an important maize insect-pest (*Spodoptera frugiperda*) and the results showed that the endophytes from maize behave as good controllers or even better than commercial entomopathogenic strains used in Brazil to control *S. frugiperda* (Fig. 13.6). The results demonstrated the importance of endophytes as entomopathogens. Even more, the same strains were tested in vitro and in vivo against the bovine tick *Rhipicephalus microplus*, an ectoparasite that causes significant losses in herds of tropical and subtropical regions of the world. To attack the tick, it was shown that endophytic strains of *Beauveria* produce several hydrolytic extracellular enzymes as proteases and chitinases suggesting that these enzymes are pathogenic determinants. Also the endophytic strains showed appressorium formation during penetration on the cuticle of the tick (Campos et al. 2005). The endophytic *Beauveria* were tested in laboratory bioassays and field conditions against the cattle tick. *Beauveria* strains tested in laboratory bioassays reduced females' egg weight and reproductive efficiency. The mortality showed that endophytic strains were equally efficient as commercial *B. bassiana* strains to kill *R. microplus* females. Field tests were carried out with cows infested with the tick. A treated group was sprayed with 3L suspension containing about one million conidia/ml and after 72 h all ticks were collected from cows and adjacent stable floor. A control group of cows were sprayed with the same amount of aqueous solution with no conidia. Field tests showed that endophytic strain was the most efficient followed by a *B. bassiana* strain collected from insects (Campos et al. 2010). Although endophytic *Beauveria* strains were isolated from maize, it is likely that they also may be found in pasture grasses which may act as controllers of cattle ticks by indirect ways as antibiosis. As far as we know, this was the first field test made in Brazil with endophytic entomopathogens fungi against ticks and the results showed an increase of 32 % mortality compared to controls. Some reports from Africa using insect isolates

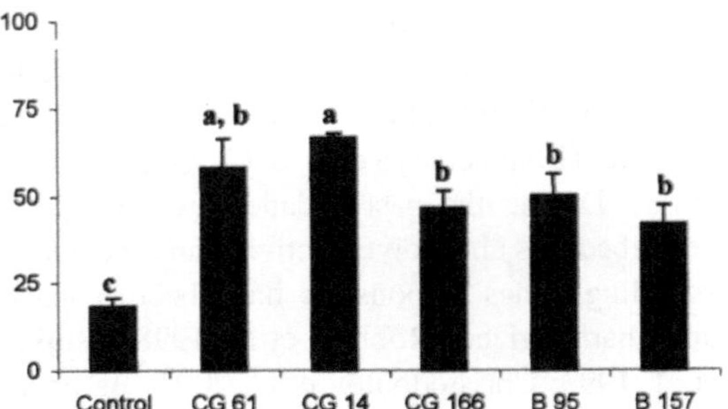

Fig. 13.6 Mortality at 20 days of *Spodoptera frugiperda* larvae treated with a conidial suspension (100,000 conidia/ml) of endophytic *Beauveria* (B95 and B157) and *Beauveria bassiana* strains (CG 61, CG 14r and CG 166). Control was treated with aqueous 0.1 % Tween 80 solution. The *bars* show standard errors. Values followed by the same letter are not significantly different from each other (Tukey test, $P > 0.05$)

entomopathogens against the African tick *R. appendiculatus* gave mortality as high as 85 % (Mwangi et al. 1995; Kaaya et al. 1996) indicating that a search for new endophytes allied to improvement of delivery conidia may increase mortality making biological control techniques able to substitute the use of synthetic compounds.

4.2.5 Endophytic Fungi and Biological Control of Plant Pathogens

The first example of biological control of an insect by an endophytic fungus already mentioned (Webber 1981) was also an indirect control of Dutch elm disease caused by the fungus *Ceratocystis ulmi*. Several other examples of endophytic fungi controlling plant diseases caused by pathogenic fungi, nematodes, and bacteria are also known as reviewed by Azevedo and Araújo (2007). The fungi *Neotyphodium (Acremonium)* and *Fusarium* are active against some *Triticum* diseases and nematodes, respectively (Pocasangre et al. 2000; Tunali et al. 2000). One type of mechanism for biocontrol is the induced systemic resistance (ISR). Thanks to the action of the endophyte, the plant is induced to produce resistant compounds as phenolic ones or increase protection by glucan and lignin formation. Other endophytes, mainly the ones with fast growth as *Trichoderma,* are able to colonize the plant inhibiting by competition or antibiosis the establishment of the pathogen. In other cases a nonpathogenic endophyte may reduce an incidence of similar pathogenic fungi. It is well known that some pathogenic fungi containing double-strand RNA (dsRNA) mycoviruses act as endophytes reducing in this way the damage caused by pathogenic strains (Agnostakis and Day 1979; Dawe et al. 2004; Deng et al. 2007; Kwon et al. 2009). Recently, our research group detected a *Colletotrichum gloeosporioides* containing dsRNA particles in cashew. This strain was hypovirulent when compared to cashew pathogenic isolates. The introduction of hypovirulent dsRNA strains could prove to be a good method to reduce cashew tree anthracnose (Figueiredo et al. 2012a, b). Other cases of endophytic behavior or mutant strains of pathogens which protect the host against disease are reported (Redman et al. 1999). In other cases very similar or even identical fungi can act as pathogenic for one host species and endophytic for other as *Guignardia* in citrus (Glienke-Blanco et al. 2002; Baayen et al. 2002) or *Moniliophthora perniciosa* in cacao (Lana et al. 2011). A good example of potential biological control of *M. perniciosa*, the causal agent of witches' broom disease, was reported by Rubini et al. (2005). From more than 30 endophytic fungi isolated from cacao, some were able to control in vitro the disease. Further in vivo tests have shown that from several endophytes which inhibited in vitro *M. perniciosa*, only one, *Gliocladium catenulatum*, significantly reduced the incidence of the pathogen. The results showed that in vitro tests must be followed by in vivo assays to show if endophytes may be used with success to control plant diseases. We are now studying the possible control of *Fusarium* in grapes (*V. labrusca*) (Brum et al. 2012) and anthracnose caused by *Colletotrichum* in an Amazonian plant, guarana (*P. cupana*), largely used as medicinal and to produce soft drinks. In these cases, several endophytic isolated fungi were active to inhibit in vitro the pathogenic fungi but in vivo tests must be performed to show the efficiency of these endophytes for biocontrol. Anyway endophytic fungi, besides protecting their plant hosts, may play a potential role to substitute chemical compounds for biocontrol of plant pathogens.

4.2.6 The Recombinant DNA Technology and Biocontrol by Endophytic Microorganisms

Recently, recombinant DNA technology has been applied to improve endophytic microorganisms, aiming to introduce new characteristics of agronomic interests, such as the biological control of insect-pests (Azevedo et al. 2000; Araújo et al. 2008). Fahey (1988) and Fahey et al. (1991) described the first work directed at the introduction of a heterologous gene in an endophytic microorganism for the purpose of insect control. As a member of the biotechnology company Crop Genetics International, he described the major steps in the construction of an endophytic bacterium for the purpose of insect control. This was achieved through the secretion of an insecticidal toxin in the host plant. He used the endophyte *Clavibacter xyli* subsp. *cynodontis*, a Gram-positive, xylem-inhabiting bacterium, capable of colonizing several plant species. This endophytic bacterium received a gene from another bacterium, *Bacillus thuringiensis*, which is able to produce the d-endotoxin active against insects, especially Lepidoptera and Coleoptera. Therefore, the genetically modified bacterium is able to secrete toxin inside the plant, protecting it against attacks by target insects (Azevedo et al. 2000). Following the work of Fahey (1988), several other researchers belonging to the same company published more detailed reports describing the construction of the insect biocontrol agent. Turner et al. (1991) showed that a plasmid carrying two copies of the *B. thuringiensis* subsp. *kurstaki cryIA*(c) d-endotoxin gene and containing a genomic DNA fragment of *C. xyli* subsp. *cynodontis* could be integrated into the chromosome of *C. xyli* subsp. *cynodontis* by homologous recombination. However, the engineered bacterium exhibited insecticidal activity in artificial diets but not *in planta*. Lampel et al. (1994) used an improved integrative vector that, although it showed some instability, resulted in toxin production *in planta*. The presence of endophytic bacteria inside the host plant may increase the plant's fitness by protecting it against pests and pathogens, improving plant growth and increasing resistance in stressful environments (Azevedo et al. 2000; Scherwinski et al. 2007). Many studies are being carried out with both natural and genetically modified microorganisms to evaluate host colonization (Germaine et al. 2004; Ferreira et al. 2008). *Methylobacterium* spp. have been described as enhancing plant systemic resistance (Madhaiyan et al. 2004), plant growth, and root formation (Senthilkumar et al. 2009). In this context, our research group decided to study the endophytic colonization of rice seedlings and *Spodoptera frugiperda* J.E. Smith larvae by the genetically modified endophytic bacterium *M. mesophilicum* in vitro (Rampelotti-Ferreira et al. 2010). The endophyte *M. mesophilicum* strain SR1.6/6 used in this work was previously isolated from *Citrus sinensis* (Araújo et al. 2002) and labeled with green fluorescent protein (*gfp*) (Gai et al. 2009). The colonization of *S. frugiperda* larvae and rice seedlings by the genetically modified endophytic bacterium *M. mesophilicum*, and also the possible transfer of this bacterium into the larva's body during consumption of the seedlings, were studied. The data obtained by bacterial reisolation and fluorescence microscopy showed that the bacteria colonized the rice seedlings and that the endophytic bacteria present in the seedlings could be acquired by the larvae. In that way, the transference of endophytic bacteria from plants to insect can be a new and important strategy in insect control using engineered endophytic bacteria.

Recombinant DNA technology in fungi is mainly restricted to the development of transformation systems. Van-Heeswijck and McDonald (1992) were probably the first to propose the use of engineering endophytic fungi to control insects and diseases of the host plant *Lolium perenne*. The use of recombinant DNA techniques as reviewed by Azevedo et al. (2000) was mainly restricted to fungi able to produce toxins, aiming to obtain more active toxin mutants to control herbivores as insects or aiming elimination of toxins which are prejudicial to domestic animals. Yunus et al. (1999) engineered the endophytic fungi *Neotyphodium lolii* to introduce an auxin growth hormone producer gene; the modified strain was able to be reintroduced into perennial ryegrass. Panaccione et al. (2001) also used *N. lolii* (Lp1) isolated from *L. perenne* in order to obtain by genetic modification a strain which was no longer able to produce the toxin ergovaline. A recent review (Mei and Flinn 2010) listed US-issued patents which relate the use of fungal and bacterial endophytes for plant-growth promotion and stress tolerance. Recombinant DNA techniques are becoming more frequently used in endophytic microorganisms and new molecular biology approaches have been introduced. For instance, fungal transformation mediated by *Agrobacterium tumefaciens* has been used for several species. Our group used *Diaporthe phaseolorum* from mangrove plants (Sebastianes et al. 2012b). This fungus is an antibiotic producer of

3-hydroxypropionic acid (Sebastianes et al. 2012a), and similar techniques may be used in endophyte-engineered fungi to produce compounds which can be used for biological control of insect-pests and diseases.

5 Symbiotic Control by Endophytic Bacteria: A Paratransgenic Approach

The strategy, paratransgenesis, was developed in order to prevent the transmission of pathogens by insect vectors to humans (Beard et al. 1998, 2001, 2002; Rio et al. 2004). The key concept in paratransgenesis is the genetic alteration of symbiotic microbes that are carried by insects (therefore, they are paratransgenic insects). The genetic alterations of the symbiotic microbes are designed to increase their competitiveness within the insect vector at the expense of the pathogen. This overall strategy of disease prevention is an example of symbiotic control and is a variation on the theme of symbiotic therapy (Ahmed 2003). The symbiotic control strategy, and therefore paratransgenesis, is to find a local candidate microbe having an existing association with the pathosystem that includes the problem or condition at hand. The local candidate microbe should occupy the same niche as, or have access to, the target pathogen or condition (Durvasula et al. 1997). The local origin of the biocontrol microbe in symbiotic differs from classical biological control, where microbes, herbivores, parasites, or predators are sought from outside of the local ecosystem for establishment in the local ecosystem to control a pest such as a plant or invertebrate (Miller 2007). In symbiotic control, all elements originate at the local site and are already coevolved with and established in the pathosystem; foreign exploration is not only unnecessary but also most likely counterproductive. Because of these strict requirements, a suitable symbiotic candidate may not always be found or may not be amenable to practical manipulation (Miller 2007). The key to symbiotic control is finding a candidate microbe having an existing association with the ecosystem that includes the problem or condition at hand and that occupies the same niche as or has access to the target pathogen (Miller 2007). In this context, endophytic microorganisms, special bacteria, have been considered as a candidate to symbiotic control strategy to control of phytopathogens (Gai et al. 2009, 2011; Ferreira Filho et al. 2012). Also, the strategy of symbiotic control employs both paratransgenic and nonrecombinant methods to control disease or health problems. In some cases these solutions may result in competitive displacement of the pathogen with a more benign microbe.

5.1 Symbiotic Control of the Phytopathogen *Xylella fastidiosa*

Citrus variegated chlorosis (CVC) is a disease of the sweet orange, *Citrus sinensis* L., which is caused by *Xylella fastidiosa* subsp. *pauca* (Hartung et al. 1994; Schaad et al. 2004), a phytopathogenic bacterium that has been shown to infect all sweet orange cultivars (Li et al. 1997). CVC was first reported in Brazil in 1987 and has rapidly become one of the most economically important diseases affecting sweet orange production in Brazil (Rossetti et al. 1990; Lee et al. 1991). CVC rapidly became widespread in most major citrus growing areas through unregulated movement of infected nursery stock due to a previous lack of certification programs and high CVC infection rates in Brazil. CVC can be found in at least 90 % of the orchards in Brazil (Lambais et al. 2000). In Brazil, CVC is responsible for losses of US $100 million per year to the citrus industry (Della-Coletta et al. 2001). Although *X. fastidiosa* subsp. *pauca* was the first plant pathogen to have its genome sequenced (Simpson et al. 2000), there is still no effective control for CVC. The pathogen is known to have an extraordinary host range among higher plants in New World ecosystems (Freitag 1951). Interestingly, within the majority of native host plants, *X. fastidiosa* does not damage the host plant and behaves as an endophyte (Purcell and Saunders 1999). In contrast, the horticultural crops that suffer from diseases caused by *X. fastidiosa* are those that have been introduced into New World ecosystems (Chen et al. 2000). The observation that a few asymptomatic trees persist in some infected orchards may lead to new approaches to the investigation of the control of CVC. These asymptomatic plants have the same genotype as diseased plants and are located in the same grove under similar climatic and edaphic conditions, suggesting that some other factor is responsible for resistance to CVC. One factor that may influence the resistance to CVC is the nature of the endophytic microbial community colonizing individual *C. sinensis* plants (Araújo et al. 2002). The key to symbiotic control is finding a candidate microbe having an existing association with the ecosystem that includes the problem or condition at hand and that occupies the same niche as or has access to the target pathogen (Miller 2007). Bacteria of the genus *Methylobacterium* are known to occupy the same niche as *X. fastidiosa* subsp. *pauca* inside citrus plants (Araújo et al. 2002; Lacava et al. 2004). During feeding, insects could acquire not only the pathogen but also endophytes from host plants. Gai et al. (2009) reported the localization of the endophytic bacterium, *M. mesophilicum*, in *C. roseus* model plant system and the transmission of this endophyte by *Bucephalogonia xanthophis,* a sharpshooter insect vector of *X. fastidiosa* subsp. *pauca*. *Methylobacterium mesophilicum*, originally isolated as an endophytic bacterium from citrus plants (Araújo et al. 2002), was genetically transformed to express *gfp* (Gai et al. 2007). The GFP-labeled strain of *M. mesophilicum* was inoculated into *C. roseus* (model plant) seedlings and was observed colonizing its xylem vessels. The transmission of *M. mesophilicum* by *B. xanthophis* was verified with insects feeding on fluids containing the GFP-labeled bacterium. Forty-five days after inoculation, the plants exhibited endophytic colonization by *M. mesophilicum*, confirming this bacterium as a nonpathogenic, xylem-associated endophyte (Gai et al. 2009). These data demonstrate that *M. mesophilicum* not only occupies the same niche as *X. fastidiosa* subsp. *pauca* inside plants but also that it may be transmitted by *B. xanthophis*. The transmission, colonization, and genetic manipulation of *M. mesophilicum* are a prerequisite to examining the potential use of paratransgenic–symbiotic control (SC) to interrupt transmission of *X. fastidiosa* subsp. *pauca*, the bacterial pathogen causing CVC, by insect vectors that propose *M. mesophilicum* as a candidate for a paratransgenic–SC strategy to reduce the spread of *X. fastidiosa* subsp. *pauca*. It is known that *X. fastidiosa* subsp. *pauca* produces a fastidian gum (da Silva et al. 2001) which may be responsible for the obstruction of xylem in affected plants (Lambais et al. 2000), so the production of endoglucanase by genetically modified endophytic bacteria may transform the endophytes into symbiotic control agents for CVC. Azevedo and Araújo (2003) have used the replicative vector pEGLA160 to produce genetically modified *Methylobacterium* expressing antibiotic resistance and endoglucanase genes. Furthermore, other strategies can be evaluated such as a production of genetically modified *Methylobacterium* to secrete soluble anti-*Xylella* protein effect in citrus, such as Lampe et al. (2006) suggested in the *Escherichia coli* α-hemolysin system for use in Axd to secrete soluble anti-*Xylella* protein effectors in grapevine. Also, Lampe et al. (2007) suggested the evaluation of proteins secreted from the grapevine bacterial symbiont *Pantoea agglomerans* for use as secretion partners of anti-*Xylella* protein effectors. One strategy that can adopt as the next step for SC control of CVC is producing a genetically modified endophytic bacterium, like *Methylobacterium*, to secrete anti-*Xylella* protein effectors.

According to Gai et al. (2011), the bacterial communities associated with vector insects and plants differ in abundance through the yearly season. Endophytic bacteria could influence disease development by reducing the insect transmission efficiency due to competition with pathogens in host plants and also in insect foreguts. In addition the bacterial communities in the foregut of insect vectors of *X. fastidiosa* subsp. *pauca* changed with time, environmental conditions, and in different insect species. However, members of the genus *Curtobacterium* were consistently detected in the sharpshooters foregut and are commonly isolated from the xylem of citrus plants (Araújo et al. 2002), and because of this, they may be candidates for biological control.

6 Siderophores from Endophytic Bacteria: Suppression of Phytopathogens

Iron is a necessary cofactor for many enzymatic reactions and is an essential nutrient for virtually all organisms. In aerobic conditions, iron exists predominantly in its ferric state (Fe^{3+}) and reacts to form highly insoluble hydroxides and oxyhydroxides that are largely unavailable to plants and microorganisms. To acquire sufficient iron, siderophores produced by bacteria can bind Fe^{3+} with a high affinity to solubilize this metal for its efficient uptake. Bacterial siderophores are low-molecular-weight compounds with high Fe^{3+} chelating affinities (Sharma and Johri 2003) responsible for the solubilization and transport of this element into bacterial cells. Some bacteria produce hydroxamate-type siderophores, and others produce catecholate types (Neilands and Nakamura 1991). In a state of iron limitation, the siderophore-producing microorganisms are also able to bind and transport the iron-siderophore complex by the expression of specific proteins (Nachin et al. 2001; Nudel et al. 2001). The production of siderophores by microorganisms is beneficial to plants because it can inhibit the growth of plant pathogens (Masclaux and Expert 1995; Nachin et al. 2001; Sharma and Johri 2003; Etchegaray et al. 2004; Siddiqui 2005). Siderophores can also induce resistance mechanisms in the plant (Schroth and Hancook 1995). Plant-growth promotion, including the prevention of the deleterious effects of phytopathogenic organisms (Sharma and Johri 2003), can be achieved by the production of siderophores (Hayat et al. 2010). Production of siderophores is a mechanism through which endophytic biocontrol agents suppress pathogens indirectly by increasing the availability of minerals to the biocontrol agent in addition to iron chelation and, thus, stimulating the biosynthesis of other antimicrobial compounds (Duffy and Defago 1999).

Endophytic bacteria colonize an ecological niche similar to that of plant pathogens, especially vascular wilt pathogens, which might favor them as potential candidates for biocontrol and growth-promoting agents (Ramamoorthy et al. 2001). Several bacterial endophytes have been reported to support plant growth by providing phytohormones, low-molecular-weight compounds, or enzymes (Lambert and Joos 1989; Frommel et al. 1991; Glick et al. 1998). Production of siderophores is another mechanism by which endophytic biocontrol agents suppress pathogens indirectly by stimulating the biosynthesis of other antimicrobial compounds by increasing availability of minerals to the biocontrol agent in addition to iron chelation (O'Sullivan and O'Gara 1992; Duffy and Defago 1999; Persello-Cartieaux et al. 2003). In this context, Vendan et al. (2010) suggested that siderophore production may be a common phenotype among endophytes. In a recent study of the diversity and potential for plant-growth promotion of endophytic bacteria isolated from ginseng (*Panax ginseng* C.A. Meyer), Vendan et al. (2010) described the siderophore production by 7 endophytic bacteria strains. These strains were classified as *Bacillus cereus*, *B. flexus*, *B. megaterium*, *Lysinibacillus fusiformis*, *L. sphaericus*, *Microbacterium phyllosphaerae*, and *Micrococcus luteus*. Siderophore production by endophytic bacteria has been investigated in only a few cases, mainly as a mechanism of certain bacteria to antagonize pathogenic fungi. Thus, it was observed that all the isolates from cotton roots having antagonistic activity, mainly *Pantoea* spp., excreted siderophores (Li et al. 2009). Also in rice, strains of the genera *Pseudomonas* and *Burkholderia* and two species of *Pantoea* (*P. ananatis* and *P. agglomerans*) having antagonistic activity excreted siderophores (Yang et al. 2008).

According to Verma et al. (2011), three endophytic actinobacteria strains isolated from the root tissues of *Azadirachta indica* plants were selected through tests for their potential as biocontrol and plant-growth-promoting agents. It was also observed that the seed treated with the spore suspension of three selected endophytic strains of *Streptomyces* significantly promoted plant growth and antagonized the growth of *Alternaria alternata*, the causal agent of early blight disease in tomato plants. It was observed that the three selected strains prolifically produce siderophores that play a vital role in the suppression of *A. alternata*. These authors concluded that these endophytic isolates have the potential to be plant-growth promoters as well as a biocontrol agent, which is a useful trait for crop production in nutrient-deficient soils. Loaces et al. (2011) described and characterized the community of endophytic, siderophore-producing bacteria (SPB) associated with *Oryza sativa*. Less than 10 % of the endophytic bacteria produced siderophores in the roots and leaves of young plants, but most of the endophytic bacteria were siderophore producers in mature plants. According to the results, 54 of the 109 endophytic SPB isolated from different plant tissues or growth stages from replicate plots of *O. sativa* were unique. The relative predominance of bacteria belonging to the genera *Sphingomonas*, *Pseudomonas*, *Burkholderia*, and *Enterobacter* alternated during plant growth, but the genus *Pantoea* was predominant in the roots at tillering and in the leaves at subsequent stages. *Pantoea ananatis* was the SPB permanently associated with all of the plant tissues of *O. sativa*. In the same study, the SPB and plant-growth-promoting bacteria (PGPB) *Azospirillum brasilense*, *A. amazonense*, and *Herbaspirillum seropedicae* were assessed using dual culture in vitro on NFbI medium to allow the simultaneous growth of PGPB and SPB. These PGPB are considered important genera of endophytic diazotrophs (Baldani and Döbereiner 1980; Baldani et al. 2000, 2003). The results indicate that the SPB *P. ananatis* is the permanent and dominant associated species and is unable to inhibit two of the relevant plant-growth-promoting bacteria, *A. brasilense* and *H. seropedicae*.

7 Concluding Remarks

Endophytic microorganisms are believed to elicit plant growth in many ways, including helping plants acquire nutrients, e.g., via nitrogen fixation, phosphate solubilization, or iron chelation; preventing infections via antifungal or antibacterial agents; out-competing pathogens for nutrients by producing siderophores; or establishing the plant's systemic resistance and producing phytohormones. However, the effects and functions of endophytes in plants have not been comprehensively defined. The challenge and goal is to be able to manage microbial communities to favor plant colonization by beneficial bacteria and fungi. This will be possible when a better knowledge of endophyte ecology and molecular interactions is attained. Although all of the approximately 300,000 plant species have been estimated to harbor one or more endophytes, few relationships between plants and these endophytes have been studied in detail; the legume–rhizobia symbiosis and associations between fungi and the root of plants (mycorrhizae) are exceptions. Additionally, there remain many barriers to commercial usage of inoculants for inducing resistance, and even more studies are necessary to permit the usage of endophytes in this way. While there is a wide diversity of endophytes to be explored, supporting the idea that the most efficient resistance inducers are still to be described, genetic transformation of bacteria should also be considered a way to group important characteristics found in different strains. The combination of inducers of systemic resistance and endophytic characteristics may affect future agricultural concepts, allowing safer production with a lower impact on the environment.

References

Adams PD, Kloepper JW (1996) Seed borne bacterial endophytes in different cotton cultivars. In: Abstract, 1996 annual meeting of American Phytopathological Society, p 97

Agnostakis SL, Day PR (1979) Hypovirulence conversion of *Endothia parasitica*. Phytopathology 69:1226–1229

Ahmed FE (2003) Genetically modified probiotics in foods. Trends Biotechnol 21:491–497

Alves SB (1998) Controle Microbiano de Insetos. Editora Fundação de Estudos Agrários Luiz de Queiroz, Piracicaba, p 1163

Andreote FD, Lacava PT, Gai CS, Araújo WL, Maccheroni W Jr, van Overbeek LS, van Elsas JD, Azevedo JL (2006) Model plants for studying the interaction between *Methylobacterium mesophilicum* and *Xylella fastidiosa*. Can J Microbiol 52:419–426

Ângelo PCS, Nunes-Silva CG, Brigido MM, Azevedo JSN, Assunção EM, Sousa ARB, Patricio FJB, Rego

MM, Peixoto JCC, Oliveira WP Jr, Freitas DV, Almeida ERP, Viana AMHA, Souza AFPN, Andrade EV, Acosta POA, Batista JS, Walter MEMT, Leomil L, Anjos DAS, Coimbra RCM, Barbosa MHN, Honda E, Pereira SS, Silva A, Pereira JO, Silva ML, Marins M, Holanda FJ, Abreu RMM, Pando SC, Goncalves JFC, Carvalho ML, Leal-Mesquita ERRBP, da Silveira MA, Batista WC, Atroch AL, Franca SC, Porto JIR, Schneider MPC, Astolfi-Filho S (2008) Guarana (*Paullinia cupana* var. *sorbilis*), an anciently consumed stimulant from the Amazon rain forest: the seeded-fruit transcriptome. Plant Cell 27:117–124

Arachevaleta M, Bacon CW, Hoveland CS, Radcliffe DE (1989) Effect of the tall fescue endophyte on plant response to environmental stress. Agron J 81:83–90

Araújo WL, Saridakis HO, Barroso PAV, Aguilar-Vildoso CI, Azevedo JL (2001) Variability and interactions between endophytic bacteria and fungi isolated from leaf tissues of citrus rootstocks. Can J Microbiol 47:229–236

Araújo WL, Marcon J, Maccheroni W Jr, Elsas JDV, Vuurde JLV, Azevedo JL (2002) Diversity of endophytic bacterial populations and interaction with *Xylella fastidiosa* in citrus plants. Appl Environ Microbiol 68:4906–4914

Araújo WL, Lacava PT, Andreote FD, Azevedo JL (2008) Interaction between endophytes and plant host: biotechnological aspects. In: Ait Barka E, Clément C (eds) Plant-microbe interactions, vol 1. Research Signpost, Kerala, pp 1–21

Araújo WL, Lacava PT, Marcon J, Lima AOS, Kuklinsky-Sobral J, Pizzirani-Kleiner AA, Azevedo JL (2010) Guia prático: Isolamento e caracterização de microrganismos endofíticos. Cop. Luiz de Queiroz, Piracicaba, 167p

Assis SMP, Silveira EB, Mariano RLR, Menezes D (1998) Bactérias endofíticas – método de isolamento e potencial antagônico no controle da podridão negra do repolho. Summa Phytopathol 24:216–220

Azevedo JL, Araújo WL (2003) Genetically modified crops: environmental and human health concerns. Mutat Res 544:223–233

Azevedo JL, Araújo WL (2007) Diversity and applications of endophytic fungi isolated from tropical plants. In: Ganguli BN, Deshmukh SK (eds) Fungi: multifaceted microbes. CRC Press, Boca Raton, pp 189–207

Azevedo JL, Maccheroni W, Pereira JO, Araújo WL (2000) Endophytic microorganisms: a review on insect control and recent advances on tropical plants. Electron J Biotechnol 3:40–65

Baayen RP, Bonants PMJ, Verkley G, Carroll GC, Van Der Aa HA, Der Weerdt M, Van Brouwershaven JR, Scutte GC, Maccheroni W Jr, Glienke-Blanco C, Azevedo JL (2002) Nonpathogenic isolates of the Citrus Black Spot fungus *Guignardia citricarpa*, identified as a cosmopolitan endophytic of woody plants *G. mangiferae (Phyllosticta capitalensis)*. Phytopathology 92:464–477

Bacon C, Hinton D (2006) Bacterial endophytes: the endophytic niche, its occupants, and its utility. In: Gnanamanickam S (ed) Plant-associated bacteria. Springer, Dordrecht, pp 155–194

Bacon CW, White JF Jr (2000) Microbial endophytes. Marcel Dekker, New York, p 487

Bacon CW, Porter JK, Robins JD, Luttrell ES (1977) *Epichloe typhi* from toxic tall fescue grasses. Appl Environ Microbiol 34:576–581

Baldani VLD, Döbereiner J (1980) Host-plant specificity in infection of cereal with *Azospirillum* spp. Soil Biol Biochem 12:433–440

Baldani VLD, Baldani JI, Döbereiner J (2000) Inoculation of rice plants with the endophytic diazotrophs *Herbaspirillum seropedicae* and *Burkholderia* spp. Biol Fertil Soils 30:485–489

Baldani JI, Baldani VLD, Döbereiner J (2003) Genus *Herbaspirillum*. In: Brenner DJ, Krieg NR, Staley JTGM (eds) Bergey's manual of determinate bacteriology, 2nd edn. Springer, New York

Barboza-Corona JE, Nieto-Mazzocco E, Velázquez-Robledo R, Salcedo-Hernandez R, Bautista M, Jiménez B, Ibarra JE (2003) Cloning, sequencing, and expression of the chitinases gene *chiA74* from *Bacillus thuringiensis*. Appl Environ Microbiol 69:1023–1029

Barzanti R, Ozino F, Bazzicalupo M, Gabbrielli R, Galardi F, Gonnelli C, Mengoni A (2007) Isolation and characterization of endophytic bacteria from the nickel hyperaccumulator plant *Alyssum bertolonii*. Microb Ecol 53:306–316

Beard CB, Durvasula RV, Richards FF (1998) Bacterial symbiosis in arthropods and the control of disease transmission. Emerg Infect Dis 4:581–591

Beard CB, Dotson EM, Pennington PM, Eichler S, Cordon-Rosales C, Durvasula RV (2001) Bacterial symbiosis and paratransgenic control of vector-borne Chagas disease. Int J Parasitol 31:621–627

Beard C, Cordon-Rosales C, Durvasula R (2002) Bacterial symbionts of the triatominae and their potential use in control of Chagas disease transmission. Annu Rev Entomol 47:123–141

Beattie GA, Lindow SE (1995) The secret life of foliar bacterial pathogens on leaves. Annu Rev Phytopathol 33:145–172

Benhamou N, Kloepper JW, Tuzun S (1998) Induction of resistance against *Fusarium* wilt of tomato by combination of chitosan with an endophytic bacterial strain: ultrastructure and cytochemistry of the host response. Planta 204:153–168

Bernardi-Wenzel J, Garcia A, Rubin-Filho CJ, Prioli AJ, Pamphile J (2010) Evaluation of foliar fungal endophyte diversity and colonization of medicinal plant *Luehea divaricata* (Martins et Zuccarini). Biol Res 43:375–385

Bhattacharya D, Nagpure A, Gupta RK (2007) Bacterial chitinases: properties and potential. Crit Rev Biotechnol 27:21–28

Bing LA, Lewis LC (1991) Suppression of *Ostrinia nubilalis* (Hubner) (Lepidoptera, Pyralidae) by endophytic *Beauveria bassiana* (Balsamo) Vuillemin. Environ Entomol 2:1207–1211

Bing LA, Lewis LC (1993) Occurrence of the entomopathogen *Beauveria bassiana* (Balsamo) Vuillemin in different tillage regimes and in *Zea mays* L. and virulence towards *Ostrinia nubilalis* (Hubner). Agric Ecosyst Environ 45:147–156

Boddey RM, Urquiaga S, Alves BJR, Reis V (2003) Endophytic nitrogen fixation in sugarcane: present knowledge and future applications. Plant Soil 252: 139–149

Boursnell JG (1950) The symbiotic seed-borne fungus in the Cistaceae. I. Distribution and function of the fungus in the seedling and in the tissues of the mature plant. Ann Bot 14:217–243

Brum MCP, Araújo WL, Maki CS, Azevedo JL (2012) Endophytic fungi from *Vitis labrusca* L. (Niagara Rosada) and its potential for the biological control of *Fusarium oxysporum*. Genet Mol Res 11:4187–4197

Burton A (2006) Dispatches – pesticides may promote Parkinson's disease. Front Ecol Environ 4:284–289

Campos RA, Arruda W, Boldo JT, Silva MV, Barros NM, Azevedo JL, Schrank A, Vainstein MH (2005) *Boophilus microplus* infection by *Beauveria amorpha* and *Beauveria bassiana*: SEM analysis and regulation of subtilisin-like proteases and chitinases. Curr Microbiol 50:257–261

Campos RA, Boldo JT, Pimentel IC, Dalfovo V, Araújo WL, Azevedo JL, Vainstein MH, Barros NM (2010) Endophytic and entomopathogenic strains of *Beauveria* sp. to control the bovine tick *Rhipicephalus* (Boophilus) *microplus*. Genet Mol Res 9:1421–1430

Cao L, Qiu Z, You J, Tan H, Zhou S (2005) Isolation and characterization of endophytic *Streptomycete* antagonists of *Fusarium* wilt pathogen from surface-sterilized banana roots. FEMS Microbiol Lett 247:147–152

Challis GL, Hopwood DA (2003) Synergy and contingency as driving forces for the evolution of multiple secondary metabolite production by *Streptomyces* species. Proc Nat Acad Sci USA 100:14555–14561

Chelius MK, Triplett EW (2001) The diversity of archaea and bacteria in association with the roots of *Zea mays* L. Microbiol Ecol 41:252–263

Chen J, Jarret RL, Qin X, Hartung JS, Chang CJ, Hopkins DL (2000) 16SrDNA analysis of *Xylella fastidiosa*. Syst Appl Microbiol 23:349–354

Cheplick GP, Clay K, Marks S (1989) Interactions between infection by endophytic fungi and nutrient limitation in the grasses *Lolium perenne* and *Festuca arundinacea*. New Phytol 111:89–97

Cherry AJ, Banito A, Djegui D, Lomer C (2004) Suppression of the stem-borer *Sesamia calamistis* (Lepidoptera, Noctuidae) in maize following seed dressing, topical application and stem injection with African isolates of *Beauveria bassiana*. Int J Pest Manag 50:67–73

Christodoulou E, Duffner F, Vorgias CE (2001) Overexpression, purification, and characterization of a thermostable chitinase (Chi40) from *Streptomyces thermoviolaceus* OPC-520. Protein Expr Purif 23:97–105

Ciampi-Panno L, Fernandez C, Bustamante P, Andrade N, Ojeda S, Conteras A (1989) Biological control of bacterial wilt of potatoes caused by *Pseudomonas solanacearum*. Am Potato J 66:315–332

Ciraulo MB (2011) Interações entre endófitos de *Coffea arabica* isolados de cultura assintomática e sintomática para a atrofia dos ramos de cafeeiro casusada por *Xylellla fastidiosa*. PhD thesis, University of Mogi das Cruzes, São Paulo, 149p

Cocolin L, Rantsiou K, Iacumin L, Cantoni C, Comi G (2002) Direct identification in food samples of *Listeria* spp. and *Listeria monocytogenes* by molecular methods. Appl Environ Microbiol 68:6273–6282

Coombs JT, Franco CMM (2003) Isolation and identification of actinobacteria from surface-sterilized wheat roots. Appl Environ Microbiol 69:5603–5608

Coombs JT, Michelsen PP, Franco CMM (2004) Evaluation of endophytic actinobacteria as antagonists of *Gaeumannomyces graminis* var. *tritici* in wheat. Biol Control 29:359–366

Cottrell MT, Kirchman DL (2000) Community composition of marine bacterioplankton determined by 16S rRNA gene clone libraries and fluorescence in situ hybridization. Appl Environ Microbiol 66:5116–5122

da Silva FR, Vettore AL, Kemper EL, Leite A, Arruda P (2001) Fastidian gum: the *Xylella fastidiosa* exopolysaccharide possibly involved in bacterial pathogenicity. FEMS Microbiol Lett 203:165–171

Dawe AL, Segers GC, Allen TD, McMains CG, Nuss DL (2004) Microarray analysis of *Cryphonectria parasitica* Ga and Gpi signaling pathways reveals extensive modulation by hypovirus infection. Microbiology 150:4033–4043

de Bary A (1866) Morphologie, Phisiologie der pilze, flechten und myxomyceten, vol I: Holmeister's handbook of physiological Botany, Leipzig

Della-Coletta FH, Takita MA, de Souza AA, Aguilar-Vildoso CI, Machado MA (2001) Differentiation of strains of *Xylella fastidiosa* by a variable number of tandem repeat analysis. Appl Environ Microbiol 67:4091–4095

Deng F, Allen TA, Nuss DL (2007) Transcription factor homologue CpdT12 is down-regulated by hypovirus infection and required for virulence and female fertility in the chestnut blight fungus *Cryphonectria parasitica*. Eukaryot Cell 6:235–244

Dong Z, Canny MJ, McCully ME, Roboredo MR, Cabadilla CF, Ortega E, Rodes R (1994) A nitrogen fixing endophyte of sugarcane stems. Plant Physiol 105:1139–1147

Dong YM, Iniguez AL, Ahmer BMM, Triplett EW (2003) Kinetics and strain specificity of rhizosphere and endophytic colonization by enteric bacteria on seedlings of *Medicago sativa* and *Medicago truncatula*. Appl Environ Microbiol 69:1783–1790

Duffy BK, Defago G (1999) Environmental factors modulating antibiotic and siderophore biosynthesis by *Pseudomonas fluorescens* biocontrol strains. Appl Environ Microbiol 65:2429–2438

Dunbar J, Takala S, Barns SM, Davis JA, Kuske CR (1999) Levels of bacterial community diversity in four arid soils compared by cultivation and 16S rRNA gene cloning. Appl Environ Microbiol 65:1662–1669

Durvasula RV, Gumbs A, Panackal A, Kruglov O, Aksoy S, Merrifield RB, Richards FF, Beard CB (1997) Prevention of insect-borne disease: an approach using transgenic symbiotic bacteria. Proc Natl Acad Sci U S A 94:3274–3278

El-Tarabily KA (2003) An endophytic chitinase-producing isolate of *Actinoplanes missouriensis*, with potential for biological control of root rot of lupine caused by *Plectosporium tabacinum*. Aust J Bot 51:257–266

El-Tarabily KA, Nassar AH, Hardy GE St. J, Sivasithamparam K (2009) Plant growth promotion and biological control of *Pythium aphanidermatum*, a pathogen of cucumber, by endophytic actinomycetes. J Appl Microbiol 106:13–26

Etchegaray A, Silva-Stenico ME, Moon DH, Tsai SM (2004) In silico analysis of nonribosomal peptide synthetases of *Xanthomonas axonopodis* pv. *citri*: identification of putative siderophore and lipopeptide biosynthetic genes. Microbiol Res 159:425–437

Fahey JW (1988) Endophytic bacteria for the delivery of agrochemicals to plants. In: Cutle HG (ed) Biologically active natural products. American Chemical Society, Washington, DC, pp 120–128

Fahey JW, Dimock MB, Tomasino SF, Taylor JM, Carlson PS (1991) Genetically engineered endophytes as biocontrol agents: a case study in industry. In: Andrews JH, Hirano SS (eds) Microbial ecology of leaves. Springer, New York, pp 401–411

Ferreira A, Quecine MC, Lacava PT, Oda S, Azevedo JL, Araújo WL (2008) Diversity of endophytic bacteria from *Eucalyptus* species seeds and colonization of seedlings by *Pantoea agglomerans*. FEMS Microbiol Lett 287:8–14

Ferreira Filho AS, Quecine MC, Bogas AC, Rossetto PB, Lima AOS, Lacava PT, Azevedo JL, Araújo WL (2012) Endophytic *Methylobacterium extorquens* expresses a heterologous β-1,4-endoglucanase A (EglA) in *Catharanthus roseus* seedlings, a model host plant for *Xylella fastidiosa*. World J Microbiol Biotechnol 28:1475–1481

Figueiredo LC, Figueiredo GS, Giancoli ACH, Tanaka FAO, Silva LAO, Kitajima EW, Azevedo JL (2012a) Detection of isometric dsRNA-containing viral particles in *Colletotrichum gloeosporioides* isolated from cashew trees. Trop Plant Pathol 37:142–145

Figueiredo LC, Figueiredo GS, Quecine MC, Cavalcanti FCN, Santos AC, Costa NT, Oliveira NT, Azevedo JL (2012b) Genetic and pathogenic diversity of *Colletotrichum gloeosporioides* the causal agent of cashew anthracnose. Indian J Fundam Appl Life Sci 2:250–259

Flärdh K, Buttner MJ (2009) *Streptomyces* morphogenetics: dissecting differentiation in a filamentous bacterium. Nat Rev Microbiol 7:36–49

Franco C, Michelsen P, Percy N, Conn V, Listiana E, Moll S, Loria R, Coombs J (2007) Actinobacterial endophytes for improved crop performance. Aust Plant Pathol 36:524–531

Freitag JH (1951) Host range of the Pierce's disease virus of grapes as determined by insect transmission. Phytopathology 41:920–934

Frommel MI, Novak J, Lazarovits G (1991) Growth enhancement and developmental modification of *in-vitro* potato (*Solanum tuberosum* ssp. *tuberosum*) as affected by nonfluorescent *Pseudomonas* sp. Plant Physiol 96:928–936

Gai CS, Lacava PT, Araújo WL, Azevedo JL (2007) Manipulation of endophytic bacteria for symbiotic control of *Xylella fastidiosa*, causal agent of citrus variegated chlorosis. In: Pierce's disease research symposium, San Diego, CA, Proceedings, 12–14 Dec 2007, pp 194–197

Gai CS, Lacava PT, Quecine MC, Auriac MC, Lopes JRS, Araújo WL, Miller TA, Azevedo JL (2009) Transmission of *Methylobacterium mesophilicum* by *Bucephalogonia xanthophis* for paratransgenic control strategy of citrus variegated chlorosis. J Microbiol 47:448–454

Gai CS, Dini-Andreote F, Andreote FD, Lopes JRS, Araújo WL, Miller TA, Azevedo JL, Lacava PT (2011) Endophytic bacteria associated to sharpshooters (Hemiptera: Cicadellidae), insect vectors of *Xylella fastidiosa* subsp. *pauca*. J Plant Pathol Microbiol 2:109. doi:10.4172/2157-7471.1000109

Garbeva P, Van Overbeek LS, Van Vuurde JWL, Van Elsas JD (2001) Analysis of endophytic bacterial communities of potato by planting and denaturing gradient gel electrophoresis (DGGE) of 16s rDNA based PCR fragments. Microb Ecol 41:369–383

Garcia A, Rhoden SA, Bernardi-Wenzel J, Orlandelli RC, Azevedo JL, Pamphile JA (2012) Antimicrobial activity of crude extracts of endophytic fungi isolated from medicinal plant *Sapindus saponaria*, L. J Appl Pharm Sci 2:35–40

Gazis R, Chavern P (2010) Diversity of fungal endophytes on leaves and stems of wild rubber trees (*Hevea brasiliensis*) in Peru. Fungal Ecol 3:240–254

Germaine K, Keogh E, Garcia-Cabellos G, Borremans B, Van der Lelie D, Barac T, Oeyen L, Vangronsveld J, Moore FP, Moore ERB, Campbell TCD, Dowling DN (2004) Colonization of poplar trees by *gfp* expressing bacterial endophytes. FEMS Microbiol Ecol 48:109–118

Glick B, Penrose D, Li J (1998) A model for the lowering of plant ethylene concentrations by plant growth-promoting bacteria. J Theor Biol 190:63–68

Glienke-Blanco C, Aguilar-Vildoso CL, Vieira MLC, Barroso PAV, Azevedo JL (2002) Genetic variability in the endophytic fungus *Guignardia citricarpa* isolated from citrus plants. Genet Mol Biol 25:251–255

Hallmann J, Quadt-Hallmann A, Mahaffee WF, Kloepper JW (1997) Bacterial endophytes in agricultural crops. Can J Microbiol 43:895–914

Hartung JS, Beretta J, Brlansky RH, Spisso J, Lee RF (1994) Citrus variegated chlorosis bacterium: axenic culture, pathogenicity, and serological relationships with other strains of *Xylella fastidiosa*. Phytopathology 84:591–597

Hawksworth DL (2001) The magnitude of fungal diversity: the 1.5 million species estimate revisited. Mycol Res 105:1422–1432

Hayat R, Ali S, Amara U, Khalid R, Ahmed I (2010) Soil beneficial bacteria and their role in plant growth promotion: a review. Ann Microbiol 60:579–598

Huang JS (1986) Ultrastructure of bacterial penetration in plants. Annu Rev Phytopathol 24:141–157

Idris R, Trifonova R, Puschenreiter M, Wenzel WW, Sessitsch A (2004) Bacterial communities associated with flowering plants of the Ni hyperaccumulator *Thlaspi goesingense*. Appl Microbiol Biotechnol 70:2667–2677

Iniguez AL, Dong Y, Carter HD, Ahmer BMM, Stone JM, Triplett EW (2005) Regulation of enteric endophytic bacterial colonization by plant defense. Mol Plant Microbe Interact 18:169–178

James EK, Olivares FL (1998) Infection and colonization of sugarcane and other graminaceous plants by endophytic diazotrophs. Crit Rev Plant Sci 17:77–119

Kaaya GP, Mwangi EN, Ouna EA (1996) Prospects for biological control of livestock ticks, *Rhipicephalus appendiculatus* and *Amblyomma variegatum* using entomogenous fungi *Beauveria bassiana* and *Metarhizium anisopliae*. J Invertebr Pathol 67:15–20

Kafur A, Khan AB (2011) Isolation of endophytic actinomycetes from *Catharanthus roseus* (L.) G. Don leaves and their antimicrobial activity. Iran J Biotechnol 9:302–306

Kibe R, Sakamoto M, Yokota H, Ishikawa H, Aiba Y, Koga Y, Benno Y (2005) Movement and fixation of intestinal microbiota after administration of human feces to germfree mice. Appl Environ Microbiol 71:3171–3178

Kim KJ, Yang YJ, Kim JG (2003) Purification and characterization of chitinase from *Streptomyces* sp. M-20. J Biochem Mol Biol 36:185–189

Kramer KJ, Muthukrishnan S (1997) Insect chitinases: molecular biology and potential use as biopesticides. Insect Biochem Mol Biol 27:887–900

Krechel A, Faupel A, Hallmann J, Ulrich A, Berg G (2002) Potato associated bacteria and their antagonistic potential towards plant pathogenic fungi and the plant parasitic nematode *Meloidogyne incognita* (kofoid and white) chitwood. Can J Microbiol 48:772–786

Kumar BSD, Dube HC (1992) Seed bacterization with a fluorescent *Pseudomonas* for enhanced plant growth and yield and disease control. Soil Biol Biochem 26:539–542

Kwon SJ, Cho SY, Lee KM, Yu J, Son M, Kim KH (2009) Proteomic analysis of fungal host factors differentially expressed by *Fusarium graminearum* infected with *Fusarium graminearum* virus-DK21. Virus Res 144:96–106

Lacava PT, Araujo WL, Marcon J, Maccheroni W Jr, Azevedo JL (2004) Interaction between endophytic bacteria from citrus plants and the phytopathogenic bacteria *Xylella fastidiosa*, causal agent of citrus-variegated chlorosis. Lett Appl Microbiol 39:55–59

Lacava PT, Li WB, Araújo WL, Azevedo JL, Hartung JS (2006) Rapid, specific and quantitative assays for the detection of endophytic bacterium *Methylobacterium mesophilicum* in inoculated plants. J Microbiol Method 65:535–541

Lacava PT, Araújo WL, Azevedo JL (2007a) Evaluation of endophytic colonization of *Citrus sinensis* and *Catharanthus roseus* seedlings by endophytic bacteria. J Microbiol 45:11–14

Lacava PT, Li WB, Araújo WL, Azevedo JL, Hartung JS (2007b) The endophyte *Curtobacterium flaccumfaciens* reduces symptoms caused by *Xylella fastidiosa* in *Catharanthus roseus*. J Microbiol 45:388–393

Lamb TG, Tonkyn DW, Kluepfel DA (1996) Movement of *Pseudomonas aureofaciens* from the rhizosphere to aerial plant tissue. Can J Microbiol 42:1112–1120

Lambais MR, Goldman MHS, Camargo LEA, Goldman GH (2000) A genomic approach to the understanding of *Xylella fastidiosa* pathogenicity. Curr Opin Microbiol 3:459–462

Lambert B, Joos H (1989) Fundamental aspects of rhizobacterial plant growth promotion research. Trends Biotechnol 7:215–219

Lampe D, Lauzon C, Miller TA (2006) Use of the *E. coli* α-hemolysin secretion system in bacteria designed for symbiotic control of pierce's disease in grapevines and sharpshooters. In: Pierce's disease research symposium, San Diego, CA, Proceedings, 27–29 Nov 2006, pp 240–241

Lampe D, Kang A, Miller TA (2007) Native secretion systems for the grapevine endophyte *Pantoea agglomerans* useful for the delivery of anti-*Xylella* effector proteins. In: Pierce's disease research symposium, San Diego, CA, Proceedings, 12–14 Dec 2007, pp 218–220

Lampel JS, Canter GL, Dimock MB, Kelly JL, Anderson JJ, Uratani BB, Foulke JS Jr, Turner JT (1994) Integrative cloning, expression, and stability of the cryIA(c) gene from *Bacillus thuringiensis* subsp. *kurstati* in a recombinant strain of *Clavibacter* xyli subsp. *cynodontis*. Appl Environ Microbiol 60:501–508

Lana TG, Azevedo JL, Pomella AWV, Monteiro RTR, Silva CB, Araújo WL (2011) Endophytic and pathogenic isolates of the cacao fungal pathogen *Moniliophthora perniciosa* (Tricholomataceae) are indistinguishable based on genetic and physiological analysis. Genet Mol Res 10:326–334

Lecuona RE (1996) Microorganismos patógenos empleados en el control microbiano de insectos plaga. Editora Lecuona, Castelar, p 338

Lee RF, Kerrick KS, Beretta MJG, Chagas CM, Rossetti V (1991) Citrus variegated chlorosis: a new destructive disease of citrus in Brazil. Citrus Ind 72:12–13

Lewis LC, Cossentine JE (1986) Season long intraplant epizootics of entomopathogens *Beauveria bassiana* and *Nosema pyrausta* in a corn agroecosystem. Entomophaga 31:36–69

Li WB, Donadio LC, Sempionato OR, Stuchi ES, Rossetti V, Beretta MJG (1997) Effect of rootstocks on the severity of citrus variegated chlorosis (CVC) of sweet orange 'Pera' (*C sinensis* L. Osbeck). Proc Int Soc Citric 1:286–289

Li C-H, Zhao M-W, Tang C-M, Li S-P (2009) Population dynamics and identification of endophytic bacteria antagonistic toward plant-pathogenic fungi in cotton root. Microb Ecol 59:344–356

Loaces I, Ferrando L, Fernandez Scavino A (2011) Dynamics, diversity and function of endophytic siderophore-producing bacteria in rice. Microb Ecol 61:606–618

Lodewyckx C, Vangronsveld J, Porteus F, Moore ERB, Taghavi S, Mezgeay M, Lelie DV (2002) Endophytic bacteria and their potential applications. Crit Rev Plant Sci 21:586–606

Madhaiyan M, Poonguzhali S, Senthilkumar M, Seshadri S, Chung H (2004) Growth promotion and induction of systemic resistance in rice cultivar Co-47 (*Oryza sativa* L.) by *Methylobacterium* spp. Bot Bull Acad Sin 45:315–324

Mahaffe WF, Kloepper JW, Van Vuurde JWL, Van Der Wolf JM, Van Den Brink M (1997) Endophytic colonization of *Phaseolus vulgaris* by *Pseudomonas fluorescens* strain 89B-27 and *Enterobactr asburiae* strain JM22. In: Ryder MH, Stephens PM, Bowen GD (eds) Improving plant productivity in rhizosphere bacteria. CSIRO, Melbourne, p 180

Mahaffee WF, Kloepper JW (1997) Temporal changes in the bacterial communities of soil, rhizosphere and endorhiza. Microb Ecol 34:210–223

Malinowski DP, Belesky DP (1999) Endophyte infection enhances the ability of tall fescue to utilize sparingly available phosphorus. J Plant Nutr 22:835–853

Mano H, Morisaki H (2008) Endophytic bacteria in the rice plant. Microbes Environ 23:109–117

Martins ES, Praça LB, Dumas VF, Silva-Werneck JO, Sone EH, Waga IC, Berry C, Monnerat RG (2007) Characterization of *Bacillus thuringiensis* isolates toxic to cotton boll weevil (*Anthonomus grandis*). Biol Control 40:65–68

Martins ES, Aguiar RWS, Martins NF, Melatti VM, Falcão R, Gomes ACMM, Ribeiro BM, Monnerat RG (2008) Recombinant Cry1Ia protein is highly toxic to cotton boll weevil (*Anthonomus grandis* Boheman) and fall armyworm (*Spodoptera frugiperda*). J Appl Microbiol 104:1363–1371

Masclaux C, Expert D (1995) Signalling potential of iron in plant–microbe interactions: the pathogenic switch of iron transport in *Erwinia chrysanthemi*. Plant J 7:121–128

Mastretta C, Taghavi S, van der Lelie D, Mengoni A, Galardi F, Gonnelli C, Barac T, Boulet J, Weyens N, Vangronsveld J (2009) Endophytic bacteria from seeds of *Nicotiana tabacum* can reduce cadmium phytotoxicity. Int J Phytoremediat 11:251–267

Mei C, Flinn BS (2010) The use of beneficial microbial endophytes for plant biomass and stress tolerance improvement. Recent Pat Biotechnol 4:81–95

Melo IS, Azevedo JL (1998) Controle biológico I. Embrapa, Jaguariuna, p 262

Mendes R, Azevedo JL (2007) Valor biotecnológico de fungos endofíticos isolados de plantas de interesse econômico. In: Abstract, 5° Congresso Brasileiro de Micologia, pp 129–140

Miller TA (2007) Symbiotic control in agriculture and medicine. Symbiosis 42:67–74

Monteiro PB, Renaudin J, Jagoueix-Eveillard S, Ayres AJ, Garnier M, Bové JM (2001) *Catharanthus roseus*, an experimental host plant for the citrus strain of *Xylella fastidiosa*. Plant Dis 85:246–251

Musson G (1994) Ecology and effects of endophytic bacteria in plant. Masters thesis, Auburn University, Auburn, AL

Mwangi eN, Kaaia GP, Essuman S (1995) Experimental infections of the tick *Rhipicephalus appendiculatus* with pathogenic fungi, *Beauveria bassiana* and *Metarhizium anisopliae*, and natural infection of some ticks with bacteria and fungi. J Afr Zool 109:151–160

Nachin L, El Hassouni M, Loiseau L, Expert D, Barras F (2001) SoxR-dependent response to oxidative stress and virulence of *Erwinia chrysanthemi*: the key role of SufC, an orphan ABC ATPase. Mol Microbiol 39: 960–972

Neilands JB, Nakamura K (1991) Detection, determination, isolation, characterization and regulation of microbial iron chelates. In: Winkelmann G (ed) CRC handbook of microbial iron chelates. CRC Press, Boca Raton, pp 1–14

Nudel C, Gonzalez R, Castaneda N, Mahler G, Actis LA (2001) Influence of iron on growth, production of siderophore compounds, membrane proteins, and lipase activity in *Acinetobacter calcoaceticus* BD 413. Microbiol Res 155:263–269

O'Sullivan DJ, O'Gara F (1992) Traits of fluorescent *Pseudomonas* spp. involved in suppression of plant root pathogens. Microbiol Rev 56:662–676

Orlandelli RC, Alberto RN, Rubin Filho CJ, Pamphile JA (2012) Diversity of endophytic fungal community associated with *Piper hispidum* Sw. (Piperaceae) leaves. Genet Mol Res 11:1575–1585

Owen NL, Hundley N (2004) Endophytes the chemical synthesizer inside plants. Sci Prog 87:79–99

Pamphile JA, Azevedo JL (2002) Molecular characterization of endophytic strains of *Fusarium verticillioides (Fusarium moniliforme)* from maize (*Zea mays L.*). World J Microbiol Biotechnol 18:391–396

Panaccione DG, Johnson RD, Wang J, Young CA, Damron K, Scott B, Schardl CL (2001) Elimination of ergovaline from a grass-*Neotyphodium* endophyte symbiosis by genetic modification of the endophyte. Proc Natl Acad Sci U S A 98:12820–12825

Persello-Cartieaux F, Nussaume L, Robaglia C (2003) Tales from the underground: molecular plant–rhizobacteria interactions. Plant Cell Environ 26:189–199

Petrini O (1986) Taxonomy of endophytic fungi in aerial plant tissues. In: Fokkema NJ, Van den Heuvel J (eds) Microbiology of the phyllosphere. Cambridge University Press, Cambridge, pp 175–187

Petrini O, Hake U, Dreifuss MM (1990) An analysis of fungal communities isolated from fruticose lichens. Mycologia 82:444–451

Pimenta PFP, Modi GB, Pereira ST, Shahabuddin M, Sacks DL (1997) A novel role for the peritrophic matrix in protecting *Leishmania* from the hydrolytic activities of the sand fly midgut. Parasitology 115:359–369

Pimentel IC (2001) Fungos endofíticos de milho (*Zea mays L.*) e soja (*Glycine max* (L.) Merril) e seu potencial valor biotecnológico no controle de pragas agrícolas. PhD thesis, University of Paraná, Curitiba, 154p

Pleban S, Ingel F, Chet I (1995) Control of *Rhizoctonia solani* and *Sclerotium rolfsii* in the greenhouse using endophytic *Bacillus* spp. Eur J Plant Pathol 101:665–672

Pocasangre L, Sikora RA, Vilich V, Schuster RP (2000) Survey of banana endophytic fungi from Central America and screening for biological control of the burrowing nematode (*Rhadopholus similis*). Informusa 9:3–5

Purcell AH, Saunders SR (1999) Fate of Pierce's disease strains of *Xylella fastidiosa* in common riparian plants in California. Plant Dis 83:825–830

Quecine MC, Araújo WL, Marcon J, Gai CS, Azevedo JL, Pizzirani-Kleiner AA (2008) Chitinolytic activity of endophytic *Streptomyces* and potential for biocontrol. Lett Appl Microbiol 47:486–491

Quecine MC, Lacava PT, Magro SR, Parra JRP, Araújo WL, Azevedo JL, Pizzirani-Kleiner AA (2011) Partial characterization of chitinolytic extract from endophytic *Streptomyces* sp. and its effects on the boll weevil. J Agric Sci Technol 5:420–427

Radji M, Sumiati A, Rachmayan R, Elya B (2011) Isolation of fungal endophytes from *Garcinia mangostana* and their antibacterial activity. Afr J Biotechnol 10:103–107

Rajkumar M, Ae N, Freitas H (2009) Endophytic bacteria and their potential to enhance heavy metal phytoextraction. Chemosphere 77:153–160

Ramamoorthy V, Viswanathan R, Raguchander T, Prakasan V, Samiyappan R (2001) Induction of systemic resistance by plant growth promoting rhizobacteria in crop plants against pests and diseases. Crop Prot 20:1–11

Ramesh R, Joshi AA, Ghanekar MP (2009) *Pseudomonas*: major endophytic bacteria to suppress bacterial wilt pathogen *Ralstonia solanacearum* in the egg plant (*Solanum melongena* L.). World J Microbiol Biotechnol 25:47–55

Rampelotti-Ferreira FT, Ferreira A, Vendramim JD, Lacava PT, Azevedo JL, Araújo WL (2010) Colonization of rice and *Spodoptera frugiperda* J.E. Smith (Lepidoptera: Noctuidae) larvae by genetically modified endophytic *Methylobacterium mesophilicum*. Neotrop Entomol 39:308–310

Redman RS, Freeman S, Clifton DR, Morrel J, Brown G, Rodriguez RJ (1999) Biochemical analysis of plant protection afforded by a nonpathogenic endophytic mutant of *Colletotrichum magna*. Plant Physiol 119:795–804

Reinhold B, Hurek T (1988) Location of diazotrophs in the interior with special attention to the kallar grass association. Plant Soil 110:259–268

Reinhold-Hurek B, Hurek T (1998) Life in grasses: diazotrophic endophytes. Trends Microbiol 6:139–144

Reis VM, Baldani JI, Baldani VLD, Döbereiner J (2000) Biological dinitrogen fixation in gramineae and palm trees. Crit Rev Plant Sci 10:227–247

Reiter B, Pfeifer U, Schwab H, Sessitsch A (2002) Response of endophytic bacterial communities in potato plants to infection with *Erwinia carotovora* subsp. *atroseptica*. Appl Environ Microbiol 68:2261–2268

Rhoden SA, Garcia A, Bongiorno VA, Azevedo JL, Pamphile JA (2012) Antimicrobial activity of crude extracts of endophytic fungi Isolated from medicinal; plant *Trichilia elegans* A. Juss. J Appl Pharm Sci 2:57–59

Rio RVM, Hu Y, Aksoy S (2004) Strategies of the home-team: symbioses exploited for vector-borne disease control. Trends Microbiol 12:325–336

Robbins PW, Allbright C, Benfield B (1998) Cloning and expression of a *Streptomyces plicatus* chitinase (chitinase 63) in *Escherichia coli*. J Biol Chem 263:443–447

Rosenblueth M, Martinez-Romero E (2006) Bacterial endophytes and their interactions with hosts. Mol Plant Microbe Interact 19:827–837

Rossetti V, Garnier M, Bove JM, Beretta MJG, Teixeira ARR, Quaggio JA, deNegri JD (1990) Presence de bacteries dans le xyleme d'orangers atteints de chlorose variegee, unenovelle maladie des agrumes au Brasil. Les Comptes Rendus de L'Académie des Sciences (Paris) 310:345–349

Rubini MR, Silva-Ribeiro R, Pomella AWV, Maki C, Araújo WL, Santos DR, Azevedo JL (2005) Diversity of endophytic fungal community of cacao (*Theobroma cacao*, L.) and biological control of *Crinipellis perniciosa* causal agent of Witches' Broom disease. Int J Biol Sci 1:24–33

Ryan RP, Germaine K, Franks A, Ryan DJ, Dowling DN (2008) Bacterial endophytes: recent developments and applications. FEMS Microbiol Lett 278:1–9

Saikkonen K, Wäli P, Helander M, Faeth SH (2004) Evolution of endophyte–plant symbioses. Trends Plant Sci 9:275–280

Sardi P, Saracchi M, Quaroni B, Borgonovi GE, Merli S (1992) Isolation of endophytic *Streptomyces* strains from surface-sterilized roots. Appl Environ Microbiol 58:2691–2693

Schaad NW, Pastnikova E, Lacey G, Fatmi M, Chang CJ (2004) *Xylella fastidiosa* subspecies: *X. fastidiosa* subsp. piercei, subsp. nov., *X. fastidiosa* subsp. multiplex subsp. nov., and *X. fastidiosa* subsp. pauca subsp. nov. Syst Appl Microbiol 27:290–300

Scherwinski K, Wolf A, Berg G (2007) Assessing the risk of biological control agents on the indigenous microbial communities: *Serratia plymuthica* HRO-C48 and *Streptomyces* sp. HRO-71 as model bacteria. BioControl 52:87–112

Schroth MN, Hancook GH (1995) Disease suppressive soil and root colonizing bacteria science. Soil Biol Biochem 24:539–542

Sebastianes FLS, Cabedo N, El-Aouad N, Valente AMMP, Lacava PT, Azevedo JL, Pizzirani-Kleiner AA (2012a) 3-Hydroxypropionic acid as an antibacterial agent from endophytic fungi *Diaporthe phaseolorum*. Curr Microbiol 65:622–632

Sebastianes LS, Lacava PT, Favaro LCL, Rodrigues MBC, Araújo WL, Azevedo JL, Pizzirani-Kleiner AA (2012b) Genetic transformation of *Diaporthe phaseolorum*, an endophytic fungus found in mangrove forests mediated by *Agrobacterium tumefaciens*. Curr Genet 58:21–33

Seghers D, Wittebolle L, Top EM, Verstraete W, Siciliano SD (2004) Impact of agricultural practices on the *Zea mays* L. endophytic community. Appl Environ Micobiol 70:1475–1482

Seipke RF, Kaltenpoth M, Hutchings MI (2012) *Streptomyces* as symbionts: an emerging and widespread theme? FEMS Microbiol Rev 36:862–876

Senthilkumar M, Madhaiyan M, Sundaram SP, Kannaiyan S (2009) Intercellular colonization and growth promoting effects of *Methylobacterium* sp. with plant-growth regulators on rice (*Oryza sativa* L. Cv CO-43). Microbiol Res 164:92–104

Sessitsch A, Reiter B, Berg G (2004) Endophytic bacterial communities of field grown potato plants and their plant growth-promoting and antagonistic abilities. Can J Microbiol 50:239–249

Sessitsch A, Hardoim P, Döring J, Weilharter A, Krause A, Woyke T, Mitter B, Hauberg-Lotte L, Friedrich F, Rahalkar M, Hurek T, Sarkar A, Bodrossy L, van Overbeek L, Brar D, van Elsas JD, Reinhold-Hurek B (2012) Functional characteristics of an endophyte community colonizing rice roots as revealed by metagenomic analysis. Mol Plant Microbe Interact 25:28–36

Sevilla M, Burris RH, Gunapala N, Kennedy C (2001) Comparison of benefit to sugarcane plant growth and N_2 incorporation following inoculation of sterile plants with *Acetobacter diazotrophicus* wild-type and *Nif*-mutant strains. Mol Plant Microbe Interact 14:358–366

Sharma A, Johri BN (2003) Growth promoting influence of siderophore-producing *Pseudomonas* strains GRP3A and PRS9 in maize (*Zea mays* L.) under iron limiting conditions. Microbiol Res 158:243–248

Sheng XF, Xia JJ, Jiang CY, He LY, Qian M (2008) Characterization of heavy metal-resistant endophytic bacteria from rape (*Brassica napus*) roots and their potential in promoting the growth and lead accumulation of rape. Environ Pollut 156:1164–1170

Sia EF (2006) Isolados endofíticos e entomopatogênicos de *Beauveria*: caracterização e importância biotecnológica. MS thesis, University of Mogi das Cruzes, São Paulo, Brazil, 55p

Siddiqui ZA (2005) PGPR: prospective biocontrol agents of plant pathogens. In: Siddiqui ZA (ed) PGPR: biocontrol and biofertilization. Springer, Dordrecht, pp 111–142

Silva KS, Rebouças TNH, Lemos OL, Bomfim MP, Bomfim AA, Esquivel GL, Barreto APP, José ARS, Dias NO, Tavares GM (2004) Patogenicidade causada pelo fungo *Colletotrichum gloesporioides* (Penz) em diferentes espécies frutíferas. Rev Bras Frutic 28:131–133

Simpson AJG, Reinach FC, Arruda P, Abreu FA, Acencio M, Alvarenga R, Alves LMC, Araya JE, Baia GS, Baptista CS et al (2000) The genome sequence of the plant pathogen *Xylella fastidiosa*. Nature 406:151–159

Smalla K, Wieland G, Buchner A, Zock A, Parzy J, Kaiser S, Roskot N, Heuer H, Berg G (2001) Bulk and rhizosphere soil bacterial communities studied by denaturing gradient gel electrophoresis: plant-dependent enrichment and seasonal shifts revealed. Appl Environ Microbiol 67:4742–4751

Strobel G, Daisy B (2003) Bioprospecting for microbial endophytes and their natural products. Microbiol Mol Biol Rev 67:491–502

Stuart RM, Romão AS, Pizzirani-Kleiner AA, Azevedo JL, Araújo WL (2010) Culturable endophytic filamentous fungi from leaves of transgenic imidazolinone-tolerant sugarcane and its non-transgenic isolines. Arch Microbiol 192:307–313

Sturz AV (1995) The role of endophytic bacteria during seed piece decay and potato tuberization. Plant Soil 175:257–263

Sturz AV, Christie BR, Matheson BG, Nowak J (1997) Biodiversity of endophytic bacteria which colonize red clover nodules, roots, stems and foliage and their influence on host growth. Biol Fertil Soils 25:13–19

Sturz AV, Christie BR, Nowak J (2000) Bacterial endophytes: potential role in developing sustainable systems of crop production. Crit Rev Plant Sci 19:1–30

Sun L, Qiu F, Zhang X, Dai X, Dong X, Song W (2008) Endophytic bacterial diversity in rice (*Oryza sativa* L.) roots estimated by 16S rDNA sequences analysis. Microb Ecol 55:415–424

Suryanarayanan TS (2011) Diversity of fungal endophytes in tropical trees. In: Prittila AM, Carolin FA (eds) Endophytes of tropical trees. Forestry sciences series 80. Springer, Netherlands, pp 67–80

Suryanarayanan TS, Murali TS, Thirumavukkarasu N, Govinda-Rajulu MB, Venkatesan G, Sukumar R (2011) Endophytic fungal communities in wood perennials of three tropical forest types of the Western Ghats, Southern India. Biodivers Conserv 20:913–928

Suryanarayanan TS, Tirunavukkarasu N, Govindarajulu MB, Gopalan V (2012) Fungal endophytes; an untapped source of biocatalysts. Fungal Divers 54:19–30

Taechowisan T, Peberdy JF, Lumyong S (2003) Isolation of endophytic actinomycetes from selected plants and their antifungal activity. World J Microbiol Biotechnol 19:381–385

Tan RX, Zou WX (2001) Endophytes: a rich source of functional metabolites. Nat Prod Rep 18:448–459

Toyota K, Kimura M (2000) Suppression of *Ralstonia solanacearum* in soil following colonization by other strains of *R. solanacearum*. Soil Sci Plant Nutr 46: 449–459

Tsujibo H, Minoura K, Miyamoto K, Endo H, Moriwaki M, Inamori Y (1993) Purification and properties of a thermostable chitinase from *Streptomyces thermoviolaceus* OPC-520. Appl Environ Microbiol 59:620–622

Tunali B, Marshall D, Royo C, Nachit MM, Fonzo NDI, Araus JL (2000) Antagonistic effect of endophytes against several non-root pathogens of wheat. In: Durum wheat improvement in the Mediterranean region: new challenges. Proceedings of the Semina, vol 40, Zaragoza, Spain, pp 381–386

Turner JT, Lampell JS, Stearmen RS, Sundin GW, Gunyuzlu UP, Anderson JJ (1991) Stability of the d-endotoxin gene from *Bacillus thuringiensis* subsp. *kurstaki* in a recombinant strain of *Clavibacter xyli* subsp. *cynodontis*. Appl Environ Microbiol 57:3522–3528

Van Der Peer R, Puente HLM, LA Weger DE, Shipper B (1990) Characterization of root surface and endorhizosphere pseudomonas in relation to their colonization of roots. Appl Environ Microbiol 56:2462–2470

Van-Heeswijck R, McDonald G (1992) *Acremonium* endophyte in perennial ryegrass and other pasture grasses in Australia and New Zealand. Aust J Agric Res 43:1683–1709

Vega FE, Posada F, Aime MC, Pava-Ripoli M, Infante F, Rhener SA (2008) Entomopathogenic fungal endophytes. Biol Control 346:72–82

Vega FE, Goettel MS, Blackwell M, Chandler D, Jackson MA, Keller S, Koike M, Maniania NK, Monzon A, Onley BH, Pell JK, Rangel DEN, Roy HE (2009) Fungal entomopathogens: new-insights on their ecology. Fungal Ecol 2:149–159

Vendan RT, Yu YJ, Lee SH, Rhee YH (2010) Diversity of endophytic bacteria in ginseng and their potential for plant growth promotion. J Microbiol 48:559–565

Verma VC, Singh SK, Prakash S (2011) Bio-control and plant growth promotion potential of siderophore producing endophytic *Streptomyces* from *Azadirachta indica* A. Juss. J Basic Microbiol 5:550–556

Wagner BL, Lewis LC (2000) Colonization of corn, *Zea mays*, by the entomopathogenic fungus *Beauveria bassiana*. Appl Environ Microbiol 66:3468–3473

Webber J (1981) A natural control of Dutch elm disease. Nature 292:449–451

Weyens N, Vander Lelie D, Taghavi S, Vangronsveld J (2009) Phytoremediation: plant-endophyte partnerships take the challenge. Curr Opin Biotechnol 20: 248–254

Wiehe W, Hecht-Bucholz C, Hoflich G (1994) Electron microscopic investigations on root colonization of *Lupinus albus* and *Pisum sativum* with two associative plant growth promoting rhizobacteria, *Pseudomonas fluorescens* and *Rhizobium leguminosarum* bv. *trifolii*. Symbiosis 86:221–224

Wilson D (1995) Endophyte – the evolution of a term and clarification of its use and definition. Oikos 72: 274–276

Wolkers H, van Bavel B, Ericson I, Skoglund E, Kovacs KM, Lydersen C (2006) Congener-specific accumulation and patterns of chlorinated and brominated contaminants in adult male walruses from Svalbard, Norway: Indications for individual-specific prey selection. Sci Total Environ 370:70–79

Yang J-H, Liu H-X, Zhu G-M, Pan Y-L, Guo J-H (2008) Diversity analysis of antagonists from rice-associated bacteria and their application in biocontrol of rice diseases. J Appl Microbiol 104:91–104

Yunus A, Kawamata S, Shimanuki T, Murakami Y, Ichinose Y, Shiraishi T, Yamada T (1999) Transformation of the endophyte *Neotyphodium* with the indole acetic acid (iaa M) gene. Ann Phytopathol Soc Jpn 55:192–196

Zinniel DK, Lambrecht P, Harris NB, Feng Z, Kuczmarski D, Higley P, Ishimaru CA, Arunakumari A, Barletta RG, Vidaver AK (2002) Isolation and characterization of endophytic colonizing bacteria from agronomic crops and prairie plants. Appl Environ Microbiol 68:2198–2208

Biocontrol and Bioremediation: Two Areas of Endophytic Research Which Hold Great Promise

14

Mary Ruth Griffin

Abstract

Research into the beneficial use of endophytic organisms has dramatically increased worldwide in recent years. Endophytes are typically bacteria or fungi which colonize the internal tissues of plant hosts without causing visible negative effects. Two areas in endophyte research, which hold tremendous positive economic and environmental potential, are biocontrol and bioremediation. Biocontrol, short for biological control is the intentional use of a specific organism or their metabolic by-products to limit the harmful impact of a plant pest. Endophytes due to their unique symbiotic relationships within their hosts have the potential to directly act antagonistically against plant pests. In addition endophytes may also act indirectly against pests, benefitting their hosts by enhancing general plant growth or plant-protection responses, such as in the case of induced systemic resistance. Bioremediation is the use of microorganisms to alter or reduce the toxic impact of pollutants through various forms of metabolic activity. Microorganisms, in part due to their short life spans, can adapt relatively fast to environmental pollutants. Endophytes with these adaptations can in some cases provide their hosts with the capability to remediate their surrounding microenvironments. In this review, we will explore recent advances made in the promising areas of biocontrol and bioremediation research.

1 Introduction

Symbiosis describes a relationship between two interacting organisms and includes a wide spectrum of resulting conditions that range from mutualistic to pathogenic. Organisms that internally colonize plants for part (facultative) or all (obligate) of their lives (Hardoim et al. 2008) occur along different areas of the symbiotic spectrum due to a plethora of reasons. Known as endophytes, these diverse organisms dwell within the internal tissues of plant structures and are found in roots, stems, leaves, flowers, and even seeds (Surette et al. 2003). As mutual or commensal symbionts, they coexist peaceably within their

M.R. Griffin (✉)
Department of Environmental and Plant Biology, Ohio University, Athens, OH 45701, USA
e-mail: maryruthgriffin@gmail.com

V.C. Verma and A.C. Gange (eds.), *Advances in Endophytic Research*,
DOI 10.1007/978-81-322-1575-2_14, © Springer India 2014

Fig. 14.1 Fungal endophyte (*Beauveria bassiana*) shown here sporulating from the flower and stem of a Venus flytrap (*Dionaea muscipula*). The fungus was inoculated via the pods and, after 6–8 weeks, moved systemically throughout the plant before sporulating from new plant tissue. As a result of the spores becoming airborne, uninoculated plants also became colonized (Photo courtesy Mary R. Griffin, unpublished data)

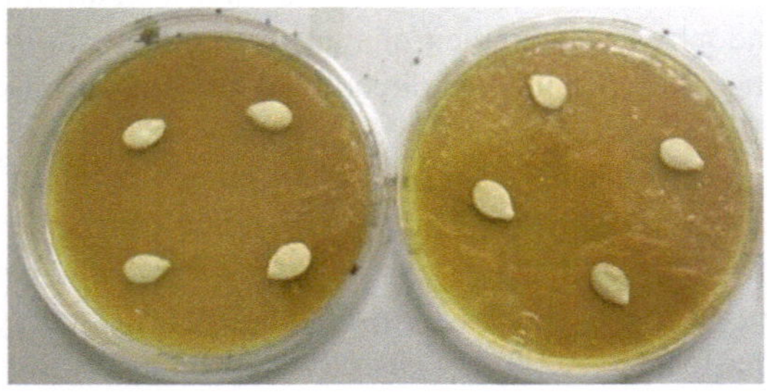

Fig. 14.2 Squash seeds plated onto plant-based media to determine if they are colonized with endophytes. Use of host-plant media (10 % juice concentrate) is useful for encouraging growth of seed endophytes in culture

plant hosts causing no visible negative effects or signs of infection (Compant et al. 2010; Ryan et al. 2008). The bacteria or fungi that become endophytes typically originate from the plant's rhizosphere (Compant et al. 2005a) or phyllosphere (Hallman et al. 1997); although some are horizontally transmitted (Fig. 14.1) and a few are vertically transmitted (Fig. 14.2) from parent to seed (Ernst et al. 2003; Müller and Krauss 2005; Canakar et al. 2005). Once internalized, endophytes may move throughout the plant and colonize additional areas (Germaine et al. 2006). Endophytes have been known about for a long time. The term endophyte was actually coined in 1884 by Heinrich Anton de Bary, who recognized that fungi and bacteria could dwell within plant tissues without causing any apparent harm. At this time however there are still ongoing discussions about what specifically constitutes a "true endophyte," because the term, as it is, only defines the location of an organism in a plant as opposed to any specific attribute (Schulz and Boyle 2005; Porras-Alfaro and Bayman 2011). As such, symbiotic titles can have fluidity and this will effect where endophytes are viewed to fall on the symbiotic spectrum.

Overall endophytes are very common worldwide and it is thought that all 300,000+ plant species identified thus far serve as hosts for at least one or multiple endophytic species (Strobel et al. 2004). Linked to their widespread presence in the plant kingdom is additional evidence that the diversity of endophytic species within ecosystems is highest among those with the highest floral diversity, such as in tropical and temperate rainforests (Strobel and Daisy 2003). Despite their commonality in plants, most endophytes were generally overlooked or even ignored in the scientific literature until recent decades. Now, however, interest in endophytes, especially mutualistic endophytes, has exploded worldwide due to the many potential benefits they may provide. Two examples of rapidly developing technologies which utilize endophytes to benefit society as a whole are biocontrol and bioremediation. Biocontrol research and technology development primarily focus on aiding plant survival by utilizing organisms to inhibit the advance of harmful pest/pathogenic organisms.

Bioremediation research and technology development focuses on the utilization of organisms which have the ability to break down or accumulate harmful toxins in the environment; to ultimately aid plant growth in a polluted environment. The use of endophytes for biocontrol and bioremediation efforts has made tremendous strides, but work is still ongoing. The primary focus of this chapter is to examine advantages and challenges related to the use of endophytes with these two technologies. It also attempts to examine many of the advances made in our understanding regarding endophytes' abilities which relate to their potential in biocontrol activities and bioremediation efforts.

1.1 Endophytes and Their Relationships with Plants: Diversity, Ecology, and Challenge

Endophytes, because of their symbiotic relationships with plant hosts, offer a novel approach to the study of plants in general, but also, specifically, they provide a unique window for examining mechanisms by which plants and other organisms interact in their environment to ultimately facilitate their own survival. Over the past several decades there have been numerous published articles and reviews calling for the importance of recognizing the potential benefits endophytes have to offer human society in the areas of biocontrol and bioremediation efforts (Strobel and Daisy 2003; Bacon and Hinton 2007; Weyens et al. 2009b; Yu et al. 2010; Khan and Doty 2011). It is now realized by researchers in plant microtechnology fields that endophytes literally represent a treasure trove of unexplored microdiversity and ecology, and as a result research literature regarding their abilities is continuing to mount worldwide.

1.2 Identifying Endophytes in a Plant: Endophyte Diversity and Its Challenges

Scientific study of the taxonomy of endophytes is really just beginning to shed light on their complex diversity and ecology (Surette et al. 2003; Tejesvi et al. 2007; Taghavi et al. 2009; Yousaf et al. 2010b). Often research focused on acquiring endophytes for future biocontrol or bioremediation efforts will begin with a survey of the endophytes currently inhabiting the plant(s) of interest. From these, selected endophytes can then be screened for specific abilities, in the hopes of

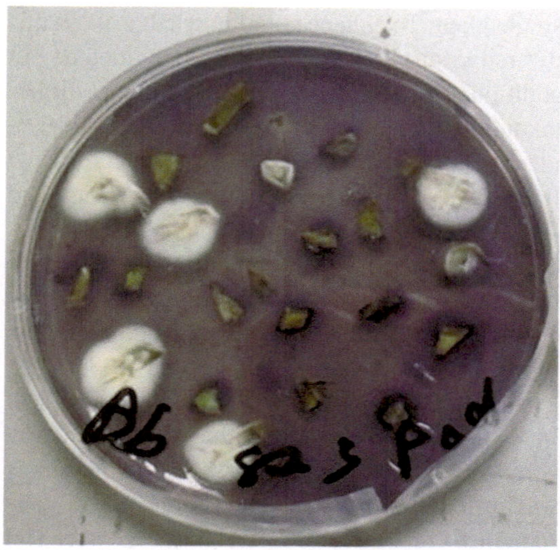

Fig. 14.3 Example results from the use of a traditional isolation culturing method using aseptic techniques. Fungal endophyte (*Beauveria bassiana*) is shown growing from surface-sterilized Venus flytrap plant material after placement on antibiotic containing media (Doberski and Tribe 1980) selective for the isolation of fungal species (Photo courtesy Mary R. Griffin, unpublished data)

isolating potential biocontrol or bioremediation agents (Bacon and Hinton 2007; Hardoim et al. 2008; Yousaf et al. 2010a). Novel endophytes with specific abilities enabling them to deter insect feeding, antagonize a plant pathogen, degrade a toxic compound, or hyperaccumulate metals internally are continually being identified through these surveys. In reality however, the results of these studies often represent only a small portion of the plant's total endophyte diversity. Currently the isolation of endophytes for identification from even well-known plants can be difficult. Traditional culturing methods, which isolate endophytes from surface-sterilized plant material plated on media, only allow for the identification of endophytes, capable of growing on that particular prepared media. It is well known that the ingredients in culture media can impact endophytic growth (Figs. 14.2, 14.3, and 14.5). In addition, endophyte surveys often overlook slow-growing or unculturable species, such as obligate biotrophs. Because of these variables, it is often difficult to know, with complete assurance, all the different endophyte species that are present in a single plant. To enhance our knowledge about this diversity, survey studies now often include, in addition to morphological identification, further taxonomic techniques such as gene sequence identification methods (16S rRNA, 16S rDNA BOX-PCR), carbon source utilization tests, and fatty acid methyl ester profile analysis. For example, in a recent study of the endophytic bacteria of the aerial parts (stems and leaves) found in a crocus wildflower (*Crocus albiflorus*), a combination of plating isolation techniques and 16S rRNA gene sequence identification of isolates was used. In this study it was found that the community composition of culturable microbial communities was different from the results found using gene sequence identification. Only three bacterial divisions were found in the culture collection, whereas six divisions were identified using gene sequence identification. Also of note, it was found that *C. albiflorus* supports diverse bacterial communities. Some of the identified communities had been described previously in association with other plants, but the *C. albiflorus* in this study also contained species that had not been described in association with plants before (Reiter and Sessitsch 2006). Obviously for any technology that would utilize beneficial endophytes, it would be important to know about the other endophytes also colonizing the plant(s) of interest. Additional endophytes could greatly impact (positive or negative) the results of a biocontrol or bioremediation effort. In past work with endophytes, this has not always been possible; however, in current studies work is now done to determine if the plant(s) of interest is colonized with endophytes that may alter the results of a study. Concerning total endophyte diversity, we've only begun to scratch the surface but innovations in survey technology capability and the mapping of genomes of many critical bacteria and fungi isolates should help alleviate many of these issues in coming years (Porras-Alfaro and Bayman 2011).

1.3 Developing a Working List of Common Endophytes

Despite the overwhelming diversity of endophytes, there is now a working list of organisms consistently identified in plant–endophyte taxonomic studies, and as such, these organisms have become associated with their ability to utilize the endophytic lifestyle. Endophytic bacterial species commonly isolated from plants often include members from the *Acetobacter*, *Arthrobacter*, *Bacillus*, *Burkholderia*, *Cellulomonas*, *Clavibacter*, *Curtobacterium*, *Enterobacter*, *Herbaspirillum*, *Pseudomonas, Streptomyces,* and *Microbacterium* genera (Lodwyckx et al. 2002). This list however is continually expanding. Endophytic fungi are also exceptionally broad in their biodiversity. These endophytes are typically from the ascomycetes and to a lesser degree the basidiomycetes.

2 Variables Which Can Impact the Success of Biocontrol or Bioremediation Efforts

As discussed earlier, all plants are thought to host one or more endophytic bacteria or fungi, with the more common being multiples of both; however, beyond this, much of our current understanding about the multitude of species that cultivate the endophytic lifestyle and their interactions between plant hosts and one another is limited. Reasons for these limitations are many but include the fact that endophytic communities, population numbers, and interspecies dynamics occurring within plants can differ depending on plant species, cultivar, plant part (leaf, stem, or root), and any number of environmental conditions, such as drought, low nutrient availability, pathogen–host interaction, or pollution presence (Gange et al. 2007; Tan et al. 2003; Yousaf et al. 2010a; Khan et al. 2011). A recent study demonstrated that even individual differences between plant genotypes can impact the type and ratio of endophytic communities colonizing plant tissues in a relatively small geographical area (Bálint et al. 2013). Some endophytes may colonize only one or a few species, while others may cultivate numerous species. In addition some endophytes may only be associated with one area of the plant, whereas others may be found throughout (Jumpponen and Trappe 1998). Because many endophytes live only part of their lives inside plants, it is likely that plants become colonized long term by endophytes capable of thriving within not only the internal plant environment but the external as well.

To complicate this, there is also increasing evidence that plants can be involved in the recruitment or shedding of endophytes (Siciliano et al. 2001). Plants also have their own attributes which limit harmful pest/pathogens (Cowan 1999; Cortesero et al. 2000) such as specific disease resistance genes which code for the production of compounds with direct activity against pest/pathogens, some of which we have exploited for agricultural gain (Martin et al. 2003) as well as numerous chemical pathways by which some species have the ability to accumulate or breakdown many harmful toxins in the environment (Peer et al. 2005).

2.1 Confusing and Sometimes Conflicting Results Can Occur When Working in Biological Systems

Work with biological systems is by its nature very difficult because scientists must account for innumerable variables which occur between interacting organisms. As a result biocontrol and bioremediation agents that have been under study for numerous years may still produce highly variable results when observed in a field study. Therefore, the large-scale practical application of these bio-agents still has numerous challenges (Weyens et al. 2009b). Knowing some but not all, of the interactions occurring in a biological system can cause unforeseen difficulties and confusion about how best to proceed. For example, in a study related to understanding the impact of long-term biocontrol efforts, endophytes from two types of corn plants were examined; control plants and corn plants selectively breed for increased production of benzoxazinoids (BXs). The BX-producing corn plants contain compounds that are known to be toxic to many microbes and insects. This toxicity to microbes and insects was assumed to be advantageous to growers because they could use fewer pesticides. However, upon comparison of the two types of corn plant–endophyte communities, it was found that they differed greatly. The corn plants with increased BX compounds were actually encouraging the selection of tolerate BX endophytes, which unfortunately happened to be *Fusarium* spp., (Saunders and Kohn 2008). *Fusarium* is a genus of well-known fungi which have been especially problematic to agriculture because of their production within plant tissues of mycotoxins that are harmful to humans and animals if consumed (Thane et al. 2004). In another study related to bioremediation processes, certain plants or bacteria have been shown to have specific genes enabling them to use toxin-degrading pathways for organic compounds. However, in some cases when the compounds are taken up by the plant, the toxins are only partially degraded. As such the compounds are typically stored within the plant's tissues and could pose a danger to herbivores if eaten. Studies such as these detail the complexity of plant's microbial ecology.

3 Changing Perceptions Regarding Fungi and Bacteria in General and Endophytes in Particular

As mentioned in the introduction, the term endophyte is under some debate particularly regarding biocontrol technology. This debate stems from the fact that the term refers primarily to the location of an organism living within a plant, without causing disease; however, as such this definition could still include latent pathogens (organisms which cause disease after a specific period) or opportunistic saprotrophs (organisms that feed on nonliving organic matter). In addition, many may limit the term "endophyte" to specific taxonomic groups primarily for the purpose of making a study or review more manageable. Lastly the terms used to describe the endophytic condition (mutualistic etc.) can also be a source of confusion for the novice as the descriptive terms can change depending on the human or plant association. Despite this, endophytic activities by fungi and bacteria have had a long history of study in biological research literature and the use of mutualistic endophytes in modern biocontrol and bioremediation efforts can be directly tied to these earlier studies.

3.1 Brief History Behind Endophytes and the Fluidity of Symbiotic Titles

Plant–bacterial and fungal relationships actually have been extensively studied for a long period of time, and the results of these have added to our overall understanding of these complex relationships. Much of what was learned about endophytes early on involved the discipline of crop science and was used primarily to improve agricultural crop yields. As a result work with bacteria and fungi that formed symbiotic internal relationships in plants examined plant–pathogen interactions. The plant science discipline, known as plant pathology, actually evolved in part due to the tragedy of the infamous Irish potato famine, which occurred in the mid-1800s. The famine which lasted over a period of several years was the result of potato blight, a condition of rapid rot caused by the plant pathogen *Phytophthora infestans*. The famine was historic because of the millions of people that died of starvation or were forced to migrate in mass to other regions of the world (Fraser 2003). Consequently at this time in history, our only understanding of plant relationships with bacteria and fungi was negative, and the idea of intentionally inoculating plants with bacteria and fungi for their benefit would take time to develop in society. However, thanks to the work in plant pathology, there is now enhanced understanding about the many organisms that cause plant disease and the mechanisms they employ (Hammond-Kosack and Jones 2000). Much of this previously acquired knowledge is now used in current biocontrol studies. In other words, information from the negative aspect of the symbiotic spectrum now provides researchers with information upon which to build a better understanding of beneficial plant symbionts. It has been suggested that what makes endophytes especially suitable as biocontrol agents is their colonization of an ecological niche similar to that of plant pathogens (Ryan et al. 2008). Furthermore, some plant pathogens may actually be considered beneficial as several have actually been sought as biocontrol agents for use against weed plants. Justification for their use is related to the phasing out or cancellation of many widespread commercial pesticides and herbicides (Charudattan 2001), several of which have through their continual use built up in the environment and are part of the many contamination site pollutants which now must be addressed through the use of bioremediation efforts.

As understanding continued to develop regarding the range of symbiotic spectrum activities and what individual symbiotic organisms in biological systems could do, recognition of beneficial microbes by scientists began to emerge. For instance some of the earliest beneficial symbionts recognized in the plant sciences were the root-colonizing endophytes known as Rhizobia bacteria. These bacteria are associated with legumes (Fabaceae) and produce nodules in the roots wherein they live and fix unusable nitrogen into a usable form for the plant. Nitrogen typically (unless artificially supplied) is a limiting factor that restricts growth in plants. Rhizobial activity within legumes is still one of the best known examples of a plant growth-promoting activity (PGPA), taught now in most basic plant science classes. This early work led to further understanding in the area of microbes and nitrogen fixation. Now we know there are many kinds of nitrogen-fixing bacteria found in soil that do not require an endophytic relationship with legumes. These organisms are known collectively as diazotrophs and help make up the beneficial rhizobacteria located in the critical rhizosphere zone of plant roots. Diazotrophs can provide protection by occupying the rhizospheric area to the exclusion of pathogens attempting to invade. Plant growth promotion using nitrogen fixation endophytes is still actively sought in research for both biocontrol and bioremediation efforts. Many diazotrophs that do not have to form nodules to fix nitrogen can also become endophytic and some provide assistance to plants through their PGPAs at contamination sites.

For instance, endophytic diazotrophs, such as *Pantoea* and *Azospirillum* sp., were recently isolated from a large diversity of plants (Verma et al. 2001; Loiret et al. 2004; Xin et al. 2009).

Also instrumental in the scientific community's increased awareness of the complexity of plant symbiotic relationships are the well-known beneficial root colonizers, arbuscular mycorrhiza fungi (AMF). These symbionts internally colonize the root cortex of plants and develop branched networks that aid plants in capturing macronutrients such as phosphorous or nitrogen as well as other micronutrients. It is understood that the type of symbiotic relationship AMF form with plants has been occurring for a very long time as evidenced by paleobotanical fossils of plant roots containing the arbuscular structures of the fungi. Today AMF continue to form relationships with a large portion of land plants, and several articles and reviews describe the demonstrated success of arbuscular mycorrhizal associations in bioremediation and biocontrol efforts (Al-Karaki and Al-Raddad 1997; Augé 2001; Denton 2007; Compant et al. 2010). Plants have prophylactically benefited from the presence of AMF by enhancing their ability to control numerous plant pathogens such as *Fusarium*, *Rhizoctonia*, *Verticillium*, *Pythium*, and *Sclerotium* (Azcón-Aguilar and Barea 1996). Typically, however, mycorrhizal fungi are generally not considered true endophytes because as a group they are identified by the specific function they provide for a plant and they are grouped in the fungal phylum Glomeromycota (Porras-Alfaro and Bayman 2011).

More generalized studies of plant–bacterial and fungal relationships also examined how rhizospheric and endophytic microbes could facilitate plant's overall ability to survive in challenging environments through PGPA other than nitrogen fixation. Now widely recognized, these PGPAs incorporate a wide variety of activities done (1) by the microbe(s) solely or (2) by the plant in response to the microbe(s) or vice versa or (3) in unison with the microbe (Strobel 2002; Tejesvi et al. 2007). Various endophytes have been shown to increase plant growth through a number of mechanisms such as drought tolerance (Bacon 1993), phosphate solubilization activity (Verma et al. 2001), or through the production of functional metabolites (Tan and Zou 2001) as in essential plant vitamin production (Pirttila et al. 2004), indole acetic acid (IAA) production (Xin et al. 2009), and production of siderophores (Rajikumar et al. 2010).

3.2 True Endophytes

The potential importance of beneficial endophytes to plants and biotechnology really did not become clear until 1975, when Charles Bacon discovered fungal endophytes in the family Clavicipitaceae growing systemically in pasture grasses (Bacon et al. 1977). Subsequent studies revealed that these endophytes actually aided plant survival through a variety of PGPAs (Arachevaleta et al. 1989; Bacon 1993; Latch 1993) and that the endophytes could be passed from parent plant to seed (Clay 1987). The reason behind the initial work performed by Charles Bacon (how his work was supported) highlights some of the greatest challenges in using endophytes for the benefit of plants. In a situation related to the issue found with high BX corn plants discussed earlier, the fungal endophytes Bacon recognized also produced alkaloids within their hosts, and as such the colonized pasture grasses were toxic to cattle, and the associated syndromes caused by their consumption had a highly negative financial impact on the livestock industry. It turned out that when the endophytes produced toxins, it was good for the plant because it deterred insect feeding (Johnson et al. 1985). Of note, there are now biocontrol commercial formulations of clavicipitaceous endophytes of grasses that protect turf grasses from insect pests (Clay 1987). By humans, clavicipitaceous endophytes are deemed beneficial in turf grasses, but negative in pasture grasses; however, in both cases, the plant would consider the fungi beneficial. Thus, symbiotic titles are defined not just by the plant but also by the humans who utilize the endophyte and plant.

For the general public however, even today bacteria and fungi are viewed as negative regardless of type or potential. For example, if questioned today about endophytes, a general layperson would likely view endophytes as potential contaminants in food, even though all food (industrial or organic) typically contains some fungi or bacteria. In contrast a layperson's response concerning bioremediation efforts and the use of endophytes will likely have a more generally positive response. This would likely arise from the assumption that food and pollutants are not associated; however, many environmental pollutants come from agricultural-related chemicals used on food.

4 Biological Control (Biocontrol) with Endophytes

The definition of a biocontrol (biological control) agent is the use of a living organism or their metabolic by-products to control pests or disease-causing pathogens. Part of the appeal of biocontrol technology is that it offers growers a natural means of control and can lessen dependency on chemical pesticides. The ecological risks of synthetic chemicals such as pesticides and herbicides used in crop protection in edible portions of foods are a cause of grave concern by the public and their governments (US National Research Council 1993). Because of this, much of the research on the use of endophytes as biocontrol agents has focused on important commercial crops which have historically been plagued with insects or pathogens that cause tremendous economic loss. In the past to ensure good yields, these high-value crops were treated with solely chemical pesticides, now however due to increasing pest and pathogen pesticide resistance, even reluctant-to-change growers are seeking workable alternatives or augments to their traditional chemical treatments (Copping and Menn 2000). Changes in agriculture growing practices are evidenced by widely enacted integrated pest management programs (IPM) and policies worldwide. As public understanding has increased regarding the multiple benefits symbiotic organisms can provide, their general use has become more common (Zhuang et al. 2007). Currently there is an amazing amount of literature regarding the utilization of bacterial or fungal–plant symbiotic partnerships for biocontrol efforts; however, many of these do not examine solely endophytes but rather also include colonizers of plant's external surrounding areas such as the rhizosphere (Weller 1988). In addition the use of endophytes for biocontrol activities is still limited for several reasons: (1) there is public concern over potential negative consequences, and (2) many of the relationships formed with plants by microbes do not necessarily require them to colonize the plant to work effectively, as such endophytes are a subgroup of the total biocontrol agents examined in the scientific literature. However, because endophytes are naturally present in all plants and within this unique niche they are sheltered from external environmental conditions, they can in some cases be potentially more effective biocontrol agents than microbes in external niches. Also considering their diversity, endophytes could be used as biocontrol agents with any crop, genetically modified or otherwise, against a multitude of pest/pathogens (bacterial, fungal, insect, mites nematodes, and others), in most environment. The problem is finding the right endophyte(s) or rather the right combination of organisms, endophyte and plant type, to grow in suitable environmental conditions. All of this relates directly back to our lack of knowledge about endophyte biodiversity, community, and population dynamics between plant hosts in addition to variable abiotic environmental influences.

4.1 Mechanisms of Action Used by Effective Biocontrol Agents

Biocontrol activity by endophytic organisms can be viewed as working indirectly or directly. Indirect action may be the simple occupation of space by an endophyte within the interior of a plant host which prevents a pathogen from establishing itself. Direct biocontrol activity emphasizes

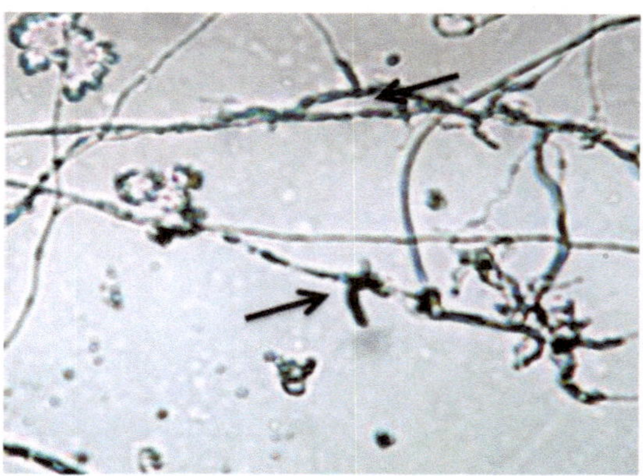

Fig. 14.4 The biocontrol mechanism of hyperparasitism in fungi is sometimes accompanied by hyphal coiling. In this figure, the biocontrol agent *Beauveria bassiana* (larger hyphae) is observed coiling around the plant pathogen *Pythium myriotylum* (Photo courtesy of Mary R. Griffin)

the interaction between the biocontrol and the pest/pathogen, such as consumption of the pathogen. There is general agreement that the mechanisms used by biocontrol agents to antagonize plant pest/pathogens can be divided into four broad categories: (1) antibiosis, (2) competition, (3) parasitism, and (4) induction of plant defense systems (Handelsman and Stabb 1996). (1) Antibiosis can specifically refer to the production of antibiotics, but it also applies more broadly to any metabolized compound capable of killing (–cidal) or inhibiting (–stasis) the growth or reproduction of organisms, such as the production of cell-wall degrading enzymes; (2) competition for space and nutrients, also known as niche exclusion, involves the inhibition of a pathogen either through substrate denial or by outcompeting the pathogen through the uptake of nutrients necessary for it to establish; (3) parasitism or hyperparasitism refers to the parasitizing of a parasite (Fig. 14.4); and (4) induction of a plant host's normal defense responses involves triggering reactions in a plant that would normally happen if the plant was in actuality being attacked by pest/pathogen. This type of response can be local or systemic throughout the plant. Examples of this type of biocontrol activity would be initiated by causing a local hypersensitive response (HR) that results in a rapid death of cells in a small area surrounding the initial infection site; this type of response will typically precede a slower systemic response known as systemic acquired resistance (SAR) or, depending on the chemical pathway used by the plant, an induced systemic resistance (ISR), in which the plant produces a similar response, but no HR response is needed for it to occur. Both SAR which uses salicylic acid and ISR which uses jasmonic acid can stimulate the production of numerous plant defense-related compounds, such as terpenoids and peroxidase, which are needed to help overcome a pest or pathogenic attack (Han et al. 2000; Ownley et al. 2008).

4.2 Surveying and Screening Biocontrol Agents Using Mechanisms of Action

Specific examples of bacterial and fungal endophytes using these mechanisms of control abound in the endophyte biocontrol literature (Yu et al. 2010). For example, hyperparasitism has been well documented for the relatively common endophytic fungi, *Trichoderma*. For this fungus the mechanism involves the secretion of chitinases and cellulases upon contact with the pathogen. Subsequent coiling of hyphae around the hyphae of the pathogen enables the biocontrol to enzymatically digest the pathogen's cell walls (Baek et al. 1999; Russo et al. 2012). Most biocontrol agents however do not antagonize plant pests/pathogens with only one mechanism of action and will often use several mechanisms to inhibit or control pest or pathogenic invaders. For example, in survey studies conducted for the purpose of finding new potential biocontrol agents, plants will often be examined for their natural endophytes for two solid reasons: (1) plants often will cultivate beneficial microbes when they are available and (2) these naturally occurring endophytes are already known to be competent, meaning they have already successfully colonized and can thrive in the plant of interest. These endophytes once isolated (Fig. 14.3) will then be screened in order to determine if the plant naturally hosts endophytes which can help defend it against

Fig. 14.5 Antibiosis challenge using an in vitro assay. In this assay the antibiosis activity of an unidentified actinomycete is examined as it grows on three different nutrient media and is challenged by a common plant pathogen. The actinomycete could survive on each substrate; however, when challenged by the fungal–plant pathogen *Aspergillus niger*, its ability to inhibit the fungus was linked to the media upon which both organisms were growing. In vitro assays are also used in bioremediation studies to determine plant and microbial tolerance of toxic pollutants (Photo courtesy of Mary R. Griffin)

harmful organisms. Screening endophytes with the use of in vitro assays are typically some of the first activities used in surveys for biocontrol or bioremediation agents (Fig. 14.5). For example, in a study conducted for the purpose of examining the endophytes which naturally colonized rape seed (*Brassica napus*), numerous bacterial and fungal specimen were isolated. Following their isolation, the endophyte's abilities were examined using in vitro assays, and all of the microorganisms isolated were found to be suppressive against the fungus *Verticillium dahliae*, a major soilborne pathogen, which causes leaf discoloration, wilting, and at times death in the crop plant. Further examination of these isolated bacteria was done, using seedlings grown in gnotobiotic assays (gnotobiotic assays are very useful when examining any potential mutualistic endophyte because the assay can be strictly limited to examine the interactions of the plant and endophyte; see Fig. 14.6), and showed that as a group the bacterial community exhibited a wide range of biocontrol mechanisms, including the ability to stimulate an HR response, as well as chitinolytic, cellulolytic, proteolytic, and phospholytic enzyme activities. In addition many of bacterial isolates examined were able to individually exhibit multiple antagonistic activities (Alström 2001).

Mechanisms utilized by biocontrol agents may also differ based on the type of pest/pathogen that is attacking the plant. In a survey study of fungal endophytes isolated from healthy *Theobroma cacao* tissues to determine potential isolates with in vitro antagonism against several major pathogens, it was found that several isolates demonstrated antagonism against *Moniliophthora roreri* (frosty pod rot), *Phytophthora palmivora* (black bud rot), and *Moniliophthora perniciosa* (witches broom). The most common antagonistic mechanism observed was substrate competition; however, some isolates did show antibiosis activity and one isolate of *Trichoderma* was parasitic on *M. roreri*. In later field work, a selected endophyte from the same study, *Colletotrichum gloeosporioides,* was shown to significantly decrease pod loss due to black pod rot (Mejía et al. 2008). In another survey study also done to screen for potential biocontrol agents for potato (*Solanum tuberosum*) plants, six fungal endophytes were selected for suppression of *Rhizoctonia solani*, an important soilborne pathogen which requires the use of large amounts of pesticides to control. Out of

Fig. 14.6 An example of a sealed gnotobiotic assay with carrot seedlings. Carrot seeds were coated with a potential mutualistic endophyte and were aseptically planted following the sterilization of test tubes, water, and vermiculite substrate. This technique can also be used to determine if a control or a colonized plant seedling can grow in a polluted system (Photo courtesy of Mary R. Griffin)

the six isolates examined, two *Trichoderma atroviride* and *Epiccum nigrum* showed significant in vitro inhibition of mycelial growth of *R. solani,* and in subsequent greenhouse work, both isolates improved potato yield significantly (Lahlali and Hijri 2010).

Other biocontrol studies may examine solely the efficacy of as specific type of endophyte such as nonpathogenic microbial isolates (harmless isolates of known pathogenic species) which are commonly found in the plants. For instance, the nematode (*Radopholus similis*) is a serious pest in commercial banana (*Musa* sp.) plants, and in a study to determine if colonization and multiplication of the nematodes could be decreased, banana plants were inoculated with four stains of nonpathogenic *Fusarium oxysporum*. Results of the study showed that nematode activity was indeed decreased in the roots and that none of the isolates caused a reduction in plant growth. The exact mechanism of action was not determined however, because the *F. oxysporum* isolates did not demonstrate any direct mechanisms of control (Niere et al. 1999). In another study with cotton (*Gossypium hirsutum*), the ISR mechanism utilized by nonpathogenic binucleate *Rhizoctonia* spp. to protect seedlings against the damping-off pathogen *R. solani* and *Alternaria macrospora*, which causes leaf spot, was found to be effective against both pathogens (Jabaji-Hare and Neate 2005).

Additionally endophytes, which have the capability of dual biocontrol activity (activity against two or more plant pathogens or against insect pests and plant pathogens), are also being sought, in part for their commercial viability. One such example is *Beauveria bassiana*, a well-known commercially available entomopathogen effective against twenty orders of insects but which can also live successfully as an endophyte and has been isolated from numerous plant species (Ownley and Griffin 2012). *Beauveria bassiana* was shown to control the bacterial pathogen *Xanthomonas malvacearum* in cotton using the ISR mechanism (Ownley et al. 2008). As discussed earlier sometimes researchers are not always sure what the exact mechanism a biocontrol agent is utilizing in every study; however, if the biocontrol agent shows promise, further work will likely continue to elucidate their mechanism(s) of action; in addition previous information collected from years of research can often help narrow the search parameters. This information is also used to help researchers select agents with "good potential." For example, the bacterial strain of

Paenibacillus polymyxa (SG-6) was shown to be effective at controlling the post-harvest disease caused by the green mold (*Penicillium digitatum*) on citrus fruit. The strain (SG-6) was originally isolated from the roots of *Sophora tonkinensis* (an herb used in traditional Chinese medicine) and identified using 16S rDNA gene analysis and Biolog tests. The bacterial strain was deemed as a likely biocontrol agent because the species has previously been shown to have antagonistic activity against several plant pathogens in greenhouse and field studies. In the study, further in vitro and in vivo assays were used to determine *P. polymyxa*'s efficacy against *P. digitatum*, and although the precise mechanism of antibiosis exhibited by this strain was not known at the time of the paper's submission, there had been previous studies which showed that the species of *P. polymyxa* had the ability to produce numerous antimicrobial compounds as well as several enzymes key to the control of even resistant plant pathogens (Lai et al. 2012).

4.3 Isolation, Identification, and Application of Biocontrol or Bioremediation Organisms

Many techniques such as traditional isolation, in vitro challenges, and gnotobiotic assays can be used in the screening of potential agents for both biocontrol and bioremediation efforts (Figs. 14.3, 14.5, and 14.6). In laboratory, greenhouse, and field settings, the intentional inoculation of plants with endophytes can occur in a variety of ways. For instance, endophytes can be incorporated (infested) into the soil as amendments and allowed to colonize plants via the roots; other ways include incorporating the endophyte in a liquid and dipping plant roots into the solution prior to planting (root dip) or by directly coating the endophyte onto plant seed all of which also encourage natural colonization of plant tissue. Other inoculation methods include direct inoculation into the plant via a syringe or for trees, "endotherapy" (trunk injection). For endophytes which enter via the phyllosphere, these are often applied as foliar sprays. Today there are a wide variety of commercial products containing bio-agents. These commercially formulized products are used for a variety of biotechnological efforts and will contain additives described as proprietary mixes. These proprietary substances are added for different reasons such as increasing general efficacy following application (helping them adhere to external plant tissues longer) or to extend the bio-agent's shelf-life expectancy.

5 Bioremediation with Endophytes

In addition to their use as biocontrol agents, endophytes are also being used in another promising area of research, the remediation of polluted soil and water. The United States Environmental Protection Agency (USEPA) defines bioremediation as a treatability technology, which uses biological activities to reduce the concentration and/or toxicity of a pollutant. Bioremediation for the reduction of toxic compounds is an issue of growing concern worldwide especially since the negative effects of many widely manufactured and released substances on human heath are more widely understood. Pollution caused by the release of toxic contaminants has become a tragic norm in our modern-day society. Many older contaminated areas are the result of contaminant releases which occurred prior to the understanding of the product's true detrimental long-term impact. Unfortunately many releases still occur today due to a variety of reasons ranging from breakdowns in oversight to sheer ignorance. Some releases (past and present) are even intentional because it is cheaper to pollute than pay for containment of the contaminant. Accidental or otherwise contaminant releases commonly occur around point sources such as industrial and agricultural sites. Following their release from a point source, these contaminants can then migrate outward causing a widespread plume as they are carried. In order to make the soil and water of these contaminated areas safe again for human-related activities, remediation efforts must be done. Unfortunately traditional (ex situ) or removal-focused remediation technologies, especially for contaminated soils, are environmentally invasive, time consuming, and financially costly. These include excavation, transport to specialized landfills, incineration, or stabilization (Weyens et al. 2009a). Because of the high cost of these traditional remediation techniques, degradation or accumulation on-site of toxic compounds in environmental soil and water by plants and their associated organisms are being actively investigated. Bioremediation (in situ) technology is generally agreed to be potentially a much cheaper alternative. Therefore, the use of microscopic organisms capable of degrading or accumulating toxic compounds in combination with specific host plants could offer an efficient, economic, and sustainable remediation technology for the future.

5.1 Types of Bioremediation

Bioremediation can generally be broken into two main types: intrinsic and engineered. Intrinsic bioremediation is also known as natural attenuation or passive bioremediation. Out of the two, intrinsic would be preferred to engineered bioremediation, primarily because its cost is much lower. Intrinsic bioremediation consists of allowing the naturally occurring organisms in the area of concern to degrade or accumulate the contaminants without implementing any additional steps to enhance the process. This type of bioremediation does naturally occur all the time worldwide without the addition of human efforts; however, before polluted land or water areas can be left to solely intrinsic mechanisms, three basic requirements must be met. The site must (1) already contain a sufficient number and type of microorganisms that can biodegrade the contaminant, (2) have good environmental conditions for growth and sufficient availability of nutrients to maintain the current populations, and (3) relate to the allowable time that the natural processes will be given to work. In other words, the land or water is essentially unusable until the natural bioremediation process is completed. Meeting all three of these requirements is often difficult for sites which have large areas of contaminations or high concentrations of toxins. Also in many cases, especially for polluted soils, if contamination levels are too high, then microbial populations tend to be decreased (Zhang et al. 2012), which would greatly affect the site's remediation timeline. Concerning the second requirement, particularly with the use of endophytes for intrinsic bioremediation efforts, environmental conditions and adequate nutrient attainment would be directly linked to overall plant survival, but many plant species are sensitive to pollutants.

Furthermore, studies have shown that even plants deemed to be tolerant to a site's contaminant usually do not grow well if pollution levels are high enough (Glick 2005; Germaine et al. 2009; Zhang et al. 2012). For example, in areas contaminated with hydrocarbons (well-known toxic compounds associated with the production of petroleum), plant growth and development are often inhibited, and due to hydrocarbon's hydrophobic properties, plants and microorganisms attempting to grow at a site have reduced ability to absorb water and nutrients from the soil (Khan et al. 2012). The final requirement for intrinsic remediation addresses the requirement of adequate time duration, which often is not feasible. Even with engineered bioremediation strategies, natural processes usually take a long time to work. Intrinsic bioremediation efforts therefore can only be used in limited situations.

Regardless of the method chosen for the remediation of a site, ex situ or in situ by intrinsic or engineered methods, there are also safety requirements which have to be met. All sites need to be monitored in order to verify that their contaminant plumes are not continuing to spread and that they and are indeed being reduced overtime. To accomplish this, networks of monitoring methods are required to determine the location and concentration of the contaminants over time, the relative numbers and types of microbes present, and other appropriate parameters.

5.2 Technologies of Bioremediation

Engineered bioremediation processes are similar to intrinsic methods in that they still rely on biotic activity to reduce contamination, but this method allows for the enhancement of the natural processes of degradation typically by modifying one or all of the three previously mentioned site requirements. Bioremediation is actually a broad descriptive term that includes several linked technologies such as phytoremediation which is used in intrinsic methods as well, but in engineered methods it involves the intentional selection and planting of specifically tolerant plants for the toxin(s) present and using them as hyperaccumulators, to take up the toxic compounds and partially degrade or store them within their internal tissues. Phytoremediation depending on the contaminants can involve such strategies as phytodegradation and phytovolatilization both of which can be used with substances such as synthetic organic compounds. Phytoextraction would be the term used to describe the removal of toxins such as heavy metals and metalloids by plants as they accumulate them internally (Peer et al. 2005). Other technologies include bioventing which uses microorganisms to degrade organic compounds in groundwater; bioleaching, which involves accumulation of metals by microorganisms; bioaugmentation which would involve the intentional release of a natural or genetically engineered microorganism to a site; or biostimulation which would involve some modification of the environment, such as the release of nutrients which can stimulate microorganisms in the area to act as better bioremediators. Bioremediation using endophytes often will link many of these technologies together. For example, with phytoremediation, the intentional planting of tolerant plant species, with extensive root systems is known to facilitate biodegradation processes. Furthermore, bioaugmentation could be used through the inoculation of soil with microorganisms known to tolerate or degrade a pollutant in the ground or through the intentional inoculation of plants with endophytes to assist them to degrade accumulating toxic compounds coming into the plant. Biostimulation technologies are also often utilized at sites of concern because introduced organisms need to be compatible with the existing conditions of the environment, such as physicochemical properties like water, oxygen, pH, and nutrient levels and temperatures. In some cases it may be possible to alter these in order to facilitate biodegradation activities.

Monitoring of engineered bioremediation sites is also critical to determine the progress of remediation. Engineered bioremediation monitoring techniques will vary based on the technologies used, for instance, in a case utilizing bioaugmentation with introduced genetically engineered endophytic remediators; monitoring gene abundance and expression along with the plant health of hosts used in phytoremediation of contaminated soils will provide evidence about gene persistence and functional activity of the applied microorganisms. Quantitative PCR (qPCR) could be also used to monitor the presence of specific organisms and functional activity of the degrading gene.

5.3 Studying Bioremediation Using Plants and Their Microorganisms

To date symbiotic interactions among microorganisms and plants have received much attention because together they have been shown to be capable of enhancing pollutant degradation. Several contributions have already been documented using organisms associated with plant rhizosphere (Franks et al. 2006; Russo et al. 2012). However, in work with endophytes, researchers have also found that many endophytes commonly dwell in the rhizosphere before moving into plant internal tissues. Therefore, plants and their associated microbial communities can have a great impact on the success of any bioremediation effort. Currently, however, the overwhelming majority of studies which involve the examination of benefits with endophytes in bioremediation have been successfully applied only in laboratory scale experiments, and large-scale field applications are still limited (Weyens et al. 2009a). The use of microbes in general and endophytes in particular for bioremediation efforts is still relatively new, but from the evidence presented in early works, it is clear that there is strong potential for mutualistic symbionts to make a significant contribution toward cheaper sustainable bioremediation efforts.

5.4 Bioremediation Success with Endophytes Is Linked to Plant Health

Plant's exterior symbiotic relationships are associated with the overall health of the plant and can greatly impact the success of any bioremediation effort with endophytes. Regarding the importance of plant health in the facilitation of biotechnologies, it is known that plants cultivate a microbial community around their roots. Growing plant roots leak relatively large amount of sugars and additional compounds such as organic acid, amino acids, and vitamins into the surrounding environment. These leaked compounds attract and feed other highly varied populations of competing microbial species. The root zone is the area where bioremediation activities mostly occur. This area is known as the "zone of effectiveness" and impacts the success of bioremediation techniques depending on the location of the contaminant in the environment and the depth it has reached. This is why not only tolerant plants are sought in areas of concern but also those with extensive root systems are deemed valuable. Thus, plant type but also overall good plant health is essential for encouraging colonization by plant symbionts. Endophytic colonization may actually be an enormous advantage to microbes. It has been suggested that the unique niche of the interior plant environment may allow the desired endophyte strain to reach larger population sizes due to reduced competition for nutrients and space and being physically protected from adverse changes in the environment.

5.5 Advantages for a Plant to Have Remediation-Capable Endophytes When Growing at a Contaminated Site

The use of bioaugmentation with endophytes for remediation of many toxic compounds found in both soil and water has been studied, and overall efficiency of remediation efforts has been shown to be associated with the survival of inoculated toxin-degrading organisms located in the rhizosphere and endosphere of the plant (Khan et al. 2012). It has also been shown that artificial inoculation of plants with desired endophytes does enhance plant resistance to contaminant stress and increases their acclimation rate and biomass formation. Also as discussed above plants through the natural release of carbon sources can enhance the external microbial population numbers in the rhizosphere as well as the internal applied microbial population numbers which would enhance their overall degradation or accumulation potential. Another important advantage of using endophytic pollutant degraders is to provide continuation of degradation efforts within plants following uptake of the contaminant. Many plant hyperaccumulators cannot completely degrade toxic compounds and may store these partially degraded contaminants internally. For instance, chlorobenzoates are toxic metabolic intermediates produced from biodegradation of a variety of compounds that are still considered environmental contaminants (McGuinness and Dowling 2009). Endophytes which could continue the degradation process internally would reduce phytotoxicity and decrease the overall toxic effect of the pollutant on any herbivorous fauna (Newman and Reynolds 2005). In addition endophytes are also helpful with some contaminants, which when only partially degraded by plants internally can become volatized and released. Endophytes have been shown to be particularly helpful in reducing evapotranspiration of pollutants from plants (Weyens et al. 2009c).

5.6 Bioremediation Efforts Facilitated with PGPAs and the Challenges of Endophytic Diversity

As discussed earlier a better understanding of all plant symbiotic partnerships is needed to fully exploit their abilities to enhance the remediation of contaminated soils and waters. Essentially any activity by an endophyte which can support a plant's efforts to survive in a contaminated area will also promote the overall bioremediation effort of the plant as well as aid other microbial organisms (epiphytic or rhizospheric) that are dependent on the plant for survival. Endophytes capable of PGPAs, like those used by biocontrol agents, have also been found at numerous sites requiring remediation efforts. These PGPAs include nitrogen fixation, production of plant growth-promoting hormones, and modification of sugar-sensing systems in plants. In addition to these there is also a particularly important activity used by some endophytes to assist their host plants in overcoming stress responses which result in elevated ethylene levels. These beneficial endophytes produce an enzyme known as ACC deaminase which can decrease harmful ethylene levels within their hosts (Glick 2005; Arshad et al. 2007, 2008). As in biocontrol efforts, PGPAs by microbial organisms are important to the success of either intrinsic or engineered bioremediation efforts; however, PGPAs in bioremediation can also include the ability to degrade toxic compounds in the environment. For example, in a study by Dashti et al. (2009), it was reported that endophytic bacteria possessing both hydrocarbon degradation and nitrogen fixation capabilities could enhance hydrocarbon degradation without adding any nitrogen source in hydrocarbon-contaminated soils. Endophytic bacteria have also been shown to indirectly improve plant growth in contaminated soils by reducing the growth and activity of pathogens through competition for nutrients and space and through stimulation of plant resistance mechanisms. A recent example of this type of biological system support was observed in an experiment with pea plants (*Pisum sativum*) inoculated with bacteria capable of degrading an herbicide. At the end of the study, the experimental pea plants had no accumulation of the toxin within their tissues and experienced little phytotoxic effects; however, control plants had accumulation of the compound internally and decreased overall biomass. The researchers also reported that a large rhizosphere population was present and were in part responsible for the enhanced degradation of the compound (Germaine et al. 2006). In a similar fashion as biocontrol surveys, bioremediation surveys are conducted in sites where plants and their microbial populations are currently under selective pressure. These studies often focus on endophyte species diversity and ecology but also seek organisms that contain unique genes which code for specific pathways to degrade environmental pollutants (Taghavi et al. 2009). For instance, in an examination of the endophytic bacteria in poplar trees (*Populus* spp.) growing at a BTEX-contaminated field site in Belgium, several survey and diversity studies were conducted. As a result many bacteria were isolated and characterized by genetic analysis, substrate utilization, and sensitivities to antibiotics and heavy metals. The work demonstrated that the bacterial communities found in poplar trees were very diverse, and from these, several endophytic strains were isolated with bioremediation capabilities of volatile organics and herbicides (Germaine et al. 2006; Porteous-Moore et al. 2006). In another survey of endophytic bacteria with hydrocarbon-degrading abilities, Italian ryegrass (*Lolium multiflorum*) was found to host a high number of endophytic bacteria belonging to diverse phylogenetic groups; however, birdsfoot trefoil (*Lotus corniculatus*) did not host such large numbers. A metagenomic study of the endophytic bacteria isolated from the roots of rice, grown in uncontaminated soil, showed hydrocarbon degradation potential (Sessitsch et al. 2012).

Often in bioremediation genetic surveys, researchers are seeking organisms which have unique genes which code for pathways by which degradation of toxic compounds can be accomplished. For instance, in a study examining the prevalence of two genes metabolically active in hydrocarbon degradation (alkB and CYP153) within the external and internal plant microbial community growing in hydrocarbon-contaminated soil, it was found that bacteria carrying these genes could colonize the rhizosphere and the plant interior (Siciliano et al. 2001; Afzal et al. 2011; Yousaf et al. 2011). In addition the genes that actually control endophytic activity within plants as well as enabling endophytic competence (ability to maintain the endophytic lifestyle overtime) within original host plants and alternate hosts are also of interest. Microbes that cannot prosper internally within a plant at a site of concern will likely not be of use long term.

Endophytic activities in original host plants versus alternate host also often vary. Variations in sensitivity and tolerance levels among different plant species or cultivars to pollutants in soils might be linked to differences in endophytic bacterial population and activities. Many studies reveal endophytic bacteria show high colonization and degrading activities in different areas of plants. For example, in a study, examining three different endophytic strains of *Enterobacter ludwigii*, it was demonstrated that different levels of gene abundance and gene-degrading expression occurred within different plants species, at different plant growth stages, and even in different compartments of the plant. In other studies examining the survival and metabolic activities of previously isolated hydrocarbon-degrading bacteria, these endophytes were found to vary distinctly in their biodegrading abilities based on the strains (endophyte), plant species, plant development, and plant region colonized.

6 Pollutants and Bioremediation Efforts

Bioremediation efforts have been shown to be beneficial in the degradation of many environmentally released contaminants. This section looks at some of the contaminants for which many bioremediation efforts attempt to alleviate. Bioremediation research is progressing at a fast rate regarding some of these pollutants; however, research with others is still sparse. Synthetic organic compounds are known to be hazardous and have been associated with numerous environmental contamination sites including polychlorinated biphenyls (PCBs), pesticides, industrial solvents, petroleum products, dioxins and furans, explosives, and brominated flame retardants. The use of many of these pollutants has been generally phased out worldwide; however, they are extremely resistant to natural breakdown processes and can remain stable for even decades. Of these many pollutants, twelve specific organic compounds were listed as persistent organic pollutants (POPS) by the Stockholm Convention on Persistent Organic Pollutants, under the United Nations Environment Program (UNEP), an international agreement enforced in 2004, and include PCBs, nine pesticides (aldrin, chlordane, dichlorodiphenyltrichloroethane also known as DDT, dieldrin, endrin, mirex, heptachlor, hexachlorobenzene, and toxaphene), and dioxins and furans. Effects of exposure to these contaminants in the environment include poisoning of plants and animals, ecosystem alteration, and human health risks, such as increased risks for cancer.

Please note the brief listing of environmental pollutants contained in this chapter is not in any way intended to be comprehensive. The high number of pollutants in our environment is difficult to quantify and beyond the scope of this chapter. Many new contaminants are being released every day for which we do not know the long impacts.

6.1 Hydrocarbons: Major Environmental Contaminants

Subgroups of the petroleum product compounds, which pose serious concerns to human and environmental health, are hydrocarbons which are released as gases, tiny particles, or droplets. Most releases of hydrocarbons into the environment are associated with the use of petrol, diesel, crude oil, and oil products in vehicles used for transportation.

Hydrocarbons can be gases (e.g., methane and propane), liquids (e.g., hexane and benzene), waxes or low-melting solids (e.g., paraffin wax and naphthalene, commercial insecticide), or polymers (e.g., polyethylene, polypropylene, and polystyrene). Also grouped within the hydrocarbons is a group of specific compounds known collectively as volatile organic compounds (VOCs). As a group VOCs can cause short- and long-term adverse health effects. A particular group of VOCs are benzene, toluene, ethylbenzene, and xylene and are collectively known as (BTEX) compounds. Acute exposure to petrol and its BTEX components have been associated with skin and sensory irritation, central nervous system depression, and effects on the respiratory system in humans (McGuinness and Dowling 2009). The presence of hydrocarbon pollutants in the environment can negatively affect plant growth and development as well as soil chemical properties and soil microorganisms' population and activities.

Bioremediation of hydrocarbons with bacterial endophytes is currently an active area of research. The first report that bacteria isolated from the root interior of plants vegetated in hydrocarbon-contaminated soils hosted genes encoding hydrocarbon degradation (Siciliano et al. 2001) stimulated much interest. Several studies also revealed hydrocarbon-degrading endophytic bacteria isolated from different plants vegetated in hydrocarbon-contaminated soils (Yousaf et al. 2010a). Bacteria possessing hydrocarbon degradation pathways and metabolic activities also have been shown to improve plant tolerance to hydrocarbon pollutants by degrading these organic compounds.

Furthermore, endophytic bacteria have been shown to produce various enzymes to degrade hydrocarbons and reduce both the phytotoxicity and evapotranspiration of hydrocarbon volatiles (Khan et al. 2012). In a bioaugmentation study, a genetically enhanced endophytic strain from poplar *Pseudomonas putida* VM1441 (pNAH7) was able to protect inoculated pea plants from the toxic effects of naphthalene. It was also shown that inoculation of plants with this strain facilitated higher (40 %) naphthalene degradation rates compared with uninoculated plants in artificially contaminated soil (Germaine et al. 2009). In another survey endophytes were isolated from hybrid poplar trees (*P. trichocarpa* × *P. deltoides*) growing on a BTEX-contaminated site in Belgium that were shown to be capable of degrading toluene, naphthalene, and the chlorinated organic herbicide (2, 4-D). In a different study a genetically engineered endophytic strain, *Burkholderia cepacia* G4, which contained the pTOM (a plasmid, which encodes a pathway for the degradation of toluene), was shown in laboratory scale experiments to increase yellow lupine plant (*Lupinus luteus*) tolerance to toluene and decrease the transpiration of toluene into the atmosphere by 50–70 %, following inoculation within the plant (Barac et al. 2004).

6.2 Additional Environmental Pollutants

Many groups of synthetic organic explosives including trinitrotoluene (TNT), hexahydrotrinitrotriazine or royal demolition explosive (RDX), and octahydro- tetranitrotetraocine or high-melting explosive (HMX) can contaminate environmental soil. All these compounds have been associated with negative ecosystem and plant, animal, or human health risks. Bioremediation effects with an endophytic *Methylobacterium* isolated from hybrid poplar trees (*Populus deltoides* × *Populus nigra* DN34) demonstrate that it was capable of degrading the explosives TNT, RDX, and HMX. Degradation was accomplished by mineralizing approximately 60 % of the RDX and HMX to carbon dioxide in approximately 2 months, suggesting that this endophyte may have potential for remediation of environmental soil containing explosive nitroaromatic compounds.

6.3 Pesticides and Herbicides Used in Agriculture

DDT was used worldwide as an insecticide from the 1940s until the 1970s, until it was banned in the USA and other countries. Other pesticides of concern include tetrachlorophenol (TCP), used as an insecticide and bactericide; pentachlorophenol (PCP), a fungicide, herbicidal defoliant, and disinfectant; and the tin-containing pesticide tributyltin (TBT), a pesticide and antifungal agent. In a study examining endophytes capable of degrading these pesticides, a genetically engineered bacteria expressing a specific bacterial glutathione-S-transferase (GST) isolated from a *Burkholderia* strain were found to be capable of degrading the toxic pesticide chlormequat chloride following inoculation into pea plants (Compant et al. 2005b). In additional work with two commonly used worldwide broadleaf herbicides are 2, 4-dichlorophenoxyacetic acid (2, 4-D) and atrazine; there has also been promising research. Both of these herbicides are listed by the USEPA as toxic and are associated with human health risks; however, they are still used on many of the world's important crops (McGuinness and Dowling 2009). In a study using pea plants (*Pisum sativum*) inoculated with endophytic *Pseudomonas*, bacteria originally isolated from hybrid poplars (*P. trichocarpa* × *P. deltoides* cv. Hoogvorst) were found to be capable of degrading the herbicide 2, 4-D. The inoculated pea plants after exposure to 2, 4-D showed no accumulation of the herbicide in their tissues and experienced little signs of phytotoxicity, whereas control plants had significant accumulation of the toxin internally and showed overall diminished plant vitality (Germaine et al. 2006).

6.4 Bioremediation Using Hyperaccumulators of Heavy Metals and Metalloids

All organisms must contend with the threat of metal and metalloid pollution in our increasingly industrialized world. Adverse effects caused by mercury, lead, nickel, cadmium aluminum, chromium, and arsenic are known and well documented. Seeking endophytes which can facilitate plant's accumulation of these compounds could help remediation efforts greatly. Unlike phyto- and biodegradation however, phytoextraction or accumulation of heavy metals in plant materials facilitates their concentration to above ground within the aerial parts of plants. These plant materials could then be collected and taken away for further remediation concentration or these compounds may even be recycled. Current bioremediation efforts with endophytes are showing much promise. In the survey study of the endophytic bacteria of the plant *Alyssum bertolonii*, a known Ni hyperaccumulator plant, it was observed that *A. bertolonii* actually harbors numerous tolerant Ni bacteria (Barzanti et al. 2007). In another survey study conducted for the purpose of seeking hyperaccumulating endophytic bacteria, four bacterial strains were isolated from surface-sterilized *Sedum alfredii*, a perennial herb plant used in phytoremediation for its tolerance and accumulating ability of zinc and cadmium. Two bacteria from this survey *Sphingomonas* sp. (SaMR10) and *Variovorax* sp. (SaNR1) were shown to significantly promote plant growth and phytoextraction of both Zn and Cd (Zhang et al. 2012). In a study of yellow lupine, plants grown on nickel-enriched substrates after being inoculated with an engineered nickel-resistant bacterium *Burkholderia cepacia* increased plant tolerance, but in addition plant roots were able to significantly increase by 30 %, their overall nickel (Ni) concentration (Lodewyckx et al. 2001). Also in another study where tomato plants (*Solanum lycopersicum*) were inoculated with *Magnaporthe oryzae* and *Burkholderia* sp., the plants had increased biomass and the roots and shoots were able to accumulate Ni and Cd from the soil (Ma et al. 2011). In an investigation of various PGPB and their siderophores within a model system contaminated with heavy metals, it was found that *Pseudomonas aeruginosa* was able to solubilize large amounts of chromium (Cr) and lead (Pb). Because *P. aeruginosa* is an opportunistic human pathogen however, regulatory agencies would never give permission for their deliberate release into the environment (Braud et al. 2009). It also has been suggested that many of these metal-resistant endophytes promote plant growth by various mechanisms such as nitrogen fixation, solubilization of minerals, production of phytohormones, siderophores, and transformation of nutrient elements (Rajikumar et al. 2010).

7 Bioengineering for Future Biocontrol and Bioremediation Efforts

Genetic engineering and metagenomic studies will help illuminate many of the issues scientists face in bioremediation and biocontrol technologies. Engineered organisms for biocontrol and bioremediation will help us to continue to advance in our global efforts for international food security and remediation of toxic land and water sites. One promising technology which may be exploited for great benefit in the future is horizontal gene transfer.

It has been demonstrated that bacterial endophytes have the ability to provide bioremediation technology efforts, an additional advantage through the utilization of a natural phenomenon known as horizontal gene transfer (Taghavi et al. 2005). Horizontal gene transfer (HGT), which is also sometimes referred to as lateral gene transfer (LGT), is any process in which an organism incorporates genetic material from another organism in a manner other than traditional reproduction. It is known that many endophytic bacteria exhibit natural competence for degrading organic contaminants and that they may also introduce contaminant-degrading genes to local bacterial populations by horizontal gene transfer (Yousaf et al. 2010a). Horizontal gene transfer (HGT) can result in a natural endophyte population acquiring the capacity to degrade environmental pollutants without the need to establish artificial or engineered inoculant strains long term. It is has been demonstrated that endophytic bacteria are capable of expressing necessary catabolic genes which could promote degradation of toxic compounds (or their metabolites) as they accumulate or translocate within the vascular tissues of the host plant and that these genes can be transferred via plasmids to other endophytes within the plant. It is thought that HGT is actually widespread among bacteria particular in environmental niches where such genes offer significant survival enhancement. Gene transfer has also been observed between two fungal pathogens where a gene encoding for a virulence factor was transferred from one species to another and allowed for the emergence of a new damaging disease of wheat (Friesen et al. 2006).

The activity of HGT by endophytic bacteria which promoted more efficient degradation of toluene in poplar plants with the degradative plasmid, pTOM-Bu61, was found to have transferred naturally to a number of different endophytes in planta (Taghavi et al. 2005). In another study using pea plants with Pseudomonas endophytes harboring the plasmids pWWO and pNAH7, it was observed that these plasmids had high rates of transfer into a range of indigenous endophytes (Ryan et al. 2008). Horizontal gene transfer is a natural phenomenon (previously observed with acquired resistance to antibiotic in sensitive population of bacteria), it has been speculated in the literature that if incompetent organisms could be used to transfer gene of value to natural endophytes, then these resulting engineered or "enabled" endophytes may not be considered genetically modified microorganisms (GMMs) and, therefore, could be exempt from current international and national GM legislation. This legal status could potentially facilitate their study in field situations at an accelerated pace (McGuinness and Dowling 2009). This approach may have tremendous practical applications in equipping the natural endophyte populations of contaminate sites worldwide with the capacity to degrade pollutants and not requiring long-term establishment of foreign inoculant strains.

In the absence of natural biodegradation ability, genetically engineered strains of endophytic and rhizospheric bacteria could be constructed and tailor-made for the desired application. This approach is considered one of the most promising new technologies for remediation of contaminated environmental sites. Molecular approaches currently in use for the isolation and characterization of bacterial endophytes and plant-associated bacteria and communities have recently been reviewed by Franks et al. (2006) and Russo et al. (2012); however, this field of study is expanding quickly.

Using biotechnology, bacterial and fungal strains can be engineered, via natural gene transfer or recombinant DNA technology, to produce specific enzymes, capable of controlling pest/pathogens or degrading toxic organic pollutants found in the environment. Biocontrol and bioremediation work will also be assisted by proteomic studies which can examine plant–pathogen interactions (Mehta et al. 2008), three-way interactions between endophyte and plant and pathogen (Maara et al. 2006), as well as three-way interactions between endophyte and plant and toxin. In the future the availability of complete genome sequences of key endophytic bacteria and fungi will become available. The identification of "endophytic genes," especially those relating to the establishment of endophytic competence (genes governing colonization and establishment) *in planta*, will greatly enhance our understanding of these unique microbes. Together this information will form the foundation for even further transcriptome and proteome analysis studies. The incorporation of this information with well-established techniques such as IVET and other "-omic" technologies will then offer scientists new abilities to search for genes on a global scale that are found to be induced or repressed during colonization of plant tissues.

8 Reaching for the Future and Addressing Potential Concerns

Research into biocontrol and bioremediation activities with endophytes will continue in the future. Much of the concerns regarding biotechnological efforts utilizing living organisms, such as endophytes, have already been discovered, discussed, and openly debated; however, the potential advantages and benefits these organisms could offer have not yet even been fully recognized. In this chapter we have discussed how our understanding about endophytes has evolved over time as well as some of the problems that can arise during attempt to implement these technologies. We also now know that many of these problems arise due to unknown variables linked to the complexity of endophyte diversity and ecology. However, despite these challenges, the goal of using mutualistic symbionts, natural organisms for biocontrol and bioremediation technologies, is sound. Encouraging nature to work for us and not against us can and will work, but the study of endophytes does force those truly interested in their use to understand that there is still a great need to know more about the multitude of interactions and dynamics occurring within plant microbial ecology before we can do what we envision on a consistently successful level.

Always in the case of any new technological advancement, there are concerns that need to be addressed, but humankind has (for better or worse) never ceased to stop advancing because of the fear of the unknown, despite the challenges. Concerns, particularly regarding biocontrol safety and innudative biocontrol activity with exotic species, have already been addressed internationally in many articles and reviews, but these concerns will need to be revisited as the technology moves forward (Simberloff and Stiling 1996; Van Lanteren et al. 2003). One of the primary concerns is that many facultative endophytes often recruited by plants come from the pool of soil and rhizospheric species and from these competent for life *in planta* may include opportunistic human/animal pathogens (Braud et al. 2009). Opportunistic pathogens pose the greatest threat to individuals that are immunocompromised, and because so many people worldwide have immune-related conditions or need to take immunosuppressive drugs, these organisms must be monitored more carefully. Research will need to continue to establish potential risks, if any, associated with the development of an endophytic niche for any biocontrol agent that would be used in widespread biotechnological applications. It is known that various opportunistic human bacterial pathogens including *Burkholderia*, *Enterobacter*, *Herbaspirillum*, *Ochrobactrum*, *Pseudomonas*, *Ralstonia*, *Staphylococcus,* and *Stenotrophomonas* have been identified as colonizers of the plant rhizosphere. The fear is that these organisms if applied at high ratios could pose human health risks. For example, the endophytic capable bacterium, *Pantoea agglomerans* (formerly *Enterobacter agglomerans*), is an opportunistic human pathogen capable of causing infections, and in a study examining infection frequencies at children's hospital in Houston, TX, USA, *P. agglomerans* was isolated a total of 53 times during a 6-year period. Infections were observed in bloodstreams, joints and bones, and urinary tracts of children (Cruz et al. 2007). *Pantoea agglomerans* has actually been widely studied in the biocontrol literature, but like the *P. aeruginosa* example mentioned earlier, it would not be released now on a large-scale basis due to potential risks. Of note, often after plant surveys, temperature tests are used. Potential opportunist can be quickly weeded out by observing if the potential biocontrol can survive/thrive at 37 °C (human body temperature).

In the development of novel biocontrols, however, whole organisms may not be needed in order to be considered effective. In some cases there is less risk in using microbial extracts to control pests and pathogens (Janisiewicz and Roitman 1988). Of course the antimicrobial structures would have to be elucidated and assessments of their effective biological activity obtained; however, the unknown variables which often can result from working with whole organisms would be reduced. For instance, *P. agglomerans* as well as other opportunistic bacteria use antibiosis as a mechanism of control; these compounds once isolated could be very useful. Actually many compounds from bio-agents have already been isolated and marketed by pesticide companies (Copping and Menn 2000). Some would argue however that this is only causing the same continuously repeating cycle wherein specific chemicals are applied to crops and the pest/pathogens develop resistance as quickly as they have with human-made synthetic compounds. Of the two technologies, bioremediation will likely have the least public resistance particularly in areas where humans can be kept away from contaminate sites. This technology will also likely increase if controlled HGT technology can be fully realized and there are few objections to their use. Valuable genes can be introduced by incorporated within incompetent bacterial strains intentionally inoculated within plants but incapable of long-term survival in the plant. These temporary organisms would theoretically live only long enough to share their unique plasmids. In conclusion, despite obvious challenges it is important to note the use of endophytes for biocontrol and bioremediation efforts is still relatively new and biotechnological development for both is really just beginning. There really is great potential for mutualistic symbionts to make significant contributions toward cheaper sustainable pest/pathogen and pollutant controls.

References

Afzal M, Yousaf S, Reichenauer TG, Kuffner M, Sessitsch A (2011) Soil type affects plant colonization, activity and catabolic gene expression of inoculated bacterial strains during phytoremediation of diesel. J Hazard Mater 186:1568–1575

Al-Karaki GN, Al-Raddad A (1997) Effects of arbuscular mycorrhizal fungi and drought stress on growth and nutrient uptake of two wheat genotypes differing in drought resistance. Mycorrhiza 7:83–88

Alström S (2001) Characteristics of bacteria from oilseed rape in relations to their biocontrol activity against *Verticillium dahlia*. J Phytopathol 149:57–64

Arachevaleta M, Bacon CW, Hoveland CS, Radcliffe DE (1989) Effects of the tall fescue endophyte on plant response to environmental stress. Agron J 81:83–90

Arshad M, Saleem M, Hussain S (2007) Perspectives of bacterial ACC deaminase in phytoremediation. Trends Biotechnol 25:356–362

Arshad M, Shaharoona B, Mahmood T (2008) Inoculation with *Pseudomonas* spp. containing ACC-deaminase partially eliminates the effects of drought stress on growth, yield and ripening of pea (*Pisum sativum* L.). Pedosphere 18:611–620

Augé RM (2001) Water relations, drought and vesicular–arbuscular mycorrhizal symbiosis. Mycorrhiza 11:3–42

Azcón-Aguilar C, Barea JM (1996) Arbuscular mycorrhizas and biological control of soil-borne plant pathogens – an overview of the mechanisms involved. Mycorrhiza 6:457–464

Bacon C (1993) Abiotic stress tolerances (moisture, nutrients) and photosynthesis in endophyte-infected tall fescue. Agri Ecosyst Environ 44:123–141

Bacon CW, Hinton DM (2007) Isolation, *in planta* detection, and uses of endophytic bacteria for plant protection. In: Hurst CJ et al (eds) Manual of environmental microbiology. ASM Press, Washington, DC

Bacon CW, Porter JK, Robbins JD, Luttrell ES (1977) *Epichlöe typhina* from toxic tall fescue grasses. Appl Environ Microbiol 34:76–81

Baek JM, Howell CR, Kenerley CM (1999) The role of an extracellular chitinase from *Trichoderma virens* Gv29-8 in the biocontrol of *Rhizoctonia solani*. Curr Genet 35:41–45

Bálint M, Tiffin P, Hallström B, O'Hara RB, Olson MS, Fankhauser JD, Piepenbring M, Schmitt I (2013) Host genotype shapes the foliar fungal microbiome of Balsam Poplar (*Populus balsamifera*). PLOS One 8(1):e53987

Barac T, Taghavi S, Borremans B, Provoost A, Oeyen L, Colpaert JV, Vangronsveld J, van der Lelie D (2004) Engineered endophytic bacteria improve phytoremediation of water-soluble, volatile, organic pollutants. Nat Biotechnol 22:583–588

Barzanti R, Ozino F, Bazzicalup M, Gabbrielli R, Galardi F, Gonnelli C, Mengoni A (2007) Isolation and characterization of endophytic bacteria from nickel hyperaccumulator plant *Alyssum bertolonii*. Microb Ecol 53:306–316

Braud A, Jézéquel K, Bazot S, Lebeau T (2009) Enhanced phytoextraction of an agricultural Cr, Hg and Pb-contaminated soil by bioaugmentation with siderophore producing bacteria. Chemosphere 74:280–286

Canakar K, Kraigher H, Ravnikar M, Rupnik M (2005) Bacterial endophytes from seeds of Norway spruce (*Picea abies* L. Karst). FEMS Microbiol Lett 244:341–345

Charudattan R (2001) Biological control of weeds by means of plant pathogens: significance for integrated weed management in modern agro-ecology. BioControl 46:229–260

Clay K (1987) Effects of fungal endophytes on seed and seedling biology of *Lolium perenne* and *Festuca arundinacea*. Oecologia 73:358–362

Compant S, Duffy B, Nowak J, Clément C, Barka EA (2005a) Use of plant growth-promoting bacteria for biocontrol of plant diseases: principles, mechanisms of action and future prospects. Appl Environ Microbiol 71:4951–4959

Compant S, Reiter B, Sessitsch A, Nowak J, Clément C, Barka EA (2005b) Endophytic colonization of *Vitis vinifera* L. by plant growth-promoting bacterium *Burkholderia* sp. Strain PsJN. Appl Environ Mirobiol 4:1685–1693

Compant S, van der Heijden MGA, Sessitsch A (2010) Climate change effects on beneficial plant-microorganism interactions. FEMS Microbiol Ecol 73:197–214

Copping LG, Menn JJ (2000) Biopesticides: a review of their action, applications and efficacy. Pest Manag Sci 56:651–676

Cortesero AM, Stapel JO, Lewis WJ (2000) Understanding and manipulating plant attributes to enhance biological control. Biol Control 17:35–49

Cowan MM (1999) Plant products as antimicrobial agents. Clin Microbiol Rev 12:564–582

Cruz AT, Cazacu AC, Allen CH (2007) *Pantoea agglomerans*, a plant pathogen causing human disease. J Clin Microbiol 45:1989–1992

Dashti N, Khanafer M, El-Nemr I, Sorkhoh N, Ali S, Radwan S (2009) The potential of oil-utilizing bacterial consortia associated with legume root nodules for cleaning oily soils. Chemosphere 74:1354–1359

Denton BD (2007) Advances in phytoremediation of heavy metals using plant growth promoting bacteria and fungi. Microbiol Mol Genet 3:1–5

Doberski JW, Tribe HT (1980) Isolation of entomogenous fungi from elm bark and soil with reference to ecology of *Beauveria bassiana* and *Metarhizium anisopliae*. Trans Br Mycol Soc 74:95–100

Ernst M, Mendgen KW, Wirsel SGR (2003) Endophytic fungal mutualists: seed-borne *Stagonospora* spp. enhance reed biomass production in axenic microcosms. Mol Plant Microbe Interact 16:580–587

Franks A, Ryan PR, Abbas A, Mark GL, O'Gara F (2006) Molecular tools for studying plant growth-promoting Rhizobacteria (PGPR). In: Molecular techniques for soil and rhizosphere microorganisms. CABI Publishing, Wallingford/Oxfordshire

Fraser EDG (2003) Social vulnerability and ecological fragility: building bridges between social and natural sciences using the Irish potato famine as a case study. Conserv Ecol 7:1–9

Friesen TL, Stukenbrock EH, Liu Z, Meinhardt S, Ling H, Faris JD, Rasmussen JB, Solomon PS, McDonald BA, Oliver RP (2006) Emergence of a new disease as a result of interspecific virulence gene transfer. Nat Genet 38:953–956

Gange AC, Dey S, Currie AF, Sutton BC (2007) Site- and species-specific differences in endophyte occurrence in two herbaceous plants. J Ecol 94:614–622

Germaine KJ, Liu X, Cabellos GG, Hogan JP, Ryan D, Dowling DN (2006) Bacterial endophyte-enhanced phytoremediation of the organochlorine herbicide 2,4-dichlorophenoxyacetic acid. FEMS Microbiol Ecol 57:302–310

Germaine KJ, Keogh E, Ryan D, Dowling DN (2009) Bacterial endophyte-mediated naphthalene phytoprotection and phytoremediation. FEMS Microbiol Lett 296:226–234

Glick BR (2005) Modulation of plant ethylene levels by the bacterial enzyme ACC deaminase. FEMS Microbiol Lett 251:1–7

Hallman J, Quadt-Hallman A, Mahaffee WF, Kloepper JW (1997) Bacterial endophytes in agricultural crops. Can J Microbiol 43:895–914

Hammond-Kosack K, Jones JDG (2000) Responses to plant pathogens. In: Buchanan B, Gruissem W, Jones R (eds) Biochemistry and molecular biology of plants. American Society of Plant Physiologists, Rockville, pp 1102–1156

Han DY, Coplin DL, Bauer WD, Hoitink HAJ (2000) A rapid bioassay for screening rhizosphere microorganisms for their ability to induce systemic resistance. Phytopathology 90:327–332

Handelsman J, Stabb EV (1996) Biocontrol of soilborne plant pathogens. Plant Cell 8:1855–1869

Hardoim PR, Van Overbeek LS, Elsas JDV (2008) Properties of bacterial endophytes and their proposed role in plant growth. Trends Microbiol 16:463–471

Jabaji-Hare S, Neate SM (2005) Nonpathogenic binucleate *Rhizoctonia* spp. and benzothiadiazole protect cotton seedlings against *Rhizoctonia* damping-off and *Alternaria* leaf spot in cotton. Biocontrol 95:1030–1036

Janisiewicz WJ, Roitman J (1988) Biological control of blue mold and grey mold on apple and pear with *Pseudomonas cepacia*. Phytopathology 78:1697–1700

Johnson MC, Dahlman DL, Siegel MR, Bush LP, Latch GCM, Potter DA, Varney DR (1985) Insect feeding deterrents in endophyte-infected tall fescue. Appl Environ Microbiol 49:568–571

Jumpponen A, Trappe JM (1998) Dark septate endophytes: a review of facultative biotrophic root-colonizing fungi. New Phytol 140:295–310

Khan Z, Doty S (2011) Endophyte-assisted phytoremediation. Curr Top Plant Biol 12:97–104

Khan T, Latif A, Hamayun M, Ahmad N, Hussain J, Kang S, Kim Y, Adnan M, Tang D, Waqas M, Radhakrishnan R, Hwang Y, Lee I (2011) Salinity stress resistance offered by endophytic fungal interaction between *Penicillium minioluteum* LHL09 and *Glycine max*. L.J. J Microbiol Biotechnol 21:893–902

Khan S, Afzal M, Iqbal S, Khan QM (2012) Plant–bacteria partnerships for the remediation of hydrocarbon contaminated soils. Chemosphere 90:1317–1332

Lahlali R, Hijri M (2010) Screening, identification and evaluation of potential biocontrol fungal endophytes against *Rhizoctonia solani* AG3 on potato plants. FEMS Microbiol Lett 311:152–159

Lai K, Chen S, Hu M, Hu Q, Geng P, Weng Q, Jia J (2012) Control of postharvest green mold of citrus fruit by application of endophytic *Paenibacillus polymyxa* strain SG-6. Postharvest Biol Technol 69:40–48

Latch GCM (1993) Physiological interactions of endophytic fungi and their hosts. Biotic stress tolerance imparted to grasses by endophytes. Agri Environ Microbiol 44:143–156

Lodewyckx C, Taghavi S, Mergeay M, Vangronsveld J, Clijsters H, van der Lelie D (2001) The effect of recombinant heavy metal resistant endophytic bacteria on heavy metal uptake by their host plant. Int J Phytoremediat 3:173–187

Lodwyckx C, Vangronsveld J, Porteous F, Moore ERB, Taghavi S, Mazgeay M, van der Leile D (2002) Endophytic bacteria and their potential applications. Crit Rev Plant Sci 21:583–606

Loiret FG, Ortega E, Kleiner D, Ortega-Rhoďes P, Roďes R, Dong Z (2004) A putative new endophytic nitrogen-fixing bacterium *Pantoea* sp. from sugarcane. J Appl Microbiol 97:504–511

Ma Y, Prasad MNV, Rajkumar M, Freitas H (2011) Plant growth promoting rhizobacteria and endophytes accelerate phytoremediation of metalliferous soils. Biotechnol Adv 29:248–258

Maara R, Ambrosimo P, Carbone V, Vinale F, Woo SL, Ruocco M, Ciliento R, Lanzuise S, Ferraioli S, Soriente I, Gigante S, Turrá D, Fogliano V, Scala F, Lorito M (2006) Study of the three-way interaction between *Trichoderma atroviride*, plant and fungal pathogens by using a proteomic approach. Curr Genet 50:307–321

Martin GB, Bogdanove AJ, Sessa G (2003) Understanding the functions of plant disease resistance proteins. Annu Rev Plant Biol 54:23–61

McGuinness M, Dowling D (2009) Plant-associated bacterial degradation of toxic organic compounds in soil. Int J Environ Res Public Health 6:2226–2247

Mehta A, Brasileiro ACM, Souza DSL, Romano E, Campos MA, Frossi-de-Śa MF, Silva MS, Franco OL, Fragoso RR, Bevitori R, Rocha TL (2008) Plant-pathogen interactions: what is proteomics telling us? FEBS J 275:3731–3746

Mejía LC, Rojas EI, Maynard Z, Van Bael S, Arnold AE, Hebbar P, Samuels GJ, Robbins N, Herre EA (2008) Endophytic fungi as biocontrol agents of *Theobroma cacao* pathogens. Biol Control 46:4–14

Müller CB, Krauss J (2005) Symbiosis between grasses and asexual fungal endophytes. Curr Opin Plant Biol 8:450–456

Newman LA, Reynolds CM (2005) Bacteria and phytoremediation: new uses for endophytic bacteria in plants. Trends Biotechnol 23:6–8

Niere BI, Speijer PR, Gold CS, Sikora RA (1999) Fungal endophytes from bananas for the biocontrol of Radopholus similis. In: Frison EA, Gold CS, Karamura EA, Sikora RA (eds) Mobilizing IPM for sustainable banana production in Africa. INIBAP, Montpellier, pp 237–245

Ownley BH, Griffin MR (2012) Dual biological control of insect pests and plant pathogens with fungi in the order Hypocreales. In: Biocontrol: management, processes and challenges. Nova Science Publishers, Inc., Hauppauge

Ownley BH, Griffin MR, Klingeman WE, Gwinn KD, Moulton JK, Pereira RM (2008) *Beauveria bassiana*: endophytic colonization and plant disease control. J Invertebr Pathol 98:267–270

Peer WA, Baxter IR, Richards EL, Freeman JL, Murphy S (2005) Phytoremediation and hyperaccumulator plants. Hort. Purdue.edu/hort/research/murphy/pdfs/metals11.pdf

Pirttila AM, Joensuu P, Pospiech H, Jalonen J, Hohtola A (2004) Bud endophytes of Scots pine produce adenine derivatives and other compounds that affect morphology and mitigate browning of callus cultures. Pysiol Plant 121:305–312

Porras-Alfaro A, Bayman P (2011) Hidden fungi, emergent properties: endophytes and microbiomes. Ann Rev Phytopathol 49:291–315

Porteous-Moore F, Barac T, Borremans B, Oeyen L, Vangronsveld J, van der Lelie D, Campbell D, Moore ERB (2006) Endophytic bacterial diversity in poplar trees growing on a BTEX-contaminated site: the characterisation of isolates with potential to enhance phytoremediation. Syst Appl Microbiol 29:539–556

Rajikumar M, Noriharu A, Prasad MNV, Freitas H (2010) Potential of siderophore-producing bacteria for improving heavy metal phytoextraction. Trends Biotechnol 28:142–149

Reiter B, Sessitsch A (2006) Bacterial endophytes of the wildflower *Crocus albiflorus* analyzed by characterization of isolates and by a cultivation-independent approach. Can J Microbiol 52:140–149

Russo A, Carrozza GP, Vettori L, Felici C, Cinelli F, Toffanin A (2012) Plant beneficial microbes and their application in plant biotechnology. In: Agbo EC (ed) Innovations in biotechnology. Intechopen, pp 57–72. doi:10.5772/2450

Ryan RP, Germaine K, Franks A, Ryan DJ, Dowling DN (2008) Bacterial endophytes: recent developments and applications. FEMS Microbiol Lett 278:1–9

Saunders M, Kohn LM (2008) Evidence for alternation of fungal endophyte community assembly by host defense compounds. New Phytol 182:229–238

Schulz B, Boyle C (2005) The endophytic continuum. Mycol Res 6:661–686

Sessitsch A, Hardoim P, Dȍring J, Weilharter A, Krause A, Woyke T, Mitter B, Hauberg-Lotte L, Friedrich F, Rahalkar M, Hurek T, Sarkar A, Bodrossy L, van Overbeek L, Brar D, van Elsas JD, Reinhold-Hurk B (2012) Functional characteristics of an endophyte community colonizing rice roots as revealed by metagenomic analysis. Mol Plant Microbe Interact 25:28–36

Siciliano SD, Fortin N, Mihoc A, Wisse G, Labelle S, Beaumier D, Ouellette D, Roy R, Whyte LG, Banks MK, Schwab P, Lee K, Greer CW (2001) Selection of specific endophytic bacterial genotypes by plants in response to soil contamination. Appl Environ Microbiol 67:1469–2475

Simberloff D, Stiling P (1996) How risky is biological control? Ecology 77:1965–1974

Strobel GA (2002) Rainforest endophytes and bioactive products. Crit Rev Biotechnol 22:315–333

Strobel G, Daisy B (2003) Bioprospecting for microbial endophytes and their natural products. Microbiol Mol Biol Rev 67:491–502

Strobel G, Daisy B, Castillo U, Harper J (2004) Natural products from endophytic microorganisms. J Nat Prod 67:257–268

Surette MA, Sturz AV, Lada RR, Nowak J (2003) Bacterial endophytes in processing carrots (*Daucus carota* L. var. *sativus*): their localization, population density, biodiversity and their effects on plant growth. Plant Soil 253:381–390

Taghavi S, Barac T, Greenberg B, Borremans B, Vangronsveld J, van der Lelie D (2005) Horizontal gene transfer to endogenous endophytic bacteria from poplar improves phytoremediation of toluene. Appl Environ Microbiol 71:8500–8505

Taghavi S, Garafola C, Monchy S (2009) Genome survey and characterization of endophytic bacteria exhibiting a beneficial effect on growth and development of poplar trees. Appl Environ Microbiol 75:748–757

Tan RX, Zou WX (2001) Endophytes: a rich source of functional metabolites. Nat Prod Rep 18:448–459

Tan Z, Hurek T, Reinhold-Hurek B (2003) Effect of N-fertilization, plant genotype and environmental conditions of *nif*H gene pools in roots of rice. Environ Microbiol 6:1009–1015

Tejesvi MV, Kini KR, Prakash HS, Subbiah V, Shetty HS (2007) Genetic diversity and antifungal activity of species of *Pestalotiopsis* isolated as endophytes from medicinal plants. Fungal Divers 24:37–54

Thane U, Adler A, Clasen P-E, Galvano F, Langseth W, Lew H, Logrieco A, Nielsen KF, Ritieni A (2004) Diversity in metabolite production by *Fusarium langsethiae*, *Fusarium poae*, and *Fusarium sporotrichioides*. Int J Food Microbiol 95:257–266

United States National Research Council, Board Agric (1993) Pesticides in the diets of infants and children. National Academy Press, Washington, DC, 408 pp

Van Lanteren JC, Babendreier D, Bigler F, Burgio G, Hokkanen HMT, Kuske S, Loomans AJM, Menzler-Hokkanen I, Van Rijn PCJ, Thomas MB, Tommasini MG, Zeng QQ (2003) Environmental risk assessment of exotic natural enemies used in innudative biological control. Biocontrol 48:3–38

Verma SC, Ladha JK, Tripathi AK (2001) Evaluation of plant growth promoting and colonization ability of endophytic diazotrophs from deep water rice. J Biotechnol 91:127–141

Weller DM (1988) Biological control of soilborne plant pathogens in the rhizosphere with bacteria. Ann Rev Phytopathol 26:379–407

Weyens N, van der Lelie D, Taghavi S, Newman L, Vangronsveld J (2009a) Exploiting plant-microbe partnerships to improve biomass production and remediation. Curr Trends Biotechnol 27:591–598

Weyens N, van der Lelie D, Taghavi S, Vangronsveld J (2009b) Phytoremediation: plant-endophyte partnerships take the challenge. Curr Opin Biotechnol 20:248–254

Weyens N, Taghavi S, Barac T, Lelie D, Jana B, Artois T, Carleer R, Vangronsveld J (2009c) Bacteria associated with oak and ash on a TCE-contaminated site: characterization of isolates with potential to avoid evapotranspiration of TCE. Environ Sci Pollut Res 16:830–843

Xin G, Zhang G, Kang JW, Staley JT, Doty SL (2009) A diazotrophic, indole-3-acetic acid-producing endophyte from wild cotton wood. Biol Fertil Soils 45:669–674

Yousaf S, Ripka K, Reichenauer TG, Andria V, Afzal M, Sessitsch A (2010a) Hydrocarbon degradation and plant colonization by selected bacterial strains isolated from Italian ryegrass and birdsfoot trefoil. J Appl Microbiol 109:1389–1401

Yousaf S, Andria V, Reichenauer TG, Smalla K, Sessitsch A (2010b) Phylogenetic and functional diversity of alkane degrading bacteria associated with Italian ryegrass (*Lolium multiflorum*) and birdsfoot trefoil

(*Lotus corniculatus*) in a petroleum oil-contaminated environment. J Hazard Mater 184:523–532

Yousaf S, Afzal M, Reichenauer TG, Brady CL, Sessitsch A (2011) Hydrocarbon degradation, plant colonization and gene expression of alkane degradation genes by endophytic *Enterobacter ludwigii* strains. Environ Pollut 159:2675–2683

Yu H, Zhang L, Li L, Zheng C, Guo L, Li W, Sun P, Qin L (2010) Recent developments and future prospects of antimicrobial metabolites produced by endophytes. Microbiol Res 165:437–449

Zhang X, Lin L, Zhu Z, Yang X, Wang Y, An Q (2012) Colonization and modulation of host growth and metal uptake by endophytic bacteria of *Sedum alfredii*. Int J Phytoremediat 15:51–64

Zhuang X, Chen J, Shim H, Bai Z (2007) New advances in plant growth-promoting rhizobacteria for bioremediation. Environ Int 33:406–413

15 Biosourcing Endophytes as Biocontrol Agents of Wilt Diseases

Adeline Su Yien Ting

Abstract

Endophytes are the centre of many investigations in the recent years, mainly for their role as biological control agents towards various pathogens. Of the many types of phytopathogens, wilt pathogens are thought to benefit the most from application of endophytes. Wilt pathogens colonize internal plant tissues, especially the vascular tissues, which are also a common colonization niche for endophytes. The pre-colonization of biocontrol endophytes has been shown to render some form of protection to the host plant, resulting in disease suppression when challenged with the pathogen. Investigations to identify potential biocontrol agents are commonly initiated by performing extensive isolation and screening of endophytes from various asymptomatic host plants. This is followed by *in vitro* assays with selected pathogens, with various mechanisms of their antagonistic interaction established. Isolates with strong biocontrol activities are subsequently tested at the glasshouse and field stage to determine biocontrol efficacy. To date, tremendous progress has been made in understanding the diversity and mechanisms of control of endophytes against wilt pathogens. Their biocontrol efficacies are evident in laboratory screenings and glasshouse trials. In field trials however, poor control efficacy is often observed, attributed to the influence of indigenous microflora in the soil and environmental conditions. To address these limitations, bioformulation of endophytes is explored. This article will discuss the endophytes identified as biocontrol agents against wilt pathogens, the typical methods for biosourcing of these biocontrol endophytes, the challenges in implementing endophytes for wilt control and strategies to address these limitations.

A.S.Y. Ting (✉)
School of Science, Monash University Sunway Campus, Jalan Lagoon Selatan, 46150 Bandar Sunway, Darul Ehsan, Selangor, Malaysia
e-mail: adeline.ting@monash.edu

V.C. Verma and A.C. Gange (eds.), *Advances in Endophytic Research*,
DOI 10.1007/978-81-322-1575-2_15, © Springer India 2014

1 Introduction

The term "endophyte" refers to microorganisms that exist and colonize tissues of its host plant (*endon* Greek for within; *phyton* for plant) without causing any visible symptoms (Petrini 1991; Wilson 1995; Stone et al. 2000). The asymptomatic nature of the association is crucial to define endophytes in recent times, as some pathogens (virulent and latent pathogens) and parasites are also endophytic (Freeman and Rodriguez 1993; Marler et al. 1999; Sturz and Nowak 2000; Sieber 2002; Sikora et al. 2008). In most studies, endophytes refer exclusively to fungi (Carroll 1988; Clay 1988) although they constitute of both bacterial and fungal origins. Endophytes have been found to infect cells inter- or intracellularly and can colonize cells locally or systematically (Stone et al. 2000). Figure 15.1 illustrates intracellular growth of endophytes in roots of wild banana. As a result of the different modes of infection and colonization, various plant organs can be colonized with the typical trend suggesting shoot colonization is often localized while root colonization is more extensive (Schulz and Boyle 2005).

Endophytes enjoy a "privileged" relationship with its host plants. They derive benefits by their mere existence within the plant tissues, obtaining nutrients and getting buffered from biotic and abiotic stresses (Schulz and Boyle 2005). The host plants also benefit from this mutualistic symbiotic relationship. Endophytic infection improves host plant response to biotic and abiotic stresses (Bacon and Hill 1996; Schardl 2001), significantly evident in the mycorrhizal-root relationship of members of Orchidaceae (Gardes 2002) and the relationship of balansiaceous endophytes with grasses (Schardl 2001). As the benefits of endophyte-host relationship become more apparent over the years, endophytes have emerged as the source of intelligent screening for biological activities to realize their potential in biological control of diseases. To date, both bacterial and fungal endophytes have shown significant potential as biocontrol agents towards wilt pathogens (M'Piga et al. 1997; Ting et al. 2010b, 2012; Sundaramoorthy et al. 2012). Improvements are, however, required to ensure that the control efficacy typically observed in laboratory and greenhouse experiments is successfully replicated in the field conditions.

Fig. 15.1 Intracellular growth and colonization of endophytes in root tissues of wild banana (×400)

2 Major Wilt Diseases in Crops

Agricultural crops succumb to both fungal and bacterial wilt diseases, with the most common wilt diseases in the tropics reported as *Fusarium* wilt, *Verticillium* wilt and bacterial wilt.

2.1 *Fusarium* Wilt

Fusarium wilt is a devastating disease affecting many valuable crops globally. The pathogen, *Fusarium oxysporum* (FO), is classified into forma specialis (f. sp.) to indicate host specificity. This disease is common in banana (Ploetz 2005), tomato (Larkin and Fravel 1998), chilies (Sundaramoorthy et al. 2012) and a variety of legumes (Cobos et al. 2005) and cucurbits (Freeman et al. 2002). Amongst the many crops, bananas are the most severely (Sturz and Nowak 2000) affected by the wilt pathogen as implications of crop loss are at a global scale and affect food supply. Therefore, in this discussion, the emphasis is on Fusarium wilt disease of banana, caused by *Fusarium oxysporum* f. sp. *cubense*

(Smith) (Snyder and Hansen) (FOC race 4), particularly the tropical strain (Ploetz 2005). FOC race 4 of the tropical strain is by far the most virulent race. This tropical strain is capable of causing the Panama disease even in resistant cultivars, devastating banana crops, causing abandonment of plantations and disrupting supply of banana as staple food. This pathogen invades the vascular tissues of the banana plant through root penetration, spreading upwards to the corm and pseudostem, resulting in blockage to the xylem vessels, leading to wilting and death of plants (chlorosis and gradually necrosis) (Nelson 1981). Early attempts to control *Fusarium* wilt include cultural improvements, chemical treatments and use of resistant cultivars (Nel et al. 2006), but were all proven futile. This is followed by investigations on the use of suppressive soil, antagonistic biocontrol agents and transgenic approaches. The use of suppressive soil has been attempted but artificial induction is difficult to sustain and replicated under various field conditions (Ting et al. 2003), especially when disease suppression is attributed to the various interactions of soil physical structures (type of soil, drainage condition, montmorillonite soils, pH), nutritional status (Ploetz 2000) and concerted effort by microbes (composition of fungi, bacteria, actinomyces) (Alabouvette et al. 1993). A more sophisticated strategy using transgenic plants carrying resistance genes has also been attempted but was difficult to manipulate. This method is also costly compared to the conventional use of antagonistic microbes. Therefore, for many years, antagonistic rhizobacteria remained a popular choice as a strategy to manage *Fusarium* wilt. Rhizobacteria showed good biocontrol potential but were susceptible to biotic and abiotic factors in the environment. This led to the emergence of endophytes as alternative biocontrol agents against wilt diseases. Endophytes are theorized as an excellent choice as they occupy similar colonization and establishment niche as the wilt pathogen. Introduction of endophytes to sterile micropropagated plants is expected to benefit disease management as pre-colonization allows early establishment of endophyte-host association and exclusion of pathogen from niche competition.

2.2 *Verticillium* Wilt

Verticillium wilt, caused by the vascular pathogens *Verticillium dahliae* or *V. albo-atrum*, is widespread in economically important crops such as cotton (Bolek et al. 2005) and vegetables such as tomatoes (Sharma and Nowak 1998). Their persistence, survival and relatively high levels of inoculum in tropical soils are attributed to their microsclerotia (Schnathorst 1981). Infected plants display similar symptoms as *Fusarium* wilt, beginning with yellowing followed by chlorosis and necrosis of leaves occurring on one or both sides of the leaf or the whole plant. Subsequently, vascular discoloration and stunting may be apparent. To date, no chemical application has been effective against *Verticillium* wilt, although crop rotation and resistant varieties have been traditionally used to achieve a certain degree of control (Uppal et al. 2008).

2.3 Bacterial Wilt

Bacterial wilt is caused by a variety of pathogens, but of the many, *Ralstonia solanacearum* (Smith) is the most extensively studied. This is due to its ability to infect over 200 plant species belonging to more than 50 families of economically important crops such as tomato, pepper, potato, tobacco, eggplant, cowpea, peanut, cashew, banana, papaya and olive (Hayward 1991; Guo et al. 2004). The virulence of *R. solanacearum* to cause disease is perpetuated by their ability to survive in various soil types and their efficient mechanism to invade host plants (Hayward 1991). Various strategies have been employed to manage bacterial wilt by *R. solanacearum,* using biofumigants (Pradhanang et al. 2003) and transgenic resistant plants (Lee et al. 2002). However, all these strategies met with limited success; thus, biological control has emerged as the new approach.

3 Endophytes as Biological Control Agents of Wilt Diseases

The role of endophytes as biological control agents originated from their antagonistic role towards feeding herbivores and insects. Balansiaceous endophytes and their anamorphs (*Neotyphodium*) were the first "biocontrol agents" as they colonize grasses and produce alkaloids that act as natural feeding deterrent towards herbivores (Petrini 1996; Schardl 2001). These endophytes occur naturally in the host plants and are nonpathogenic. Jumpponen and Trappe (1998) found that roots of plants in the natural habitat are often colonized by nonpathogenic endophytes. This leads to the hypothesis that naturally occurring endophytes may have potential to suppress diseases, especially those relevant to the host plant. To the scientific community, endophytes present untapped potential as biological control agents for wilt diseases. They are highly desirable because they can colonize the rhizosphere, enter into the plant tissues to survive and proliferate endophytically in the plants. Endophytes are often also suitable as biocontrol agents as they are non-host specific. For example, endophytes isolated from oil palm roots can express biocontrol activity towards FOC race 4 when inoculated into banana plants (Mohd Fishal et al. 2010). Endophytes produce substances that have immeasurable value for exploitation, not just for agricultural applications but pharmaceutical and industrial applications as well.

3.1 Antagonistic Bacterial Endophytes

Existing literatures would reveal that a variety of bacterial endophytes can demonstrate biocontrol activity against various wilt pathogens. Key isolates include *Bacillus* sp. and *Pseudomonas* sp. which contribute to disease suppression through induction of host resistance (M'Piga et al. 1997; Nejad and Johnson 2000; Sundaramoorthy et al. 2012). Actinobacteria are also found to display traits of endophytic infection and colonization (Smith et al. 1990; El-Abyad et al. 1993). However, the rate of isolation of Actinobacteria is rather low as they were not found in any of the 54 plants (fruit trees, ornamentals, weeds, medicinal) sampled by Ting et al. (2009d). In roots of wild bananas, only four of the 341 endophytes were Actinobacteria (Ting, unpublished). Contrarily, recent developments by Tan et al. (2011) found a variety of Actinobacteria *Streptomyces*, *Nesterenkonia*, *Arthrobacter*, *Microbacterium*, *Cellulomonas* and *Propionibacterium* as natural colonizers of tomato roots. Amongst the isolates, two of the isolates, *S. virginiae* isolate Y30 and E36, showed good biocontrol potential towards *Ralstonia solanacearum* by producing siderophores and ACC deaminase activity (Tan et al. 2011). This strongly suggests the endophytic nature of Actinobacteria and their potential as biocontrol agents towards wilt pathogens. The inclusion of rhizobacteria strains with bacterial endophytes has shown ability to enhance biocontrol activity (Sundaramoorthy et al. 2012). In their study, application of endophytic *B. subtilis* (EPCO16, EPC5) with the rhizobacterium *P. fluorescens* (Pf1) effectively reduced Fusarium wilt incidence in chilies. Treated plants recorded enhanced activities of peroxidase (PO), polyphenol oxidase (PPO) (peroxidase), phenylalanine ammonia lyase (PAL), β-1,3-glucanase, chitinase and phenolics, which demonstrates disease suppression via induced host resistance. This multi-strain approach is suitable for endophyte-rhizobacteria mixtures, but may not be as effective for endophyte-endophyte combinations. Ting et al. (2008, 2009a, 2010b) found that banana plantlets inoculated with *P. aeruginosa* (UPM13B8) and *Serratia marcescens* (UPM39B3) did not enhance disease reduction rather the reverse. This was also seen by Mohd Fishal et al. (2010) where combination of *Pseudomonas* sp. (UPMP3) and *Burkholderia* sp. (UPMB3) produced only intermediate effect, inferior to single treatment with *Pseudomonas* UPMP3. In both examples, host defence mechanisms were induced but disease suppression by multi-strain applications did not lead to exceptional biocontrol.

In addition to eliciting host defence mechanisms, infection by bacterial endophytes also improved growth and vigour of host plants. Improved growth benefited the host plants as survival of wilt-infected plants is magnified and the onset of symptoms is significantly delayed. In Ting et al. (2008), treatment with *S. marcescens* (UPM39B3) improved plant height, pseudostem diameter, root mass and total number of leaves per plant and was able to delay onset of symptoms by 7–10 days compared to plantlets without endophytes. In addition, plantlets pretreated with endophytes did not collapse even after 49 days, while plantlets without endophytes died by day 42.

3.2 Beneficial Fungal Endophytes

The beneficial fungal endophytes studied for biocontrol properties against wilt pathogens are usually non-balansiaceous fungi. They include a variety of nonpathogenic isolates of *Fusarium oxysporum* (FO), as well as *Penicillium* spp., *Cladosporium* spp. and mycorrhizal fungi. Nonpathogenic *Fusarium oxysporum* isolates are the most extensively studied fungal endophytes. Ting et al. (2008, 2009c) discovered two nonpathogenic FO isolates (UPM31P1 and UPM31F4) from roots of wild bananas capable of inducing host defence mechanisms to suppress wilt development and improving growth and vigour. Other fungal endophytes with biological control properties include *Penicillium citrinum* (BTF08) (Ting et al. 2012), *Cladosporium* sp. (BTF21, ALF01), *Phomopsis* sp. (MIF21, WAA02, WAA03) and *Nigrospora* sp. (BTF05, BTF07) (Ting et al. 2009d). The antagonistic nature of these endophytes is new for some isolates such as *P. citrinum* (Ting et al. 2012). However, for some isolates like *Phomopsis,* their biocontrol activity towards FO is rather established (Yu et al. 2010). In addition to nonpathogenic fungi, arbuscular mycorrhizal fungi (AMF) are also useful biocontrol agents. However, their application is limited as bioactivity is dependent on host response to AMF infection. In a study by Garmendia et al. (2004), three *Glomus* spp. were investigated for biocontrol activity. *G. intraradices* was not effective as higher disease incidence (DI) was recorded. *G. mosseae* improved plant growth but was not able to suppress disease development. Only *G. deserticola* improved yield in both healthy and diseased plants, attributed to higher P intake which diminishes the deleterious effects of pathogen on yield difference.

The biocontrol activities by endophytes have now been exploited for the management of other wilt-related diseases. These diseases such as viral outbreak and nematode infestation are common in wilt-infected plants as the plant health is compromised. For bananas, endophytes have been used to control Banana Bunchy Top Virus (BBTV) of banana and nematode infestations through induction of pathogenesis-related proteins to suppress disease development. In BBTV, *Pseudomonas* (Pf1) and *Bacillus* strains (EPB22) have been successfully inoculated into banana plants, to reduce incidence of BBTV by 52–80 % (Harish et al. 2009). Nonpathogenic *Fusarium oxysporum* endophytes (V5w2) have been able to reduce *Radopholus similis* infestation in East African highland cooking bananas (Paparu et al. 2007). For cotton and cucumbers, seeds treated with *Enterobacter asburiae* (JM22) and *Pseudomonas fluorescens* (89B-61) have shown effective control against *Meloidogyne incognita* (Hallmann et al. 1998). All this propel the beneficial use of endophytes as strategies in pest and disease management.

3.3 Mechanisms of Disease Suppression

Endophytes have various mechanisms to suppress growth and spread of pathogens. The main mechanisms include antimicrobial activity, improved plant growth and vigour and induced resistance. Inhibition via production of antimicrobial compounds is by far the simplest to detect and quantify and can be achieved using plate assays and biochemical analysis. Disease suppression via improved plant growth and vigour and host-induced resistance is more complex and takes into account the soil and environmental factors. However, it is the interaction of these factors-soils, environment and host plant with endophytes, which provides a better understanding on the extent of the bioactivity of endophytes under field conditions.

3.3.1 Inhibition Via Production of Antimicrobial Compounds

Endophytes produce a variety of antimicrobial compounds. Numerous literatures would show peptides, quinones, phenols, alkaloids, steroids, terpenoids and flavonoids (Ezra et al. 2004), as well as a host of enzymes such as hydrolases, chitinases, laminarinases and glucanases (Chernin and Chet 2002), as main antimicrobial compounds responsible for pathogen inhibition. These antimicrobial compounds affect the hyphae of pathogen, rendering cellular abnormality interfering with growth (Fig. 15.2). The antimicrobial compounds studied are usually non-volatile compounds. Occasionally, some novel metabolites are discovered and elucidated for further development as bioagents. One example is the production of two new antibiotics pyrrocidines A and B from *Acremonium zeae*, an endophyte of maize, which is able to suppress *Aspergillus flavus* and the wilt fungi *Fusarium verticillioides* (Wicklow et al. 2005). For bacterial endophytes, important inhibitory molecules have also been identified such as bacisubin from *Bacillus subtilis* (Liu et al. 2007) and pyocyanin, siderophores and antibiotics from *Pseudomonas* spp. (Gupta et al. 2001). One crucial advantage of the antimicrobial compounds produced by the endophytes is the non-host specificity of the compounds (Schulz et al. 2002). This allows the use of endophytes against various pathogens and in a rather wide range of hosts. However, when introduced into field, the exact nature of endophytes and their antimicrobial production is not known. There were hypotheses indicating that the direct synthesis of antimicrobial metabolites seldom occurs *in planta,* unlike in cultures (Chaurasia et al. 2005). In fact, rugulosin is one of the few that has been detected as synthesized in the host plant (*Pinus sylvestris*) by a non-balansiaceous endophyte (Miller et al. 2002).

The potential of endophytes in producing volatile inhibitory metabolites is less explored, presumably due to the limited application of volatiles as a strategy for disease management. Unlike non-volatiles where the extracts are easily extracted, purified, characterized and amenable for further innovation (Liu et al. 2007), analysis of volatiles requires high-end instrumentation such as gas chromatography–mass spectrometry (GCMS) to detect and quantify amount produced. The analysis is laborious as the profiles of inhibitory volatile compounds were identified based on comparisons with controls (non-inhibitory isolates, agar plugs). The technique itself also does not discriminate inhibitory volatiles and non-inhibitory volatiles. Ting et al. (2010a, 2011a) have tested several bacterial and fungal endophytes against FOC race 4. The volatiles produced by the bacterial and fungal endophytes

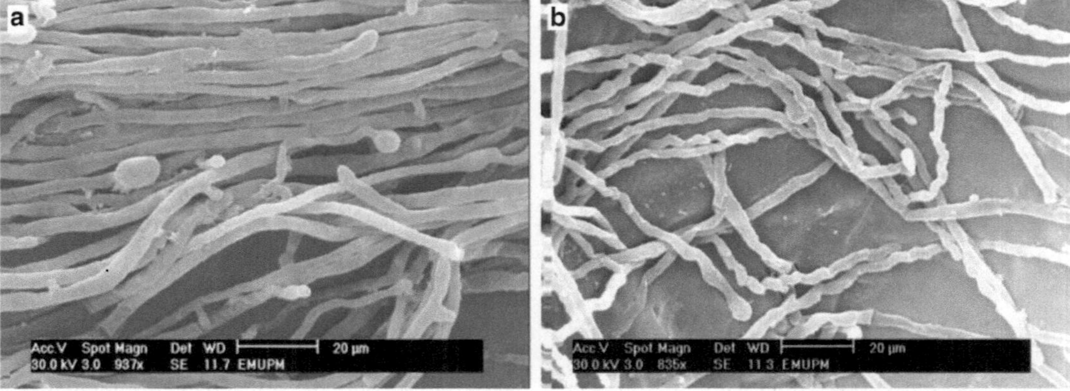

Fig. 15.2 (**a**) Scanning electron micrographs (*SEM*) of normal structures of FOC race 4 hyphae in control plate and (**b**) abnormal hyphae structures upon exposure to endophytes

were entrapped and extracted with the Solid-Phase Microextraction (SPME) technique using the SPME syringe equipped with glass fibre. Bacterial endophytes produced between 24 and 52 volatile compounds, while fungal endophytes produced 15–47 volatile compounds. There were three notable inhibitory compounds they found in bacterial endophytes: methanethiol, 2-pentanone 3-methyl and 3-undecene. These metabolites were consistently detected in *Pseudomonas aeruginosa* (LCB01 and AVA02) which showed 20.3 and 1.4 % inhibition towards FOC race 4. These compounds have not been associated with antifungal activities, but their derivatives are antimicrobial (El-Shazly et al. 2002). For fungal endophytes, volatile inhibitors were identified as 1-butanol, 3-methyl, B-butyrolactone and 1-propanol 2-methyl, produced by *Nigrospora* sp., *Penicillium citrinum*, *Cladosporium* and *Phomopsis* (Ting et al. 2010a). The production of 1-butanol, 3-methyl and 1-butanol 2-methyl has also been detected in *Gliocladium* spp. which has a role in inhibiting *Pythium ultimum* and *Verticillium dahliae* (Stinson et al. 2003). Strobel et al. (2001, 2008) also found butyrolactone produced by *Gliocladium roseum* and *Aspergillus terreus* against *Botrytis cinerea* (Cazar et al. 2005). The production of multiple volatile compounds does not necessary translate to better inhibitory effect. As observed by Ting et al. (2010a, 2011a), *Nigrospora* sp. (BTF05) may have 29 inhibitory volatile compounds but only showed 8.57 % inhibition. Contrary, *Penicillium citrinum* (BTF08) produced lesser metabolites (13 volatile compounds) but demonstrated 31.43 % inhibition towards FOC race 4 (Ting et al. 2010a). In both studies, comparison on which metabolite was the most effective was not further evaluated.

3.3.2 Improved Plant Growth and Vigour

Plants benefit from endophytic infection as when growth and vigour is significantly improved, disease suppression was also noticeably more effective. Endophytes promote growth of plants by producing phytohormones and growth-promoting substances (Tudzynski and Sharon 2002). For bacterial endophytes, growth promotion is promoted via mechanisms such as phosphate solubilization, secretion of growth-inducing phytohormones, enhancing nitrogen fixation and supplying nutrients to the host plant (Wakelin et al. 2004; Compant et al. 2005). Other benefits of plant-growth improvements due to bacterial endophytic infection include improved physiological characteristics (osmotic regulation, changes in stomatal, adjustment to root size and morphology), modification of N accumulation and metabolism and increased uptake of certain minerals essential for growth (Compant et al. 2005). With improved growth and vigour, host plants have better tolerance to disease. Bacterial endophytes in host tissues are also known to produce the enzyme ACC deaminase that utilizes the plant compound 1-aminocyclopropane-1-carboxylate (ACC). ACC deaminase cleaves to α-ketobutyrate and ammonia (the ethylene precursors), reducing the ethylene levels in host plants (Sessitsch et al. 2002; Glick et al. 2007). With reduced ethylene levels, plants do not undergo senescence rapidly, thus delaying onset of symptoms and prolonging survival and tolerance to pathogenic infection. This may explain the prolonged survival of endophytically infected plants despite succumbing to pathogenic infection (Ting et al. 2008). ACC-possessing bacteria are, therefore, generally accepted as far more superior than as non-ACC bacteria in promoting growth of host plant (Rashid et al. 2012). Host plants have also been found to respond differently to various endophytic isolates. In a study conducted on banana plantlets of the Berangan-type cv. Intan, it was found that the plantlets responded better to bacteria compared to fungal endophytes (Ting et al. 2008). Endophytic infection with *Bacillus* sp. (UPM14B1), *S. marcescens* (UPM39B3) and *Pseudomonas aeruginosa* (UPM13B8) recorded higher growth values than treatments with fungal endophytes. Combinations of endophytic isolates have been attempted as well to determine their effect on growth. It was demonstrated that combinations may not necessarily be beneficial to promote plant growth. Inoculation with nonpathogenic FO with *S. marcescens* and a cocktail of *Bacillus*, *P. aeruginosa* and *S. marcescens* were all inferior to single application of nonpathogenic FO and *S. marcescens*, reflected in root mass (Ting et al. 2008).

3.3.3 Induced Host Resistance

Induced systemic resistance (ISR) in host plants is the most common mechanism expressed upon treatments with endophytes. ISR can be induced by viable cells or dried mycelium (Dong et al. 2006). Occurrence of ISR is reflected by the elevated levels of key enzymes such as peroxidase (PO), polyphenol oxidase (PPO), phenylalanine ammonia lyase (PAL), lignothioglycolic acid (LTGA), chitinase and β-1,3-glucanase (Cabral et al. 1993; Vidhyasekaran 1997; Ting et al. 2009c, 2010b, 2012; Mohd Fishal et al. 2010). The role of enzymes in plant defence is closely associated with the synthesis of phytoalexins which are antimicrobial in nature (Sundaramoorthy et al. 2012). Some of these enzymes are also involved in the synthesis of lignin, a valuable substance in cell walls that forms a protective barrier to further penetration by pathogens (Pankhurst et al. 1979). Higher LTGA levels have been found to correlate to formation of lignified cell walls and the subsequent tolerance to pathogen infection (Ting et al. 2009c, 2010b). Enzyme activities upon endophytic infection are often high, and this is sustained throughout even with the subsequent pathogenic infection. In studies by Ting et al. (2009c, 2010b, 2012), enzyme levels in plantlets treated with the endophyte are not significantly different from plantlets subsequently challenged with FOC race 4. All these are evidences that suggest endophytic infection serves as the primary defence and sensitizes the plant to pathogenic infection. The enzymatic activities are higher in root tissues compared to leaf tissues as root tissues are the entry sites for endophytes (Ting et al. 2009c, 2010b). Levels of PO, PPO and PAL are consistently higher in root tissues while phenol levels are typically higher in leaf tissues. Disease suppression achieved via induced host resistance can at least prolong plant survival and remained symptomless for up to 21 days after infection by FOC race 4.

4 Biosourcing Endophytes

The quest to source for beneficial endophytes for wilt management begins with the few basic steps. Firstly is to isolate the endophytes and identify the potential biocontrol agents, secondly to screen for mechanisms of antagonism and thirdly to determine their host plant interactions upon application and introduction to host plants.

4.1 Isolation and Identification of Endophytes

To initiate the isolation process, selection of host plants is important. An appropriate host would increase the success of discovering endophytes that are compatible to the target plants. Schulz et al. (2002) and Strobel (2003) have proposed that selection of endophytes for the control of wilt diseases should be kept within plants in the tropics as metabolites produced may vary with biotope from which they are isolated. The most conventional approach is to isolate endophytes from close species of plants of interest, members within same families, and gradually progressing to host plants consisting of wild or resistant cultivars and nonrelated random host plants (Ting et al. 2008, 2009d). Ting et al. (2008, 2009d) have embarked on several studies on endophyte isolation from a variety of host plants. They discovered that higher numbers of nonpathogenic FO were obtained from wild bananas (a related host species to commercial banana cultivars) compared to any other host plant species. From wild bananas, 69 nonpathogenic FOs (56 % of total isolates) were recovered with 61 of these isolates antagonistic towards FOC race 4 (Ting, unpublished). This clearly validates that the approach of isolating antagonistic endophytes from known host plants such as from wild bananas (Ting et al. 2008) is strategic. It is also apparent that their observations are similar to Gloer (1997) that the production of secondary metabolites by the endophytes corresponds to respective taxon and ecological niche. Recent developments have proposed the isolation of endophytes from plants growing in special habitats (diseased areas, suppressive soil) as well as from nonrelated random host plants. Isolation of endophytes from resistant or asymptomatic plants in disease-infected areas and in suppressive soils is strategic as endophytes in these plants would have produced antimicrobial substances to resist infection (Nel et al. 2006). Forsyth et al. (2006) isolated three nonpathogenic FO isolates from roots of banana grown in Fusarium wilt suppressive soils. One of the isolates (BRIP29089) significantly reduced severity of symptom development in FOC race 1- and 4-infected ladyfinger and cavendish, respectively. Adaptability of plants to grow in these areas is therefore the consequence of the presence of endophytes in the tissues. For nonrelated host plants, random selections of medicinal plants, ornamentals, weeds and fruit trees have produced interesting results. Ting et al. (2009d) found that medicinal plants have the highest number of endophytes effective against FOC race 4, followed by weeds, ornamental and fruit trees. The rate of endophyte recovery per plant was 3.0, 2.3, 1.8 and 1.3 isolate plant^{-1} in which 1.5, 1.5, 0.5 and 0.4 isolate plant^{-1} from medicinal, weeds, ornamental and fruit trees, respectively, were antagonistic towards FOC race 4 (Ting et al. 2009d). Investigations using nonrelated hosts revolutionized the approach to source for beneficial endophytes, particularly with the exploitation of medicinal plants for endophyte isolation (Yu et al. 2010).

Endophyte isolation is conducted almost immediately upon sampling to avoid colonization of saprophytes and epiphytes which complicates the recovery of endophytes (Ting, unpublished). Surface sterilization using ethanol, Tween or sodium hypochlorite is typically employed to sterilize the tissues (Silvani et al. 2008; Ting et al. 2008). Sterilizing agents are selected carefully as tissues from different plant organs or age of plant have varying susceptibility to the sterilants (Schulz et al. 1998). The tissues are then plated on appropriate growth media such as tryptic soy agar and malt extract agar for isolation of bacterial and fungal endophytes, respectively (Silvani et al. 2008). The effectiveness of sterilization method to isolate endophytes has been challenged by the possible growth of epiphytes rather than endophytes. Epiphytes may be present as spores (Petrini 1984) and are protected from sterilizing agents by plant surface tissue. Therefore, the imprinting method has been used where tissues were imprinted on agar, and when no growth is observed, the sterilization process is considered successful (Schulz et al. 1998). In some studies, endophyte detection is important, especially studies with ecological implications and host colonization studies. The conventional and typical method to detect endophytes is through the use of histological observation. This is performed by staining with trypan blue solution. Tissue observations however are inadequate to identify the taxon and prevalence of certain hyphae morphology (Cabral et al. 1993). In recent years, more elegant method has been introduced such as the use of green fluorescent protein (*gfp*) for detection of endophytes (Mikkelsen et al. 2001). Molecular methods are to detect and identify non-sporing endophytes based on non-specific 5.8S gene, ITS1 and ITS2 regions (Wirsel et al. 2001). 16S rRNA analysis is performed for bacterial endophytes (Sessitsch et al. 2002).

4.2 Screening for Biocontrol Properties

The typical screening procedure to detect bioactivity of endophytes is by co-inoculation of endophytes with test pathogens to detect growth inhibition (dual-culture test) (Rihakova et al. 2002). This technique is appropriate for rapid screening of large number of isolates that produce non-volatile compounds as their primary mode of antagonism. This dual-culture test can be employed for both fungal and bacterial endophytes (Ting et al. 2009d). In some studies, crude extracts containing antimicrobial metabolites are further extracted using solvents like ethyl acetate, where residues are redissolved in methanol for antifungal assays through paper disk diffusion test, agar dilution assay, disk diffusion assay or mycelial radial growth test (Wicklow et al. 2005;

Yu et al. 2010). After bioactivities of substances have been tested, a separation test to identify bioactive compounds is conducted. This can be performed using thin-layer chromatography (TLC) and bioautography where the TLC plates are inoculated or seeded with test pathogens and the inhibition detected. Ultimately, active components can be isolated through precipitation-thin-layer chromatography, liquid preparation chromatography and column chromatography (Hu et al. 2010). For volatile compounds, detection is based on the Solid-Phase Microextraction (SPME) technique. In this technique, volatiles are entrapped in a glass syringe with glass fibre and the compounds eluted via gas chromatography-mass spectrophotometry (GCMS) (Strobel et al. 2001; Ting et al. 2010a, 2011a). Although this method allows for the detection of volatile compounds, distinguishing antimicrobials from non-antimicrobial compounds is difficult and laborious. It has been noted that the number of volatile compounds produced does not necessarily reflect that the isolate is more potent (Ting et al. 2010a, 2011a). At the end of the screening exercise, endophytic isolates that provide the best possible bioactivity is selected. This is typically achieved based on evidences collated from various parameters assessed, primarily the antibiosis assays. Although researchers have universally adopted this approach, it has raised concerns that preliminary laboratory tests may not reflect the true potential of endophytes as the approach lacks in mathematical or statistical analysis to link and substantiate the data. A recent study by Cavaglieri et al. (2004) attests to the reliability of antibiosis assays to select biocontrol candidates. They successfully correlated screening methods to the selection of best candidate for the control of *Fusarium verticillioides* in maize based on Pearson correlation coefficient analysis. The correlation analysis determined the effectiveness of screening method which includes niche overlap index (NOI), indices of dominance, growth rate, lag phase, antibiosis and fumonisin production, with results and bioactivity expression by four different bacteria in the greenhouse. Results based on this correlation test clearly showed that antibiosis test correlated significantly with greenhouse conditions.

4.3 Endophyte-Host Plant Interactions

The interaction of endophytes, particularly artificially inoculated endophytes with its host plants, must be examined to determine the implications of endophytic infection to their new hosts and the extent of their colonization *in planta* (Garmendia et al. 2004; Mohd Fishal et al. 2010). This would ultimately highlight the compatibility of the introduced endophyte as biocontrol agent and their effectiveness in rendering beneficial association with host and in disease control. The endophyte-host plant interaction is often conducted at the greenhouse stage where occurrence of symptoms (if any), growth promotion and disease suppression can be observed (Ting et al. 2008). Selected endophytes are inoculated to target plants, and the subsequent response of host plants to endophytic infection is recorded. Vegetative growth parameters, such as root mass, plant height, pseudostem diameter, number of leaves plant per plant, shoot weight and shoot diameter, are few examples of key indicators of growth response of host plants to endophytic infection (Pillay and Nowak 1997; Yates et al. 1997; Ting et al. 2008). Occurrence of pathogenic relationship is evident with symptom appearance and the decline in plant growth. Contrary, mutualistic association is asymptomatic with noticeable improved growth of plants. Ideally, the endophytes must be able to enhance plant growth without causing any disease or interfere with the growth and well-being of plants. At this stage, effectiveness of multi-strain applications can also be investigated to provide useful information for improvements to the application of endophytes in the field trials (Mohd Fishal et al. 2010; Sundaramoorthy et al. 2012). The interaction between endophytes and its host plants should also include investigations on the extent of endophyte colonization *in planta*. This interaction has become increasingly important in recent years due to their influence on disease suppression. It has been hypothesized that the nature of colonization of some endophytes in host tissues diluted their antimicrobial effect on

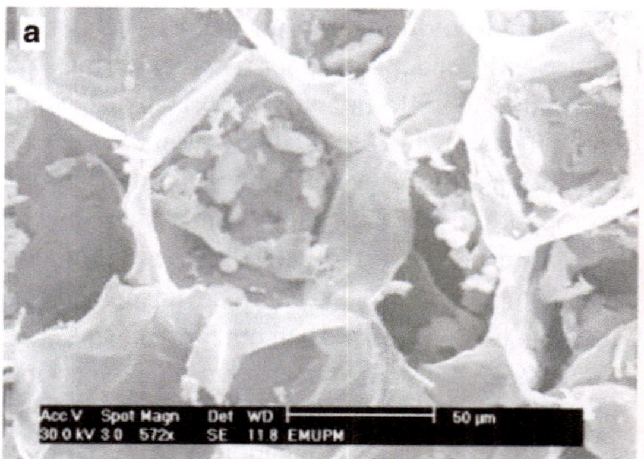
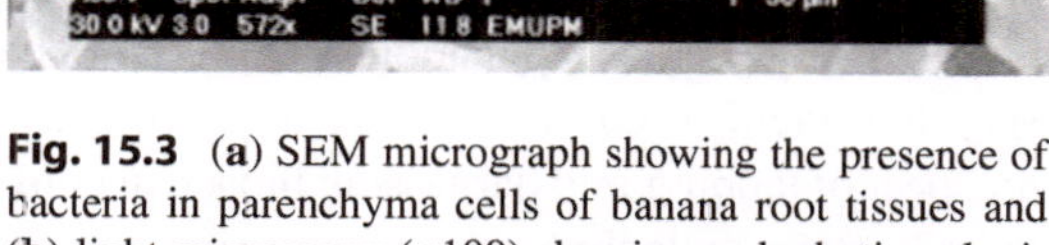

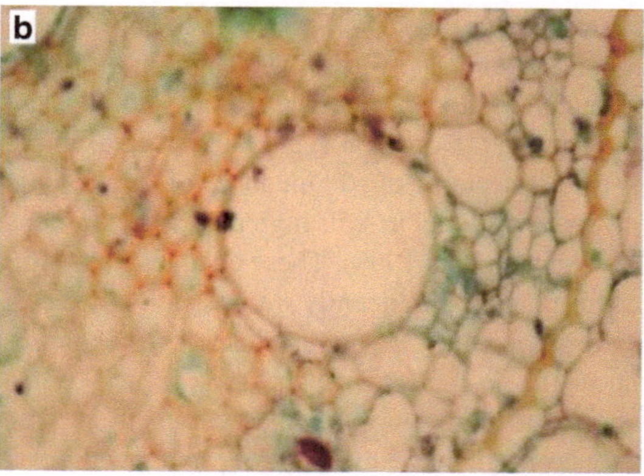

Fig. 15.3 (**a**) SEM micrograph showing the presence of bacteria in parenchyma cells of banana root tissues and (**b**) light microscopy (×100) showing endophytic colonization of fungal endophyte (stained with neutral red stain) in xylem vessels and adjacent angular parenchyma cells

pathogens (Boyle et al. 2001). The extent of endophytic colonization is determined by both the host plant and the endophyte. Endophytes colonize tissues by secreting extracellular enzymes (proteases, amylases, lipases, laccases, cellulases, xylanase) that degrade plant surfaces and cell walls to enable intracellular or intercellular colonization (Boyle et al. 2001).

In response, host plants produce host defence enzymes that may subsequently lead to limited or confined endophytic colonization to a single cell (Stone et al. 2000; Boyle et al. 2001; Deckert et al. 2001). This delicate balance of endophytes overcoming host defences, and for the host plant to allow endophyte colonization to progress in the tissues, determines the extent of endophyte colonization in both above-ground and below-ground parts of the plant. Figure 15.3 shows the colonization of bacterial and fungal endophyte in host tissues. In addition, the extent of colonization by endophytes is also motivated by anatomical differences in tissues of the plant, source-sink relationships and permeability of nutrients (Schulz and Boyle 2005). As such, the practice of providing adequate nutrients (fertilizers, soil amendments) that enhance plant growth can produce better disease suppression as well (Ting et al. 2003).

5 Challenges to the Use of Endophytes as Biocontrol Agents

The use of endophytes for the control of wilt diseases is rather limited in spite of the discoveries of several excellent endophytic isolates as biocontrol agents. The main challenge is that the biocontrol mechanism in laboratory or greenhouse is poorly translated to the field. In the field, endophytes are susceptible to environmental influences and may suffer from incompatibility with new host plants. In addition, some endophytes may also not be amenable for development as bioformulations for field application. Interactions of these factors with one another ultimately lead to the poor survival of endophytes and the expression of their bioactivity.

5.1 Environmental Influences

Environmental influences include factors natural or anthropogenic in nature. In natural soils, the soil microbiota is composed of various indigenous species which could ultimately interact with the introduced endophytes and form antagonistic or synergistic relationships. For soils exposed to anthropogenic activities such as addition of lime and fertilizers, these soils experience shifts in the relative abundance of the soil microbiota due to changes in the substrate availability and physical environment (Donnison et al. 2000). One example is the selection of dominance of endophytic mycorrhizal fungi in soils high in P concentrations (Sylvia and Neal 1990). These changes to the soil environment consequently impact endophyte survival significantly as endophytes are usually introduced to host plants via the root tissues. When applied to the roots, they face competition from the heterogenous diversity of indigenous soil microflora and the varying degree of their population densities. If the density of the introduced endophyte is diminished, the host plants may end up "acquiring" other endophytic isolates present in the soil into its host tissues, or none at all. Therefore, to mitigate this, early establishment of endophyte community on root rhizosphere as well as in host tissues is highly recommended.

5.2 Host Factors

In biological control practices, endophytes are inoculated to micropropagated plantlets to foster early colonization of endophytes to generate benefits (Sturz and Nowak 2000). In a compatible host plant, a fine balance between progressive endophytic colonization and host response to endophyte virulence is achieved (Bishop 2002). Good colonization will ensure improved host-ecological adaptability, leading to better growth of endophytes (Boyle et al. 2001), and the subsequent effective production and distribution of antimicrobial metabolites (Miller et al. 2002; Schulz et al. 2002). It is recommended that the relationship and host compatibility of introduced endophytes with every intended host plants is examined in greenhouse or field trials as different endophytes have different host specificity and may elicit different response due to plant tissue sensitivity (Sturz et al. 1999).

5.3 Endophyte Amenability

Some endophytes, in spite of their excellent bioactivity in laboratory and greenhouse trials, are not developed as bioformulations for field application. This could be due to several factors such as the nature of the endophyte, poor viability during storage and toxicity to nontargeted organisms. Some endophytes, such as *P. aeruginosa* and *S. marcescens* (Gyaneshwar et al. 2001; Ting et al. 2008), are opportunistic pathogens towards humans and alfalfa, respectively (Goto 1992). The antimicrobial compounds produced are so unspecific that they can be toxic towards all other organisms. This limits their use as biocontrol agents due to possible health risks. Some endophytes are also not amenable to large-scale application as they produce very low yield in cultures (Yu et al. 2010). The prospect of tapping into harnessing bioactive compounds from endophytes for application is progressing slowly as many of these processes are poorly understood, with missing links in biosynthesis and regulation of the antimicrobial products and their intermediates (Yu et al. 2010).

6 Improvements and Innovations to Enhance Use of Endophytes

6.1 Bioformulation

Bioformulation is the development of inoculum that allows storage while maintaining the bioactivity of the isolate when applied to the field. This approach is essential in biocontrol practices to maintain the inoculum level so that efficacy is improved (Spadaro and Gullino 2005). The conventional approach involves bioformulation into two main formulations: the dry and liquid formulations. These formulations can be adopted for endophytic isolates. Dry formulations are essentially preferred as high viability and effective disease suppression is achieved. Sporulating filamentous fungi benefits the most as dry conidia can be effectively stored with 80 % viability (Sabuquillo et al. 2010). The dried mycelium of *P. chrysogenum* collected as waste products of the pharmaceutical industry has also demonstrated effective control of *FO vasinfectum* and *V. dahliae* in cotton via induced host resistance (Dong et al. 2006). For bacteria, talc-based formulations appear to be suitable for *B. atrophaeus* S2BC-2 and mixture with

Burkholderia cepacia to control *FO gladioli* (Shanmugam and Kanoujia 2011). Talc-based preparation can also be used for fungi as in the case of nonpathogenic *Fusarium* sp. for the control of *Fusarium* wilt on basil (Minuto et al. 1997). Although dry forms are better, liquid-based formulations are useful too, especially when modernization of agro-techniques necessitates development of liquid inoculants for easy delivery. Trehalose, polyvinylpyrrolidone and glycerol are key liquid-based amendments that can act as liquid inoculants. In a study by Manikandan et al. (2010), glycerol amendment was the most effective in maintaining viability up to 6 months of storage. In the recent years, the formulation of beneficial microbes has been further studied to include enrichment materials or stabilizers. The aim is to enhance survivability of cells leading to good colonization, dominance in field over indigenous microflora and the subsequent bioactivity. Stabilizers, such as sodium alginate, glucose, sucrose, sorbitol, molasses and glycerol, have been found to benefit bioformulations (Jin and Custis 2011). *P. frequentans* have better germinability when 1.5 % sodium alginate or 7.5 % glucose is added to the conidia prior to drying (Guijarro et al. 2007). Shelf life and biocontrol efficacy of *P. oxalicum* was enhanced with the incorporation of sodium alginate (1.5 %), glycerol (20 %), sucrose (5 %) and sorbitol (5 %) (Sabuquillo et al. 2010). Ting et al. (2009b, 2011b) attempted the bioformulation of bacterial endophyte *S. marcescens* (UPM39B3) using clay-based materials enriched with starch, non-fat skimmed milk and para-aminobenzoic acid. The efficacy of the stabilizers was tested to determine viability and bioefficacy in response to UV irradiation. It was found that stabilizers improved viability of *S. marcescens* during storage, but were ineffective to protect cells from UV irradiation.

6.2 Optimization of Endophyte Application Procedure

Application of endophytes to the host plants must be studied and employed correctly to promote pre-colonization and early adaptation of endophytes in host plants. Pre-colonization encourages biotization, the metabolic response of an *in vitro* plant to microbial inoculant which leads to developmental and physiological changes (growth promotion), resulting in enhanced resistance to biotic and abiotic stress. The early adaptation by endophytes in host tissues also benefits disease control as it allows the exclusion of pathogens from niche competition. The key to effective endophyte application is to understand how and when best to introduce selected endophytes to the host plants. Endophytes can be introduced through several techniques. They can be inoculated to seeds via bacterization and to seedlings or plantlets via root dip which have all shown good control (Manikandan et al. 2010). Some application techniques require direct introduction of endophytes to the soil, which is also typically conducted for micropropagated plants. The introduction of endophytes at this stage allows the plantlets to undergo the "biopriming" (or biohardening) stage so that plantlets are strengthened against biotic and abiotic stress (Nowak 1998). Often, increased plant growth is also observed as a positive influence of "biopriming" with endophytes (Jie et al. 2009). Therefore, introduction to host plants at the early stage especially on micropropagated plants is highly recommended (Nowak 1998). The approach on how to apply endophytes is further exemplified by Manikandan et al. (2010). They revealed that a combination of seed treatment, seedling dip and soil drenching of liquid formulation was the most effective as minimum disease incidence (DI) of *Fusarium* wilt on tomato was achieved in greenhouse (17.33 %) and field (4.81 %) conditions. This observation, however, cannot be used to generalize that all endophytic applications will benefit from multiple techniques of application. The economics of implementing combination of techniques must also be considered. Similarly, multi-strain applications do not necessarily benefit from differences in techniques of application. The timing of when to introduce the multi-strains is irrelevant as similar bioefficacy is achieved when applied in a mixed or spatially separated method (Martinuz et al. 2012).

6.3 Integrated Management to Enhance Biocontrol

Integrated management adopts the combination of biological methods and chemical applications to control diseases (Elmer and McGovern 2004). In recent developments, plant-based products have been explored and applied in combination with bacterial antagonists. This approach is less toxic and organic and was first tested on antagonistic rhizobacteria. Leaf extracts of *Datura metel* (botanical formulation) mixed with *P. fluorescens* (Pf1) and *B. subtilis* (TRC54) effectively reduced wilt incidence in bananas by 64 and 75 % under greenhouse and field conditions (Akila et al. 2011). For endophytes, integrated management offers an attractive alternative to improve biocontrol activity because treatment with endophytes alone is often inadequate to sustain control. However, research on integrated management of endophytes is limited due to the complexity of the endophyte-host-environment relationship. As such, while the conventional integrated approach of implementing chemical pesticide, soil disinfestations, agronomical practices and mixtures of antagonists may favour most antagonist, benefits to endophytes remain to be seen (Spadaro and Gullino 2005).

7 Conclusions

To conclude, endophytes have excellent potential as biocontrol agents for wilt pathogens, primarily due to their strong mechanisms of antagonism, as well as their ability to render benefits to host plants (growth promotion, induced host resistance). Improvements on the delivery of endophytes and introduction to host plants can be made so that the advantage of endophyte precolonization and adaptation in host plants is fully exploited. Biotechnological innovations in bioformulations, in optimization of application and in elucidating useful bioactive compounds from endophytes can all contribute to strengthening the role of endophytes for the control of wilt diseases.

Acknowledgments The author extends her gratitude to the mentoring of Professor Dr. Sariah Meon and Associate Professor Dr. Jugah Kadir from Universiti Putra Malaysia for their invaluable insights on the field of biological control. The author also acknowledges Monash University Sunway Campus for the opportunity to continue with the pursuit of research in endophytes. Last but not least, the author is indebted to the Malaysian Ministry of Science, Technology and Innovation and the Malaysian Ministry of Agriculture, for the fellowship and funding that enabled the publication of key results by the author discussed here.

References

Akila R, Rajendran L, Harish S, Saveetha K, Raguchander T, Samiyappan R (2011) Combined application of botanical formulations and biocontrol agents for the management of *Fusarium oxysporum* f. sp. *cubense* (Foc) causing Fusarium wilt in banana. Biol Control 57:175–183

Alabouvette C, Lemanceau P, Steinberg C (1993) Recent advances in the biological control of *Fusarium* wilts. Pestic Sci 37:365–373

Bacon CW, Hill NS (1996) Symptomless grass endophytes products of coevolutionary symbioses and their role in the ecological adaptations of infected grasses. In: Redlin SC, Carris LM (eds) Endophytic fungi in grasses and woody plants. American Phytopathological Society Press, St Paul, pp 155–178

Bishop DL (2002) Gene expression of a vacuolar peroxidase with stress-induced pathogenesis in wheat sheaths. Physiol Mol Plant Pathol 61:65–71

Bolek Y, Elzik KM, Pepper AE, Bell AA, Magill CW, Thaxtoon PM, Umesh O, Reddy K (2005) Mapping of *Verticillium* wilt resistance genes in cotton. Plant Sci 168:1581–1590

Boyle C, Gotz M, Dammann-Tugend U, Schulz B (2001) Endophyte-host interactions III. Local vs. systemic colonization. Symbiosis 31:259–281

Cabral D, Stone JK, Carroll GC (1993) The internal mycoflora of Juncus spp.: microscopic and cultural observations of infection patterns. Mycol Res 97:367–376

Carroll GC (1988) Fungal endophytes in stems and leaves: from latent pathogen to mutualistic symbiont. Ecology 69:2–9

Cavaglieri LR, Passone A, Etcheverry MG (2004) Correlation between screening procedures to select root endophytes for biological control of *Fusarium verticillioides* in *Zea mays* L. Biol Control 31:259–267

Cazar ME, Schmeda-Hirschmann G, Astudillo L (2005) Antimicrobial butyrolactone I derivatives from the Ecuadorian soil fungus *Aspergillus terreus* Thorn. var *terreus*. World J Microbiol Biotechnol 21:1067–1075

Chaurasia B, Pandey A, Palni LMS, Tiveldi P, Kumar B, Colvin N (2005) Diffusible and volatile compounds produced by an antagonistic *Bacillus subtilis* strain cause structural deformations in pathogenic fungi *in-vitro*. Microbiol Res 160:75–81

Chernin L, Chet I (2002) Microbial enzymes in the biocontrol of plant pathogens and pests. In: Burns RG, Dick RP (eds) Enzymes in the environment: activity, ecology, and applications. Marcel-Dekker Inc., New York, USA, pp 171–226

Clay K (1988) Fungal endophytes of grasses: a defensive mutualism between plants and fungi. Ecology 69:10–16

Cobos MJ, Fernandez MJ, Rubio J, Kharrat M, Moreno MT, Gil J, Millan T (2005) A linkage map of chickpea (*Cicer arietinum* L.) based on populations from Kabuli × Desi crosses: location of genes for resistance to *Fusarium* wilt race 0. Theor Appl Genet 110:1347–1353

Compant S, Duffy B, Nowak J, Clement C, Barka EA (2005) Use of plant growth-promoting bacteria for biocontrol of plant diseases: principles, mechanisms of action, and future prospects. Appl Environ Microbiol 71:4951–4959

Deckert RJ, Melville L, Peterson RL (2001) Structural features of a *Lophodermium* endophyte during the cryptic lifecycle in the foliage of *Pinus strobus*. Mycol Res 105:991–997

Dong H, Zhang X, Choen Y, Zhou Y, Li W, Li Z (2006) Dry mycelium of *Penicillium chrysogenum* protects cotton plants against wilt diseases and increases yield under field conditions. Crop Prot 25:324–330

Donnison LM, Griffith GS, Hedger J, Hobbs PJ, Bardgett RD (2000) Management influences on soil microbial communities and their function in botanically diverse hay meadows of northern England and Wales. Soil Biol Biochem 32:253–263

El-Abyad MS, El-Sayed MA, El-Shanshoury AR, El-Batanouny NH (1993) Inhibitory effects of UV mutants of *Streptomyces corchorusii* and *Streptomyces spiroverticillatus* on bean and banana wilt pathogens. Can J Bot 71:1080–1086

Elmer WH, McGovern RJ (2004) Efficacy of integrating biologicals with fungicides for the suppression of *Fusarium* wilt of cyclamen. Crop Prot 23:909–914

El-Shazly A, Doral G, Wink M (2002) Chemical composition and biological activity of the essential oils of *Senecio aegyptiacus* var. *discoideus* Boiss. Z Naturforsch 57c:434–439

Ezra D, Castillo UF, Strobel GA, Hess WM, Porter H, Jensen JB, Condron MAM, Teplow DB, Sears J, Maranta M, Hunter M, Weber B, Yaver D (2004) Coronamycins, peptide antibiotics produced by a verticillate *Streptomyces* sp. (MSU-2110) endophytic on *Monstera* sp. Microbiology 150:785–793

Forsyth LM, Smith LJ, Aitken EAB (2006) Identification and characterization of non-pathogenic *Fusarium oxysporum* capable of increasing and decreasing Fusarium wilt severity. Mycol Res 110:929–935

Freeman S, Rodriguez RJ (1993) Genetic conversion of a fungal plant pathogen to a nonpathogenic, endophytic mutualist. Science 260:75–78

Freeman S, Zveibil A, Vintal H, Maymon M (2002) Isolation of nonpathogenic mutants of *Fusarium oxysporum* f. sp. *melonis* for biological control of *Fusarium* wilt in cucurbits. Phytopathology 92:164–168

Gardes M (2002) An orchid-fungus marriage: physical promiscuity, conflict and cheating. New Phytol 154:4–7

Garmendia I, Goicoechea N, Aguirreolea J (2004) Effectiveness of three *Glomus* species in protecting pepper (*Capsicum annuum* L.) against *Verticillium* wilt. Biol Control 31:296–305

Glick BR, Cheng Z, Czarny J, Duan J (2007) Promotion of plant growth by ACC deaminase-producing soil bacteria. Eur J Plant Pathol 119:329–339

Gloer JB (1997) Applications of fungal ecology in the search for new bioactive natural products. In: Wicklow DT, Soderstrom BE (eds) The mycota, vol IV, Environmental and microbial relationships. Springer, New York, pp 249–268

Goto M (1992) Fundamentals of bacterial plant pathology. Academic, San Diego

Guijarro B, Melgarejo P, De Cal A (2007) Effect of stabilizers on the shelf-life of *Penicillium frequentans* conidia and their efficacy as a biological agent against peach brown rot. Int J Food Microbiol 113:117–124

Guo JH, Qi HY, Guo YH, Ge HL, Gong LY, Zhang LX, Sun PH (2004) Biocontrol of tomato wilt by plant growth-promoting rhizobacteria. Biol Control 29:66–72

Gupta C, Dubey R, Kang S, Maheswari D (2001) Antibiosis-mediated necrotrophic effect of *Pseudomonas* GRC2 against two fungal pathogens. Curr Sci 81:91–94

Gyaneshwar P, James EK, Mathan N, Reddy PM, Reinhold-Hurek B, Ladha JK (2001) Endophytic colonization of rice by a diazotrophic strain of *Serratia marcescens*. J Bacteriol 183:2634–2645

Hallmann J, Quadt-Hallman A, Rodroaguez-Kaabana R, Kloepper JW (1998) Interactions between *Meloidogyne incognita* and endophytic bacteria in cotton and cucumber. Soil Biol Biochem 30:925–937

Harish S, Kavino M, Kumar N, Balasubramaniam P, Samiyappan R (2009) Induction of defense-related proteins by mixtures of plant growth promoting endophytic bacteria against Banana bunchy top virus. Biol Control 51:16–25

Hayward AC (1991) Biology and epidemiology of bacterial wilt caused by *Pseudomonas solanacearum*. Ann Rev Phytopathol 29:65–87

Hu HQ, Li XS, He H (2010) Characterization of an antimicrobial material from a newly isolated *Bacillus amyloliquefaciens* from mangrove for biocontrol of Capsicum bacterial wilt. Biol Control 54:359–365

Jie L, Wang Z, Cao L, Tan H, Inderbitzin P, Jiang Z, Zhou S, Shining J (2009) Artificial inoculation of banana tissue culture plantlets with indigenous endophytes originally derived from native banana plants. Biol Control 51:427–434

Jin X, Custis D (2011) Microencapsulating aerial conidia of *Trichoderma harzianum* through spray drying at elevated temperatures. Biol Control 56:202–208

Jumpponen A, Trappe JM (1998) Performance of *Pinus contorta* inoculated with two strains of root endophytic fungus, *Phialocephala fortinii*: effects of synthesis system and glucose concentration. Can J Bot 76:1205–1213

Larkin RP, Fravel DR (1998) Efficacy of various fungal and bacterial biocontrol organisms for control of *Fusarium* wilt of tomato. Plant Dis 82:1022–1028

Lee TJ, Coyne DP, Clemente TE, Mitra A (2002) Partial resistance to bacterial wilt in transgenic tomato plants expressing antibacterial lactoferrin gene. J Am Soc Hortic Sci 127:158–164

Liu YF, Chen ZY, Ng TB, Zhang J, Zhou MG, Song FP, Lu F, Liu YZ (2007) Bacisubin, an antifungal protein with ribonuclease and hemagglutinating activities from *Bacillus subtilis* strain B-916. Peptides 28:553–559

M'Piga P, Belanger RR, Paulitz TC, Benhamou N (1997) Increased resistance to *Fusarium oxysporum* f. sp. *radicis lycopersici* in tomato plants treated with the endophytic bacterium *Pseudomonas fluorescens* strain 63–28. Physiol Mol Plant Pathol 50:301–320

Manikandan R, Saravanakumar D, Rajendran L, Raguchander T, Samiyappan R (2010) Standardization of liquid formulation of *Pseudomonas fluorescens* Pf1 for its efficacy against *Fusarium* wilt of tomato. Biol Control 54:83–89

Marler M, Pedersen D, Mitchell OT, Callaway RM (1999) A polymerase chain reaction method for detecting dwarf mistletoe infection in Douglas fir and western larch. Can J For Res 29:1317–1321

Martinuz A, Schouten A, Menjivar RD, Sikora RA (2012) Effectiveness of systemic resistance toward *Aphis gossypii* (Hom., Aphididae) as induced by combined applications of the endophytes *Fusarium oxysporum* Fo162 and *Rhizobium etli* G12. Biol Control 62:206–212

Mikkelsen L, Roulund N, Lubeck M, Funck-Jensen D (2001) The perennial ryegrass endophyte *Neotyphodium lolii* genetically transformed with the green fluorescent protein gene (gfp) and visualization in the host plant. Mycol Res 105:644–650

Miller JD, Mackenzie S, Foto M, Adams GW, Findlay JA (2002) Needles of white spruce inoculated with rugulosin producing endophytes contain rugulosin reducing spruce budworm growth rate. Mycol Res 106:471–479

Minuto G, Migheli Q, Mocioni M, Gullino ML (1997) Effect of antagonistic *Fusarium* spp. and of different commercial biofungicide formulations on Fusarium wilt of basil (*Ocimum basilicum* L.). Crop Prot 16:765–769

Mohd Fishal EM, Meon S, Wong MY (2010) Induction of tolerance to *Fusarium* wilt and defense-related mechanisms in the plantlets of susceptible Berangan banana pre-inoculated with *Pseudomonas* sp. (UPMP3) and *Burkholderia* sp. (UPMB3). Agri Sci China 9:1140–1149

Nejad P, Johnson P (2000) Endophytic bacteria induce growth promotion and wilt disease suppression in oilseed tape and tomato. Biol Control 18:208–215

Nel B, Steinberg C, Labuschagne N, Viljoen A (2006) The potential of nonpathogenic *Fusarium oxysporum* and other biological control organisms for suppressing *Fusarium* wilt of banana. Plant Pathol 55:217–223

Nelson PE (1981) Life cycle and epidemiology of *Fusarium oxysporum*. In: Mace M, Bell E, Beckman CH (eds) Fungal wilt diseases of plants. Academic, New York, pp 51–77

Nowak J (1998) Benefits of *in-vitro* "biotization" of plant tissue cultures with microbial inoculants. In Vitro Cell Dev Biol Plant 34:122–130

Pankhurst CE, Craig AG, Jones WT (1979) Effectiveness of root nodules rhizobia. J Exp Bot 30:1085–1093

Paparu P, Dubois T, Coyne D, Viljoen A (2007) Defense-related gene expression in susceptible and tolerant bananas (*Musa* spp.) following inoculation with non-pathogenic *Fusarium oxysporum* endophytes and challenge with *Radopholus similis*. Physiol Mol Plant Pathol 71:149–157

Petrini O (1984) Endophytic fungi in British Ericaceae: a preliminary study. Trans Br Mycol Soc 83:510–512

Petrini O (1991) Fungal endophytes of tree leaves. In: Andrews J, Hirano S (eds) Microbial ecology of leaves. Springer, New York, pp 179–197

Petrini O (1996) Ecological and physiological aspects of host specificity in endophytic fungi. In: Redlin SC, Carris LM (eds) Endophytic fungi in grasses and woody plants. American Phytopathological Society Press, St Paul, pp 87–100

Pillay VK, Nowak J (1997) Inoculum density, temperature and genotype effects on in vitro growth promotion and epiphytic and endophytic colonization of tomato (*Lycopersicum esculentum* L.) seedlings inoculated with a pseudomonad bacterium. Can J Microbiol 43:354–361

Ploetz RC (2000) Panama disease: a classic and destructive disease of banana. Plant Health Prog (Online). doi:10.1094/PHP-2000-1204-01-HM

Ploetz RC (2005) Panama disease, an old nemesis rears its ugly head: Part 1, the beginnings of the banana export trades. Plant Health Prog (Online). doi:10.1094/PHP-2005-1221-01-RV

Pradhanang PM, Momol MT, Olson SM (2003) Effects of plant essential oils on *Ralstonia solanacearum* population density and bacterial wilt incidence in Tomato. Plant Dis 87:423–427

Rashid S, Charles TC, Glick BR (2012) Isolation and characterization of new plant growth-promoting bacterial endophytes. Appl Soil Ecol 61:217–224

Rihakova Z, Filip V, Plockova M, Smidrkal J, Cervenkova R (2002) Inhibition of *Aspergillus niger* DMF0801 by monoacylglycerols prepared from coconut oil. Czech J Food Sci 20:48–52

Sabuquillo P, DeCal A, Melgarejo P (2010) Development of a dried *Penicillium oxalicum* conidial formulation for use as a biological agent against *Fusarium* wilt of tomato: selection of optimal additives and storage conditions for maintaining conidial viability. Biol Control 54:221–229

Schardl CL (2001) *Epichloë festucae* and related mutualistic symbionts of grasses. Fungal Genet Biol 33:69–82

Schnathorst WC (1981) Life cycle and epidemiology of *Verticillium*. In: Mace M, Bell AA, Beckman CH (eds) Fungal wilt diseases of plants. Academic Press, San Diego, USA, pp 81–111

Schulz B, Boyle C (2005) The endophytic continuum. Mycol Res 109:661–686

Schulz B, Guske S, Dammann U, Boyle C (1998) Endophyte host interactions II. Defining symbiosis of the endophyte-host interaction. Symbiosis 25: 213–227

Schulz B, Boyle C, Draeger S, Rommert AK, Krohn K (2002) Endophytic fungi: a source of biologically active secondary metabolites. Mycol Res 106:996–1004

Sessitsch A, Reiter B, Pfeifer U, Wilhelmet E (2002) Cultivation-independent population analysis of bacterial endophytes in three potato varieties based on eubacterial and Actinomycetes-specific PCR of 16S rRNA genes. FEMS Microbiol Ecol 39:23–32

Shanmugam V, Kanoujia N (2011) Biocontrol of vascular wilt and corm rot of gladiolus caused by *Fusarium oxysporum* f. sp. *gladioli* using plant growth promoting rhizobacterial mixture. Crop Prot 30:807–813

Sharma VK, Nowak J (1998) Enhancement of *Verticillium* wilt resistance in tomato transplants by in vitro coculture of seedlings with a plant growth promoting rhizobacterium (*Pseudomonas* sp. strain. PsJN). Can J Microbiol 44:528–536

Sieber TN (2002) Fungal root endophytes. In: Waisel Y, Eshel A, Kafkafi U (eds) The hidden half. Marcel Dekker, New York, pp 887–917

Sikora R, Pocasangre L, Felde A, Niere B, Vu TT, Dababat AA (2008) Mutualistic endophytic fungi and in-planta suppressiveness to plant parasitic nematodes. Biol Control 46:15–23

Silvani V, Fernandez L, Pergola M, Godeas A (2008) A simple method to obtain endophytic microorganisms from field-collected roots. Soil Biol Biochem 40:1259–1263

Smith J, Putnam A, Nair M (1990) *In-vitro* control of *Fusarium* diseases of *Asparagus officinalis* L. with a *Streptomyces* or its polyene antibiotic, faerifungin. J Agri Food Chem 38:1729–1733

Spadaro D, Gullino ML (2005) Improving the efficacy of biocontrol agents against soil borne pathogens. Crop Prot 24:601–613

Stinson M, Ezra D, Hess WM, Sears J, Strobel G (2003) An endophytic *Gliocladium* sp. of *Eucryphia cordifolia* producing selective volatile antimicrobial compounds. Plant Sci 165:913–922

Stone JK, Bacon CW, White JF Jr (2000) An overview of endophytic microbes: endophytism defined. In: Bacon CW, White JF Jr (eds) Microbial endophytes. Marcel Dekker, New York, pp 3–29

Strobel GA (2003) Endophytes as sources of bioactive products. Microbes Infect 5:535–544

Strobel GA, Dirkse E, Sears J, Markworth C (2001) Volatile antimicrobials from *Muscodor albus*, a novel endophytic fungus. Microbiology 147: 2943–2950

Strobel GA, Knighton B, Kluck K, Ren Y, Livinghouse T (2008) The production of mycodiesel hydrocarbons and their derivatives by the endophytic fungus *Gliocladium roseum* (NRRL50072). Microbiology 154:3319–3328

Sturz AV, Nowak J (2000) Endophytic communities of rhizobacteria and the strategies required to create yield enhancing associations with crops. Appl Soil Ecol 15:183–190

Sturz AV, Christie BR, Matheson BG, Arsenault WJ, Buchanan NA (1999) Endophytic bacterial communities in the periderm of potato tubers and their potential to improve resistance to soil-borne plant pathogens. Plant Pathol 48:360–369

Sundaramoorthy S, Raguchander T, Ragupathi N, Samiyappan R (2012) Combinatorial effect of endophytic and plant growth promoting rhizobacteria against wilt disease of *Capsicum annum* L. caused by *Fusarium solani*. Biol Control 60:59–67

Sylvia DM, Neal LH (1990) Nitrogen affects the phosphorus response of VA mycorrhiza. New Phytol 115:303–310

Tan H, Zhou S, Deng Z, He M, Cao L (2011) Ribosomal-sequence-directed selection for endophytic *Streptomycete* strains antagonistic to *Ralstonia solanacearum* to control tomato bacterial wilt. Biol Control 59:245–254

Ting ASY, Meon S, Jugah K, Anuar AR (2003) Effect of artificially induced suppressive soil on *Fusarium* wilt. INFOMUSA 12:33–34

Ting ASY, Meon S, Kadir J, Radu S, Singh G (2008) Endophytic microorganisms as potential growth-promoters of banana. Biocontrol 53:541–553

Ting ASY, Meon S, Kadir J, Radu S, Singh G (2009a) Field evaluation of non-pathogenic *Fusarium oxysporum* isolates UPM31P1 and UPM39B3 for the control of *Fusarium* wilt in 'Pisang Berangan' (*Musa*, AAA). Acta Hortic 828:139–144

Ting ASY, Fang MT, Tee CS (2009b) Assessment on the effect of formulative materials on the viability and efficacy of *Serratia marcescens*-a biocontrol agent against *Fusarium oxysporum* f.sp. *cubense* race 4. Am J Agri Biol Sci 4:283–288

Ting ASY, Meon S, Kadir J, Radu S, Singh G (2009c) Induced host resistance by non-pathogenic *Fusarium* endophyte as a potential defense mechanism in *Fusarium* wilt management of banana. Pest Technol 3:67–72

Ting ASY, Mah SW, Tee CS (2009d) Prevalence of endophytes antagonistic towards *Fusarium oxysporum* f. sp. *cubense* race 4 in various plants. Am Eur J Sustain Agri 3:399–406

Ting ASY, Mah SW, Tee CS (2010a) Identification of volatile metabolites from fungal endophytes with biocontrol potential towards *Fusarium oxysporum* f.sp. *cubense* race 4. Am J Agri Biol Sci 5:177–182

Ting ASY, Meon S, Kadir J, Radu S, Singh G (2010b) Induction of host defense enzymes by the endophytic bacterium *Serratia marcescens* in banana plantlets. Int J Pest Manag 56:183–188

Ting ASY, Mah SW, Tee CS (2011a) Detection of potential volatile inhibitory compounds produced by endobacteria with biocontrol properties towards *Fusarium oxysporum* f.sp. *cubense* race 4. World J Microbiol Biotechnol 27:229–235

Ting ASY, Fang MT, Tee CS (2011b) Efficacy of clay-based formulated *Serratia* in reducing inoculum of *Fusarium oxysporum* f.sp. *cubense* race 4. Acta Hortic 897:421–426

Ting ASY, Mah SW, Tee CS (2012) Evaluating the feasibility of induced host resistance by endophytic isolate *Penicillium citrinum* BTF08 as a control mechanism for Fusarium wilt in banana plantlets. Biol Control 61:155–159

Tudzynski B, Sharon A (2002) Biosynthesis, biological role and application of fungal phytohormones. In: Osiecwacz HD (ed) The mycota, vol X, Industrial applications. Springer, Berlin, pp 183–212

Uppal AK, El-hadrami A, Adam LR, Tenuta M, Daayf F (2008) Biological control of potato *Verticillium* wilt under controlled and field conditions using selected bacterial antagonists and plant extracts. Biol Control 44:90–100

Vidhyasekaran P (1997) Fungal pathogenesis in plants and crops. Marcel Dekker, New York

Wakelin SA, Warren RA, Harvey PR, Ryder MH (2004) Phosphate solubilization by *Penicillium* spp. closely associated with wheat roots. Biol Fertil Soils 40:36–43

Wicklow DT, Roth S, Deyrup ST, Gloer JB (2005) A protective endophyte of maize: *Acremonium zeae* antibiotics inhibitory to *Aspergillus flavus* and *Fusarium verticillioides*. Mycol Res 109:610–618

Wilson D (1995) Endophyte-the evolution of a term, and clarification of its use and definition. Oikos 73:274–276

Wirsel SG, Liebinger W, Ernst M, Mendgen K (2001) Genetic diversity of fungi closely associated with common reed. New Phytol 149:589–598

Yates IE, Bacon CW, Hinton DM (1997) Effects of endophytic infection by *Fusarium moniliforme* on corn growth and cellular morphology. Plant Dis 81:723–728

Yu H, Zhang L, Li L, Zheng C, Guo L, Li W, Sun P, Qin L (2010) Recent developments and future prospects of antimicrobial metabolites produced by endophytes. Microbiol Res 165:437–449

Ecology and Functional Potential of Endophytes in Bioremediation: A Molecular Perspective

16

Sohail Yousaf, Muhammad Afzal, Mariam Anees, Riffat Naseem Malik, and Andrea Campisano

Abstract

Hazardous waste sites around the world result from the manufacturing, storage, use, or disposal of compounds such as petroleum hydrocarbons, nitroaromatics, organohalogens, pesticides, and metals. Traditional remediation options are expensive and environmentally invasive. In last two decades, bioremediation has emerged as a more suitable alternative, mainly for the remediation of large polluted sites. Endophytic bacteria and fungi have been the subject of considerable study to explore their potential for improving the remediation of polluted environments. In case of phytoremediation of inorganic pollutants, endophytic bacteria can reduce the phytotoxicity and increase the mobilization and accumulation of heavy metals in aboveground plant biomass. The competency of several endophytes to degrade organic pollutants and their resistance to heavy metals probably originates from their exposure to these compounds, when present in the plant/soil niche. A wide range of molecular techniques have been applied to illustrate the ecology, diversity, composition, and role of endophytes in bioremediation. Fingerprinting techniques such as terminal restriction fragment length polymorphism (T-RFLP), denaturing gradient gel electrophoresis (DGGE), real-time PCR, microarrays, and metagenomics are being used to characterize the metal-resistant and organic pollutant-degrading endophytes.

S. Yousaf (✉) • R.N. Malik
Department of Environmental Sciences, Quaid-i-Azam University, Islamabad, Pakistan
e-mail: syousaf@qau.edu.pk

M. Afzal
Environmental Biotechnology, National Institute for Biotechnology and Genetic Engineering (NIBGE), Faisalabad, Pakistan

M. Anees
Department of Biochemistry, Quaid-i-Azam University, Islamabad, Pakistan

A. Campisano
Sustainable Agro-ecosystems and Bioresources Department, IASMA Research and Innovation Centre, Fondazione Edmund Mach, Via E. Mach 1, 38010 San Michele all'Adige (TN), Italy

V.C. Verma and A.C. Gange (eds.), *Advances in Endophytic Research*,
DOI 10.1007/978-81-322-1575-2_16, © Springer India 2014

1 Introduction

Contamination of soil, surface and groundwater, and ultimately food with organic and inorganic contaminants (such as petroleum hydrocarbons, polycyclic aromatic hydrocarbons, pesticides, salts, and heavy metals) is becoming one of the sternest environmental problems all over the world. Higher levels of these toxic contaminants in the environment have been associated with human health risks including cancer (McGuinness and Dowling 2009). Substantial efforts are being made to remediate contaminated environments. In the light of the high cost of site remediation, it is important to develop and refine innovative, low-cost, and environment-friendly methods for cleaning polluted environments. During the last two decades, bioremediation has emerged as a potential tool to clean the contaminated environments. The remediation of polluted soil and water by the use of biological agents such as microorganisms or plants is termed bioremediation. Phytoremediation (a type of bioremediation), the use of plants and their associated microorganisms for the detoxification of soil and water, is a relatively new and promising technique. It is a low-cost technique as compared to expensive and destructive mechanical methods (Sung et al. 2003; Germaine et al. 2006; Yousaf et al. 2011; Afzal et al. 2013).

Since traditional remediation options currently available are expensive and environmentally invasive, phytoremediation turns out to be a more suitable alternative, mainly for the remediation of large polluted sites with diffuse contamination (Weyens et al. 2009a). One of the major drawbacks of this technology is that plants are sensitive to higher concentration of pollutants (Glick 2003). Toxicity of organic pollutants or their toxic end products inhibit plant growth and biomass production and consequently cannot support effective degradation and sequestration of pollutants (Puschenreiter et al. 2001; Rajkumar et al. 2009). Moreover, the metals at elevated levels are generally toxic to most of the plants, impairing their metabolism and reducing plant growth (Sheoran et al. 1990). The toxicity of organic and inorganic pollutants can be reduced by the inoculation of plants with pollutant-degrading and/or plant growth-promoting microorganisms (Weyens et al. 2009a; Glick 2010; Yousaf et al. 2011; Afzal et al. 2012; Ahmad et al. 2012; Khan et al. 2013a, b).

The plant–microbe interactions that enhance plant growth have been studied widely. Recently, many studies have been performed to explore the potential of plant-associated microorganisms for increasing the remediation of polluted soil and water. Though plant growth-promoting rhizobacteria have been in use for a long time as inoculants for improving phytoremediation activity (Gentry et al. 2004; Thompson et al. 2005; Lebeau et al. 2008), endophytic bacteria and fungi show an even higher potential for improving the remediation of polluted environment (Weyens et al. 2009b; Yousaf et al. 2011). Besides the production of many plant growth-promoting chemicals, endophytic bacteria have often demonstrated a natural competency for pollutant degradation, either directly or (as vectors) by carrying degradative traits (Weyens et al. 2009a).

In case of phytoremediation of inorganic pollutants, endophytic bacteria can reduce the phytotoxicity and increase the mobilization and accumulation of heavy metals in aboveground plant biomass. The competency of several endophytes to degrade organic pollutants and their resistance to heavy metals probably originates from their elevated level of exposure to the compounds present in the plant/soil niche. This natural potential of endophytic bacteria is being explored with regard to enhanced phytoremediation activity (Weyens et al. 2010; Yousaf et al. 2011; Afzal et al. 2012). Different molecular techniques have been used to illustrate the ecology, diversity, composition, and role of endophytes in bioremediation (Yousaf et al. 2011; Afzal et al. 2012). This book chapter describes the ecology and functional potential of endophytes to improve phytoremediation of organic xenobiotics and toxic metals in contaminated environment.

2 Ecology of Endophytes

Endophyte can be defined as microorganisms (bacteria and fungi) that colonize the internal tissues of the plant without causing disease in their host (Schulz and Boyle 2006), and every plant sampled so far has shown to host at least one endophytic bacterium and/or fungus (Ryan et al. 2008; Li et al. 2012). The host plant tissues are at least transiently symptomless, and the microbial colonization inside plant tissues can be observed through histological means, by isolation from surface-sterilized plant tissues or through direct amplification of bacterial or fungal nuclear DNA from plant tissues (Stone et al. 2000). Endophytes have been studied in different geographic and ecological regions and were found to be ubiquitous within all tested plants (Li et al. 2012). Endophytes consist of endophytic bacteria, endophytic fungi, and actinomycetes (Raghukumar 2008). However, endophytic bacteria and fungi were extensively studied and applied to improve plant growth and phytoremediation activity.

Recently, colonization and metabolic activity of endophytic bacteria have been demonstrated in rhizo- and endosphere of different plants (Weyens et al. 2009a; Afzal et al. 2012). During phytoremediation of contaminated soil, endophytes, able to degrade and resist pollutants, may colonize within the plant tissues that are less toxic than soil. From this perspective, the heavy metal-resistant and alkane-degrading endophytic bacterial community has been investigated in the roots and shoots of different plants vegetated in contaminated soils. Recently, high numbers of endophytic bacteria from Italian ryegrass and birdsfoot trefoil having alkane monooxygenase (*alkB*) and hydroxylase (CYP153) genes with the potential to degrade hydrocarbons (Yousaf et al. 2010a) have been isolated.

The endophytic microorganisms are secluded from environmental changes due to their colonization inside the plant tissues. Endophytes usually colonize the intercellular spaces and have been isolated from shoot, root, and seed (Posada and Vega 2005; Compant et al. 2010; Afzal et al. 2012). Moreover, they have been found in woody tree species, such as pear and oak, herbaceous crop plants such as maize and sugar beet, and grasses such as ryegrass. Endophytes have also been isolated from different plants showing tolerance/resistance to different pollutants. For example, endophytic bacteria were isolated from poplar trees, vegetated in hydrocarbon contaminated soil, with the potential to degrade different hydrocarbon compounds (Porteous-Moore et al. 2006). In another study, Dashti et al. (2009) isolated endophytic bacteria from nodules of legume crops (*Vicia faba* and *Lupinus albus*), and these bacteria were possessing hydrocarbon degradation activities. Pollutants can shape the microbial community of endophytes naturally present in the host plant. Even such synthetic chemicals as antifungal compounds used in agriculture appear to affect the diversity of endophytic microbial species (Pancher et al. 2012).

Heavy metal-resistant endophytic bacteria were also isolated from plants vegetated in metal-polluted soil (Idris et al. 2004; Chen et al. 2010). These studies demonstrated that within the diverse endophytic bacterial communities, several endophytic bacterial strains have the potential to increase phytoremediation of organic and inorganic pollutants. Differences in the resistance levels among diverse plant species and even cultivars to organic and inorganic pollutants in soil and water might relate to variations in their endophytic microbial population and activities (Weyens et al. 2009c; Khan et al. 2013a).

Endophytic bacteria are thought to enter plant tissues mainly from roots or at sites of wounding, but some phyllosphere bacteria may also be the source of endophytes (Compant et al. 2010). Plants are very specific to soil microorganisms and prefer successful, competent, and beneficial endophytes while selecting (Sessitsch et al. 2002; Hardoim et al. 2008; Chen et al. 2010). Many endophytes can efficiently colonize the rhizosphere as well as endosphere. Recently several studies demonstrated that inoculated endophytic bacteria efficiently colonize the rhizosphere as well as plant interior tissues (Yousaf et al. 2011; Afzal et al. 2012). As compared to rhizosphere and phyllosphere bacteria, endophytes are likely to interact more closely with their host. In these very close plant–bacteria interactions, plants provide nutrients and residency for endophytic bacteria; in return endophytic bacteria can enhance plant growth and development. The composition of bacterial endophytes in various plants was different, with many strains closely related to common soil bacteria representative of genera such as *Enterobacter*, *Pseudomonas*, *Bacillus*, *Arthrobacter*, *Burkholderia*, and *Methylobacterium* (Lodewyckx et al. 2002; Yousaf et al. 2010a; Luo et al. 2011a, b).

Different molecular techniques have been applied to study the ecology of endophytic bacteria. Clone libraries, fingerprinting techniques, terminal restriction fragment length polymorphism (T-RFLP), real-time PCR, microarrays, and metagenomics are all being used to characterize the metal-resistant and organic pollutant-degrading endophytes. For example, the endophytic bacterial community in the roots of *Cyperus rotundus* L. was investigated by culture-dependent and molecular approaches (Jurelevicius et al. 2010). PCR–DGGE analysis of the 16S rRNA gene showed that the alkane-degrading and nitrogen-fixing bacterial population in rhizosphere and root samples had a high degree of similarity, indicating that rhizobacteria are source of endophytes. In another metagenomics study, Sessitsch et al. (2012) proposed that high more diverse endophytic communities have higher potential for plant growth promotion, improvement of plant stress resistance, biocontrol against pathogens, and bioremediation, regardless of their cultivability. Some endophytes possessed alkane-degrading genes (*alkB*), indicating their potential application in bioremediation. Recently, endophytes were observed in roots, stems, and leave of *Sedum alfredii*, vegetated in heavy metal-contaminated soil, with a significantly higher density in roots, followed by leave and stems (Xinxian et al. 2011). These endophytic bacteria were closely related phylogenetically to *Pseudomonas*, *Bacillus*, *Stenotrophomonas*, and *Acinetobacter* by 16S rRNA sequence analysis.

Pollution in soil not only delays plant growth but also causes changes in the size, composition, and activity of plant-associated microbial communities. Numerous studies have demonstrated the effect of different pollutants on endophytic bacterial diversity, biomass, and activity (Doelman 1986; Kamnev et al. 2005). However, endophytes isolated from plants growing in polluted soil are tolerant to high concentration of pollutants than those isolated from plants vegetated in uncontaminated soil (Lodewyckx et al. 2002; Idris et al. 2004). For instance, the metal-resistant endophytic bacteria have been isolated from different plants such as *Alyssum bertolonii*, *Thlaspi caerulescens*, *Thlaspi goesingense*, and *Nicotiana tabacum* (Lodewyckx et al. 2002; Idris et al. 2004; Barzanti et al. 2007; Mastretta et al. 2009).

The diversity of endophytic fungi from six dominant plant species growing in a Pb–Zn-contaminated soil was investigated (Li et al. 2012). Higher endophytic fungi colonization was observed in stems than leaves in each plant species. Furthermore, it was observed that among the isolated endophytic fungi, *Phoma*, *Alternaria*, and *Peyronellaea* were the main genera and the relative frequencies were 39.6, 19.0, and 20.4 %, respectively. Some were showing sensitivity to metals and did not grow on the media containing 3.6 mM Pb^{2+} or 11.5 mM Zn^{2+}. However, growth of some endophytic fungi was stimulated in the presence of tested metals. The results indicated that heavy metal-resistant fungal population in interior tissues of the plants growing in Pb–Zn-polluted soil was moderately abundant and some of the isolates have a marked adaptation to Pb^{2+} and Zn^{2+} metals, which has a potential application in phytoremediation. Similarly, the highest diversity of fungal genotypes was observed in the roots of *Thlaspi praecox* (Brassicaceae) vegetated in soil containing Cd, Zn, and Pb (Pongrac et al. 2008). The sequences obtained corresponded to *Glomus* species (Glomeromycota) to putative dark septate endophytes *Phialophora verrucosa* and *Rhizoctonia* sp. and to some other fungi from Asco- and Basidiomycota. This was the first report of dark septate endophytes occurrence in roots of hyperaccumulating *T. praecox*, a promising candidate for phytoextraction.

3 Role of Endophytes in Enhanced Bioremediation

There is an increasing interest in developing the potential biotechnological applications of endophytes for improving bioremediation of contaminated soil and water as well as the sustainable production of nonfood crops for biomass and biofuel production (Barac et al. 2004; Ryan et al. 2008). Despite this interest, details on the metabolic cross talk between endophytic microorganism (with potential to contribute to the plant bioremediation effect) and the host plant are far from being fully explored. Moreover, for efficient phytoremediation of organic pollutants, plants have to host an efficient pollutant-degrading microflora. The colonization and metabolic activity of inoculated alkane-degrading endophytic bacteria were determined in the root and shoot of Italian ryegrass vegetated in diesel-contaminated soil by using quantitative PCR (Andria et al. 2009). Endophytic bacteria efficiently colonized the rhizosphere and particularly plant interior and also showed higher levels of expression of alkane-degrading genes (*alkB*) in the rhizosphere, shoot and root interior. Similarly, Afzal et al. (2011) and Yousaf et al. (2011) demonstrated the colonization and catabolic activity of alkane-degrading endophytic bacteria during phytoremediation of diesel-contaminated soil. They found that endophytic bacteria showed higher levels of gene abundance and expression in the root and shoot of Italian ryegrass (*Lolium multiflorum* var. Taurus), birdsfoot trefoil (*Lotus corniculatus* var. Leo), and alfalfa (*Medicago sativa* var. Harpe). These findings indicate that endophytic bacteria can efficiently degrade the organic pollutants *in planta* and therefore can reduce both the phytotoxicity and evapotranspiration of organic pollutants after their uptake by the plant.

Some endophytes can also reduce the toxicity of organic and inorganic pollutants through production of different chemicals and enzymes such as iron chelators, siderophores, organic acids, 1-aminocyclopropane-1-carboxylate (ACC) deaminase, and various degrading enzymes (Sheng et al. 2008a; Soleimani et al. 2010a; Li et al. 2012). For instance, Germaine et al. (2006) demonstrated the effect of bacterial endophytes, able to degrade herbicides, on phytoremediation of the organochlorine herbicide 2,4-dichlorophenoxyacetic acid. They found that the inoculated endophytic bacterium efficiently colonized the roots and few or no symptoms of toxicity were observed. In another study, Yousaf et al. (2011) found that the endophytic *Enterobacter ludwigii* strains, possessing alkane degradation and plant growth-promoting ACC deaminase activity, were more efficient in plant growth enhancement and hydrocarbon degradation as compared to the strains possessing only alkane degradation activity. Weyens et al. (2010) observed that the engineered endophyte, *Burkholderia cepacia* VM1468 possessing (a) the pTOM-Bu61 plasmid, coding for constitutive trichloroethylene (TCE) degradation, and (b) the *ncc–nre* Ni resistance/sequestration system, enhanced plant biomass production and decreased phytotoxicity of both Ni and trichloroethylene. In another study, Madhaiyan et al. (2007) demonstrated the influence of endophytic bacteria *Methylobacterium oryzae* and *Burkholderia* sp. inoculation to tomato plants vegetated in Ni- and Cd-contaminated soil. They found that endophytes reduced the toxicity of Ni and Cd in tomato plants and improved plant growth under gnotobiotic and pot culture experiments.

Endophytes produce different phytohormones, small signaling molecules essential for plant growth, development, and defense, and these play an important role in accelerating phytoremediation activity. Among them, bacterial ACC deaminase reduces soil contamination-induced plant stress. An ethylene precursor, ACC, is degraded into ammonia and α-ketobutyrate, resulting in decrease of ethylene biosynthesis (Hardoim et al. 2008). Some endophytes can improve plant growth indirectly by producing ACC deaminase to modulate the ethylene levels in plants (Chen et al. 2010; Ma et al. 2011a, b; Zhang et al. 2011). Moreover, some endophytic fungi can produce enzymes to degrade the phenolic acid allelochemicals released by decomposing foliage, which have negative impacts on the growth of plants and microbes in soil, thus potentially alleviating the effects of the ecological suppression via allelochemicals (Chen et al. 2011; Li et al. 2012).

In general the endophytes can improve plant growth and enhance plant adaptation in contaminated soil by various mechanisms. These include pollutant detoxification, nitrogen fixation, phosphate solubilization, indole acetic acid production, and the production of siderophores (Verma et al. 2001; Muthukumarasamy et al. 2002; Lee et al. 2004). Endophytes can also improve plant growth and health by improving mineral nutrition or increasing resistance or tolerance to contaminants (Ryan et al. 2008; Khan et al. 2013a). In addition, different beneficial effects of endophytic microorganisms on plant growth have been observed including osmotic adjustment, stomatal regulation, modification of root morphology, improved uptake of minerals, and alteration of nitrogen accumulation and metabolism (Compant et al. 2005). Enhanced phytoremediation of organic pollutants has also been observed by using engineered endophytes; pollutant-degrading and/or resistance genes were introduced into bacteria by genetic engineering (Barac et al. 2004; Newman and Reynolds 2005; Doty 2008; Soleimani et al. 2010a).

More recently, the increasing interest of the researchers for endophytes has opened their application for the remediation of heavy metal-contaminated soils (Chen et al. 2010; Glick and Stearns 2011). Metal-resistant and/or plant growth-promoting endophytes are reported to be present in various plants growing in metal-contaminated soils and play an important role in successful survival and growth of plants (Rajkumar et al. 2009), and endophyte-assisted phytoremediation has been documented as a promising technology for in situ remediation of metal-contaminated soils (Li et al. 2012). Table 16.1 shows examples of successful applications of plant–endophyte partnerships for the remediation of contaminated soil.

4 Endophytic Bacteria and Phytoremediation

During phytoremediation of organic pollutants, plants can benefit from their associated endophytic bacteria possessing pollutant degradation pathways and metabolic activities, leading to the reduction of both phytotoxicity and evapotranspiration of volatile contaminants (Weyens et al. 2009b).

4.1 Endophytic Bacteria-Assisted Phytoremediation of Organic Pollutants

Several findings reveal that plants draw organic pollutants into their rhizosphere to varying extents via the transpiration stream (Harvey et al. 2002). Subsequently organic compound degradation may occur in the rhizosphere or in the plant or both. However, some organic compounds can be taken up into the root symplast and translocated via the xylem (apoplast) to the shoot, in the transportation stream. In the shoot the pollutant can be taken up into the shoot symplast, where it may be sequestered or degraded by the endophytic bacteria (Weyens et al. 2009b). Although plants can sequester or degrade organics, they do not rely on organic compounds as a source of energy and carbon. So in order to get more efficient mineralization of these organic pollutants, plants rely on their associated endophytic microorganisms, mainly bacteria and fungi. Endophytic bacteria can efficiently colonize in internal tissues of plants as well as being metabolically active in organic pollutant degradation (Weyens et al. 2009b; Yousaf et al. 2011; Khan et al. 2013a).

Many studies have reported that endophytes can improve plant's adaptation and growth while growing in polluted soil by the virtue of their plant growth-promoting ACC deaminase activity. This bacterial activity plays an important role in alleviation of different types of stress in plants, including the stress induced by the presence of organic pollutants in soil (Glick et al. 1998; Arshad et al. 2007; Weyens et al. 2009b). Endophytic bacteria with appropriate catabolic pathways, oftenly encoded on plasmids or transposons, offer more suitable candidates for efficient plant–endophyte partnership for the remediation of organic pollutants. The mobile elements for such catabolic pathways are believed to transfer across endophytic bacteria by horizontal gene transfer (Taghavi et al. 2005; Weyens et al. 2009b; Yousaf et al. 2010a).

Table 16.1 Examples of successful application of plant–endophyte partnership for the remediation of contaminated soil

Plant	Endophyte	[a]EB/EF	Contaminant	References
Lolium multiflorum	*Pantoea* sp. ITSI10, *Pseudomonas* sp. *Rhodococcus* sp. ITRI43, *Enterobacter ludwigii*	EB	Diesel	Yousaf et al. (2010b, 2011), Afzal et al. (2011, 2012)
Pisum sativum	*Pseudomonas putida* strain POPHV6	EB	2,4-Dichlorophenoxy acetic acid	Germaine et al. (2006)
Pisum sativum	*Pseudomonas putida* VM1441 (pNAH7)	EB	Naphthalene	Germaine et al. (2009)
Thlaspi goesingense	*Methylobacterium* sp. V3, *Sphingomonas* sp. pfB27, *Curtobacterium* sp. VKM, *Curtobacterium* sp. VKM,	EB	Ni	Idris et al. (2004, 2006)
Alyssum bertolonii	*Microbacterium* O1, *Pseudomonas* B7, *Curtobacterium* C2, *Staphylococcus* A3, *Bacillus* B3, *Arthrobacter* F3B	EB	Ni	Barzanti et al. (2007)
Brassica napus	*Pseudomonas fluorescens* G10, *Microbacterium* sp. G16.	EB	Pb	Sheng et al. (2008a)
Thlaspi caerulescens	*Sphingomonas* sp., *Methylobacterium* sp.	EB	Zn and Cd	Lodewyckx et al. (2002)
Lycopersicon esculentum	*Methylobacterium oryzae* strain CBMB20 and *Burkholderia* sp.	EB	Cd	Madhaiyan et al. (2007)
Ricinus communis	*Pseudomonas* sp. M6, *Pseudomonas jessenii* M15	EB	Ni, Cu, Zn	Rajkumar and Freitas (2008)
Brassica juncea	*Enterobacter aerogenes*, *Rahnella aquatilis*	EB	Ni, Cr	Kumar et al. (2009)
Orychophragmus violaceus	*Flavobacterium* sp.		Zn	He et al. (2010)
Festuca arundinacea, *Festuca pratensis*	*Neotyphodium coenophialum*	EF	Cd	Soleimani et al. (2010a)
Festuca arundinacea, *Festuca pratensis*	*Neotyphodium coenophialum* and *Neotyphodium uncinatum*	EF	Polyaromatic hydrocarbons	Soleimani et al. (2010b)
Brassica juncea	*Acacia auriculaeformis*	EF	Cd and Ni	Jiang et al. (2008)
Rape	*Mucor* sp. CBRF59	EF	Cd and Pb	Deng et al. (2013)
Triticum aestivum, *Vigna radiata*, and *Solanum melongena*	Glomus mosseae	EF	Poly aromatic hydrocarbons	Rabie 2005
Rice	*Phomopsis* sp. B3	EF	Phenanthrene	Tian et al. (2007)
Festuca arundinacea	*Lewia* sp.	EF	Poly aromatic hydrocarbons	Cruz-Hernández et al. (2013)

[a]*EB* phytoremediation endophytic bacteria, *EF* endophytic fungi

The beneficial traits of endophytic bacteria can improve plant growth and thereby contribute to enhance phytoremediation activity. Consequently, isolation and characterization of beneficial traits of endophytic bacteria in pollution-tolerant plants become frequent practice (Luo et al. 2011a, b; Yousaf et al. 2010a, 2011). Several endophytic bacteria isolated from different plants have the potential to degrade organic pollutants and have been proven to be useful in improving phytoremediation activity. Siciliano et al. (2001) were the first to report that endophytes isolated from plants vegetated in hydrocarbon-contaminated soil would be naturally rich in alkane-degrading genes and would enhance phytoremediation of organic pollutants. Lodewyckx et al. (2001) observed that endophytes of yellow pine enhanced the phytoremediation activity of inoculated plant. Latter on endophytic bacteria isolated from poplar trees (*Populus deltoides* × *Populus nigra* DN34) were found capable of degrading nitro-aromatic compounds (Van Aken et al. 2004a). The phytoremediation of explosive compounds is of great interest, and an endophytic *Methylobacterium* sp. strain (isolated from poplar plant) showed the ability to degrade 2,4,6-trinitrotoluene, hexahydro-1,3,5-trinitro-1,3,5-triazine, and ocyahydro-1,3,5,7-tetranitro-1,3,5-tetrazocine (Van Aken et al. 2004b). Barac et al. (2004) reported that the inoculation of plants with an endophytic *B. cepacia*, capable of degrading toluene, reduced the volatilization of toluene and also the phytotoxic effect of toluene on inoculated plant. It may be possible that this endophyte played a major role in the mineralization of explosive compounds present in poplar. Germaine et al. (2006) inoculated pea (*Pisum sativum*) plant with an endophyte naturally possessing the ability to mineralize the 2,4-dichlorophenoxyacetic acid. The inoculated strain actively colonized in the rhizo- and endosphere of the plant and reduced the accumulation of 2,4-dichlorophenoxyacetic acid in the aerial tissues. These studies reveal the usefulness of bacterial endophytes to enhance the phytoremediation of herbicide-contaminated soil and to reduce levels of toxic organic pollutants in the crop plants. Porteous-Moore et al. (2006) looked at the diversity of endophytic bacteria associated with poplar trees vegetated in benzene, toluene, ethylbenzene, and xylene (BTEX)-contaminated soil. They found that most of the endophytic bacteria possessed the ability to degrade the BTEX. Later on Barac et al. (2009) observed on the same site that after remediation, when the BTEX concentration decreased below the detection limit, the degradation capacity of the endophytic bacteria disappeared, resembling more that of a natural situation. In another study, Germaine et al. (2009) found that an endophytic bacterial strain, *Pseudomonas putida* VM1441, efficiently colonized both the rhizosphere and interior root tissues. Inoculation with this endophytic strain resulted in the protection of the host plant from the phytotoxic effects of naphthalene. Furthermore, inoculation facilitated higher (40 %) naphthalene degradation as compared to uninoculated plants. Recently, Weyens et al. (2010) observed that poplar cuttings inoculated with *P. putida* W619-TCE promoted plant growth and reduced TCE phytotoxicity and the amount of TCE present in the leaves. In another study, they observed that inoculation with *P. putida* W619 (wild type) enhanced plant growth, reduced activities of antioxidative defense-related enzymes, and decreased stomatal resistance (Weyens et al. 2012).

Weyens et al. (2009b) discussed the advantages of use of endophytic bacteria in bioremediation and emphasized that plant–endophyte partnership can be applied to improve plant biomass production even on marginal land. In a more recent review, Khan et al. (2013a) mentioned some benefits of using endophytes in phytoremediation of hydrocarbon-polluted soil. These can be (a) inoculated endophytic bacteria possessing pollutant degradation potential that efficiently colonized the plant; (b) endophytic bacteria showing higher levels of abundance and expression of bioremediation genes *in planta*, indicating that toxic pollutants taken up by plant may be degraded inside the plant tissues by such endophytic bacteria; and (c) inoculation method and soil type affecting endophytic bacterial colonization and activity.

4.2 Endophytic Bacteria-Assisted Phytoremediation of Heavy Metals

Heavy metals found in nature are the main component of a variety of enzymes, transcription factors, and other proteins. However, their higher concentration in environment is thought of as a contaminant. Soil–heavy metals cannot be degraded biologically; they can only be transformed into organic complexes. To stimulate their remediation, fast-growing plants with high metal uptake and biomass production are required. However, metal availability, metal uptake, and phytotoxicity for the plants are the main limiting factors for the application of phytoextraction. To improve phytoremediation of heavy metals-contaminated soil, plant–endophyte partnerships are considered a promising approach. During the phytoremediation of toxic metals, the metal-resistant plant growth-promoting endophytic bacteria can improve the plant health and development, reduce metal toxicity, and influence metal translocation and accumulation in different plant tissues. Endophytic bacteria with sequestration activity can also decrease metal phytotoxicity and affect metal translocation to the aboveground plant biomass (Weyens et al. 2009b).

Many endophytic bacteria isolated from different plants growing well in metal-contaminated soils have been found to be metal resistant; the presence of high amounts of heavy metals in the plant might directly select endophytes able to resist to contaminated environmental conditions. It is also possible that plants accumulating high concentration of heavy metals may be colonized by heavy metal-resistant endophytic bacteria (Idris et al. 2004; Rajkumar et al. 2009; Ma et al. 2011a). These bacteria have been found in different plants, such as *Alyssum bertolonii*, *Alnus firma*, *Brassica napus*, *Nicotiana tabacum*, *Thlaspi caerulescens*, and *Solanum nigrum*. Recently, the potential use of plant growth-promoting endophytes to accelerate phytoremediation of metalliferous soils has been reviewed by Ma et al. (2011b). During phytoremediation of heavy metal-contaminated soil, in some cases, endophytic bacteria may confer to the plants a higher tolerance to heavy metal stress. Recently, four heavy metals-resistant endophytic bacteria, *Serratia nematodiphila* LRE07, *Enterobacter aerogenes* LRE17, *Enterobacter* sp. LSE04, and *Acinetobacter* sp. LSE06, were isolated from *Solanum nigrum* L. vegetated in metal-polluted soil (Chen et al. 2010). Their plant growth-promoting activities such as production of ACC deaminase, indole-3-acetic acid (IAA), siderophores, and phosphate-solubilizing acids were determined. When these endophytic bacteria were inoculated to *S. nigrum*, all of these enhanced plant growth and Cd extraction from soil. All four inoculated strains colonized the rhizosphere and even the plant interior tissues. It was demonstrated that plant growth-promoting endophytic bacteria are a valuable resource which could be exploited to enhance phytoremediation activity. Truyens et al. (2012) demonstrated changes in the population of endophytic bacteria present in seeds of transgenerationally Cd-resistant *Arabidopsis thaliana*. Their data support the hypothesis that certain endophytes are selected for transmission to the next generation and that their presence might be important for subsequent germination and early seedling development. Similarly, Mastretta et al. (2009) demonstrated that Cd-resistant seed endophytes improve plant growth and decrease metal toxicity when inoculated to *Nicotiana tabacum*.

Plant-associated bacteria can improve phytoremediation efficiency by increasing the solubility, availability, and transport of heavy metals and nutrients through the production of organic acids, release of chelators, or redox changes (Puente et al. 2009; Shin et al. 2012; Zhang et al. 2011). Some endophytic bacteria can enhance heavy metal mobilization through the production of low-molecular-mass organic acids. For example, Sheng et al. (2008b) found that the water-soluble Pb significantly increased, along with a decrease in pH, in a suspension with endophytic bacterial growth and suggested that this might be due to the production of organic acids by endophytic bacteria. Similarly, it was observed that the release of 5-ketogluconic acid by the endophytic bacterium, *Gluconacetobacter diazotrophicus*, increased the solubility of different Zn sources such as ZnO, $ZnCO_3$, and $Zn_3(PO_4)$, thus increasing the availability of Zn for plant uptake (Saravanan et al. 2007). Kuffner et al. (2010) also reported that some endophytic bacteria could produce metal-mobilizing organic compounds into the contaminated soil. However, compared to rhizobacteria-assisted phytoremediation, fewer studies were performed to explore plant–endophyte partnerships for the remediation of heavy metal-contaminated soil (Chen et al. 2010).

The effectiveness of phytoremediation of metal-polluted soil depends mainly on metal uptake and accumulation in aboveground biomass. Several studies have shown that endophytic bacteria possessing heavy metal-resistant and plant growth-promoting activities can enhance metal uptake and accumulation in plant. For instance, plant growth-promoting endophytic bacterium, *Pseudomonas* sp. A3R3, significantly enhanced Ni concentration in *Alyssum serpyllifolium* (Ma et al. 2011b). Similarly, inoculation of *Solanum nigrum* with heavy metal-resistant endophytic bacteria increased Cd uptake and accumulation in root, shoot, and leaf tissues (Chen et al. 2010). In another study, Mastretta et al. (2009) demonstrated that the application of Cd-resistant endophyte *Sanguibacter* sp. into *Nicotiana tabacum* enhanced Cd accumulation in shoot tissues.

Plant–endophytic bacteria partnership can also be exploited for the remediation of soil contaminated with both organics and heavy metals. Generally phytoremediation of such contaminated soil is complicated. The presence of hazardous metals potentially decreases microbial activities including degradation of organic contaminants (Sandrin and Maier 2003). A very promising simple strategy for the phytoremediation of contaminated soil with mixed waste is the use of endophytic bacteria that are able to (1) mineralize organic pollutants and (2) enhance the translocation of toxic metals from soil to aboveground plant biomass (Weyens et al. 2009b).

5 Endophytic Fungi and Bioremediation

Although microbial-assisted bioremediation of organic and inorganic pollutants from soils has been extensively studied, there is limited information about the effect of inoculation of endophytic fungi on plants for the remediation of polluted sites.

5.1 Endophytic Fungi-Assisted Phytoremediation of Organic Pollutants

Plant–fungi partnership can also be useful for the remediation of soil contaminated with organic pollutants. For instance, Escalante-Espinosa et al. (2005) demonstrated that mutual benefits between *Cyperus laxus* and inoculated hydrocarbon-degrading microorganisms (including endophytic fungi) enhanced phytoremediation of hydrocarbon-contaminated soil. Phenological characteristics of inoculated plants were improved as compared to non-inoculated plants. The rhizospheric bacteria and fungi counts were higher for planted treatments (inoculated and non-inoculated) than for unplanted pots. The maximum phytoremediation rate (0.51 mg of TPH g^{-1} of dry plant d^{-1}) for inoculated plants was attained at day 60 of experiment and was two times higher than non-inoculated plants. Similarly, the effect of endophytic fungi on remediation efficacy of wheat, mung bean, and eggplant grown in soil spiked with hydrocarbons was assessed (Rabie 2005). Fungal inoculation significantly increased degradation of hydrocarbons in planted soil compared to uninoculated planted soil. Moreover, physiological data indicated that plant growth and tolerance increased with endophytic fungi inoculation. As consequence of the treatment with fungi, the plants provide a greater sink for the contaminants since they survive and grow better. Later on Tian et al. (2007) demonstrated the degradation of phenanthrene by endophytic fungi, *Phomopsis* sp., with rice plant. The degradation rate of phenanthrene was enhanced by fungal inoculation. In addition, the presence of fungi decreased the injury to rice under the condition of phenanthrene stress. In another study, endophytic fungi improved phytoremediation of petroleum oil-contaminated soil (Soleimani et al. 2010b). In this study, the effect of inoculation of two grass species (*Festuca arundinacea* Schreb. and *Festuca pratensis* Huds.) by endophytic fungi (*Neotyphodium coenophialum* and *Neotyphodium uncinatum*) on the remediation of soil polluted with petroleum hydrocarbons was assessed. Endophytic fungi inoculation enhanced root and shoot biomass and resulted in higher activity levels of water-soluble phenols and dehydrogenase in the soil. Significantly higher hydrocarbon degradation was observed in the rhizosphere of plants inoculated with these endophytic fungi. This study concluded that inoculation of grasses with endophytic fungi could be an efficient approach for the remediation of soils, polluted with hydrocarbons. Recently, Cruz-Hernández et al. (2013) demonstrated that *Lewia* sp. (endophytic fungus) improved the efficiency of polyaromatic hydrocarbon removal by *Festuca arundinacea*, on both perlite and soil, stimulating pyrene accumulation in roots. Inoculation with *Lewia* sp. stimulated (100 %) root growth in spiked perlite. Inoculated plants exhibited higher phenanthrene degradation (100 %) as compared to non-inoculated plants in perlite and soil.

5.2 Endophytic Fungi-Assisted Phytoremediation of Heavy Metals

Fungal endophytes have been shown to ameliorate metal toxicity for their plant hosts by restricting the uptake of toxic metals and by improving the supply of essential elements. As effective metal phytoremediation strategies depend on the ability of the plant to tolerate and accumulate metals from the environment, the wide prevalence of endophytic fungi and their potential to modulate metal speciation, toxicity, and mobility make them a key component of any remediation effort (Likar 2011). In polluted soil, endophytic fungi restored plant biomass despite higher Cu and Zn accumulation in plant organs, especially roots. Endophytic fungi can also enhance expression of certain genes in plants vegetated in polluted soil. For example, inoculation with the endophytic fungi caused an overall induction of *PaMT1*, *PaMT2*, *PaMT3*, *PaSPDS1*, *PaSPDS2*, and *PaADC* gene expression, together with increased free and conjugated polyamine levels in plants grown on polluted soil, but not in those grown on nonpolluted soil (Cicatelli et al. 2010).

Endophytic fungi may increase host plant tolerance to biotic and abiotic stresses. Soleimani et al. (2010a) demonstrated the effect of inoculation of endophytic fungi on cadmium (Cd) tolerance, accumulation, and translocation in grasses. Plants inoculated with fungi exhibited higher biomass production (12–24 %) and higher potential to accumulate Cd in roots (6–16 %) and shoots (6–20 %) than fungi-free plants. Maximum photochemical efficiency of photosystem II (F_v/F_m) revealed that Cd stress was significantly reduced in fungi-infected plants compared to noninfected ones. In another study, effect of inoculation of *Acacia auriculaeformis*-associated endophytic fungi on the growth of mustard [*Brassica juncea* (L.) Coss. var. foliosa Bailey] vegetated in Cd- and Ni-contaminated soils was assessed for improving phytoremediation activity (Jiang et al. 2008). Endophytic *Trichoderma* H8 and rhizosphere *Aspergillus* G16 were applied for rhizoremediation of Cd-, Ni-, and Cd–Ni combination-contaminated soils through association with *B. juncea*. Compared with the non-inoculated control plants, inoculation with *Trichoderma* H8 produced 109 %, 41 %, and 167 % more fresh weight (FW) in the Cd-, Ni-, and Cd–Ni-contaminated soils, respectively ($P<0.05$). The inoculation also increased the translocation factors and metal bioconcentration factors. The study suggested that the use of plant–fungi association may be a promising strategy to remediate metal-contaminated soils.

Recently, a fungal endophyte was isolated from rapeseed roots grown in a heavy metal-contaminated soil and characterized to determine its potential in improving phytoremediation of heavy metals from soil (Deng et al. 2011). The isolate CBRF59 was identified as *Mucor* sp. based on morphological characteristics and phylogenetic analysis. The addition of active mycelia of CBRF59 significantly increased the availability of soil Pb and Cd. The results showed that the endophytic fungus was potentially applicable for the decontamination of metal-polluted media.

Endophytic fungi can also improve plant adaptation and growth in heavily contaminated soil, including those contaminated with heavy metals. More recently, Deng et al. (2013) isolated protoplasts from endophytic fungi to carry out self-fusion of protoplasts for their improvement of metal tolerance. Self-fusant CBRF59T3 with resistance to 25 mM Cd(II) was constructed by self-fusion of inactivated protoplasts from *Mucor* sp. CBRF59. The dry weight of rapeseed inoculated with CBRF59 and CBRF59T3 was higher than that of the uninoculated rapeseed. Inoculation of CBRF59T3 further increased the dry weight of rapeseed by 62 % than CBRF59 in the higher Cd(II) + Pb(II)-contaminated soil. Compared with CBRF59, CBRF59T3 inoculation increased the concentration of Cd(II) in rapeseed shoots by 35–189 % in Cd(II)- and Cd(II) + Pb(II)-contaminated soils. The inoculation of CBRF59T3 also enhanced the translocation of Cd(II) from roots to shoots and increased the amount of extracted Cd(II) from rapeseed. These results proposed that the mutant constructed by protoplast fusion is a feasible and efficient method to enhance stress tolerance of uncharacterized fungi for phytoremediation of heavy metal-contaminated soils.

Deram et al. (2011) studied the effect of non-mycorrhizal (dark septate fungi) and mycorrhizal (arbuscular mycorrhizae) fungi in order to assess the most efficient utilization of each type in relation to heavy metal uptake and tolerance. Mycorrhizal infestation (hyphae, arbuscules, and vesicles) was adversely affected by soil pollution almost to exclusion. The intensity of colonization with non-mycorrhizal was very low in the presence of arbuscular mycorrhizal in non-contaminated soils but higher in polluted soils. Recently, during endophytic fungi-assisted phytoremediation of heavy metal-contaminated soil, genes belonging to different functional categories, plus other genes related to heavy metal stress (metallothioneins, phytochelatin synthase, glutathione synthase, arginine decarboxylase), were analyzed by quantitative (q)RT-PCR (Cicatelli et al. 2012). The levels of gene expression were generally downregulated, or unaffected, in polluted soil compared with controls, the main exceptions being phytochelatin synthase and clathrin, and strongly upregulated in the presence of arbuscular mycorrhizae fungi, especially *Glomus mosseae*.

6 Molecular Tools Used in Endophytic Bioremediation

Bioremediation requires a good understanding of the physicochemical characteristics of the contaminated environment, as well as a thorough description of the microbial communities involved in key physiological processes. The assessment of the microbial communities, their capabilities to degrade the target contaminants, and the resilience of these abilities can often be the most relevant aspects to take into consideration in the design and implementation of a bioremediation application. For such assessments to be as complete and momentous as possible, microbial communities need to be characterized in terms of structure, phenotypic potential, function, and interactions with the environment (Rittmann et al. 2006). While 90–99 % of microbes living in the environment defy conventional cultivation in the laboratory on synthetic solid or liquid media (Amann et al. 1995), a major methodological revolution in microbial ecology occurred in the 1990s which made possible the application of culture-independent molecular tools to study the diversity and dynamics of microbial communities in fine detail. Over the past few years, such powerful tools are enabling the qualitative (e.g., fingerprinting techniques) and quantitative (e.g., dot blot and fluorescence in situ hybridization, real-time PCR, pyrosequencing based) depiction of environmental microbial communities and are helpful in identifying new catabolic operons of xenobiotics in environmental bacteria.

A number of culture-independent molecular techniques presently used to study complex endophytic microbial communities are compatible with a high-throughput setup. These include fingerprinting, real-time PCR, microarrays, metagenomics, metatranscriptomics, metaproteomics, and metabolomics. These techniques are discussed below.

6.1 Fingerprinting Techniques

Genetic fingerprinting techniques provide a definite pattern or profile of a given microbial community. They are based on the separation of amplicons after PCR amplification of phylogenetic (e.g., 16S rRNA) or functional genes using universal or specific primers. Some of these fingerprinting techniques have the prospective for high-throughput design, such as terminal restriction fragment length polymorphism (T-RFLP), denaturing gradient gel electrophoresis (DGGE), or automated ribosomal intergenic spacer analysis (ARISA).

6.1.1 Terminal Restriction Fragment Length Polymorphism (T-RFLP)

T-RFLP separates fragments obtained by enzymatic restriction of PCR amplicons according to their size. The use of labeled primers allows a rapid, automated, and high-throughput detection of polymorphic terminal fragments. The separated fragments are visualized by an automated DNA sequencer as a pattern of peaks on an electropherogram. In addition, the identification of T-RFLP peaks can be directly obtained by comparing them to databases (Marsh et al. 2000). Yousaf et al. (2010b) studied the hydrocarbon degradation potential, colonization, and community composition of alkane-degrading endophytic bacteria in diesel-contaminated soils by using T-RFLP analysis. The diversity of indigenous alkane-degrading endophytes was investigated on the basis of cytochrome P450-type and *alkB* alkane hydroxylase endosphere of Italian ryegrass and birdsfoot trefoil. The effect of compost amendment on hydrocarbon degradation and alkane-degrading communities during phytoremediation of diesel fuel was also examined. T-RFLP analysis of alkane degrader showed that both uninoculated plants hosted different communities carrying the *alkB* gene. The *alkB* genes were detected in the root interior of Italian ryegrass but not in the endosphere of birdsfoot trefoil. Cultivation-independent analysis revealed that Italian ryegrass and birdsfoot trefoil, both sampled at flowering, hosted distinct alkane-degrading communities. In association with both plants, *alkB*- as well as CYP153-containing microorganisms were detected, and different subtypes of alkane degradation genes were encountered.

T-RFLP analysis was used to study the effect of inoculation method (seed imbibement and soil inoculation) on endophytic bacterial colonization, plant growth promotion, and hydrocarbon degradation (Afzal et al. 2012). This study demonstrated that the inoculation of hydrocarbon-degrading microorganisms decreased the potential toxic effects of hydrocarbons. The endophytic strain ITSI10 exhibiting alkane degradation as well as ACC deaminase activities was highly efficient in enhancing plant biomass (especially root biomass) and consequently hydrocarbon degradation and performed better than strain MixRI75 lacking ACC deaminase activity. Plant growth and hydrocarbon degradation were correlated with bacterial colonization, and T-RFLP analysis confirmed the presence of ITSI strain in the endosphere and rhizosphere of ryegrass.

6.1.2 Automated Ribosomal Intergenic Spacer Analysis (ARISA)

ARISA is a method of microbial community analysis which provides an estimation of microbial diversity and community composition without the bias imposed by culture-based approaches or the labor and expense involved with 16S rRNA gene clone library construction (Fisher and Triplett 1999). This method has been successfully used to evaluate the microbial diversity of both bacteria and fungi in environmental samples (Ranjard et al. 2001). ARISA provides a community-specific profile, with each peak ideally corresponding to one kind of organism in the original environmental sample. ARISA has been successfully used to determine the microbial diversity of both bacteria and fungi in marine, freshwater, and soil environments (Ranjard et al. 2001; Brown et al. 2005; Kennedy et al. 2005).

Torzilli et al. (2006) compared fungal communities from four salt marsh plants through ARISA. By using a semiquantitative transformation of ARISA data (individual peak heights/total of peak heights from the entire community), they were able to distinguish among the fungal communities associated with four different salt marsh plants, two of which (*D. spicata* and *S. perennis*) had not been examined previously with molecular techniques.

6.1.3 Denaturing Gradient Gel Electrophoresis (DGGE)

DGGE analysis demonstrated the potential to improve phytoremediation of aromatic pollutants by inoculating functional endophytic bacterial strains. The endophytic bacterial strain *Achromobacter xylosoxidans* F3B, which was able to utilize aromatic compounds as a sole carbon source, was inoculated into vetiver grass. The results showed that the endophytic bacteria strain F3B could maintain a stable population in plant roots without largely interfering with the diversity of native endophytes. Furthermore, the strain F3B could protect plants against toluene stress and maintain chlorophyll content of leaves. A 30 % reduction of evapotranspiration through vetiver leaves was observed. This research showed the potential of the endophytic bacterium *A. xylosoxidans* F3B in reducing phytotoxicity and improving phytoremediation (Ho et al. 2013).

The DGGE analysis from soybean roots revealed the effect of glyphosate herbicide application on some endophytic groups, not observed by isolation. These endophytes were exclusive for plants cultivated in soil with preplanting glyphosate application, such as *Herbaspirillum* sp., and other groups in plants that were cultivated in soil without glyphosate, such as *Xanthomonas* sp. and *Stenotrophomonas maltophilia*. Furthermore, only two bacterial species *Pseudomonas oryzihabitans* and *Burkholderia gladioli* were recovered from soybean plants by glyphosate enrichment isolation and showed different sensibility profiles to the glyphosate. These results suggest that the application at preplanting of the glyphosate herbicide may interfere with the endophytic bacterial community's equilibrium. A more complete comprehension of the interaction between herbicides and plant-associated bacterial communities is an important factor for more effective crop management. The results from this study indicated that preplanting application of glyphosate herbicide influenced endophytic bacterial communities in soybean plants. Increased application of glyphosate may change endophytic populations, such as latent pathogens and plant growth-promoting bacteria, which could result in changes in plant production (Sobral et al. 2005).

6.2 Real-Time PCR

Real-time PCR monitors the progress of a PCR reaction based on the detection and quantification of a fluorescent reporter molecule that binds to the target PCR template. From the amount of fluorescence emitted at each cycle in the exponential phase, it is possible to calculate the initial amount of target template. Real-time PCR is highly sensitive, down to a detection limit of 1–2 genome copies (Inglis and Kalischuk 2004). Real-time PCR does not require any tedious post-PCR steps for the quantification of amplicons, as their amount is monitored in real time. Therefore, this is a high-throughput technique with superior analytical sensitivity for the detection and quantification of specific genes in environmental samples (Harms et al. 2003).

A quantitative and real-time PCR enabled determination of microbial abundance and expression of alkane monooxygenase (*alkB*) genes in rhizosphere, shoot and root interior of Italian ryegrass (*Lolium multiflorum* L.) (Andria et al. 2009). To assess the role of endophytes in alkane degradation, Italian ryegrass (*Lolium multiflorum* L.) was grown in sterile soil with 0, 1, or 2 % diesel and inoculated with alkane-degrading endophytes. Plant colonization potential of these strains as well as the abundance and expression of alkane monooxygenase (*alkB*) genes in rhizosphere, shoot and root interior, was examined. Results showed that the endophyte strain better colonized the plant, particularly the plant interior, and also showed higher expression of *alkB* genes suggesting a more efficient degradation of the pollutant. This study suggested that endophytes have a high potential to be used in phytoremediation applications.

7 Emerging Technologies in Endophytic Bioremediation

In recent years, a number of technological advancements have overcome some of the above constraints leading to improved reliability, cost efficacy, and speed of bioremediation. These methods range from mere monitoring and improvement of intrinsic bioremediation to novel ideas of genetically engineering the functional genes for bioremediation application.

7.1 DNA Microarray

Microarrays (or microchips) are based on the property of a single-stranded DNA or RNA molecule ("target molecule") to hybridize to a complementary molecule ("probe") attached to a solid support (Zhou 2003). Compared to conventional nucleic acid membrane hybridization, microarrays offer the advantage of efficiency (thousands of probes can be spotted on a slide), high sensitivity, and rapid ("real-time") detection (Eyers et al. 2004). In environmental genomics, three main classes of microarrays have been developed: (i) phylogenetic oligonucleotide arrays (POAs), which contain oligonucleotide probes targeting taxonomic genes (e.g., 16S rRNA gene); (ii) functional gene arrays (FGAs), where probes target genes encoding key enzymes involved in particular processes; and (iii) community genome arrays (CGAs), which are constructed from whole genomic DNA of many different strains or species (Zhou 2003). To explore the structure of environmental microbial communities, successful hybridization of microbial community DNA amplified using the phi29 DNA polymerase was performed, and its application to groundwater samples containing sub-nanogram quantities of microbial DNA was demonstrated (Wu et al. 2006). When the objective is to examine not just existing but transcribed, i.e., functioning genes (mRNA based analysis), a T7 polymerase-based linear amplification approach using fusion primers provides ample and representative amounts of mRNAs for functional analysis of microbial communities (Gao et al. 2007). This suggests great application potential of microarrays to investigate the endophytic microbial communities involved in bioremediation of pollutants.

7.2 Pyrosequencing-Based Metagenomics

Environmental microbes represent a central source of genetic material with biotechnological interest and applications across all key industries including bioremediation. More than 99 % of microbes are uncultivable under existing laboratory regime, which prevents access to the vast variety of their products which have the potential for industrial utilization. Metagenomics promises continuous source of novel pollutant-degrading genes for increased effectiveness and service of transgenic (microbes and plants) technologies for direct use in bioremediation sectors. Additionally, the technology can be used to manufacture novel degrading enzymes from uncultivable bacteria for improved enzymatic remediation technology. In recent years, metagenomic approaches have started yielding some novel industrial products including bioremediation gene/enzyme from uncultivable microbes. Using such an approach, Fan et al. (2012) isolated a novel thermo stable pyrethroid-hydrolyzing enzyme which could be used in the detoxification of pyrethroids. Following a similar metagenomic approach in cow rumen, a novel gene responsible for the degradation of 3,5,6-trichloro-2-pyridinol, a persistent and toxic metabolite of the insecticide chlorpyrifos, was isolated (Renukaradhya and Shah 2010). Exploring the microbial community structure by using DNA-dependent molecular and metagenomic techniques is helping to better understand the role of these endophytes in bioremediation. Further analysis of sequenced genomes, the characterization of yet unknown genes, and the identification of genes expressed during degradation of different organic and/or inorganic pollutants will help to improve our understanding of endophytes and their role in bioremediation. It will not be surprising if some new factors, functions, as well as genes required for endophytic lifestyle of microorganisms will be identified in the near future.

References

Afzal M, Yousaf S, Reichenauer TG, Kuffner M, Sessitsch A (2011) Soil type affects plant colonization, activity and catabolic gene expression of inoculated bacterial strains during phytoremediation of diesel. J Hazard Mater 186:1568–1575

Afzal M, Yousaf S, Reichenauer TG, Sessitsch A (2012) The inoculation method affects colonization and performance of bacterial inoculant strains in the phytoremediation of soil contaminated with diesel oil. Int J Phytoremediat 14:35–47

Afzal M, Yousaf S, Reichenauer TG, Sessitsch A (2013) Ecology of alkane-degrading bacteria and their interaction with the plant. In: de Bruijn FJ (ed) Molecular microbial ecology of the rhizosphere, vols 1–2. Wiley-Blackwell, Hoboken, pp 975–989

Ahmad F, Iqbal S, Anwar S, Afzal M, Islam E, Mustafa T, Khan QM (2012) Enhanced remediation of chlorpyrifos from soil using ryegrass (*Lollium multiflorum*) and chlorpyrifos-degrading bacterium *Bacillus pumilus* C2A1. J Hazard Mater 237–238:110–115

Amann RI, Ludwig W, Schleifer KH (1995) Phylogenetic identification and in situ detection of individual microbial cells without cultivation. Microbiol Rev 59:143–169

Andria V, Reichenauer TG, Sessitsch A (2009) Expression of alkane monooxygenase (alkB) genes by plant-associated bacteria in the rhizosphere and endosphere of Italian ryegrass (*Lolium multiflorum* L.) grown in diesel contaminated soil. Environ Pollut 157:3347–3350

Arshad M, Saleem M, Hussain S (2007) Perspectives of bacterial ACC deaminase in phytoremediation. Trends Biotechnol 25:356–362

Barac T, Taghavi S, Borremans B, Provoost A, Oeyen L, Colpaert JV, Vangronsveld J, van der Lelie D (2004) Engineered endophytic bacteria improve phytoremediation of water-soluble, volatile, organic pollutants. Nat Biotechnol 22:583–588

Barac T, Weyens N, Oeyen L, Taghavi S, van der Lelie D, Dubin D, Spliet M, Vangronsveld J (2009) Fiels note: hydraulic containment of a BTEX plum using poplar trees. Int J Phytoremediat 11:416–424

Barzanti R, Ozino F, Bazzicalupo M, Gabbrielli R, Galardi F, Gonnelli C, Mengoni A (2007) Isolation and characterization of endophytic bacteria from the nickel hyperaccumulator plant *Alyssum bertolonii*. Microb Ecol 53:306–316

Brown MV, Schwalbach MS, Hewson I, Fuhrman JA (2005) Coupling 16SITS rDNA clone libraries and automated ribosomal intergenic spacer analysis to show marine microbial diversity: development and application to a time series. Environ Microbiol 7:1466–1479

Chen L, Luo S, Xiao X, Guo H, Chen J, Wan Y, Li B, Xu T, Xi Q, Rao C (2010) Application of plant growth-promoting endophytes (PGPE) isolated from *Solanum nigrum* L. for phytoextraction of Cd-polluted soils. Appl Soil Ecol 46:383–389

Chen Y, Peng Y, Dai CC, Ju Q (2011) Biodegradation of 4-hydroxybenzoic acid by *Phomopsis liquidambari*. Appl Soil Ecol 51:102–110

Cicatelli A, Lingua G, Todeschini V, Biondi S, Torrigiani P, Castiglione S (2010) Arbuscular mycorrhizal fungi restore normal growth in a white poplar clone grown on heavy metal-contaminated soil, and this is associated with upregulation of foliar metallothionein and polyamine biosynthetic gene expression. Ann Bot 106:791–802

Cicatelli A, Lingua G, Todeschini V, Biondi S, Torrigiani P, Castiglione S (2012) Arbuscular mycorrhizal fungi modulate the leaf transcriptome of a *Populus alba* L. clone grown on a zinc and copper-contaminated soil. Environ Exp Bot 75:25–35

Compant S, Reiter B, Sessitsch A, Nowak J, Clement C, Ait Barka E (2005) Endophytic colonization of *Vitis vinifera* L. by plant growth-promoting bacterium *Burkholderia* sp. strain PsJN. Appl Environ Microbiol 71:1685–1693

Compant S, Clément C, Sessitsch A (2010) Plant growth-promoting bacteria in the rhizo- and endosphere of plants: their role, colonization, mechanisms involved and prospects for utilization. Soil Biol Biochem 42:669–678

Cruz-Hernández A, Tomasini-Campocosio A, Pérez-Flores LJ, Fernández-Perrino FJ, Gutiiérrez-Rojas M (2013) Inoculation of seed-borne fungus in the rhizosphere of *Festuca arundinacea* promotes hydrocarbon removal and pyrene accumulation in roots. Plant Soil 362:261–270

Dashti N, Khanafer M, El-Nemr I, Sorkhoh N, Ali N, Radwan S (2009) The potential of oil-utilizing bacterial consortia associated with legume root nodules for cleaning oily soils. Chemosphere 74:1354–1359

Deng Z, Cao L, Huang H, Jiang X, Wang W, Shi Y, Zhang R (2011) Characterization of Cd- and Pb-resistant fungal endophyte *Mucor* sp. CBRF59 isolated from rapes (*Brassica chinensis*) in a metal-contaminated soil. J Hazard Mater 185:717–724

Deng Z, Zhang R, Shi Y, Hu L, Tan H, Cao L (2013) Enhancement of phytoremediation of Cd- and Pb-contaminated soils by self-fusion of protoplasts from endophytic fungus *Mucor* sp. CBRF59. Chemosphere 91(1):41–47

Deram A, Languereau F, Haluwyn CV (2011) Mycorrhizal and endophytic fungal colonization in *Arrhenatherum elatius* L. roots according to the soil contamination in heavy metals. Soil Sediment Contam 20:114–127

Doelman P (1986) Resistance of soil microbial communities to heavy metals. In: FEMS symposium, Elsevier, Copenhagen, 33(4):369–384

Doty SL (2008) Enhancing phytoremediation through the use of transgenics and endophytes. New Phytol 179:318–333

Escalante-Espinosa E, Gallegos-Martínez ME, Favela-Torres E, Gutiérrez-Rojas M (2005) Improvement of the hydrocarbon phytoremediation rate by *Cyperus laxus* Lam. inoculated with a microbial consortium in a model system. Chemosphere 59:405–413

Eyers L, George I, Schuler L, Stenuit B, Agathos SN, El Fantroussi S (2004) Environmental genomics: exploring the unmined richness of microbes to degrade xenobiotics. Appl Microbiol Biotechnol 66:123–130

Fan X, Liu X, Huang R, Liu Y (2012) Identification and characterization of a novel thermostable pyrethroid-hydrolyzing enzyme isolated through metagenomic approach. Microb Cell Fact 11:33–37

Fisher MM, Triplett EW (1999) Automated approach for ribosomal intergenic spacer analysis of microbial diversity and its application to freshwater bacterial communities. Appl Environ Microbiol 65:4630–4636

Gao H, Yang ZK, Gentry TJ, Wu L, Schadt CW, Zhou J (2007) Microarray-based analysis of microbial community RNAs by whole-community RNA amplification. Appl Environ Microbiol 73:563–571

Gentry T, Rensing C, Pepper I (2004) New approaches for bioaugmentation as a remediation technology. Crit Rev Environ Sci Technol 34:447–494

Germaine KJ, Liu X, Cabellos GG, Hogan JP, Ryan D, Dowling DN (2006) Bacterial endophyte-enhanced phytoremediation of the organochlorine herbicide 2,4-dichlorophenoxyacetic acid. FEMS Microbiol Ecol 57:302–310

Germaine KJ, Keogh E, Ryan D, Dowling DN (2009) Bacterial endophyte-mediated naphthalene phytoprotection and phytoremediation. FEMS Microbiol Lett 296:226–234

Glick BR (2003) Phytoremediation: synergistic use of plants and bacteria to clean up the environment. Biotechnol Adv 21:383–393

Glick BR (2010) Using soil bacteria to facilitate phytoremediation. Biotechnol Adv 28:367–374

Glick BR, Stearns JC (2011) Making phytoremediation work better: maximizing a plant's growth potential in the midst of adversity. Int J Phytoremediat 13:4–16

Glick BR, Penrose DM, Li J (1998) A model for the lowering of plant ethylene concentrations by plant growth-promoting bacteria. J Theor Biol 190:63–68

Hardoim PR, van Overbeek LS, Elsas JDV (2008) Properties of bacterial endophytes and their proposed role in plant growth. Trends Microbiol 16:463–471

Harms G, Layton AC, Dionisi HM, Gregory IR, Garrett VM, Hawkins SA (2003) Real-time PCR quantification of nitrifying bacteria in a municipal wastewater treatment plant. Environ Sci Technol 37:343–351

Harvey PJ, Campanella BF, Castro PML, Harms H, Lichtfouse E, Schäffner AR, Smrcek S, Werck-Reichhart D (2002) Phytoremediation of polyaromatic hydrocarbons, anilines and phenols. Environ Sci Pollut Res 9:29–47

He CQ, Tan G, Liang X, Du W, Chen Y, Zhi G, Zhu Y (2010) Effect of Zn-tolerant bacterial strains on growth and Zn accumulation in *Orychophragmus violaceus*. Appl Soil Ecol 44:1–5

Ho YN, Hsieh JL, Huang CC (2013) Construction of a plant-microbe phytoremediation system: combination of vetiver grass with a functional endophytic bacterium, *Achromobacter xylosoxidans* F3B, for aromatic pollutants removal. Bioresour Technol. doi:http://dx.doi.org/10.1016/j.biortech.2013.02.051

Idris R, Trifonova R, Puschenreiter M, Wenzel WW, Sessitsch A (2004) Bacterial communities associated with flowering plants of the Ni hyperaccumulator *Thlaspi goesingense*. Appl Environ Microbiol 70:2667–2673

Idris R, Kuffner M, Bodrossy L, Puschenreiter M, Monchy S, Wenzel WW, Sessitsch A (2006) Characterization of Ni-tolerant methylobacteria associated with the hyperaccumulating plant *Thlaspi goesingense* and description of *Methylobacterium goesingense* sp. nov. Syst Appl Microbiol 29:634–644

Inglis GD, Kalischuk LD (2004) Direct quantification of *Campylobacter jejuni* and *Campylobacter lanienae* in feces of cattle by real-time quantitative PCR. Appl Environ Microbiol 70:2296–2306

Jiang M, Cao L, Zhang R (2008) Effects of Acacia (*Acacia auriculaeformis* A. Cunn)-associated fungi on mustard (*Brassica juncea* (L.) Coss. var. foliosa Bailey) growth in Cd- and Ni-contaminated soils. Lett Appl Microbiol 47:561–565

Jurelevicius D, Korenblum E, Casella R, Vital RL, Seldin L (2010) Polyphasic analysis of the bacterial community in the rhizosphere and roots of *Cyperus rotundus* L. grown in a petroleum-contaminated soil. J Microbiol Biotechnol 20:862–870

Kamnev AA, Tugarova AV, Antonyuk LP, Tarantilis PA, Polissiou MG, Gardiner PHE (2005) Effects of heavy metals on plant-associated rhizobacteria: comparison of endophytic and non-endophytic strains of *Azospirillum brasilense*. J Trace Elem Med Biol 19:91–95

Kennedy N, Edwards S, Clipson N (2005) Soil bacterial and fungal community structure across a range of unimproved and semi-improved upland grasslands. Microb Ecol 50:463–473

Khan S, Afzal M, Iqbal S, Khan QM (2013a) Plant-bacteria partnerships for the remediation of hydrocarbon contaminated soils. Chemosphere 90:1317–1332

Khan S, Afzal M, Iqbal S, Mirza MS, Khan QM (2013b) Inoculum pretreatment affects bacterial survival, activity and catabolic gene expression during phytoremediation of diesel contaminated soil. Chemosphere 91:663–668

Kuffner M, De Maria S, Puschenreiter M, Fallmann K, Wieshammer G, Gorfer M, Strauss J, Rivelli A, Sessitsch A (2010) Culturable bacteria from Zn- and Cd-accumulating *Salix caprea* with differential effects on plant growth and heavy metal availability. J Appl Microbiol 108:1471–1484

Kumar KV, Srivastava S, Singh N, Behl H (2009) Role of metal resistant plant growth promoting bacteria in ameliorating fly ash to the growth of *Brassica juncea*. J Hazard Mater 170:51–57

Lebeau T, Braud A, Jézéquel K (2008) Performance of bioaugmentation-assisted phytoextraction applied to metal contaminated soils: a review. Environ Pollut 153:497–522

Lee S, Flores-Encarnacion M, Contreras-Zentella M, Garcia-Flores L, Escamilla J, Kennedy C (2004) Indole-3-acetic acid biosynthesis is deficient in Gluconacetobacter diazotrophicus strains with mutations in cytochrome c biogenesis genes. J Bacteriol 186:5384–5391

Li H-Y, Wei D-Q, Shen M, Zhou Z-P (2012) Endophytes and their role in phytoremediation. Fungal Divers 54:11–18

Likar M (2011) Dark septate endophytes and mycorrhizal fungi of trees affected by pollution. In: Maria A, Frank AC (eds) Endophytes of forest trees. Springer, Dordrecht, Springer Netherlands, pp 189–201

Lodewyckx C, Taghavi S, Mergeay M, Vangronsveld J, Clijsters H, van der Lelie D (2001) The effect of recombinant heavy metal-resistant endophytic bacteria on heavy metal uptake by their host plant. Int J Phytoremediation 3:173–187

Lodewyckx C, Vangronsveld J, Porteous F, Moore ERB, Taghavi S, Mezgeay M, van der Lelie D (2002) Endophytic bacteria and their potential applications. Crit Rev Plant Sci 21:583–606

Luo S, Wan Y, Xiao X, Guo H, Chen L, Xi Q, Zeng G, Liu C, Chen J (2011a) Isolation and characterization of endophytic bacterium LRE07 from cadmium hyperaccumulator *Solanum nigrum* L. and its potential for remediation. Appl Microbiol Biotechnol 89:1637–1644

Luo S-L, Chen L, Chen J-L, Xiao X, Xu T-Y, Wan Y, Rao C, Liu C-B, Liu Y-T, Lai C, Zeng G-M (2011b) Analysis and characterization of cultivable heavy metal-resistant bacterial endophytes isolated from Cd-hyperaccumulator *Solanum nigrum* L. and their potential use for phytoremediation. Chemosphere 85:1130–1138

Ma Y, Prasad M, Rajkumar M, Freitas H (2011a) Plant growth promoting rhizobacteria and endophytes accelerate phytoremediation of metalliferous soils. Biotechnol Adv 29:248–258

Ma Y, Rajkumar M, Luo YM, Freitas H (2011b) Inoculation of endophytic bacteria on host and non-host plants-effects on plant growth and Ni uptake. J Hazard Mater 195:230–237

Madhaiyan M, Poonguzhali S, Sa T (2007) Metal tolerating methylotrophic bacteria reduces nickel and cadmium toxicity and promotes plant growth of tomato (*Lycopersicon esculentum* L.). Chemosphere 69:220–228

Marsh TL, Saxman P, Cole J, Tiedje J (2000) Terminal restriction fragment length polymorphism analysis program, a web-based research tool for microbial community analysis. Appl Environ Microbiol 66:3616–3620

Mastretta C, Taghavi S, van der Lelie D, Mengoni A, Galardi F, Gonnelli C, Barac T, Boulet J, Weyens N, Vangronsveld J (2009) Endophytic bacteria from seeds of *Nicotiana tabacum* can reduce cadmium phytotoxicity. Int J Phytoremediat 11:251–267

McGuinness M, Dowling D (2009) Plant-associated bacterial degradation of toxic organic compounds in soil. Int J Environ Res Pub Health 6:2226–2247

Muthukumarasamy R, Revathi G, Seshadri S, Lakshminarasimhan C (2002) *Gluconacetobacter diazotrophicus* (syn. *Acetobacter diazotrophicus*), a promising diazotrophic endophyte in tropics. Curr Sci 83:137–145

Newman LA, Reynolds CM (2005) Bacteria and phytoremediation: new uses for endophytic bacteria in plants. Trends Biotechnol 23:6–8

Pancher M, Ceol M, Corneo PA, Longa CMO, Yousaf S, Pertot I (2012) Fungal endophytic communities in grapevines (*Vitis vinifera* L.) respond to crop management. Appl Environ Microbiol 78:4308–4317

Pongrac P, Sonjak S, Vogel-Miku K, Kump P, Neiemer M, Regvar M (2008) Roots of metal hyperaccumulating population of *Thlaspi Praecox* (*Brassicaceae*) harbour arbuscular mycorrhizal and other fungi under experimental conditions. Int J Phytoremediat 11:347–359

Porteous-Moore F, Barac T, Borremans B, Oeyen L, Vangronsveld J, van der Lelie DCD, Moore E (2006) B. Endophytic bacterial diversity in poplar trees growing on a BTEX-contaminated site: the characterization of isolates with potential to enhance phytoremediation. Syst Appl Microbiol 29:539–556

Posada F, Vega FE (2005) Establishment of the fungal entomopathogen *Beauveria bassiana* (Ascomycota: Hypocreales) as an endophyte in cocoa seedlings (*Theobroma cacao*). Mycologia 97:1195–1200

Puente ME, Li CY, Bashan Y (2009) Endophytic bacteria in cacti seeds can improve the development of cactus seedlings. Environ Exp Bot 66:402–408

Puschenreiter M, Stoger G, Lombi E, Horak O, Wenzel WW (2001) Phytoextraction of heavy metal contaminated soils with *Thlaspi goesingense* and *Amaranthus hybridus*: rhizosphere manipulation using EDTA and ammonium sulfate. J Plant Nutr Soil Sci 164:615–621

Rabie GH (2005) Role of arbuscular mycorrhizal fungi in phytoremediation of soil rhizosphere spiked with poly aromatic hydrocarbons. Mycobiology 33:41–50

Raghukumar C (2008) Marine fungal biotechnology: an ecological perspective. Fungal Divers 31:19–35

Rajkumar M, Freitas H (2008) Effects of inoculation of plant-growth promoting bacteria on Ni uptake by Indian mustard. Bioresour Technol 99:3491–3498

Rajkumar M, Noriharu A, Freitas H (2009) Endophytic bacteria and their potential to enhance heavy metal phytoextraction. Chemosphere 77:153–160

Ranjard L, Poly F, Lata JC, Mougel C, Thioulouse J, Nazaret S (2001) Characterization of bacterial and fungal soil communities by automated ribosomal intergenic spacer analysis fingerprints: biological and methodological variability. Appl Environ Microbiol 67:4479–4487

Renukaradhya M, Shah AI (2010) Isolation of a novel gene encoding a 3,5,6-trichloro-2-pyridinol degrading enzyme from a cow rumen metagenomic library. Biodegradation 21:565–573

Rittmann BE, Hausner M, Löffler F, Love NG, Muyzer G, Okabe S, Oerther DB, Peccia J, Raskin L,

Wagner M (2006) A vista for microbial ecology and environmental biotechnology. Environ Sci Technol 40:1096–1103

Ryan RP, Germaine K, Franks A, Ryan DJ, Dowling DN (2008) Bacterial endophytes: recent developments and applications. FEMS Microbiol Lett 278:1–9

Sandrin TR, Maier RM (2003) Impact of metals on the biodegradation of organic pollutants. Environ Health Perspect 111:1093–1198

Saravanan V, Madhaiyan M, Thangaraju M (2007) Solubilization of zinc compounds by the diazotrophic, plant growth promoting bacterium *Gluconacetobacter diazotrophicus*. Chemosphere 66:1794–1798

Schulz B, Boyle C (2006) What are endophytes? In: Schulz BJE, Boyle CJC, Sieber TN (eds) Microbial root endophytes. Springer, Berlin/Heidelberg, pp 1–13

Sessitsch A, Reiter B, Pfeifer U, Wilhelm E (2002) Cultivation-independent population analysis of bacterial endophytes in three potato varieties based on eubacterial and actinomycetes-specific PCR of 16S rRNA genes. FEMS Microbiol Ecol 39:23–32

Sessitsch A, Hardoim P, Döring J, Weilharter A, Krause A, Woyke T, Mitter B, Hauberg-Lotte L, Friedrich F, Rahalkar M (2012) Functional characteristics of an endophyte community colonizing rice roots as revealed by metagenomic analysis. Mol Plant Microbe Interact 25:28–36

Sheng X, Chen X, He L (2008a) Characteristics of an endophytic pyrene-degrading bacterium of *Enterobacter* sp. 12J1 from *Allium macrostemon* Bunge. Int Biodeter Biodegr 62:88–95

Sheng XF, Xia JJ, Jiang CY, He LY, Qian M (2008b) Characterization of heavy metal-resistant endophytic bacteria from rapeseed (*Brassica napus*) roots and their potential in promoting the growth and lead accumulation of rapeseed. Environ Pollut 156:1164–1170

Sheoran I, Singal H, Singh R (1990) Effect of cadmium and nickel on photosynthesis and the enzymes of the photosynthetic carbon reduction cycle in pigeon pea (*Cajanus cajan* L.). Photosynth Res 23:345–351

Shin M-N, Shim J, You Y, Myung H, Bang K-S, Cho M, Kamala-Kannan S, Oh B-T (2012) Characterization of lead-resistant endophytic *Bacillus* sp. MN3-4 and its potential for promoting lead accumulation in metal hyperaccumulator *Alnus firma*. J Hazard Mater 200:314–320

Siciliano SD, Fortin N, Mihoc A, Wisse G, Labelle S, Beaumier D, Ouellette D, Roy R, Whyte LG, Banks MK, Schwab P, Lee K, Greer CW (2001) Selection of specific endophytic bacterial genotypes by plants in response to soil contamination. Appl Environ Microbiol 67:2469–2475

Sobral JK, Welington LA, Rodrigo M, Aline AP-K, Azevedo JL (2005) Isolation and characterization of endophytic bacteria from soybean (*Glycine max*) grown in soil treated with glyphosate herbicide. Plant Soil 273:91–99

Soleimani M, Afyuni M, Hajabbasi MA, Nourbakhsh F, Sabzalian MR, Christensen JH (2010a) Phytoremediation of an aged petroleum contaminated soil using endophyte infected and non-infected grasses. Chemosphere 81:1084–1090

Soleimani M, Hajabbasi MA, Afyuni M, Mirlohi A, Borggaard OK, Holm PE (2010b) Effect of endophytic fungi on cadmium tolerance and bioaccumulation by *Festuca arundinacea* and *Festuca pratensis*. Int J Phytoremediat 12:535–549

Stone JK, Bacon CW, White J (2000) An overview of endophytic microbes: endophytism defined. In: Bacon CW, White J (eds) Microbial endophytes. Marcel Dekker, New York, pp 29–33

Sung K, Munster CL, Rhykerd R, Drew MC, Corapcioglu MY (2003) The use of vegetation to remediate soil freshly contaminated by recalcitrant contaminants. Water Res 37:2408–2418

Taghavi S, Barac T, Greenberg B, Borremans B, Vangronsveld J, van der Lelie D (2005) Horizontal gene transfer to endogenous endophytic bacteria from poplar improves phytoremediation of toluene. Appl Environ Microbiol 71:8500–8505

Thompson IP, Van Der Gast CJ, Ciric L, Singer AC (2005) Bioaugmentation for bioremediation: the challenge of strain selection. Environ Microbiol 7:909–915

Tian L, Dai C, Zhao Y, Zhao M, Yong Y, Wang X (2007) The degradation of phenanthrene by endophytic fungi *Phomopsis* sp. single and co-cultured with rice. China Environ Sci 27:757–763

Torzilli AP, Masoumeh S, David C, Patrick MG (2006) A comparison of fungal communities from four salt marsh plants using automated ribosomal intergenic spacer analysis (ARISA). Mycologia 98:690–698

Truyens S, Weyens N, Cuypers A, Vangronsveld J (2012) Changes in the population of seed bacteria of transgenerationally Cd-exposed Arabidopsis thaliana. Plant Biol. doi:10.1111/j.1438-8677.2012.00711.x

Van Aken B, Peres CM, Doty SL, Yoon JM, Schnoor JL (2004a) *Methylobacterium populi* sp. nov., a novel aerobic, pink-pigmented, facultatively methylotrophic, methane-utilizing bacterium isolated from poplar trees (*Populus deltoides x nigra* DN34). Int J Syst Evol Microbiol 54:1191–1196

Van Aken B, Yoon JM, Schnoor JL (2004b) Biodegradation of nitro-substituted explosives 2,4,6-trinitrotoluene, hexahydro-1,3,5-trinitro-1,3,5-triazine, and octahydro-1,3,5,7-tetranitro-1,3,5-tetrazocine by a phytosymbiotic *Methylobacterium* sp. associated with poplar tissues (*Populus deltoides* x nigra DN34). Appl Environ Microbiol 70:508–517

Verma SC, Ladha JK, Tripathi AK (2001) Evaluation of plant growth promoting and colonization ability of endophytic diazotrophs from deep water rice. J Biotechnol 91:127–141

Weyens N, van der Lelie D, Artois T, Smeets K, Taghavi S, Newman L, Carleer R, Vangronsveld J (2009a) Bioaugmentation with engineered endophytic bacteria improves contaminant fate in phytoremediation. Environ Sci Technol 43:9413–9418

Weyens N, van der Lelie D, Taghavi S, Newman L, Vangronsveld J (2009b) Exploiting plant-microbe partnerships to improve biomass production and remediation. Trends Biotechnol 27:591–598

Weyens N, van der Lelie D, Taghavi S, Vangronsveld J (2009c) Phytoremediation: plant-endophyte partnerships take the challenge. Curr Opin Biotechnol 20:248–254

Weyens N, Truyens S, Dupae J, Newman L, Taghavi S, van der Lelie D, Carleer R, Vangronsveld J (2010) Potential of the TCE-degrading endophyte *Pseudomonas putida* W619-TCE to improve plant growth and reduce TCE phytotoxicity and evapotranspiration in poplar cuttings. Environ Pollut 158:2915–2919

Weyens N, Boulet J, Adriaensen D, Timmermans J-P, Prinsen E, Van Oevelen S, D'Haen J, Smeets K, van der Lelie D, Taghavi S, Vangronsveld J (2012) Contrasting colonization and plant growth promoting capacity between wild type and a *gfp*-derivative of the endophyte *Pseudomonas putida* W619 in hybrid poplar. Plant Soil 356:217–230

Wu L, Liu X, Schadt CW, Zhou J (2006) Microarray-based analysis of subnanogram quantities of microbial community DNAs by using whole-community genome amplification. Appl Environ Microbiol 72:4931–4941

Xinxian L, Xuemei C, Yagang C, Woon-Chung W, Zebin W, Qitang W (2011) Isolation and characterization endophytic bacteria from hyperaccumulator *Sedum alfredii* hance and their potential to promote phytoextraction of zinc polluted soil. World J Microbiol Biotechnol 27:1197–1207

Yousaf S, Andria V, Reichenauer TG, Smalla K, Sessitsch A (2010a) Phylogenetic and functional diversity of alkane degrading bacteria associated with Italian ryegrass (*Lolium multiflorum*) and birdsfoot trefoil (*Lotus corniculatus*) in a petroleum oil-contaminated environment. J Hazard Mater 184:523–532

Yousaf S, Ripka K, Reichenauer T, Andria V, Afzal M, Sessitsch A (2010b) Hydrocarbon degradation and plant colonization by selected bacterial strains isolated from Italian ryegrass and birdsfoot trefoil. J Appl Microbiol 109:1389–1401

Yousaf S, Afzal M, Reichenauer TG, Brady CL, Sessitsch A (2011) Hydrocarbon degradation, plant colonization and gene expression of alkane degradation genes by endophytic *Enterobacter ludwigii* strains. Environ Pollut 159:2675–2683

Zhang Y, He L, Chen Z, Wang Q, Qian M, Sheng X (2011) Characterization of ACC deaminase-producing endophytic bacteria isolated from copper-tolerant plants and their potential in promoting the growth and copper accumulation of *Brassica napus*. Chemosphere 83:57–62

Zhou J (2003) Microarrays for bacterial detection and microbial community analysis. Curr Opin Microbiol 6:288–294

Ecological Aspects of Endophyte-Based Biocontrol of Forest Diseases

17

Johanna Witzell, Juan A. Martín, and Kathrin Blumenstein

Abstract

Recent studies have shown that the asymptomatic fungal endophytes may influence the outcome of forest trees' interactions with pathogens and herbivores, raising a promise that endophytes might be utilized as biocontrol agents in integrated pest and disease management. However, practical applications for forest protection based on endophytes are still rare, in particular in the case of the economically and ecologically important large trees and their diseases. A better understanding of the ecological and biological background of the protection provided by endophytes may help to design new forest protection strategies that utilize endophytes in control of tree diseases. More information is also needed regarding the effects of silvicultural methods on endophyte communities at the level of single trees and forest stands. In this chapter, we discuss the motivation for continued research on endophyte-based biocontrol of forest tree diseases and some ecological aspects related to the topic.

1 Introduction

Asymptomatic infections by fungal endophytes have been found to be present in different parts of forest trees, including leaves, bark, wood, seeds, and roots (Carroll 1988, 1995; Petrini 1991; Petrini and Fisher 1990; Danti et al. 2002; Ganley and Newcombe 2006; Sieber 2007; Saikkonen 2007). Endophyte communities of forest trees seem to be highly diverse: in several studies, saturating species accumulation curves have not been obtained, and thus only a fraction of the endophyte diversity in forest trees has probably been described so far (Unterseher 2011). Because of their omnipresence and apparently high diversity in trees (Arnold et al. 2000), infections by endophytic fungi have a potential to influence the physiology, metabolism, and ecological interactions of the trees in various ways. However, the functions of tree endophytes are in many cases as poorly known as their diversity and community

J. Witzell (✉) • K. Blumenstein
Southern Swedish Forest Research Centre, Swedish University of Agricultural Sciences, Box 49, 23053 Alnarp, Sweden
e-mail: johanna.witzell@slu.se

J.A. Martín
Departamento de Silvopascicultura, Universidad Politécnica de Madrid, Ciudad Universitaria s/n, 28040 Madrid, Spain

V.C. Verma and A.C. Gange (eds.), *Advances in Endophytic Research*,
DOI 10.1007/978-81-322-1575-2_17, © Springer India 2014

structure. Particularly little is known about the temporal and spatial variations in endophyte communities in large, long-lived forest trees (Jumpponen and Jones 2010). Moreover, the xylem-bound endophyte communities have generally received less attention in research, as compared to the endophyte communities in leaves and bark (Rodriguez et al. 2011).

Prompted by the studies that demonstrated how the seed-transmitted fungal endophytes of grasses protect their hosts from herbivory (e.g., Clay 1988, 1996; Clay and Holah 1999), also many tree-endophyte studies have set out to test the hypothesis that endophytes protect their hosts against natural enemies, herbivores, and pathogens. Indeed, the results of several studies indicate that the presence, diversity, or frequency of tree endophytes may be linked to patterns of defense or expression of resistance against natural enemies (Bettucci and Alonso 1997; Arnold et al. 2003; Gennaro et al. 2003; Ragazzi et al. 2003; Clay 2004; Santamaría and Diez 2005; Ganley et al. 2008; Mejia et al. 2008; Albrectsen et al. 2010). These results raise an attractive prospect of using endophytes as tools in forest protection (Newcombe 2011).

Active research on endophytic fungi of forest trees has been carried out during last decades, and at least in a few cases, there seems to be adequate evidence demonstrating the *proof of principle* for endophyte-based biocontrol of pests and pathogens of forest trees (e.g., Webber 1981; Dvorák et al. 2006; Martín et al. 2013). So far, however, the contribution of endophytes to innovative solutions for practical forest protection has been modest, in particular when it comes to diseases of large trees that often have high economic and ecological importance. Furthermore, some studies have emphasized the ecologically and evolutionarily non-static nature of the host plant-endophyte interactions and the conditioning effect of environment on the outcome of these interactions and their cost to the host plants (e.g., Saikkonen et al. 1998; Faeth and Fagan 2002; Lehtonen et al. 2005). Thus, it is maybe timely to revisit the rationale behind and need for continued research and development in the topic and to unearth the different factors that can either delay or promote the progress with innovations and practical solutions.

In a recent review, Newcombe (2011) analyzed some of the main challenges that are holding back the use of endophytes in forest management, considering not only the potential of endophytes in disease control but also in growth promotion and stress tolerance. Here, we aim to contribute to the discussion about the application potential of tree endophytes by bringing up some additional aspects. We first consider the fundamental rationale behind the research and development of endophyte-based biocontrol of tree diseases. Then we will examine some ecological aspects related to the mechanisms of protection. We limit our main scope to the potential of endophytes in biocontrol of diseases in mature, large-sized trees that are important in production and recreation forests and also have high ecological values. Finally, we discuss the possibility to engineer endophyte communities at forest stand or landscape level, using silvicultural actions.

2 The Rationale Behind Research on Endophyte-Based Biocontrol of Forest Diseases

2.1 Pros and Cons of Biocontrol in Forestry

Although the acceptance and enthusiasm has varied during the years, the concept of biological control (use of living organisms to control pests; Waage and Greathead 1988) appears to be rather well established in agricultural systems today (Pal and McSpadden Gardener 2006) and accepted as a part of integrated pest management (IPM) strategies (Bale et al. 2008). The success stories of biological control come mainly from control of weeds and insects, including also some examples in forest environments, e.g., the use of entomophthoralean fungus *Entomophaga maimaiga* Humber, Shimazu, and Soper in control of gypsy moth (*Lymantria dispar*) (Hajek 1997; Hajek et al. 1997; Lacey et al. 2001) and the use of the saprophytic *Phlebiopsis gigantea* fungus to control *Heterobasidion* root and butt rot (Holdenrieder and Greig 1998; Pratt et al. 1998).

The anticipated benefits of biocontrol include the safety to people and animals and the possibility to reduce the use of broadscale fungicides and pesticides in our environment (Pal and McSpadden Gardener 2006). In large-scale production, however, the cost-effectiveness of biocontrol may easily appear inferior to other control methods. For instance, chemical treatments have a history of being both fast and reliable (in the absence of developed resistance), and because the production and distribution of biocontrol agents is often a bottleneck, it is generally much easier to get hold of the chemical products in volumes and at time points needed (Bale et al. 2008). Resistance breeding, although slow in its traditional form, is attractive because of its sustainability. Boosted by the recent advances in genetic engineering, it can also be rather fast (Strauss et al. 2004). In recent years, also a method that was first met with great skepticism, i.e., the utilization of chemically or biologically induced plant responses for increased resistance (Solla and Gil 2003; Hubbes 2004; Blodgett et al. 2007; Schiebe et al. 2012), has gained increasing interest as an environmentally sound way to suppress pests and pathogens of forest trees and may appear as an attractive alternative to biocontrol in cases where the more natural methods are preferred.

The potential nontargeted effects, and the difficulty in predicting them, are obviously one of the major ecological concerns in the use of any biocontrol. Louda et al. (2003) reviewed ten cases of released biocontrol agents, which had been studied for the nontargeted effects. Although the included case studies represented biocontrol of weeds and insects control, some of the conclusions should be relevant also for biocontrol of diseases and be at least to some extent transferable to forest tree systems. Louda et al. (2003) found that the closely related species had the highest risk of getting non-intentionally targeted. They also point out that the nontarget effects of biocontrol species can be indirect. Through indirect effects, even highly specific biocontrol agents may influence other than the targeted species (see also Pearson and Callaway 2005).

A related ecological concern is the host specificity of endophytes or, rather, the lack of it that would make uncontrolled host shifts by nonspecialized endophytes possible. Reports regarding the host specificity of endophytes have been conflicting (Arnold and Lutzoni 2007 and references within). Clearly, many of the common, readily isolated tree endophytes are to be considered generalists (e.g., *Phomopsis*, *Xylaria*, *Colletotrichum*, *Fusarium*, and *Botryosphaeria*), while more specialized interactions may be found in the slow growing or unculturable fractions (Arnold and Lutzoni 2007). In its classical form, biocontrol is based on specialist enemies of the pest to be controlled (Müller-Schärer et al. 2004), which minimizes the probability of nontargeted effects. For this reason, a high degree of specificity could be a desirable trait also in endophytes that are selected to be utilized in biocontrol of forest tree diseases. Yet, from the pragmatic point of view, it is the endophytes that are readily culturable in standard conditions that would be most useful for the biocontrol purposes, because their maintenance and multiplication would be cheap and easy.

Another common concern in biological control, its slowness (Bale et al. 2008), is not likely to be as disturbing in the long-lived trees as it can be, e.g., in greenhouse environments. However, in the specific case of large trees, little is known about the duration and spatial extent of the potential protection provided by endophytes.

2.2 Challenges of Endophyte-Based Biocontrol of Forest Diseases

Obviously, the cryptic lifestyle of endophytes may bring about extra difficulties in the use of these fungi in biological control, in terms of stability, intensity, and reliability. In fact, the mere rationale of research investments on endophyte-based biocontrol of forest diseases is often questioned also by members of researcher community. Frequent criticism includes statements that tree breeding, especially with the current possibilities to enhance tree resistance using gene technologies (Harfouche et al. 2011), outcompetes biocontrol because of its power and robustness; that forest disease control based on mechanisms where fungi control fungi has been successful enough to be commercialized and to be applied in practical forestry only in few exceptional cases, such as the saprophytic *Phlebiopsis gigantea* that controls *Heterobasidion* root and butt rot; and that the use of any endophyte-derived fungicidal or antifungal chemicals will bring about the same environmental concerns as any other chemicals. Considering all these reservations, it seems warranted to carefully revise the motivation for continued investments into research and development activities that specifically target fungal endophytes as biological control against forest diseases.

It is likely that the lack of breakthroughs with endophyte-based biocontrol applications in forestry reflects the fact that our knowledge about the fundamental biology and environmental regulation of endophytes and their assemblages in trees is still rudimentary. This, in turn, is likely to be partly caused by the practical difficulties in studying the fungal communities in vivo, inside the living, large-sized trees (Albrectsen and Witzell 2012). However, the rapid development of molecular methods such as pyrosequencing to study fungal communities (Margulies et al. 2005; Hamady et al. 2008; Amend et al. 2010; Mardis 2011) and phenotype microarrays to explore the substrate utilization profiles of the fungi (Garland 2006; Borglin et al. 2012), together with bioinformatics to effectively mine the high-throughput data, have opened up new possibilities also for tree-endophyte research. While it is conceivable that proper identification of the sterile endophyte isolates will be a struggle for some more years (Nilsson et al. 2011; Seifert 2009), our possibilities to study the community-level responses and functional consequences of endophyte infections are significantly better today than they were only 10 years ago, thanks to the advances in molecular and physiological techniques and know-how and the solid basic research that has been carried out using mainly cultivation-dependent approaches during the past decades. Therefore, it seems reasonable to expect that the scientific knowledge base covering the biology and regulation of endophytes will be considerably broadened only in the very the near future (Albrectsen and Witzell 2012). This development is likely to remove some obstacles on the way towards practical applications in forestry.

Newcombe (2011) draws attention to the difficulties in selection-based assays, to the importance of microbial community interactions, and to the interdependency between tree genotype and endophytes. He also reminds of the possibility that nontargeted, unintentional effects, such as host shifting or invasiveness, may occur if endophytes are introduced into an area. Introduction of nonindigenous endophyte species or enrichment of certain existing endophyte species might also alter the tree-associated microbial communities, including the beneficial symbionts such as mycorrhiza (cf. Louda et al. 2003). In fungi, the taxonomic relations are not as straightforward as in insects and plants, which complicates the analysis of nontargeted effects of endophyte-based biocontrol on microbial communities. To avoid such effects, it would be important to map the characteristic microbial profiles in the healthy individuals of the tree species of interest, with special emphasis on the rare microbes that inhabit these trees, and to test as rigorously as possible how the putative biocontrol agent (endophyte) interacts with these microbes. This task alone is a huge challenge for researchers.

2.3 Future Prospects for Endophyte-Based Biocontrol of Forest Diseases

Despite the obviously many and extensive challenges, there are also many reasons to continue studies on endophyte-based biocontrol of forest diseases. One of the most important and pragmatic motivation for continued investigations is that the character of many forest disease problems is changing, because of the alterations in the general operational environment for forestry (Foley et al. 2005; Schröter et al. 2005; Millar et al. 2007; Santini et al. 2013). The old forest protection methods, designed for the earlier prevailing conditions, may not be functional or effective enough in the new situations. A central factor driving this development is the changing climate that may increase the frequency and intensity of disturbances (Dale et al. 2001), some of which may directly or indirectly involve forest diseases. The directional changes, such as increasing temperatures or humidity, may alter the distribution patterns of pest and pathogens (Brasier and Scott 1994). For instance, warmer winters were identified as a major factor behind the outbreak of mountain pine beetles in western North America (Kurz et al. 2008) and the populations of fungal associates of the beetles, e.g., *Ophiostoma* sp. and *Leptographium* sp. (Lee et al. 2006) benefitted from this development.

The future operational environment for forestry is also characterized by intensification, globalization, and diversification (cf. Anderson et al. 2004), which all call for new forest protection solutions. For example, the threats by alien invasive species in forests are expected to further increase in the future, due to intensive global trade that provides rapid dispersal routes for pathogens to new areas and due to the changing climate that may create opportunities for pathogens to establish and thrive in areas that have been climatically unfavorable for them earlier (Sturrock et al. 2011; Santini et al. 2013). Dispersal of alien pathogens to forests increases the probability of hybridizations between alien and native species, possibly creating progenies that are more aggressive than the parents as forest pathogens. Cases where hybridization of pathogens brought together by human activity has resulted in the emergence of new pathotypes are, for example, the causal agents of Dutch elm disease, *Ophiostoma novo-ulmi* races EAN and NAN, and the emergence of a new *Phytophthora* species, pathogenic of alder (*Alnus* spp.) in Europe (Brasier et al. 1999; Brasier 2001). Anderson et al. (2004) conclude that emergence of new pathogen strains owing to hybridization between agents that are not naturally sympatric is a repeating phenomenon behind emerging infectious diseases (EIDs) of plants. Characteristic traits to these diseases are increased incidence (geographic or host range), changed pathogenesis, or newly evolved causal agents. EIDs can lead to extinction of host species and thus pose a concrete threat to biodiversity (Anderson et al. 2004; Fisher et al. 2012).

At the same time when the pathogen pollution is spreading, there is interest in forestry to gain superior growth or quality yields by growing tree species outside their natural habitats (exotic or introduced trees), which further promotes the global biological homogenization in time and space. Introduced trees can be less troubled by diseases in the new habitat because they escape the specialist pests from the earlier habitat (enemy release hypothesis), but they may get affected by new generalist pests against which they have not developed tolerance (biotic resistance hypothesis) (Morrison and Hay 2011). Intensified forestry is further characterized, e.g., by interest in plantation forestry, sometimes using monoclones, shorter rotation times, and fertilization (Martín-García et al. 2011; Edenius et al. 2012), all of which have the potential to affect the biodiversity associated with forest trees, as well as the vulnerability of trees to diseases. The new operational conditions for forestry and control of forest diseases also include the increasing environmental concerns from the society that imposes limits for use of chemical treatments (Bazoche et al. 2012) and the ongoing afforestation processes where forests are established to areas not classified as forest lands (Smith 2002). In order to meet all the new challenges, it seems highly justified to revise and update the tool box for forest protection and to complete it by exploring new and innovative solutions, such as the potential endophyte-based biocontrol.

Newcombe (2011) concludes his review by pointing out the importance of information transfer from research communities to end users and by suggesting that the adoption of endophytes as tools in forest management could be promoted by a better inclusion of endophytology into the forestry curriculum. In addition, also other relevant professional groups, e.g., arborists, landscape engineers, and nature conservationists, could be informed more actively in order to promote the recognition of endophytes as a functional layer in tree and forest protection. Because commercial actors are the usual channel for the production and distribution of possible products, they should also be included in the dialogue as early as possible. In fact, professional analyses of the commercial prospects, revenues, and cost-effectiveness scenarios in cases where the *proof of principle* for endophyte-based biocontrol is convincing could create powerful arguments for practical implementations in forestry. Clearly, the ethical aspects of commercialization process need to be carefully considered on case-to-case basis. However, before innovations and practical forest protection solutions, based on endophytes, can be made available, more basic research is needed, for example, regarding the stability and reliability of endophytes as potential forest protection tools.

2.4 Endophyte Research: Frontier of Biological Research

A further argument for continued studies on the endophytic biodiversity is that rather than being outdated by the progress of gene technology in tree breeding, biocontrol using endophytes is currently in the very front line of biological research. The interest in endophytes as regulators of tree fitness is congruent with the increasing, general scientific interest in microbiome as an epigenetic domain affecting the functions and health of an organism (Cho and Blaser 2012). For several decades, the research has focused on the genome of the organisms as the ultimate regulator of traits and performance. However, the recently emerging view seems to be that the joint metagenome of the organism and its associated microbiome may be crucial for many functions and interactions. In particular the intestinal microbiome of humans has received attention in this context and is sometimes referred to as our additional or forgotten organ, or our second genome (Bruls and Weissenbach 2011; Qin et al. 2010), to emphasize the intimate and functional relation between microbes and human body. Evidence supporting the importance of microbes as a functional interface between different trophic levels is accumulating from other systems as well: for instance, Becher et al. (2012) found that the attraction, oviposition, and development of fruit flies (*Drosophila melanogaster*) are in fact not regulated by volatile signals from the fermenting fruits, but by yeasts that inhabit these fruits. Thus, there is an increasing scientific interest to continue endophytological studies, including the study systems that involve forest trees and their health. One of the most important aspects to be studied is the mechanisms that allow endophytes to suppress tree diseases in different cases. A proper understanding of these mechanisms is the key for design of successful biocontrol strategies. In the following section, we consider the ecological basis of the mechanisms that have been described for endophytes in literature and discuss potential strategies that are based on these mechanisms and aim at biological control of forest tree diseases.

3 Biocontrol of Tree Diseases by Endophytes: Ecological Considerations

3.1 Mechanisms of Protection Provided by Endophytes in Forest Trees

It has been suggested that endophytes can shape their hosts' resistance against pathogens through several mechanisms that may act simultaneously or in concert in a certain plant-endophyte interaction (Gao et al. 2010). Some of the described mechanisms are based on the direct interaction between the endophyte and pathogen, such as mycoparasitism (endophyte feeding on the pathogen), competitive exclusion by differential ability to utilize the substrates, or inhibition of pathogen by extracellular chemicals produced by the endophyte (Rodriguez and Redman 1997; Strobel and Daisy 2003; Arnold 2007). On the other hand, the endophytes may affect the trees' resistance through their indirect effect on the host tree, e.g., by stimulating the defensive metabolism (White and Torres 2010). Webber (1981) described a mechanism where an endophyte (*Phomopsis oblonga*) provides the host trees, elms, protection against Dutch elm disease through negative effects on the vector insects, bark beetles. Existence of even more complex indirect interactions with protective outcomes cannot be excluded.

Perhaps the best characterized mechanisms of protection by endophytes are derived from studies with grasses. The clavicipitalean endophytes of grasses are primarily vertically transmitted (in seeds) and infect the hosts systematically (Saikkonen et al. 2002; Rodriguez et al. 2009). Their protective action has been coupled to the toxic or deterring metabolites, mainly alkaloids, which they produce inside their hosts (Bush et al. 1997; Clay and Schardl 2002). Interestingly, alkaloids have been detected in low concentrations in bark samples of forest trees such as Norway spruce (*Picea abies*), and variation between samples was found to be high (Schiebe et al. 2012). Norway spruce is otherwise known to rely heavily on carbon-based phenolics and terpenoids in its defensive chemistry (Witzell and Martín 2008; Schiebe et al. 2012). In another conifer, Sitka spruce (*Picea sitchensis*), piperidine alkaloids were studied by Gerson and Kelsey (2002) who found that the total alkaloid concentration, as well as diversity of individual alkaloid compounds, was higher in bark than in needles of Sitka spruce. It cannot be excluded that fungal endophytes might contribute to the observed patchy patterns of alkaloids in trees: for example, piperine has been detected in an endophytic fungus (Verma et al. 2011), and endophyte assemblage can show tissue-specific variation (e.g., Sun et al. 2011; Martín et al. 2013). More targeted research is needed to explore the overlapping chemical domains of endophytes and their host trees. In general, however, the potential role of endophytes as a factor causing variation in "plant chemistry" has been neglected, despite the accumulating evidence showing how some endophytes have the ability to produce the same or similar bioactive compounds as those originated from their host plants (Stierle et al. 1993; Eyberger et al. 2006; Kusari et al. 2008; Zhao et al. 2011). Thus, many bioactive metabolites that have been regarded as plant products in fact can be partly or completely fungal products. The contribution of endophytes to plant defensive chemistry should thus be better incorporated in future studies on chemical ecology also in the case of forest trees.

In contrast to the vertically transmitted grass endophytes, tree endophytes generally spread horizontally from the environment with wind and rain, and the colonizations may be localized (Saikkonen et al. 1998, 2004; Saikkonen 2007). Thus, while a lot of the information gathered from grass-endophyte studies is likely to be relevant also for the tree-endophyte interactions, it is important to keep in mind the special traits of the trees that may affect the expression and functionality of the endophyte activities. As long-lived, large-sized plants, trees are likely to be exposed to attacks by pathogens (and pest) continuously and simultaneously during their whole life time. Diseases like stem rots or cankers often develop under long time periods, in the beginning often without external symptoms. Therefore, the spatial and temporal scales for the protection differ considerably between large forest trees and annual plants and perennial herbs. Even within the tree, the mechanisms of protection by endophytes are likely to differ. Nutritious leaf tissues are readily exposed to environmental inocula and may support generalist endophytes better than the less nutritive woody tissues. The xylem of the trees is protected by the bark and tends to support an endophyte flora with less and more specialized species that can cross the anatomical borders (Baum et al. 2003). The woody parts of trees are known for their ability to compartmentalize damages like stem rot or cankers by formation of tyloses, strengthening of cell walls, and buildup of reaction zones (Shigo 1984). Whether endophyte infections affect these processes is not known. Endophyte communities in wood may, however, contribute to the wood degradation in senescing and dead wood (Schwarze et al. 2000; Baum et al. 2003).

In addition to the ecological dependencies of endophyte-based biocontrol of tree diseases, we also need to better understand the anatomical, physiological, and molecular mechanisms behind it. Upon entrance and during spread and growth inside the trees, endophytic and pathogenic fungi are likely to encounter the same or similar tree defensive mechanisms of the host. Because endophytes and pathogens seem to possess many of the same virulence factors (Schulz and Boyle 2006), their microbe-associated molecular patterns (MAMPs, molecular signatures typical of whole classes of microbes; Boller and Felix 2009) may be recognized in a similar manner by the host plant. In coevolved plant-endophyte interactions, the virulence of the fungus and the defensive responses of the host tree are in a balance that does not result in development of disease (Schulz and Boyle 2006). However, if, for example, the defensive mechanisms affecting pathogenic fungi are enhanced by resistance breeding or by forest management actions, such as use of fertilizers, the conditions for beneficial endophyte infections might become less favorable. For instance, elm (*Ulmus minor*) trees with low susceptibility against Dutch elm disease also harbored a poorer endophyte diversity and frequency in xylem tissues than the conspecific trees that were known to be more susceptible to the disease (Martín et al. 2013). Thus, it is possible that inoculations with endophytes are less successful or effective in tree genotypes that show high resistance to pathogens.

3.2 Biocontrol Strategies

In order to utilize the abovementioned mechanisms in biological control of tree diseases, two general approaches can be envisioned: (1) application of endophytes, mixtures of endophytes, or bioactive products from endophytes on trees and (2) management of the endophyte communities at the habitat level, e.g., within a forest stand. Of these two general strategies, the first is best fitted for smaller spatial scales, e.g., urban forest stands with limited number of trees to be treated, while the second could be applied also on larger areas with higher number of trees to be treated, e.g., production forests or larger recreation forests. For the first approach, two temporal strategies can be employed: (a) preventive application, before the targeted pathogen attacks, and (b) therapeutic application, after the disease is discovered. The second option, habitat level management, would be likely to require a longer period to reach a full effect and could be seen mainly as a preventive method.

Some endophytes exist as latent pathogens or dormant saprophytes (Saikkonen et al. 2004). Triggered by environmental or tree intrinsic signals, some endophytes may switch to a saprophytic or pathogenic lifestyle, and, for example, the wood-inhabiting endophytes may start to decompose wood as the tree gets senescent (Saikkonen 2007; Sieber 2007; Rodriguez et al. 2011). This possibility needs to be carefully taken into consideration in planning biocontrol strategies based on endophytes. If alien endophytes would be introduced, as in the *classical biocontrol* strategy (Waage and Greathead 1988; Eilenberg et al. 2001), they might become invasive and show higher virulence towards their hosts in the introduced conditions, as compared to their native conditions (cf. Keane and Crawley 2002). Therefore, the classical version of biological control cannot be considered as the first choice strategy in endophyte-based biocontrol of forests. Effective use of another established biocontrol strategy, *inundation*, defined by Eilenberg et al. (2001) as "the use of living organisms to control pests when control is achieved exclusively by the released organisms themselves" (i.e., the progeny of the released population is not counted for and the desired effect is corrective) is obscured by the spatial and temporal continuum of fungal generations. In contrast, biocontrol based on *inoculation*, "the intentional release of a living organism as a biological control agent with the expectation that it will multiply and control the pest for an extended period, but not permanently" (Eilenberg et al. 2001), appears as an applicable strategy for forest tree protection using endophytes. Yet another applicable strategy, *conservation*, is discussed in the end of this chapter.

From the ecological point of view, the operational basis of biocontrol, which relies on the activity of a microbe or microbes (e.g., endophytic fungus or fungi) and is targeted against other microbes (e.g., disease-causing fungi or oomycetes), may differ considerably from the classical biocontrol of weeds and herbivores. In the classical biocontrol, the *top-down* regulation (natural enemies, predators) controls the population (Bale et al. 2008). Except for direct mycoparasitism, fungal interactions leading to biocontrol are more likely to involve *bottom-up* regulation, e.g., the quality of the wood as a substrate for the endophyte and the pathogen. This kind of regulation might include higher nonspecificity than the prey-predator or parasite-host interactions. For instance, it is likely that the niches of some beneficial fungi overlap with that of the pathogen (Fodor 2011). An endophyte that effectively competes with a pathogen might therefore also outcompete several other fungi, some of which might be beneficial for the host tree or functionally important for other partners in the forest ecosystem (e.g., as a wood-degrading entity). Yet another difference between endophyte-based and classical biocontrol systems could be that the density-dependent feedbacks, which regulate the populations of some biocontrol agents (Bale et al. 2008), may not function in a similar manner in fungal populations. Nevertheless, it is known that fungi can show developmental transitions induced by environmental factors. For instance, Hornby et al. (2004) found that *Ceratocystis* (*Ophiostoma*) *ulmi*, the causal agent of Dutch elm disease, went through a developmental switch between yeast and filamentous type in response to density-dependent extracellular signal (possibly lipophilic isoprenoids), which did not cross-react with another fungus (*Candida albicans*). The quorum-sensing mechanisms in dimorphic fungi are also an essential research field for future studies.

4 Engineering of Endophyte Communities for Improved Tree Health?

Several environmental factors may have an impact on the outcome of the interactions between trees, endophytes, and pathogens. The effect of environment can operate through direct effects on any of the partners. For instance, temperature and humidity can strongly regulate the development of pathogen or endophyte populations. Indirect effects, e.g., fertilization of trees that alters their chemical quality (Edenius et al. 2012), could have consequences for the microbial activity within the trees. For example, the alkaloid production of endophytes may respond to the nitrogen status of their hosts (Lehtonen et al. 2005 and references therein), potentially resulting in altered endophyte-based defense in fertilized trees. Interestingly, this kind of endophyte-mediated mechanisms could explain some of the ambiguous results that have been obtained in studies testing the effects of nitrogen fertilization on herbivory in the context of plant defense theories and carbon-based metabolites (Witzell and Martín 2008). If unexplored, such dependencies between endophyte infections, environment, and tree genotypes may profoundly destabilize and reduce the effectiveness of endophyte-based forest protection solutions. On the other hand, such dependencies may also provide possibilities to manipulate the available endophyte inoculum in a way that promotes tree health.

Different silvicultural actions may alter the quality of woody tissues as a substrate and habitat for endophytes. Silvicultural actions, e.g., removal of slash (Bernhold et al. 2008) or utilization of nurse plants (Jensen et al. 2012), may also modify the surrounding vegetation that is a source of inoculum for both pathogens and endophytes and regulate the microclimate that is important for microbes. Helander et al. (2006) investigated the effects of silviculture and local environmental variables on endophyte frequencies in silver birch (*Betula pendula* Roth) leaves, sampled from seedling stands, managed mature forest, and old natural forest. They found that the sapling stands had the highest endophyte frequency, possibly because of a high availability of spores or favorable microclimate. The managed forest had the lowest total infection frequency, and the old natural forest tended to have the most diverse identified fungal species community, but the difference was not statistically significant (Helander et al. 2006).

In a study with endophytes in plantation-cultured poplars, Martín-García et al. (2011) emphasize that silvicultural factors such as rotation length, site quality, and possible fertilization regime may affect the endophytic fungi. They found that fungal species richness and relative isolation frequency were higher in young stands than in adult stands and that the lowest richness levels were observed in adult stands located in poor sites.

Biocontrol through *conservation* method is defined as "modification of the environment or existing practices to protect and enhance specific natural enemies or other organisms to reduce the effect of pests," and unlike the other methods, it does not include actual release of organisms (Eilenberg et al. 2001). Treatment of forest trees with selected endophytes is not always a practically or ecologically feasible method, and therefore the conservation method, applied in a landscape perspective (Tscharntke et al. 2007), could be a better option for disease control in larger forests. However, in order to engineer the endophytic diversity in forests so that it would support the vitality and disease resistance of trees, we need more information about the baseline endophyte communities in healthy trees. It can be of special interest to characterize the endophytic microbiome of tree individuals that remain vital in areas that otherwise are severely affected. Similar to the discovery of "suppressive soils," rich with plant growth-promoting bacteria (Compant et al. 2005) that promote plant resistance, it might be possible to discover endophytic communities that suppress forest tree diseases.

References

Albrectsen BR, Witzell J (2012) Disentangling functions of fungal endophytes in forest trees. In: Paz Silva A, Sol M (eds) Fungi: types, environmental impact and role in disease. Nova, Hauppauge, pp 235–246

Albrectsen BR, Bjorken L, Varad A et al (2010) Endophytic fungi in European aspen (*Populus tremula*) leaves – diversity, detection, and a suggested correlation with herbivory resistance. Fungal Divers 41: 17–28

Amend AS, Seifert KA, Bruns TD (2010) Quantifying microbial communities with 454 pyrosequencing: does read abundance count? Mol Ecol 19:5555–5565

Anderson PK, Cunningham AA, Patel NG et al (2004) Emerging infectious diseases of plants: pathogen pollution, climate change and agrotechnology drivers. TREE 19:535–543

Arnold AE (2007) Understanding the diversity of foliar endophytic fungi: progress, challenges, and frontiers. Fungal Biol Rev 21:51–66

Arnold AE, Lutzoni F (2007) Diversity and host range of foliar fungal endophytes: are tropical leaves biodiversity hotspots? Ecology 88:541–549

Arnold AE, Maynard Z, Gilbert GS et al (2000) Are tropical fungal endophytes hyperdiverse? Ecol Lett 3:267–274

Arnold AE, Mejía LC, Kyllo D et al (2003) Fungal endophytes limit pathogen damage in a tropical tree. PNAS 100:15649–15654

Bale JS, van Lenteren JC, Bigler F (2008) Biological control and sustainable food production. Philos Trans R Soc B 36:761–776

Baum S, Sieber TN, Schwarze FWMR, Fink C (2003) Latent infections of *Fomes fomentarius* in the xylem of European beech (*Fagus sylvatica*). Mycol Prog 2:141–148

Bazoche P, Bunte F, Combris P et al (2012) Willingness to pay for pesticides' reduction in E.U.: nothing but organic? ALISS working paper 2012–01. http://www7.paris.inra.fr/aliss/publications_working_papers/working_papers/aliss_working_papers. Accessed 5 Mar 2013

Becher PG, Flick G, Rozpędowska E et al (2012) Yeast, not fruit volatiles mediate *Drosophila melanogaster* attraction, oviposition and development. Funct Ecol 26:822–828

Bernhold A, Witzell J, Hanson P (2008) Effect of slash removal on *Gremmeniella abietina* incidence on *Pinus sylvestris* after clear-cutting in northern Sweden. Scand J For Res 21:489–495

Bettucci L, Alonso R (1997) A comparative study of fungal populations in healthy and symptomatic twigs of *Eucalyptus grandis* in Uruguay. Mycol Res 101:1060–1064

Blodgett JT, Eyles A, Bonello P (2007) Organ-dependent induction of systemic resistance and systemic susceptibility in *Pinus nigra* inoculated with *Sphaeropsis sapinea* and *Diplodia scrobiculata*. Tree Physiol 27:511–517

Boller T, Felix G (2009) A renaissance of elicitors: perception of microbe-associated molecular patterns and danger signals by pattern-recognition receptors. Annu Rev Plant Biol 60:379–406

Borglin S, Joyner D, DeAngelis KM et al (2012) Application of phenotypic microarrays to environmental microbiology. Curr Opin Biotechnol 23:41–48

Brasier CM (2001) Rapid evolution of introduced plant pathogens via interspecific hybridization. Bioscience 51:123–133

Brasier CM, Scott JK (1994) European oak declines and global warming: a theoretical assessment with special reference to the activity of *Phytophthora cinnamomi*. EPPO Bull 24:221–232

Brasier CM, Cooke DE, Duncan JM (1999) Origin of a new *Phytophthora* pathogen through interspecific hybridization. Proc Natl Acad Sci U S A 96:5878–5883

Bruls T, Weissenbach J (2011) The human metagenome: our other genome? Hum Mol Genet 20:R142–R148

Bush LP, Wilkinson HH, Schardl CL (1997) Bioprotective alkaloids of grass-fungal endophyte symbioses. Plant Physiol 114:1–7

Carroll GC (1988) Fungal endophytes in stems and leaves: from latent pathogen to mutualistic symbiont. Ecology 69:2–9

Carroll G (1995) Forest endophytes: pattern and process. Can J Bot 73:1316–1324

Cho I, Blaser MJ (2012) The human microbiome: at the interface of health and disease. Nat Rev Genet 13:260–270

Clay K (1988) Fungal endophytes of grasses: a defensive mutualism between plants and fungi. Ecology 69:10–16

Clay K (1996) Interactions among fungal endophytes, grasses and herbivores. Res Popul Ecol 38:191–201

Clay K (2004) Fungi and the food of the gods. Nature 427:401–402

Clay K, Holah J (1999) Fungal endophyte symbiosis and plant diversity in successional fields. Science 285:1742–1745

Clay K, Schardl C (2002) Evolutionary origins and ecological consequences of endophyte symbiosis with grasses. Am Nat 160(Suppl 4):S99–S127

Compant S, Duffy B, Nowak J et al (2005) Use of plant growth-promoting bacteria for biocontrol of plant diseases: principles, mechanisms of action, and future prospects. Appl Environ Microbiol 71:4951–4959

Dale VH, Joyce LA, McNulty S et al (2001) Climate change and forest disturbances. Bioscience 51:725–734

Danti R, Sieber TN, Sanguineti G (2002) Endophytic mycobiota in bark of European beech (*Fagus sylvatica*) in the Apennines. Mycol Res 106:1343–1348

Dvorák M, Palovcíková D, Jankovský L (2006) The occurrence of endophytic fungus *Phomopsis oblonga* on elms in the area of southern Bohemia. J For Sci 52:531–535

Edenius L, Mikusinski G, Witzell J, Bergh J (2012) Effects of repeated fertilization of young Norway spruce on phenolics and arthropods: implications for insectivorous birds food resources. For Ecol Manage 277:38–45

Eilenberg J, Hajek A, Lomer C (2001) Suggestions for unifying the terminology in biological control. BioControl 46:387–400

Eyberger AL, Condapati R, Porter JRJ (2006) Endophyte fungal isolates from *Podophyllum peltatum* produce podophyllotoxin. J Nat Prod 69:1121–1124

Faeth SH, Fagan WF (2002) Fungal endophytes: common host plant symbionts but uncommon mutualists. Integr Comp Biol 42:360–368

Fisher MC, Henk DA, Briggs CJ et al (2012) Emerging fungal threats to animal, plant and ecosystem health. Nature 484:186–194

Fodor E (2011) Ecological niche of plant pathogens. Ann For Res 54:3–21

Foley JA, DeFries R, Asner GP et al (2005) Global consequences of land use. Science 309:570–574

Ganley RJ, Newcombe G (2006) Fungal endophytes in seeds and needles of *Pinus monticola*. Mycol Res 110:318–327

Ganley RJ, Sniezko RA, Newcombe G (2008) Endophyte-mediated resistance against white pine blister rust in *Pinus monticola*. For Ecol Manage 255:2751–2760

Gao F, Dai C, Liu X (2010) Mechanisms of fungal endophytes in plant protection against pathogens. Afr J Microbiol Res 4:1346–1351

Garland JL (2006) Analysis and interpretation of community-level physiological profiles in microbial ecology. FEMS Microbiol Ecol 24:289–300

Gennaro M, Gonthier P, Nicolotti G (2003) Fungal endophytic communities in healthy and declining *Quercus robur* L. and *Q. cerris* L. trees in northern Italy. J Phytopathol 151:529–534

Gerson EA, Kelsey RG (2002) Piperidine alkaloids in sitka spruce with varying levels of resistance to white pine weevil (Coleoptera: Curculionidae). J Econ Entomol 95:608–613

Hajek AE (1997) Fungal and viral epizootics in gypsy moth (Lepidoptera: Lymantriidae) populations in central New York. Biol Control 10:58–68

Hajek AE, Elkinton JS, Humber RA (1997) Entomopathogenic hyphomycetes associated with gypsy moth (Lepidoptera: Lymantriidae) larvae. Mycologia 89:825–829

Hamady M, Walker JJ, Harris JK et al (2008) Error-correcting barcoded primers for pyrosequencing hundreds of samples in multiplex. Nat Method 5:235–237

Harfouche A, Meilan R, Altman A (2011) Tree genetic engineering and applications to sustainable forestry and biomass production. Trends Biotechnol 29:9–17

Helander M, Wäli P, Kuuluvainen T, Saikkonen K (2006) Birch leaf endophytes in managed and natural boreal forests. Can J For Res 36:3239–3245

Holdenrieder O, Greig BJW (1998) Biological methods of control. In: Woodward S, Stenlid J, Karjalainen R, Hutterman A (eds) *Heterobasidion annosum*: biology, ecology, impact and control. CAB International, Wallingford, pp 235–259

Hornby JM, Jacobitz-Kizzier SM, McNeel DJ (2004) Inoculum size effect in dimorphic fungi: extracellular control of yeast-mycelium dimorphism in *Ceratocystis ulmi*. Appl Environ Microbiol 70:1356–1359

Hubbes M (2004) Induced resistance for the control of Dutch elm disease. Invest Agrar Sist Recur For 13:185–196

Jensen AM, Löf M, Witzell J (2012) Effects of competition and indirect facilitation by shrubs on *Quercus robur* saplings. Plant Ecol 231:535–543

Jumpponen A, Jones KL (2010) Seasonally dynamic fungal communities in the *Quercus macrocarpa* phyllosphere differ between urban and nonurban environments. New Phytol 186:496–513

Keane RM, Crawley MJ (2002) Exotic plant invasions and the enemy release hypothesis. TREE 17:164–170

Kurz WA, Dymond CC, Stinson G et al (2008) Mountain pine beetle and forest carbon feedback to climate change. Nature 452:987–990

Kusari S, Lamshöft M, Zuhlke S, Spiteller M (2008) An endophytic fungus from *Hypericum perforatum* that produces hypericin. J Nat Prod 71:159–162

Lacey LA, Frutos R, Kaya HK, Vail P (2001) Insect pathogens as biological control agents: do they have a future? Biol Control 21:230–248

Lee S, Kim JJ, Breuil C (2006) Diversity of fungi associated with the mountain pine beetle, *Dendroctonus ponderosae* and infested lodgepole pines in British Columbia. Fungal Divers 22:91–105

Lehtonen P, Helander M, Saikkonen K (2005) Are endophyte-mediated effects on herbivores conditional on soil nutrients? Oecologia 142:38–45

Louda SM, Pemberton RW, Johnson MT, Follett PA (2003) Nontarget effects – the Achilles' heel of biological control? Retrospective analyses to reduce risk associated with biocontrol introductions. Annu Rev Entomol 48:365–396

Mardis ER (2011) A decade's perspective on DNA sequencing technology. Nature 470:198–203

Margulies M, Egholm M, Altman WE et al (2005) Genome sequencing in microfabricated high-density picolitre reactors. Nature 437:376–380

Martín J, Witzell J, Blumenstein K et al (2013) Resistance to Dutch elm disease reduces xylem endophytic fungi presence in elms (*Ulmus* spp.). PLoS One 8(2):e56987

Martín-García J, Espiga E, Pando V, Diez JJ (2011) Factors influencing endophytic communities in poplar plantations. Silva Fenn 45:169–180

Mejía LC, Rojas EI, Maynard Z et al (2008) Endophytic fungi as biocontrol agents of *Theobroma cacao* pathogens. Biol Control 46:4–14

Millar CI, Stephenson NL, Stephens SL (2007) Climate change and forests of the future: managing in the face of uncertainty. Ecol Appl 17:2145–2151

Morrison WE, Hay ME (2011) Herbivore preference for native vs. exotic plants: generalist herbivores from multiple continents prefer exotic plants that are evolutionarily naïve. PLoS One 6(3):e17227

Müller-Schärer H, Schaffner U, Steinger T (2004) Evolution in invasive plants: implications for biological control. TREE 19:417–422

Newcombe G (2011) Endophytes in forest management: four challenges. In: Pirttilä AM, Frank AC (eds) Endophytes of forest trees: biology and applications, Forestry sciences 80. Springer, Berlin/Heidelberg/New York, pp 251–262

Nilsson HR, Tedersoo L, Lindahl BD et al (2011) Towards standardization of the description and publication of next-generation sequencing datasets of fungal communities. New Phytol 191:314–318

Pal KK, McSpadden Gardener B (2006) Biological control of plant pathogens. Plant Health Instr. Available via APSnet. http://www.apsnet.org/edcenter/advanced/topics/Documents/PHI-BiologicalControl.pdf. Accessed 27 Feb 2013

Pearson DE, Callaway RM (2005) Indirect nontarget effects of host-specific biological control agents: implications for biological control. Biol Control 35:288–298

Petrini O (1991) Fungal endophytes of leaves. In: Andrew JH, Hirano SS (eds) Microbial ecology of leaves. Springer, Berlin, pp 179–197

Petrini O, Fisher PJ (1990) Occurrence of fungal endophytes in twigs of *Salix fragilis* and *Quercus robur*. Mycol Res 94:1077–1080

Pratt JE, Johansson M, Hütterman A (1998) Chemical control of *Heterobasidion annosum*. In: Woodward S, Stenlid J, Karjalainen R, Hutterman A (eds) *Heterobasidion annosum*: biology, ecology, impact and control. CAB International, Wallingford, pp 259–282

Qin J, Li RQ, Raes J et al (2010) A human gut microbial gene catalogue established by metagenomic sequencing. Nature 464:59–65

Ragazzi A, Moricca S, Capretti P et al (2003) Differences in composition of endophytic mycobiota in twigs and leaves of healthy and declining *Quercus* species in Italy. For Pathol 33:31–38

Rodriguez RJ, Redman RS (1997) Fungal lifestyles and ecosystem dynamics: biological aspects of plant pathogens, plant endophytes, and saprophytes. Adv Bot Res 24:169–193

Rodriguez RJ, White JF, Arnold AE, Redman RS (2009) Fungal endophytes: diversity and functional roles. New Phytol 182:214–220

Rodríguez J, Elissetche JP, Valenzuela S (2011) Tree endophytes and wood biodegradation. In: Pirttilä AM, Frank AC (eds) Endophytes of forest trees: biology and applications, Forestry sciences 80. Springer, Berlin/Heidelberg/New York, pp 81–93

Saikkonen K (2007) Forest structure and fungal endophytes. Fungal Biol Rev 21:67–74

Saikkonen K, Faeth SH, Helander M, Sullivan TJ (1998) Fungal endophytes: a continuum of interaction with host plants. Annu Rev Ecol Syst 29:319–343

Saikkonen K, Ion D, Gyllenberg M (2002) The persistence of vertically transmitted fungi in grass metapopulations. Proc R Soc B Biol Sci 269:1397–1403

Saikkonen K, Wäli P, Helander M, Faeth SH (2004) Evolution of endophyte-plant symbioses. Trends Plant Sci 9:275–280

Santamaría O, Diez JJ (2005) Fungi in leaves, twigs and stem bark of *Populus tremula* from northern Spain. For Pathol 35:95–104

Santini A, Ghelardini L, De Pace C et al (2013) Biogeographical patterns and determinants of invasion by forest pathogens in Europe. New Phytol 197:238–250

Schiebe C, Hammerbacher A, Birgersson G et al (2012) Inducibility of chemical defences in Norway spruce bark is correlated with unsuccessful mass attacks by the spruce bark beetle. Oecologia 170:183–198

Schröter D, Cramer W, Leemans R et al (2005) Ecosystem service supply and vulnerability to global change in Europe. Science 310:1333–1337

Schulz B, Boyle C (2006) What are endophytes? In: Schulz B, Boyle C, Sieber TN (eds) Microbial root endophytes, Soil biology 9. Springer, Heidelberg, pp 1–13

Schwarze FWMR, Engels J, Matcheck C (2000) Fungal strategies of wood decay in trees. Springer, Berlin, 185pp

Seifert K (2009) Progress towards DNA barcoding of fungi. Mol Ecol Res 9(Suppl 1):83–89

Shigo AL (1984) Compartmentalization: a conceptual framework for understanding how trees grow and defend themselves. Annu Rev Phytopathol 22:189–214

Sieber TN (2007) Endophytic fungi in forest trees: are they mutualists? Fungal Biol Rev 21:75–89

Smith J (2002) Afforestation and reforestation in the clean development mechanism of the Kyoto protocol: implications for forests and forest people. Int J Global Environ Issue 2:322–343

Solla A, Gil L (2003) Evaluating *Verticillium dahliae* for biological control of *Ophiostoma novo-ulmi* in *Ulmus minor*. Plant Pathol 52:579–585

Stierle A, Strobel G, Stierle D (1993) Taxol and taxane production by *Taxomyces andreanae*, an endophytic fungus of Pacific yew. Science 260:214–216

Strauss SH, Brunner AM, Busov VB et al (2004) Ten lessons from 15 years of transgenic *Populus* research. Forestry 77:457–465

Strobel G, Daisy B (2003) Bioprospecting for microbial endophytes and their natural products. Microbiol Mol Biol Rev 67:491–502

Sturrock RN, Frankel SJ, Brown AV et al (2011) Climate change and forest diseases. Plant Pathol 60:133–149

Sun X, Guo LD, Hyde KD (2011) Community composition of endophytic fungi in *Acer truncatum* and their role in decomposition. Fungal Divers 47:85–95

Tscharntke T, Bommarco R, Clough Y et al (2007) Conservation biological control and enemy diversity on a landscape scale. Biol Control 43:294–309

Unterseher M (2011) Diversity of fungal endophytes in temperate forest trees. In: Pirttilä AM, Frank CA (eds) Endophytes of forest trees: biology and applications. Springer, Heidelberg, pp 31–46

Verma VC, Lobkovsky E, Gange AC et al (2011) Piperine production by endophytic fungus *Periconia* sp. isolated from *Piper longum* L. J Antibiot 64: 427–431

Waage JK, Greathead DJ (1988) Biological control: challenges and opportunities. Philos Trans R Soc Lond B 318:111–128

Webber JF (1981) A natural biological-control of Dutch elm disease. Nature 292:449–451

White JF, Torres MS (2010) Is plant endophyte-mediated defensive mutualism the result of oxidative stress protection? Physiol Plant 138:440–446

Witzell J, Martín JA (2008) Phenolic metabolites in the resistance of northern forest trees to pathogens-past experiences and future prospects. Can J For Res 38:2711–2727

Zhao J, Shan T, Mou Y, Zhou L (2011) Plant-derived bioactive compounds produced by endophytic fungi. Mini Rev Med Chem 11:159–168

18 Endophyte-Mediated Biocontrol of Herbaceous and Non-herbaceous Plants

Orna Liarzi and David Ezra

Abstract

All plants in natural ecosystems appear to be symbiotic with endophytes. This includes many economically important agricultural products as well as essential food crops. The endophytes confer fitness benefits to their hosts in various and variable aspects such as growth enhancement and increased reproductive success and confer tolerance to biotic and abiotic stresses. In this chapter we will focus on the biocontrol activity of endophytes, i.e., the biological effects of endophytes on herbaceous or non-herbaceous host plants and the mechanisms, if known, by which the endophytes increase the fitness of their hosts.

1 Introduction

All plants in natural ecosystems appear to be symbiotic with fungal endophytes (Rodriguez and Redman 2008; Rodriguez et al. 2009b; Singh et al. 2011a). It is estimated that 20–30 % of grass species worldwide, including many economically important forage and turf grasses, are associated with endophytic fungi (Leuchtmann 1992). Throughout its motionless life, the plant is exposed to various biotic and abiotic stresses, from which it can either escape or mitigate (Rodriguez and Redman 2008). However, since grasses lack the biosynthetic capacity for the production of secondary metabolites, which are useful in the long-term survival strategy (Kuldau and Bacon 2008), their dependence on microorganisms that produce secondary metabolites is more pronounced.

Symbiosis is defined as "the permanent association between two or more specifically distinct organisms, at least during a part of the life cycle" (de Bary 1879). The association of fungal symbionts with plants can be in a form of endophyte or as mycorrhizal fungi (Singh et al. 2011a). Unlike mycorrhizal fungi that colonize plant roots and grow into the rhizosphere, endophytes reside entirely within plant tissues and may grow within roots, stems, and/or leaves, emerging to sporulate at plant or host-tissue senescence, and their presence in the plant tissue causes no symptoms of disease (Sherwood and Carroll 1974; Hallmann et al. 1997; Carroll 1988; Stone et al. 2004). Similarly, a variety of bacteria have been reported to maintain endophytic lifestyle

O. Liarzi • D. Ezra (✉)
Department of Plant Pathology, ARO the Volcani Center, P.O. Box 6, Bet Dagan 50250, Israel
e-mail: dezra@volcani.agri.gov.il

V.C. Verma and A.C. Gange (eds.), *Advances in Endophytic Research*,
DOI 10.1007/978-81-322-1575-2_18, © Springer India 2014

in plants (Reinhold-Hurek and Hurek 2011; Mei and Flinn 2010; Dudeja et al. 2012; Rosenblueth and Martinez-Romero 2006). It is generally assumed that many bacterial endophyte communities are the product of a colonizing process initiated in the root zone (McInroy and Kloepper 1995; Sturz et al. 2000; Welbaum et al. 2004). However, they may also originate from other sources such as the phyllosphere, the anthosphere, or the spermosphere (Hallmann et al. 1997). In contrast to extensively studied grass endophytes, endophytes associated with woody angiosperms are poorly known (Arnold et al. 2003). Even though examples of endophytes and woody plants associations are published extensively in the last few years.

In this chapter, we will focus on the biocontrol activity of endophytes (both fungi and bacteria), i.e., the biological effects of the endophytes on herbaceous or non-herbaceous host plants. Endophytes confer fitness benefits to their hosts in various and variable aspects. These aspects include growth enhancement and tolerance to biotic and abiotic stresses. The mechanisms of each of these aspects and their significance are detailed below.

2 Herbaceous Plants

Herbaceous plants from the family Poaceae (known as "true grasses") contribute to the development of humankind since there are several important species that serve as essential food crops (e.g., wheat, rice, maize, and barley), as forage of livestock, and for recreational and conservation purposes (Kuldau and Bacon 2008). Other herbaceous plants such as Fabaceae and Asteraceae families also comprise a valuable source of food. Here, we review the biocontrol effects of endophytes on herbaceous host plants.

In general, endophytic fungi of grasses can be classified into two main groups, constitutive mutualists and inducible mutualists (Carroll 1988). It is generally accepted that the former represents clavicipitaceous endophytes while the latter represents nonclavicipitaceous endophytes (Singh et al. 2011a). These two groups are different in evolutionary relatedness, taxonomy, plant hosts, and ecological functions (Rodriguez et al. 2009b). The clavicipitaceous endophytes represent a small number of phylogenetically related clavicipitaceous species that exhibit a narrow host range (limited only to some cool- and warm-season grasses) and extensively colonize plant shoot and rhizome, mainly with one dominant fungal isolate/genotype (Bischoff and White 2005; Rodriguez et al. 2009b; Wille et al. 1999). The transmission of clavicipitaceous endophytes is mainly vertical, i.e., maternal plants passing fungi onto offspring via seed infections (Saikkonen et al. 2002). The clavicipitaceous endophytes may further be subdivided into three types (Clay and Schardl 2002); however, in this review we will discuss the biocontrol effects of type III only, because of their symptomless behavior. On the other hand, the nonclavicipitaceous endophyte group represents a broad host range – these endophytes can be recovered from asymptomatic tissues of every major lineage of land plants (Rodriguez et al. 2008a; Higgins et al. 2007), and from all terrestrial ecosystems (Arnold and Lutzoni 2007), and considered to be the largest group of fungal symbionts (Petrini 1996). These endophytes are capable of forming limited to extensive colonization in the host tissue, their transmission pattern can be either vertical or horizontal, and they are able to colonize both above- and below-ground tissues (Rodriguez et al. 2009b; Singh et al. 2011a). The nonclavicipitaceous endophytes can be further subdivided into three functional classes based on host colonization pattern, mechanism of transmission between host generations, *in planta* biodiversity levels, and ecological function (Rodriguez et al. 2009b), yet, in this article, we generalize and refer to the nonclavicipitaceous endophyte as one functional group.

2.1 Increased Plant Growth

2.1.1 Clavicipitaceous Endophytes

Endophyte infection increases growth rate of perennial ryegrass and tall fescue (Clay 1988). Controlled environmental studies conducted on single cultivars and natural ecotypes of the grasses tall fescue (*Festuca arundinacea*), meadow fescue (*Lolium pratense*), and perennial ryegrass (*Lolium perenne* L.) suggest that their epichloe endophytes (*Neotyphodium coenophialum*, *N. uncinatum*, and *N. lolii*, respectively) enhance biomass production, tiller numbers and survivor, seed production, and root growth (Belesky et al. 1989; de Battista et al. 1990b; Funk et al. 1993; Latch et al. 1985b; Clay 1987; Joost 1995). In tall fescue and meadow fescue, the endophytes increase root growth, extend root hairs, and decrease root diameter (Malinowski et al. 1999b; Malinowski and Belesky 2000). An increase in the rate of growth and herbage yield may be due to physiological response of the grass from an increase in endogenous levels of plant hormones, which may be an additive effect from the fungal endophyte or from an increase in the water and nutrient content (Kuldau and Bacon 2008). Enhanced plant growth observed in endophyte-infected grasses is attributed to either or both production of synthetic growth hormones or phytohormones, such as indole acetic acid (IAA), which controls tillering in grasses (Joost 1995), and has been demonstrated to accumulate in vitro in cultures of *N. coenophialum* (de Battista et al. 1990a) and related species (Porter et al. 1985). The finding of small molecular weight indole compounds such as indole-3-acetic acid, indole-3-ethanol, and several indole glycerols in in vitro cultures of endophytes suggests that these chemicals may serve as synthetic growth hormones secreted by the endophyte and promote plant growth (Porter et al. 1977, 1985). In addition, it has been suggested that loline alkaloids, secondary metabolites secreted by the endophyte, serve as allelochemicals responsible for allelopathy phenomenon found in plants, particularly rosaceous species, grown in soils planted previously with endophyte-infected tall fescue (Petroski et al. 1990). The results of such a phenomenon produce a competitive edge for infected grasses, resulting in an increase in population density. It was found that the influence of *N. coenophialum* on the growth of tall fescue begins at germination because endophyte-infected seeds exhibit higher germination rates than endophyte-free seeds in half of the genotypes tested (Pinkerton et al. 1990), probably in a mechanism that involves reduction of the relative water gain during imbibitions (Rice et al. 1990). Lower germination and seedling vigor of endophyte-free tall fescue grass caused 20 % reduction in ground cover relative to endophyte-infected plants (Joost 1995). Recent evidence suggests that morphological changes in the endophyte *Neotyphodium lolii* affect its host (*Lolium perenne*) phenotype: old mycelium-induced dwarf symptoms in a non-mutational manner (Simpson et al. 2012). It should be noted that the growth promotion activity of the endophyte might be dependent on environmental conditions. For example, in symbiotic *Agrostis perennans* plants, the enhanced inflorescence occurs under conditions of restricted light availability (Davitt et al. 2010).

To conclude, the exact biochemical basis for endophyte-induced growth is still obscure, but it is suggested that production of indole acetic acid and/or other phytohormones may play a role in plant growth alternations (de Battista et al. 1990a; Yue et al. 2000).

2.1.2 Nonclavicipitaceous Endophytes

Most nonclavicipitaceous endophytes increase host shoot and/or root biomass, probably in a mechanism that involves induction of plant hormones by the host and/or biosynthesis of plant hormones by the fungi (Rodriguez et al. 2009a; Tudzynski and Sharon 2002). Nevertheless, the exact nature of plant growth promotion effects is not always clear (Druege et al. 2007; Pham et al. 2004). The plant-root-colonizing basidiomycete fungus *Piriformospora indica* has been obtained from the rhizosphere soils of the woody shrubs *Prosopis juliflora* and *Zizyphus nummularia* in the sandy soils of the arid Thar Desert (Verma et al. 1998). This fungus exhibits strong promotion of vegetative growth during its symbiosis with species from various plant families, including cereal crops, rice, wheat, and barley as well as Dicotyledoneae (reviewed in Franken 2012), and this growth-promoting effect can occur also under abiotic stress conditions (Shahabivand et al. 2012). *P. indica* is also important due to its ability to increase tomato fruit biomass under indoor production conditions (Fakhro et al. 2010). *P. indica* promotes the initial stages of plant development and therefore an earlier switch to generative stages (Barazani et al. 2005; Achatz et al. 2010; Andrade-Linares et al. 2012). The mechanism for this promotion is probably based on acceleration of root development (Waller et al. 2005; Baltruschat et al. 2008) and earlier expression of age-dependent genes (Waller et al. 2008). In addition, *P. indica* produces auxin (Sirrenberg et al. 2007) and induces auxin-regulated genes in barley (Schäfer et al. 2009) and Chinese cabbage (Lee et al. 2011), which cause strong growth-promoting effect of the roots. Yet, evidence suggests that indole-3-acetic acid is not required for growth promotion in barley, but is involved in the establishment of biotrophic colonization in the roots (Hilbert et al. 2012). Moreover, evidence suggests that *P. indica* inhibits ethylene signaling and thereby contributes to plant growth promotion (Barazani et al. 2007; Schäfer et al. 2009). Additional phytohormones synthesized or manipulated by *P. indica* include cytokinins (Vadassery et al. 2008), gibberellins, brassinosteroids, and abscisic acid (Schäfer et al. 2009). The latter is proposed to enhance plant growth via calcium (Vadassery et al. 2009), phosphoinositide, and protein kinases (Camehl et al. 2011). Thus, a wide set of phytohormones and their signaling networks appear to be involved in increasing early root growth promotion, which leads to greater biomass. *P. indica* also activates nitrate reductase that plays a role in nitrate acquisition and also a starch-degrading enzyme glucan water dikinase involved in early events of starch degradation in the plants as tobacco and *Arabidopsis* (Sherameti et al. 2005).

More examples for nonclavicipitaceous endophytes that promote plant growth are the endophytic fungi *Phoma glomerata* LWL2 and *Penicillium* sp. LWL3 that both promote host-cucumber plant growth probably in a mechanism that involves the phytohormones gibberellins and indole acetic acid that they secrete (Waqas et al. 2012). Additional strain of *Penicillium* (*Penicillium citrinum* KACC43900) is a growth promotion fungal endophyte isolated from the roots of the sand dune flora (*Ixeris repenes* (L.) Gray) and exerts gibberellin-producing capacity (Khan et al. 2008). *Epicoccum nigrum* is an important sugarcane endophyte fungus that increases root system biomass (Fávaro et al. 2012). This phenomenon could be the outcome of plant hormone production by the endophyte, previously observed in culture medium (Rowan and Latch 1994). The culture filtrate of *Paecilomyces formosus* LHL10, an endophytic fungus isolated from cucumber roots, significantly increases the growth of gibberellin-deficient mutant rice cultivars. Analysis of this culture filtrate revealed the presence of both gibberellins and indole acetic acid phytohormones (Khan et al. 2012a). Another fungal endophyte from cucumber, *Exophiala* sp. LHL08, also promotes its host growth by production of gibberellins (Khan et al. 2011b). The endophyte PGP-HSF, isolated from *Mentha piperita*, enhances its vegetative growth (Mucciarelli et al. 2002), and it is localized in peppermint green tissues and improves peppermint metabolic and photosynthetic apparatus (Mucciarelli et al. 2003). In rice (*Oryzae sativa*), the growth and development is regulated epigenetically by a fungal endophyte: in symbiotic plants the resources are preferentially allocated into root growth until root hairs are established, therefore increasing the rate of expansion of the symbiotic roots (Rodriguez et al. 2009a). Also, the rice endophyte *Pantoea agglomerans* YS19 promotes host plant growth probably due to endophytic production of phytohormones, enhancing the transportation of photosynthetic assimilation products from the flag leaves (source) to stachys (sink) (Feng et al. 2006).

Not only fungal endophytes promote growth, but also bacterial endophytes exert plant growth effects (Sturz et al. 2000; Welbaum et al. 2004; Compant et al. 2005a; Dudeja et al. 2012). *Burkholderia phytofirmans* strain PsJN is a plant-promoting bacterial endophyte, isolated from surface-sterilized onion roots (Sessitsch et al. 2005) exhibiting a broad host range including potatoes, tomatoes, grape vines, rice, and the bioenergy crop candidate switchgrass (*Panicum virgatum* L.) cv. Alamo (Compant et al. 2005b; Barka et al. 2000; Mattos et al. 2008; Conn et al. 1997; Nowak et al. 2003; Kim et al. 2012). This endophyte produces high level of 1-aminocyclopropane-1-carboxylic acid (ACC) deaminase (Sessitsch et al. 2005) that catalyzes the cleavage of 1-aminocyclopropane-1-carboxylic, the immediate precursor of ethylene, and therefore lowers the ethylene levels in host plants, leading to stimulation of plants growth (Glick et al. 1998). It is also capable to produce auxin that may have a stimulatory effect on plant development (Mattos et al. 2008). Evidence suggests that the activity of quinolinate phosphoribosyltransferase plays a role in the signal pathway for promotion of plant growth by the endophyte *Burkholderia* sp. strain PsJN (Wang et al. 2006). The endophytic bacteria *Gluconacetobacter diazotrophicus* isolate IS100 can colonize sugarcane roots and significantly improve plant growth and nutrient uptake (Suman et al. 2005; Saravanan et al. 2008). Additional examples for plant growth promotion and biomass yield increase by microbial endophytes can be found in the reviews by Mei and Flinn (2010) and Hardoim et al. (2008).

The major sources of nitrogen for agricultural soils are from mineral fertilizers and biological nitrogen fixation (Sturz et al. 2000). Rhizosphere-based diazotrophic bacteria tend to retain the products of nitrogen fixation for their own use, and therefore the benefit to the crop is only realized after the bacterial death (van Berkum and Bohlool 1980; Okon 1985). In contrast, endophytic nitrogen-fixing bacteria are believed to be capable of contributing directly to the nitrogen requirements of their host in sugarcane (Boddey et al. 1995; Sevilla et al. 2001), rice (Ladha and Reddy 1995; Yanni et al. 1997; Hurek and Reinhold-Hurek 2003), and wheat (Webster et al. 1997). Additional examples for endophytic nitrogen-fixing bacteria can be found in the interior roots of rice, maize, and grasses as reviewed in Sturz et al. (2000). For more general modes of actions for plant growth enhancement by bacterial endophyte, see Sturz et al. (2000).

2.2 Stress Tolerance

Fitness benefits conferred by the endophyte contribute to or are responsible for plant adaptation to stress (Stone et al. 2000; Rodriguez et al. 2004). Therefore, the endophytes are acting as biological triggers that control the activation of host stress response (Rodriguez et al. 2004). We differentiate between biotic and abiotic stress. Biotic stress could be the outcome of interspecific competition, invertebrate pests, herbivory of mammals, and diseases caused by phytopathogens. Abiotic stress could be the outcome of heavy metal pollutions, drought, salinity, and temperature stresses. Abiotic stress is the primary cause of crop loss worldwide, reducing average yields for most major crop plants by more than 50 % (Boyer 1982; Bray et al. 2000; Sturz et al. 2000; Singh et al. 2011a). The mechanisms by which endophytes confer stress tolerance are described below.

2.2.1 Biotic Stress

Interspecific Competition

Centaurea stoebe is an invasive forb in North America. The presence of Alternaria endophyte enhances its competitive ability without increasing its size. The mechanism by which the endophyte increases its host competiveness is unknown, but it is not related to increased growth (Aschehoug et al. 2012). It is suggested that the mechanisms for interspecific competition involve increased clonal growth and lateral spread, production of allelochemicals, increased seedling vigor and seed yield (Kuldau and Bacon 2008; Bush et al. 1997; Malinowski et al. 1999a), as well as increased number of tillers, greater leaf elongation rate, and altered root architecture (Malinowski and Belesky 2000). In field experiments, infected tall fescue suppressed other grasses and forbs relative to uninfected fescue (Clay and Schardl 2002). The following evidence emphasizes the competitive advantage of endophyte-infected grasses: the number of white clover (*Trifolium repens* L.) plants declined in pastures dominated by endophyte-infected, compared with noninfected, perennial ryegrass (Percival and Duder 1983; Sutherland and

Hoglund 1989). An example for involvement of allelochemicals is the finding that seed extracts of endophyte-infected tall fescue inhibit germination of *Trifolium* spp. (Springer 1997). It is suggested that loline alkaloids enhance the competitive ability of endophyte-infected grasses by retarding the establishment of competitors in a sward. This is based on the finding that loline alkaloids are the only group of endophyte-related alkaloids shown to reduce germination rate of monocot and dicot seeds (Petroski et al. 1990). It should be noted that endophyte infection seems to improve the ability of tall fescue plants to survive environmental stresses, whereas no difference in the plant survival observed under temperate locations (Bouton et al. 1993; Joost 1995). On the other hand, it was shown that the presence of the fungal endophyte *Neotyphodium schardlii* in its host grass (*Cinna arundinacea*) results in reduction of host survival but increases regeneration, and therefore the negative effect on plant survival is overwhelmed by the beneficial effect on regeneration (Rudgers et al. 2012). Recently, it was demonstrated that hybridization of symbiotic *Neotyphodium* endophytes may increase the competitive potential of its host (*Festuca arizonica*) in stressful environments, and this may enable niche expansion of Arizona fescue in the environments with low resources (Saari and Faeth 2012).

Invertebrate Pests

Endophytic fungi belonging to the genus *Neotyphodium* confer resistance to infected host grasses against insect pests. This can be achieved by the production of endophyte-related alkaloids or alkaloid groups, peramine, lolitrem B, ergovaline, and the lolines. For example, the corn flea beetle (*Chaetocnema pulicaria* Melsheimer) feeding and survival is reduced by infection of tall fescue with *N. coenophialum*, and the suggested mechanism is antixenosis (Ball et al. 2011). Seed predation is lower in endophyte-containing tall fescue grasses by cocksfoot moth (Saari et al. 2010). Similarly, in a native grass experiment, herbivores show a significant preference for endophyte-free plant material, and the presence of endophyte reduces the performance of the Orthoptera *Schistocerca americana* (Crawford et al. 2010). Endophyte in ryegrass reduces Argentine stem weevil oviposition, feeding, and also larval survival (Barker et al. 1984a, b). This is achieved by the production of secondary metabolite by the endophyte such as peramine (being the most potent (Tanaka et al. 2005)), ergovaline (Popay et al. 1990), and loliterm B (Prestidge and Gallagher 1985). Endophyte presence in ryegrass deters the adult black beetle (*Heteronychus arator*) resulting in fewer eggs and larvae (Ball and Prestidge 1992), and this effect is mediated by ergovaline (Ball et al. 1997). The presence of endophyte in ryegrass reduces mealy bug (*Balanococcus poae*) numbers (Pennell et al. 2005), probably due to failure of the dispersing crawler stage of the bug to establish on plants (Pearson 1989). In perennial ryegrass, *Lolium perenne* L., the hairy chinch bug (*Blissus leucopterus hirtus* Montandon) damages and population density decreases linearly as the proportion of *Neotyphodium* endophyte infection increases (Richmond and Shetlar 2000). The deterrent activities of the lolines and peramine against sucking insects may also help to reduce infections by plant viruses vectored by those insects (Mahmood et al. 1993). It should be noted that loline alkaloids represent an inducible defense in the symbiotum since its level increases in response to clipping (Bush et al. 1997; Craven et al. 2001). Similarly, endophyte genes for secondary metabolite biosynthesis are only expressed in the plant and under conditions of restricted growth (Tanaka et al. 2012).

In general, *Neotyphodium*- and *Epichloë*-infected grasses deter approximately 45 species of invertebrate pests belonging to the following families: Aphididae, Chrysomelidae, Cicadidae, Curculionidae, Gryllidae, Lygaeidae, Miridae, Noctuidae, Pyralidae, Scarabaeidae, and Tenebrionidae (Kuldau and Bacon 2008). Studies suggest that secondary metabolites of the fungal endophyte contribute to insect toxicity, especially pyrrolopyrazine alkaloid peramine (Ball et al. 1995; Rowan et al. 1986), and for a broader range ergot alkaloid ergovaline, pyrrolizidine loline alkaloids (Siegel et al. 1990; Wilkinson et al. 2000; Riedell et al. 1991), and janthitrems

(Tapper and Lane 2004). A detailed list of insect and nematodes pests deterred by *Neotyphodium/Epichloë* species can be found in Kuldau and Bacon (2008).

In addition to the grass species, the endophyte *Fusarium oxysporum* strain 162 confers resistance to the nematodes *Meloidogyne incognita* in tomato (Dababat and Sikora 2007) and *Radopholus similis* in banana in combined application with the egg pathogen *Paecilomyces lilacinus* strain 251 and the antagonistic bacteria *Bacillus firmus* (Mendoza and Sikora 2009). Interestingly, *N. coenophialum* is localized in the aboveground tissues of tall fescue, whereas the root-knot nematodes (*Meloidogyne marylandi*) attack the roots. It is suggested that the endophyte-induced structural changes in the root, i.e., thickening of endodermal cell walls, might reduce the ability of the nematode to penetrate the stele (Gwinn and Bernard 1993; Kimmons et al. 1990). However, the migratory nematode *Pratylenchus scribneri* is able to penetrate the roots of *N. coenophialum-infected* tall fescue; yet, its reproduction is inhibited by the endophyte presence in an unknown mechanism (Kimmons et al. 1990).

It should be noted that the effects of endophyte-infected grasses on the preference and performance of phytophagous insects may be variable (Clement et al. 2011; Tintjer and Rudgers 2006). It was found that for the black cutworm *Agrotis ipsilon* (Hufnagel), *Neotyphodium* endophyte-mediated resistance is based mainly on *N*-acetyl norloline and peramine, whereas ergovaline exhibits smaller effect (Baldauf et al. 2011). Recently, it was demonstrated that the insect-pathogenic fungus *Metarhizium robertsii* (Clavicipitaceae) is not only rhizosphere competent but also displays endophytic association with *Panicum virgatum* and *Phaseolus vulgaris* roots and this association results in the proliferation of root hairs (Sasan and Bidochka 2012).

Herbivory of Mammals

In the 1930s, tall fescue (*Lolium arundinaceum*) was bred and widely used in the United States, due to its improved characteristics such as considerable longevity, stress tolerance, and its capacity to prevent soil erosion. However, in the mid-1970s, the problem of fescue toxicosis in cattle and the livestock had been recognized, with symptoms resembling ergot poisoning caused by *Claviceps purpurea* (Schardl et al. 2004). A relationship between health disorders in cattle and a high level of endophyte infestation in tall fescue from toxic pastures in the United States was first shown by Bacon et al. (1977) (Bacon et al. 1977), and of particular importance is ergovaline (Lyons et al. 1986). *Neotyphodium lolii* is identified as the ryegrass endophyte causing staggers (Fletcher and Harvey 1981). Lolitrems are indole diterpene alkaloids produced by *N. lolii* that cause tremor-inducting neurotropic activity (Gallagher et al. 1984; Rowan 1993; Tor-Agbidye et al. 2001). Similarly to the resistance against invertebrates, the ergot alkaloid levels greatly increases in *Festuca rubra*-endophyte symbiosis upon clipping, suggesting that an epichloe metabolite can represent an inducible plant defense (Bazely et al. 1997). Several examples for the effect of endophyte-infected grasses on livestock are decrease in productivity (Burke and Rorie 2002; Cross et al. 1995; Paterson et al. 1995; Porter and Thompson 1992), increase in systemic relaxin in pregnant pony mares (Ryan et al. 2001), alteration in hemograms and serum biochemical analytes of steers (Oliver et al. 2000), lower phagocytic activity (Saker et al. 1998), and abdominal lipomatosis in deer (Wolfe et al. 1998). Also, the presence of *Neotyphodium coenophialum* endophyte in tall fescue (*Festuca arundinacea* Schreb.) lowers the copper concentrations in the plant, and this may contribute to lower copper status in animals and therefore to the etiology of fescue toxicity (Dennis et al. 1998). Additional examples for the effect of the secondary metabolites produced by the endophyte and livestock health can be found in Powell and Petroski (1992), Bacon (1995), Malinowski and Belesky (2000), Schardl (2001), and Rodriguez et al. (2008a), as well as strategies to control their effects in di Menna et al. (2012).

It should be noted that since perennial ryegrass is the most important pasture species in New Zealand, it has been estimated that the total cost of ryegrass staggers to the New Zealand livestock industry is more than 40$ million (Prestidge et al. 1991). Similarly, tall fescue (*Festuca arundinacea*) is the most important cool-season grass in the United States, providing the primary ground cover on some 35 million acres, and therefore the estimated cost to beef producers, due to decreases in productivity, is more than 600$ million annually (Porter and Thompson 1992; Paterson et al. 1995).

Additional endophyte-infected grasses that cause related symptoms are drunken horse grass (*Achnatherum inebrians*) in Asia and sleepygrass (*Achnatherum robustum*) in North America, which are associated with ergot alkaloids that induce stupor and aversion to future grazing (Miles et al. 1998; Petroski et al. 1992). Dronkgras (*Melica decumbens*) in South Africa and *Poa huecu* (causing the lethal heucú toxicosis) in Argentina are associated with tremors due to the effects of indole diterpene (Coetzer et al. 1985; Moon et al. 2002; Pimilio et al. 1989). Endophyte affects also populations of small mammals. For example, endophyte-infected tall fescue reduces vole reproduction (Fortier et al. 2000). *Neotyphodium* presence in *Lolium multiflorum* reduces seed removal by rodents (Uchitel et al. 2011). Rabbit weight gain and intake is reduced by feeding endophyte-infected tall fescue seed diets (Filipov et al. 1998).

Locoweeds are toxic to mammals due to the presence of the alkaloid swainsonine, produced by the endophytic fungus *Undifilum oxytropis*. There is a correlation between swainsonine concentration and the proportion of endophytic DNA in plant nonreproductive tissues (Achata Böttger et al. 2012). Also, there is a correlation between swainsonine produced by the endophyte *Embellisia* sp. and dinitrogen fixation by *Rhizobium* in the perennial legume *Oxytropis sericea*, with the latter increasing the production of swainsonine of the former (Valdez Barillas et al. 2007).

Plant Disease

Clavicipitaceous endophytes suppress plant pathogens in both in vitro and field experiments, in a mechanism that involves production of degradative enzymes and antibiotics by the endophyte (Siegel and Latch 1991; White and Cole 1985). Specific chemicals, such as several indole compounds sesquiterpene, diacetamide, and unidentified volatile compounds (Yue et al. 2000, 2001), have been associated with *Epichloë* species for resistance to leaf spot, *Cladosporium phlei*, and stem rust, *Puccinia graminis*, on infected *Phleum pratense* (Koshino et al. 1988, 1989; Yoshihara et al. 1985), and the mitigation effect of these diseases by the endophyte was further confirmed in field studies (Greulich et al. 1999; Welty et al. 1993). The maize endophyte *Acremonium zeae* produces antibiotics that augment host defenses against a variety of pathogen, whereas protective endophytes, including mycoparasites which grow asymptomatically within healthy maize tissues, show little sensitivity to these antibiotics (Wicklow et al. 2005; Wicklow and Poling 2009). Additional fungal pathogens that are controlled to some level by endophyte infections are reviewed in Kuldau and Bacon (2008) and Schardl (2001).

Yet, the fungal disease suppression in endophyte-infected grasses is not always clear (Kuldau and Bacon 2008): There is a resistance to *Sclerotinia homoeocarpa* in chewing fescue, hard fescue, and strong creeping red fescue, but there is an increase in disease incidence to *Pythium* blight in *N. coenophialum*-infected tall fescue (Clarke et al. 2006; Blank 1992). Mixed results also obtained with the ryegrass endophyte *Neotyphodium lolii*. This endophyte inhibits the growth of *Colletotrichum graminicola*, *Limonomyces rosipellis*, and *Rhizoctonia zeae* but did not affect *Bipolaris sorokiniana*, *Pythium aphanidermatum*, *Sclerotinia homoeocarpa*, *Gaeumannomyces graminis*, *Rhizoctonia solani*, *Leptosphaeria korrae*, *Phialophora graminicola*, and *Magnaporthe poae* (Siegel and Latch 1991). In vitro, this endophyte exerts only limited antibiotic effects against fungal pathogens (Siegel and Latch 1991; White and Cole 1985). Furthermore, there is no correlation between the presence of *Neotyphodium lolii* in Tasmanian pastures and barley yellow dwarf virus (Guy 1992) nor effect on the virus vector *Rhopalosiphum padi* (Latch et al. 1985a). In contrast, Mahmood et al. (1993) demonstrated that the presence of the endophyte significantly reduces the virus indirectly due to reduction of virus spread by controlling the aphid vector (Mahmood et al. 1993). In accordance, evidence suggests that this endophyte manages to increase plant's growth in the presence of the virus (Hesse and Latch 1999) and that in symbiotic *Lolium pratense*, endophyte infection protects its host by reducing the percentage of barley yellow dwarf virus infection as well as the number of its aphid vector (Lehtonen et al. 2006).

Many nonclavicipitaceous endophytes protect their hosts against fungal pathogens (Danielsen and Jensnen 1999; Narisawa et al. 2002; Campanile et al. 2007; Rodriguez et al. 2009a; Mei and Flinn 2010). This protection might be achieved via the production of secondary metabolites harmful to plant pathogens (Schulz et al. 1999), fungal parasitism (Samuels et al. 2000), induction of systemic resistance (Vu et al. 2006; Waller et al. 2005; Compant et al. 2005a), activation of host defenses upon exposure to virulent pathogens (Redman et al. 1999), and competition between the endophyte and the pathogen for resources or niche space (Rodriguez et al. 2009a; Combès et al. 2012).

Several mechanisms are proposed for disease control by microbial endophytes in plants (Kuldau and Bacon 2008; Reinhold-Hurek and Hurek 2011): (A) The endophyte induces plants' systemic resistance (Chen et al. 1995; Kloepper and Beauchamp 1992; Kunkel et al. 2004; Roberts et al. 1992; Waller et al. 2005; Serfling et al. 2007; Waller et al. 2008). (B) Niche exclusion by the endophyte – the epiphyllous mycelia nets found in several clavicipitaceous endophyte-grass associations may oppress the pathogen (Moy et al. 2000). Niche exclusion has been demonstrated also for bacterial endophyte (Cook and Baker 1983) and for dark septate endophytic fungus (Khastini et al. 2012). (C) The endophyte fortifies plant cell wall strength; for example, *P. fluorescens* WCS417r causes thickening tomato cell walls (Duijff et al. 1997) and dark septate endophytic fungus causes thickening of barley cell walls and thus limits the ingress of the *Verticillium longisporum* pathogen into adjacent cells (Narisawa et al. 2004). (D) Endophyte-induced accumulation of pathogenesis-related proteins (M'Piga et al. 1997; Seo et al. 2012). (E) Additional biocontrol mechanisms such as production of antifungal or antibacterial agents, nutrient competition (reviewed in Sturz et al. 2000), siderophore (Khastini et al. 2012), down-regulation of the activity of antioxidant enzymes and thereby reduction of disease severity (Kumar et al. 2009), and production of chitinase by the biocontrol endophyte (*Trichoderma virens* 223) against the pineapple disease pathogen (*Ceratocystis paradoxa*) in sugarcane (Romão-Dumaresq et al. 2012).

The protection conferred by the endophytic fungal isolate *Fusarium solani* to tomato plants against the root pathogen *Fusarium oxysporum* f. sp. *radicis-lycopersici* is mediated through ethylene signaling pathway, whereas the jasmonic acid pathway is not essential for the biocontrol activity (Kavroulakis et al. 2007). However, the exact pathway for conferring resistance is not always known. For example, in rice, the endophytic bacteria *Azospirillum* sp. B510 enhances the resistance against the virulent rice blast fungus *Magnaporthe oryzae* and the virulent bacterial pathogen *Xanthomonas oryzae*, without salicylic acid accumulation or expression of pathogenesis-related genes in the rice plants. Therefore, it is suggested that this endophyte is able to induce disease resistance in rice by activating unknown resistance mechanism independent of salicylic acid-mediated defense signaling (Yasuda et al. 2009).

The ability to act as bioprotectants by triggering induced systemic resistance (ISR) has been demonstrated also for bacterial endophytes (Compant et al. 2005a; Kloepper and Ryu 2006). There is no compelling evidence for an overall ISR signal produced by bacteria (Haas et al. 2002). However, possible candidates are bacterial traits (i.e., flagellation and production siderophores and lipopolysaccharides), volatile organic compounds secreted by the endophyte, triggering salicylic acid-dependent signaling pathway or activating independent salicylic acid pathway involving jasmonate and ethylene signals (reviewed in Compant et al. 2005a; Sturz et al. 2000).

Several examples for disease protection by bacterial endophytes: ISR is triggered by *Burkholderia phytofirmans* PsJN against *Botrytis cinerea* on grapevine, in a mechanism that involves disrupting cellular membrane and inducing cell death (Barka et al. 2000, 2002), and *Verticillium dahliae* on tomato (Sharma and Nowak 1998). *Pseudomonas denitrificans* 1–15 and *Pseudomonas putida* 5–48 against *Fusarium oxysporum* f. sp. *radicis-lycopersici* on tomato (M'Piga et al. 1997) and *Pythium ultimum* and *Fusarium oxysporum* f. sp. *pisi* on pea roots (Benhamou et al. 1996a). *Bacillus pumilus* SE34 against *Fusarium oxysporum* f. sp. *pisi* on pea roots (Benhamou et al. 1996b) and *Fusarium oxysporum* f. sp. *vasinfectum* on cotton roots (Conn et al. 1997). Pre-inoculation of *Arabidopsis* with the endophytic bacteria Streptomyces isolate IFB-A03 activates the salicylic acid-mediated plant defense response upon exposure to pathogenic *Streptomyces scabies*, and it is suggested that the endophyte acts at the upstream of salicylic acid accumulation in the defense signaling pathway associated with systemic acquired resistance (Lin et al. 2012). Endophytic actinobacteria prime the defense pathways by inducing low-level expression of systemic acquired resistance and jasmonic acid/ethylene genes. Upon pathogen infection, the defense genes are strongly upregulated and exhibit high-defense gene expression (Conn et al. 2008). A strain of *Bacillus subtilis* (Lu144) reduces disease incidence of bacterial wilt of mulberry if inoculated prior to the pathogen (Ji et al. 2008). Systemic-induced resistance can act also against nematodes. For example, the bacterial endophyte *Rhizobium etli* strain G12 (and also the fungal endophyte *Fusarium oxysporum* strain (Fo162)) has been shown to systemically induce resistance in tomato plants toward *Meloidogyne incognita* (Martinuz et al. 2012).

It should be noted that the positive effect of the endophyte on its host is not only restricted to suppress plant's disease but also to promote growth under pathogen attack. Recent evidence demonstrates that the endophyte *Fusarium verticillioides* modulates the growth of the pathogen *Ustilago maydis* in maize and decreases its aggressiveness toward the plant, by interfering with early infection process (Lee et al. 2009a), as well as the coinfected plant growth was similar to that which could be gained in the absence of the pathogen (Rodriguez Estrada et al. 2012).

Interestingly, in some cases, the protective effect of an endophyte requires interplay between the endophyte and the plant. For example, the *Verticillium dahliae* Kleb. nonhost isolate Dvd-E6 protects Craigella tomatoes (*Lycopersicon esculentum* Mill.) from pathogenic *V. dahliae* Kleb. race 1 (Vd1), only in *in planta* context, whereas culturing Dvd-E6 and Vd1 (separate or together), the growth rates remain similar and neither is inhibitory to the other (Shittu et al. 2009a) and the protection conferred by Dcd-E6 is range restricted (Shittu et al. 2009b). Similarly, the endophytic bacteria *Methylobacterium* spp. strains have varying effects on plant disease resistance against *Pectobacterium atrosepticum*, *Phytophthora infestans*, and *Pseudomonas syringae* pv. tomato DC3000 in potato (*Solanum tuberosum* L.). These effects are modulated through endophyte community of the host (Ardanov et al. 2012).

Endophyte can also confer local disease resistance, i.e., the disease resistance is localized to tissues that the endophyte has colonized and is not systemic. For example, a nonpathogenic *Colletotrichum* mutant confers disease resistance in watermelon and cucumber in a rapid and strong activation of biochemical processes that confer resistance such as peroxidase and phenylalanine ammonia lyase activity and lignin deposition (Redman et al. 1999; Rodriguez and Redman 2008).

An alternative approach for biocontrol is engineering the endophyte in order to improve its biocontrol abilities. For example, introduction of the gene of the major chitinase of *Serratia marcescens*, ChiA, to the endophytic strain of *Pseudomonas fluorescens* resulted in effective biocontrol properties against the phytopathogenic fungus *Rhizoctonia solani* on bean seedlings under growth chamber conditions (Downing and Thomson 2000). Another example is the expression of *N*-acyl-homoserine lactonase gene (inhibiting production of quorum-sensing signals) from

Bacillus thuringiensis in the bacterial endophyte *Burkholderia* sp. KJ006, and this reduces the disease incidence of rice seedling rot caused by the pathogenic *Burkholderia glumae* in situ, since the latter is controlled in a population-dependent manner (Cho et al. 2007).

2.2.2 Abiotic Stress

Heavy Metals Pollution

Evidence suggests that the presence of endophyte enhances heavy metal tolerance to their hosts: aluminum tolerance of fine fescues (*Festuca* spp.) (Zaurov et al. 2001), *Neotyphodium lolii* induces tolerance to zinc stress in *Lolium perenne* (Fabien et al. 2001), and the endophyte *Neotyphodium gansuense* improves infected *Achnatherum inebrians* plant growth under high cadmium concentration, in a mechanism that involves anti-oxidative enzyme activities (Zhang et al. 2010b). Endophyte infection enhances perennial ryegrass tillering ability and reduces leaf elongation under cadmium stress condition by alleviating the detrimental effects of cadmium (Ren et al. 2006). In the presence of cadmium, the root endophyte *Piriformospora indica* reduces cadmium content in the shoot of *Triticum aestivum* cv. Sardari39 plants as well as increases its growth parameters (Shahabivand et al. 2012). Endophyte infection of *Lolium arundinaceum* increases tiller number and biomass under both control and cadmium stress conditions (Ren et al. 2011). Similarly, under cadmium stress, the presence of Neotyphodium endophytes in the grasses *Festuca arundinacea* and *Festuca pratensis* results in higher biomass production and higher potential to accumulate cadmium in roots and shoots (Soleimani et al. 2010a, b).

The endophytic fungus Sordariomycetes sp., which was isolated from leaves of *Suaeda salsa* and introduced into rice (*Oryza sativa*), improves rice growth under moderate lead levels in a mechanism that involves enhancement of photosynthesis and antioxidant activity (Li et al. 2012b). The endophyte *Neotyphodium coenophialum* confers aluminum tolerance to tall fescue in a mechanism that involves aluminum sequestering on root surfaces and root tissues. This sequestration is capable probably due to increased exudation of phenolic-like compounds from roots of endophyte-infected plants that chelate the aluminum (Malinowski and Belesky 1999). Another example is of the dark septate endophyte *Exophiala pisciphila* H93 that promotes the roots and shoots growth of maize under heavy metals (lead, zinc, and cadmium) stress conditions and the improve tolerance of maize to the heavy metals pollutant is archived by restricting the translocation of the heavy metal ions from roots to shoots (Li et al. 2011).

Drought and Water-Stress Tolerance

Clavicipitaceous endophytes such as *Neotyphodium* sp., *Acremonium* sp., *Phialophora* sp., and *Curvularia* sp. confer drought tolerance in grasses (Bacon and Hill 1996; Bacon 1993; West 1994; Joost 1995; Singh et al. 2011a). Additional examples for enhanced drought tolerance of endophyte-infected species can be found in Kuldau and Bacon (2008).

Drought tolerance is achieved by osmoregulation and stomatal regulation (Bacon and Hill 1996), as well as accumulation of drought-protective osmolytes in the grass tissues (Richardson et al. 1992). The production of loline alkaloids affects the osmotic potential and therefore reduces the effects of drought stress (Bush et al. 1997). The level of these alkaloids increases in response to heat or drought, and if their level is sufficient to affect the osmotic balance, they might protect macromolecules from denaturation and/or scavenge reactive oxygen species associated with drought stress (Malinowski and Belesky 2000). However, there is no significant correlation between symbiotically conferred stress tolerance and increase osmotic potential or abscisic acid (Rodriguez et al. 2008b). Another possible mechanism is the involvement of dehydrin proteins (Carson et al. 2004). In addition, since endophyte-infected grasses exhibit increase rate and length of root growth (Richardson et al. 1990; Richardson et al. 1993), it can be expected to play a role also in drought protection (Kuldau and Bacon 2008). Under water stress conditions, clavicipitaceous endophytes are associated with increasing cell wall elasticity (White et al. 1992), as well as effecting root physiology: increasing root growth, prolonging root hairs, and decreasing root diameter (Malinowski et al. 1997, 1999b). Reduced stomatal conductance is associated with water conservation in *Festuca arizonica-Neotyphodium* sp. interactions (Morse et al. 2002).

In their review, Malinowski and Belesky (2000) suggested that the adaptations conferred by the clavicipitaceous endophytes for drought are: (A) Drought avoidance through morphological adaptations to maintain favorable water status. This can be achieved by improving water uptake from the soil by an extensive root system, reducing transpiration losses and maintaining higher water storage in plant tissues. (B) Drought tolerance through physiological and biochemical adaptations that enable plant tissues to withstand water deficits. This can be achieved by accumulation and translocation of assimilates, osmotic adjustment, and maintenance of cell wall elasticity. (C) Recovery from drought by increasing water-use efficiency (Malinowski and Belesky 2000).

Increasing amounts of evidence demonstrate the involvement of nonclavicipitaceous endophytes in conferring drought tolerance (reviewed in Singh et al. 2011a). The finding that symbiotic plants (e.g., rice and tomato) consume less water than nonsymbiotic plants regardless of the colonizing endophyte, taken together with the fact that these symbiotic plants achieve increase biomass levels, suggests that symbiotic plants exhibit increase water-use efficiency that may provide a unique mechanism for symbiotically conferred drought tolerance (Singh et al. 2011a). The endophytic fungus *Piriformospora indica* improves plant drought tolerance (Sherameti et al. 2008), which is in accordance with its natural desert origin (Verma et al. 1998). In *Arabidopsis*, *P. indica* confers drought tolerance in a mechanism that is associated with priming of expression of stress-related genes in the leaves (Sherameti et al. 2008; Oelmüller et al. 2009). In Chinese cabbage, drought tolerance conferred by *P. indica* is achieved in three targets: antioxidant enzyme activities (such as peroxidases, catalases, and superoxide dismutases) in the leaves, upregulating drought-related genes (such as DREB2A, CBL1, ANAC072, and RD29A), and increasing the amount of CAS protein (Sun et al. 2010). In cucumber, the presence of the endophyte fungi *Phoma glomerata* LWL2 and *Penicillium* sp. LWL3 increases the plant biomass under drought conditions (Waqas et al. 2012). Yet, the beneficial effect of the endophyte is dependent on the host genotype (Cheplick 2004).

Bacterial endophytes also increase drought resistance (Sturz et al. 2000). *Pseudomonas strain* PsJN improves stomata function and reduces transplanting shock through improved water management (Nowak et al. 1995). Wheat seedlings cocultured with *Azospirillum brasilense* strain SP245 exhibit improved water relations under osmotic stress (Creus et al. 1998).

Salinity

Soil salinization is an extensive threat to crop productivity (Singh et al. 2011a). Approximately 7 % of the global land surface is covered with saline soils (Ruiz-Lozano et al. 1996), and 5 % of the cultivated land is affected with excess salt content (Munn et al. 1999). In barley, the endophyte *Piriformospora indica* eliminates the salt stress effects, probably due to elevation of the metabolic activity in the leaves and therefore compensating the salt-induced inhibition of the leaf metabolic activity, induction of changes in the fatty acid composition in the leaves of the host plant, and upregulation of the activity of antioxidant enzymes (Baltruschat et al. 2008). Additional mechanism involves enhancement of the ratio of reduced to oxidized ascorbate and induction of DHAR activity (Waller et al. 2005). Moreover, *P. indica* abolishes the detrimental effect of moderate salt stress and increases barley biomass (Waller et al. 2005). *P. indica* also induces ethylene biosynthesis in barley roots, and ethylene signaling may be required for plant salt tolerance (Cao et al. 2006).

Symbiotic plants containing root endophytes activate stress response systems more quickly and strongly than nonsymbiotic plants (Rodriguez et al. 2004). The newly isolated endophytic fungus *Paecilomyces formosus* LHL10 from cucumber enhances its host shoot length under salinity stress.

This could be achieved by accumulation of proline and antioxidants and maintaining plant water potential and consequently reducing the electrolytic leakage and membrane damage of the host (Khan et al. 2012a). Similarly, the endophytic fungi *Phoma glomerata* LWL2 and *Penicillium* sp. LWL3 increase the plant biomass and enhance assimilation of essential nutrient under salinity conditions. The symbiotic association mitigates stress by compromising the activities of reduced glutathione, catalase, peroxidase, and polyphenol oxidase. In addition, under stress conditions, the presence of the endophyte modulates stress in a mechanism that involves downregulation of abscisic acid, alters jasmonic acid, and elevates salicylic acid contents (Waqas et al. 2012). Recently, two different strains of *Penicillium* (*Penicillium minioluteum* LHL09 and *Penicillium funiculosum* LHL06) isolated from *Glycine max.* L. promote growth characteristics under salinity stress conditions. These endophytes ameliorate the effects of the salt stress by influencing biosynthesis of the plant's hormones and flavonoids (Khan et al. 2011a). In addition, *Exophiala* sp. LHL08, an endophyte isolated from cucumber, confers salinity stress tolerance by elevating the level of salicylic acid (Khan et al. 2011b). Additional examples for nonclavicipitaceous endophyte-induced tolerance for salt stress can be found in a review by Singh and colleagues (Singh et al. 2011a).

The symbiotically conferred stress tolerance is a habitat-specific phenomenon: introduction of nonclavicipitaceous endophytes into stress-sensitive commercial rice varieties achieved tolerance to salt, drought, and cold stresses, as well as increase growth characteristics (Redman et al. 2011). Another example is the dunegrass *Leymus mollis* that colonize the endophyte *Fusarium culmorum*, which confers salt tolerance in a habitat-specific manner (Rodriguez et al. 2008b).

Heat and Cold Stresses

Endophytes protect host plants from extreme temperature damages. For example, inoculation of *Dichanthelium lanuginosum* plants with the fungal endophyte *Curvularia* enabled them to tolerate high soil temperature in which non-inoculated plants are dead (Redman et al. 2002), and this tolerance is dependent on viral infection of the endophytic fungus (Márquez et al. 2007). Since only endophytes isolated from geothermal plants confer heat tolerance (Rodriguez et al. 2008b), it is suggested that this ability is a habitat-adapted phenomenon (Rodriguez and Redman 2008). A possible mechanism for heat tolerance involves osmoprotectants such as trehalose, glycine betaine, and taurine in the heat response, as well as the fungal pigment melanin and heat shock protein (Morsy et al. 2010). In contrast, *Cucumis sativus* plants inoculated with the endophytic fungus *Paecilomyces formosus* LHL10 display higher plant growth only under high-temperature stress. This suggests that the endophyte has varying effects in response to temperature stress (Khan et al. 2012b). Similarly, introduction of the endophytic bacterium *Clavibacter* sp. strain Enf12 into *Chorispora bungeana* plantlets improves their tolerance to chilling stress through enhancement of the antioxidant defense system (Ding et al. 2011).

3 Trees and Woody Plants

As mentioned above, endophytes, both fungi and bacteria, can be found in most, if not all, plants in nature (Petrini 1986; Rodriguez et al. 2008a). Nonclavicipitaceous endophytes can be recovered from ferns, conifers, and seed plants from the arctic tundra to the tropics (Strobel 2006; Arnold and Herre 2003; Arnold and Lutzoni 2007; Rodriguez et al. 2008a; Aly et al. 2010). In contrast to extensively studied grass endophytes, endophytes associated with woody angiosperms are poorly known (Arnold et al. 2003). Very few published reviews discuss trees and woody plant endophytes and their role in the host protection and benefit (Albrectsen and Witzell 2012; Pirttilä 2001; Pirttilä and Frank 2011). Two major methods for the analysis of the diversity within host plant are currently used: isolation and cultivation of fungi and bacteria from the plant tissue and DNA-based techniques. It has been estimated that less than 1 % of bacterial species and less than 5 % of fungal species are currently known, suggesting that millions of microbial species remain to be discovered (Gunatilaka 2006). The biodiversity of endophytes in woody plants can be remarkable and varies from single species (Arnold et al. 2000, 2003; Arnold and Herre 2003; Arnold 2008; and many more) to more than 100 taxa per plant (González and Tello 2011). Mechanisms by which endophytes benefit plants (discussed above in details) may overlap – a single endophyte can employ several of them (Porras-Alfaro and Bayman 2011). As an example, many plant-growth-promoting bacteria (PGPB) promote plant growth, support nitrogen fixation, and prime plants' induced systemic resistance (Compant et al. 2005b).

In this subchapter we will try to review the knowledge present for the involvement of endophytes in biocontrol of trees and woody plants.

3.1 Increased Plant Growth

Induction and/or synthesis of plant-growth-promoting phytohormones (auxins, cytokinins), N_2 fixing, synthesis of enzymes/peptides that provide nutrient availability (phosphatases, siderophores, etc.), improvement of nutrient, water uptake, and tolerance to various types of stresses are among the direct mechanisms through which endophytes promote plant growth (Hanada et al. 2010; Barka et al. 2000). Growth stimulation may also act indirectly through the biocontrol of phytopathogens in the root zone and induction of phytohormone synthesis by the plant (Sturz et al. 2000).

Some liverworts and mosses, hornworts, the fern genus *Azolla* de Lamarck, Cycads, the angiosperm *Gunnera* L., and various orchids evolved symbiotic relations with endophytic cyanobacteria. The primary benefit to the plant in these relations is fixed nitrogen originating from N_2-fixation of the cyanobacteria (Krings et al. 2009). Only few examples of endophytes associated with woody plants with the ability to fix nitrogen are present: the nitrogen-fixing bacteria associated with woody plants are *Frankia* sp. – filamentous bacteria that convert atmospheric N_2 gas into ammonia. *Frankia* fixes nitrogen while living in root nodules on "actinorhizal plants," which are plants that colonize soils that are low in combined nitrogen (Benson and Silvester 1993). Some Acacia koa trees form a symbiosis with *Rhizobium* and *Bradyrhizobium* sp. that not only fix nitrogen in the root but can also form "canopy nodulation" in which they fix nitrogen (Leary et al. 2004). *Acetobacter diazotrophicus*, a nitrogen-fixing Acetobacteria, was isolated from coffee plants surface-sterilized stems and roots (Jimenez-Salgado et al. 1997). Some abundant bacteria with the ability to fix nitrogen were isolated from poplar (*Populus trichocarpa*) and willow (*Salix sitchensis*) (Doty et al. 2009; Xin et al. 2009b). It will not be an exaggeration to assume that there are many more endophytic microorganisms involved in nitrogen fixation in woody plants, waiting to be discovered.

Barka et al. (2000) demonstrated that grapevine plantlet roots (*Vitis vinifera* L. cultivar "Chardonnay") cocultured with *Burkholderia* sp. strain PsJN grow faster and had significantly more secondary roots compared to the uninoculated control (Barka et al. 2000; Compant et al. 2005b). Fungi and bacteria have been illustrated to produce and secrete plant hormones as part of their pathogenicity process or manipulate the plant for their benefit (Baca and Elmerich 2003). Plant hormone-secreting endophytes are described in many herbaceous plants (Jacobson et al. 1994; Dai et al. 2008; Khan and Doty 2009 and many more); however, the involvement of phytohormones in growth promoting of woody plants is much less demonstrated. In a study by Pirttilä (2001) of endophytes from scots pine (*Pinus sylvestris* L.), the endophytes *Methylobacterium extorquens*, *Pseudomonas synxantha*, and *Rhodotorula minuta* were examined for possible plant hormone production (gibberellins, auxins, or cytokinins); none were detected. On the other hand, *Rhizoctonia* sp. isolated from scots pine and Norway spruce were found to enhance the growth of inoculated seedlings of scots pine. The researchers concluded that this growth is due to the production of plant growth regulators (Doty 2011; Gronberg et al. 2006). Additional growth-promoting endophyte bacterium is *Enterobacter* sp. 638, which genome was sequenced lately, isolated from *Populus trichocarpa* × *deltoides* cv. H11-11. This bacterium improves poplar growth and development through the production of the phytohormones indole acetic acid, acetoin, and 2,3-butanediol. Interestingly, the production of the two latter phytohormones is induced by the presence of sucrose, the major plant sugar, therefore linking between the availability of sucrose in the host plant and the synthesis of plant-growth-promoting phytohormones by the endophytic bacterium (Taghavi et al. 2010).

Another example is of endophytic yeasts identified as *Rhodotorula graminis* and *R. mucilaginosa*, both isolated from Populus, Cottonwood. These endophytes convert plant-derived L-tryptophan to phytoactive IAA. The presence of the endophyte on the trees' seedlings was expressed as very rapid growth of roots (Xin et al. 2009a).

3.2 Stress Tolerance

Recent studies indicate that fitness benefits conferred by mutualistic fungi contribute to or are responsible for plant adaptation to stress. Collectively, mutualistic fungi may confer tolerance to drought, salinity, metals, and extreme temperatures. It has become apparent that at least some plants are unable to tolerate habitat imposed abiotic and biotic stresses in the absence of fungal endophytes (Rodriguez et al. 2004, 2008b; Redman et al. 2002; reviewed by Singh et al. 2011a and many more). The term "induced systemic tolerance" (IST) for plant-growth-promoting rhizobacteria (PGPR) was proposed by Yang et al. (2009) to PGPR-induced physical and chemical changes in plants that result in enhanced tolerance to abiotic stress, including salinity and temperature. This term may be applicable to endophytic bacteria and fungi as they activate many of the same mechanisms of induction in nonwoody plants (Yang et al. 2009).

On the other hand, only few examples of such contribution to woody plants are published. *Trichoderma hamatum* (DIS 219b), isolated from cacao (*Theobroma cacao*), was found to induce several cacao-expressed sequence tags (ESTs) sharing homology with genes reported to function in plant responses to environmental stresses and abiotic stresses such as drought and to biotic stresses including plant disease (Bailey et al. 2006; Bae et al. 2009). These findings led to the assumption that this endophyte may contribute to drought tolerance. Bae et al. (2009) found that colonization of the cacao seedlings by *T. hamatum* DIS 219b enhances root growth, resulting in improve water acquisition and increase water content. The roots of colonized seedlings perceive the dry soils and respond while the leaves take advantage of the increased water availability through the roots, resulting in a delayed drought response. Khan et al. (2012c) inoculated herbal crop varieties with endophytic bacteria and yeasts isolated from poplar and willow trees, finding them to promote growth and increase plant tolerance to drought in some of the crops (Khan et al. 2012c). This finding does not necessarily prove the ability of these endophytes to act for abiotic tolerance in poplar and willow trees or any other woody plant, but it demonstrates the possibility of existence of abiotic stress-tolerance-inducing endophytes in tress waiting to be revealed. The bacterial endophyte *Burkholderia phytofirmans* strain PsJN increases grapevine growth at low temperature and improves its host ability to endure cold stress (Barka et al. 2006).

The role of endophytes as bioremediators will not be discussed in this review. For information about endophytes and bioremediation, see Li et al. (2012a), Ryan et al. (2008), Ma et al. (2011), or Soleimani et al. (2010b) for woody and herbaceous plants.

3.3 Mechanisms of Biocontrol

Mechanisms of inhibition of plant pathogens by endophytes may be of several means, including direct effect, indirect effects, and ecological effects (Gao et al. 2010). Direct effect involves direct interaction between the endophytes and the pathogens. The most commonly reported mechanism of biological control is antagonism. Antagonism includes the more specific mechanisms of antibiosis – the secretion of secondary metabolites or enzymes by the endophytes to their surrounding area, competition over food source and nutrient, and mycoparasitism – direct parasitism on the hyphae of the pathogen (Kloepper and Ryu 2006; Meji'a et al. 2008; Gao et al. 2010).

Many of the natural products occurring in endophytes have been shown to have antimicrobial activity. In many cases this activity is implicated in protecting the host against pathogens (Gunatilaka 2006; Aly et al. 2010). Yet, in most cases, evidence for the production of these metabolites *in planta*, in woody plants, is still absent (Aly et al. 2010).

Gonzalez and Tello (2011) described the isolation of fungal endophytes from several varieties of grapevines (*Vitis vinifera*). Many of which possess antifungal properties that are useful against a number of plant pathogens. For example, *Fusarium proliferatum* has been employed to control grapevine downy mildew caused by *Plasmopara viticola* (Bakshi et al. 2001; Falk et al. 1996). In these studies *F. proliferatum* is considered a mycoparasitic, cold-tolerant fungus, capable of controlling the development of *P. viticola* via secretion of extracellular glucanolytic enzymes (Bakshi et al. 2001; Pancher et al. 2012). *Epicoccum nigrum* represents another promising biocontrol agent, capable to produce secondary metabolites with antibiotic activity (Martini et al. 2009). Some authors have indicated the biocontrol properties of *E. nigrum* against pathogens such as *Monilinia* (Larena et al. 2005), as well as other several grapevine pathogens like *Plasmopara viticola* (Kortekamp 1997) or *Botrytis cinerea* (Fowler et al. 1999). Another fungus with potential for its use as microbial antagonist is *Aureobasidium pullulans*. This taxon is known to possess activity against a wide range of grapevine pathogens, including postharvest fungi (Schena et al. 1999, 2003). *A. pullulans* was also found in apples from organic orchards considered to be a potential biocontrol for apple storage pathogens (Granado et al. 2008). *Phaeosphaeria nodorum* isolated from plums (*Prunus domestica*) was found to secrete to its growth medium inhibitory substances suppressing the growth of *Monilinia fructicola* and *Colletotrichum gloeosporioides*, two pathogens causing brown rot, blossom blight, twig blight, and anthracnose on plum (Pimenta et al. 2012). In wild banana (*Musa acuminata*), two endophytes (*Cordana* sp. and *Nodulisporium* sp.) exhibit potential activity against *Colletotrichum* (Nuangmek et al. 2008). Pirttilä and colleagues have studied endophyte involvement and influence on scots pine growth and disease resistance in a number of publications. One example is the antagonistic activity of endophytes isolated from scots pine, which was studied in vitro. In this study, *Hormonema dematioides* was found to inhibit the growth of two strains of *Gremmeniella abietina* (HR3 and KR), a fungal pathogen responsible for the Brunchorstia disease of coniferous and *Hymenoscyphus ericae*. Other endophytes in this study produced antagonistic substances toward *H. dematioides* (Pirttilä 2001).

Only few examples of endophytes from trees, shown to be active both in vitro and in vivo, have been published to date. One example of such is a complex of endophytes, isolated from cacao trees (*Theobroma cacao*) in Panama. When inoculated into leaf tissues, the endophytes significantly reduce damage by an important foliar pathogen – *Phytophthora* sp. The anti-pathogen defense is localized to endophyte-infected tissues (Arnold et al. 2003). This suggests that direct or indirect interactions between endophytes and the *Phytophthora* pathogen are responsible for limiting the pathogen's spread (Arnold 2008). Another example is of *Trichoderma martiale*, an endophytic *Trichoderma* sp. isolated from cacao tree in Brazil, when spore inoculum (of about 5×10^7 ml^{-1}) in a formulation of vegetable oil and sucrose was sprayed on cacao pods, it reduced black pod disease severity for at least 30 days post-inoculation (Hanada et al. 2009). The most likely mode of action of *Trichoderma* in this case is parasitism on the pathogen. A second possible mode of biocontrol action is stimulation of resistance reaction in the host toward the parasites (Hanada et al. 2008, 2009). Two other important pathogens of cacao *Moniliophthora roreri* and *Moniliophthora perniciosa* are controlled by endophytes isolated from a healthy cacao tree (Meji'a et al. 2008). An endophytic *Gliocladium catenulatum* reduces up to 70 % incidence of witches' broom disease in cacao

(Rubini et al. 2005). Campanile et al. (2007) examined the ability of endophytic fungi (*Trichoderma viride*, *Epicoccum nigrum*, *Fusarium tricinctum*, *Alternaria alternata*, *Sclerotinia sclerotiorum and Cytospora* teleomorph: *Valsa* sp.) to control *Diplodia corticola*, an oak dieback disease agent in the Mediterranean, both in vitro and *in planta*. Their result indicates that although in vitro dual culture tests indicated strong inhibition of the pathogen by few of the inspected endophytes, *in planta* results were not in full agreement. Moreover, the distance between endophyte and pathogen inoculation affects disease severity. While close inoculation (3 cm) is effective in reducing pathogen development, 6 cm distance indicates no differences from the control. This finding is in accordance to local influence as found by Arnold et al. (2003).

A few examples of endophytic bacterial isolates of *Bacillus*, *Erwinia*, and *Pseudomonas* from oak have been proven biologically active against the oak wilt pathogen (*Ceratocystis fagacearum*) (Brooks et al. 1994). *Pseudomonas* sp. introduced into Elm trees to control Dutch elm disease (Myers and Strobel 1983), and *Bacillus* sp. against *Verticillium* wilt of maple trees (Hall et al. 1986) display antibiosis biological activity in vitro and reduction of disease and trees mortality in vivo. Arau'jo et al. (2002) found a higher frequency of *Curtobacterium flaccumfaciens* in asymptomatic citrus plants suggesting a role for this bacterium in the resistance of *Citrus sinensis* to Citrus variegated chlorosis (CVC), a disease caused by *Xylella fastidiosa*. Another example in citrus is the attempt to use isolate of *Pseudomonas* sp. as endophytic bacteria for the control of *Phoma tracheiphila*, a pathogenic fungus causing the Mal secco disease of lemons and other citrus in the Mediterranean basin (Lima et al. 1994; Coco et al. 2004; Migheli et al. 2009).

In addition, some endophytic bacteria from trees have been demonstrated to secrete antibacterial (Guan et al. 2005; Castillo et al. 2002, 2003), antifungal (Li et al. 2007), antiviral (Guo et al. 2000), and other biological active metabolites against human parasites, i.e., malaria (Castillo et al. 2002; Ezra et al. 2004a) and anticancer metabolites (Strobel et al. 1993). Mejia et al. (2008) argue that endophyte isolates that outcompete or displace pathogens by outgrowing them tend to be those that are commonly isolated from cacao in their field survey. In contrast, endophytes showing antibiosis tend to be slower growing and are relatively less abundant. At the same time, less abundant endophytes in cacao tissues in those surveys are relatively poor. The outcome of these arguments for biocontrol strategies is that if the researcher choose isolates that show in vitro antibiosis activity against a particular pathogen, one needs to recognize that effectively introducing and keeping them inside the tree tissues may be more challenging than assumed. It must be taken into account that even when a specific endophyte displays antibiosis activity against a particular pathogen in in vitro experiments, when transferred into the host plant, the presence of fast growing insensitive to its active metabolites might outgrow the biocontrol endophyte, preventing it from protecting the plant against pathogen attack (Meji'a et al. 2008).

Another aspect is the possibility of introducing an endophyte, proven active by in vitro experiments, from one host into another, expecting it to establish and protect the new host against pathogens. While in some cases the endophyte would not establish in the new host, in others the introduced endophyte will not provide protection to the new host (Wäli et al. 2006). Another possibility is that the new endophyte, although established in the host, causing no visible symptoms, when the pathogens are introduced to the host plant, disease severity is expressed in a much higher degree than demonstrated on control plants lacking the endophyte in it. In this case it is obvious that the endophyte exerts a negative influence on host's capability to resist the pathogen. An example for the latter possibility is given by Ardanov et al. (2011). In vitro-grown potato cultivar Blue Congo was inoculated at high and low inoculation densities with *Methylobacterium* sp. IMBG290 against *Pectobacterium atrosepticum*. Low inoculation density resulted in resistance, while high density led to susceptibility toward the pathogen, but no obvious mechanism behind the phenomenon was identified. The researchers report that the inoculation of the plants (potato and pine) with a *Methylobacterium* sp. caused a change in the population structures of innate endophytic communities, which correspond with plant responses toward pathogens (Ardanov et al. 2011).

Genetically modified endophytes (GME) for the protection of their hosts against pests and plant diseases and the promotion of plant growth are one more possible use of endophytes for biological control application. Introduction of heterologous genes in endophytic bacteria may confer new characteristics useful in biocontrol of diseases and pests that harm the host plant. For example, the endophytic bacterium *Clavibacter xyli* subsp. *ynodontis*, a xylem colonizer of different plant species, was genetically modified to express the gene *cry*A(c) from *Bacillus thuringiensis* which its protein product controls the larvae of *Ostrinia nubilalis* (Lampel et al. 1994). Andreote et al. (2004) observed that the bacterial community, endophytically associated with citrus seedlings, is affected by the introduction of GMEs expressing different heterologous genes. This study was done as part of a future effort to use GME in citrus for the control of CVC (Andreote et al. 2004).

3.4 Induction of Plant Resistance

An important mechanism for biological control is that of fungal and bacterial presence or their metabolites, affecting the plant by increasing the plant's resistance to pathogens, a process termed induced systemic resistance (ISR) (Kloepper and Ryu 2006; Compant et al. 2005a). Endophytic fungi or bacteria can induce systemic resistance in plants against pathogens after actively penetrating and colonizing the host, promoting the synthesis of biologically active compounds or causing changes in plant morphology and/or physiology (Hanada et al. 2010). Resistance can also be elicited in plants by the application of chemicals or necrosis-producing pathogens, and this process is termed systemic acquired resistance (SAR). Pieterse et al. (1998) proposed that ISR and SAR can be differentiated not only by the elicitor but also by the signal transduction pathways that are elicited within the plant. Accordingly, ISR is elicited by rhizobacteria or other nonpathogenic microorganisms, while SAR is elicited by pathogens or chemical compounds (Kloepper and Ryu 2006). Endophyte-mediated resistance in forest trees was demonstrated for Western white pine (*Pinus monticola*). Fungal endophytes isolated from Western white pine increase the survival of seedlings against *Cronartium ribicola*, the White pine blister rust pathogen, and this effect persists over time. It is not indicated how the endophytes function to reduce disease symptoms and delay mortality in the host but the researchers concluded according to their results that the endophytes are involved in the host defense (Ganley et al. 2008). Significant reductions in fusiform rust disease caused by *Cronartium quercuum* f. sp. *fusiforme* was reported by Enebak and Carey (2000). The potential elicitation of ISR by *Bacillus sphaericus* SE56 and *B. pumilus* strains INR7, SE34, SE49, and SE52 against a suspension of *C. quercuum* basidiospores was demonstrated in this study. Bacteria were applied at seeding, and *C. quercuum* basidiospores were sprayed onto the pine seedlings at five different times. Disease symptoms were evaluated 6 months after last application of basidiospores. All strains except SE49 resulted in significant reductions in disease incidence (Kloepper and Ryu 2006). Recently, Ardanov et al. (2012) reported that *Methylobacterium extorquens* DSM13060 induces the expression of plant defense genes in pine.

Endophytes involved in the protection of cacao and cupuacu (*Theobroma grandiflorum*) trees against *Phytophthora palmivora* constitute an indirect example of plant-induced resistance, as the mode of control was undefined, but strong antagonistic activity was not found for the isolated endophytes described (Hanada et al. 2010).

Secondary metabolites produced by trees and woody plants for their protection against plant pathogens are well known and have been studied. Among these metabolites are phytoalexins including flavonoid and terpenoids (Smith 1996). Plants also produce for their protection polyphenols and defense-related enzymes including phenylalanine ammonia lyase, peroxidase, catalase, and superoxide dismutase. Endophytes may promote the plant production of these molecules. Gao et al. (2011) reported that addition of an elicitor produced by an endophytic isolate of *Fusarium* sp. from *Euphorbia pekinensis* to the plant suspension cultures causes cell stimulation to produce polyphenol, terpenoids, and antioxidant enzymes including peroxidase, superoxide dismutase, and catalase that are all in higher concentration in the presence of the elicitor than in the control. However, the activity of these enzymes is much lower in intensity for long period than indicated for induction by pathogens. Elicitors such as lipopolysaccharides, polysaccharides, and glycoprotein stimulate plant defense and plant secondary metabolites and suppress pathogen attack efficiently (Gao et al. 2010, 2011). This may indicate the involvement of endophytes in the induction of plant defense by stimulation of induced resistance in the host, making the plant prepared for a "real" attack by phytopathogens, resembling the "priming" effect achieved by inoculation of plants by non-pathogenic rhizobacteria (Ardanov et al. 2012; van Loon 2007).

Compant et al. (2005a) observed a localized accumulation of phenolic compounds in several cortical cells of grapevine plantlet following inoculation and colonization by *Burkholderia* sp. strain PsJN. They concluded that this bacterium can induce a host defense response in the plants root. The colonization of grapevine by the bacterium causes strengthening of hosts cell walls (Compant et al. 2005a). The thickening of the cell wall is due to the deposition of callose and the accumulation of phenolic compounds at the site of pathogen attack (Benhamou et al. 1998). Moreover, vines inhabited by the bacteria expressed resistance to grapevines' gray mold disease caused by *Botrytis cinerea* (Barka et al. 2000, 2002).

Endophytes may also produce secondary metabolites that directly inhibit insects or pathogens or produce elicitors that stimulate the plant to produce this type of secondary metabolites. Furthermore, a single endophyte may offer protection from both fungal pathogens and insects. For example, *Beauveria bassiana* inhibits both fungal pathogens and insects, mostly by production of secondary metabolites. *Lecanicillium* spp. and *Trichoderma* spp. are both mycoparasites and insect parasites, although they also produce inhibitory metabolites (Ownley et al. 2010; Porras-Alfaro and Bayman 2011). Another example is of fungal endophytes isolated from *Picea rubens* (red spruce) needles showing toxicity against *Choristoneura fumiferana*, the eastern spruce budworm (Sumarah et al. 2010; Porras-Alfaro and Bayman 2011). *Meira geulakonigii*, an endophytic fungus isolated from grapefruit peel, has been shown to reduce populations of citrus rust mite (CRM; *Phyllocoptruta oleivora*) on citrus leaves and fruits, both in field and laboratory (Paz et al. 2007). Examples of endophytes controlling pathogenic nematodes have been reviewed by Sikora et al. (2008). Mainly inoculation with *Fusarium oxysporum* and, to a lesser extent, species of *Trichoderma* reduces populations of nematodes in roots of banana and tomato plants (Sikora et al. 2008).

3.5 Ecological Niche Occupation

Given that biological populations of an ecosystem interact with one another, positive interactions (commensalism, mutualism, and synergism) may enable some populations to function as a community within this habitat. Viewed as such, one mechanism by which systemic acquired resistance operates may be through the utilization of the so-called system-level effects among established endophyte communities. For example, positive interactions among autochthonous populations are usually better developed in mature communities than in newly established communities. Thus, the new colonist (invader or pathogen) will encounter severe negative interactions with autochthonous populations (Sturz et al. 2000). In essence, the invading population is prevented from becoming established by the dynamics of the ecosystem it is trying to invade, a form of defensive mutualism (Clay 1988). Thus, the network of connections among species in a mature (established) ecosystem protects those member species from outside competition, which benefits the plant if the putative colonist is a phytopathogen (Sturz et al. 2000). Fungal endophytes are generally thought to protect a plant by rapid colonization and thereby exhausting the limited available substrates so that none would be available for pathogens to grow (Pal and Gardener 2006). It is reasonable to assume that the colonization of tree tissues occurs over long periods of time. In these tissues, the endophyte diversity may be mainly limited by competition from resident fungi that occupy the relatively poor nutrient niche or microhabitat. This effect may limit invasion of the tissue by other microorganisms, endophytes, or pathogens (Albrectsen and Witzell 2012).

3.6 Volatile Emitting Endophytes

Another yet very unique application of endophytes for biological control of plants and plant products is volatile organic compounds (VOCs) emitting fungi and bacteria. Fungal VOCs have been used as part of biological control strategies to prevent the growth of plant pathogens. In addition, VOCs' plant-growth-promoting effects have created increasing interest recently (Morath et al. 2012). One known example is *Muscodor albus*, an endophytic fungi from certain tropical trees and vine species mainly from Central and South America, Australia, and Thailand (Atmosukarto et al. 2005; Woropong et al. 2001, 2002; Sopalun et al. 2003; Ezra et al. 2004b and many more) that have been used by Mercier and Manker (2005) to control soilborne diseases. Addition of *M. albus* to soil mixtures provides control to pathogens as *Rhizoctonia solani*, which causes damping-off of broccoli, and *Phytophthora capsici*, which causes root rot of bell pepper.

Although not recorded as endophytic fungi, soilborne VOCs emitting fungi may benefit plants by activating defense responses and priming of plants against future pathogen attack (Morath et al. 2012). Mixtures of bacterial VOCs induce a defense response in plants (Ryu et al. 2003). For example, exposure of *Arabidopsis thaliana* to 1-octen-3-ol ("mushroom alcohol"), a major fungal VOC, causes defense genes upregulation and provides protection against *Botrytis cinerea* (Kishimoto et al. 2007). Although not described as a phenomenon in woody trees, a very interesting finding by Ryu et al. (2003, 2004) describes VOCs secreted by the endophyte *Bacillus amyloliquefaciens* IN937a that elicit plant growth promotion (Ryu et al. 2003) and ISR (Ryu et al. 2004).

Examination of *Muscodor albus* for biocontrol application against plant pathogens demonstrated that *M. albus* produces VOCs that inhibit and kill plant pathogenic fungi and bacteria (Strobel et al. 2001), as well as grain pathogens (Goates and Mercier 2009). Additionally, the VOCs produced by *Muscodor yucatanensis*, *Muscodor fengyangensis*, and a second isolate of *M. albus* all inhibit pathogenic species of bacteria, fungi, and oomycota (Atmosukarto et al. 2005; Macias-Rubalcava et al. 2010; Zhang et al. 2010a). *Muscodor crispans* was found to produce a mixture of VOCs that inhibits a wide range of plant pathogens, including the fungi *Mycosphaerella fijiensis* (the black sigatoka pathogen of banana) and the serious bacterial pathogen of citrus, *Xanthomonas axonopodis* pv. *citri* (Mitchell et al. 2010). Other fungi were found to emit active VOCs as well. Pimenta et al. (2012) isolated an endophytic fungus, *Phaeosphaeria nodorum*, from plums (*Prunus domestica*), producing inhibitory VOCs to *Monilinia fructicola.* A *Phoma* sp. isolated from creosote bush emits VOCs that may contribute to this shrub survival in harsh desert habitats (Strobel et al. 2011). This *Phoma* sp. produces a unique mixture of VOCs that inhibit or kill a range of plant pathogens, including *Verticillium*, *Ceratocystis*, *Cercospora*, and *Sclerotinia* (Strobel et al. 2011).

The term "mycofumigation" was given to the use of volatile emitting fungi and their VOCs for the control of other organisms, pathogenic on fruit, vegetables, and food products in postharvest (Stinson et al. 2003). VOCs of *M. albus* are toxic to the peach pathogens, *Penicillium expansum*, *B. cinerea*, and *Monilinia fructicola*, as determined in vitro. Furthermore, the volatiles prevent fungal contamination of postharvest peaches over 7 day of storage (Mercier and Jimenez 2004). Muscodor™ a product based on *M. albus* was introduced by AgraQuest for use in postharvest of different fruits, nuts, seed, soil treatment, and more (AgraQuest 2005). VOCs of *Oxyporus latemarginatus* EF069, an endophyte isolated from red peppers, reduce postharvest decay of apples caused by *B. cinerea* and *Rhizoctonia* root rot of moth orchid (Lee et al. 2009b). *Phomopsis* spp., *Nodulisporium* spp., and *Hypoxylon* spp. produce volatile compounds that control fruit decay as well (Park et al. 2010; Tomsheck et al. 2010; Singh et al. 2011b). Suwannarach et al. (2013) reported on the isolation of a *Nodulisporium* spp. CMU-UPE34 with the ability to control green mold decay on *Citrus limon* caused by *Penicillium digitatum* and blue mold decay of *Citrus aurantifolia* and *Citrus reticulata* caused by *Penicillium expansum*.

Not only fungi are able to produce and emit VOCs but some bacterial strains, such as *Bacillus subtilis*, *Pseudomonas chlororaphis*, *Pseudomonas fluorescens*, *Serratia odorifera*, and *Stenotrophomonas maltophilia*, all non-endophytic from nonwoody plants (Kai et al. 2007; Athukorala et al. 2010), have also been shown to produce volatile organic compounds active against *Alternaria*, *Colletotrichum*, *Fusarium*, *Rhizoctonia*, *Pythium*, *Sclerotium*, *and Verticillium* (Suwannarach et al. 2013). It is very reasonable that endophytic bacteria with the ability to produce and emit biologically active VOCs will be found in woody and non-woody plants in the future.

Insecticidal properties of fungal VOCs are also being investigated. *Muscodor* spp. producing nitrosoamide have been demonstrated to kill insects (Strobel et al. 2010). *M. vitigenus* produces naphthalene, an effective insect repellent (Daisy et al. 2002). The volatile mixture produced by strain CZ-620 has some nematicidal and insecticidal activities (Lacey and Neven 2006; Riga et al. 2008). In addition, VOCs' profiles correlate with varying levels of pathogenicity of entomopathogenic fungi *Beauveria bassiana* and *Metarhizium anisopliae* studied for their potential as biocontrol agents to reduce termite populations (Hussain et al. 2010).

4 Concluding Remarks and Future Prospects

- Microbial endophytes appear to be symbiotic with all plants in natural ecosystems and have profound impacts on the survival and fitness of plants.
- Microbial endophytes benefit plants by promoting plant growth, and thereby increase crop yields, and confer tolerance to both biotic (interspecific competition, invertebrate pests, herbivory of mammals, and diseases caused by phytopathogens) and abiotic (heavy metal pollutions, drought, salinity, and temperature) stresses. Endophytes also produce novel substances that may have significance to human health. These characteristics establish endophytes as good candidates for both biocontrol and bioremediation agents. Thus, if such endophytes can be identified and confer benefits in mechanized, agricultural systems, they would be increasingly important in agricultural production.
- The use of microbial endophytes for biocontrol holds much promise for the reasons discussed above. However, there are challenges due to the complexity of the system – the endophyte-host associations are highly plastic (Malinowski and Belesky 2000). In order to achieve the biotechnological potential of these microbes, understanding of the mechanisms enabling endophytes to interact with plants is needed (Jalgaonwala and Mahajan 2011; Kuldau and Bacon 2008). The complexity of interactions among endophytes, pathogens, insects, and plants demonstrates the difficulty of predicting the outcomes for plant protection by the endophytes. In addition, there likely remain much undescribed endophytes, especially regarding woody plants and trees (Arnold and Herre 2003).
- To date, although an enormous amount of literature on the possible use of endophytes as biological control and growth-promoting agents was published (reviewed by Albrectsen and Witzell 2012; Mei and Flinn 2010), it has not become an established part of most pest management systems. This may be due to great expectation by researchers, as well as the

consequence of simplified screening systems, where candidate biocontrol agents are tested using newly recovered isolates, under controlled environmental conditions, against one disease and on one crop (Sturz et al. 2000). For example, the application of an endophyte, exhibiting in vitro antibiosis properties against a systemic pathogen, into the tree's xylem, where chemical control is less efficient, may potentially play a role in the pathogens control. Yet, at least for mal secco disease of citrus, none of the attempts were proven to be useful (Lima et al. 1994; Coco et al. 2004; Ezra et al. unpublished).

- The question whether the introduction of an endophyte into a new environment might have an influence on the ecosystem should be evaluated by research. The possibility that endophytes could be pathogenic to other members of the native forest cannot be ruled out. It should be of a practice to confirm that the biocontrol agents exert no pathogenic effects on both the target host and on other plant species that are part of agro systems, polycultures, or native vegetation associated with the target host (Meji'a et al. 2008).
- Future plans for endophytes include the introduction of native and novel or transformed endophytes expressing specific and desirable characteristics for plant improvement. For example, in grasses the uses of transformed endophytes are for delivery of pesticides, delivery of genes for enhanced biotic and abiotic stress resistance, accelerated seedling emergence and subsequent plant development, increasing or improving nutritional qualities, and increase herbage yield (Kuldau and Bacon 2008). In addition, endophytic fungi have the potential for use as vectors for transformation of useful products that can be expressed *in planta* (Kuldau and Bacon 2008).
- Endophytes are, undoubtedly, an integrated part of the plants environment, yet most of the microbial endophytes are still obscure. Also, their impact on the host and their relations with it are still vague. However, endophytes' potential benefits for human kind are huge. They may contain the solution to the world's food shortage, by increasing crop yield, as well as to climate changes, by increasing the plant's tolerance to stresses. Therefore, it is research duty in the future to discover and utilize the full potential they still pose.

References

Achata Böttger J, Creamer R, Gardner D (2012) Seasonal changes in *Undifilum* colonization and swainsonine content of locoweeds. J Chem Ecol 38:486–495

Achatz B, von Ruden S, Andrade D, Neumann E, Pons-Kuhnemann J, Kogel KH, Franken P, Waller F (2010) Root colonization by *Piriformospora indica* enhances grain yield in barley under diverse nutrient regimes by accelerating plant development. Plant Soil 333:59–70

AgraQuest (2005) http://mbao.org/2005/MBAO%20 2005%20pdfs/Preplant/11/Walgenbach.pdf

Albrectsen BR, Witzell J (2012) Disentangling functions of fungal endophytes in forest trees. In: Silva AP, Sol M (eds) Fungi types, environmental impact and role in disease. Nova Science Publishers Inc, New York

Aly AH, Debbab A, Kjer J, Proksch P (2010) Fungal endophytes from higher plants: a prolific source of phytochemicals and other bioactive natural products. Fungal Divers 41:1–16

Andrade-Linares DR, Müller A, Fakhro A, Schwarz D, Franken P (2012) Impact of *Piriformospora indica* on tomato. In: Varma A, Kost G, Oelmuller R (eds) *Piriformospora indica*. Springer, Berlin

Andreote FD, Gullo MJ, de Souza Lima AO, Junior WM, Azevedo JL, Araujo WL (2004) Impact of genetically modified *Enterobacter cloacae* on indigenous endophytic community of *Citrus sinensis* seedlings. J Microbiol 42:169–173

Araujo WL, Marcon J, Maccheroni W, van Elsas JD, van Vuurde JWL, Azevedo JL (2002) Diversity of endophytic bacterial populations and their interaction with *Xylella fastidiosa* in *Citrus* plants. Appl Environ Microbiol 68:4906–4914

Ardanov P, Ovcharenko L, Zaets I, Kozyrovska N, Pirttilä AM (2011) Endophytic bacteria enhancing growth and disease resistance of potato (*Solanum tuberosum* L.). Biol Control 56:43–49

Ardanov P, Sessitsch A, Häggman H, Kozyrovska N, Pirttilä AM (2012) *Methylobacterium*-induced endophyte community changes correspond with protection of plants against pathogen attack. PLoS One 7(10):e46802. doi:10.1371/journal.pone.0046802

Arnold AE (2008) Endophytic fungi: hidden components of tropical community ecology. In: Schnitzer S, Carson W (eds) Tropical forest community ecology. Blackwell Scientific, Inc, Hoboken, pp 254–271

Arnold AE, Herre EA (2003) Canopy cover and leaf age affect colonization by tropical fungal endophytes: ecological pattern and process in *Theobroma cacao* (Malvaceae). Mycologia 95:388–398

Arnold AE, Lutzoni F (2007) Diversity and host range of foliar fungal endophytes: are tropical leaves biodiversity hotspots? Ecology 88:541–549

Arnold AE, Maynard Z, Gilbert G, Coley PD, Kursar TA (2000) Are tropical fungal endophytes hyperdiverse? Ecol Lett 3:267–274

Arnold AE, Mejía L, Kyllo D, Rojas E, Maynard Z, Herre EA (2003) Fungal endophytes limit pathogen damage in a tropical tree. Proc Natl Acad Sci U S A 100:15649–15654

Aschehoug ET, Metlen KL, Callaway RM, Newcombe G (2012) Fungal endophytes directly increase the competitive effects of an invasive forb. Ecology 93:3–8

Athukorala SNP, Dilantha-Fernando WG, Rashid KY, Kievit TD (2010) The role of volatile and non-volatile antibiotics produced by *Pseudomonas chlororaphis* strain PA23 in its root colonization and control of *Sclerotinia sclerotiorum*. Biocontrol Sci Technol 20:875–890

Atmosukarto I, Castillo U, Hess WM, Sears J, Strobel G (2005) Isolation and characterization of *Muscodor albus* I-41.3s, a volatile antibiotic producing fungus. Plant Sci 169:854–861

Baca BE, Elmerich C (2003) Microbial production of plant hormones. In: Elmerich C, Newton WE (eds) Associative and endophytic nitrogen-fixing bacteria and cyanobacterial associations. Kluwer Academic, Dordrecht

Bacon CW (1993) Abiotic stress tolerances (moisture, nutrients) and photosynthesis in endophyte-infected tall fescue. Agric Ecosyst Environ 44:123–141

Bacon CW (1995) Toxic endophyte-infected tall fescue and range grasses: historic perspectives. J Anim Sci 73:861–870

Bacon CW, Hill NS (1996) Symptomless grass endophytes: products of coevolutionary symbioses and their role in the ecological adaptations of grasses. In: Redkin SC, Carris LM (eds) Endophytic fungi in grasses and woody plants. APS Press, St. Paul, pp 155–178

Bacon CW, Porter JK, Robbins JD, Luttrell ES (1977) *Epichloe typhina* from toxic tall fescue grasses. Appl Environ Microbiol 34:576–581

Bae H, Sicher RC, Kim MS, Kim SH, Strem MD, Melnick RL et al (2009) The beneficial endophyte *Trichoderma hamatum* isolate DIS 219b promotes growth and delays the onset of the drought response in *Theobroma cacao*. J Exp Bot 60:3279–3295

Bailey BA, Bae H, Strem MD, Roberts DP, Thomas SE, Samuels GJ, Choi Y, Holmes KA (2006) Fungal and plant gene expression during the colonization of cacao seedlings by endophytic isolates of four *Trichoderma* species. Planta 224:1449–1464

Bakshi S, Sztejnberg A, Yarden O (2001) Isolation and characterization of a cold-tolerant strain of *Fusarium proliferatum*, a bio-control agent of grape downy mildew. Phytopathology 91:1062–1068

Baldauf MW, Mace WJ, Richmond DS (2011) Endophyte-mediated resistance to black cutworm as a function of plant cultivar and endophyte strain in tall fescue. Environ Entomol 40:639–647

Ball OJ-P, Prestidge RA (1992) The effect of the endophytic fungus *Acremonium lolii* on adult black beetle (*Heteronychus arator*) feeding. In: Proceedings of the 45th New Zealand plant protection conference, Wellington, 1992, pp 201–204

Ball OJ-P, Prestidge RA, Sprosen JM (1995) Interrelationships between *Acremonium lolii*, peramine and lolitrem B in perennial ryegrass. Appl Environ Microbiol 61:1527–1533

Ball OJ-P, Miles CO, Prestidge RA (1997) Ergopeptine alkaloids and *Neotyphodium lolii*-mediated resistance in perennial ryegrass against adult *Heteronychus arator* (Coleoptera: Scarabaeidae). J Econ Entomol 90:1382–1391

Ball OJ, Gwinn KD, Pless CD, Popay AJ (2011) Endophyte isolate and host grass effects on *Chaetocnema pulicaria* (Coleoptera: Chrysomelidae) feeding. J Econ Entomol 104:665–672

Baltruschat H, Fodor J, Harrach BD, Niemczyk E, Barna B, Gullner G, Janeczko A, Kogel KH, Schäfer P, Schwarczinger I, Zuccaro A, Skoczowski A (2008) Salt tolerance of barley induced by the root endophyte *Piriformospora indica* is associated with a strong increase in antioxidants. New Phytol 180:501–510

Barazani O, Benderoth M, Groten K, Kuhlemeier C, Baldwin IT (2005) *Piriformospora indica* and *Sebacina vermifera* increase growth performance at the expense of herbivore resistance in *Nicotiana attenuata*. Oecologia 146:234–243

Barazani O, Von Dahl CC, Baldwin IT (2007) *Sebacina vermifera* promotes the growth and fitness of *Nicotiana attenuata* by inhibiting ethylene signaling. Plant Physiol 144:1223–1232

Barka EA, Belarbi A, Kachet C, Nowak J, Audran JC (2000) Enhancement of *in vitro* growth and resistance to gray mould of *Vitis vinifera* cocultured with plant growth-promoting rhizobacteria. FEMS Microbiol Lett 186:91–95

Barka EA, Gognies S, Nowak J, Audran JC, Belarbi A (2002) Inhibitory effect of endophyte bacteria on *Botrytis cinerea* and its influence to promote the grapevine growth. Biol Control 24:135–142

Barka EA, Nowak J, Clément C (2006) Enhancement of chilling resistance of inoculated grapevine plantlets with a plant growth-promoting rhizobacterium, *Burkholderia phytofirmans* strain PsJN. Appl Environ Microbiol 72:7246–7252

Barker GM, Pottinger RP, Addison PJ, Prestidge RA (1984a) Effect of *Lolium* endophyte fungus infections on behaviour of adult Argentine stem weevil. NZ J Agric Res 27:271–277

Barker GM, Pottinger RP, Addison PJ (1984b) Effect of *Lolium* endophyte fungus infections on survival of larval Argentine stem weevil. NZ J Agric Res 27:279–281

Bazely DR, Vicaru M, Emmerich S, Filip L, Lin D, Inman A (1997) Interactions between herbivores and endophyte-infected *Festuca rubra* from the Scottish islands of St. Kilda, Benbecula and Rum. J Appl Ecol 34:847–860

Belesky DP, Stringer WC, Hill NS (1989) Influence of endophytes and water regime upon tall fescue accessions. I. Growth characteristics. Ann Bot 63:495–503

Benhamou N, Belanger RR, Paulitz TC (1996a) Induction of differential host responses by *Pseudomonas fluorescens* in Ri T-DNA-transformed pea roots after challenge with *Fusarium oxysporum* f. sp. *pisi* and *Pythium ultimum*. Phytopathology 86:114–178

Benhamou N, Kloepper JW, Quadt-Hallmann A, Tuzun S (1996b) Induction of defense-related ultrastructural modifications in pea root tissues inoculated with endophytic bacteria. Plant Physiol 112:919–929

Benhamou N, Kloepper JW, Tuzun S (1998) Induction of resistance against *Fusarium* wilt of tomato by combination of chitosan with and endophytic bacterial strain: ultrastructure and cytochemistry of the host response. Planta 204:153–168

Benson DR, Silvester WB (1993) Biology of Frankia strains, actinomycete symbionts of actinorhizal plants. Microbiol Rev 57:293–319

Bischoff JF, White JF Jr (2005) Evolutionary development of the Clavicipitaceae. In: Dighton J, White JF, Oudemans P (eds) The fungal community: its organization and role in the ecosystem. Taylor & Francis, Boca Raton, pp 505–518

Blank CA (1992) Interactions of tall fescue seedlings infected with *Acremonium coenophialum* with soilborne pathogens. MS thesis, University of Tennessee, Knoxville

Boddey RM, de Oliveira OC, Urquiaga S, Reis VM, de Oliveira FL, Baldani VKD, Döbereiner J (1995) Biological nitrogen fixation associated with sugarcane and rive: contributions and prospects for improvement. Plant Soil 174:195–209

Bouton JH, Gates RN, Belesky DP, Owsley M (1993) Yield and persistence of tall fescue in the southeastern Coastal Plain after removal of its endophyte. Agron J 85:52–55

Boyer JS (1982) Plant productivity and environment. Science 218:443–448

Bray EA, Bailey-Serres J, Weretilnyk E (2000) Responses to abiotic stresses. In: Gruissem W, Buchannan B, Jones R (eds) Biochemistry and molecular biology of plants. American Society of Plant Physiologists, Rockville, pp 1158–1249

Brooks DS, Gonzalez CF, Appel DN, Filer TH (1994) Evaluation of endophytic bacteria as potential biological control agents for oak wilt. Biol Control 4:373–381

Burke JM, Rorie RW (2002) Changes in ovarian function in mature beef cows grazing endophyte infected tall fescue. Theriogenology 57(6):1733–1742

Bush LP, Wilkinson HH, Schardl CL (1997) Bioprotective alkaloids of grass-fungal endophyte symbioses. Plant Physiol 114:1–7

Camehl I, Drzewiecki C, Vadassery J, Shahollari B, Sherameti I, Forzani C, Munnik T, Hirt H, Oelmuller R (2011) The OXI1 kinase pathway mediates *Piriformospora indica*-induced growth promotion in *Arabidopsis*. PLoS Pathog 7:e1002051

Campanile F, Ruscelli A, Luisi N (2007) Antagonistic activity of endophytic fungi towards *Diplodia corticola* assessed by in vitro and in planta tests. Eur J Plant Pathol 117:237–246

Cao WH, Liu J, He XJ, Mu RL, Zhou HL, Chen SY, Zhang JS (2006) Modulation of ethylene responses affects plant salt-stress responses. Plant Physiol 140:707–719

Carroll G (1988) Fungal endophytes in stems and leaves – from latent pathogen to mutualistic symbionts. Ecology 69:2–9

Carson RD, West CP, Reyes BDL, Rajguru S, Guerber CA (2004) Endophyte effects on dehydrin protein expression and membrane leakage in tall fescue. In: Kallenbach R, Rosenkrans CT, Ryan L (eds) 5th international symposium on *Neotyphodium*/grass interactions, University of Arkansas, Fayetteville, Abstract #202

Castillo UF, Strobel GA, Ford EJ et al (2002) Munumbicins, wide-spectrum antibiotics produced by *Streptomyces* NRRL30562, endophytic on *Kennedia nigriscans*. Microbiology 148:2675–2685

Castillo U, Harper JK, Strobel GA et al (2003) Kakadumycins, novel antibiotics from *Streptomyces* sp. NRRL 30566, an endophyte of *Grevillea pteridifolia*. FEMS Microbiol Lett 224:183–190

Chen C, Bauske EM, Musson G, Rodriguez-Kabana R, Kloepper JW (1995) Biological control of *Fusarium* wilt of cotton by use of endophytic bacteria. Biol Control 6:83–91

Cheplick GP (2004) Recovery from drought stress in *Lolium perenne* (Poaceae): are fungal endophytes detrimental? Am J Bot 91:1960–1968

Cho HS, Park SY, Ryu CM, Kim JF, Kim JG, Park SH (2007) Interference of quorum sensing and virulence of the rice pathogen *Burkholderia glumae* by an engineered endophytic bacterium. FEMS Microbiol Ecol 60:14–23

Clarke BB, White JF Jr, Hurley RH, Torres MS, Sun S, Huff DF (2006) Endophyte-mediated suppression of dollar spot disease in fine fescues. Plant Dis 90:994–998

Clay K (1987) Effects of fungal endophytes on the seed and seedling biology of *Lolium perenne* and *Festuca arundinacea*. Oecologia (Heidelb) 73:358–362

Clay K (1988) Fungal endophytes of grasses: a defensive mutualism between plants and fungi. Ecology 69:10–16

Clay K, Schardl CL (2002) Evolutionary origins and ecological consequences of endophyte symbiosis with grasses. Am Nat 160:S99–S127

Clement SL, Hu J, Stewart AV, Wang B, Elberson LR (2011) Detrimental and neutral effects of a wild grass-fungal endophyte symbiotum on insect preference and performance. J Insect Sci 11:1–13

Coco V, Grimaldi V, Licciardello G, Cirvilleri G, Grasso S, Catara A (2004) Inhibition of *Phoma tracheiphila* by *pseudomonads* in citrus seedlings. Proc Int Soc Citriculture 2:729–732

Coetzer JAW, Kellerman TS, Naude TW (1985) Neurotoxicoses of livestock caused by plants and fungi in southern Africa, Report 199, Technical

Commununication, Department of Agriculture, South Africa, Onderstepoort

Combès A, Ndoye I, Babce C, Bruzaud J, Djediat C, Dupont J, Nay B, Prado S (2012) Chemical communication between the endophytic fungus *Paraconiothyrium variabile* and the phytopathogen *Fusarium oxysporum*. PLoS One 7(10):e47313

Compant S, Duffy B, Nowak J, Clément C, Barka EA (2005a) Use of plant growth-promoting bacteria for bio-control of plant diseases: principles, mechanisms of action and future prospects. Appl Environ Microbiol 71:4951–4959

Compant S, Reciter B, Sessitsch A, Nowak J, Clement C, Barka EA (2005b) Endophytic colonization of *Vitis vinifera* L. by plant growth-promoting bacterium *Burkholderia* sp. strain PsJN. Appl Environ Microbiol 71:1685–1693

Conn KL, Nowak J, Lazarovits G (1997) A gnotobiotic bioassay for studying interactions between potatoes and plant growth-promoting rhizobacteria. Can J Microbiol 43:801–808

Conn VM, Walker AR, Franco CMM (2008) Endophytic actinobacteria induce defense pathways in *Arabidopsis thaliana*. Mol Plant Microbe Interact 21:208–218

Cook RJ, Baker KF (1983) The nature and practice of biological control of plant pathogens. Am Soc, St. Paul

Craven KD, Blankenship JD, Leuchtmann A, Hignight K, Schardl CL (2001) Hybrid fungal endophytes symbiotic with the grass *Lolium pratense*. Sydowia 53:44–73

Crawford KM, Land JM, Rudgers JA (2010) Fungal endophytes of native grasses decrease insect herbivore preference and performance. Oecologia 164:431–444

Creus CM, Sueldo RJ, Barassi CA (1998) Water relations in *Azospirillum*-inoculated wheat seedlings under osmotic stress. Can J Bot 76:238–244

Cross DL, Redmond LM, Strickland JR (1995) Equine fescue toxicosis: signs and solutions. J Anim Sci 73:899–908

Dababat AEFA, Sikora RA (2007) Induced resistance by the mutualistic endophyte, *Fusarium oxysporum* strain 162, toward *Meloidogyne incognita* on tomato. Biocontrol Sci Technol 17:969–975

Dai CC, Yu BY, Li X (2008) Screening of endophytic fungi that promote the growth of *Euphorbia pekinensis*. Afr J Biotechnol 7:3505–3510

Daisy BH, Strobel GA, Castillo U, Ezra D, Sears J, Weaver DK, Runyon JB (2002) Naphthalene, an insect repellent, is produced by *Muscodor vitigenus*, a novel endophytic fungus. Microbiology 148:3737–3741

Danielsen S, Jensnen DF (1999) Fungal endophytes from stalks of tropical maize and grasses: isolation, identification, and screening for antagonism against *Fusarium verticillioides* in maize stalks. Biocontrol Sci Technol 9:545–553

Davitt AJ, Stansberry M, Rudgers JA (2010) Do the costs and benefits of fungal endophyte symbiosis vary with light availability? New Phytol 188:824–834

de Bary A (1879) Die erschenung symbiose. In: Trubner KJ (ed) Vortarg auf der versammlung der naturforscher und artze zu cassel, Strassburg, 1879, pp 1–30

de Battista JP, Bacon CW, Severson R, Plattner RD, Bouton JH (1990a) Indole acetic acid production by the fungal endophyte of tall fescue. Agron J 82:878–880

de Battista JP, Bouton JH, Bacon CW, Siegel MR (1990b) Rhizome and herbage production of endophyte-removed tall fescue clones and populations. Agron J 82:651–654

Dennis SB, Allen VG, Saker KE, Fontenot JP, Ayad JY, Brown CP (1998) Influence of *Neotyphodium coenophialum* on copper concentration in tall fescue. J Anim Sci 76:2687–2693

di Menna ME, Finch SC, Popay AJ, Smith BL (2012) A review of the *Neotyphodium lolii*/*Lolium perenne* symbiosis and its associated effects on animal and plant health, with particular emphasis on ryegrass staggers. N Z Vet J 60:315–328

Ding S, Huang CL, Sheng HM, Song CL, Li YB, An LZ (2011) Effect of inoculation with the endophyte *Clavibacter* sp. strain Enf12 on chilling tolerance in *Chorispora bungeana*. Physiol Plant 252:141–151

Doty SL (2011) Growth-promoting endophytic fungi of forest trees. In: Pirttila AM, Carolin CA (eds) Endophytes of forest trees: biology and applications. Springer, New York

Doty SL, Oakley B, Xin G, Kang JW, Singleton G, Khan Z, Vajzovic A, Staley JT (2009) Diazotrophic endophytes of native black cottonwood and willow. Symbiosis 47:27–33

Downing KJ, Thomson JA (2000) Introduction of the *Serratia marcescens* ChiA gene into an endophytic *Pseudomonas fluorescens* for the bio-control of phytopathogenic fungi. Can J Microbiol 46:363–369

Druege U, Baltruschat H, Franken P (2007) *Piriformospora indica* promotes adventitious root formation in cuttings. Sci Hortic 112:465–477

Dudeja SS, Giri R, Saini R, Suneja-Madan P, Kothe E (2012) Interaction of endophytic microbes with legumes. J Basic Microbiol 52:248–260

Duijff BJ, Gianinazzi-Pearson V, Lemanceau P (1997) Involvement of the outer membrane lipopolysaccharides in the endophytic colonization of tomato roots by bio-control *Pseudomonas fluorescens* strain WCS417r. New Phytol 135:325–334

Enebak SA, Carey WA (2000) Evidence for induced systemic protection to fusiform rust in loblolly pine by plant growth-promoting rhizobacteria. Plant Dis 84:306–308

Ezra D, Castillo UF, Strobel GA et al (2004a) Coronamycins, peptide antibiotics produced by a verticillate *Streptomyces* sp. (MSU-2110) endophytic on *Monstera* sp. Microbiology 150:785–793

Ezra D, Hess WM, Strobel GA (2004b) New endophytic isolates of *Muscodor albus*, a volatile antibiotic producing fungus. Microbiology 150:4023–4031

Fabien M, Nathalle V, Adnan H, Alain C, Huguette S (2001) Endophytic *Neotyphodium lolii* induced tolerance to Zn stress in *Lolium perenne*. Physiol Plant 113:557–563

Fakhro A, Andrade-Linares DR, von Bargen S, Bandte M, Büttner C, Grosch R, Schwarz D, Franken P (2010)

Impact of *Piriformospora indica* on tomato growth and on interaction with fungal and viral pathogens. Mycorrhiza 20:191–200

Falk SP, Pearson RC, Gadoury DM, Seem RC, Sztejnberg A (1996) *Fusarium proliferatum* as a bio-control agents against grape downy mildew. Phytopathology 86:1010–1017

Fávaro LC, Sebastianes FL, Araújo WL (2012) *Epicoccum nigrum* P16, a sugarcane endophyte, produces antifungal compounds and induces root growth. PLoS One 7(6):e36826

Feng Y, Shen D, Song W (2006) Rice endophyte *Pantoea agglomerans* YS19 promotes host plant growth and affects allocations of host photosynthates. J Appl Microbiol 100:938–945

Filipov NM, Thompson FN, Hill NS, Dawe DL, Stuedemann JA, Price JC, Smith CK (1998) Vaccination against ergot alkaloids and the effect of endophyte-infected seed-based diets on rabbits. J Anim Sci 76:2456–2463

Fletcher LR, Harvey IC (1981) An association of *Lolium* endophyte with ryegrass staggers. N Z Vet J 28: 185–186

Fortier GM, Bard N, Jansen M, Clay K (2000) Effects of tall fescue endophyte infection and population density on growth and reproduction in prairie voles. J Wildl Manage 64:122–128

Fowler SR, Jaspers MV, Walter M, Steward A (1999) Suppression of overwintering *Botrytis cinerea* inoculum on grape rachii using antagonistic fungi. In: Proceedings of the 52nd New Zealand plant protection conference, Auckland, 1999, pp 141–147

Franken F (2012) The plant strengthening root endophyte *Piriformospora indica*: potential application and the biology behind. Appl Microbiol Biotechnol 96: 1455–1464

Funk CR, White RH, Breen J (1993) Importance of *Acremonium* endophytes in turfgrass breeding and management. Agric Ecosyst Environ 44:215–232

Gallagher RT, Hawkes AD, Steyn PS, Vleggaar R (1984) Tremorgenic neurotoxins from perennial ryegrass causing ryegrass staggers disorder of livestock: structure elucidation of lolitrem B. J Chem Soc Chem Commun 1984:614–616

Ganley RJ, Sniezko RA, Newcombe G (2008) Endophyte-mediated resistance against white pine blister rust in *Pinus monticola*. Forest Ecol Manage 255:2751–2760

Gao F, Dai C, Liu X (2010) Mechanisms of fungal endophytes in plant protection against pathogens. Afr J Microbiol Res 4:1346–1351

Gao F, Yong Y, Dai C (2011) Effects of endophytic fungal elicitor on two kinds of terpenoids production and physiological indexes in *Euphorbia pekinensis* suspension cells. J Med Plants Res 5:4418–4425

Glick BR, Penrose DM, Li J (1998) A model for the lowering of plant ethylene concentrations by plant growth promoting bacteria. J Theor Biol 190:63–68

Goates BJ, Mercier J (2009) Effect of biofumigation with volatiles from *Muscodor albus* on the viability of *Tilletia* spp. teliospores. Can J Microbiol 55:203–206

González V, Tello ML (2011) The endophytic mycota associated with *Vitis vinifera* in central Spain. Fungal Divers 47:29–42

Granado J, Thürig B, Kieffer E, Petrini L, Flieβbach A, Tamm L, Weibel FP, Wyss GS (2008) Culturable fungi of stored 'golden delicious' apple fruits: a one-season comparison study of organic and integrated production systems in Switzerland. Microb Ecol 56:720–732

Greulich F, Horio E, Shimanuki T, Yoshihara T (1999) Field results confirm natural plant protection by the endophytic fungus *Epichloë typhina* against the pathogenic fungus *Cladosporium phlei* on timothy leaves. Ann Phytopathol Soc Jpn 65:454–459

Gronberg H, Kaparakis G, Sen R (2006) Binucleate *Rhizoctonia* (*Ceratorhiza* spp.) as non-mycorrhizal endophytes alter *Pinus sylvestris* L. seedling root architecture and affect growth of rooted cuttings. Scand J Forest Res 21:450–457

Guan SH, Sattler I, Lin WH, Guo DA, Grabley S (2005) p-Aminoacetophenonic acids produced by a mangrove endophyte: *Streptomyces griseus* subspecies. J Nat Prod 68:1198–1200

Gunatilaka AAL (2006) Natural products from plant-associated microorganisms: distribution, structural diversity, bioactivity, and implications of their occurrence. J Nat Prod 69:509–526

Guo B, Dai JR, Ng S, Huang Y, Leong C, Ong W, Carte BK (2000) Cytonic acids A and B: novel tridepside inhibitors of hCMV protease from the endophytic fungus *Cytonaema* species. J Nat Prod 63:602–604

Guy PL (1992) Incidence of *Acremonium lolii* and lack of correlation with barley yellow dwarf viruses in Tasmanian perennial ryegrass pastures. Plant Pathol 41:29–34

Gwinn KD, Bernard EC (1993) Interactions of endophyte-infected grasses with the nematodes *Meloidogyne marylandi* and *Pratylenchus scribneri*. In: Proceedings of 2nd international symposium *Acremonium*/grass interact. Plenary Papers, Plamerston North, 1993

Haas D, Keel C, Reimmann C (2002) Signal transduction in plant-beneficial rhizobacteria with bio-control properties. Antonie von Leeuwenhoek 81:385–395

Hall TJ, Schreiber LR, Leben C (1986) Effects of xylem-colonizing *Bacillus* spp. on verticillium wilt in maples. Plant Dis 70:521–524

Hallmann J, Quadt-Hallmann A, Mahaffee WF, Kloepper JW (1997) Endophytic bacteria in agricultural crops. Can J Microbiol 43(10):895–914

Hanada RE, de Jorge ST, Pomella AWV, Hebbar KP, Pereira JO, Ismaiel A, Samuels GJ (2008) *Trichoderma martiale* sp. nov., a new endophyte from sapwood of *Theobroma cacao* with a potential for biological control. Mycol Res 112:1335–1343

Hanada RE, Pomella AWV, Soberanis W, Loguercio LL, Pereia JO (2009) Biocontrol potential of *Trichoderma martiale* against the black-pod disease (*Phytophthora palmivora*) of cacao. Biol Control 50:143–149

Hanada RE, Pomella AWV, Costa HS, Bezerra JL, Loguercio LL, Pereira JO (2010) Endophytic fungal diversity in *Theobroma cacao* (cacao) and

T. grandiflorum (cupuaçu) trees and their potential for growth promotion and bio-control of black-pod disease. Fungal Biol 114:901–910

Hardoim PR, van Overbeek LS, van Elsas JD (2008) Properties of bacterial endophytes and their proposed role in plant growth. Trends Microbiol 16:463–471

Hesse U, Latch GCM (1999) Influence of *Neotyphodium lolii* and barley yellow dwarf virus, individually and combined, on the growth of *Lolium perenne*. Aust Plant Pathol 28:240–247

Higgins KL, Arnold AE, Miadlikowska J, Sarvate SD, Lutzoni F (2007) Phylogenetic relationships, host affinity, and geographic structure of boreal and arctic endophytes from three major plant lineages. Mol Phylogenet Evol 42:543–555

Hilbert M, Voll LM, Ding Y, Hofmann J, Sharma M, Zuccaro A (2012) Indole derivative production by the root endophyte *Piriformospora indica* is not required for the growth promotion but for biotrophic colonization of barley roots. New Phytol 196:520–534

Hurek T, Reinhold-Hurek B (2003) *Azoarcus* sp. strain BH72 as a model for nitrogen-fixing grass endophytes. J Biotechnol 106:169–178

Hussain A, Tian MY, He YR, Lei YY (2010) Differential fluctuation in virulence and VOC profiles among different cultures of entomopathogenic fungi. J Invertebr Pathol 104:166–171

Jacobson CB, Pasternak JJ, Glick BR (1994) Partial purification and characterization of 1-amino-cyclopropane-1-carboxylate deaminase from the plant growth promoting rhizobacterium *Pseudomonas putida* GR 12–2. Can J Microbiol 40:1019–1025

Jalgaonwala RE, Mahajan RT (2011) A review: bacterial endophytes and their bioprospecting. J Pharm Res 4:795–799

Ji X, Lu G, Gai Y, Zheng C, Mu Z (2008) Biological control against bacterial wilt and colonization of mulberry by an endophytic *Bacillus subtilis* strain. FEMS Microbiol Ecol 60:565–573

Jimenez-Salgado T, Fuentes-Ramirez E, Tapia-Hernandez A, Mascarua-Esparza MA, Martinez-Romero E, Caballero-Mellado J (1997) *Coffea arabica* L., a new host plant for *Acetobacter diazotrophicus*, and isolation of other nitrogen-fixing acetobacteria. Appl Environ Microbiol 63:3676–3683

Joost RE (1995) *Acremonium* in fescue and ryegrass: boon or bane? A review. J Anim Sci 73:881–888

Kai M, Effmert U, Berg G, Piechulla B (2007) Volatile of bacterial antagonists inhibit mycelial growth of the plant pathogen *Rhizoctonia solani*. Arch Microbiol 187:351–360

Kavroulakis N, Ntougias S, Zervakis GI, Ehaliotis C, Haralampidis K, Papadopoulou KK (2007) Role of ethylene in the protection of tomato plants against soil-borne fungal pathogens conferred by and endophytic *Fusarium solani* strain. J Exp Bot 58:3853–3864

Khan Z, Doty SL (2009) Characterization of bacterial endophytes of sweet potato plants. Plant Soil 322: 197–207. doi:10.1007/s11104-009-9908-1

Khan SA, Hamayun M, Yoon H, Kim HY, Suh SK, Hwang SK, Kim JM, Lee IJ, Choo YS, Kong WS, Lee BM, Kim JG (2008) Plant growth promotion and *Penicillium citrinum*. BMC Microbiol 8:231–240

Khan AL, Hamayun M, Ahmad N, Hussain J, Kang SM, Kim YM, Adnan M, Tang DS, Waqas M, Radhakrishnan R, Hwang YH, Lee IJ (2011a) Salinity stress resistance offered by endophytic fungal interaction between *Penicillium minioluteum* LHL09 and *Glycine max* L. J Microbiol Biotechnol 21:893–902

Khan AL, Hamayun M, Ahmad N, Waqas M, Kang SM, Kim YH, Lee IJ (2011b) *Exophiala* sp. LHL08 reprograms *Cucumis sativus* to higher growth under abiotic stresses. Physiol Plant 143:329–343

Khan AL, Hamayun M, Kim YH, Kang SM, Lee IJ (2011c) Ameliorative symbiosis of endophyte (*Penicillium funiculosum* LHL06) under salt stress elevated plant growth of *Glycine max*. L. Plant Physiol Biochem 49:852–861

Khan AL, Hamayun M, Kang SM, Kim YH, Jung HY, Lee JH, Lee IJ (2012a) Endophytic fungal association via gibberellins and indole acetic acid can improve plant growth under abiotic stress: an example of *Paecilomyces formosus* LHL10. BMC Microbiol 12:3. doi:10.1186/1471-2180-12-3

Khan AL, Hamayun M, Radhakrishnan R, Waqas M, Kang SM, Kim YH, Shin JH, Choo YS, Kim JG, Lee IJ (2012b) Mutualistic association of *Paecilomyces formosus* LHL10 offers thermotolerance to *Cucumis sativus*. Antonie von Leeuwenhoek 101:267–279

Khan Z, Guelich G, Phan H, Redman R, Doty S (2012c) Bacterial and yeast endophytes from poplar and willow promote growth in crop plants and grasses. ISRN Agronomy 2012:Article ID 890280. doi:10.5402/2012/890280

Khastini RO, Ohta H, Narisawa K (2012) The role of dark septate endophytic fungus, *Veronaeopsis simplex* Y34, in *Fusarium* disease suppression in Chinese cabbage. J Microbiol 50:618–624

Kim S, Lowman S, Hou G, Nowak J, Flinn B, Mei C (2012) Growth promotion and colonization of switchgrass (*Panicum virgatum*) cv. Alamo by bacterial endophyte *Burkholderia phytofirmans* strain PsJN. Biotechnol Biofuels 5:37

Kimmons CA, Gwinn KD, Bernard EC (1990) Nematode reproduction on endophyte-infected and endophyte-free tall fescue. Plant Dis 74:757–761

Kishimoto K, Matsui K, Ozawa R, Takabayashi J (2007) Volatile 1-octen-3-ol induces a defensive response in *Arabidopsis thaliana*. J Gen Plant Pathol 73:35–37

Kloepper JW, Beauchamp CJ (1992) A review of issues related to measuring colonization of plant roots by bacteria. Can J Microbiol 38:1219–1232

Kloepper JW, Ryu CM (2006) Bacterial endophytes as elicitors of induced systemic resistance. In: Schulz B, Boyle C, Sieber TN (eds) Microbial root endophytes. Springer, Berlin/Heidelberg, pp 33–52

Kortekamp A (1997) *Epicoccum nigrum* link: a biological control, agent of *Plasmopara viticola* (Berk. et Curt.) Berl. et De Toni? Vitis 36:215–216

Koshino H, Terada S, Yoshihara T, Sakamura S, Shimanuki T, Sato T, Tajimi A (1988) Three phenolic acid derivatives from stromata of *Epichloë typhina* on *Phleum pratense*. Phytochemistry 27:1333–1338

Koshino H, Yoshihara T, Sakamura S, Shimanuki T, Sato T, Tajimi A (1989) A ring B aromatic sterol from stromata of *Epichloë typhina*. Phytochemistry 28:771–772

Krings M, Hass H, Kerp H, Taylor TN, Agerer R, Dotzler N (2009) Endophytic cyanobacteria in a 400-million-yr-old land plant: a scenario for the origin of a symbiosis? Rev Palaeobot Palyetalnol 153:62–69

Kuldau G, Bacon C (2008) Clavicipitaceous endophytes: their ability to enhance resistance of grasses to multiple stresses. Biol Control 46:57–71

Kumar M, Yadav V, Tuteja N, Johri AK (2009) Antioxidant enzyme activities in maize plants colonized with *Piriformospora indica*. Microbiology 155:780–790

Kunkel BA, Grewal PS, Quigley MF (2004) A mechanism of acquired resistance against an entomopathogenic nematode by *Agrotis ipsilon* feeding on perennial ryegrass harboring a fungal endophyte. Biol Control 29:100–108

Lacey LA, Neven LG (2006) The potential of the fungus, *Muscodor albus*, as a microbial control agent of potato tuber moth (Lepidoptera: Gelechiidae) in stored potatoes. J Invertebr Pathol 91:195–198

Ladha JK, Reddy PM (1995) Extension of nitrogen fixation to rice – necessity and possibilities. Geo J 35:363–372

Lampel JS, Canter GL, Dimock MB, Kelly JL, Anderson JJ, Uratani BB, Foulke JS Jr, Turner JT (1994) Integrative cloning, expression, and stability of the *cry*IA(c) gene from *Bacillus thuringiensis* subsp. *kurstati* in a recombinant strain of *Clavibacter xyli* subsp. *cynodontis*. Appl Environ Microbiol 60:501–508

Larena I, Torres R, De Cal A, Liñán M, Melgarejo P, Domenichini P, Bellini A, Mandrin JF, Lichou J, Ochoa de Eribe X, Usall J (2005) Biological control of postharvest brown rot (*Monilinia* spp.) of peaches by field applications of *Epicoccum nigrum*. Biol Control 32:305–310

Latch GCM, Christensen MJ, Gaynor DL (1985a) Aphid detection of endophyte infection in tall fescue. N Z J Agric Res 28:129–132

Latch GCM, Hunt WT, Musgrave DR (1985b) Endophytic fungi affect growth of perennial ryegrass. N Z J Agric Res 28:165–168

Leary JJK, Singleton PW, Borthakur D (2004) Canopy nodulation of the endemic tree legume acacia koa in the mesic forests of Hawaii. Ecology 85:3151–3157

Lee K, Pan JJ, May G (2009a) Endophytic *Fusarium verticillioides* reduces disease severity caused by *Ustilago maydis* on maize. FEMS Microbiol Lett 299:31–37

Lee SO, Kim HY, Cho GJ, Lee HB, Jang KS, Choi YH, Kim JC (2009b) Mycofumigation with *Oxyporus latemarginatus* EF069 for control of postharvest apple decay and *Rhizoctonia* root rot on moth orchid. J Appl Microbiol 106:1213–1219

Lee YC, Johnson JM, Chien CT, Sun C, Cai DG, Lou BG, Oelmuller R, Yeh KW (2011) Growth promotion of Chinese cabbage and *Arabidopsis* by *Piriformospora indica* is not stimulated by mycelium-synthesized auxin. Mol Plant Microbe Interact 24:421–431

Lehtonen PT, Helander M, Siddiqui SA, Lehto K, Saikkonen K (2006) Endophytic fungus decreases plant virus infections in meadow ryegrass (*Lolium pretense*). Biol Lett 2:620–623

Leuchtmann A (1992) Systematics, distribution, and host specificity of grass endophyte. Nat Toxins 1:150–162

Li J, Beatty PK, Shah S, Jensen SE (2007) Use of PCR-targeted mutagenesis to disrupt production of Fusaricidin-type antifungal antibiotics in *Paenibacillus polymyxa*. Appl Environ Microbiol 73:3480–3489

Li T, Liu MJ, Zhang XT, Zhang HB, Sha T, Zhao ZW (2011) Improved tolerance of maize (*Zea mays* L.) to heavy metals by colonization of a dark septate endophyte *Exophiala pisciphila*. Sci Total Environ 409:1069–1074

Li HY, Wei DQ, Wei M, Zhou ZP (2012a) Endophytes and their role in phytoremediation. Fungal Divers 54:11–18

Li X, Bu N, Li Y, Ma L, Xin S, Zhang L (2012b) Growth, photosynthesis and antioxidant responses of endophyte infected and non-infected rice under lead stress conditions. J Hazard Mater 213–214:55–61

Lima G, Ippolito A, Nigro F, Salerno M (1994) Attempts in the biological control of citrus mal secco (*Phoma tracheiphila*) using endophytic bacteria (in Italian). Dif Piante 17:43–49

Lin L, Ge HM, Yan T, Qin YH, Tan RX (2012) Thaxtomin A-deficient endophytic *Streptomyces* sp. enhances plant disease resistance to pathogenic *Streptomyces scabies*. Planta 236:1849–1861

Lyons PC, Evans JJ, Bacon CW (1986) Occurrence of peptide and clavinet ergot alkaloids in tall fescue grass. Science 232:487–489

Ma Y, Prasad MNV, Rajkumar M, Freitas H (2011) Plant growth promoting rhizobacteria and endophytes accelerate phytoremediation of metalliferous soils. Biotechnol Adv 29:248–258

Macias-Rubalcava ML, Hernandez-Bautista BE, Oropeza F, Duarte G, Gonzalez MC, Glenn AE, Hanlin RT, Anaya AL (2010) Allelochemical effects of volatile compounds and organic extracts from *Muscodor yucatanensis*, a tropical endophytic fungus from *Bursera simaruba*. J Chem Ecol 36:1122–1131

Mahmood T, Gergerich RC, Milius EA, West CP, D'Arcy CJ (1993) Barley yellow dwarf virus in wheat, endophyte-infected and endophyte-free tall fescue and other hosts in Arkansas. Plant Dis 77:225–228

Malinowski DP, Belesky DP (1999) Tall fescue aluminum tolerance is affected by *Neotyphodium coenophialum* endophyte. J Plant Nutr 22(8):1335–1349

Malinowski DP, Belesky DP (2000) Adaptations of endophyte-infected cool-season grasses to environmental stresses: mechanism of drought and mineral stress tolerance. Crop Sci 40:923–940

Malinowski D, Leuchtmann A, Schmidt D, Nösberger J (1997) Symbiosis with *Neotyphodium uncinatum* endophyte may increase the competitive ability of meadow fescue. Agron J 89:833–839

Malinowski DP, Belesky DP, Fedder JM (1999a) Endophyte infection may affect the competitive ability of tall fescue grown with red clover. J Agron Crop Sci 183:91–102

Malinowski DP, Brauer DK, Belesky DP (1999b) *Neotyphodium coenophialum*-endophyte affects root morphology of tall fescue grown under phosphorous deficiency. J Agron Crop Sci 183:53–60

Márquez LM, Redman RS, Rodriguez RJ, Roossinck MJ (2007) A virus in a fungus in a plant: three-way symbiosis required for thermal tolerance. Science 315:513–515

Martini M, Musetti R, Grisan S, Polizzotto R, Borselli S, Pavan F, Osler R (2009) DNA-dependent detection of the grapevine fungal endophytes *Aureobasidium pullulans* and *Epicoccum nigrum*. Plant Dis 93:993–998

Martinuz A, Schouten A, Sikora RA (2012) Systemically induced resistance and microbial competitive exclusion: implications on biological control. Phytopathology 102:260–266

Mattos KA, Pádua VLM, Romeiro A, Hallack LF, Neves BC, Ulisses TMU, Barros CF, Todeschini AR, Previato JO, Mendonça-Previato L (2008) Endophytic colonization of rice by the diazotrophic bacterium *Burkholderia kururiensis* and its ability to enhance plant growth. An Acad Bras Cienc 80:477–493

McInroy JA, Kloepper JW (1995) Population dynamics of endophytic bacteria in field-grown sweet corn and cotton. Can J Microbiol 41:895–901

Mei C, Flinn BS (2010) The use of beneficial microbial endophytes for plant biomass and stress tolerance improvement. Recent Pat Biotechnol 4:81–95

Meji'a LC, Rojas EI, Maynard Z, Van Bael S, Arnold AE, Hebbar P, Samuels GJ, Robbins N, Herre EA (2008) Endophytic fungi as bio-control agents of *Theobroma cacao* pathogens. Biol Control 46:4–14

Mendoza AR, Sikora RA (2009) Biological control of *Radopholus similis* in banana by combined application of the mutualistic endophyte *Fusarium oxysporum* strain 162, the egg pathogen *Paecilomyces lilacinus* strain 251 and the antagonistic bacteria *Bacillus firmus*. Biocontrol 54:263–272

Mercier J, Jimenez JI (2004) Control of fungal decay of apples and peaches by the biofumigant fungus *Muscodor albus*. Postharvest Biol Technol 31:1–8

Mercier J, Manker D (2005) Bio-control of soil-borne diseases and plant growth enhancement in greenhouse soilless mix by the volatile-producing fungus *Muscodor albus*. Crop Prot 24:355–362

Migheli Q, Cacciola SO, Balmas V, Pane A, Ezra D, di San Lio Magnano G (2009) Mal secco disease caused by *Phoma tracheiphila*: a potential threat to lemon production worldwide. Plant Dis 93:852–867

Miles CO, Di Menna ME, Jacobs SWL, Garthwaite I, Lane GA, Prestidge RA, Marshall SL, Wilkinson HH, Schardl CL, Ball OJ-P, Larch GCM (1998) Endophytic fungi in indigenous Australasian grasses associated with toxicity to livestock. Appl Environ Microbiol 64:601–606

Mitchell AM, Strobel GA, Moore E, Robison R, Sears J (2010) Volatile antimicrobials from *Muscodor crispans*, a novel endophytic fungus. Microbiology 156:270–277

Moon CD, Miles CO, Jarlfors U, Schardl CL (2002) The evolutionary origins of three new *Neotyphodium* endophytes species from grasses indigenous to the Southern Hemisphere. Mycologia 94:694–711

Morath SU, Hung R, Bennett JW (2012) Fungal volatile organic compounds: a review with emphasis on their biotechnological potential. Fungal Biol Rev 26:73–83

Morse LJ, Day TA, Faeth SH (2002) Effect of *Neotyphodium* endophyte infection on growth and leaf gas exchange of Arizona fescue under contrasting water regimes. Environ Exp Bot 48:257–268

Morsy MR, Oswald J, He J, Tang Y, Roossinck MJ (2010) Teasing apart a three-way symbiosis: transcriptome analyses of *Curvularia protuberata* in response to viral infection and heat stress. Biochem Biophys Res Commun 401:225–230

Moy M, Belanger F, Duncan R, Freehoff A, Leary C, Meyer W, Sullivan R, White JF Jr (2000) Identification of epiphyllous mycelial nets on leaves of grasses infected by Clavicipitaceous endophytes. Symbiosis 28:291–302

M'Piga P, Belanger RR, Paulitz TC, Benhamou N (1997) Increased resistance to *Fusarium oxysporum* f. sp. *radicis-lycopersici* in tomato plants treated with the endophytic bacterium *Pseudomonas fluorescens* strain 63–28. Physiol Mol Plant Pathol 50:301–320

Mucciarelli M, Scannerini S, Bertea CM, Maffei M (2002) An ascomycetous endophyte isolated from *Mentha piperita* L.: biological features and molecular studies. Mycologia 94:28–39

Mucciarelli M, Scannerini S, Bertea CM, Maffei M (2003) *In vitro* and *in vivo* peppermint (*Mentha piperita*) growth promotion by nonmycorrhizal fungal colonization. New Phytol 158:579–591

Munn R, Cramer GR, Ball MC (1999) Interactions between rising CO_2, soil salinity and plant growth. In: Luo Y, Mooney HA (eds) Carbon dioxide and environmental stress. Academic, London, pp 139–167

Myers DF, Strobel GA (1983) *Pseudomonas syringae* as a microbial antagonist against *C. ulmi* in the apoplast of American elms. Trans Br Mycol Soc 80:389–394

Narisawa K, Kawamata H, Currah US, Hashiba T (2002) Suppression of *Verticillium* wilt in eggplant by some fungal root endophytes. Eur J Plant Pathol 108:103–109

Narisawa K, Usuki F, Hashiba T (2004) Control of *Verticillium* yellows in Chinese cabbage by the dark septate endophyte fungus LtVB3. Phytopathology 94:412–418

Nowak J, Asiedu SK, Lazarovits G, Pillay V, Stewart A, Smith C, Liu Z (1995) Enhancement of in vitro growth and transplant stress tolerance of potato and vegetable plantlets co-cultured with a plant growth promoting pseudomonad bacterium. In: Carre F, Chagcardieff P (eds) Proceedings of the international symposium on ecophysiology and photosynthetic in vitro cultures, Aix-en-Provence, 1995, pp 173–180

Nowak J, Asiedu SK, Lazarovits G (2003) Enhancement of in vitro growth and transplant stress tolerance of potato and vegetable plants co-cultured with a plant growth promoting rhizobacterium. In: Carre F, Chagvardieff P (eds) Ecophysiology and photosynthetic in vitro cultures. CEA, Aix-en-Provence, pp 173–180

Nuangmek W, Mc Kenzie EHC, Lumyong S (2008) Endophytic fungi from wild banana (*Musa acuminata* Colla) works against anthracnose disease caused by *Colletotrichum musae*. Res J Microbiol 3:358–374

Oelmüller R, Sherameti I, Tripathi S, Varma A (2009) *Piriformospora indica*, a cultivable root endophyte with multiple biotechnological applications. Symbiosis 49:1–17

Okon YPG (1985) *Azospirillum* as a potential inoculants for agriculture. Trends Biotechnol 3:223–228

Oliver JW, Schulze AE, Rohrbach BW, Fribourd HA, Ingle T, Waller JC (2000) Alterations in hemograms and serum biochemical analytes of steers after prolonged consumption of endophyte-infected tall fescue. J Anim Sci 78:1029–1035

Ownley BH, Gwinn KD, Vega FE (2010) Endophytic fungal entomopathogens with activity against plant pathogens: ecology and evolution. Biocontrol 55:113–128

Pal KK, Gardener BM (2006) Biological control of plant pathogens. The plant health instructor. APS Net. doi:10.1094/PHI-A-2006-1117-02

Pancher M, Ceol M, Corneo PE, Oliveira Longa CM, Yousaf S, Pertot I, Campisano A (2012) Fungal endophytic communities in grapevines (*Vitis vinifera* L.) respond to crop management. Appl Environ Microbiol 78:4308–4317

Park MS, Ahn J, Choi GJ, Choi YH, Jang KS, Kim JC (2010) Potential of the volatile-producing fungus *Nodulisporium* sp. CF016 for the control of postharvest diseases of apple. Plant Pathol J 26:253–259

Paterson J, Forcherio C, Larson B, Samford M, Kerley M (1995) The effects of Fescue toxicosis on beef cattle productivity. J Anim Sci 73:889–898

Paz Z, Burdman S, Gerson U, Sztejnberg A (2007) Antagonistic effects of the endophytic fungus *Meira geulakonigii* on the citrus rust mite *Phyllocoptruta oleivora*. J Appl Microbiol 103:2570–2579

Pearson WD (1989) The pasture mealy bug *Balanococcus poae* (Maskell) in Canterbury: a preliminary report. In: Stahle PP (ed) Proceedings of the 5th Australasian conference of grassland invertebrate ecology. D&D Printing, Victoria, pp 297–303

Pennell CGL, Popay AJ, Ball OJ-P, Hume DE, Baird DB (2005) Occurrence and impact of pasture mealybug *Balanococcus poae* and root aphid (*Aploneura lentisci*) on ryegrass (*Lolium* spp.) with and without infection by *Neotyphodium* fungal endophytes. N Z J Agric Res 48:329–338

Percival NS, Duder FR (1983) A comparison of perennial grasses under sheep grazing on the Central Plateau. Proc N Z Grassl Assoc 44:81–90

Petrini O (1986) Taxonomy of endophytic fungi of aerial plant tissues. In: Fokkema NJ, van den Huevel J (eds) Microbiology of the phyllosphere. Cambridge University Press, Cambridge, UK, pp 175–187

Petrini O (1996) Ecological and physiological aspects of host-specificity in endophytic fungi. In: Redlin SC, Carris LM (eds) Endophytic fungi in grasses and woody plants. APS Press, St. Paul, pp 87–100

Petroski RJ, Dornbos DL Jr, Powell RG (1990) Germination and growth inhibition of annual ryegrass (*Lolium multiflorum* L.) and alfalfa (*Medicago sativa*) by loline alkaloids and synthetic N-acetyl loline derivatives. J Agric Food Chem 38:1716–1718

Petroski R, Powell RG, Clay K (1992) Alkaloids of *Stipa robusta* (sleepygrass) infected with an *Acremonium* endophyte. Nat Toxins 1:84–88

Pham GH, Kumari R, Singh AN, Sachdev M, Prasad R, Kaldorf M (2004) Axenic cultures of *Piriformospora indica*. In: Varma A, Abbott L, Werner D, Hampp R (eds) Plant surface microbiology. Springer, Germany, pp 593–616

Pieterse CMJ, van Wees SCM, van Pelt JA, Knoester M, Laan R, Gerrits H, Weisbeek PJ, van Loon LC (1998) A novel signaling pathway controlling induced systemic resistance in *Arabidopsis*. Plant Cell 10:1571–1580

Pimenta SR, Moreira JF, Silva DA, Buyer JS, Janisiewicz WJ (2012) Endophytic fungi from plums (*Prunus domestica*) and their antifungal activity against *Monilinia fructicola*. J Food Prot 75:1883–1889

Pimilio AB, Rofi RD, Gambino MP, Mazzini CA, Debenedetti de Langenheim RT (1989) The lethal principle of *Poa huecu* (coirón blanco): a plant indigenous to Argentina. Toxicon 27:1251

Pinkerton BS, Rice JS, Undersander DJ (1990) Germination in *Festuca arundinacea* as affected by the fungal endophyte, *Acremonium coenophialum*. In: Quisenberry SS, Joost RE (eds) Proceedings of international symposium on *Acremonium*/grass interactions, Louisiana Agricultural Experimental Station, Baton Rouge, p 176. Agriculture ecosystem and environment, vol 44, p 283

Pirttilä AM (2001) Endophytes in the buds of Scots pine (*Pinus sylvestris* L.) Oulu University Press, Oulu. ISBN 951-42-6444-4. http://herkules.oulu.fi/isbn9514264444/

Pirttilä AM, Frank AC (eds) (2011) Endophytes of forest trees: biology and applications. Springer, Dordrecht. ISBN: 978-94-007-1598-1

Popay AJ, Prestidge RA, Rowan DD, Dymock JJ (1990) The role of *Acremonium lolii* mycotoxins in insect resistance of perennial ryegrass (*Lolium perenne*). In: Quisenberry SS, Joost RE (eds) Proceedings of international symposium on *Acremonium*/grass interactions, Baton Rouge, 1990, pp 44–48

Porras-Alfaro A, Bayman P (2011) Hidden fungi, emergent properties: endophytes and microbiomes. Annu Rev Phytopathol 49:291–315

Porter JK, Thompson FN Jr (1992) Effects of Fescue toxicosis on reproduction in livestock. J Anim Sci 70:1594–1603

Porter JK, Bacon CW, Robbins JD, Himmelsbach DS, Higman HC (1977) Indole alkaloids from *Balansia epichloe* (Weese). J Agric Food Chem 25:88–93

Porter JK, Bacon CW, Cutler HG, Arrendale RF, Robbins JD (1985) *In vitro* auxin production by *Balansia epichloë*. Phytochemistry 25:88–93

Powell RG, Petroski RJ (1992) Alkaloid toxins in endophyte-infected grasses. Nat Toxins 1:163–170

Prestidge RA, Gallagher RT (1985) Lolitrem B – a stem weevil toxin isolated from *Acremonium*-infected ryegrass. In: Proceedings of the 38th New Zealand weed and pest control conference, Hastings, 1985, pp 38–40

Prestidge RA, Barker GM, Pottinger RP (1991) The economic cost of Argentine stem weevil in pastures in New Zealand. In: Proceedings of the 44th New Zealand weed and pest control conference, Palmerston North, 1991, p 165

Redman RS, Freeman S, Clifron DR, Morrel J, Brown G, Rodriguez RJ (1999) Biochemical analysis of plant protection afforded by a nonpathogenic endophytic mutant of *Colletotrichum magna*. Plant Physiol 119:795–804

Redman RS, Sheehan KB, Stout RG, Rodriguez RJ, Henson JM (2002) Thermotolerance generated by plant/fungal symbiosis. Science 298:1581

Redman RS, Kim YO, Woodward CJ, Greer C, Espino L, Doty SL, Rodriguez RJ (2011) Increased fitness of rice plants to abiotic stress via habitat adapted symbiosis: a strategy for mitigating impacts of climate change. PLoS One 6:e14823

Reinhold-Hurek B, Hurek T (2011) Living inside plants: bacterial endophytes. Curr Opin Plant Biol 14:435–443

Ren AZ, Gao YB, Zhang L, Xie F (2006) Effects of cadmium on growth parameters of endophyte-infected and endophyte-free ryegrass. J Plant Nutr Soil Sci 169:857–860

Ren A, Li C, Gao Y (2011) Endophytic fungus improves growth and metal uptake of *Lolium arundinaceum* Darbyshire ex. Schreb Int J Phytorem 13(3):233–243

Rice JS, Pinkerton BW, Undersander DJ (1990) Imbibition of tall fescue as affected by *Acremonium coenophialum*. In: Quisenberry SS, Joost RE (eds) Proceedings of international symposium of *Acremonium*/grass interactions, Louisiana Agricultural Experiment Station, Baton Rouge, 1990, p 185

Richardson MD, Hill NS, Hoveland CS (1990) Rooting patterns of endophyte infected tall fescue grown under drought stress. Agron Abs 81:129

Richardson MD, Chapman GW Jr, Hoveland CS, Bacon CW (1992) Sugar alcohols in endophyte-infected tall fescue under drought. Crop Sci 32:1060–1061

Richardson MD, Bacon CW, Hill NS, Hinton DM (1993) Growth and water relations of *Acremonium coenophialum*. In: Hume DE, Latch GCM, Easton HS (eds) Proceeding of the second international symposium on *Acremonium*/grass interactions. AgResearch, Grassland Research Centre, Palmerston North, pp 181–184

Richmond DS, Shetlar DJ (2000) Hairy chinch bug (Hemiptera: Lygaeidae) damage, population density, and movement in relation to the incidence of perennial ryegrass infected by *Neotyphodium* endophyte. J Econ Entomol 93:L1167–L1172

Riedell WE, Kieckhefer RE, Petroski RJ, Powell RG (1991) Naturally occurring and synthetic loline alkaloids derivatives: insect feeding behavior modification and toxicity. J Entomol Sci 26:122–129

Riga E, Lacey L, Guerra N (2008) *Muscodor albus*, a potential bio-control agent against plant-parasitic nematodes of economically important vegetable crops in Washington State, USA. Biocontrol 45:380–385

Roberts CA, Marek SM, Niblack TL, Karr AL (1992) Parasitic *Meloidogyne* and *Acremonium* increase chitinase in tall fescue. J Chem Ecol 18:1007–1116

Rodriguez R, Redman R (2008) More than 400 million years of evolution and some plants still can't make it on their own: plant stress tolerance via fungal symbiosis. J Exp Bot 59:1109–1114

Rodriguez RJ, Redman RS, Henson JM (2004) The role of fungal symbioses in the adaptation of plants to high stress environments. Mitig Adapt Strat Glob Chang 9:261–272

Rodriguez RJ, White JF, Arnold AE, Redman RS (2008a) Fungal endophytes: diversity and functional roles. New Pathol 182:314–330

Rodriguez RJ, Henson K, Van Volkenburgh E, Hoy M, Wright L, Beckwith F, Kim YO, Redman RS (2008b) Stress tolerance in plants via habitat-adapted symbiosis. ISME J 2:404–416

Rodriguez RJ, Freeman DC, McArthur ED, Kim YO, Redman RS (2009a) Symbiotic regulation of plant growth, development and reproduction. Commun Integr Biol 2:141–143

Rodriguez RJ, White JF Jr, Arnold AE, Redman RS (2009b) Fungal endophyte: diversity and functional roles. New Phytol 182:314–330

Rodriguez Estrada AE, Jonkers W, Kistler HC, May G (2012) Interactions between *Fusarium verticillioides*, *Ustilago maydis*, and *Zea mays*: and endophyte, a pathogen, and their shared plant host. Fungal Genet Biol 49:578–587

Romão-Dumaresq AS, de Araújo WL, Talbot NJ, Thornton CR (2012) RNA interference of endochitinases in the sugarcane endophyte *Trichoderma virens* 223 reduces its fitness as a bio-control agent of pineapple disease. PLoS One 7(10):e47888

Rosenblueth M, Martinez-Romero E (2006) Bacterial endophytes and their interactions with hosts. Mol Plant Microbe Interact 19:827–837

Rowan DD (1993) Lolitrems, paxilline and peramine: mycotoxins of the ryegrass/endophyte interaction. Agric Ecosyst Environ 44:103–122

Rowan DD, Latch GCM (1994) Utilization of endophyte-infected perennial ryegrasses for increased insect resistance. In: Bacon CW, White JF Jr (eds) Biotechnology of endophytic fungi of grasses. CRC Press, Boca Raton, pp 169–183

Rowan DD, Hunt MB, Gaynor DL (1986) Peramine, a novel insect feeding deterrent from ryegrass infected with the endophyte *Acremonium lolii*. J Chem Soc Chem Commun 1986:935–936

Rubini MR, Silva-Ribeiro RT, Pomella AWV, Maki CS, Ara'ujo WL et al (2005) Diversity of endophytic fungal community of cacao (*Theobroma cacao* L.) and biological control of *Crinipellis perniciosa*, causal agent of witches' broom disease. Int J Biol Sci 1:24–33

Rudgers JA, Miller TEX, Ziegler SM, Craven KD (2012) There are many ways to be mutualist: endophytic fungus reduces plant survival but increases population growth. Ecology 93:565–574

Ruiz-Lozano JM, Azcon R, Palma JM (1996) Superoxide dismutase activity in arbuscular mycorrhizal *Lactuca sativa* plants subjected to drought stress. New Phytol 13:327–333

Ryan PL, Bennett-Wimbush K, Vaala WE, Bagnell CA (2001) Systemic relaxin in pregnant pony mares grazed on endophyte-infected fescue: effects and fluphenazine treatment. Theriogenology 56:471–483

Ryan RP, Germaine K, Franks A, Ryan DJ, Dowling DN (2008) Bacterial endophytes: recent developments and applications. FEMS Microbiol Lett 278:1–9

Ryu CM, Farag MA, Hu CH, Reddy MS, Wei H-X, Pare PW, Kloepper JW (2003) Bacterial volatiles promote growth in *Arabidopsis*. Proc Natl Acad Sci U S A 100:4927–4932

Ryu CM, Farag MA, Hu CH, Munagala SR, Kloepper JW, Pare PW (2004) Bacterial volatiles induce systemic resistance in *Arabidopsis*. Plant Physiol 134:1017–1026

Saari S, Faeth SH (2012) Hybridization of *Neotyphodium* endophytes enhances competitive ability of the host grass. New Phytol 195:231–236

Saari S, Helander M, Faeth SH, Saikkonen K (2010) The effects of endophytes on seed production and seed predation of tall fescue and meadow fescue. Microb Ecol 60:928–934

Saikkonen K, Ion D, Gyllenberg M (2002) The persistence of vertically transmitted fungi in grass metapopulations. Proc R Soc B Biol Sci 269:1397–1403

Saker KE, Allen VG, Kalnitsky J, Thatcher CD, Swecker WS Jr, Fontenot JP (1998) Monocyte immune cell response and copper status in beef steers that grazed endophyte-infected tall fescue. J Anim Sci 76:2694–2700

Saravanan VS, Madhaiyan M, Osborne J, Thangaraju M, Sa TM (2008) Ecological occurrence of *Gluconacetobacter diazotrophicus* and nitrogen-fixing *Acetobacteraceae* members: their possible role in plant growth promotion. Microb Ecol 55:130–140

Sasan RK, Bidochka MJ (2012) The insect-pathogenic fungus *Metarhizium robertsii* (Clavicipitaceae) is also an endophyte that stimulates plant root development. Am J Bot 99:101–107

Schäfer P, Pfiffi S, Voll LM, Zajic D, Chandler PM, Waller F, Scholz U, Pons-Kuhnemann J, Sonnewald S, Sonnewald U, Kogel KH (2009) Manipulation of plant innate immunity and gibberellin as factor of compatibility in the mutualistic association of barley roots with *Piriformospora indica*. Plant J 59:461–474

Schardl CL (2001) *Epichloë festucae* and related mutualistic symbionts of grasses. Fungal Genet Biol 33:69–82

Schardl CL, Leuchtmann A, Spiering M (2004) Symbioses of grasses with seedborne fungal endophytes. Annu Rev Plant Biol 55:315–340

Schena L, Ippolito A, Zahavi T, Cohen L, Nigro F, Droby S (1999) Genetic diversity and bio-control activity of *Aureobasidium pullulans* isolates against postharvest rots. Postharvest Biol Technol 17:189–199

Schena L, Nigro F, Pentimone I, Ligorio A, Ippolito A (2003) Control of postharvest roots of sweet cherries and table grapes with endophytic isolates of *Aureobasidium pullulans*. Postharvest Biol Technol 30:209–220

Schulz B, Rommert AK, Dammann U, Aust HJ, Strack D (1999) The endophyte-host interaction: a balanced antagonism? Mycol Res 10:1275–1283

Seo DJ, Nguyen DM, Song YS, Jung WJ (2012) Induction of defense response against *Rhizoctonia solani* in cucumber plants by endophytic bacterium *Bacillus thuringiensis* GS1. J Microbiol Biotechnol 22:407–415

Serfling A, Wirsel SG, Lind V, Deising HB (2007) Performance of the bio-control fungus *Piriformospora indica* on wheat under greenhouse and field conditions. Phytopathology 97:523–531

Sessitsch A, Coenye T, Sturz AV, Vandamme P, Barka EA, Salles JF, Elsas JD, Faure D, Reiter B, Glick BR, Wang-Pruski G, Nowak J (2005) *Burkholderia phytofirmans* sp. nov., a novel plant-associated bacterium with plant-beneficial properties. Int J Syst Evol Microbiol 55:1187–1192

Sevilla M, Burris RH, Gunapala N, Kennedy C (2001) Comparison of benefit to sugarcane plant growth and $15N_2$ incorporation following inoculation of sterile plants with *Acetobacter diazotrophicus* wild-type and Nif-mutants stains. Mol Plant Microbe Interact 14:358–366

Shahabivand S, Maivan HZ, Goltapeh EM, Sharifi M, Aliloo AA (2012) The effects of root endophyte and arbuscular mycorrhizal fungi on growth and cadmium accumulation in wheat under cadmium toxicity. Plant Physiol Biochem 60:53–58

Sharma VK, Nowak J (1998) Enhancement of *Verticillium* wilt resistance in tomato transplants by in vitro coculture of seedlings with a plant growth-promoting rhizobacterium (*Pseudomonas* sp. strain PsJN). Can J Microbiol 44:528–536

Sherameti I, Shahollari B, Venus Y, Altschmied L, Varma A, Oelmüller R (2005) The endophytic fungus *Piriformospora indica* stimulates the expression of nitrate reductase and the starch-degrading enzyme glucan-water dikinase in tobacco and *Arabidopsis* roots through a homeodomain transcription factor which binds to a conserved motif in their promoters. J Biol Chem 280:2641–2647

Sherameti I, Tripathi S, Varma A, Oelmüller R (2008) The root-colonizing endophyte *Piriformospora indica* confers drought tolerance in *Arabidopsis* by stimulating the expression of drought stress-related genes in leaves. Mol Plant Microbe Interact 21:799–807

Sherwood M, Carroll G (1974) Fungal succession on needles and young twigs of old-grown Douglas fir. Mycologia 66:499–506

Shittu HO, Castroverde DC, Nazar RN, Robb J (2009a) Plant-endophyte interplay protects tomato against a virulent *Verticillium*. Planta 229:415–426

Shittu HO, Shakir AS, Nazar RN, Robb J (2009b) Endophyte-induced *Verticillium* protection in tomato is range-restricted. Plant Signal Behav 4:160–161

Siegel MR, Latch GCM (1991) Expression of antifungal activity in agar culture by isolates of grass endophytes. Mycologia 83:529–537

Siegel MR, Latch GCM, Bush LP, Fannin FF, Rowan DD, Tapper BA, Bacon CW, Johnson MC (1990) Fungal endophyte-infected grasses: alkaloid accumulation and aphid response. J Chem Ecol 16:3301–3315

Sikora RA, Pocasangre L, zum Felde A, Niere B, Vu TT, Dababat AA (2008) Mutualistic endophytic fungi and *in planta* suppressiveness to plant parasitic nematodes. Biol Control 46:15–23

Simpson WR, Schmid J, Singh J, Faville MJ, Johnson RD (2012) A morphological change in the fungal symbionts *Neotyphodium lolii* induces dwarfing in its host plant *Lolium perenne*. Fungal Biol 116:234–240

Singh LP, Gill SS, Tuteja N (2011a) Unraveling the role of fungal symbionts in plant abiotic stress tolerance. Plant Sig Behav 6:175–191

Singh SK, Stroble GA, Knighton B, Geary B, Sears J, Ezra D (2011b) An endophytic *Phomopsis* sp. bioactivity and fuel potential with its volatile organic compounds. Microb Ecol 61:729–739

Sirrenberg A, Goebel C, Grond S, Czempinski N, Ratzinger A, Karlovsky P, Santos P, Feussner I, Pawlowski K (2007) *Piriformospora indica* affects plant growth by auxin production. Physiol Plant 131:581–589

Smith CJ (1996) Tansley Review No. 86 accumulation of phytoalexins: defence mechanism and stimulus response system. New Phytol 132:1–45

Soleimani M, Hajabbasi MA, Afyuni M, Mirlohi A, Borggaard OK, Holm PE (2010a) Effect of endophytic fungi on cadmium tolerance and bioaccumulation by *Festuca arundinacea* and *Festuca pratensis*. Int J Phytoremediation 12:535–549

Soleimani M, Afyuni M, Hajabbasi MA, Nourbakhsh F, Sabzalian MR, Christensen JH (2010b) Phytoremediation of an aged petroleum contaminated soil using endophyte infected and non-infected grasses. Chemosphere 81:1084–1090

Sopalun K, Strobel GA, Hess WM, Worapong J (2003) A record of *Muscodor albus*, an endophyte from *Myristica fragrans*, in Thailand. Mycotaxon 88:239–247

Springer WC (1997) Allelopathic effects of tall fescue. In: Brink GE (ed) Proceedings of southern forage crop improvement conference 53rd, Fort Worth, 1997, pp 25–33

Stinson M, Ezra D, Hess WM, Sears J, Strobel G (2003) An endophytic *Gliocladium* sp. of *Eucryphia cordifolia* producing selective volatile antimicrobial compounds. Plant Sci 165:913–922

Stone JK, Bacon CW, White JF (2000) An overview of endophytic microbes: endophytism defined. In: Bacon CW, White JF (eds) Microbial endophytes. Marcel Dekker, Inc, New York, pp 3–30

Stone JK, Polishook JD, White JRJ (2004) Endophytic fungi. In: Mueller G, Bills GF, Foster MS (eds) Biodiversity of fungi: inventory and monitoring methods. Elsevier, Burlington, pp 241–270

Strobel GA (2006) *Muscodor albus* and its biological promise. J Ind Microbiol Biotechnol 33:514–522

Strobel GA, Stierle A, Stierle D, Hess WM (1993) Taxomyces andreanae a proposed new taxon for a bulbilliferous hyphomycete associated with Pacific yew. Mycotaxon 47:71–78

Strobel GA, Dirkse E, Sears J, Markworth C (2001) Volatile antimicrobials from *Muscodor albus* a novel endophytic fungus. Microbiology 147:2943–2950

Strobel G, Manker D, Mercier J (2010). United States Patent 7,754,203-endophytic fungi and methods of use. AgraQuest, Inc., Davis

Strobel G, Singh SK, Riyaz-Ul-Hassan S, Mitchell AM, Geary B, Sears J (2011) An endophytic/pathogenic *Phoma* sp. from creosote bush producing biologically active volatile compounds having fuel potential. FEMS Microbiol Lett 320:87–94

Sturz AV, Christie B, Nowak J (2000) Bacterial endophytes: potential role in developing sustainable system of crop production. Crit Rev Plant Sci 19:1–30

Suman A, Gaur A, Shrivastava AK, Yadav RL (2005) Improving sugarcane growth and nutrient uptake by inoculating *Gluconacetobacter diazotrophicus*. Plant Growth Regul 47:155–162

Sumarah MW, Puniani E, Sørensen D, Blackwell BA, Miller JD (2010) Secondary metabolites from anti-insect extracts of endophytic fungi isolated from *Picea rubens*. Phytochemistry 71:760–765

Sun C, Johnson JM, Cai D, Sherameti I, Oelmüller R, Lou B (2010) *Piriformospora indica* confers drought tolerance in Chinese cabbage leaves by stimulating antioxidant enzymes, the expression of drought-related genes and the plastid-localized CAS protein. J Plant Physiol 167:1009–1017

Sutherland BL, Hoglund JH (1989) Effect of ryegrass containing the endophyte (*Acremonium lolii*) on the performance of associated white clover and subsequent crops. Proc N Z Grassl Assoc 50:265–269

Suwannarach N, Kumla J, Bussaban B, Nuangmek W, Matsui K, Lumyong S (2013) Biofumigation with the endophytic fungus *Nodulisporium* spp. CMU-UPE34 to control postharvest decay of citrus fruit. Crop Prot 45:63–70

Taghavi S, van der Lelie D, Hoffman A, Zhang YB, Walla MD, Vangronsveld J, Newman L, Monchy S (2010) Genome sequence of plant growth promoting endophytic bacterium *Enterobacter* sp. 638. Plos Genet 6:e1000943

Tanaka A, Tapper BA, Popay AJ, Parker EJ, Scott B (2005) A symbiosis expressed non-ribosomal peptide synthetase from mutualistic fungal endophyte of perennial ryegrass confers protection to the symbiotum from insect herbivory. Mol Microbiol 57:1036–1050

Tanaka A, Takemoto D, Chujo T, Scott B (2012) Fungal endophytes of grasses. Curr Opin Plant Biol 15: 462–468

Tapper B, Lane GA (2004) Janthitrems found in a *Neotyphodium* endophyte of perennial ryegrass. In: Kallenbach R, Rosenkrans CF Jr, Lock TR (eds) 5th international endophyte symposium, University of Arkansa, Fayetteville

Tintjer T, Rudgers JA (2006) Grass-herbivore interactions altered by strains of a native endophyte. New Phytol 170:513–521

Tomsheck AR, Stroble GA, Booth E, Geary B, Spakowicz D, Knighton B, Floerchinger C, Sears J, Liarzi O, Ezra D (2010) Hypoxylon sp., an endophyte of *Persea indica*, producing 1,8-cineole and other bioactive volatiles with fuel potential. Microb Ecol 60:903–914

Tor-Agbidye J, Blythe LL, Craig AM (2001) Correlation of endophyte toxins (ergovaline and lolitrem B) with clinical disease: fescue foot and perennial ryegrass staggers. Vet Hum Toxicol 43:140–146

Tudzynski B, Sharon A (2002) Biosynthesis, biological role and application of fungal phyto-hormones. In: Osiewacz HD (ed) The mycota X. Industrial applications. Springer, Berlin, pp 183–212

Uchitel A, Omacini M, Chaneton EJ (2011) Inherited fungal symbionts enhance establishment of an invasive annual grass across successional habitats. Oecologia 165:465–475

Vadassery J, Ritter C, Venus Y, Camehl I, Varma A, Shahollari B, Novak O, Strand M, Ludwig-Muller J, Oelmuller R (2008) The role of auxins and cytokinins in the mutualistic interaction between *Arabidopsis* and *Piriformospora indica*. Mol Plant Microbe Interact 21:1371–1383

Vadassery J, Ranf S, Drzewiecki C, Mithofer A, Mazars C, Scheel D, Lee J, Oelmuller R (2009) A cell wall extract from the endophytic fungus *Piriformospora indica* promotes growth of *Arabidopsis* seedlings and induces intracellular calcium elevation in roots. Plant J 59:193–206

Valdez Barillas JR, Paschke MW, Ralphs MH, Child RD (2007) White locoweed toxicity is facilitated by a fungal endophyte and nitrogen-fixing bacteria. Ecology 88:1850–1856

van Berkum P, Bohlool BB (1980) Evaluation of nitrogen fixation by bacteria in association with roots of tropical grasses. Microbiol Rev 44:491–517

van Loon LC (2007) Plant responses to plant growth-promoting rhizobacteria. Eur J Plant Pathol 119:243–254

Verma S, Varma A, Rexer KH, Hassel A, Kost G, Sarbhoy A, Bisen P, Bütenhorn B, Franken P (1998) *Piriformospora indica*, gen. et sp. nov., a new root-colonizing fungus. Mycologia 90:896–903

Vu T, Hauschild R, Sikora RA (2006) *Fusarium oxysporum* endophytes induced systemic resistance against *Radopholus similis* on banana. Nematology 8:847–852

Wäli PR, Helander M, Nissinen O, Saikkonen K (2006) Susceptibility of endophyte-infected grasses to winter pathogens (snow molds). Can J Bot 84:1043–1051

Waller F, Achatz B, Baltruschat H, Fodor J, Becker K, Fisher M, Heier T, Huckelhoven R, Neumann C, von Wettstein D, Franken P, Kogel KH (2005) The endophytic fungus *Piriformospora indica* reprograms barley to salt-stress tolerance, disease resistance, and higher yield. Proc Natl Acad Sci U S A 102:13386–13391

Waller F, Mukherjee K, Deshmukh SD, Achatz B, Sharma M, Schäfer P, Kogel KH (2008) Systemic and local modulation of plant responses by *Piriformospora indica* and related *Sebacinales* species. J Plant Physiol 165:60–70

Wang K, Conn K, Lazarovits G (2006) Involvement of quinolinate phophoribosyl transferase in promotion of potato growth by a *Burkholderia* Strain. Appl Environ Microbiol 72:760–768

Waqas M, Khan AL, Kamran M, Hamayun M, Kang SM, Kim YH, Lee IJ (2012) Endophytic fungi produce gibberellins and indoleacetic acid and promotes host-plant growth during stress. Molecules 17: 10754–10773

Webster G, Gough C, Vasse J, Batchelor CA, O'Callaghan KJ, Kothari SL, Davey MR (1997) Interactions of rhizobia with rice and wheat. Plant Soil 194:115–122

Welbaum G, Sturz AV, Dong Z, Nowak J (2004) Managing soil microorganisms to improve productivity of agro-ecosystems. Crit Rev Plant Sci 23:175–193

Welty RE, Barker RE, Azevedo MD (1993) Response of field-grown tall fescue infected by *Acremonium coenophialum* to *Puccinia graminis* subsp. *Graminicola.* Plant Dis 77:574–575

West CP (1994) Physiology and drought tolerance of endophyte-infected grasses. In: Bacon CW, White JF Jr (eds) Biotechnology of endophytic fungi of grasses. CRC Press, Boca Raton, pp 87–99

White JF Jr, Cole GT (1985) Endophyte-host associations in forage grasses. III. In vitro inhibition of fungi by *Acremonium coenophialum*. Mycologia 77:487–489

White RH, Engelke MC, Morton SJ, Johnson-Cicalese JM, Ruemmele BA (1992) *Acremonium* endophyte effects on tall fescue drought tolerance. Crop Sci 32:1392–1396

Wicklow DT, Poling SM (2009) Antimicrobial activity of pyrrocidines from *Acremonium zeae* against endophytes and pathogens of maize. Phytopathology 99:109–115

Wicklow DT, Roth S, Deyrup ST, Gloer JB (2005) A protective endophyte of maize: *Acremonium zeae* antibiotics inhibitory to *Aspergillus flavus* and *Fusarium verticillioides*. Mycol Res 109:610–618

Wilkinson HH, Siegel MR, Blankenship JD, Mallory AC, Bush LP, Schardl CL (2000) Contribution of fungal loline alkaloids to protection from aphids in grass-endophyte mutualism. Mol Plant Microbe Interact 13:1027–1033

Wille PA, Aeschbacher RA, Boller T (1999) Distribution of fungal endophyte genotypes in doubly infected host grasses. Plant J 18:349–358

Wolfe BA, Bush M, Monfort SL, Mumford SL, Pessier A, Montali RJ (1998) Abdominal lipomatosis attributed to tall fescue toxicosis in deer. J Am Vet Med Assoc 213(12):1783–1786

Woropong J, Strobel GA, Ford EJ, Li JY, Baird G, Hess WM (2001) *Muscodor albus* anam. nov. an endophyte from *Cinnamomum zeylanicum*. Mycotaxon 79:67–79

Woropong J, Strobel GA, Daisy B, Castillo U, Baird G, Hess WM (2002) *Muscodor roseus* anna. nov. an endophyte from *Grevillea pteridifolia*. Mycotaxon 81:463–475

Xin G, Glawe D, Doty SL (2009a) Characterization of three endophytic indole-3-acetic acid-producing yeasts occurring in Populus trees. Mycol Res 113:973–980

Xin G, Zhang G, Kang JW, Staley JT, Doty SL (2009b) A diazotrophic, indole-3-acetic acid-producing endophyte from wild cottonwood. Biol Fertil Soils 45:669–674

Yang J, Kloepper JW, Ryu CM (2009) Rhizosphere bacteria help plants tolerate abiotic stress. Trends Plant Sci 14:1–4

Yanni YG, Rizk RY, Corich V, Squartini A, Ninke K, Philip-Hollingsworth S, Orgambide F, de Bruijn F, Stolzfus J, Buckley D, Schmidt TM, Mateos PF, Dazzo FB (1997) Natural endophytic association between *Rhizobium leguminosarum* bv. *trifolii* and rice roots and assessment of its potential to promote rice growth. Plant Soil 194:99–114

Yasuda M, Isawa T, Shinozaki S, Minamisawa K, Nakashita H (2009) Effects of colonization of a bacterial endophyte, *Azospirillum* sp. B510, on disease resistance in rice. Biosci Biotechnol Biochem 73(12):2595–2599

Yoshihara T, Togiya S, Koshino H, Sakamura S, Shimanuki T, Sato T, Tajimi A (1985) Three fungitoxic sesquiterpenes from stromata of *Epichloë typhina*. Tetrahedron Lett 26:5551–5554

Yue Q, Miller CJ, White JF Jr, Richardson MD (2000) Isolation and characterization of fungal inhibitors from *Epichloë festucae*. J Agric Food Chem 48:4687–4692

Yue Q, Wang CL, Gianfagna TJ, Meyer WA (2001) Volatile compounds of endophyte-free and infected tall fescue (*Festuca arundinacea* Schreb). Phytochemistry 58:935–941

Zaurov DE, Bonos S, Murphy JA, Richardson M, Belanger FC (2001) Endophyte infection can contribute to aluminum tolerance in fine fescues. Crop Sci 41:1981–1984

Zhang CL, Wang GP, Mao LJ, Komon-Zelazowska M, Yuan ZL, Lin FC, Druzhinina IS, Kubicek CP (2010a) *Muscodor fengyangensis* sp. nov. from southeast China: morphology, physiology and production of volatile compounds. Fungal Biol 114:797–808

Zhang XX, Li CJ, Nan ZB (2010b) Effects of cadmium stress on growth and anti-oxidative systems in *Achnatherum inebrians* symbiotic with *Neotyphodium gansuense*. J Hazard Mater 175:703–709

Part VI

Endophytes and Cancer

19 Implication of Endophytic Metabolite and Their Derivatives in Cancer Chemotherapy: A Prospective Study

Pragya Mishra, Raghvendra Raman Mishra, Mallika Tiwari, Parjanya Shukla, Archana Singh, and Hari S. Shukla

Abstract

Incidence of cancer keep increasing worldwide may be due to genetic aberration, environmental effect, diet, socioeconomic factors, and various types of infections. Our previous studies revealed an association of *Helicobacter pylori* (*H. pylori*) and their species, *Salmonella* Typhi (*S.* Typhi), and *Mycobacterium* with various gastrointestinal tract (GI) cancers including oral, oropharyngeal, esophageal, gastric, gallbladder, pancreatic, and anal-canal cancers. We experience that poor cure rate is reported due to failure of conventional medicine, drug resistance, and failure to know the exact cause. As we are a group of oncologist and basic researcher, it is our experience that surgical procedures and chemotherapy are better adjuvant therapeutic option for cancer treatment. The area of chemotherapy is enhanced, but basic foundation is devised from natural products, which are used directly or as synthetic derivatives as stand-alone or in different combinations. Microorganisms, either bacteria or fungi that live inside plant tissue (endophytes) system, are big source of natural antimicrobial compound. It is known that endophytic alkaloids, taxoids, podophyllotoxins, etc., have an antineoplastic activity. Keeping these facts in mind, this chapter

P. Mishra
Ranjan Physiology and Biochemistry Laboratory, Department of Botany, Centre of Food Technology, University of Allahabad, Allahabad 211002, India

R.R. Mishra
Department of Microbiology, Institute of Medical Sciences, Banaras Hindu University, Varanasi 221005, Uttar Pradesh, India

M. Tiwari • H.S. Shukla (✉)
Department of Surgical Oncology, Institute of Medical Sciences, Banaras Hindu University, Varanasi 221005, Uttar Pradesh, India
e-mail: harishukla@usa.net

P. Shukla
Department of Pharmaceutical Sciences, Faculty of Health Science, Sam Higginbottom Institute of Agriculture Technology and Sciences – Deemed University, Allahabad, India

A. Singh
Department of Botany, Faculty of Science, Banaras Hindu University, Varanasi, Uttar Pradesh, India

V.C. Verma and A.C. Gange (eds.), *Advances in Endophytic Research*,
DOI 10.1007/978-81-322-1575-2_19, © Springer India 2014

points out the active exploration and implication of endophytic metabolite and their derivatives in cancer chemotherapy in near future. Obtained data were analyzed and result showed that the endophytic metabolites may be potential source of newer cancer chemotherapeutic drugs. It concluded that in the field of cancer chemotherapy, search for novel drugs from endophytic origin is still a priority.

1 Introduction

More than hundred types of cancers exist and they can affect any part of the body. In the present scenario approximately 70 % of all cancer deaths occur in low- and middle-income countries. Worldwide, the five most common types of cancers associated with lung, stomach, liver, colorectal, and esophagus are responsible for major death cases. The curative procedure is either chemotherapy-based surgery or radiotherapy. The cancer chemotherapy based on natural origin is used as a potential therapeutic application for cancer treatment from prehistoric time till date. Natural products from several sources have been used for cancer treatment as alone as well as in combination to combat cancer with the emergence of taxol (Fig. 19.1). The main sources of these successful compounds are microbes such as bacteria, fungi, and plants of different ecosystem. These microbes give out a major source of natural products with cancer therapeutic activity. Plant alkaloids, taxoids, and podophyllotoxins are biologically active metabolites of endophytes, which can be used effectively in cancer treatment. The search for novel cancer chemotherapeutic drugs is still a priority due to the rapid development of resistance pattern. In addition, the high toxicity, usually associated with some cancer chemotherapy drugs, and their undesirable side effects increase the demand for novel antitumor drugs active against untreatable tumor, with fewer side effects and with greater therapeutic efficiency. In this context there are many research works and reviews available, but among them only a few are relevant to its implication with cancer chemotherapy.

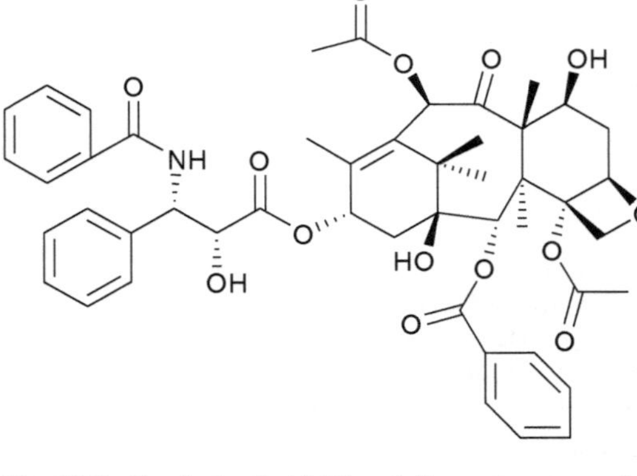

In our experience the study area, Varanasi, U. P., India, had a higher load of cancer patients (Fig. 19.2) and low rate of therapeutics (Tewari et al. 2008a, b). This chapter points out recent update which focuses in the identification, production, and implication of the cancer chemotherapeutic compounds in the future. We will also try to correlate the present cancer burden and role of endophytic metabolite and their derivatives in cancer chemotherapy.

Fig. 19.1 Taxol: the first billion dollar anticancer molecule of endophytic origin

2 Cancer Burden and Etiology

Cancer is multifactorial and may occur due to result of multistep pathway. It seems a high load of new cases and deaths will be added due to cancer alone (Table 19.1) (Parkin et al. 2001). The variation in the incidence rate of cancer is influenced by environmental and socioeconomic factors with genetic

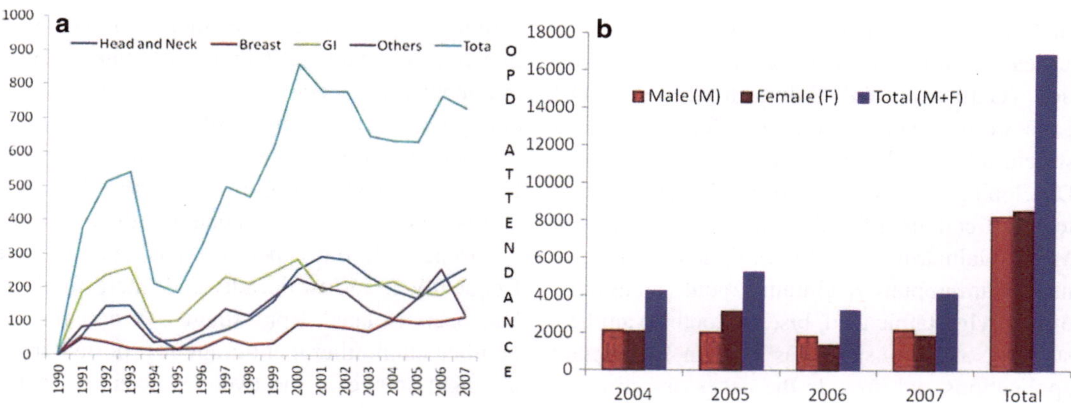

Fig. 19.2 (a) Estimated burden and observation status of four common cancers in Varanasi region of north India since 1990–2007. (b) Male and female distribution of Out Patient Department, Sir Sunder Lal Hospital, Institute of Medical Sciences, Surgical Oncology Wing

Table 19.1 Fifteen common cancers of worldwide with their estimated numbers of new cases and deaths in male and female under thousand

S. no	Site of cancer	Male ($n = 1{,}000$) Incidence	Male Mortality	Female ($n = 1{,}000$) Incidence	Female Mortality
1	Lung	902	810	337	293
2	Breast	–	–	1,050	370
3	Colorectal	499	255	446	234
4	Stomach	558	405	318	241
5	Liver	398	384	166	165
6	Prostrate	543	204	–	–
7	Cervix	–	–	471	233
8	Esophagus	279	227	133	111
9	Bladder	260	99	76	33
10	Non-Hodgkin lymphoma	167	93	121	68
11	Leukemia	144	109	113	86
12	Oral cavity	170	81	97	47
13	Pancreas	116	112	101	101
14	Kidney	119	57	71	34
15	Ovary	–	–	192	114

After Parkin et al. (2001)

aberration (Maurya et al. 2010), epigenetic changes (Tewari et al. 2013a, b), overexpression/suppress expression of gene (Tewari et al. 2013a, b), and various types of infection such as *Helicobacter pylori* (*H. pylori*) and their species (Mishra et al. 2010, 2011), *Salmonella* Typhi (*S.* Typhi), *Mycobacterium* (Tewari et al. 2009, 2010), etc.

3 Conventional and Newer Approaches for Cancer Treatment

Conventional cancer chemotherapy has the limitation of multidrug resistance (MDR) caused by overexpression of integral membrane transporters, such as P-gp, which can efflux intracellular anticancer drugs thus decreasing drug accumulation (Galmarini and Galmarini 2003). MDR cells are resistant to cytotoxic effects of various structurally unrelated chemotherapeutic agents. Developing new anticancer drugs that are efficient to MDR cells is a feasible strategy to overcome MDR (Galmarini et al. 2012). Folic acid antagonist, 4-aminopteroyl-glutamic acid, antagonist of pteroylglutamic acid, bis(2-haloethyl) amines, sulfides, and "mustard gas" are in therapeutic applications and provide the necessary stimulus for the development of other anticancer agents (Farber and Diamond 1948; Gilman 1946; Law 1951; Philips 1950; Seeger et al. 1947; Stevens et al. 1950). Few newer approaches such as dendritic cell therapy (Tewari et al. 2012) and endophytic metabolite-based cancer therapy are also under consideration (Philips 1950).

Since half of the century, natural products are only compounds used to serve us in cancer therapy. The main sources of these successful compounds are microbes and plants from various parts of the ecosystem. The major source of natural products with antitumor activity evaluated with the emergence of antibiotics. Few alkaloids, taxoids, and podophyllotoxins are also biological metabolites that can be obtained from the plant systems which can be used for effective cancer treatment. Few polymer conjugated to the anticancer protein neocarzinostatin is accumulated more in tumor tissues than did neocarzinostatin. This tumoritropic accumulation was studied with radioactive (51Cr-labeled) proteins of various molecular sizes (M, 12,000–160,000) and other properties in combination of dye-complexed serum albumin to visualize the accumulation in tumors of tumor-bearing animal model indicating macromolecular therapeutics approach in cancer chemotherapy is established in spite of others (Matsumura and Maeda 1986).

4 Recognition and Emergence of Endophytic Metabolite

Endophytes reside inside the leaf, stem, or roots of higher plants. They are the source of secondary metabolites with promising chemotherapeutic activity. Fewer cancer treatment options with higher side effects and high cost make cancer treatment very difficult. Endophytic fungi are an emerging potential source of chemotherapeutic compounds from different chemical classes. From endophytic origin more than hundred compounds have been evaluated with significant cytotoxicity. In our experience paclitaxel is used frequently form last decade, and interestingly it has been isolated from fungus. Vincristine is another chemotherapeutic agent reported from a fungal source in the twentieth century. After that interest on endophytic-based chemotherapy is enhanced, nowadays we have approximately ten anticancer and hundreds of compounds with significant cytotoxic activity (Kharwar et al. 2009).

Endophytic bacteria have been studied; it improves the biomass production and the carbon sequestration potential from *Populus* spp. It is reported that 78 bacterial endophytes isolated from willow tree (*Salix* sp.). *Gammaproteobacteria* dominated in the above collection, which include *Enterobacter* spp. strain 638 and *Stenotrophomonas maltophilia*. It draws design and strategies for improvement of biomass production with the interactions between endophytic bacteria and their host plants (Taghavi et al. 2009). Endophytic microbes may promote plant growth and confer enhanced resistance to various pathogens. In the case of *Oryza sativa*, endophytes isolated and used as the test plant produced two types of interactions; first are biofilms (bacteria attached to mycelia) and mixed cultures with no such attachments. Indoleacetic acid-like substances (IAAS) of biofilms were observed higher in fungi or bacteria. In vitro production and application of beneficial biofilm inoculation of endophytes are important for improved plant production in agroecosystem (Bandara et al. 2006). In 2009, 34 bacterial endophytes were isolated and characterized from stem of *Chelidonium majus* L. and evaluated for antifungal activity (Goryluk et al. 2009).

Biodiversity of endophytes may yield products of great use to humans (Smith et al. 2008). Recently, many endophytic bioactive metabolites known as well as new substances, possessing a wide variety of biological activities as antibiotic, antitumor, anti-inflammatory, antioxidant, etc., have been identified. Different bioactive secondary metabolites produced by endophytic microorganisms as well as microbial sources of these metabolites and their host plants are well categorized (Firakova et al. 2007).

Penicillin (source *Penicillium notatum*) opened discovery of novel bioactives from microbial metabolites. Bioactive natural products from endophytic microbes have enormous potential as the source of new medicinal products. The cloning of the genes of endophytic metabolites has begun to open up attractive screening possibilities for the direct identification of endophytic strains (Zhang et al. 2001). It seems that endophytes are the group of poorly investigated microorganisms. They have reliable source of bioactive and chemically novel compounds with potential for exploitation in a wide variety of medical importance (Strobel and Daisy 2003). Endophytes of ethnomedicinal plants could be a good source of antibacterial substances. Endophytic fungi were isolated from surface-sterilized leaves and small branches of *Garcinia mangostana* plant found in Indonesia. The crude extracts of ethyl acetate (EtOAc) of the 24 fermentation broths from 24 endophytic fungi were tested for their antibacterial activity against *Staphylococcus aureus* ATCC 25923, *Bacillus subtilis* ATCC 6633, *Escherichia coli* ATCC 25922, *Pseudomonas aeruginosa* ATCC 27853, *Salmonella typhi* ATCC 14028, and *Micrococcus luteus* (ATCC 10240), opening newer future prospects of endophytes. The minimum inhibitory concentration (MIC) of the crude ethyl acetate extracts of isolate RGM-02 inhibited *S. aureus* (MIC 25 μg/ml), *B. subtilis* (MIC 50 μg/ml), *M. luteus* (MIC 25 μg/ml), *E. coli* (MIC 200 μg/ml), *S. typhi* (MIC 200 μg/ml), and *P. aeruginosa* (MIC 100 μg/ml), respectively. The molecular identification revealed that the isolate RGM-02 represented *Microdiplodia hawaiiensis* CZ315 (Radji et al. 2011).

In *Prunus mume*, *Rosaceae*, an endophytic bacterial strain ZZ120 was isolated from stems and identified as *Bacillus subtilis*. ZZ120 culture filtrate contains *n*-butanol extract which is a strong growth inhibitor against disease and phytopathogens including *Fusarium graminearum*, *Alternaria alternata*, *Rhizoctonia solani*, *Cryphonectria parasitica*, and *Glomerella glycines*. Antifungal compounds were isolated from *n*-butanol extract as a mixture of its iturins which had strong antifungal activity for *B. subtilis* ZZ120, and its bioactive components might provide an alternative agent for the biocontrol of replant diseases (Li et al. 2012). Brazilian mangrove plant *Laguncularia racemosa* (L.) used for identification of 70 endophytic fungal strains. These fungal metabolites were active towards bacteria, *Staphylococcus aureus*, *Bacillus subtilis*, *Enterococcus faecalis*, *Micrococcus luteus*, *Escherichia coli*, and *Pseudomonas aeruginosa*. Among 70, thirty-four (48.6 %) endophytic fungi strains were identified to produce secondary metabolites having an antimicrobial activity. Crude extracts of *Aspergillus niger*, *Curvularia pallescens*, *Guignardia bidwellii*, *Paecilomyces variotii*, and *Mycelia sterilia* presented good results (Silva et al. 2011).

Rigorous studies on animal microbial world and plant system show that these harbor a wide range of diverse bacteria. Depending on the colonized compartment, these bacteria are rhizospheric (root colonizers), endophytic (colonizing the endosphere, the bulk of internal tissues), and phyllospheric (Pini et al. 2012). Roots of *Pongamia glabra* from Jalgaon, Maharashtra (India), were used for the identification of endophytic bacteria under different environmental conditions (Jalgaonwala and Mahajan 2011). In total 3, 16 endophytic antibacterial products were isolated; few are possessing great antifungal activity. In spite of extensive research, scientist mainly focused on endophytic fungi for producing plant-derived bioactive compounds such as paclitaxel, podophyllotoxin, camptothecine, vinblastine, hypericin, and diosgenin (Zhao et al. 2010).

From the roots of *Oryza sativa* L., 192 positive clones in the 16S rDNA library of endophytes and 52 OTUs (Operational Taxonomic Units) were identified based on the similarity of the amplified ribosomal DNA restriction analysis (ARDRA), banding profiles. *Betaproteobacteria* (27.08 % of the total clones) traced, and the most dominant genus was *Stenotrophomonas*. More than 14.58 % were uncultured bacteria and may be member of endophytic bacterial community

(Sun et al. 2008). Another genome sequence of strain RR-10 from endophyte *Stenotrophomonas maltophilia* which was isolated from a rice root in a rice field of China (Zhu et al. 2012). 853 endophytic strains were isolated from aerial tissues of four agronomic crop species and 27 prairie plant species. Among them endophytes exhibiting the most promising levels of colonization and an ability to persist were identified as *Cellulomonas*, *Clavibacter*, *Curtobacterium*, and *Microbacterium* isolates by 16S rRNA gene sequence, fatty acid, and carbon source utilization analyses. *Microbacterium testaceum* defines for the first time the endophytic nature which is useful for biocontrol and other applications (Zinniel et al. 2002).

Endophytes are divers and useful for biological control of pathogens and plant growth promotion. Twenty-one isolates of endophytes had been identified, belonging to 11 genera (*Alternaria*, *Bipolaris*, *Colletotrichum*, *Glomerella*, *Guignardia*, *Lasiodiplodia*, *Marasmius*, *Phlebia*, *Phoma*, *Phomopsis*, and *Schizophyllum*); one isolate was identified only to the order level (Diaporthales). The phylogenetic analysis confirmed the molecular identification of some isolates to genus level, while for others it was confirmed at the species level (Orlandelli et al. 2012). Root, stem, petiole, leaf, and seed are main parts of host plants where endophytes can reside. *Panax notoginseng* is an example which is evaluated for antagonistic activity against *Fusarium oxysporum*, *Ralstonia* sp., and *Meloidogyne hapla*. Thousands of endophytic bacterial strains were evaluated in vitro; 104 strains exhibited antagonistic properties against at least one of these three pathogens belonging to four clusters: *Firmicutes*, *Proteobacteria*, *Actinobacteria*, and *Bacteroidetes/Chlorobi* (Ma et al. 2012).

5 Endophytic Metabolite in Cancer Chemotherapy

Cancer is increasing with alarming rate due to changing lifestyle, nutrition, global warming, genetic aberration/changes, and infections (Maurya et al. 2010; Tewari et al. 2009, 2010, 2013a, b; Mishra et al. 2010, 2011). The cancer therapy is costly with several side effects and immunomodulations (Mishra et al. 2013). So the assumption is that the natural products derived from medicinal plants have gained significance in cancer chemotherapy. Natural products and their derivatives are used from prehistoric in clinical uses. The National Cancer Institute (NCI) of the United States of America (USA) has screened more than lacks extracts from an estimated about 35,000 plant samples against a number of tumor systems (Cragg and Boyd 1996). In the twentieth century approximately 92 anticancer drugs were commercially available, approximately 62 % can be related to natural origin (Cragg et al. 1997), and few are under uses (Boopathy and Kathiresan 2010). In the Table 19.2, plant-derived natural products, paclitaxel and camptothecin were estimated to account for nearly one third of the global anticancer market (Oberlies and Kroll 2004). Over the last few decades, significant efforts have been made, by both pharmaceutical companies and academic institutions, to isolate and identify new natural products, especially from fungal species (Jimeno et al. 2004).

With the all aspects of the ecology of bacterial endophytes evaluated for the potential use of bacterial endophytes for beneficial purposes (Lodewyckx et al. 2002). Other applications in

Table 19.2 Ten common anticancer compounds derived from natural resources used in chemotherapy

S. no.	Source of origin	Anticancer compound
1	*Catharanthus roseus*	Vinblastine and vincristine
2	*Podophyllum peltatum*	Epipodophyllotoxin, an isomer of podophyllotoxin
3	*Taxus baccata*	Paclitaxel
4	*T. brevifolia*	Paclitaxel
5	*T. canadensis*	Paclitaxel
6	*Camptotheca acuminata*	Camptothecin
7	*Cephalotaxus harringtonia* var. *drupacea*	Homoharringtonine
8	*Bleekeria vitensis*	Elliptinium
9	*Dysoxylum binectariferum*	Flavopiridol
10	*Ipomoea batatas*	Ipomeanol

industry may also be discovered among the novel products produced by endophytic microbes are under consideration (Strobel et al. 2004). Anticancer activity of 14 anthracenedione derivatives separated from the secondary metabolites of the mangrove endophytic fungi *Halorosellinia* sp. (No. 1403) and *Guignardia* sp. (No. 4382) had been reported (Zhang et al. 2010). One hundred thirty endophytic fungi that were isolated from 12 Chinese traditional medicinal plants were tested for antitumor and antifungal activities by MTT assay on human gastric tumor cell line BGC-823 and the growth inhibition test against 7 phytopathogenic fungi. The results showed that fermentation broths from 9.2 % of the isolates exhibited antitumor activity and 30 % exhibited antifungal activity; moreover, some of them exhibited broad-spectrum antifungal activity. The active isolates were identified to 32 taxa. The results indicate that the endophytic fungi of Chinese traditional medicinal plants are promising sources of novel bioactive compounds (Li et al. 2005a, b). Taxol is an active agent derived from plant *Taxomyces andreanae* of plant association *Taxus brevifolia* confirmed as anticancer (Stierle et al. 1993).

6 Evaluation of Anticancer Activity of Endophytic Metabolites

Several endophytic metabolites were evaluated till now, among which we identified few compounds (Ryan et al. 2008) with their structure, source of origin, and endophyte with their activity on various cell lines (Verma 2012) (Table 19.3). *Xylaria* sp. produces benzoquinone metabolites, and penicillenols B1, B2, C1, and C2, [17, 18, 19, 20] were identified from *Penicillium* sp. GQ-7, from fungus *Aegiceras cornice* (Xu et al. 2008). Leaf of *Kandelia candel* was evaluated to isolate cyclic depsipeptides, 1962A and 1962B, from the mangrove endophytic fungus (No. 1962). The MTT bioassay, 1962A, showed weak activity against human breast cancer MCF-7 cells (Huang et al. 2007). Preussomerin EG2 [21] and preussomerin EG3 [22] and palmarumycin CP2 were isolated from the mycelium of *Edenia gomezpompae* present on the leaves of *Callicarpa acuminata* (Verbenaceae). Three other polyke-tides, penicillenone [23], arugosin I [24], and 9-demethyl FR-901235 [25], were isolated from the *Penicillium* sp. JP-1, an endophytic fungus isolated from *Aegiceras corniculatum.* Tyrosol [26] was isolated from *Glomerella cingulata*, the most common endophytic fungus associated with *V. arenaria*, but this was inactive. Metabolites such as phomopsin B [27] and C [28] were isolated from the mangrove endophytic fungus *Phomopsis* sp. ZSUH76 (Xu et al. 2008). Mangrove endophytic fungus *Xylaria* sp. is a potent resource of metabolites. Xyloketal B [29] and Xyloketal J [30], Xyloester A [31], and Xyloallenolid B [32] were isolated from the *Xylaria* sp. (Xu et al. 2008). Five new metabolites, (+)-(5*S*, 10*S*)-4′-hydroxymethylcyclozonarone [33], phyllospinarone [34], 3-ketotauranin [36], 3α-hydroxytauranin [37], and 12-hydroxytauranin [38], together with tauranin [35], were isolated from *Phyllosticta spinarum*, a fungal endophyte in *Platycladus orientalis*. All these natural products were evaluated for in vitro antiproliferative activity against a panel of five sentinel cancer cell lines, NCI-H460 (non-small cell lung), MCF-7 (breast), SF-268 (CNS glioma), PC-3 M (prostate), and MIA Pa Ca-2 (pancreatic). Only tauranin [35] showed antiproliferative activity against the cancer cell lines tested (Kithsiri Wijeratne et al. 2008; Ryan et al. 2008).

Five new metabolites, (+)-(5*S*, 10*S*)-4′-hydroxymethylcyclozonarone [31] phyllospinarone [32] together with tauranin [33] 3-ketotauranin [34] 3α-hydroxytauranin [35], 12 hydroxytauranin [36], were isolated from *Phyllosticta spinarum*, a fungal endophyte in *Platycladus orientalis*. All these natural products were evaluated for *in vitro* antiproliferative activity against a panel of five sentinel cancer cell lines, NCI-H460 (non-small cell lung), MCF-7 (breast), SF-268 (CNS glioma), PC-3 M (prostate), and MIA Pa Ca-2 (pancreatic). Only tauranin [37] showed anti proliferative activity against the cancer cell lines tested (Kithsiri Wijeratne et al. 2008; Ryan et al. 2008).

Table 19.3 Metabolites isolated from the defined endophytic origin with their structure

Metabolites with structure [no.]	Source	Endophyte	Cytotoxic test	Reference
2-chloro-5-methoxy-3-methylcyclohexa-2, 5-diene-1,4-dione [1] and xylariaquinone A [2]	*Sandoricum koetjape*	*Xylaria* sp.	African green monkey kidney fibroblasts (Vero cells)	Tansuwan et al. (2007)
Penicillenols A1 and B1 [3-4]	*Sandoricum koetjape*	*Xylaria* sp.	HL-60 cell line	Lin et al. (2008a)
Nigerasperone A-B [5-6]	*Aegiceras corniculatum*	*Penicillium* sp. GQ-7	A549 and SMMC- 7721 tumor cell lines	Zhang et al. (2007)
Nigerasperone C [7]	*Aegiceras corniculatum*	*Penicillium* sp. GQ-7,	Candida albicans	Zhang et al. (2007)
Excelsione [8]	*Knightia excelsa*	Unidentified fungus	P388 murine leukemia cells	Lang et al. (2007)
Preussomerin EG1[9]	*Callicarpa acuminata*	*Edenia gomezpompae*	–	Macías-Rubalcava et al. (2008)
Leptosphaerone C [10]	*Aegiceras corniculatum*	*Penicillium* sp. JP-1	A-549 cells	Lin et al. (2008b)
Nectriapyrone [11]	*V. arenaria*	*Glomerella cingulata*	JURKAT T leukemia cells and B16F10 melanoma cells	Guimarães et al. (2008)
Phomopsin A [12]	South China Sea plants	*Phomopsis* sp. ZSUH76	–	Huang et al. (2008)
Phyllospinarone [13]	*Platycladus orientalis*	*Phyllosticta spinarum*	NCI-H460 (non-small cell lung), MCF-7 (breast), SF-268 (CNS glioma), PC-3 M (prostate), MIA Pa Ca-2 (pancreatic)	KithsiriWijeratne et al. (2008)
Dipeptide trichodermamide A–C [14–16]	Culture broth	*Eupenicillium* sp.	Human colorectal carcinoma HCT116	Davis et al. (2008)

7 Prevention of *Helicobacter pylori* Infection Using Endophytic Metabolites Opens Preventive Strategy of Gastric Cancer

H. pylori is Gram-negative, spiral-shaped bacterium that colonizes the gastrointestinal tract of human (Marshall and Windsor 2005; Mishra et al. 2010, 2011) and other zooans (Wadstrom and Hanninen 1999). Several case-control studies have shown significant association between *H. pylori* infection and the risk gastric cancer (Munoz et al. 1968). For example, intestinal-type tumors predominate in countries with high prevalence of *H. pylori*, e.g., East Asia, while diffuse-type tumors have more uniform geographic distribution (Nomura et al. 1995). Prospective studies have also supported the association between *H. pylori* infection and gastric cancer risk (Konturek et al. 2002). Perhaps the most compelling evidence for establishing the link between *H. pylori* infection and gastric cancer comes from prospective studies on 1526 Japanese *H. pylori*-infected patients during a 7-year study. It was observed that 2.9 % of infected patients developed cancer, whereas other subject remained safe. In *H. pylori*, 4.7 % of patients with non-ulcer dyspepsia developed gastric cancer (Uemura et al. 2002). Many antimicrobial agents are poorly secreted in the mucosa or are inactivated in the acid environment of the stomach. Because of this, sometimes *H. pylori*

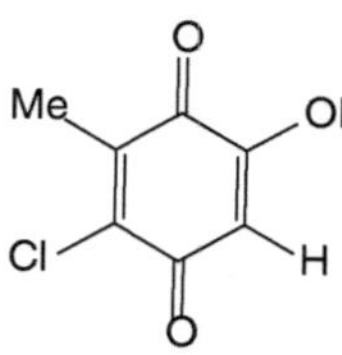

[1]

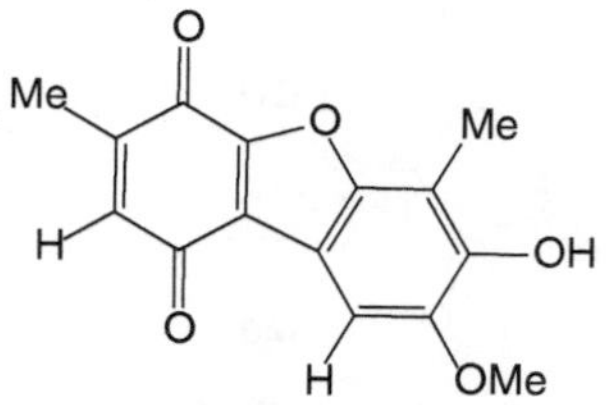

Xylariaquinone [2]

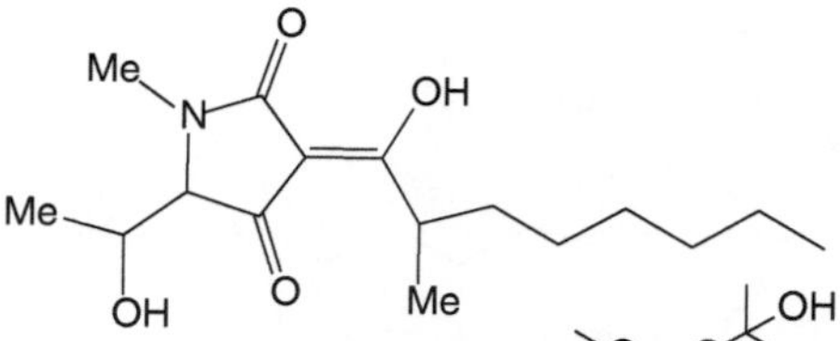

PenicillenolsA$_1$ [3]

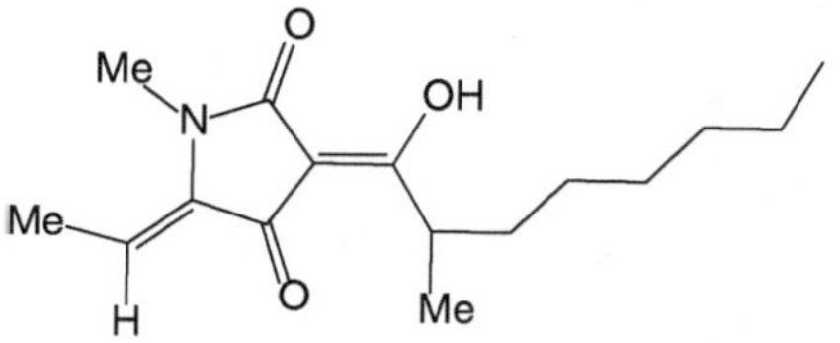

Penicillenols B$_1$ [4]

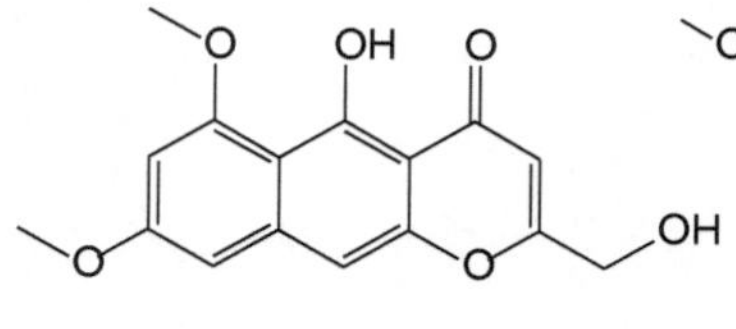

Nigerasperone A [5]

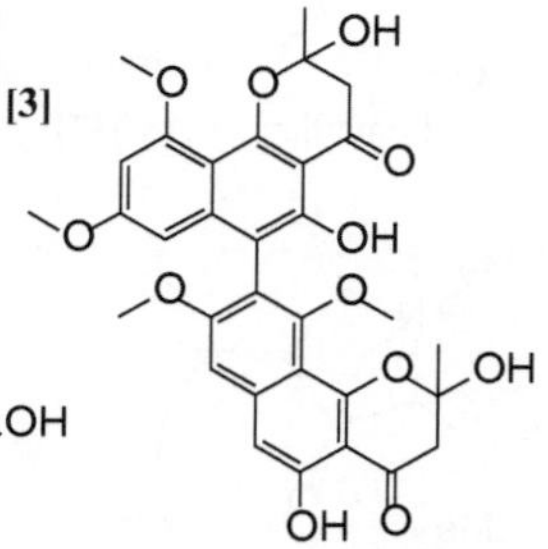

Nigerasperone B [6]

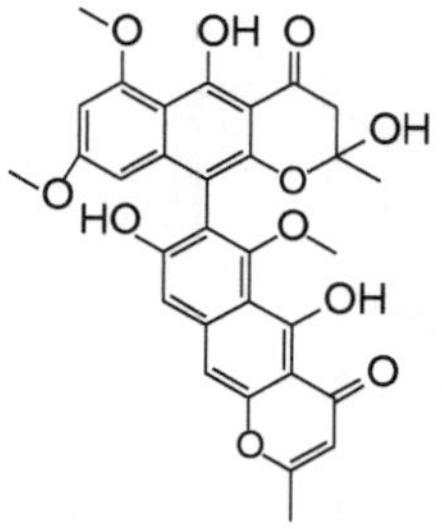

Nigerasperone C [7]

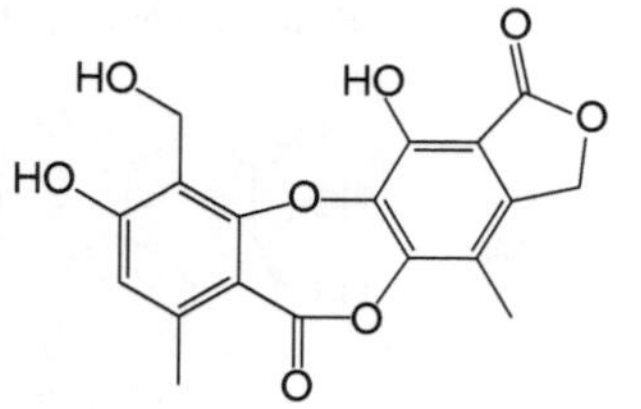

Excellsione [8]

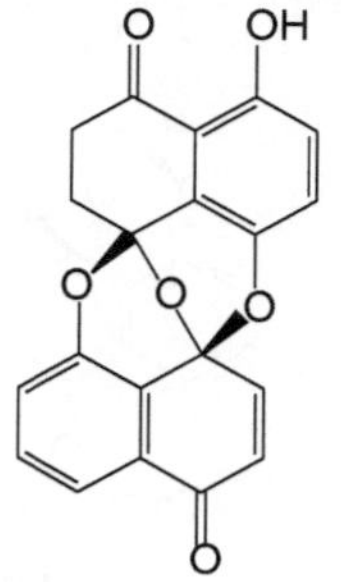

Preussomerin EG$_1$ [9]

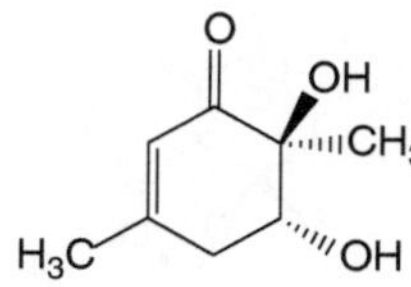

Leptosphaerone C [10]

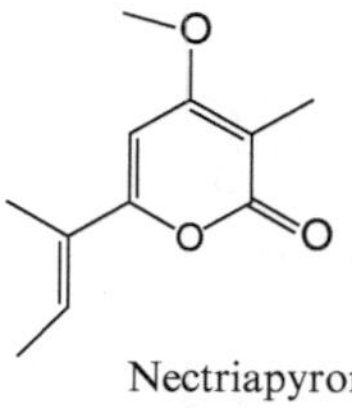

Nectriapyrone [11]

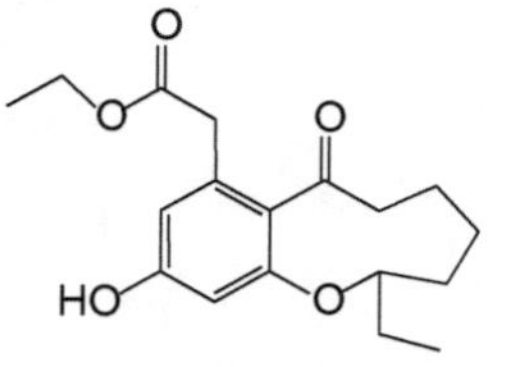

Phomopsin A [12]

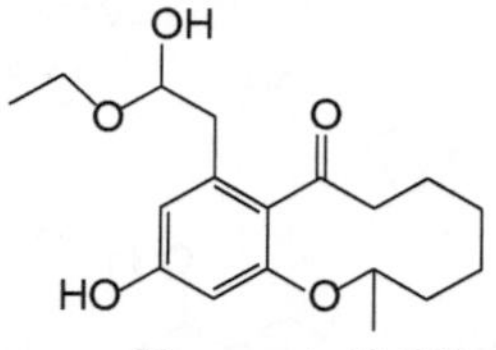

Phomopsin B [13]

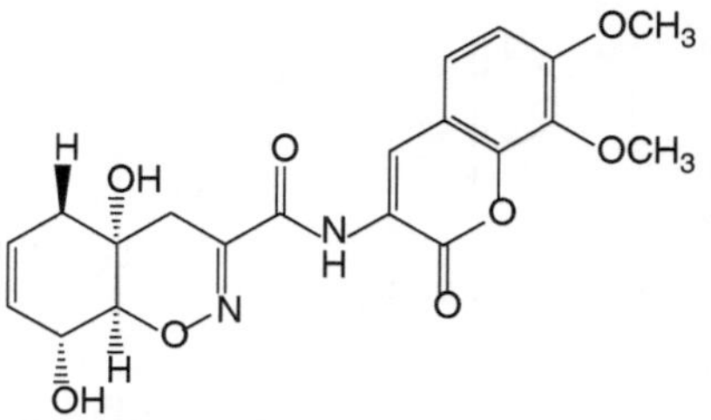

Trichodermamide A and B
[14] R=OH
[15] R=Cl

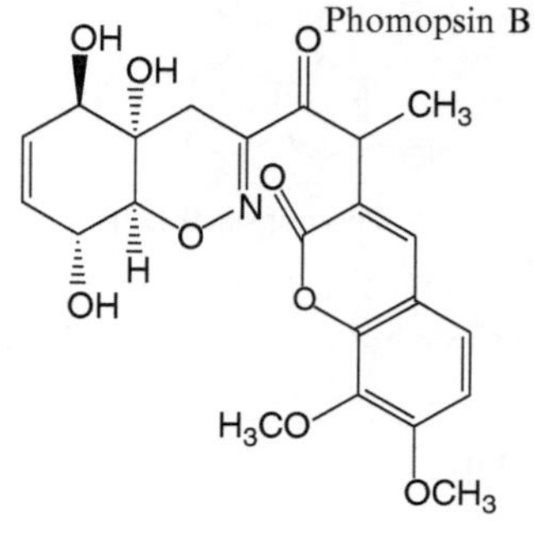

Trichodermamide C [16]

show susceptibility to antibiotics in vitro but has proved difficult to eradicate in vivo. Aminopenicillins, macrolides, tetracyclines, nitroimidazole, and proton pump inhibitor are commonest agents that are generally used for the treatment of *H. pylori*. The therapy of *H. pylori* should be planned as monotherapy or dual therapy or triple therapy or quadruple therapy (Table 19.4) (Hunt 1997). Surgical procedure and active chemotherapy in combination with various anti-*H. pylori*

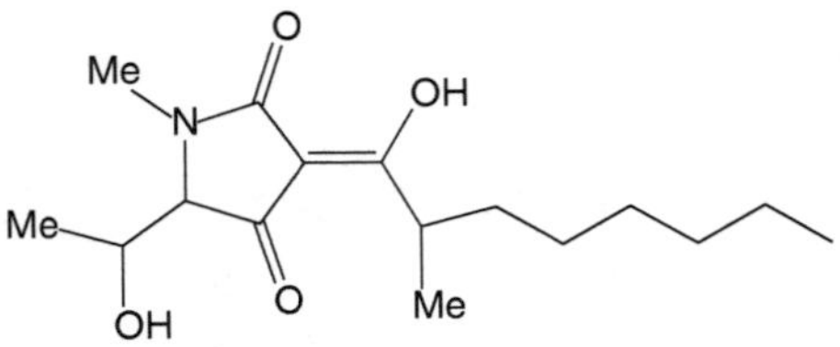

Penicillenols A_2 **[17]**

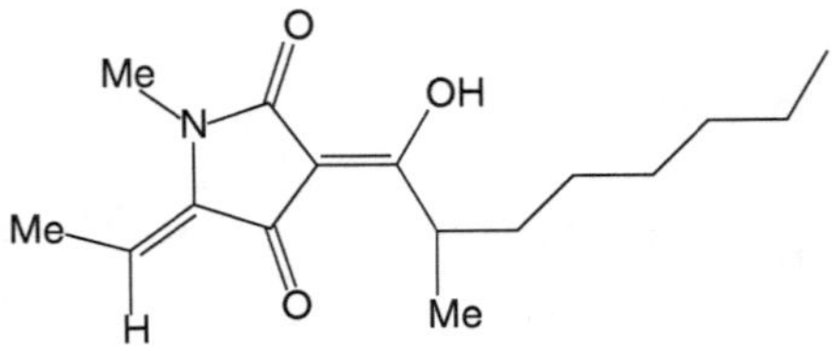

Penicillenols B_2 **[18]**

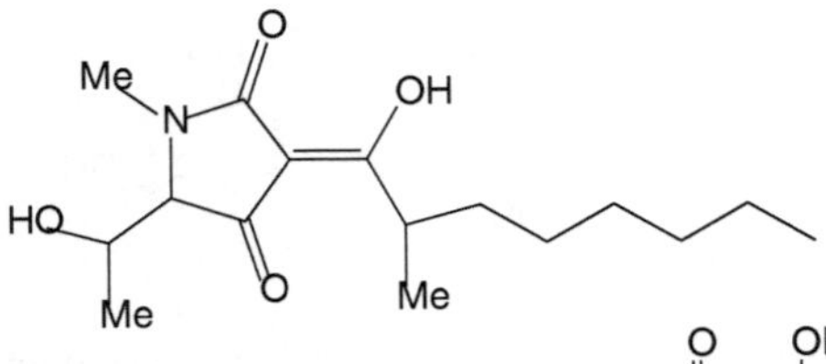

Penicillenols C_1 **[19]**

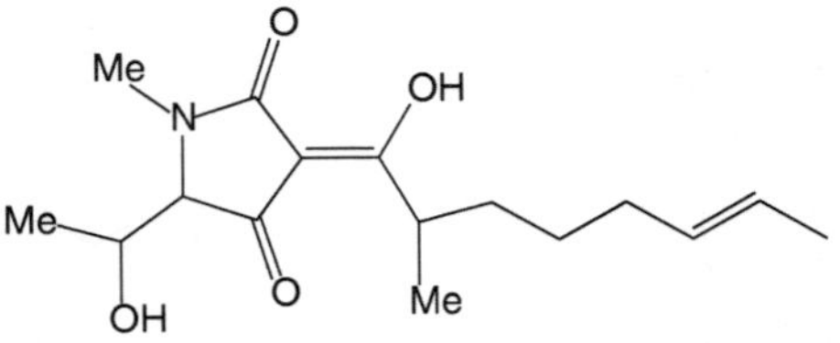

Penicillenols C_2 **[20]**

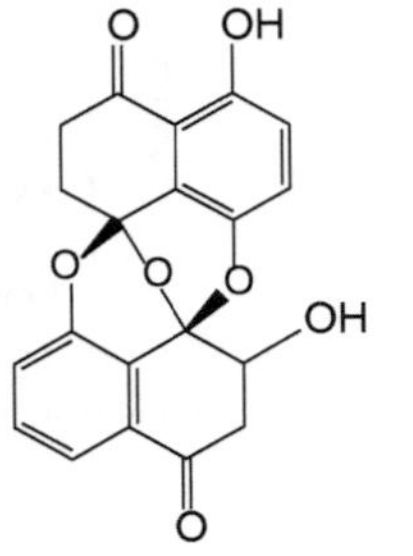

Preussomerin EG_2 **[21]**

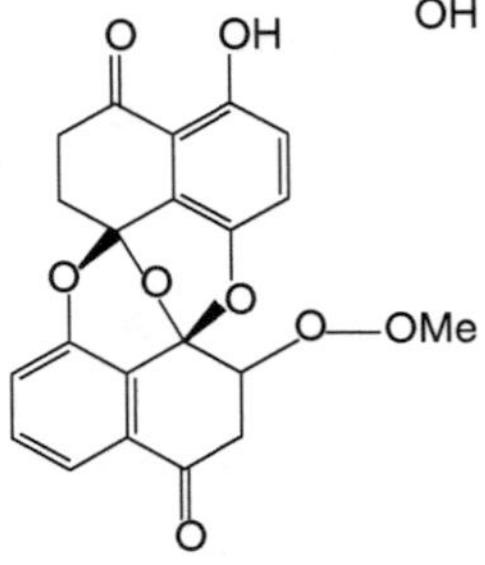

Preussomerin EG_3 **[22]**

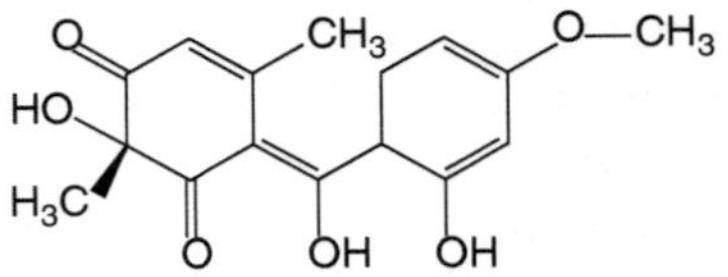

Penicillenone **[23]**

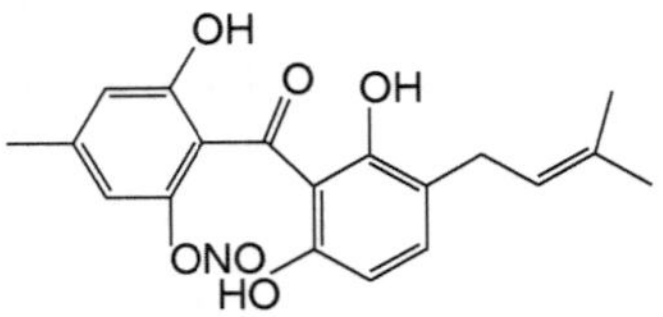

Arugesin I **[24]**

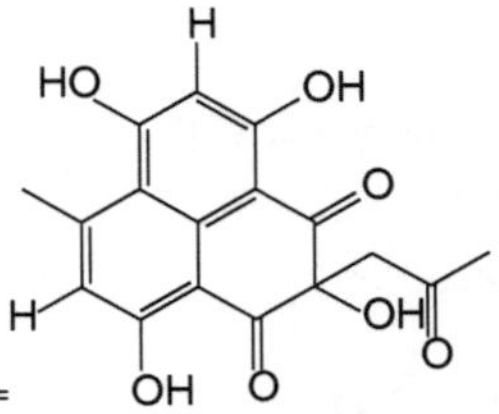

FR-901235 **[25]**

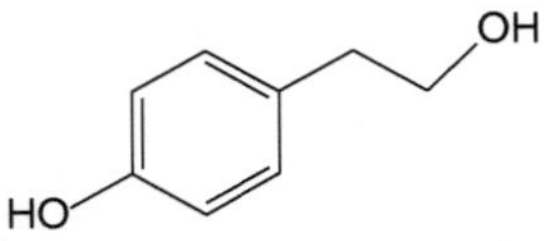

Tyrosol **[26]**

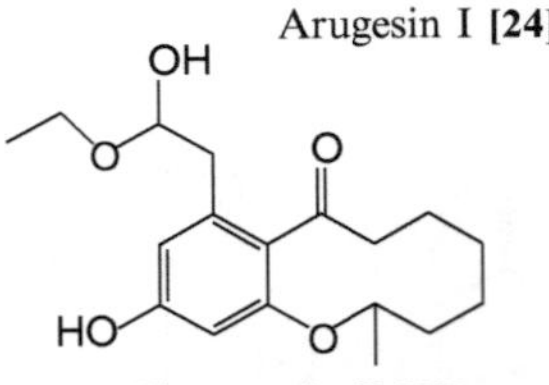

Phomopsin B **[27]**

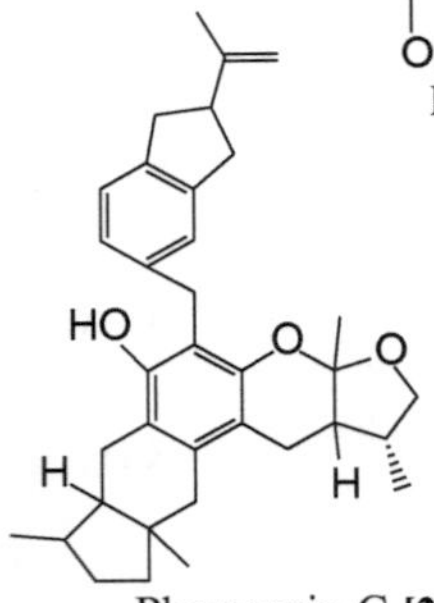

Phomopsin C **[28]**

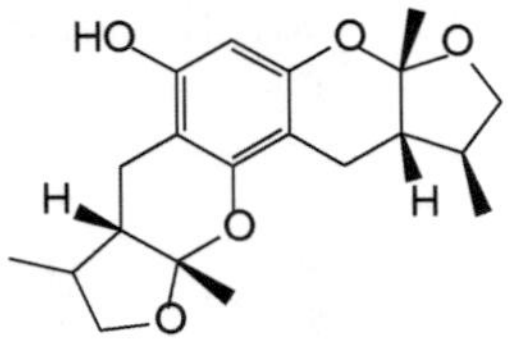

Xyloketal B **[29]**

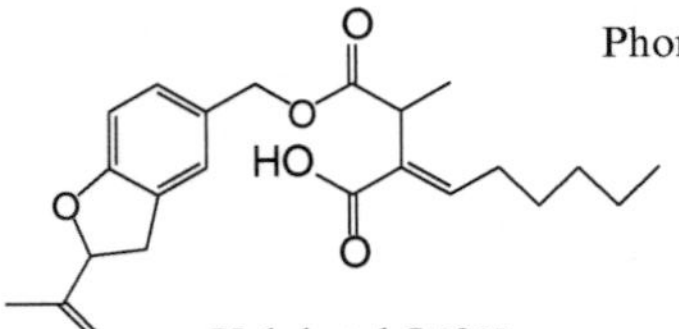

Xyloketal J **[30]**

drugs are the therapeutic option for its treatment. In our experience poor cure rate is reported in *H. pylori* infection due to failure of conventional medicine that may be due to antibiotic drug resistance. This problem enhanced area of emergence of chemotherapy against *H. pylori* infection. Natural products, which are used directly or as synthetic derivatives as alone, as well as in combination, are an alternative for its prospective therapy, because it appears that endophytic alkaloids, taxoids, and podophyllotoxins have an antineoplastic activity with full antimicrobial potentials. Recently *Rhizoctonia* sp. (Cy064), an endophytic fungus in the leaf of *Cynodon*

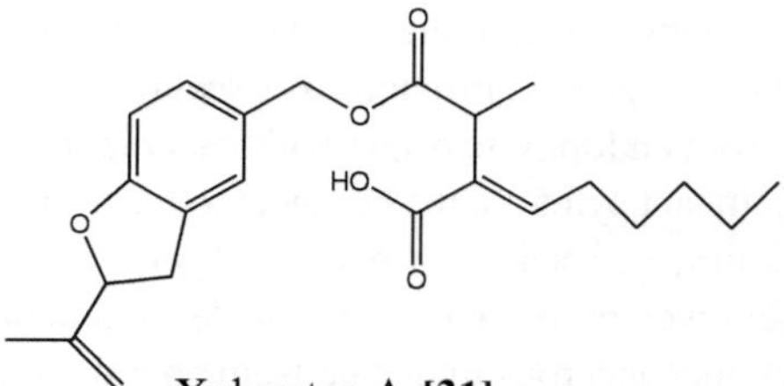

Xyloester A [31]

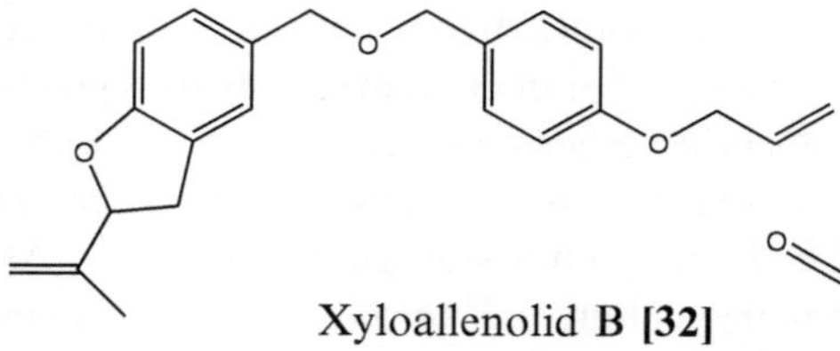

Xyloallenolid B [32]

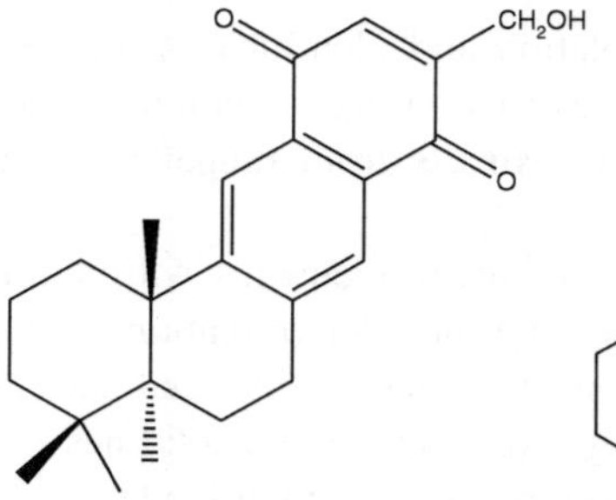

[33]

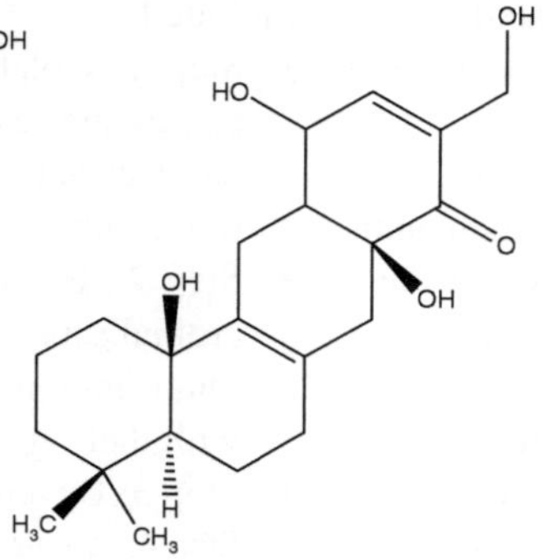

[34]

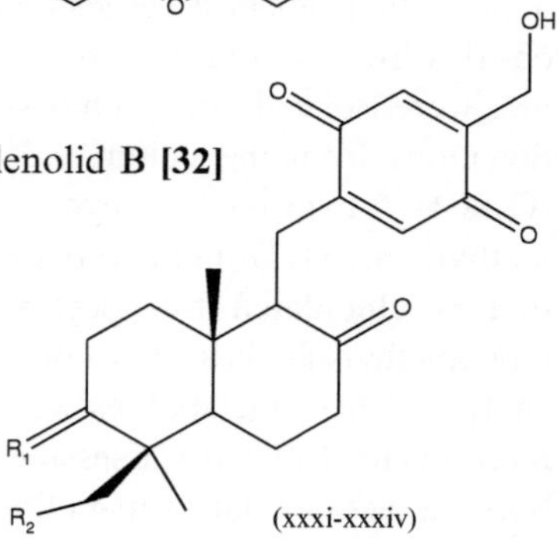

(xxxi-xxxiv)

	R1	R2
[35]	H	H
[36]	O	H
[37]	OH	H
[38]	H	OH

Table 19.4 The following agents are generally used for the treatment of *H. pylori*

Therapy	Antibiotics/combination of antibiotics	Eradication rate (%)
Monotherapy	Amoxicillin	20
	Erythromycin	20
	Clarithromycin	54
	Ranitidine bismuth citrate	20
Dual therapy	Tripotassium dicitratobismuthate + Amoxicillin	40
	Tripotassium dicitratobismuthate + Metronidazole	80
	Clarithromycin + Omeprazole	58–83
	Colloidal bismuth subcitrate + Omeprazole	30–40
	Amoxicillin + Omeprazole	72–84
	Colloidal bismuth subcitrate + Erythromycin	40–60
	Ranitidine bismuth citrate + Amoxicillin	40–60
	Ranitidine bismuth citrate + Clarithromycin	44–82
Triple therapy	Omeprazole + Amoxicillin + Clarithromycin	96.9 (12 days)
	Omeprazole + Clarithromycin + Tinidazole	82.2 (6 days)
	Lansoprazole + Amoxicillin + Clarithromycin	92–94
	Pantoprazole + Amoxicillin + Metronidazole	85
	Lansoprazole + Azithromycin + Tinidazole (Ultrashort therapy for 3 days)	82
	Lansoprazole + Azithromycin + Rebamipide	75
	Omeprazole + Amoxicillin + Plaunotol	83.4 (4 weeks)
Quadruple therapy	Omeprazole + Colloidal bismuth subcitrate (CBS), Tripotassium dicitratobismuthate (TDB) + Tetracycline + Metronidazole	98
	Omeprazole + Amoxicillin + Clarithromycin + Metronidazole	96
	Omeprazole + Amoxicillin + TDB + Metronidazole	95

dactylon, opens opportunity for detection of newer therapeutic metabolites. It contains rhizoctonic acids (benzophenone) in combination with monomethylsulochrin, ergosterol, and 3b, 5a, 6b-trihydroxyergosta-7, 22-diene. These metabolites were isolated through bioassay-guided fractionations from the culture of *Rhizoctonia* sp. (Cy064). 5-hydroxy-2-(2-hydroxy-6-methoxy-4-methylbenzoyl)-3-methoxybenzoic acid structure is elucidated by spectral analysis, and monomethylsulochrin was confirmed by $^{13C\text{-}NMR}$ analysis. These metabolites were subjected to a more detailed in vitro assessment of their antibacterial action against five clinical isolates and one reference (ATCC 43504) *H. pylori* strains (Maa et al. 2004).

It proves endophytic metabolites have a versatile capacity of antimicrobial agents. Few endophytes have been shown to possess superior biosynthetic capabilities owing to their presumable gene recombination with the host, while residing and reproducing inside the healthy plant tissues. In another study 32 endophytic fungi were isolated from the medicinal herb *Cynodon dactylon* (Poaceae). The ethyl acetate extracts of the cultures were examined in vitro for the anti-H. *pylori* activity. It was reported among 32, sixteen endophyte culture extracts have potent anti-*H. pylori* activities. Four metabolites, helvolic acid, monomethylsulochrin, ergosterol, and 3b-hydroxy-5a, 8a-epidioxy- ergosta-6, 22-diene, were identified with good minimum inhibitory concentration (Li et al. 2005a, b). The antimicrobial spectrum of helvolic acid is most active against *H. pylori.* The study strategies (Maa et al. 2004; Li et al. 2005a, b) opening newer pathway to detect endophytic metabolites in the prevention of *H. pylori* infection may open preventive strategy of gastric cancer.

8 Future Prospect

It is established that an endophyte may be a good source of secondary metabolites with promising cancer chemotherapeutic activity. It is very tough to search new anticancer agents frequently because sources are limited. Paclitaxel, the well-defined fungal endophytic metabolite, initiated us to work in this field. Apart from paclitaxel several compounds from endophytic origin had been reported with significant cytotoxicity, but their active implication is limited because of poor work plan.

As per given proposal plan for the development of newer anticancerous molecule from endophytic origin (Fig. 19.3), the following steps will be included:

- Step 1: Isolation and identification of endophytes and extraction and isolation of crude ethyl acetate extracts from fungal fermentation broths
- Step 2: Test of microorganisms with antimicrobial screenings and determination of minimum inhibitory concentration and analysis with their cytotoxic activity on cell lines
- Step 3: Development of animal model
- Step 4: Result analysis using statistical parameter and consideration of newer cancer chemotherapeutics with their active trial

9 Conclusion

The lifestyle, global warming, malnutrition, genetic aberration/epigenetic changes, and various environmental factors increase the incidences of cancer. Cancer is a global problem with less curative rate and indication of poor survival with various side effects. Natural derived compounds can be assumed to play an important role to prevent the cancer incidences. Our ecosystems, including aerial, terrestrial, and marine, have a potential source of anticancer compounds, but they are least explored till now. In this chapter we mainly focus on such limited compounds. Owing to a diverse chemical ecology, it is concluded that endophytic metabolites have a great promise for providing potent, inexpensive, and safer anticancer drugs, which deserve an extensive investigation in near future.

Acknowledgements The authors are thankful to Dr. V. C. Verma (Postdoctoral Fellow) and Prof. Gopal Nath (Head), Department of Microbiology, Banaras Hindu University, Varanasi, for his encouragement to write about such newer aspects of medical and basic science. Financial

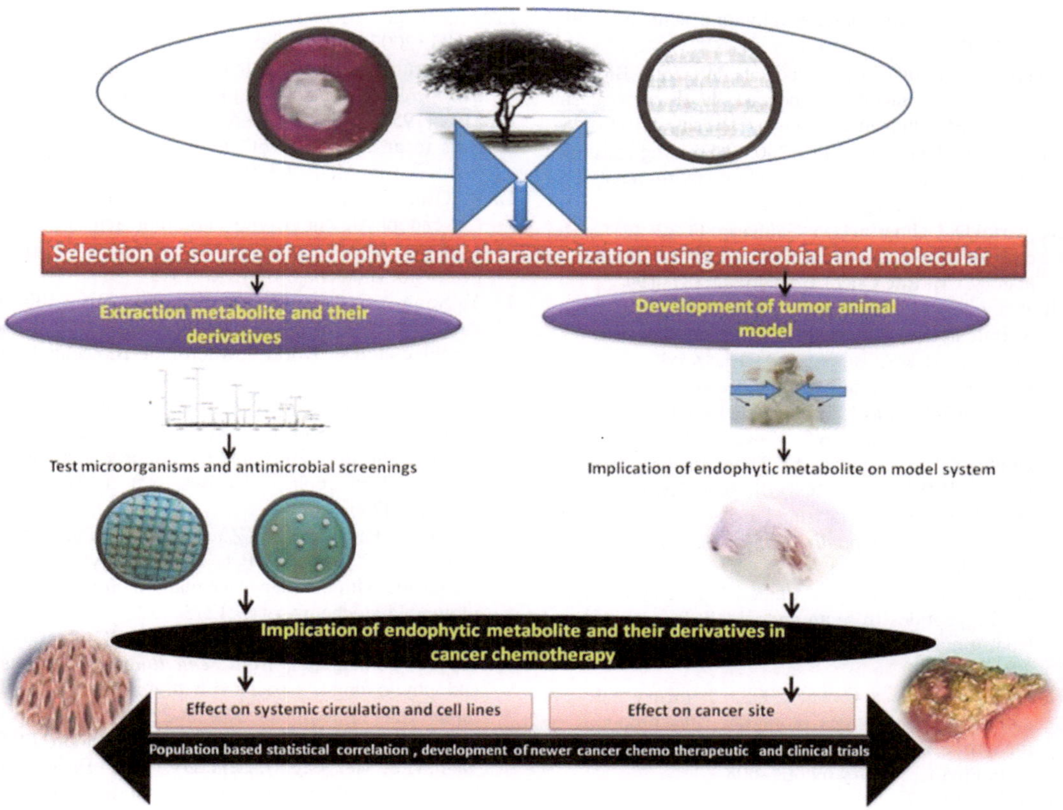

Fig. 19.3 A layout of future bio-prospection of microbial endophytes for search of new anticancer molecules

assistance from the Council of Scientific and Industrial Research, New Delhi, India (file no. 9/13(306)/2010-EMR-I), as Research associateship to RRM is gratefully acknowledged for the years 2010–2013.

Dedication

This article is dedicated to late Er. Shivendra Ranjan Mishra (1982–2012), younger brother of Raghvendra Raman Mishra, who lost his life in an accident of explosion in piped natural gas (PNG) at Kanpur, Uttar Pradesh, India.

References

Bandara WMMS, Seneviratne G, Kulasooriya SA (2006) Interactions among endophytic bacteria and fungi: effects and potentials. J Biosci 31:645–650

Boopathy NS, Kathiresan K (2010) Anticancer drugs from Marine flora: an overview. J Oncol 2010:1–18

Cragg GM, Boyd MR (1996) Drug discovery and development at the National Cancer Institute: the role of natural products of plant origin. In: Balick MJ, Elisabetsky E, Laird SA (eds) Medicinal plant resources of the tropical forest. Columbia University Press, New York, pp 101–136

Cragg GM, Newman DJ, Snader KM (1997) Natural products in drug discovery and development. J Nat Prod 60:52–60

Davis RA, Longden J, Avery VM et al (2008) The isolation, structure determination and cytotoxicity of the new fungal metabolite, trichodermamide C. Bioorg Med Chem Lett 18:2836–2839

Farber S, Diamond LK (1948) Temporary remissions in acute leukemia in children produced by folic acid antagonist, 4-aminopteroyl-glutamic acid. N Engl J Med 238:787–793

Firakova S, Sturdikova M, Muckova M (2007) Bioactive secondary metabolites produced by microorganisms associated with plants. Biologia (Bratislava) 3:251–257

Galmarini CM, Galmarini FC (2003) Multidrug resistance in cancer therapy: role of the microenvironment. Curr Opin Investig Drugs 4:1416–1421

Galmarini D, Galmarini CM, Galmarini FC (2012) Cancer chemotherapy: a critical analysis of its 60 years of history. Crit Rev Oncol Hematol 84:181–199

Gilman A (1946) Therapeutic applications of chemical warfare agents. Fed Proc 5:285–292

Goryluk A, Rekosz-Burlaga H, Blaszczyk M (2009) Isolation and characterization of bacterial endophytes of *Chelidonium majus* L. Polish J Microbiol 58:355–361

Guimarães DO, Borges WS, Kawano CY, Ribeiro PH, Goldman GH, Nomizo A, Thiemann OH, Oliva G, Lopes NP, Pupo MT (2008) Biological activities from extracts of endophytic fungi isolated from *Viguiera arenaria* and *Tithonia diversifolia*. FEMS Immun Med Microbiol 52:134–144

Huang H, She Z, Lin Y, Vrijmoed LLP, Lin W (2007) Cyclic peptides from an endophytic fungus obtained from a mangrove leaf (*Kandelia candel*). J Nat Prod 70:1696–1699

Huang Z, Cai X, Shao C et al (2008) Chemistry and weak antimicrobial activities of phomopsins produced by mangrove endophytic fungus *Phomopsis* sp. ZSU-H76. Phytochemistry 69:604–608

Hunt RH (1997) Peptic ulcer disease: defining the treatment strategies in the era of *Helicobacter pylori*. Am J Gastroenterol 92:36S–40S

Jalgaonwala RE, Mahajan RT (2011) Isolation and characterization of endophytic bacteria from roots of *Pongamia glabra* vent. Int J Pharma Biosci 1:280–287

Jimeno J, Faircloth G, Fern'andez Sousa-Faro JM, Scheuer P, Rinehart K (2004) New marine derived anticancer therapeutics—a journey from the sea to clinical trials. Mar Drugs 2:14–29

Kharwar RN, Verma VC, Kumar A, Gond SK, Harper JK, Hess WM, Lobkovsky E, Ma C, Ren Y, Strobel GA (2009) Javanicin, an antibacterial naphthoquinone from an endophytic. Fungus of neem, *Chloridium* sp. Curr Microbiol 58:233–238

Kithsiri Wijeratne EM, Paranagama PA, Marron MT, Gunatilaka MK, Arnold AE, Gunatilaka AAL (2008) Sesquiterpene quinones and related metabolites from *Phyllosticta spinarum*, a fungal strain endophytic in *Platycladus orientalis* of the Sonoran Desert. J Nat Prod 71:218–222

Konturek SJ, Starzynska T, Konturek PC, Karczewska E, Marlicz K, Lawniczak M, Jaroszewicz-Heigelman H, Bielanski W, Hartwich A, Ziemniak A, Hahn EG (2002) *Helicobacter pylori* and CagA status, serum gastrin, interleukin-8 and gastric acid secretion in gastric cancer. Scand J Gastroenterol 37:891–898

Lang G, Cole ALG, Blunt JW, Robinson WT, Munro MHG (2007) Excelsione, a depsidone from an endophytic fungus isolated from the New Zealand endemic tree *Knightia excelsa*. J Nat Prod 70:310–311

Law LW (1951) Response of a resistant variant of leukemic cells to an antagonist of pteroylglutamic acid. Proc Soc Exp Biol Med 77:340–344

Li H, Qing C, Zhao Z, Zhang Y (2005a) Screening for endophytic fungi with antitumor and antifungal activities from Chinese medicinal plants. World J Microbiol Biotechnol 21:1515–1519

Li Y, Song YC, Liu JY, Ma YM, Tan RX (2005b) Anti-Helicobacter *pylori* substances from endophytic fungal cultures. World J Microbiol Biotechnol 21: 553–558

Li H, Wang X, Han M, Zhao Z, Wang M, Tang Q, Liu C, Kemp B, Gu Y, Shuang J, Xue Y (2012) Endophytic *Bacillus subtilis* ZZ120 and its potential application in control of replant diseases. Afr J Biotechnol 11:231–242

Lin Z, Zhu T, Fang Y, Gu Q, Zhu W (2008a) Polyketides from *Penicillium* sp. JP-1, an endophytic fungus associated with the mangrove plant *Aegiceras corniculatum*. Phytochemistry 69:1273–1278

Lin ZJ, Lu ZY, Zhu TJ, Fang YC, Gu QQ, Zhu WM (2008b) Penicillenols from *Penicillium* sp. GQ-7, an endophytic fungus associated with *Aegiceras corniculatum*. Chem Pharm Bull 56:217–221

Lodewyckx C, Vangronsveld J, Porteous F, Moore ERB, Taghavi S, Mezgeay M, Lelie DV (2002) Endophytic bacteria and their potential applications. Crit Rev Plant Sci 21:583–606

Ma L, Cao YH, Cheng MH, Huang Y, Mo MH, Wang Y, Yang JZ, Yang FX (2012) Phylogenetic diversity of bacterial endophytes of *Panax notoginseng* with antagonistic characteristics towards pathogens of root-rot disease complex. Antonie Van Leeuwenhoek 102(3):407–408

Maa YM, Lia Y, Liua JY (2004) Anti-*Helicobacter pylori* metabolites from *Rhizoctonia* sp. Cy064, an endophytic fungus in *Cynodon dactylon*. Fitoterapia 75:451–456

Macías-Rubalcava ML, Hernández-Bautista BE, Jiménez-Estrada M, González MC, Glenn AE, Hanlin RT, Hernández-Ortega S, Saucedo-García A, Muria-González JM, Anaya AL (2008) Naphthoquinone spiroketal with allelochemical activity from the newly discovered endophytic fungus *Edenia gomezpompae*. Phytochemistry 69:1185–1196

Marshall BJ, Windsor HM (2005) The relation of Helicobacter pylori to gastric adenocarcinoma and lymphoma: pathophysiology, epidemiology, screening, clinical presentation, treatment, and prevention. Med Clin North Am 89:313–344

Matsumura Y, Maeda H (1986) A new concept for macromolecular therapeutics in cancer chemotherapy: mechanism of tumoritropic accumulation of proteins and the antitumor agent smancs. Cancer Res 46: 6387–6392

Maurya SK, Tewari M, Mishra RR, Shukla HS (2010) Genetic aberrations in gallbladder cancer. Surg Oncol 634:1–7

Mishra RR, Tewari M, Shukla HS (2010) *Helicobacter* species and pathogenesis of gallbladder cancer. Hepatobiliary Pancreat Dis Int 9:129–134

Mishra RR, Tewari M, Shukla HS (2011) *Helicobacter pylori* and pathogenesis of gallbladder cancer. J Gastroenterol Hepatol 26:260–266

Mishra RR, Tewari M, Shukla HS (2013) Association of *Helicobacter pylori* infection with inflammatory cytokine expression in patients with gallbladder cancer. Indian J Gastroenterol 32(4):232–235. doi:10.1007/s12664-013-0321-6

Munoz N, Correa P, Cuello C, Duque E (1968) Histological types of gastric carcinoma in high and now risk areas. Int J Cancer 3:809–818

Nomura AM, Stemmermann GN, Chyou PH (1995) Gastric cancer amount the Japanese in Hawaii. Jpn J Cancer Res 86:916–923

Oberlies NH, Kroll DJ (2004) Camptothecin and taxol: historic achievement in natural products research. J Nat Prod 67:129–135

Orlandelli RC, Alberto RN, Rubin Filho CJ, Pamphile JA (2012) Diversity of endophytic fungal community associated with *Piper hispidum* (Piperaceae) leaves. Genet Mole Res 11:1575–1585

Parkin DM, Bray FI, Devesa SS (2001) Cancer burden in the year 2000. The global picture. Eur J Cancer 37:S4–S66

Philips FS (1950) Recent contributions to the pharmacology of bis (2-haloethyl) amines and sulfides. J Pharmacol Exp Ther 99:281–323

Pini F, Frascella A, Santopolo L, Bazzicalupo M, Biondi EG, Scotti C, Mengoni A (2012) Exploring the plant-associated bacterial communities in *Medicago sativa* L. BMC Microbiol 12:78

Radji M, Sumiati A, Rachmayani R, Elya B (2011) Isolation of fungal endophytes from *Garcinia mangostana* and their antibacterial activity. Afr J Biotechnol 10:103–107

Ryan RP, Germaine K, Franks A, Ryan DJ, Dowling DN (2008) Bacterial endophytes: recent developments and applications. FEMS Microbiol Lett 278:1–9

Seeger DR, Smith JM Jr, Hultquist ME (1947) Antagonist for pteroylglutamic acid. J Am Chem Soc 69:2567

Silva MRO, Almeida AC, Arruda FVF, Gusmão N (2011) Endophytic fungi from Brazilian mangrove plant *Laguncularia racemosa* (L.) Gaertn. (Combretaceae): their antimicrobial potential. In: Mendez Vilas A (ed) Science against microbial pathogens: communicating current research and technological advances. Formatex Research Center, Badajoz, pp 1260–1266

Smith SA, Tank DC, Boulanger L-A, Bascom-Slack CA, Eisenman K, Kingery D, Babbs B, Fenn K, Greene JS, Hann BD, Keehner J, Swift EGK, Kembaiyan V, Lee SJ, Li P, Light DY, Lin EH, Ma C, Moore E, Schorn MA, Vekhter D, Nunez PV, Strobel GA, Donoghue MJ, Strobel SA (2008) Bioactive endophytes warrant intensified exploration and conservation. PLoS ONE 8:3052

Stevens CM, Mylorie A, Auerbach C (1950) Biological action of 'mustard gas' compounds. Nature 166: 1019–1021

Stierle A, Stierle D, Stierle D (1993) Taxol and taxane production by *Taxomyces andreanae*, an endophytic fungus of Pacific yew. Science 260:214–216

Strobel G, Daisy B (2003) Bioprospecting for microbial endophytes and their natural products. Microbiol Mole Biol Rev 67:491–502

Strobel G, Daisy B, Castillo U, Harper J (2004) Natural products from endophytic microorganisms. J Nat Prod 67:257–268

Sun L, Qiu F, Zhang X, Dai X, Dong X, Song W (2008) Endophytic bacterial diversity in rice (*Oryza sativa* L.) roots estimated by 16S rDNA sequence analysis. Microb Ecol 55:415–424

Taghavi S, Garafola C, Monchy S, Newman L, Hoffman A, Weyens N, Barac T, Vangronsveld J, Lelie DV (2009) Genome survey and characterization of endophytic bacteria exhibiting a beneficial effect on growth and development of poplar trees. Appl Environ Microbiol 75:748–757

Tansuwan S, Pornpakakul S, Roengsumran S, Petsom A, Muangsin N, Sihanonta P, Chaichit N (2007) Antimalarial benzoquinones from an endophytic fungus, *Xylaria* sp. J Nat Prod 70:1620–1623

Tewari M, Mishra RR, Kumar V, Kar AG, Shukla HS (2008a) Isolated tuberculosis of the ampulla of vater masquerading as periampullary carcinoma: a case report. J Pancreas 10:16001–16003

Tewari M, Kumar V, Mishra RR, Kumar M, Shukla HS (2008b) Is there a role for cholecystectomy in gallbladder carcinoma discovered to be unrespectable for cure at laparotomy? World J Surg 32:2683–2687

Tewari M, Kumar V, Mishra RR, Shukla HS (2009) Gallbladder carcinoid masquerading as gallbladder carcinoma: a case report. Hepato Pancreat Dis Int 8:121–123

Tewari M, Mishra RR, Shukla HS (2010) *Salmonella typhi* and gallbladder carcinoma: report from an endemic region. Hepatobiliary Pancreat Dis Int 9:458–464

Tewari M, Sahai S, Mishra RR, Shukla SK, Shukla HS (2012) Dendritic cell therapy in advanced gastric cancer: a promising new hope? Surg Oncol 21:164–171

Tewari M, Agarwal A, Mishra RR, Meena RN, Shukla HS (2013a) Epigenetic changes in carcinogenesis of gallbladder. Indian J Surg Oncol. doi:10.1007/s13193-013-0240-0,1-10

Tewari M, Srivastava AK, Mishra RR, Kumar M, Shukla HS (2013) HER2 expression in gastric and gastroesophageal cancer: report from a tertiary care hospital in North India. Indian J Surg. doi:10.1007/s12262-013-0871-y, 1–5

Uemura N, Okamoto S, Yamamoto S (2002) *Helicobacter pylori* infection and the development of gastric cancer. N Engl J Med 345:784–789

Verma VC (ed) (2012) Microbial endophyte of neem: a biotechnological perspective. LAP Lambert Academic Publishing Gmbh & Co Kg, Heinrich-Bocking–strr. 6–8, 66121, Saarbrücken, Germany

Wadstrom T, Hanninen ML (1999) *Helicobacter* in the digestive tract. Curr Opin Gastroenterol 15:S53–S56

Xu F, Zhang Y, Wang J, Pang J, Huang C, Wu X, She Z, Vrijmoed LLP, Jones EBG, Lin Y (2008) Benzofuran derivatives from the mangrove endophytic fungus *Xylaria* sp. J Nat Prod 71:1251–1253

Zhang HW, Song YC, Tan RX (2001) Biology and chemistry of endophytes. Nat Prod Rep 18:448

Zhang Y, Ming LX, Wang BG (2007) Nigerasperones A-C, new monomeric and dimeric naphtho-γ-pyrones from a marine alga-derived endophytic fungus *Aspergillus niger* EN-13. J Antibiot 60:204–210

Zhang J, Tao L, Liang Y, Chen L, Mi Y, Zheng L, Wang F, She Z, Lin Y, To KKW, Fu L (2010) Anthracenedione derivatives as anticancer agents isolated from secondary metabolites of the mangrove endophytic fungi. Mar Drugs 8:1469–1481

Zhao J, Zhou L, Wang J, Shan T, Zhong L, Liu X, Gao X (2010) Méndez-Vilas A (Ed) FORMATEX Madrid (Spain) Endophytic fungi for producing bioactive compounds originally from their host plants. Curr Res Technol Edu Topics Appl Microbiol Microb Biotechnol 1:567–576

Zhu B, Liu H, Tian WX, Fan XY, Li B, Zhou XP, Jin GL, Xiea GL (2012) Sequence of *Stenotrophomonas maltophilia* RR-10, isolated as an endophyte from rice root. J Bacteriol 194:1280–1281

Zinniel DK, Lambrecht P, Harris NB, Feng Z, Kuczmarski D, Higley P, Ishimaru CA, Arunakumari A, Barletta RG, Vidaver AK (2002) Isolation and characterization of endophytic colonizing bacteria from agronomic crops and prairie plants. Appl Environ Microbiol 68:2198–2208

20 Endophytic Fungi: Novel Sources of Anticancer Molecules

Vinod Kumar, Shalini Rai, Pallavi Gaur, and Tayaba Fatima

Abstract

Cancer is an enormous global health burden, touching every region of the world. The term "anticancer" is based on the assessment of the authors of the paper of the cytotoxicity of each compound against specific cancer cell lines. Endophytic fungi, fungi that are residing asymptomatically in internal tissues of all higher plants, are of growing interest as promising sources of biologically active agents. Endophytic fungi are one of the most creative groups of secondary metabolite producers that play important biological roles for human life. This chapter mainly focuses on endophytic fungi, which produce anticancer bioactive compounds such as paclitaxel, podophyllotoxin, camptothecin, vinblastine, hypericin, and diosgenin. The uniqueness of the endophytic community of fungi is stressed as a promising source of novel compounds with anticancer activity. Endophytes represent a dependable source of specific secondary metabolites and can be manipulated both physicochemically and genetically to increase yields of desired metabolites and to produce novel analogues of active metabolites.

1 Introduction

There is a need to search for new antimicrobial agents because infectious diseases are still a global problem because of the development and spread of drug-resistant pathogens (Pillay and Zambon 1998; Espinel et al. 2001). Novel anticancer drugs are also required due to the high worldwide mortality (Pisani et al. 1999). Cancer is a group of diseases that can affect various organs of the body and is characterized by the uncontrolled growth of abnormal cells and invasion into normal tissue. Cancer cells can also spread to other parts of the body and produce new tumors. If the spread of cells becomes uncontrolled, it can lead to death. Today, cancer accounts for one in every eight deaths worldwide – more than HIV/AIDS, tuberculosis, and malaria

V. Kumar (✉) • S. Rai • P. Gaur • T. Fatima
National Bureau of Agriculturally Important Microorganisms, Kushmaur, Post Box No. 6, Kaithauli, Mau Nath Bhanjan 275 101, Uttar Pradesh, India
e-mail: drvinod119@gmail.com

V.C. Verma and A.C. Gange (eds.), *Advances in Endophytic Research*, DOI 10.1007/978-81-322-1575-2_20, © Springer India 2014

combined. In 2008, there were an estimated 12.7 million cases of cancer diagnosed and 7.6 million deaths from cancer around the world. More than 60 % of all cancer deaths lack the medical resources and health systems to support the disease burden. Moreover, the global cancer burden is growing at an alarming pace; in 2030 alone, about 21.4 million new cancer cases and 13.2 million cancer deaths are expected to occur, simply due to the growth and aging of the population, adoption of behaviors, as well as lifestyle and environmental influences including smoking, poor diet, physical inactivity, and reproductive patterns (American Cancer Society 2013). The treatment of the disease is very difficult due to limited number of cancer chemotherapies, their deleterious side effects, and high cost of the drugs.

The discovery of new chemotherapeutic agents is a key goal for natural product and medicinal chemists because many existing therapies do not effectively treat certain cancers and multidrug-resistant tumors exacerbate treatment challenges. Secondary metabolites (natural products) have played an important role in the discovery and development of medicinal agents. Many important anticancer drugs have been isolated from plant sources. These compounds include the vinca alkaloids such as vinblastine and vincristine, which were isolated from the Madagascar periwinkle, *Catharanthus roseus* (Noble et al. 1958; Johnson et al. 1959) and paclitaxel (Wani et al. 1971), which is currently used for the treatment of breast cancer. Unfortunately, plant-derived natural products, being potent cytotoxic metabolites, are often produced in very low quantities by the source organisms. For instance, paclitaxel constitutes only 0.01–0.03 % of the dry phloem weight of *Taxus* (Cragg and Newman 2005). Supply is one of the serious issues because; if a source plant is endangered or has been collected in a politically quixotic part of the world. Many of the anticancer chemotherapeutics widely prescribed today – including tubulin inhibitors, alkylating agents, and compounds that target DNA topoisomerases I and II – are cytotoxic (cell-killing) agents. These secondary metabolite compositions make the re-isolation of a desired compound problematic because of environmental variations. Many researchers have looked to endophytic microorganism's natural products as a source of new compounds to combat the complex of diseases called cancer. Most of these bioactive compounds interact with enzyme targets and help the organism survive against a wide array of challenges. Each new microbe has the potential for yielding as yet undiscovered compounds with bioactivity that can be adapted for medicinal purposes. It has been estimated by Demain (2000) and others that fewer than 16 % of the fungal species that have been described have been cultured and studied. These described species probably represent fewer than 5 % of the total fungal species that await exploration. The focus of this report is anticancer agents produced by fungal endophytes.

2 Endophytic Fungi

2.1 Definition of an Endophyte

Endophytes are microorganisms that internally infect living plant tissues without causing any visible manifestation of disease and live in mutualistic association with plants for at least a part of their life cycle (Bacon and White 2000). The term "endophyte" (Gr. endon, within; phyton, plant) was first contrived by de Bary (1866). All types of microorganisms (fungi, bacteria, and actinomycetes) have been discovered as endophytes. The most frequently encountered endophytes are fungi (Staniek et al. 2008). Fungal endophytes constitute an inexplicably diverse group of polyphyletic fungi ubiquitous in plants and maintain an indiscernible dynamic relationship with their hosts for at least a part of their life cycle. The existence of fungi inside the tissues of asymptomatic plants has been known since the end of the nineteenth century (Guerin 1898). Evidence of plant-associated microorganisms found in the fossilized tissues of stems and leaves has revealed that endophyte-plant associations may have evolved from the time higher plants first appeared on the earth (Redecker et al. 2000). However, except for some infrequent studies, it was not until the end of the twentieth century that fungal endophytes began to receive more attention from scientists. Since endophytes were first described in *Lolium persicum* (Freeman 1904), various investigators have isolated endophytes from different plant species. These discoveries led to a worldwide search for novel endophytes for the better understanding and applicability of such a promising group of microorganisms. On the one hand, the ecological aspects of endophytic fungi such as host range, evolutionary relatedness, infection, colonization, transmission patterns, tissue specificity, and mutualistic fitness benefits have been investigated relating to a plethora of plants (Arnold et al. 2003, 2007; Arnold 2005, 2007; Stone et al. 2004; Schulz and Boyle 2005; Rodriguez et al. 2009). Several endophyte-derived natural products have shown potential as antimicrobial, insecticidal, cytotoxic, and anticancer activity agents against a variety of plant and human pathogens (Verma et al. 2009).

2.2 Collection and Isolation Techniques of Endophytes

A number of methods for isolation of endophytes are described in literatures. After a plant is selected for study, it is identified, and its location is plotted using a global positioning device. Small stem pieces are cut from the plant and placed in sealed plastic bags after excess moisture is removed. Every attempt is made to store the materials at 4 °C until isolation procedures can begin (Shrestha et al. 2001; Strobel et al. 2002). In the laboratory, plant materials are thoroughly surface treated with 70 % ethanol, sometimes they are flamed, and ultimately they are air dried under a laminar flow hood. This is done in order to eliminate surface-contaminating microbes. Then, with a sterile knife blade, outer tissues are removed from the samples, and the inner tissues are carefully excised and placed on water agar plates. After several days of incubation, hyphal tips of the fungi are removed and transferred to potato dextrose agar. The endophytes are encouraged to sporulate on specific plant materials and are eventually identified via standard morphological and molecular biological techniques and methods. Eventually, when an endophyte is acquired in pure culture, it is tested for its ability to be grown in shake or still culture by the use of various media and growth conditions. It is also placed in storage under various conditions, including 15 % glycerol at −70 °C. Ultimately, once appropriate growth conditions are found, the microbe is fermented and extracted, and the bioactive compound(s) is isolated and characterized. Virtually all of the common and advanced procedures for product isolation and characterization are utilized in order to acquire the product(s) of interest. Central to the processes of isolation is the establishment of one or more bioassays that will guide the compound purification processes. One cannot put too much emphasis on this point since the ultimate success of any natural-product isolation activity is directly related to the development or selection of appropriate bioassay procedures. These can involve target organisms, enzymes, tissues, or model chemical systems that relate to the purpose for which the new compound is needed (Strobel and Daisy 2003).

2.3 Biodiversity and Distribution of Fungal Endophytes

Endophytes have been found in every plant studied to date. There are over 300,000 higher plant species, and it can be assumed that each of these species hosts a complex community of endophytic microbes (Saikkonen et al. 1998). Fungal endophytes are a diverse and versatile group of microorganisms that colonize plants in the Arctic and Antarctic, and in geothermal soils, deserts, oceans, rainforests, mangrove swamps, and coastal forests (Fisher et al. 1995; Regina et al. 2002; Strobel 2002; Bashyal et al. 2005; Suryanarayanan et al. 2005; Wang et al. 2006; Lin et al. 2008a, b; Rosa et al. 2009). They have been isolated from the root complexes and aerial parts of a diverse range of hosts including algae, bryophytes, pteridophytes, gymnosperms, and angiosperms (Swatzell et al. 1996; Wang et al. 2006; Kralj et al. 2006; Gond et al. 2007; Silvia et al. 2008; Hoffman and Arnold 2008; Kharwar et al. 2008). Fungal endophytes are an important component of microbial biodiversity. Endophytic fungal symbionts can have profound effects on plant ecology, fitness, and evolution. Diverse group of this organism are able to produce number of bioactive agents (Brundrett 2006). The fossil record indicates that plants have been associated with endophytic fungi, for >400 million year, and were likely associated when plants colonized land, thus having an important role in driving the evolution of life on land. Clavicipitaceous endophytes are class I endophytes that represent a small number of phylogenetically related Clavicipitaceous species that are fastidious in culture and limited to some cool and warm season grasses (Stone et al. 2004; Bischoff and White 2005). Transmission of class I endophytes is primarily vertical, with maternal plants passing fungi on to offspring via seed infections (Saikkonen et al. 2002). The benefits conferred by these fungi appear to depend on the host species, host genotype, and environmental conditions (Faeth et al. 2006). Diversity of class II endophytes in individual host plants is quite limited. Class II endophytes comprise a diversity of species, all of which are members of the Dikarya (Ascomycota or Basidiomycota). They have ability to confer habitat-specific stress tolerance to host plants (Rodriguez et al. 2008). Researchers proposed that Clavicipitaceous endophytes are defensive mutualists of host grasses, and this hypothesis gets widely accepted on endophytes natural history, evolution, ecology, and physiology and followed by a number of researchers (Lane et al. 2000; Panaccione 2005; Panaccione et al. 2006; Koulman et al. 2007). Class III endophytes are distinguished on the basis of their occurrence and horizontal transmission. This includes vascular, nonvascular plants, woody, and herbaceous angiosperms in tropical forest and Antarctic communities (Davis et al. 2003; Higgins et al. 2007; Murali et al. 2007; Davis and Shaw 2008). Class III endophytes are especially known for their great diversity within individual host tissues, plants, and populations. Individual leaves may harbor up to one isolate per 2 μM of leaf tissue and contain a number of species. A single plant may harbor hundreds of different endophytic fungi. Class IV endophytes have darkly melanized septa and restricted to plant roots. They are generally Ascomycetes fungi which are conidial or sterile and that form melanized structures like inter- and intracellular hyphae and microsclerotia in the roots. This class of endophytes is found in host plants like nonmycorrhizal from Antarctic, Arctic, alpine, subalpine, temperate zones and tropical ecosystems (Jumpponen 2001).

To investigate the secondary metabolites of microorganisms from unusual or specialized niches may increase the chances of finding novel compounds. Scientists often focus their efforts on fungi that cause problems either as animal or plant pathogens. Plant endophytes are more subtle, rarely causing problems, coexisting with their hosts under most circumstances. They are generally nonpathogenic in nature but may produce secondary metabolites that enable them to survive in the competitive world of plant interstitial space. An overview of recent literature indicated that 51 % of bioactive substances isolated from endophytic fungi were previously unknown, compared to 38 % from soil fungi. Since the discovery of a fungus that produced more than 100 compounds with demonstrated anticancer activity have been isolated from endophytic fungi including several compounds originally found in other higher plants. This chapter will describe each of these compounds in terms of their source microorganism, plant host, and biological activity.

3 Cytotoxic Secondary Metabolites from Fungal Endophytes

3.1 Contribution of Endophytic Fungi to the Discovery of Novel Anticancer Molecules

The first chemotherapeutic agent was discovered quite by accident over 50 years ago. "During World War I, mustard gas (1,5-dichloro-3-thiapentane)" that decreases white blood cells was used as a chemical warfare agent (Goodman et al. 1946). Scientists reasoned that an agent that damaged the rapidly growing white blood cells might have a similar effect on certain cancers of the blood. Many of the anticancer chemotherapeutics widely prescribed today – including antimetabolites, tubulin inhibitors, alkylating agents, and compounds that target DNA topoisomerases I and II – are cytotoxic (cell-killing) agents. These compounds are designed to kill cancer cells more effectively than normal cells because they generally target the more rapidly dividing cancer cells. However, this is not always the case. Bone marrow cells, hair follicles, and epithelial cells such as those lining the GI tract also divide rapidly and are often the targets of side effects that can range from unpleasant to seriously debilitating. Despite the problems associated with the use of cytotoxic agents, cytotoxicity assays using a wide array of cancer cell types have played an important role in the discovery of compounds like paclitaxel, camptothecin, and the vinca alkaloids that target cancer cells (Thurston 2007; Wu 2006). In this report anticancer activity is generally associated with the cytotoxicity of the compounds described.

The anticancer drugs show nonspecific toxicity to proliferating normal cells, possess enormous side effects, and are not effective against many forms of cancer (Gangadevi and Muthumary 2008; Pasut and Veronese 2009). Thus, the cure of cancer has been enhanced mainly due to diagnosis improvements which allow earlier and more precise treatments (Pasut and Veronese 2009). There are some evidences that bioactive compounds produced by endophytes could be alternative approaches for discovery of novel drugs, since many natural products from plants, microorganisms, and marine sources were identified as anticancer agents (Fir et al. 2007). The anticancer properties of several secondary metabolites from endophytes have been investigated recently. Endophytic fungi have received less attention than their more pathogenic relatives because they reside within plant tissue. Studies of these organisms indicate that they are prolific producers of compounds that can be exploited as both agrochemical and medicinal agents. The search for new compounds is certainly important. Of equal importance, however, has been the discovery that some endophytes produce compounds that have been exclusively isolated from higher plants (Stierle et al. 1993). Since the initial report of the production of anticancer compound paclitaxel from a Northwest Pacific yew endophyte in 1993, several other important anticancer agents from fungal endophytes including camptothecin and several analogues, vincristine, and podophyllotoxin have been reported by researchers (Stierle et al. 1993; Zhang et al. 2000; Lingqi et al. 2000; Yang et al. 2004; Puri et al. 2005, 2006; Eyberger et al. 2006; Rehman et al. 2008; Kusari et al. 2009). More than a hundred of anticancer compounds belonging to 19 different chemical classes with activity against 45 different cell lines have been isolated from over 50 different fungal species belonging to 6 different endophytic fungal groups. Of the total compounds isolated from endophytic fungi, 57 % were novel or were analogues of known compounds. There has been a significant increase in the number of anticancer compounds isolated from endophytic fungi following the first report of the production of paclitaxel by a fungus (Stierle et al. 1993). In this report, compounds will be listed by chemical classification, although some compounds could be assigned to multiple chemical classes.

3.2 Anticancer Compounds from Endophytic Fungi

3.2.1 Alkaloids

Alkaloids are naturally occurring chemical compounds and have been studied as potential anticancer agents secreted by both host plant and endophytic fungi (Table 20.1, Fig. 20.1). Endophytes have usually been associated with a host organism that has also been reported to produce the compound of interest. Camptothecin (CPT) is a pentacyclic quinoline alkaloid that inhibits topoisomerase I (topo I), an enzyme involved in DNA replication. The compound exerts its cytotoxic effect by inhibiting the dissociation of the DNA–topoisomerase I complex during replication (Ling-Hua et al. 2003; Pommier 2006). Camptothecin was initially isolated from the wood of *Camptotheca acuminata* (Nyssaceae) called “xi shu” or the “happy tree,” which is

Table 20.1 Alkaloids isolated from endophytic fungi

Compound	Activity	Cell line/target enzyme	Host	Fungal endophyte	Reference
Penochalasin A	40.0 μM[a]	KB cell line	*Imperata cylindrica*	*Chaetomium globosum*	Ding et al. (2006)
9-Deacetoxy fumigaclavine	3.10 μM[a]	K562	*Cynodon dactylon*	*Aspergillus fumigatus*	Ge et al. (2009)
Emindole DA	5.5 μg/mL[a]	36 human tumor	*Mediterranean green alga*	*Emericella nidulans*	Kralj et al. (2006)
Cytochalasin 1	3.91, 15.6, 3.91 μg/mL[b]	A2780S, HCT-116, SW-620	*Tripterygium wilfordii*	*Rhinocladiella* sp.	Lee (1995)
Cytochalasin 2	15.6, 62.5, 15.6 μg/mL[b]	A2780S, HCT-116, SW-620	*Tripterygium wilfordii*	*Rhinocladiella* sp.	Lee (1995)
Cytochalasin 3	3.91, 15.6 μg/mL[b]	A2780S, SW-620	*Tripterygium wilfordii*	*Rhinocladiella* sp.	Lee (1995)
Cytochalasin E	<0.015, 0.98 μg/mL[b]	A2780S, HCT-116	*Tripterygium wilfordii*	*Rhinocladiella* sp.	Lee (1995)
	0.244 μg/mL[b]	SW-620	*Tripterygium wilfordii*	*Rhinocladiella* sp.	Lee (1995)
Cytoglobosin C	2.26 μM[a]	A549	*Ulva pertusa*	*Chaetomium globosum*	Cui et al. (2010)
Cytoglobosin D	2.55 μM[a]	A549	*Ulva pertusa*	*Chaetomium globosum*	Cui et al. (2010)
Chaetominine	21.0, 28.0 nM[a]	K562, SW1116	*Adenophora axilliflora*	*Chaetomium* sp. IFB-E015	Jiao et al. (2006)
Chaetoglobosin	3.125, 6.25 μg/mL[a]	H22, MFC	*Curcuma wenyujin*	*Chaetomium globosum* L18	Wang et al. (2012)
Chaetoglobosin U	16.0 μM[a]	KB cell line	*Imperata cylindrical*	*Chaetomium globosum*	Ding et al. (2006)
Chaetoglobosin C	34.0 μM[a]	KB cell line	*Imperata cylindrical*	*Chaetomium globosum*	Ding et al. (2006)
Chaetoglobosin F	52.0 μM[a]	KB cell line	*Imperata cylindrical*	*Chaetomium globosum*	Ding et al. (2006)
Chaetoglobosin E	48.0 μM[a]	KB cell line	*Imperata cylindrical*	*Chaetomium globosum*	Ding et al. (2006)
Camptothecin	–	A549, HEP-2	*Nothapodytes foetida*	*Entrophospora infrequens*	Puri et al. (2005)
Camptothecin	–	–	*Camptotheca acuminata*	*Neurospora crassa*	Rehman et al. (2008)
Camptothecin	–	OVCAR-5	*Camptotheca acuminata*	*Fusarium solani*	Kusari et al. (2009)
9-Methoxy camptothecin	–	–	*Camptotheca acuminata*	*Fusarium solani*	Kusari et al. (2009)
10-Hydroxy camptothecin	–	–	*Camptotheca acuminata*	*Fusarium solani*	Kusari et al. (2009)
Vincristine	–	–	*Catharanthus roseus*	*Fusarium oxysporum*	Zhang et al. (2000), Yang et al. (2004)

[a]IC_{50}

[b]IC_{100}

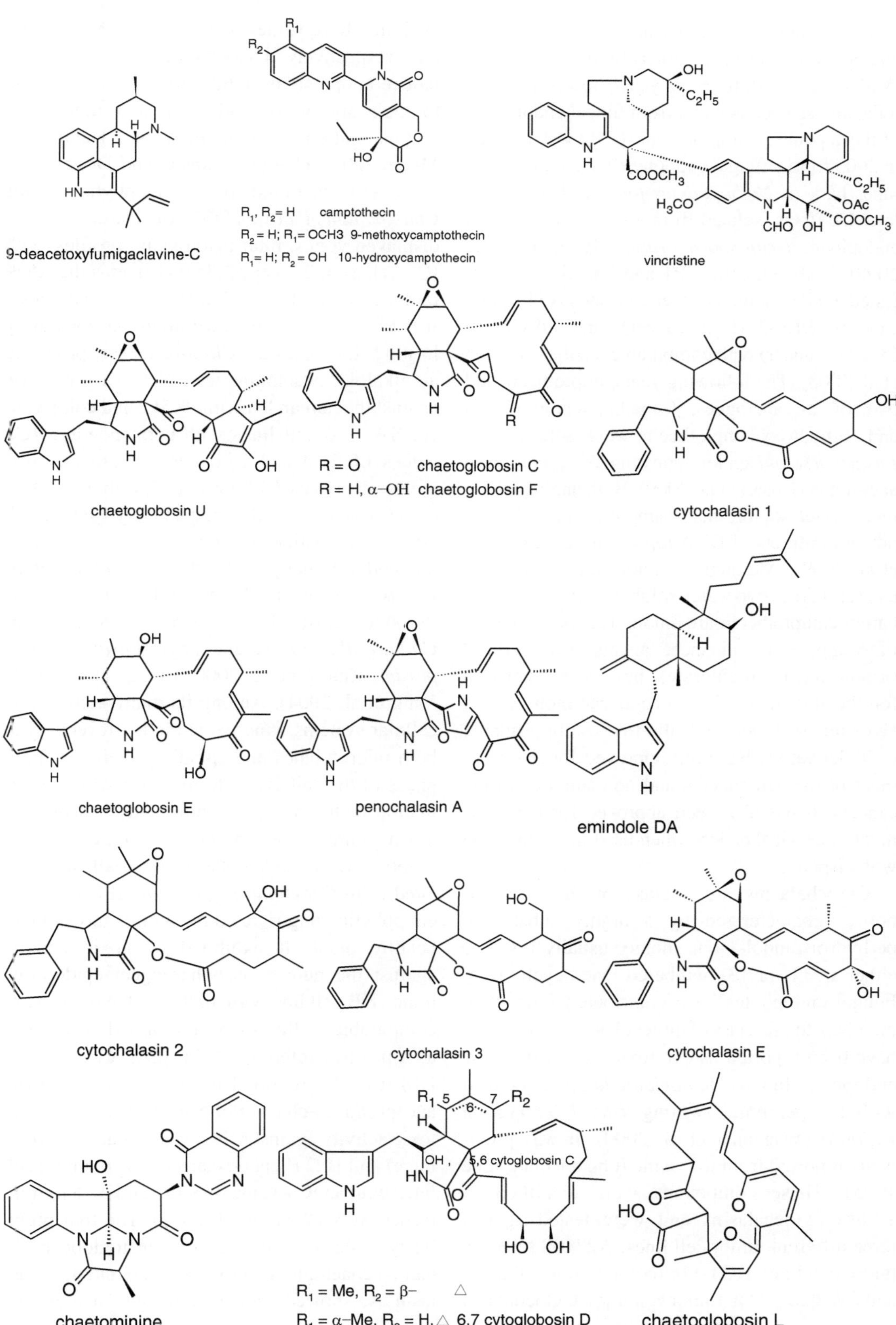

Fig. 20.1 Chemical structure of alkaloids isolated from endophytic fungi

native to mainland China, and exhibited potent antileukemic and antitumor activities in animals (Wall et al. 1966). In recent years, however, camptothecin has been isolated from fungal endophytes of these plants. Camptothecin (CPT) was isolated in 2005 from a fungal endophyte isolated from the inner bark of *Nothapodytes foetida*. Three years later CPT was isolated from a *C. acuminata* seed endophyte, *Neurospora crassa* (Rehman et al. 2008). Both authentic CPT and fungal CPT were tested against human cancer cell lines A549 (lung cancer), HEP-2 (liver cancer), and OVCAR-5 (ovarian cancer) with comparable results (Rehman et al. 2008). The following year camptothecin and two of its analogues, 9-methoxycamptothecin and 10-hydroxycamptothecin, were isolated from *Fusarium solani*, endophytic fungi of Camptotheca acuminata (Kusari et al. 2009). Both analogues are more water soluble than camptothecin and more potent inhibitors of DNA topoisomerase I (Kusari et al. 2009). Although camptothecin itself is not used as a drug, two water-soluble derivatives of the parent camptothecin are among the most recently FDA-approved anticancer agents. Camptosar® (irinotecan hydrochloride) has been approved for the treatment of colorectal carcinomas, and Hycamtin® (topotecan), the first orally available CPT derivative, has been approved for the treatment of ovarian cancers and non-small cell lung cancers. It has also been approved for the treatment of cervical cancer when used in conjunction with cisplatin.

Cytochalasins are a class of fungal metabolites characterized by a highly substituted perhydroisoindol-1-one moiety usually fused to either an 11- or 13-membered macrocyclic ring. Fungal endophytes have contributed four novel members to this class of molecules. Cytochalasins have been reported as cytotoxic agents from the endophytic fungus *Rhinocladiella* sp. associated with the perennial twining vine *Tripterygium wilfordii* (Wagenaar et al. 2000). It was previously reported from this same fungal isolate (Lee 1995). These compounds were identified as 22-oxa cytochalasins and were tested against three different tumor cell lines: A2780S (ovarian tumor cell line), HCT-116 (colon tumor cell line), and SW-620 (colon tumor cell line). Cytochalasin exhibited IC_{100} values of 3.91, 15.6, and 3.91 μg/mL, respectively. Cytochalasins are known to induce apoptosis by inhibiting cell division due to their ability to bind with, and inhibit the polymerization of, actin filaments (Haidle and Myers 2004). Cytoglobosins C and D were isolated and identified from endophytic fungus *Chaetomium globosum* QEN-14. The compounds displayed very similar cytotoxicity profiles, with IC_{50} values of 2.26 and 2.55 μM against the A549 tumor cell line (Cui et al. 2010). *Chaetomium* sp. IFB-E015, an endophytic fungus on apparently healthy *Adenophora axilliflora* leaves, produced an alkaloid, chaetominine, which was cytotoxic against the human leukemia K562 and colon cancer SW1116 cell lines with corresponding IC_{50} values of 21.0 and 28.0 nM. Its potency was greater than that of 5-fluorouracil, with IC_{50} values of 33.0 and 76.0 nM, respectively (Jiao et al. 2006). Vincristine, or leurocristine, is a vinca alkaloid originally isolated from *Catharanthus roseus*, a member of the family Apocyanaceae (Svoboda 1961). It has been isolated from the *Catharanthus roseus* endophyte *Fusarium oxysporum* (Zhang et al. 2000; Lingqi et al. 2000; Yang et al. 2004). Among its many activities in cellular systems, vincristine binds irreversibly to both microtubules and spindle proteins in the S phase of the cell cycle. It interferes with the formation of the mitotic spindle and consequently arrests tumor cells in the metaphase. Chaetoglobosin U is a cytochalasin-based alkaloid isolated from *Chaetomium globosum* IFB-E019, an endophytic fungus residing within the stem of healthy plant. It exhibited cytotoxic activity against the human nasopharyngeal epidermoid tumor KB cell line with an IC_{50} value of 16.0 μM comparable to that of 5-fluorouracil co-assayed as a positive reference (14.0 μM). *C. globosum* L18 was also reported from *Curcuma wenyujin* and produces chaetoglobosin. It exhibited cytotoxic activity against MFC (gastric cancer cells in mice) and H22 (hepatic cancer cells in mice) cell lines with an IC_{50} value of 3.125 and 6.25 μg/mL, respectively (Wang et al. 2012). The four previously isolated analogues of chaetoglobosin U, named chaetoglobosins C, F, and E and penochalasin A, showed moderate activity against the same cell line, with IC_{50} values of 34.0, 52.0, 48.0, and 40.0 μM, respectively (Ding et al. 2006). 9-Deacetoxyfumigaclavine C was isolated from the endophyte *Aspergillus fumigatus*, which was obtained from a healthy stem of *Cynodon dactylon*. It exhibited potent cytotoxicity against human leukemia cells (K562) with an IC_{50} value of 3.1 μM, which was similar to that of doxorubicin hydrochloride (1.2 μM), a drug which is currently used for the treatment of leukemia (Ge et al. 2009). Indole alkaloid emindole DA was isolated from *Emericella nidulans* var. *acristata*, an endophyte of unspecified Mediterranean green alga (Kralj et al. 2006). It exhibited antitumor activity against 36 human tumor cell lines representing 11 different tumor types, with a mean IC_{50} value of 5.5 μg/mL compared to the reference compound Adriamycin tested in parallel in the same assays with an IC_{50} value of 0.16 μg/mL (Kralj et al. 2006).

3.2.2 Terpenes

Periconicin B is a fusicoccane diterpene (Table 20.2, Fig. 20.2) isolated from the endophytic fungus *Periconia atropurpurea*, associated with *Xylopia aromatica* (Teles et al. 2006). Periconicin B exhibited potent cytotoxic activity against the two mammalian cell lines, HeLa (cervical cancer) and CHO (Chinese hamster ovary). It decreased cell viability of HeLa cells and CHO cells with an IC_{50} of 8.0 μM, showing potency similar to that of cisplatin, a well-known antineoplastic agent (IC_{50} 5.0 μM) used as a cytotoxic positive control (Teles et al. 2006).

It could be reasonably argued that no other secondary metabolite has had such a dramatic effect on cancer chemotherapy as Taxol (paclitaxel) (Kingston 2005; Cragg and Newman 2005). This highly functionalized diterpene is the prototypical taxane, isolated from the bark of the Northwest Pacific yew tree *Taxus brevifolia* for the first time by Wani et al. (1971). Unfortunately, as paclitaxel garnered more attention for its unique mode of action and potential as a chemotherapeutic agent, it also gained attention because of problems associated with the supply issue. Early estimates suggested that the population of Northwest Pacific yew trees could not adequately supply the projected demands for paclitaxel. Alternative sources were considered for the compound including total synthesis, semi-synthesis, and tissue culture (Kingston 2005). Stierle et al. (1993) took another approach and reported the isolation of a fungal endophyte from the needles of *T. brevifolia* that produced paclitaxel independently of the tree. The fungus had not been previously described and was designated *Taxomyces andreanae* in honor of its discoverer (Stierle et al. 1993). Stierle later reported the discovery of paclitaxel by a second fungus, *Penicillium raistrickii*, isolated from the inner bark of a yew tree. Kumaran also reported this compound from *Taxus cuspidata* and isolated from two different endophytic fungi *Pestalotiopsis neglecta* and *Pestalotiopsis versicolor* (Kumaran et al. 2010). Several other scientists have since reported the isolation of paclitaxel from different endophytic fungi associated not only with *Taxus* sp. but with other host plants as well. Strobel reported the production of paclitaxel from *Pestalotiopsis microspora* isolated from *Taxus wallichiana* (Strobel et al. 1996) and a second isolate of *Pestalotiopsis microspora* from bald cypress, *Taxodium distichum* (Li et al. 1996). Paclitaxel also reported from *Morinda citrifolia* isolated from *Lasiodiplodia theobromae* (Pandi et al. 2011). It has been reported from *P. pausiceta* isolated from *Cardiospermum helicacabum* (Gangadevi et al. 2008) and from *Pestalotiopsis terminaliae*, an endophytic fungus of *Terminalia arjuna* (Gangadevi and Muthumary 2008). It has also been reported from *Chaetomella raphigera*, a second endophytic paclitaxel producer reported from *Terminalia arjuna* (Gangadevi and Muthumary 2009a, b). The same scientists also reported the production of paclitaxel by *Bartalinia robillardoides*, an endophyte of *Aegle marmelos* (Gangadevi and Muthumary 2009a). This is not a comprehensive list of paclitaxel-producing endophytes, and more producers are reported every year. Fungal paclitaxel has been tested by apoptotic assay against a number of different cancer cell lines, including BT220, H116, HLK210, HL251, and INT-407. As the paclitaxel concentration increased from 0.005 to 0.05 μM, paclitaxel induced cell death through apoptosis

Table 20.2 Terpenes isolated from endophytic fungi

Compound	Activity (IC_{50})	Cell line/target enzyme	Host	Fungal endophyte	Reference
Periconicin B	8.0, 8.0 μM	HeLa and CHO	*Xylopia aromatica*	*Periconia atropurpurea*	Teles et al. (2006)
Paclitaxel	–	–	*Taxus brevifolia*	*Taxomyces andreanae*	Stierle et al. (1993)
	–	–	*Taxus wallichiana*	*Pestalotiopsis microspora*	
	0.005–5 μM	BT220, H116, INT-407, HL251, HLK210	*Terminalia arjuna*	*Pestalotiopsis terminaliae*	Gangadevi and Muthumary (2008)
	–	–	*Aegle marmelos*	*Bartalinia robillardoides*	Gangadevi and Muthumary (2009a)
	11.4 μg/mL	MCF-7	*Morinda citrifolia*	*Lasiodiplodia theobromae*	Pandi et al. (2011)
	0–5 μM	BT 220, HLK 210, HL 251	*Taxus cuspidata*	*Pestalotiopsis neglecta*	Kumaran et al. (2010)
	0–5 μM	BT 220, HLK 210, HL 251		*Pestalotiopsis versicolor*	Kumaran et al. (2010)
ent-4(15)-eudes men-11-ol-1-one	11, 20 μM	NCI-H187, MCF7	*Etlingera littoralis*	*Eutypella sp.*	Isaka et al. (2009)
	32, 32 μM	KB and Vero			
8-Deoxy-tricho thecin	0.88, 1.48 μM	BC-1, NCI-H187	*Knema laurina*	*KLAR 5*	Chinworrungsee et al. (2008)
7α-Hydroxy trichodermol 91	2.37, 1.73 μM	BC-1, NCI-H187			
Trichothecolone	12.90, 10.06 μM	KB, BC-1			
	11.31 μM	NCI-H187			
7α-Hydroxy scirpene	8.47, 21.53 μM	KB, BC-1			
	27.76 μM	NCI-H187			
Tauranin	4.3, 3.5 μM	NCI-H460, PC-3 M	*Platycladus orientalis*	*Phyllosticta spinarum*	Wijeratne et al. (2008)
	1.5, 1.8 μM	MCF-7, SF-268			
	2.8 μM	MIA Pa Ca-2			
Merulin A	4.98, 4.84 μg/mL	BT474, SW620	*Xylocarpus granatum*	XG8D (a basidiomycete)	Chokpaiboon et al. (2010)
Merulin C	1.57, 4.11 μg/mL	BT474, SW620			
Eremophilanolides 1–3	3.8–21 μM	KB, MCF-7, NCI-H187	*Licuala spinosa*	*Xylaria* sp.	Isaka et al. (2010)

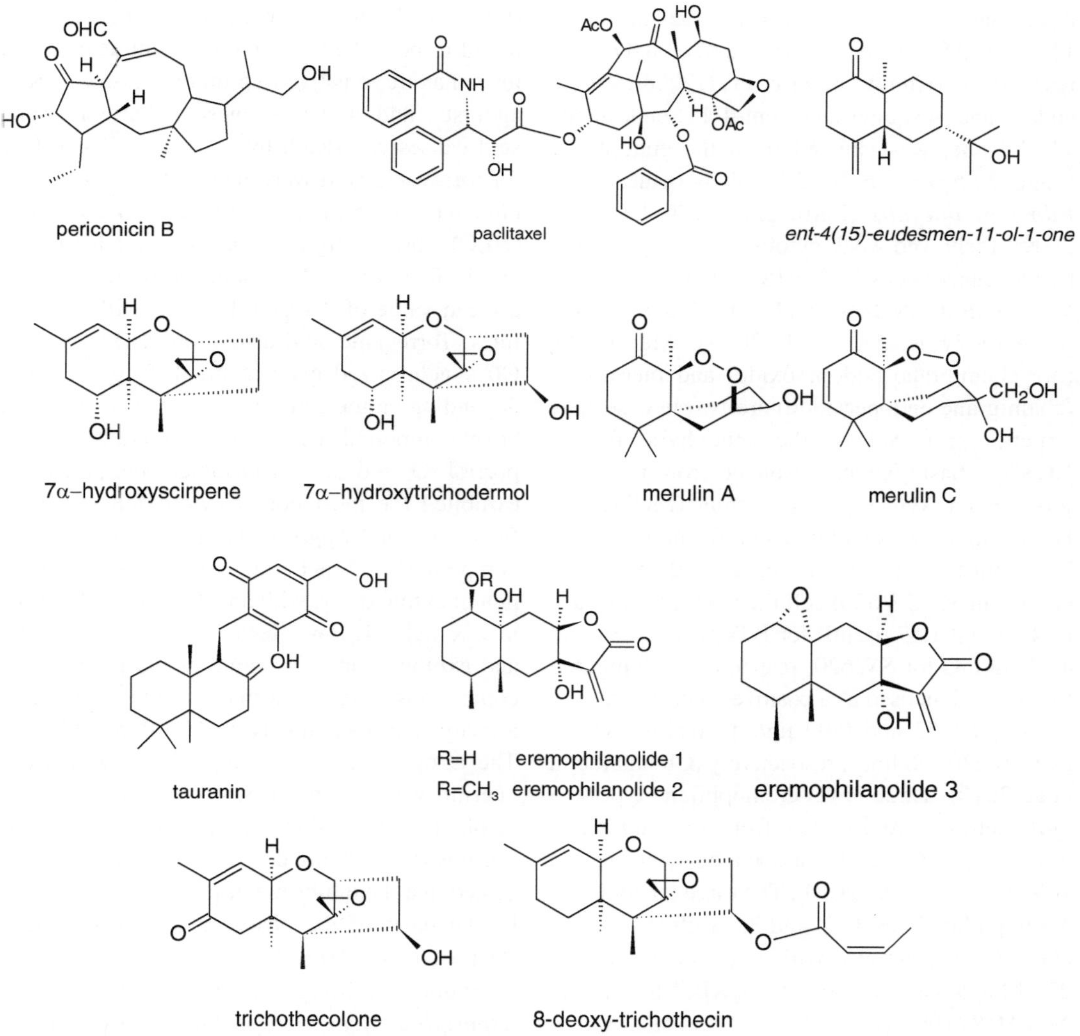

periconicin B paclitaxel *ent-4(15)-eudesmen-11-ol-1-one*

7α–hydroxyscirpene 7α–hydroxytrichodermol merulin A merulin C

tauranin R=H eremophilanolide 1 R=CH_3 eremophilanolide 2 eremophilanolide 3

trichothecolone 8-deoxy-trichothecin

Fig. 20.2 Chemical structure of terpenes isolated from endophytic fungi

increased accordingly, but the level of cell death only increased slightly with a further increase to 0.5 μM, while a further increase to 5 μM resulted in a dramatic decrease in cell death (Gangadevi and Muthumary 2008).

Four cytotoxic sesquiterpene compounds, 8-deoxytrichothecin, trichothecolone, 7α-hydroxytrichodermol, and 7α-hydroxyscirpene, were isolated from fungal isolate KLAR 5, which the authors identified as a "sister taxon of *Acremonium crotocinigenum*," a mitosporic Hypocreales found in a healthy twig of the Thai medicinal plant *Knema laurina* (Chinworrungsee et al. 2008). Trichothecolone and 7α-hydroxyscirpene exhibited selective activity against BC-1 (human breast cancer cells), with effective IC_{50} values of 0.88 and 2.37, respectively, and against NCI-H187 (human small-cell lung cancer cells) with IC_{50} values of 1.48 and 1.73 μM, respectively, compared to the standard drug ellipticine that exhibited an IC_{50} value of 0.63 μM against the BC-1 cell line. These compounds were not active against the KB cell line (human epidermoid cancer of the mouth) (Chinworrungsee et al. 2008). Compounds trichothecolone and 7 α-hydroxyscirpene were moderately active against all three cancer cell lines with IC_{50} values of 10.06, 11.31, 12.90 μM and 21.53, 27.76, 8.47 μM, respectively (Chinworrungsee et al. 2008). A new eudesmane sesquiterpene, ent-4(15)-eudesmen-11-ol-1-one, was isolated from the endophytic fungus *Eutypella* sp. BCC 13199 associated with *Etlingera littoralis* (Earth ginger) (Isaka et al. 2009). It showed weak cytotoxic activity against human cancer cells NCI-H187, MCF-7, KB, and Vero, with IC_{50} values of 11, 20, 32, and 32 μM, respectively (Isaka et al. 2009). Merulin A (nor-chamigrane endoperoxide) and merulin C (chamigrane endoperoxide) are two new sesquiterpenes produced by the endophytic fungus XG8D, a basidiomycete isolated from the mangrove plant *Xylocarpus granatum* (Meliaceae). Both compounds exhibited significant cytotoxicity against human breast cancer (BT474) and colon cancer (SW620) cell lines with IC_{50} values of 4.98 and 1.57 μg/mL for BT474 and 4.84 and 4.11 μg/mL for SW620, respectively, compared to doxorubicin used as a positive control with IC_{50} values of 0.53 and 0.09 μg/mL against BT474 and SW620 cell lines, respectively (Chokpaiboon et al. 2010). Three novel eremophilane-type sesquiterpenes were isolated from the endophyte *Xylaria* sp. BCC 21097 associated with *Licuala spinosa* (Isaka et al. 2010). The three compounds, eremophilanolides 1, 2, and 3, exhibited moderate cytotoxic activity with IC_{50} values of 3.8–21 μM against cancer cell lines KB, MCF-7, and NCI-H187 (Isaka et al. 2010). *Phyllosticta spinarum* was isolated from *Platycladus orientalis*, a plant of the Sonoran Desert (Wijeratne et al. 2008). Although the fungus produced a series of compounds, only tauranin exhibited cytotoxic activity against several cancer cell lines: NCI-H460 (non-small cell lung cancer), MCF-7 (breast cancer), SF-268 (CNS cancer – glioma), PC-3 M (metastatic prostate cancer), and MIA Pa Ca-2 (pancreatic carcinoma) at values of 4.3, 1.5, 1.8, 3.5, and 2.8 μM, respectively (Wijeratne et al. 2008).

3.2.3 Quinone

Torreyanic acid is unusual dimeric quinine (Table 20.3, Fig. 20.3) isolated from *Pestalotiopsis microspora*, an endophyte of *Torreya taxifolia* (Lee et al. 1996). In general, torreyanic acid was found to be 5–10 times more potent against cell lines that are sensitive to protein kinase C (PKC) agonists, and it was suggested that torreyanic acid causes cell death by apoptosis. IC_{50} values for torreyanic acid were between 3.5 μg/mL for human colorectal neuroendocrine cell carcinoma (NEC) and 45 μg/mL for human adenocarcinomic alveolar basal epithelial cells (A549), with a mean value of 9.4 μg/mL for 25 different cell lines. Torreyanic acid also showed G1 arrest of G0 synchronized cells at the 1–5 μg/mL level depending on the cell line (Lee et al. 1996). Five novel compounds an unresolved mixture of alterporriol G and its atropisomer alterporriol H exhibited the most potent cytotoxicity, with an EC_{50} value of 2.7 μg/mL. The previously reported compound 6-*O*-methylalaternin also exhibited potent cytotoxicity, with an EC_{50} value of 4.2 μg/mL. Kahalalide F was tested as a positive control and exhibited an EC_{50} value of 6.3 μg/mL. The compounds were also tested for kinase inhibitory activity in an assay involving 24 different kinases. The atropisomers and compound 6-*O*-methylalaternin were the most potent kinase inhibitors, displaying EC_{50} values between 0.64 and 1.4 μg/mL towards individual kinases. The authors suggested that the inhibition of protein kinases could be the basis of the observed cytotoxic activity (Debbab et al. 2009a, b).

Endophytic fungi *Alternaria alternata* and *Aspergillus niger* were isolated from *Tabebuia argentea* and produce lapachol (Sadananda et al. 2011). Later on endophytic fungi *Alternaria* sp. also isolated from *Aegiceras corniculatum* produced bianthraquinone derivatives (alterporriol K, L) and other endophytic fungi *Mycosphaerella* sp. were isolated from *Psychotria horizontalis* produced cercosporin, showed cytotoxicity against different cell lines (Huang et al. 2011; Moreno et al. 2011). Another endophytic fungi *Chaetomium* sp. was isolated from the stem of *Salvia officinalis*, produced two cytotoxic compound cochliodinol and isocochliodinol, tested for cytotoxicity against L5178Y mouse lymphoma cells (Sekita 1983; Debbab et al. 2009a). Compound cochliodinol was an order of magnitude more potent than its isomer, with an

Table 20.3 Quinones isolated from endophytic fungi

Compound	Activity	Cell line/targetenzyme	Host	Fungal endophyte	Reference
Alterporriol K	13.1 μM[a]	MDA-MB-435	*Aegiceras corniculatum*	*Alternaria* sp. ZJ9-6B	Huang et al. (2011)
Alterporriol L	29.1 μM[a]	MCF-7			
Cercosporin	4.68 and 3.56 μM[a]	Vero, MCF7	*Psychotria horizontalis*	*Mycosphaerella* sp.	Morenoa et al. (2011)
Lapachol	–	–	*Tabebuia argentea*	*Alternaria alternata*	Sadananda et al. (2011)
Lapachol	–	–	*Tabebuia argentea*	*Aspergillusniger*	Sadananda et al. (2011)
Torreyanic acid	3.5, 45.0 μg/mL[a]	NEC, A549	*Torreyataxifolia*	*Pestalotiopsis microspora*	Lee et al. (1996)
	9.5 μg/mL (mean)[a]	25 tumorcell lines			
Mixture of alterporriol G and alterporriol H	2.7 μg/mL[b]	L5178Y	*Mentha pulegium*	*Stemphylium globuliferum*	Debbab et al. (2009b)
6-*O*-Methylalaternin	4.2 μg/mL[b]	L5178Y			
Cochliodinol	7.0 μg/mL[b]	L5178Y	*Salviaofficinalis*	*Chaetomium* sp.	Debbab et al. (2009a)
Isocochliodinol	71.5 μg/mL[b]	L5178Y			
2-Chloro-5-methoxy-3-methylcyclohexa-2, 5-diene-1,4-dione	1.35 μM[a]	Vero cells	*Sandoricum koetjape*	*Xylaria* sp.	Tansuwan et al. (2007)
Xylariaquinone A	>184 μM[a]	Vero cells			
Anthracenedione 1	57.32, 90.86 μM[a]	KB, KBv200	Mangroveplant	*Halorosellinia* sp.	Zhang et al. (2010)
Anthracenedione 5	86.45 μM[a]	KBv200			
Anthracenedione 6	3.17, 3.21 μM[a]	KB, KBv200			
Anthracenedione 7	56.56 μM[a]	KB			
Anthracenedione 9	38.05, 34.64 μM[a]	KB, KBv200			
Anthracenedione 14	68.39 μM[a]	KB			

[a]IC_{50}
[b]EC_{50}

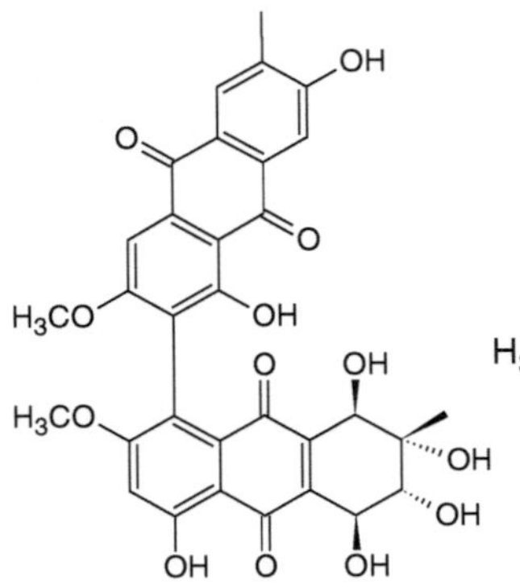

alterporriol G and H

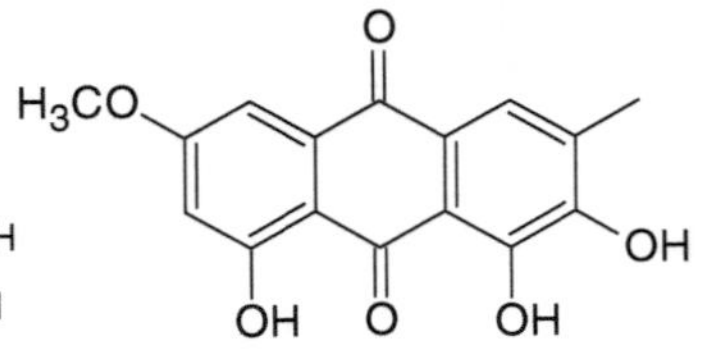

6-O-methylalaternin

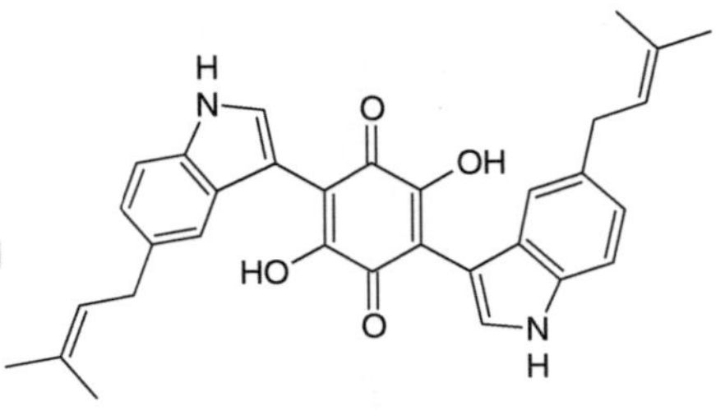

cochliodinol

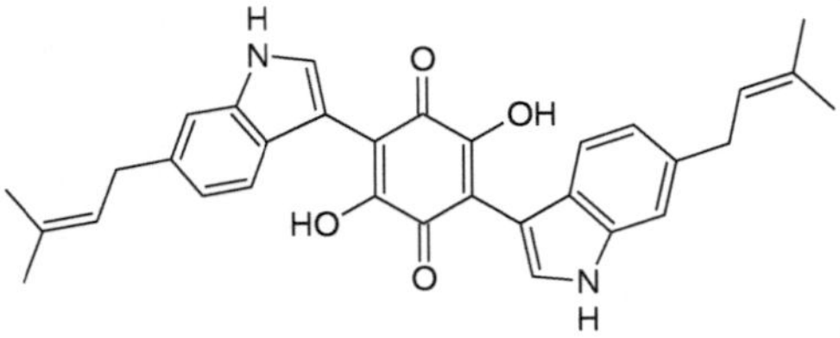

isocochliodinol

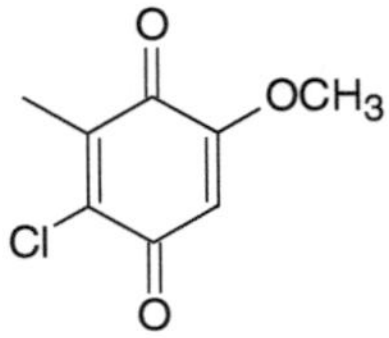

2-chloro-5-methoxy-3-methyl
cyclohexa-2,5-diene-1,4-dione

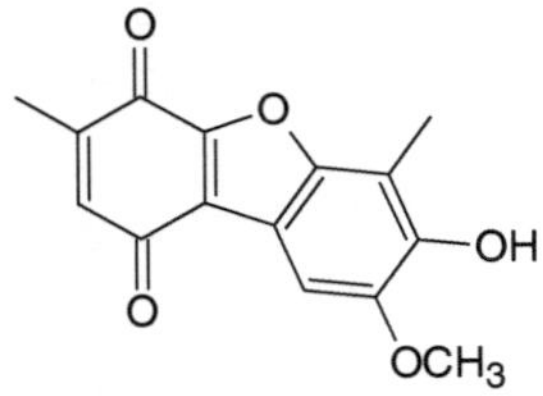

xylariaquinone A

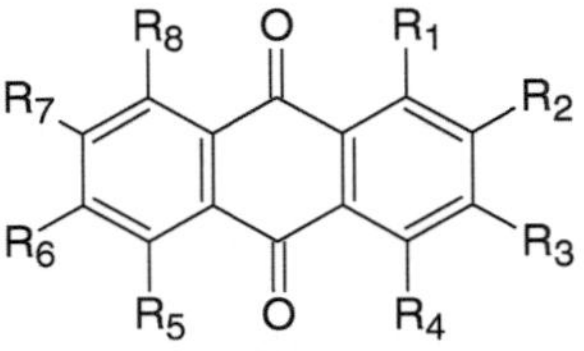

anthracenedione derivative1	R1=R3=OCH3, R6 =CH3
anthracenedione derivative5	R1=R4=OH, R7 =CH3
anthracenedione derivative6	R1=OH,R3=CH3
anthracenedione derivative7	R1=R8=OH
anthracenedione derivative 9	R1=R3=OH, R6 =OCH3,R8=CH3
anthracenedione derivative14	R1=OH, R2=R4 =OCH3,R7=CH3
undefined R=H	

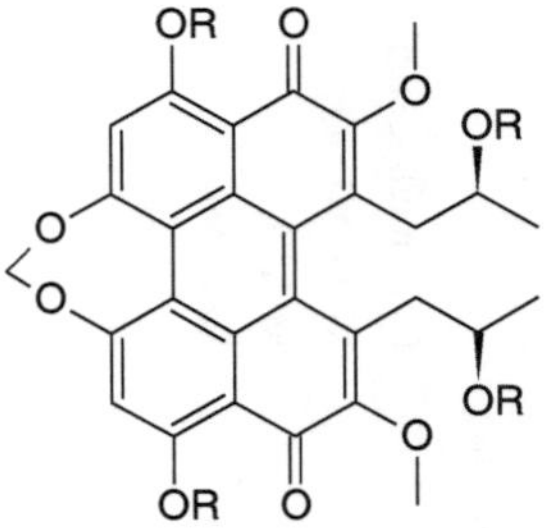

cercosporin 1 R= H
cercosporin 1 R = Ac

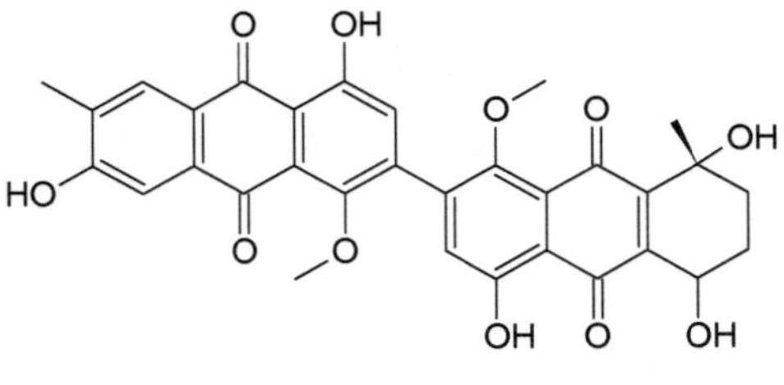

alterporriol K

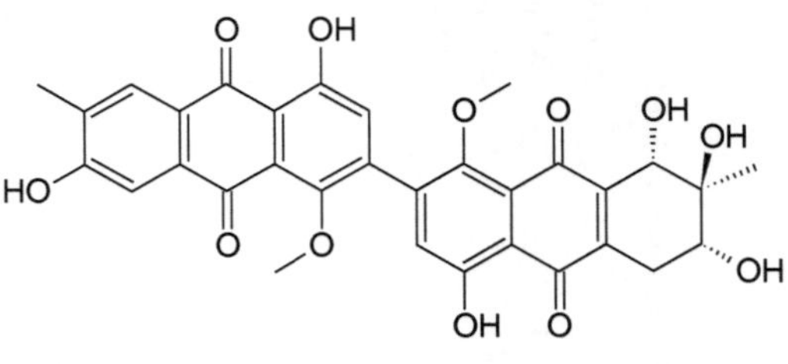

alterporriol L

Fig. 20.3 Chemical structure of quinones isolated from endophytic fungi

EC_{50} of 7.0 µg/mL, compared to 71.5 µg/mL for compound isocochliodinol (Debbab et al. 2009a). Two novel benzoquinone derivatives, 2-chloro-5-methoxy-3-methylcyclohexa-2, 5-diene-1,4-dione and xylariaquinone A, were isolated from *Xylaria* sp., an endophytic fungus of *Sandoricum koetjape*. These compounds showed potent cytotoxicity against African green monkey kidney fibroblasts (Vero cells) with an IC_{50} value of 1.35 µM compared to the positive control ellipticine, with an IC_{50} value of 2.03 µM (Tansuwan et al. 2007).

Five novel and eight known compounds were isolated from *Stemphylium globuliferum*, an endophyte of the Egyptian medicinal plant *Mentha pulegium* (Debbab et al. 2009a, b). Each of the compounds isolated from this fungus was tested for cytotoxicity against L5178Y mouse lymphoma cells. Of the 14 previously reported anthracenedione derivatives were isolated from *Halorosellinia* sp. (no. 1403) and *Guignardia* sp. (no. 4382), fungal endophytes of an unspecified mangrove plant (Zhang et al. 2010). All fourteen compounds exhibited some degree of cytotoxicity, but six anthracenedione compounds 1, 5, 6, 7, 9, and 14 exhibited the greatest potency. These compounds, anthracenedione derivatives, exhibited cytotoxicity towards KB and KBv200 cell lines, with IC_{50} values between 3.7 and 70 µM. Anthracenedione 6 compound was the most potent, with an IC_{50} value of 3.17 µM (KB) and 3.21 (KBv200). Anthracenediones 1, 5, and 9 also exhibited cytotoxicity against KBv200, with IC_{50} values between 3.21 and 91 µM. The literature suggests that both the number and location of hydroxyl groups play a key role in cytotoxicity. Anthracenedione 6 has a single hydroxyl group and is the most potent cytotoxic agent against both cell lines (Zhang et al. 2010).

3.2.4 Polyketides

The novel oblongolides Y and Z were isolated from *Phomopsis* sp. BCC 9789 associated with *Musa acuminate* (wild banana) (Taridaporn et al. 2010). Oblongolide Y showed cytotoxic activity against human breast cancer cell line BC with an IC_{50} value of 48 µM, while oblongolide Z exhibited cytotoxicity against KB (human oral epidermoid cancer), BC, and NCI-H187 (small-cell lung cancer), and nonmalignant (Vero) cell lines with IC_{50} values of 37, 26, 32, and 60 µM, compared to doxorubicin as a positive control, which had IC_{50} values of 0.24 µM (KB), 0.30 µM (BC), and 0.08 µM (NCI-H187) (Taridaporn et al. 2010). An endophytic *Alternaria* sp., isolated from the Egyptian medicinal plant *Polygonum senegalense*, produced several tricyclic lactone polyketides including the known alternariol, alternariol 5-*O*-sulfate, and alternariol 5-*O*-methyl ether. Alternariol 5-*O*-sulfate has not been previously reported (Aly et al. 2008). These compounds were cytotoxic to L5178Y mouse lymphoma cells with EC_{50} values of 1.7, 4.5, and 7.8 µg/mL, respectively, compared to the positive control kahalalide F, which had an EC_{50} value of 6.3 µg/mL (Table 20.4, Fig. 20.4).

The same fungal endophyte also produced two bicyclic acid derivatives – the known altenusin and the novel desmethylaltenusin (Aly et al. 2008). These compounds also exhibited significant cytotoxic activity against L5178Y cells, with EC_{50} values of 6.8 and 6.2 µg/mL, respectively, compared to the positive control kahalalide F (Aly et al. 2008). Leptosphaerone C and penicillenone are novel polyketides isolated from *Penicillium* sp. JP-1, an endophytic fungus associated with the mangrove plant *Aegiceras corniculatum* (Lin et al. 2008a, b). Leptosphaerone C showed activity against A549 cells (adenocarcinomic human alveolar basal epithelial) with an IC_{50} value of 1.45 µM, and penicillenone exhibited cytotoxicity against P388 leukemia cells with an IC_{50} value of 1.38 µM (Lin et al. 2008a, b). Another mangrove endophyte *Phomopsis* sp. ZSU-H76wasthesourceof2-(7′-hydroxyoxooctyl)-3-hydroxy-5-methoxybenzeneacetic acid ethyl ester, a new polyketide. The endophyte was isolated from the stem of *Excoecaria agallocha* from Dong Zhai, Hainan, China (Huang et al. 2009). This compound exhibited cytotoxicity towards Hep-2 and HepG2 cell lines, with IC_{50} values of 25 and 30 µg/mL (Huang et al. 2009). Arugosins A and B are benzophenone polyketides isolated from *Emericella nidulans* var. *acristata*, an endophyte of a Mediterranean green alga (Kralj et al. 2006). Both compounds showed

Table 20.4 Polyketide isolated from endophytic fungi

Compound	Activity	Cell line/target enzyme	Host	Fungal endophyte	Reference
Oblongolide Y	48 μM[a]	BC	*Musa acuminata*	*Phomopsis* sp.	Taridaporn et al. (2010)
Oblongolide Z	37.0, 26.0 μM[a]	KB, BC,			
	32.0 μM[a]	NCI-H187			
	60 μM[a]	Vero cells			
Alternariol	1.7 μg/mL[b]	L5178Y	*Polygonum senegalense*	*Alternaria* sp.	Aly et al. (2008)
Alternariol 5-*O*-sulfate	4.5 μg/mL[b]	L5178Y			
Alternariol 5-*O*-methyl ether	7.8 μg/mL[b]	L5178Y			
Altenusin	6.8 μg/mL[b]	L5178Y	*Polygonum senegalense*	*Alternaria* sp.	
Desmethylaltenusin	6.2 μg/mL[b]	L5178Y			
Leptosphaerone C 5	1.45 μM[a]	A549	*Aegiceras corniculatum*	*Penicillium* sp.	Lin et al. (2008a, b)
Penicillenone	1.38 μM[a]	P388			
2-(7′-Hydroxyoxooctyl)-3-hydroxy-5-methoxybenzene acetic acid ethylester	25 and 30 μg/mL[a]	HEp-2 and HepG2	*Excoecaria agallocha*	*Phomopsis* sp.	Huang et al. (2009)
Arugosin A	10 μg/mL[a]	7 out of 36	Mediterranean green alga	*Emericella nidulans* var. *acristata*	Kralj et al. (2006)
Arugosin B	–	Human tumor			
Bikaverin	0.43 μM[a]	NCI-H460	*Cylindropuntia echinocarpa*	*Fusarium oxysporum*	Zhan et al. (2007)
	0.26 μM[a]	MIA Pa Ca-2			
	0.42, 0.38 μM[a]	MCF-7, SF-268			

Sequoiatones A	4–10 μM[c]	BC	*Sequoia sempervirens*	*Aspergillus parasiticus*	Stierle et al. (1999, 2003)
Sequoiatones B	4–10 μM[c]	BC			
Sequoiamonascin A	1 % cell growth[d]	MCF7			
	1 % cell growth[d]	NCI-H460			
	2 % cell growth[d]	SF-268 (CNS)			
Sequoiamonascin B	19 % cell growth[d]	MCF7			
	4 % cell growth[d]	NCI-H460			
	15 % cell growth[d]	SF-268 (CNS)			
Kasanosin A	27.3 μM, 35.0 μM[a]	DNA pol β, γ	Seaweed	*Talaromyces* sp.	Kimura et al. (2008)
Kasanosin B	60.1 μM, 72.9 μM[a]				
Hypericin Emodin 69	fungal extract, cell viability-1.0 % (light)	THP-1	*Hypericum perforatum*	*Thielavia subthermophila*	Kusari et al. (2009)
Rubrofusarin B	4.5 μg/mL[a]	SW1116	*Cynodon dactylon*	*Aspergillusniger*	Song et al. (2004)
Botryorhodines A	96.97 μM[a]	HeLa	*Botryosphaeria rhodina*	*Bidens pilosa*	Abdou et al. (2010)
Botryorhodines B	36.41 μM[a]	HeLa			
Penicitides A and B	27.3 μM[a]	HepG2	*Laurencia*	*Penicillium chrysogenum*	Gao et al. (2011)

[a]IC_{50}
[b]EC_{50}
[c]IG_{50}
[d]10 mM

moderate antitumor activity against 7 out of 36 human tumor cell lines at a concentration of 10 μg/mL (Kralj et al. 2006). The reference compound Adriamycin, tested in parallel in the same assays, was more potent, with an IC_{50} value of 0.016 μg/mL.

Bikaverin, a polyketide isolated from *Fusarium oxysporum* strain CECIS, an endophyte of *Cylindropuntia echinocarpa*, exhibited cytotoxicity against a panel of four sentinel cancer cell lines, NCI-H460 (non-small cell lung), MIA Pa Ca-2 (pancreatic), MCF-7 (breast), and SF-268 (CNS glioma) with IC_{50} values of 0.43, 0.26, 0.42, and 0.38 μM, respectively. It was compared to the standard compound doxorubicin, which exhibited IC_{50} values of 0.01, 0.05, 0.07, and 0.04 μM, respectively (Zhan et al. 2007).

Two novel polyketides, sequoiatone A and B, were isolated from the endophyte *Aspergillus parasiticus* from the bark of *Sequoia sempervirens* (Stierle et al. 1999). The compounds showed moderate and somewhat selective inhibition of human tumor cells, with greatest efficacy against breast cancer cell lines. Most of the GI_{50} values were between 4 and 10 μM, with LC_{50} values >100 μM (Stierle et al. 1999). Botryorhodines A and B are benzophenone polyketides isolated from *Bidens pilosa*, an endophyte of *Botryosphaeria rhodina*

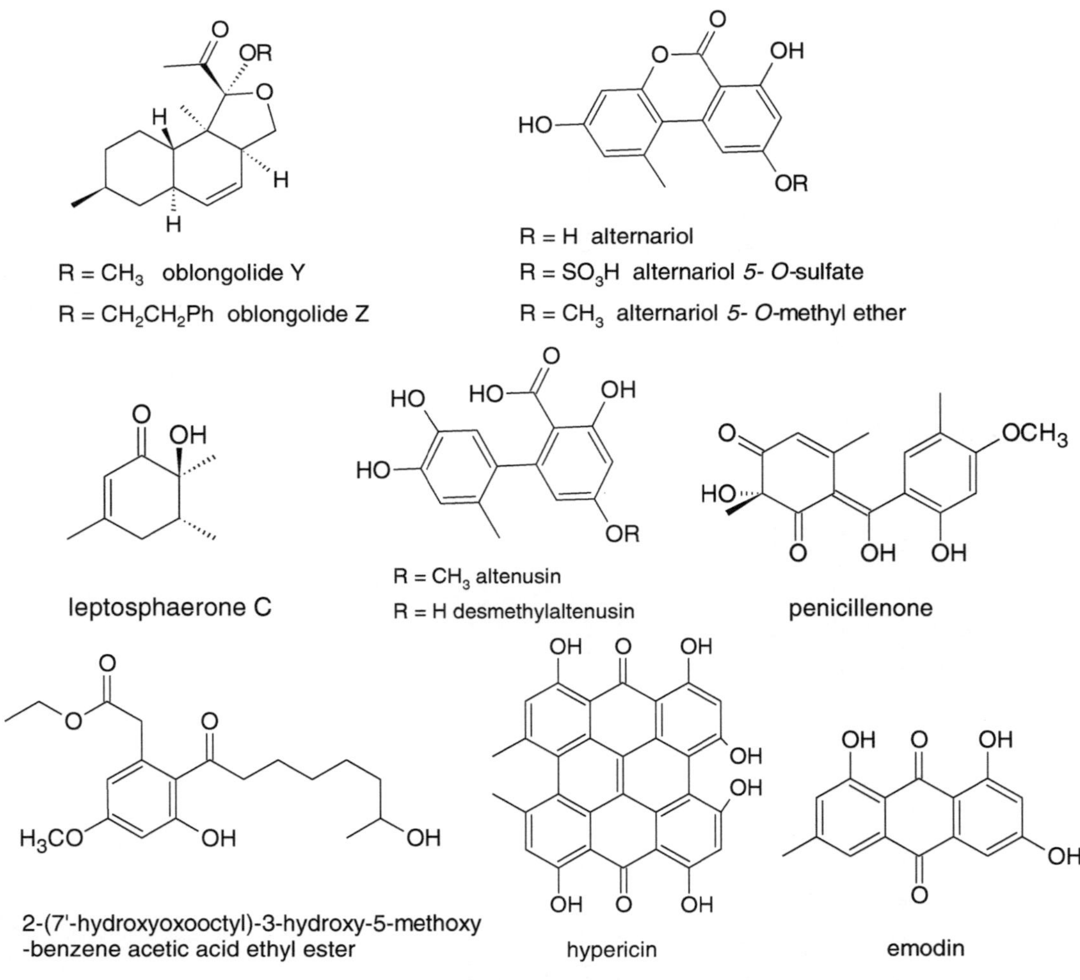

Fig. 20.4 Chemical structure of polyketide isolated from endophytic fungi

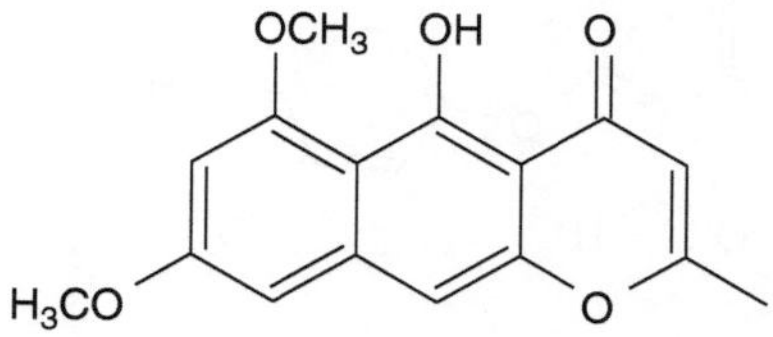

Fig. 20.4 (continued)

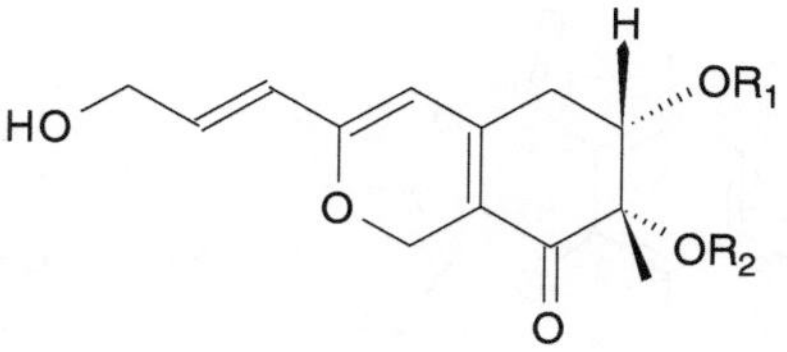

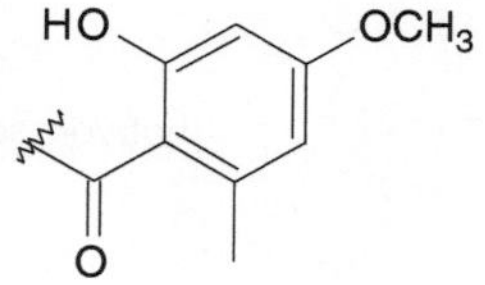

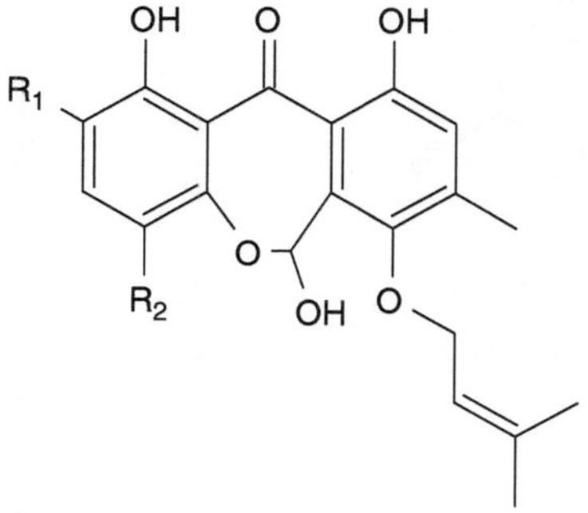

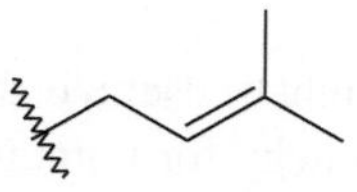

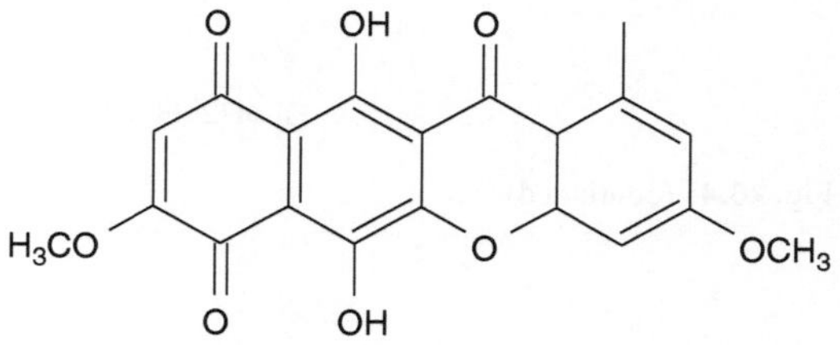

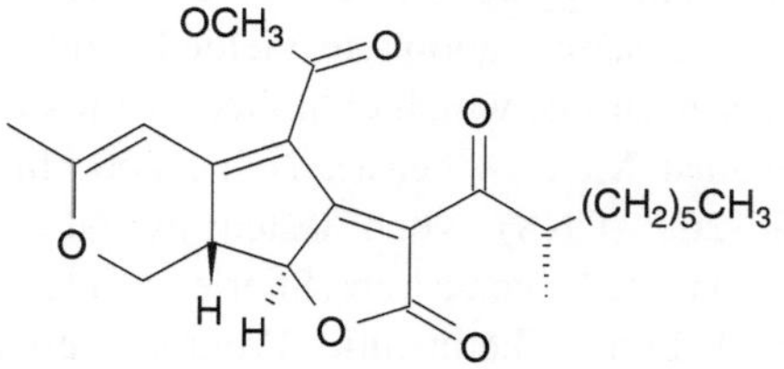

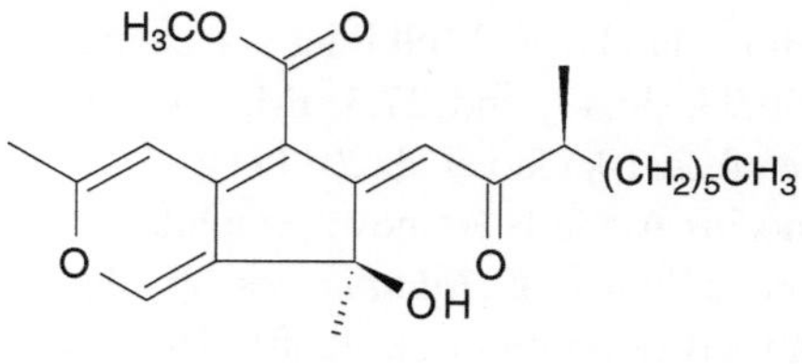

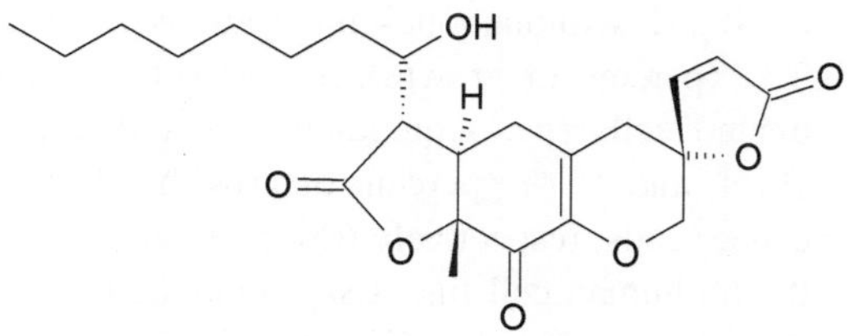

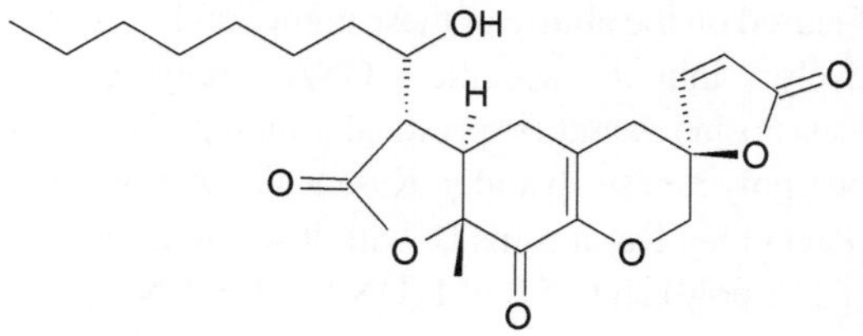

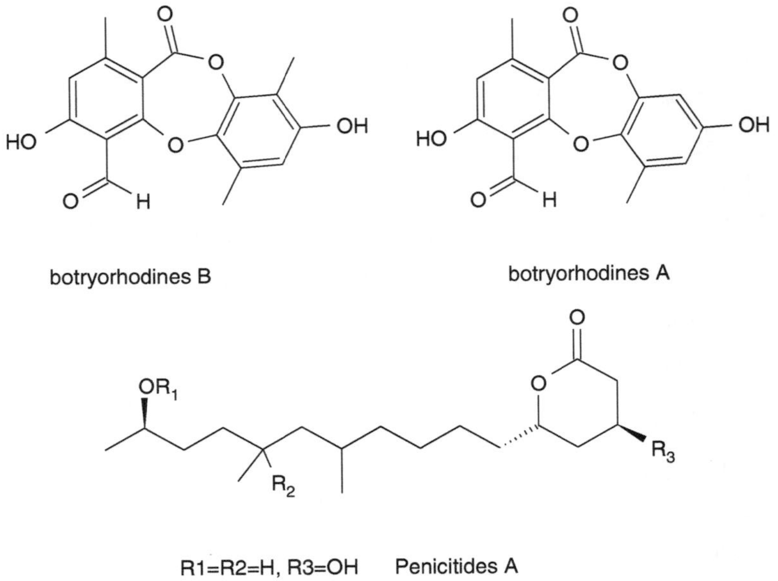

botryorhodines B

botryorhodines A

R1=R2=H, R3=OH Penicitides A

Fig. 20.4 (continued)

(Abdou et al. 2010), and penicitides A and B isolated from *Penicillium chrysogenum*, an endophyte of *Laurencia* (Gao et al. 2011). Both compounds showed moderate antitumor activity against HeLa and HepG2 cell lines at a concentration of 96.97, 36.41, and 27.3 μM, respectively (Abdou et al. 2010; Gao et al. 2011).

Kasanosins A and B are novel azaphilones isolated from cultures of *Talaromyces* sp. derived from seaweed (Kimura et al. 2008). These compounds were not evaluated for cytotoxicity against specific cancer cell lines. Instead, the authors focused on the ability of these compounds to selectively inhibit specific DNA polymerases. Kasanosins A and B specifically inhibited eukaryotic polymerases β and γ. Kasanosins A was more potent than Kasanosins B, with IC_{50} values of 27.3 (DNA polβ) and 35.0 μM (DNA polγ). DNA polymerases are important target molecules of antitumor agents, especially for antimetabolite nucleosides that include 1-β-D-arabinofuranosylcytosine (araC) and 2′-deoxy-2′,2′-difluorocytidine (gemcitabine) (Miura and Izuta 2004). Kasanosins A and B have very high specificity for families of DNA polymerases, which might be useful in the development of a drug design strategy for immunosuppressive and/or anticancer chemotherapy agents (Kimura et al. 2008).

The same endophyte yielded sequoiamonascins A and B, which exhibited cytotoxic activity against MCF7 (breast), NCI-H460 (lung), and SF-268 (CNS) when tested by NCI in their human cell line screen (Stierle et al. 2003). The NCI Drug Therapeutic Program reported the activity in terms of percent of growth of treated cells compared to untreated cells; values below 32 % were considered active. At concentrations of 10 μM, sequoiamonascins A allowed 1, 1, and 2 % (percent of growth), respectively, for each treated cell type; sequoiamonascins B allowed 19, 4, and 15 % (percent of growth) of treated cancer cells, respectively (Stierle et al. 2003). In the 60-human cell line assay, sequoiamonascins A had a median log GI_{50} of –5.00, below the potency threshold established by NCI to warrant further study. Sequoiamonascins A showed selective activity towards all six leukemia cell lines, one breast cancer cell line, and two melanoma cell lines, with median log GI_{50} values approaching –6.00 (Stierle et al. 2003).

Hypericin, along with emodin, was isolated from a stem endophyte similar to *Chaetomium globosum* of *Hypericum perforatum* harvested in India (Kusari et al. 2008). The organism was ultimately identified as *Thielavia subthermophila* (Kusari et al. 2009). The fungal extract containing both compounds exhibited photodynamic cytotoxicity against the human acute monocytic leukemia cell line (THP-1) in two different assays. In the resazurin-based assay, dark vs. light cell viability was 92.7 vs. 4.9 %, and in the ATPlite assay, dark vs. light cell viability was 91.1 vs. 1.0 % (Kusari et al. 2009). The known naphtha-γ-pyrone rubrofusarin B was isolated from *Aspergillus niger* IFB-E003, an endophyte of *Cynodon dactylon*. It was cytotoxic to colon cancer cell line SW1116, with an IC_{50} value of 4.5 μg/mL, compared to the positive control 5-fluorouracil at 5 μg/mL (Song et al. 2004). Rubrofusarin B also reversed multidrug resistance of human epidermal KB carcinoma cells (Song et al. 2004).

3.2.5 Some Other Anticancer Compounds Produced by Endophytic Fungi

Chromones

Pestalotiopsone F (5-carbomethoxy-methyl-7-hydroxy-2-pentylchromone) is a novel chromone isolated from the fungus *Pestalotiopsis* sp., an endophyte of the Chinese mangrove plant *Rhizophora mucronata* (Xu et al. 2009). Compound displayed moderate cytotoxicity against the murine cancer cell line L5178Y, with an EC_{50} value of 8.93 μg/mL (Xu et al. 2009). Four novel isoprenylated chromone derivatives, pestalofíciol I-L (heterodimer), were isolated from *Pestalotiopsis fici*, a fungal endophyte of *Camellia sinensis*. The IC_{50} values of the 4 compounds ranged between 8.7 and >136.1 μM for HeLa cells and between 17.4 and >153.8 μM for MCF7 cells, compared to the positive control 5-fluorouracil with IC_{50} values of 10.0 and 15.0 μM, respectively (Ling et al. 2009). It exhibited the most potent cytotoxicity, with IC_{50} values of 8.7 and 17.4 μM, respectively (Table 20.5, Fig. 20.5).

Benzo[j]fluoranthenes

Daldinone C and daldinone D were isolated from *Hypoxylon truncatum* IFB-18, an endophyte of *Artemisia annua*. Both compounds exhibited potent cytotoxicity against SW1116 cells (human colorectal cancer cell line), with IC_{50} values of 49.5 and 41.0 μM, respectively, comparable to that of 5-fluorouracil (37.0 μM) (Gu et al. 2007) (Table 20.5, Fig. 20.5).

Cyclohexanones

The known compound epiepoxydon (Nagasawa et al. 1978; Nagata et al. 1992; Iwamoto et al. 1999) was isolated from a marine endophyte, *Apiospora montagnei* of the North Sea alga *Polysiphonia violacea* (Klemke et al. 2004). In the brine shrimp assay, the compound was strongly cytotoxic. It exhibited an LC_{50} of 3.6 μg/mL for the breast adenocarcinoma cell line MCF7 and GI_{50} concentrations of 0.7 μg/mL for the human gastric carcinoma HM02, 0.75 μg/mL for the human liver carcinoma HepG2, and 0.8 μg/mL for MCF7. Total growth inhibition (TGI) for these cell lines was also determined and was found to be 1.0 μg/mL for HM02, 4.6 μg/mL for HepG2, and 1.5 μg/mL for MCF7. In the case of HM02 and HepG2 cells, the LC_{50} of compound27 was >10 μg/mL. Epiepoxydon was previously reported to have an ED_{50} of 0.2 μg/mL towards the P388 lymphocytic leukemia cell line (Iwamoto et al. 1999) (Table 20.5, Fig. 20.5).

Depsidones

Depsidone 1 was isolated from an endophytic fungus of the order *Pleosporales* (BCC 8616) that was isolated from an unidentified leaf of the Hala-Bala evergreen forest (Pittayakhajonwut et al. 2006). Its exhibited weak cytotoxic activity against KB and BC cell lines, with IC_{50} values of 6.5 and 4.1 μg/mL, respectively (Pittayakhajonwut et al. 2006) (Table 20.5, Fig. 20.5).

Depsipeptides

Beauvericin is a depsipeptide isolated from *Fusarium oxysporum* EPH2RAA, an endophytic fungus of the Sonoran Desert plant *Ephedra fasciculata* (Zhan et al. 2007). It has previously been

Table 20.5 Anticancer compounds isolated from endophytic fungi

Chemical nature	Compound	Activity	Cell line/target enzyme	Host	Fungal endophyte	Reference
Depsipeptide	Beauvericin	1.41 mM[a]	NCI-H460	*Ephedra fasciculata*	*Fusarium oxysporum*	Zhan et al. (2007)
		1.66 mM[a]	MIA Pa Ca-2			
		1.81, 2.29 mM[a]	MCF-7, SF-268			
Ergochrome	Ergoflavin	1.9, 0.1 mM[a]	TNF-a	*Mimusops elengi*	PM0651480	Deshmukh et al. (2009)
		1.2, 0.3 mM[a]	IL-6			
		1.2, 4.0 mM[a]	ACHN, H-460			
		2.4, 8.0 mM[a]	Panc1, HCT116			
		1.5 mM[a]	Calu1			
	Dicerandrol A	7.0, 7.0 mg/mL[b]	A549, HCT-116	*Dicerandra frutescens*	*Phomopsis longicolla*	Wagenaar and Clardy (2001)
	Dicerandrol B	1.8, 1.8 mg/mL[b]	A549, HCT-116			
	Dicerandrol C	1.8, 7.0 mg/mL[b]	A549, HCT-116			
	Secalonic acid D	0.38, 0.43 mM[a]	HL-60, K562	Mangrove plant	ZSU44	Zhang et al. (2009)
Ester	Globosumone A	6.50, 21.30 mM[a]	NCI-H460, MCF-7	*Ephedra fasciculata*	*Chaetomium globosum*	Bashyal et al. (2005)
		8.80, 13.00 mM[a]	SF-268, WI-38			
		10.60 mM[a]	MIA Pa Ca-2			
	Globosumone B	24.80, 21.90 mM[a]	NCI-H460, MCF-7			
		29.10, 14.20 mM[a]	SF-268, WI-38			
		30.20 mM[a]	MIA Pa Ca-2			
Lactone	Brefeldin A	10.0 ng/mL[a]	HL-60	*Taxus mairei*	*Aspergillus clavatus*	Wang et al. (2002)
		9.0 ng/mL[a]	KB	*Torreya grandis*	*Paecilomyces* sp.	
		2.0 ng/mL[a]	MCF-7			
		1.0 ng/mL[a]	Spc-A-1			
		1.8 ng/mL[a]	HeLa			
		0.18, 0.04 mM[a]	KB, BC-1	*Knema laurina*	*Acremonium* sp.	Chinworrungsee et al. (2008)
		0.11 mM[a]	NCI-H187			
	Radicicol	0.03 mM[a]	MCF-7	*Ephedra fasciculata*	*Chaetomium chiversii*	Turbyville et al. (2006)
	Photinides A–F	10 mg/mL[c]	MDA-MB-231	*Roystonea regia*	*Pestalotiopsis photiniae*	Ding et al. (2009)
	Eutypellin A	12, 84 mM[a]	NCI-H187, MCF-7	*Etlingera littoralis*	*Eutypella* sp.	Isaka et al. (2009)
		38, 88 mM[a]	KB, Vero			

Lignan	Podophyllotoxin	–	Topoisomerase I	*Podophyllum hexandrum*	*Trametes hirsuta*	Puri et al. (2006)
		–	Topoisomerase I	*Podophyllum peltatum*	*Phialocephala fortinii*	Eyberger et al. (2006)
Peptide	Leucinostatin A	2.0 nM[d]	BT-20	*Taxus baccata*	*Acremonium* sp.	Strobel and Hess (1997)
Spirobis naphthalene	Spiromamakone A	0.33 mM[a]	P388	*Knightia excelsa*	Mycelia sterilia	van der Sar et al. (2006)
	Spiropreussione A	2.4 mM[a]	A2780	*Aquilaria sinensis*	*Preussia* sp.	Chen et al. (2009)
		3.0 mM[a]	BEL-7404			
Xanthone	Phomoxanthone A	0.99, 0.51, 1.4 mg/mL[a]	KB, BC-1, Vero	*Tectona grandis*	*Phomopsis* sp.	Isaka et al. (2001)
	Phomoxanthone B	4.1, 0.70, 1.8 mg/mL[a]	KB, BC-1, Vero			
Aldehyde	Chaetopyranin	15.4 mg/mL[a]	HMEC	*Polysiphonia urceolata*	*Chaetomium globosum*	Wang et al. (2006)
		28.5 mg/mL[a]	SMMC-7721			
		39.1 mg/mL[a]	A549			
Benzo[j]fluoranthene	Daldinone C	49.5 mM[a]	SW1116	*Artemisia annua*	*Hypoxylon truncatum*	Gu et al. (2007)
	Daldinone D	41.0 mM[a]	SW1116			
Chromone	Pestalotiopsone F	8.93 mg/mL[e]	L5178Y	*Rhizophora mucronata*	*Pestalotiopsis* sp.	Xu et al. (2009)
	Pestalofliciol I	>136.1, 136.1 mM[a]	HeLa, MCF7	*Camellia sinensis*	*Pestalotiopsis fici*	Ling et al. (2009)
	Pestalofliciol J	21.2, >153.8 mM[a]	HeLa, MCF7			
	Pestalofliciol K	99.3, >132.5 mM[a]	HeLa, MCF7			
	Pestalofliciol L	8.7, 17.4 mM[a]	HeLa, MCF7			
Cyclohexanone	Epiepoxydon	0.70 mg/mL[f]	HM02	*Polysiphonia violacea*	*Apiospora montagnei*	Klemke et al. (2004)
		0.75 mg/mL[f]	HepG2			
		0.8 mg/mL[f]	MCF7			
Depsidone	Depsidone 1	6.5, 4.1 mg/mL[a]	KB, BC	Leaf, Hala-Bala Forest	BCC 8616	Pittayakhajonwut et al. (2006)

[a] IC_{50}
[b] IC_{100}
[c] IG_{25}
[d] LD_{50}
[e] EC_{50}
[f] IG_{50}
[g] 10 mM

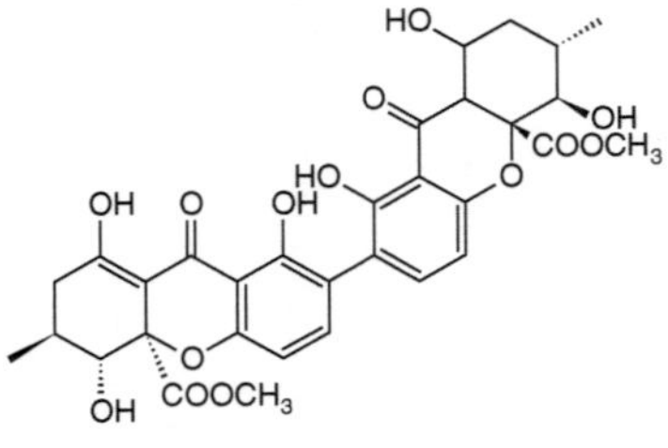

secaionic acid D

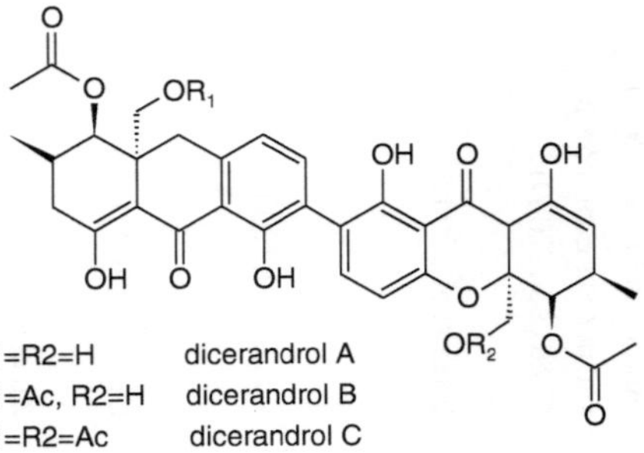

R1=R2=H dicerandrol A
R1=Ac, R2=H dicerandrol B
R1=R2=Ac dicerandrol C

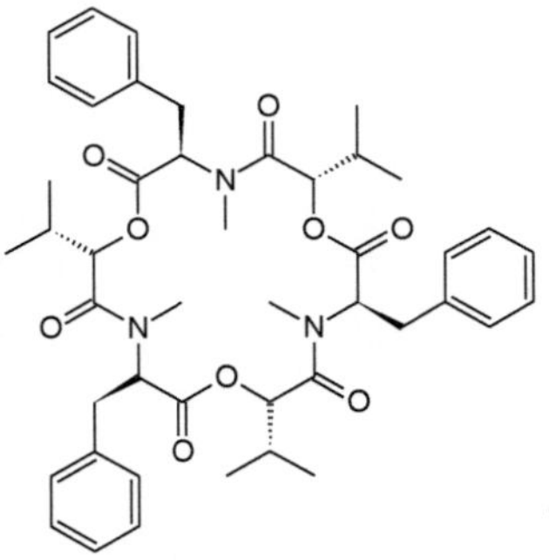

beauvericin

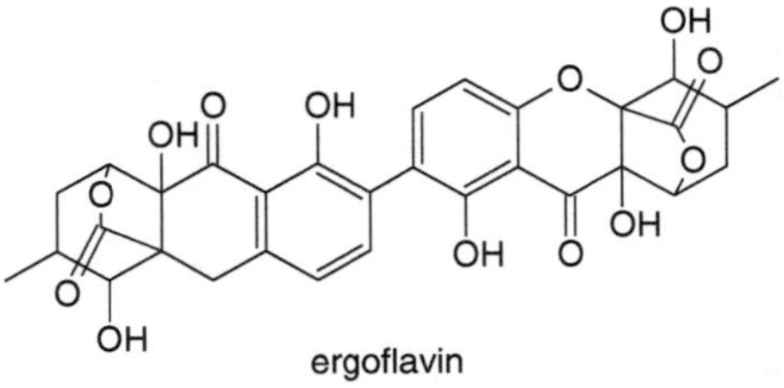

ergoflavin

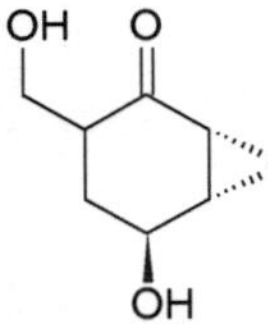

(+) epiepoxydon

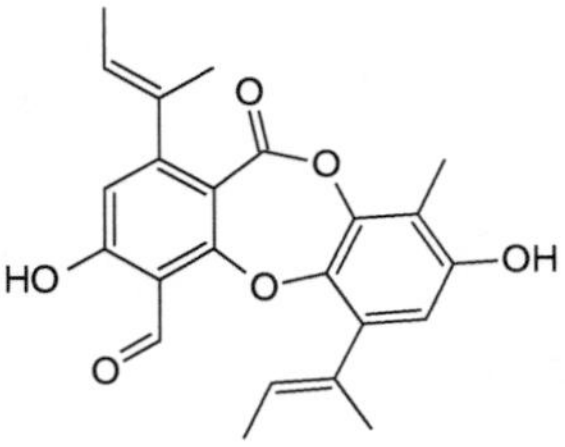

depsidone 1

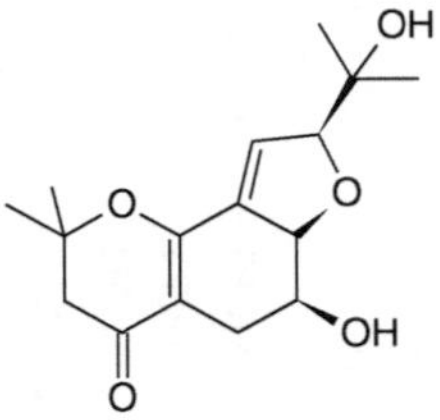

pestaloficiol I

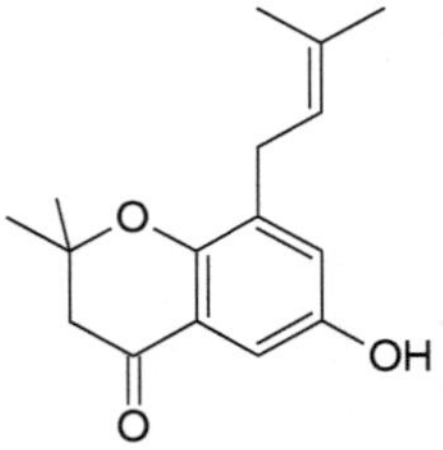

pestaloficiol J

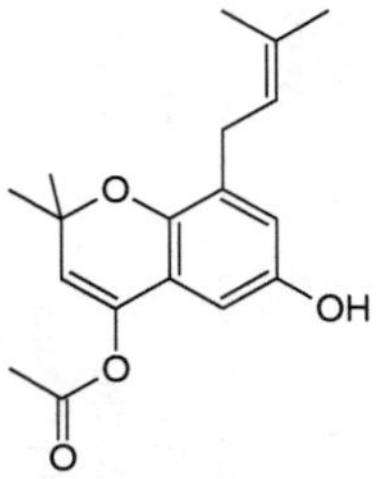

pestaloficiol K

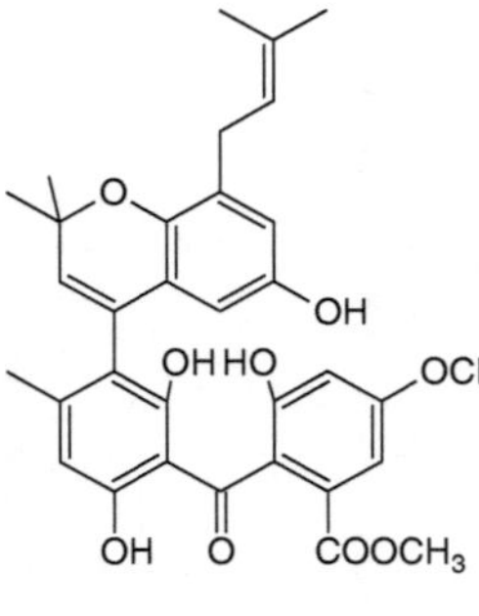

pestaloficiol L

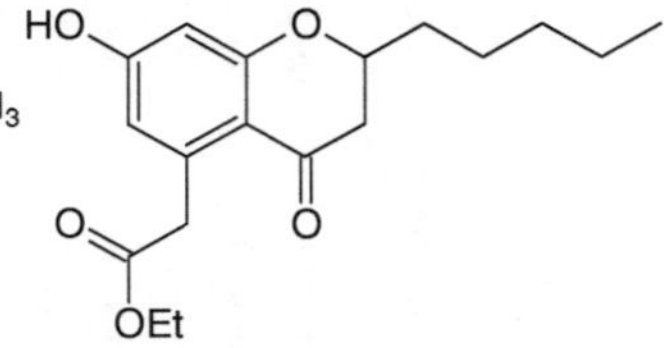

pestalotiopsone F

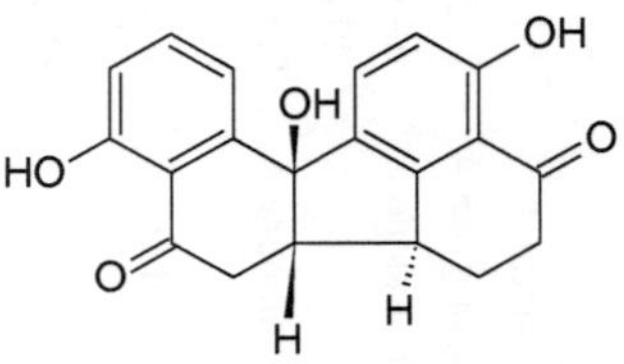

daldinone C

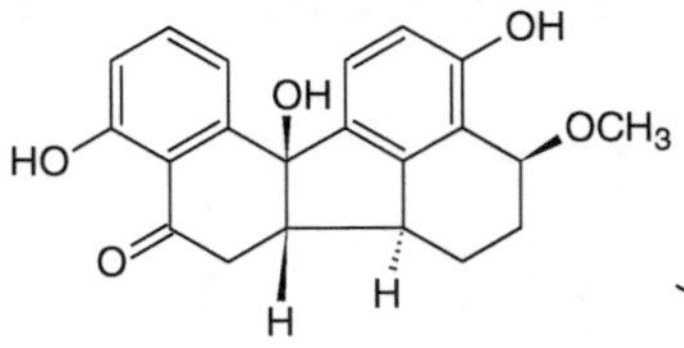

daldinone D

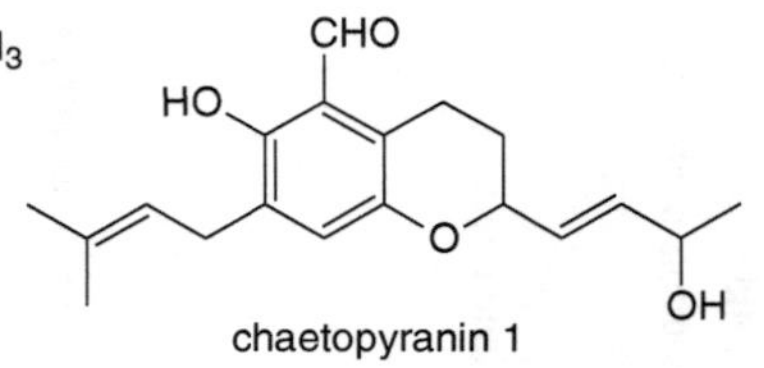

chaetopyranin 1

Fig. 20.5 Chemical structure of anticancer compounds isolated from endophytic fungi

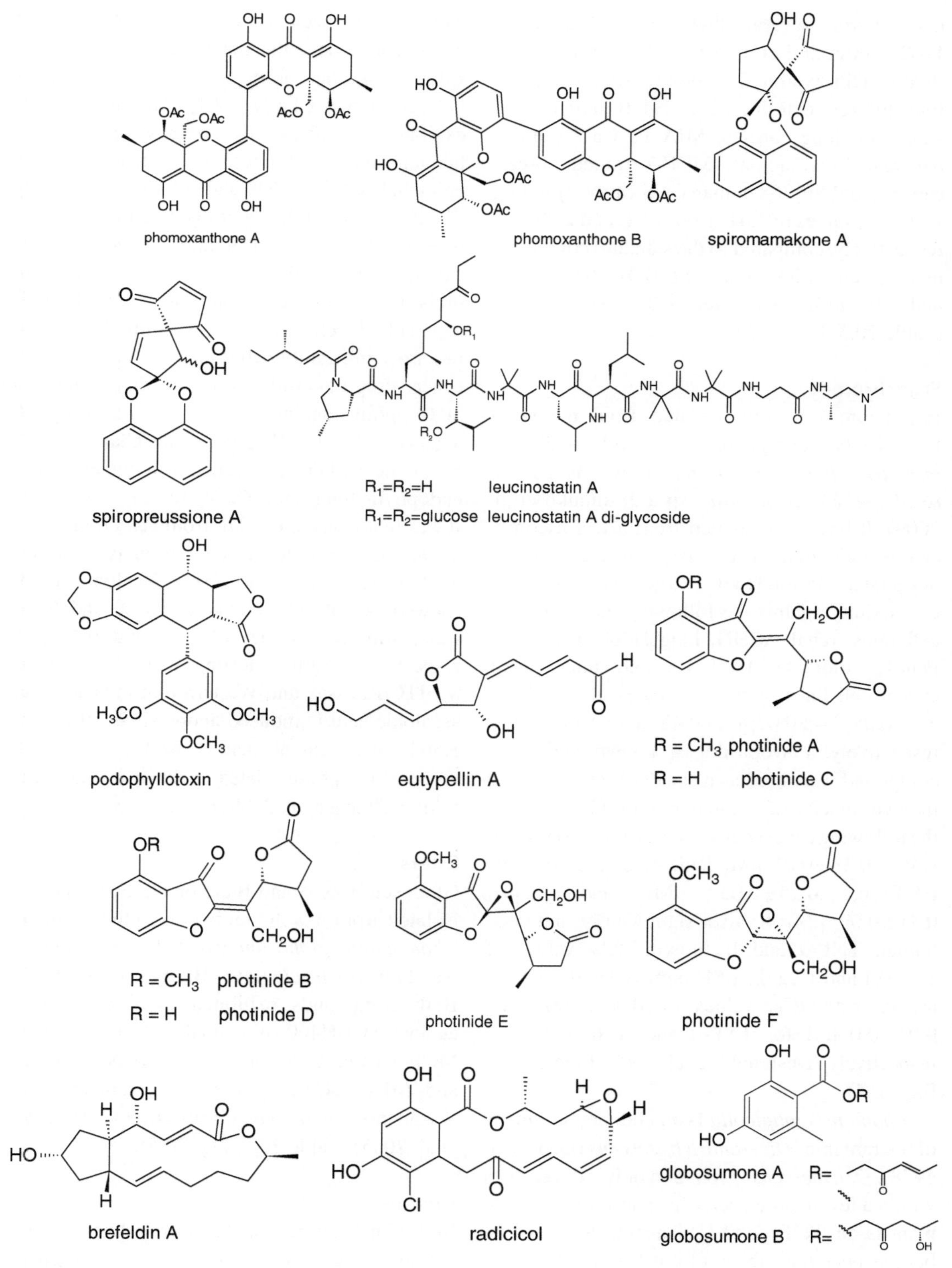

phomoxanthone A

phomoxanthone B

spiromamakone A

spiropreussione A

$R_1=R_2=H$ leucinostatin A

$R_1=R_2=$glucose leucinostatin A di-glycoside

podophyllotoxin

eutypellin A

$R = CH_3$ photinide A

$R = H$ photinide C

$R = CH_3$ photinide B

$R = H$ photinide D

photinide E

photinide F

brefeldin A

radicicol

globosumone A

globosumone B

Fig. 20.5 (continued)

isolated from several other fungi (Hamill et al. 1969; Bernardini et al. 1975; Deol et al. 1978). Beauvericin exhibited cytotoxic activity against four different cell lines, NCI-H460 (human non-small cell lung cancer), MIA Pa Ca-2 (human pancreatic carcinoma), MCF-7 (human breast cancer), and SF-268 human CNS cancer (glioma) with IG_{50} values of 1.41, 1.66, 1.81, and 2.29 μM, respectively, compared to the standard compound doxorubicin with values of 0.01, 0.05, 0.07, and 0.04 μM, respectively (Zhan et al. 2007) (Table 20.5, Fig. 20.5).

Ergochromes

Ergoflavin were first isolated from the ergot fungus *Claviceps purpurea*, as well as *Phoma terrestris*, *Pyrenochaeta terrestris*, *Penicillium oxalicum*, and *Aspergillus* sp. (Deshmukh et al. 2009). It has been isolated from a leaf ascomycetous endophyte of *Mimusops elengi* ('Bakul') designated PM0651480. Ergoflavin exhibited cytotoxicity against the following human cancer cell lines: renal ACHN, lung H460, pancreatic Panc1, colorectal HCT116, and lung Calu1 cancer cell lines, with IC_{50} values of 1.2 ± 0.20, 4.0 ± 0.08, 2.4 ± 0.02, 8.0 ± 0.45, and 1.5 ± 0.21 μM, respectively. Flavopiridol, a known anticancer compound, was used as a standard for evaluating the cytotoxicity of ergoflavin with IC_{50} values in the following cancer cells: ACHN, 0.84 ± 0.03 μM; H460, 0.38 ± 0.01 μM; Panc-1, 0.23 ± 0.07 μM; HCT116, 0.25 ± 0.03 μM; and Calu1, 0.41 ± 0.09 μM. It also significantly inhibited human TNF-α and IL-6, with IC_{50} values of 1.9 ± 0.1 and 1.2 ± 0.3 μM compared to dexamethasone, with IC_{50} values of 0.06 ± 0.007 and 0.01 ± 0.0 μM for TNF-α and IL-6 inhibition, respectively (Deshmukh et al. 2009) (Table 20.5, Fig. 20.5).

Phomopsis longicolla is an endophytic fungus of the rare mint *Dicerandra frutescens* (Deol et al. 1978). *D. frutescens* is found in only a dozen sites within a few hundred acres in central Florida. The plant is on the Federal Endangered Species List but has been the subject of much study due to its rich chemistry (Eisner et al. 1990; McCormick et al. 1993). The fungal endophyte produced three compounds designated dicerandrols A, B, and C, which have been classified as ergochromes because they have the same tricyclic C_{15} system with a similar arrangement of substituents (Wagenaar and Clardy 2001). The dicerandrols exhibited significant cytotoxicity against two human cancer cell lines, lung adenocarcinoma epithelial cell line A549 and colorectal HCT-116. The IC_{100} value of dicerandrols A against both cell lines and the value of dicerandrols C against HCT-116 was 7.0 μg/mL. The IC_{100} value of dicerandrols C against A549 and of dicerandrols B against both cell lines was 1.8 μg/mL. These values are significantly better than the standard anticancer drug etoposide, which has IC_{100} values of 30.0 μg/mL against A549 and 125.0 μg/mL against HCT-116 (Wagenaar and Clardy 2001). Secalonic acid D was isolated from the mangrove endophytic fungus no. ZSU44 (Zhang et al. 2009). It was first isolated in 1970 from *Penicillium oxalicum* and was found to be extremely toxic and teratogenic (Zhang et al. 2009). Secalonic acid D showed potent cytotoxicity to HL60 and K562 cells, with IC_{50} values of 0.38 and 0.43 μM, respectively. Further testing with the Annexin V-FITC/PI assay and Western blot indicated that secalonic acid D induced apoptosis in HL60 and K562 cells. Secalonic acid D also led to cell cycle arrest of G1 phase related to down regulation of c-Myc (Zhang et al. 2009) (Table 20.5, Fig. 20.5).

Esters

Globosumones A and B are orsellinic acid esters isolated from a well-studied endophytic fungus, *Chaetomium globosum* isolated from Mormon tea, *Ephedra fasciculata* (Bashyal et al. 2005). Both compounds exhibited cytotoxic activity against NCI-H460 (non-small cell lung cancer), MCF-7 (breast cancer), SF-268 (CNS glioma), and MIA PaCa-2 (pancreatic carcinoma) and WI-38 normal human fibroblast cells (Bashyal et al. 2005) (Table 20.5, Fig. 20.5).

Lactones

Brefeldin A has been isolated from several fungal species including *Curvularia*, *Alternaria*, *Ascochyta*, *Phyllosticta*, *Penicillium*, and *Cercospora* (Wang et al. 2002). The compound has antifungal, anticancer, and antiviral activities.

Brefeldin A was isolated from two different endophytic fungi, *Aspergillus clavatus* and *Paecilomyces* sp., which were isolated from the tissues of Chinese *Taxus mairei* and *Torreya grandis*. This compound showed strong cytotoxicity against HL-60, KB, HeLa, MCF-7, and Spc-A-1 cell lines, with IC_{50} values of 10.0, 9.0, 1.8, 2.0, and 1.0 ng/mL, compared to the standard anticancer compound paclitaxel, which had IC_{50} values of 1.2, 0.16, 1.8, 5.0, and 0.8 ng/mL, respectively (Wang et al. 2002). Brefeldin A was also isolated from a new species of *Acremonium* which was isolated from a healthy twig of the Thai medicinal plant *Knema laurina* (Chinworrungsee et al. 2008). In this study, brefeldin A showed potent activity against the following human cancer cell lines: KB (epidermoid cancer of the mouth), BC-1 (breast cancer), and NCI-H187 (small-cell lung cancer), with IC_{50} values of 0.18, 0.04, and 0.11 μM, respectively (Chinworrungsee et al. 2008) (Table 20.5, Fig. 20.5).

The known compound radicicol was isolated from *Chaetomium chiversii*, an endophytic fungus of *Ephedra fasciculata*, as part of an ongoing investigation of the endophytes of Sonoran Desert plants and of inhibitors of HSP90 (heat shock protein). Hsp90 may play a critical role in the cancer phenotype and thus provide an effective target for cancer chemotherapy. Cancer cells frequently express high levels of Hsp90, presumably in response to the stress conditions within the tumor microenvironment. Radicicol also exhibited antiproliferative activity against breast cancer cell line MCF-7, with an IC_{50} value of 0.03 μM (Turbyville et al. 2006). Six novel benzofuranone-derived γ-lactones, photinides A–F, were isolated from *Pestalotiopsis photiniae*, an endophyte of *Roystonea regia* (Ding et al. 2009). All six γ-lactones exhibited cytotoxicity against breast cancer cell line MDA-MB-231 with inhibitory rates of 24.4, 24.2, 23.1, 24.4, and 24.6 %, respectively, at a concentration of 10 μg/mL. Eutypellin A is a γ-lactone that exhibited cytotoxic activity against NCI-H187 (human small-cell lung cancer cells), MCF-7, KB, and nonmalignant Vero cells with IC_{50} values of 12, 84, 38, and 88 μM compared to the standard ellipticine, which exhibited IC_{50} values of 3.6, 2.5, and 5.5 μM, respectively (Isaka et al. 2009). Eutypellin A was isolated from the endophytic fungus *Eutypella* sp. BCC 13199, itself isolated from *Etlingera littoralis* (Earth ginger) (Isaka et al. 2009).

Lignans

The aryltetralin lignan podophyllotoxin is an important natural product which was originally isolated in 1950 from the higher plant *Podophyllum emodi* (Leiter et al. 1950). Podophyllotoxin is the precursor to three anticancer drugs, the topoisomerase I inhibitors etoposide, teniposide, and etoposide phosphate (Eyberger et al. 2006; Puri et al. 2006). Endophytes capable of producing podophyllotoxin and related analogues (Eyberger et al. 2006; Puri et al. 2006). Puri isolated *Trametes hirsute* from the dried rhizomes of *Podophyllum hexandrum* collected from the Himalayan region, India (Puri et al. 2006), while Porter and colleagues isolated two different strains of *Phialocephala fortinii* from rhizomes of *Podophyllum peltatum* (Eyberger et al. 2006) (Table 20.5, Fig. 20.5).

Peptides

Leucinostatin A was isolated almost 40 years ago from cultures of *Penicillium lilacinum* (Arai et al. 1973). Scientists have found that it inhibits prostate cancer growth through the reduction of insulin-like growth factor-I expression in prostate stromal cells (Kawada et al. 2010). *Acremonium* sp. isolated from *Taxus baccata* was also shown to produce Leucinostatin A when grown in liquid culture (Strobel and Hess 1997). The fungal endophyte also produced leucinostatin A di-O-β-glucoside, a glycosylated analogue of leucinostatin A which had an LD_{50} of >25 nM against breast cancer cell line BT-20, compared to leucinostatin A, which had an LD_{50} of 2 nM (Strobel and Hess 1997) (Table 20.5, Fig. 20.5).

Spirobisnaphthalenes

The spirobisnaphthalenes are a relatively new class of compounds that was first reported in 1990. They possess two naphthalene-derived C_{10} units bridged through a spiroketal linkage. Spiromamakone A was isolated from an unspecified nonsporulating endophytic fungus

(Mycelia sterilia) isolated from the native New Zealand tree *Knightia excelsa* (rewarewa) (van der Sar et al. 2006). Spiromamakone A exhibited potent cytotoxic activity against P388 (murine leukemia cell line), with an IC_{50} value of 0.33 μM. The compound also exhibited potent antimicrobial activity (van der Sar et al. 2006). This endophytic fungus *Preussia* sp. was isolated from a mature stem of *Aquilaria sinensis* (Thymelaeaceae), collected from Guangxi Medicinal Arboretum (Chen et al. 2009). It produced a series of novel spirobisnaphthalenes, one of which, spiropreussione A, exhibited in vitro cytotoxicity against the A2780 human ovarian carcinoma cell line and the BEL-7404 human liver carcinoma cell line, with IC_{50} values of 2.4 and 3.0 μM, respectively. Spiropreussione A was inactive (IC_{50}> 10 μM) against the HCT-8 (colon carcinoma), BGC-823 (gastric carcinoma), and A549 (lung adenocarcinoma) human cancer cell lines. None of the other novel compounds exhibited cytotoxicity in these assays at the concentrations tested (Chen et al. 2009) (Table 20.5, Fig. 20.5).

Xanthones

Phomoxanthones A and B, two novel xanthone dimers, were isolated from the fungus *Phomopsis* sp. BCC 1323, an endophyte of *Tectona grandis*. Both compounds exhibited impressive cytotoxic activity against KB cells, BC-1 cells, and nonmalignant Vero cells. Phomoxanthone A had IC_{50} values of 0.99, 0.51, and 1.4 μg/mL, respectively, while phomoxanthone B had IC_{50} values of 4.1, 0.70, and 1.8 μg/mL, compared to the standard compound ellipticine, which had IC_{50} values of 0.46 μg/mL for KB cells and 0.60 μg/mL for BC-1 cells, respectively (Isaka et al. 2001) (Table 20.5, Fig. 20.5).

Aldehydes

Chaetopyranin was evaluated for its radical scavenging abilities using DPPH (1,1-diphenyl-2-picrylhydrazyl). This compound is a benzaldehyde derivative isolated from the endophytic fungus *Chaetomium globosum* associated with the marine red alga *Polysiphonia urceolata* (Wang et al. 2006). Chaetopyranin exhibited moderate or weak cytotoxic activities against three human tumor cell lines: HMEC (human microvascular endothelial cells), SμMC-7721 (hepatocellular carcinoma cells), and A549 (human lung epithelial cells) with IC_{50} values of 15.4, 28.5, and 39.1 μg/mL, respectively. This compound showed moderate activity with an IC_{50} value of 35 μg/mL, compared to an IC_{50} value of 18 μg/mL for the positive control BHT (butylated hydroxytoluene) (Wang et al. 2006) (Table 20.5, Fig. 20.5).

4 Conclusion

Endophytic fungi are prolific producers of secondary metabolites, in particular, are of considerable interest to researchers and pharmacists due to their ability to synthesize a wide range of economically important bioactive molecules. This chapter highlights the importance of endophytic fungi – those hidden, subtle inhabitants of the interstitial spaces in plants – as a source of secondary metabolites with promising anticancer activity. Access to a limited number of cancer chemotherapies, their serious side effects, and high cost make treatment particularly challenging. In addition, many therapies do not effectively treat certain cancers, and multidrug-resistant tumors exacerbate treatment complexity. The search for new anticancer agents and for new sources of potent plant-derived compounds is critical, considering the number of deaths associated with cancers on an annual basis and the likelihood that this number will increase in the future. The discovery of new chemotherapeutic agents is therefore a key goal for natural product and medicinal chemists. Many of the compounds discussed in this chapter had IC_{50} values comparable to those of the standard reference drugs, making the search for anticancer compounds isolated from endophytic fungi a promising one. In the past 10 years, more than 100 compounds with significant cytotoxicity were reported from endophytic fungi, and the isolation of anticancer compounds has been increasing over 5 year intervals – it is interesting to note from 1990 to 1995, only a single novel anticancer agent was reported from endophytic fungi. This discovery spurred interest not only in fungal endophytes as a source of novel anticancer agents but also in endophytes as an alternative source of valuable higher-plant metabolites. A fungal source of a desired anticancer agent is of particular value, as fungal fermentation provides a virtually inexhaustible source of desired metabolites. As natural products chemists turn their attention to endophytic fungi, the number of new compounds isolated should increase over the next 5 years. Novel compounds or previously isolated compounds are readily available and accessible to whatever specific anticancer screens researchers use for isolation and evaluation. As our understanding of the mechanisms associated with the onset and metastasis of cancers increases, our ability to use this knowledge to select for ever more potent and selective compounds should increase commensurately. Endophytic fungi will continue to provide a fertile arena for these quests.

Acknowledgment The authors are thankful to the Director, NBAIM, for providing the necessary facilities and support for this work.

References

Abdou R, Scherlach K, Dahse HM, Sattler I, Hertweck C (2010) Botryorhodines A–D, antifungal and cytotoxic depsidones from *Botryosphaeria rhodina*, an endophyte of the medicinal plant *Bidens pilosa*. Phytochemistry 71:110–116

Aly AH, Edrada-Ebel RA, Indriani ID, Wray V, Müller WEG, Totzke F, Zirrgiebel U, Schächtele C, Kubbutat MHG, Lin WH, Proksch P, Ebel R (2008) Cytotoxic metabolites from the fungal endophytic *Alternaria* sp. and their subsequent detection in its host plant Polygonum senegalense. J Nat Prod 71:972–980

Arai T, Mikami Y, Fukushima K, Utsumi T, Yazawa K (1973) A new antibiotic, leucinostatin, derived from *Penicillium lilacinum*. J Antibiot 26:157–161

Arnold AE (2005) Diversity and ecology of fungal endophytes in tropical forests. In: Deshmukh D (ed) Current trends in mycological research. Oxford & IBH Publishing Co. Pvt. Ltd., New Delhi, pp 49–68

Arnold AE (2007) Understanding the diversity of foliar endophytic fungi: progress, challenges, and frontiers. Fungal Biol 21:51–66

Arnold AE, Mejia LC, Kyllo D, Rojas EI, Maynard Z, Robbins N, Herre EA (2003) Fungal endophytes limit pathogen damage in a tropical tree. Proc Natl Acad Sci U S A 100:15649–15654

Arnold AE, Henk DA, Eells RA, Lutzoni F, Vilgalys R (2007) Diversity and phylogenetic affinities of foliar fungal endophytes in loblolly pine inferred by culturing and environmental PCR. Mycologia 99:185–206

Bacon CW, White JF Jr (2000) Physiological adaptation in the evolution of endophytism in the Clavicipitaceae. In: Bacon CW, White JF Jr (eds) Microbial endophytes. Marcel Dekker, Inc., New York, pp 237–261

Bary AD (1866) Morphologie und physiologie der plize, Flechten, und Myxomyceten. In: Hofmeister's handbook of physiological botany, vol 2. Engelmann, Leipzig

Bashyal BP, Wijeratne EMK, Faeth SH, Gunatilaka AAL (2005) Globosumones AC, cytotoxic orsellinic acid esters from the Sonoran Desert endophytic fungus *Chaetomium globosum*. J Nat Prod 68:724–728

Bernardini M, Carilli A, Pacioni G, Santurbano B (1975) Isolation of Beauvericin from *Paecilomyces fumosoroseus*. Phytochemistry 14:1865

Bischoff JF, White JF Jr (2005) Evolutionary development of the Clavicipitaceae. In: Dighton J, White JF, Oudemans P (eds) The fungal community: its organization and role in the ecosystem. Taylor & Francis, Boca Raton, pp 505–518

Brundrett MC (2006) Understanding the roles of multifunctional mycorrhizal and endophytic fungi. In: Schulz BJE, Boyle CJC, Sieber TN (eds) Microbial root endophytes. Springer, Berlin, pp 281–293

Cancer Facts & Figures, American Cancer Society (2013) P-49 (Please visit http://www.cancer.org/docroot/NWS/content/NWS)

Chen Y, Guo H, Du Z, Liu XZ, Che Y, Ye X (2009) Ecology-based screen identifies new metabolites from a *Cordyceps*-colonizing fungus as cancer cell proliferation inhibitors and apoptosis inducers. Cell Prolif 42(6):838–847

Chinworrungsee M, Wiyakrutta S, Sriubolmas N, Chuailua P, Suksamrarn A (2008) Cytotoxic activities of trichothecenes isolated from an endophytic fungus belonging to order hypocreales. Arch Pharm Res 31:611–616

Chokpaiboon S, Sommit D, Teerawatananond T, uangsin N, Bunyapaiboonsri T, Pudhom K (2010) Cytotoxic nor-chamigrane and chamigrane endoperoxides from a basidiomycetous fungus. J Nat Prod 73: 1005–1007

Cragg GM, Newman DJ (2005) Plants as source of anticancer agents. J Ethnopharmacol 100:72–79

Cui CM, Li XM, Li CS, Proksch P, Wang BG (2010) Cytoglobosins A–G, cytochalasans from a marine-derived endophytic fungus, *Chaetomium globosum* QEN-14. J Nat Prod 73:729–733

Davis EC, Shaw AJ (2008) Biogeographic and phylogenetic patterns in diversity of liverwort-associated endophytes. Am J Bot 95(8):914–924

Davis EC, Franklin JB, Shaw AJ, Vilgalys R (2003) Endophytes *Xylaria* among liverworts and angiosperm: phylogenetic, distribution and symbiosis. Am J Bot 40:1661–1667

Debbab A, Aly AH, Edrada-Ebel RA, Muller WEG, Mosaddak M, Hakiki A, Ebel R, Proksch P (2009a) Bioactive secondary metabolites from the endophytic fungus *Chaetomium* sp. isolated from *Salvia officinalis* growing in Morocco. Biotechnol Agron Soc Environ 13:229–234

Debbab A, Aly AH, Edrada-Ebel RA, Wray V, Müller WEG, Totzke F, Zirrgiebel U, Schächtele C, Kubbutat MHG, Lin WH, Mosaddak M, Hakiki A, Proksch P, Ebel R (2009b) Bioactive metabolites from endophytic fungus *Stemphylium globuliferum* isolated from *Mentha pulegium*. J Nat Prod 72:626–631

Demain AL (2000) Microbial natural products: a past with a future. In: Wrigley SK, Hayes MA, Thomas R, Chrystal EJT, Nicholson N (eds) Biodiversity: new leads for pharmaceutical and agrochemical industries. The Royal Society of Chemistry, Cambridge, pp 3–16

Deol BS, Ridley DD, Singh P (1978) Isolation of cyclodepsipeptides from plant pathogenic fungi. Aust J Chem 31:1397–1399

Deshmukh SK, Mishra PD, Kulkarni-Almeida A, Verekar S, Sahoo MR, Periyasamy G, Goswami H, Khanna A, Balakrishnan A, Vishwakarma RA (2009) Anti-inflammatory and anticancer activity of Ergoflavin isolated from an endophytic fungus. Chem Biodivers 6:784–789

Ding G, Song YC, Chen JR, Xu C, Ge HM, Wang XT, Tan RX (2006) Chaetoglobosin U, a cytochalasan alkaloid from endophytic *Chaetomium globosum* IFB-E019. J Nat Prod 69:302–304

Ding G, Zheng Z, Liu S, Zhang H, Guo L, Che Y (2009) Photinides A-F, cytotoxic benzofuranone-derived γ-lactones from the plant endophytic fungus Pestalotiopsis photiniae. J Nat Prod 72:942–945

Eisner T, McCormick KD, Sakaino M, Eisner M, Smedley SR, Aneshansley DJ, Deyrup M, Myers RL, Meinwald J (1990) Chemical defense of a rare mint plant. Chemoecology 1:30–37

Espinel MA, Laszlo A, Simonsen L, Boulahbal F, Kim SJ, Reniero A, Hoffner S, Rieder HL, Binkin N, Dye C, Williams R, Raviglione MC (2001) Global trends in resistance to antituberculosis drugs. N Eng J Med 344:1294–1303

Eyberger AL, Dondapati R, Porter JR (2006) Endophyte fungal isolates from *Podophyllum peltatum* produces podophyllotoxin. Nat Prod 69:1121–1124

Faeth SH, Gardner DR, Hayes CJ, Jani A, Wittlinger SK, Jones TA (2006) Temporal and spatial variation in alkaloid levels in *Achnatherum robustum*, a native grass infected with the endophyte *Neotyphodium*. J Chem Ecol 32:307–324

Fir áková S, Šturdíková M, Múčková M (2007) Bioactive secondary metabolites produced by microorganisms associated with plants. Biologia 62(3):251–257

Fisher PJ, Graf F, Petrini LE, Sutton BC, Wookey PA (1995) Fungal endophytes of *Dryas octopetala* from a high arctic polar semidesert and from the Swiss Alps. Mycologia 87:319–323

Freeman EM (1904) The seed-fungus of *Lolium temulentum* L. the Darnel. Philos Trans R Soc B 196:1–27

Gangadevi V, Muthumary J (2008) Taxol, an anticancer drug produced by an endophytic fungus *Bartalinia robillardoides* Tassi, isolated from a medicinal plant, *Aegle marmelos* Correa ex Roxb. W J Microbiol Biotechnol 24:717–724

Gangadevi V, Muthumary J (2009a) Taxol production by *Pestalotiopsis terminaliae*, an endophytic fungus of *Terminalia arjuna* (arjun tree). Biotechnol Appl Biochem 52:9–15

Gangadevi V, Muthumary J (2009b) A novel endophytic Taxol-producing fungus *Chaetomella raphigera* isolated from a medicinal plant, *Terminalia arjuna*. Appl Biochem Biotechnol 158:675–684

Gangadevi V, Murugan M, Muthumary J (2008) Taxol determination from *Pestalotiopsis pauciseta*, a fungal endophyte of a medicinal plant. Chin J Biotechnol 24:1433–1438

Gao SS, Li XM, Du FY, Li CS, Proksch P, Wang BG (2011) Secondary metabolites from a Marine-derived endophytic fungus *Penicillium chrysogenum* QEN-24S. Mar Drugs 9:59–70

Ge HM, Yu ZG, Zhang J, Wu JH, Tan RX (2009) Biphenylquinolizidine alkaloids from *Lagerstroemia indica*. J Nat Prod 72:753–775

Gond SK, Verma VC, Kumar A, Kumar V, Kharwar RN (2007) Study of endophytic fungal community from different parts of *Aegle marmelos* Correae (Rutaceae) from Varanasi (India). W J Microbiol Biotchnol 23:1371–1375

Goodman LS, Wintrobe MM, Dameshek W, Goodman MJ, Gilman A (1946) Use of methyl-Bis(Beta-Chloroethyl)amine hydrochloride and Tris(Beta-Chloroethyl)amine Hydrochloride for Hodgkin's Disease, Lymphosarcoma, leukemia and certain allied and miscellaneous disorder. J Am Med Assoc 132:126–132

Gu W, Ge HM, Song YC, Ding H, Zhu HL, Zhao XA, Tan RX (2007) Cytotoxic benzo[j]fluoranthene metabolites from *Hypoxylon truncatum* IFB-18, an endophyte of *Artemisia annua*. J Nat Prod 70:114–117

Guerin P (1898) Sur la presenced'un champignon dans l'ivraie. J Bot 12:230–238

Haidle AM, Myers AG (2004) An enantioselective, modular, and general route to the cytochalasins: synthesis of L-696,474 and cytochalasin B. Proc Natl Acad Sci 101:12048–12053

Hamill RL, Higgens CE, Boaz HE, Gorman M (1969) The structure of beauvericin, a new depsipeptide antibiotic toxic to *Artemia salina*. Tetrahedron Lett 10:4255–4258

Higgins KL, Arnold AE, Miadlikowska J, Sarvate SD, Lutzoni F (2007) Phylogenetic relationships, host affinity, and geographic structure of boreal and arctic endophytes from three major plant lineages. Mol Phylogenet Evol 42:543–555

Hoffman MT, Arnold AE (2008) Geographic locality and host identity shape fungal endophyte communities in cupressaceous trees. Mycol Res 112:331–344

Huang WY, Cai YZ, Surveswaran S, Hyde KD, Corke H, Sun M (2009) Molecular phylogenetic identification of endophytic fungi isolated from three *Artemisia* species. Fungal Divers 36:69–88

Huang CH, Pan JH, Chen B, Yu M, Huang HB, Zhu X, Lu YJ, She ZG, Lin YC (2011) Three Bianthraquinone derivatives from the mangrove endophytic fungus *Alternaria* sp. ZJ9-6B from the South China Sea. Mar Drugs 9:832–843

Isaka M, Jaturapat A, Rukseree K, Danwisetkanjana K, Tanticharoen M, Thebtaranonth Y (2001) Phomoxanthones A and B, Novel xanthone dimers from the endophytic fungus *Phomopsis* Species. J Nat Prod 64:1015–1018

Isaka M, Palasarn S, Lapanun S, Chanthaket R, Boonyuen N, Lumyong S (2009) Gamma-lactones and ent-eudesmane sesquiterpenes from the endophytic fungus *Eutypella* sp. BCC 13199. J Nat Prod 72:1720–1722

Isaka M, Chinthanom P, Boonruangprapa T, Rungjindamai N, Pinruan U (2010) Eremophilane-type sesquiterpenes from the fungus *Xylaria* sp. BCC 21097. J Nat Prod 73:683–687

Iwamoto C, Minoura K, Oka T, Ohta T, Hagishita S, Numata A (1999) Absolute stereostructures of novel cytotoxic metabolites, penostatins A–E, from a *Penicillium* species separated from an Enteromorpha alga. Tetrahedron 55:14353–14368

Jiao RH, Xu S, Liu JY, Ge HM, Ding H, Xu C, Zhu HL, Tan RX (2006) Chaetominine, a cytotoxic alkaloid produced by endophytic *Chaetomium* sp. IFB-E015. Org Lett 8:5709–5712

Johnson IS, Wright HF, Svoboda GH (1959) Experimental basis of clinical evaluation of anti-tumor principles from *Vinca rosea* Linn. J Lab Clin Med 54:830

Jumpponen A (2001) Dark septate endophytes – are they mycorrhizal? Mycorrhiza 11:207–211

Kawada M, Inoue H, Ohba SI, Masuda T, Momose I, Ikeda D (2010) Leucinostatin A inhibits prostate cancer growth through reduction of insulin-like growth factor-I expression in prostate stromal cells. Int J Cancer 1:810–818

Kharwar RN, Verma VC, Strobel G, Ezra D (2008) The endophytic fungal complex of *Catharanthus roseus* (L.) G. Don. Curr Sci 95:228–233

Kimura T, Nishida M, Kuramochi K, Sugawara F, Yoshida H, Mizushina Y (2008) Novel azaphilones, kasanosins A and B, which are specific inhibitors of eukaryotic DNA polymerases β and λ from *Talaromyces* sp. Bioorg Med Chem 16:4594–4599

Kingston DGI (2005) In: Cragg GM, Kingston DG, Newman DJ (eds) Anticancer agents from natural products. Taylor & Francis Group/CRC Press, Boca Raton, pp 89–120

Klemke C, Kehraus S, Wright AD, König GM (2004) New secondary metabolites from the marine endophytic fungus *Apiospora montagnei*. J Nat Prod 67:1058–1063

Koulman A, Lane GA, Christensen MJ, Fraser K, Tapper BA (2007) Peramine and other fungal alkaloids are exuded in the guttation fluid of endophyte-infected grasses. Phytochemistry 68:355–360

Kralj A, Kehrarus S, Krick A, Eguereva E, Kelter G, Maurer M, Wortmann A, Fiebig HH, König GM (2006) Arugosins G and H: prenylated polyketides from the marine-derived fungus *Emericella nidulans* var. *acristata*. J Nat Prod 69:995–1000

Kumaran RS, Kim HJ, Hur BK (2010) Taxol promising fungal endophyte, Pestalotiopsisspecies isolated from Taxus cuspidate. J Biosci Bioeng 110(5):541–546

Kusari S, Lamshöft M, Zühlke S, Spiteller M (2008) An endophytic fungus from *Hypericum perforatum* that produces hypericin. J Nat Prod 71:159–162

Kusari S, Lamshoft M, Spiteller M (2009) *Aspergillus fumigatus* Fresenius, an endophytic fungus from *Juniperus communis* L. Horstmann as a novel source of the anticancer prodrug deoxypodophyllotoxin. J Appl Microbiol 107:1019–1030

Lane GA, Christenses MJ, Miles CO (2000) Coevolution of fungal endophytes with grasses: the significance of secondary metabolites. In: Bacon CW, White JF (eds) Microbial endophytes. Marcel Dekker, New York, pp 341–388

Lee JC (1995) Isolation and structure determination of bioactive compounds from endophytic and insect-associated fungi. Ph.D. thesis, Cornell University, New York, pp 9–51

Lee JC, Strobel GA, Lobkovsky E, Clardy JC (1996) Torreyanic acid: a selectively cytotoxic quinone dimer from the endophytic fungus *Pestalotiopsis microspora*. J Org Chem 61:3232–3233

Leiter J, Downing V, Hartwell J et al (1950) Damage induced in sarcoma 37 with podophyllin, podophyllotoxin, alpha-peltatin, beta-peltatin and quercetin. J Natl Cancer Inst 10:1273–1293

Li Y, Strobel GA, Sidhu RS, Hess WM, Ford EJ (1996) Endophytic taxol producing fungi from bald cypress *Taxodium distichum*. Microbiology 142:2223–2226

Lin Z, Zhu T, Fang Y, Gu Q, Zhu W (2008a) Polyketides from *Penicillium* sp. JP-1, an endophytic fungus associated with the mangrove plant *Aegiceras corniculatum*. Phytochemistry 69:1273–1278

Lin ZJ, Lu ZY, Zhu TJ, Fang YC, Gu QQ, Zhu WM (2008b) Penicillenols from *Penicillium* sp. GQ-7, an endophytic fungus associated with *Aegiceras corniculatum*. Chem Pharm Bull 56:217–221

Ling L, Liu S, Niu S, Guo L, Chen X, Che Y (2009) Isoprenylated chromone derivatives from the plant endophytic fungus *Pestalotiopsis fici*. J Nat Prod 72:1482–1486

Ling-Hua M, Zhi-Yong L, Pommier Y (2003) Non-camptothecin DNA Topoisomerase I Inhibitors in cancer therapy. Curr Top Med Chem 3:305–320

Lingqi Z, Guo B, Li H, Zeng S, Shao H, Gu S, Wei R (2000) Preliminary study on the isolation of endophytic fungus of *Catharanthus roseus* and its fermentation to produce products of therapeutic value. Chin Tradit Herbal Drugs 31:805–807

McCormick KD, Deyrup MA, Menges ES, Wallace SR, Meinwald J, Eisner T (1993) Relevance of chemistry to conservation of isolated populations: the case of volatile leaf compounds of Dicerandra mints. Proc Natl Acad Sci 90:7701–7705

Miura S, Izuta S (2004) DNA polymerases as targets of anticancer nucleosides. Curr Drug Targets 5:191–195

Moreno E, Varughese T, Spadafora C, Arnold AE, Coley PD, Kursar TA, Gerwick WH, Rios LC (2011) Chemical constituents of the new endophytic fungus *Mycosphaerella* sp. nov. and their anti-parasitic activity. Nat Prod Commun 6:835–840

Murali TS, Suryanarayanan TS, Venkatesan G (2007) Fungal endophyte communities in two tropical forests of southern India: diversity and host affiliation. Mycol Prog 6:191–199

Nagasawa H, Suzuki A, Tamura S (1978) Isolation and structure of (+)- desoxyepiepoxydon and (+)-epiepoxy-don, phytotoxin fungal metabolites. Agric Biol Chem 42:1303–1304

Nagata T, Ando Y, Hirota A (1992) Phytotoxin from tea gray blight fungi, *Pestalotiopsis longiseta* and *Pestalotiopsis theae*. Biosci Biotechnol Biochem 56:810–811

Noble RL, Beer CT, Cutts JH (1958) Role of chance observations in chemotherapy: *Vinca rosea*. Ann N Y Acad Sci 76:882–894

Panaccione DG (2005) Origins and significance of ergot alkaloid diversity in fungi. FEMS Microbiol Lett 251:9–17

Panaccione DG, Cipoletti JR, Sedlock AB, Blemings KP, Schardl CL, Machado C, Seidel GE (2006) Effects of ergot alkaloids on food preference and satiety in rabbits, as assessed with gene knockout endophytes in perennial ryegrass (*Lolium perenne*). J Agric Food Chem 54:4582–4587

Pandi M, Kumaran RS, Choi YK, Kim HJ, Muthumary J (2011) Isolation and detection of taxol, an anticancer drug produced from *Lasiodiplodia theobromae*, an endophytic fungus of the medicinal plant *Morinda citrifolia*. Afr J Biotechnol 10:1428–1435

Pasut G, Veronese FM (2009) PEGylation for improving the effectiveness of therapeutic biomolecules. Adv Drug Deliv Rev 61:1177–1188

Pillay D, Zambon M (1998) Antiviral drug resistance. Br Med J 317:660–662

Pisani P, Parkin DM, Bray F, Ferlay J (1999) Estimates of the worldwide mortality from 25 cancers in 1990. Int J Cancer 83:870–873

Pittayakhajonwut P, Dramae A, Madla S, Lartpornmatulee N, Boonyuen N, Tanticharoen M (2006) Depsidones from the endophytic fungus BCC 8616. J Nat Prod 69:1361–1363

Pommier Y (2006) Topoisomerase I inhibitors: camptothecins and beyond. Nat Rev Cancer 6:789–802

Puri SC, Verma V, Amna T, Qazi GN, Spiteller M (2005) An endophytic fungus from *Nothapodytes foetida* that produces camptothecin. J Nat Prod 68:1717–1719

Puri SC, Nazir A, Chawla R, Arora R, Riyaz-ul-Hassan S, Amna T, Ahmed B, Verma V, Singh S, Sagar RK, Sharma AK, Kumar R, Sharma RK, Qazi GN (2006) The endophytic *Trametes hirsuta* as a novel alternative source of podophyllotoxin and related aryl tetralin lignans. J Biotechnol 122:494–510

Redecker D, Kodner R, Graham LE (2000) Glomalean fungi from the Ordovician. Science 289:1920–1921

Regina RS, Sheehan KB, Stout RG, Rodriguez RJ, Henson JM (2002) Thermotolerance conferred to plant host and fungal endophyte during mutualistic symbiosis. Science 298:1581

Rehman S, Shawl AS, Kour A, Andrabi R, Sudan P, Sultan P, Verma V, Qazi GN (2008) An endophytic *Neurospora* sp. from *Nothapodytes foetida* producing camptothecin. Appl Biochem Microbiol 44:203–209

Rodriguez RJ, Henson J, Van Volkenburgh E, Hoy M, Wright L, Beckwith F, Kim Y, Redman RS (2008) Stress tolerance in plants via habitat-adapted symbiosis. Int Soc Microb Ecol 2:404–416

Rodriguez RJ, White JFJ, Arnold AE, Redman RS (2009) Fungal endophytes: diversity and functional roles. New Phytol 182:314–330

Rosa LH, Vaz ABM, Caligiorne RB, Campolina S, Rosa CA (2009) Endophytic fungi associated with the Antarctic grass *Deschampsia antarctica* Desv. (Poaceae). Polar Biol 32:161–167

Sadananda TS, Nirupama R, Chaithra K, Govindappa M, Chandrappa CP, Vinay Raghavendra B (2011) Antimicrobial and antioxidant activities of endophytes from Tabebuia argenteaand identification of anticancer agent (lapachol). J Med Plant Res 5(16):3643–3652

Saikkonen K, Faeth SH, Helander M, Sullivan TJ (1998) Fungal endophytes: a continuum of interactions with host plants. Annu Rev Ecol Syst 29:319–343

Saikkonen K, Ion D, Gyllenberg M (2002) The persistence of vertically transmitted fungi in grass metapopulations. Proc R Soc B Biol Sci 269:1397–1403

Schulz BJE, Boyle CJC (2005) The endophytic continuum. Mycol Res 109:661–687

Sekita S (1983) Isocochliodinol and neocochliodinol, bis (3-indolyl)- benzoquinones from *Chaetomium* spp. Chem Pharm Bull 31:2998–3001

Shrestha K, Strobel GA, Prakash S, Gewali M (2001) Evidence for paclitaxel from three new endophytic fungi of Himalayan yew of Nepal. Planta Med 67:374–376

Silvia P, Roberto L, Duckett JG, Davis EC (2008) A novel ascomycetous endophytic association in the rhizoids of the leafy liverwort family, Schistochilaceae (Jungermanniidae, Hepaticopsida). Am J Bot 95:531–541

Song YC, Li H, Ye YH, Shan CY, Yang YM, Tan RX (2004) Endophytic napthopyrone metabolites are co-inhibitors of xanthine oxidase, SW1116 cell and some microbial growths. FEMS Microbiol Lett 241:67–72

Staniek A, Woerdenbag HJ, Kayser O (2008) Endophytes: exploiting biodiversity for the improvement of natural product-based drug discovery. J Plant Interact 3:75–93

Stierle A, Strobel G, Stierle D (1993) Taxol and taxane production by *Taxomyces andreanae*, an endophytic fungus of Pacific yew. Science 260:214–218

Stierle A, Stierle D, Bugni T (1999) Sequoiatones A and B: novel antitumor metabolites of a redwood endophyte. J Org Chem 64:5479–5484

Stierle D, Stierle A, Bugni T (2003) Sequoiamonascins A-D: novel anticancer metabolites isolated from a redwood endophyte. J Org Chem 68:4966–4969

Stone JK, Polishook JD, White JF Jr (2004) Endophytic fungi. In: Mueller G, Bills GF, Foster MS (eds) Biodiversity of fungi: inventory and monitoring methods. Elsevier, Burlington, pp 241–270

Strobel GA (2002) Rainforest endophytes and bioactive products. Crit Rev Biotechnol 22:315–333

Strobel GA, Daisy B (2003) Bioprospecting for microbial endophytes and their natural products. Microbiol Mol Biol 67:491–502

Strobel GA, Hess WM (1997) Glucosylation of the peptide leucinostatin A, produced by an endophytic fungus of European yew, may protect the host from leucinostatin toxicity. Chem Biol 4:529–536

Strobel GA, Sears J, Kramer R, Sidhu RS, Hess WM (1996) Taxol from *Pestalotiopsis microspora* an endophytic fungus of *Taxus wallichiana*. Microbiology 142:435–440

Strobel G, Ford E, Worapong J (2002) Isopestacin, an isobenzofuranone from *Pestalotiopsis microspora*, possessing antifungal and antioxidant activities. Phytochemistry 60:179–183

Suryanarayanan TS, Wittlinger SK, Faeth SH (2005) Endophytic fungi associated with cacti in Arizona. Mycol Res 109:635–639

Svoboda GH (1961) Alkaloids of *Vinca rosea* IX. Extraction and characterization of leurosidine and leurocristine. Lloydia 24:173–178

Swatzell LJ, Powell MJ, Kiss JZ (1996) The relationship of endophytic fungi to the gametophyte of the fern *Schizaea pusilla*. Int J Plant Sci 157:53–62

Tansuwan S, Pornpakakul S, Roengsumran S, Petsom A, Muangsin N, Sihanonta P, Chaichit N (2007) Antimalarial benzoquinones from an endophytic fungus. Xylaria sp. J Nat Prod 70:1620–1623

Taridaporn B, Seangaroon Y, Prasert S, Kitlada S, Saisamorn L (2010) Oblongolides from the endophytic fungus *Phomopsis* sp. BCC 9789. J Nat Prod 73:55–59

Teles HL, Sordi R, Silva GH, Castro-Gamboa I, Bolzani Vda S, Pfenning LH, de Abreu LM, Costa-Neto CM, Young MC, Araújo AR (2006) Aromatic compounds produced by *Periconia atropurpurea*, an endophytic fungus associated with *Xylopia aromatica*. Phytochemistry 67:2686–2690

Thurston DE (2007) Chemistry and pharmacology of anticancer drugs. Taylor & Francis Group/CRC Press, Boca Raton, pp 13–16

Turbyville TJ, Wijeratne EMK, Liu MX, Burns AM, Seliga CJ, Luevano LA, David CL, Faeth SH, Whitesell L, Gunatilaka AAL (2006) Search for Hsp90 inhibitors with potential anticancer activity: isolation and SAR studies of radicicol and monocillin I from two plant-associated fungi of the Sonoran desert. J Nat Prod 69:178–184

Van der Sar SA, Blunt JW, Munro MHG (2006) Spiro-Mamakone A: a unique relative of the Spirobisnaphthalene class of compounds. Org Lett 8:2059–2061

Verma VC, Kharwar RN, Strobel GA (2009) Chemical and functional diversity of natural products from plant associated endophytic fungi. Nat Prod Commun 4:1511–1532

Wagenaar MM, Clardy J (2001) Dicerandrols, new antibiotic and cytotoxic dimers produced by the fungus *Phomopsis longicolla* isolated from an endangered mint. J Nat Prod 64:1006–1009

Wagenaar MM, Corwin J, Strobel G, Clardy J (2000) three new cytochalasins produced by an endophytic fungus in the genus Rhinocladiella. J Nat Prod 63:1692–1695

Wall ME, Wani MC, Cook CE, Palmer KH, Mcphail AT, Sim GA (1966) Plant antitumor agents. I. The isolation and structure of camptothecin, a novel alkaloidal leukemia and tumor inhibitor from *Camptotheca acuminata*. J Am Chem Soc 88:3888–3890

Wang J, Huang Y, Fang M, Zhang Y, Zheng Z, Zhao Y, Su W (2002) Brefeldin A, a cytotoxin produce by *Paecilomyces* sp. and *Aspergillus clavatus* isolated from *Taxus mairei* and *Torreya grandis*. FEMS Immunol Med Microbiol 34:51–57

Wang S, Li XM, Teuscher F, Li DL, Diesel A, Ebel R, Proksch P, Wang BG (2006) Chaetopyranin, a benzaldehyde derivative, and other related metabolites from *Chaetomium globosum*, an endophytic fungus derived from the marine red alga Polysiphonia urceolata. J Nat Prod 69:1622–1625

Wang Y, Xu L, Ren W, Zhao D, Zhu Y, Wu X (2012) Bioactive metabolites from *Chaetomium globosum* L18, an endophytic fungus in the medicinal plant Curcuma wenyujin. Phytomedicine 19:364–368

Wani MC, Taylor HL, Wall ME, Goggon P, McPhail AT (1971) Plant antitumor agents, VI. The isolation and structure of taxol, a novel antileukemic and antitumor agent from *Taxus brevifolia*. J Am Chem Soc 93:2325–2327

Wijeratne EM, Paranagama PA, Marron MT, Gunatilaka MK, Arnold AE, Gunatilaka AA (2008) Sesquiterpene quinones and related metabolites from *Phyllosticta spinarum*, a fungal strain endophytic in *Platycladus orientalis* of the Sonoran Desert. J Nat Prod 71:218–222

Wu XZ (2006) A new classification system of anticancer drugs—based on cell biological mechanisms. Med Hypotheses 5:883–887

Xu J, Kjer J, Sendker J, Wray V, Guan H, Edrada RA, Lin WH, Wu J, Proksch P (2009) Chromones from the endophytic fungus *Pestalotiopsis* sp. isolated from the Chinese mangrove plant *Rhizophora mucronata*. J Nat Prod 72:662–665

Yang X, Zhang L, Guo B, Guo S (2004) Preliminary study of a vincristine-producing endophytic fungus isolated from leaves of *Catharanthus roseus*. Chin Tradit Herbal Drugs 35:79–81

Zhan J, Burns AM, Liu MX, Faeth SH, Gunatilaka AAL (2007) Search for cell motility and angiogenesis inhibitors with potential anticancer activity: beauvericin and other constituents of two endophytic strains of *Fusarium oxysporum*. J Nat Prod 70:227–232

Zhang LB, Gou LH, Zeng SV (2000) Preliminary study on the isolation of endophytic fungus of *Catharanthus roseus* and its fermentation to produce product of therapeutic value. Chin Tradit Herbal Drugs 11:805–807

Zhang JY, Tao LY, Liang YJ, Yan YY, Dai CL, Xia XK, She ZG, Lin YC, Fu LW (2009) Secalonic acid D induced leukemia cell apoptosis and cell cycle arrest of G(1) with involvement of GSK-3beta/beta-catenin/c-Myc pathway. Cell Cycle 8:2444–2450

Zhang X, Shi Y, Wang X, Zhang W, Lou K (2010) Isolation, identification and insecticidal activity of endophyte from *Achnatherum inebrians*. Wei Sheng Wu Xue Bao 50:530–536

Part VII

Future Challenges

A Functional View of Plant Microbiomes: Endosymbiotic Systems That Enhance Plant Growth and Survival

21

James F. White Jr., Mónica S. Torres, Holly Johnson, Ivelisse Irizarry, and Mariusz Tadych

Abstract

Over the past several decades, it has become clear that numerous nonpathogenic or weakly pathogenic microbes inhabit plants both internally and externally. The challenge for plant biologists who study endophytism lies not only in the discovery of endophytes in plants but also in articulating the precise mechanisms whereby these endophytes function to support the growth and survival of their plant hosts. In this chapter, we discuss the phenomenon of microbial endophytism from a functional perspective. We propose that endophytic microbes in plants comprise a critical part of the plant's functional systems. We propose three broad categories of endosymbiotic systems, including (1) *Defensive Endosymbiotic Systems*, (2) *Stress Tolerance Endosymbiotic Systems*, and (3) *Nutritional Endosymbiotic Systems*. We will also consider potential interactions between endosymbiotic organisms of plants and relativity of function of endosymbionts. A particular endophyte may serve multiple functions in the ecology of its host plant, and predominant functions of an endophyte may change depending on the ecological circumstances affecting its host. Only now are we beginning to realize how important endophytic microbes are to plants. Much research is needed to elucidate the mechanisms of action and the roles that endophytes play in modulating host plant ecology and enhancing plant growth and survival.

1 Introduction

Since the development of the germ theory, most scientists, working with plant- and animal-infecting microbes, have focused their research efforts on microbes that cause disease (Ainsworth 1981). Endophytes (fungi and/or bacteria that live within tissues of plants) largely appear to be the inverse of plant pathogens since generally

J.F. White Jr. (✉) • M.S. Torres
• I. Irizarry • M. Tadych
Department of Plant Biology and Pathology, School of Environmental and Biological Sciences, Rutgers University, New Brunswick, NJ 08901, USA
e-mail: white@rci.rutgers.edu

H. Johnson
Department of Computer Science, Central Washington University, Ellensburg, WA, USA

V.C. Verma and A.C. Gange (eds.), *Advances in Endophytic Research*,
DOI 10.1007/978-81-322-1575-2_21, © Springer India 2014

they represent cryptic benign infections of healthy plants. Although we cannot cite any statistics, experience with endophytic microbial populations in plants suggests that they outnumber plant pathogens many times to one (Bills 1996; Suryanarayanan et al. 1998; Arnold et al. 2001). Recent diversity studies of microbial endophytes would seem to support this view (Arnold et al. 2001; West et al. 2010; Lucero et al. 2011; Porras-Alfaro and Bayman 2011; Zimmerman and Vitousek 2012). In fact, endophytism is so common among microbes (as compared to pathogenicity) that it may well be that endophytism is the normal state for most plant-infecting microbes and pathogenicity is the "out-of-balance" condition (Schulz and Boyle 2005). In this chapter, we discuss the current state of knowledge in the domain of endophyte biology and highlight those areas that we believe represent future fertile ground for expanding our understanding of the functioning of "endosymbiotic systems" and their roles in biology or ecology of host plants. We further posit that analyzing the phenomenon of endophytism from a functional systems perspective will permit us to develop a better understanding of the ecological context in which these endosymbioses function.

Endophytic microbes have for much of history been perceived as having limited function. Scientists have viewed fungal endophytes in particular as parasites, weak pathogens, or saprophytes that enter plants but cannot function until the host is weakened or senescent (Saikkonen et al. 2004). Other endophytes are speculated to be degenerate pathogens whose life cycles have been curtailed by genetic phenomena or partial host incompatibility resulting in microbes that are trapped in hosts and unable to reproduce or evolve (White 1988; Schardl and Phillips 1997; Schardl and Craven 2003; Moon et al. 2004). If these models of endophyte functionality were, in fact, correct, endophytes would be largely nonfunctional and perhaps have only negative impacts on host plants. However, there exists a body of empirically anchored research that indicates that the reverse appears to be correct. Endophytic microbes are increasingly being found to have positive impacts on host plant fitness, with infections frequently resulting in greater growth, fecundity, herbivore deterrence, disease resistance, abiotic stress tolerance, etc. (Cheplick and Clay 1988; Clay 1988; Bashan et al. 1989; Wilkinson et al. 2000; Malinowski et al. 2005; Malinowski and Belesky 2006; Waller et al. 2005; Clarke et al. 2006; Feng et al. 2006; Ortíz-Castro et al. 2008; Puente et al. 2009; Álvarez-Loayza et al. 2011; Bacon and Hinton 2011). Although we previously proposed that fungal endophytes were trapped in host plants, we have since discovered that they actually possess cryptic conidial states on surfaces of plants where they disseminate horizontally (White et al. 1996; Moy et al. 2000; Dugan et al. 2002; Tadych et al. 2012). An increasing body of research further suggests that endophytes are not functionless at all but instead have definable functions in plants and ecosystems (Clay 1988; Puente and Bashan 1994; Saikkonen et al. 1996; White et al. 2001; Rudgers et al. 2004, 2005; Kuldau and Bacon 2008). It is becoming evident that endophytes are adapted to hosts, express different life cycle stages at distinct stages of host development, and transmit with seeds to succeeding generations of the hosts (Latch et al. 1987; White 1987; White et al. 1991; Afkhami and Rudgers 2008; Rodriguez et al. 2009b; Álvarez-Loayza et al. 2011; White et al. 2012a, b). The adaptation to hosts and seed transmission aspects of many endophytes both emphasizes the importance of endophytes to their plant hosts. Endophytic microbes, whether bacteria or fungi, inhabit niches within plants that result in enhancements of host fitness and the subsequent ability to enable hosts to colonize and reproduce in a particular ecological niche within a larger ecosystem (Matthews and Clay 2001; Rudgers et al. 2005; Rodriguez et al. 2009b). Rodriguez et al. (2009a, b) demonstrated that endophytic associations with plants were habitat-adaptive symbioses, serving to enable host plants to survive and reproduce in habitats where hosts could not otherwise grow (see also Redman et al. 2011). Thus, it has become clear that a more accurate and appropriate view of endophytic microbe has emerged: They constitute endosymbiotic systems of plants that enable plants to thrive in particular ecological niches. We believe that one important course for future lines of investigation will be to better define the endosymbiotic systems and elucidate how they function to enable plants to compete with other species and adapt to their environments. In this chapter, we will provide evidence to support the existence and functioning of three broad categories of endosymbiotic systems, including (1) *Defensive Endosymbiotic Systems*, (2) *Stress Tolerance Endosymbiotic Systems*, and (3) *Nutritional Endosymbiotic Systems*. Lastly, we look to move the study of endophytes beyond individual species associations in order to recognize the complexity of interactions between endophytes and effects of microbial endophyte consortia on plants.

2 Defensive Endosymbiotic Systems

There exists a large body of work that supports a defensive function for certain fungal endophytes (Cheplick and Clay 1988; Lewis et al. 1993; White et al. 1993; Azevedo et al. 2000; Clay and Schardl 2002; Arnold et al. 2003; Schardl et al. 2004; Spiering et al. 2005; Álvarez-Loayza et al. 2011). Much of this work focuses on various clavicipitaceous fungal endophytes found within grasses. In cool-season and some warm-season grasses, species of the clavicipitaceous fungal genera *Epichloë* (including *Neotyphodium*) and *Balansia* systemically colonize plants and produce alkaloids (and possibly other metabolites) that reduce or alter herbivory (Panaccione 2005; Panaccione et al. 2006; Clay and Cheplick 1989).

Certain species of morning glories (*Ipomoea* spp.; Convolvulaceae) have long been regarded as toxic and avoided by herbivores due to possession of high levels of ergot alkaloids (Austin 1973). In recent years, it has been shown that ergot alkaloids that toxify certain species of morning glories are produced by clavicipitaceous endophytes or epiphytes (Steiner et al. 2006; Markert et al. 2008; Leistner and Steiner 2009). Even though much of the work on endophytes in Convolvulaceae has yet to be done, it is difficult to ascribe any function other than defense to these symbionts.

While some controversy arises with regard to defining fungal endophytes universally as anti-herbivorous (Faeth et al. 1999; Saikkonen et al. 1999), it seems apparent that some do appear to function in this capacity and thus may constitute an herbivory defensive system (Clay 1988). Herbivory defensive systems may function through production of toxins by the endophytes themselves or through endophyte-induced upregulation of host defensive compounds. Clavicipitaceous fungal endophytes in grasses have been shown to have a "reprogramming effect" on host plants. Grasses bearing these particular endophytes have increased levels of phenolics and other potential anti-feeding compounds (Waller et al. 2005; Sullivan et al. 2007; Kumar et al. 2009; White and Torres 2010; Torres et al. 2012). This "reprogramming effect" is a topic of current interest since it may be key to understanding the mechanism of endophyte-host interactions. For any given endophyte-plant association, observed anti-herbivory could be the result of (1) endophyte-induced anti-herbivore compounds, (2) endophyte produced anti-herbivore compounds, or (3) a combination of both.

Another specific example of a defensive endosymbiotic system is the *Diplodia*-palm association. *Diplodia mutila* (Botryosphaeriaceae; Ascomycota) is an endophyte of the neotropical palm *Iriartea deltoidea* (Álvarez-Loayza et al. 2011). *Diplodia* is an asymptomatic endophyte in the leaves and stems of mature palm populations. The fruits of the palm transmit the fungus, which often forms a black carbonaceous mycelium on the surface of fruits. Mature plants containing *Diplodia* were also found to be resistant to stem borer insects. While mature palm plants show no symptoms of infection by *Diplodia*, the fungus may be mortally pathogenic to seedlings of the palm under certain environmental conditions. In high sunlight conditions (e.g., under the gaps in the rainforest canopy), *Diplodia* expresses a pathogenic phase where the fungus causes extensive necrosis of seedling tissues to such an extent that many seedlings in high light areas do not survive. If seedlings bearing the endophyte grow in the shaded areas of the forest understory, the fungus does not cause disease. Instead, it remains as an asymptomatic endophytic defensive mutualist of the plant (Álvarez-Loayza et al. 2011). Studies conducted on cultures of *Diplodia* provide a possible explanation for how light affects the expression of the pathogenic phase of the endophyte. Cultures of *Diplodia* that were exposed to high light were shown to have enhanced secretion of hydrogen peroxide (H_2O_2). Hydrogen peroxide acts as an important defense signal molecule that triggers a hypersensitive response in some plants that results in cell/tissue death (Heath and Packer 1968). The extent to which endophytes produce hydrogen peroxide in plant tissues appears to determine whether the fungus remains a defensive mutualist or becomes a pathogen capable of killing plant tissues, colonizing them, and later producing conidia and ascospores on the necrotic tissues.

Defensive protection that results from endophyte infection may include pathogen protection. Bacterial endophytes are frequently found to protect hosts from fungal pathogens (Cho et al. 2007). Clavicipitaceous endophytes in grasses have been shown to protect grass hosts from certain fungal diseases including dollar spot, caused by *Sclerotinia homoeocarpa*, and red thread disease caused by *Laetisaria fuciformis* (Clarke et al. 2006; Bonos et al. 2005). Defensive mechanisms against pathogens are generally not clear, but could involve antibiosis-like antagonisms (White and Cole 1985), physical exclusion phenomena (White et al. 1996), or physical colonization of the pathogens by endosymbiotic microbes. We have observed numerous instances in which fungal hyphae within plant tissues become colonized by endophytic bacteria (unpublished data). What interactions occur between fungi and bacteria in such circumstances is completely unknown.

3 Stress Tolerance Endosymbiotic Systems

An increasing body of work suggests that many endophytes enhance host plant tolerance to abiotic stresses; however, the mechanism that facilitates this enhancement is not clear (Zhang and Nan 2007; Kuldau and Bacon 2008; Hamilton et al. 2012). This endosymbiotic system has a parallel in animal biology where it has been shown that intestinal microbes may enhance the ability of animals to cope with stress, but the facilitating mechanism also remains unclear (Bravo et al. 2011). It is likely that these endosymbiotic systems in plants increase oxidative stress resistance and enhance host tolerance to soils having high salinity, heavy metals, extreme heat, extremely arid conditions, and various biotic and abiotic assaults to plants that manifest as increased oxidative stress. In animals or plants, enhanced oxidative stress resistance could stem from nutrients that the host obtains directly from the endophyte (Bravo et al. 2011). This is a logical hypothesis to explain increased stress tolerance in plants and cannot be discarded; unfortunately, we have little evidence for this mechanism at the present time.

Some research suggests that fungal endophytes of plants may produce antioxidants that could modulate oxidative stress through the scavenging of reactive oxygen generated during biotic or abiotic stress events. Endophytic fungi have been shown to be the producers of numerous antioxidant compounds that may play a role in enhancing stress tolerance in host plants (Schulz et al. 2002; Rasmussen et al. 2008). Huang et al. (2007) examined the total antioxidant capacity and total phenolic content of 292 endophytic fungal isolates and demonstrated a high correlation between phenolic content and antioxidant capacity, suggesting that the endophytes themselves may be producing phenolic antioxidants. These investigators identified phenolic acids, flavonoids, tannins, hydroxyanthraquinones, and phenolic terpenoids as potential antioxidants. From the endophyte *Pestalotiopsis microspora*, the potent antioxidants pestacin and isopestacin have been identified. These compounds scavenge superoxide and hydroxyl free radicals (Strobel and Daisy 2003).

Fungal endophytes may also produce carbohydrate compounds that have antioxidant capacity. The fungal sugar alcohol mannitol has been shown to have antioxidant activity (Jennings et al. 1998). Richardson et al. (1992) reported higher concentrations of mannitol and other potential fungal carbohydrates with antioxidant activity in the apoplasts of tall fescue grass tissues infected by the endophytic fungus *Neotyphodium coenophialum*. Mannitol is used by fungi as a common storage sugar, and it has been hypothesized that it functions as an osmoprotectant in plants that also produce it. Mannitol is produced by the endophytic pathogen *Alternaria alternata*. Some scientists have suggested that mannitol suppresses reactive oxygen species (ROS)-mediated plant defense responses in Alternaria's tobacco host (Jennings et al. 1998). Fungal antioxidants of all forms may contribute to enhance overall oxidative stress tolerance in plants. The "habitat-adapted symbiosis" phenomenon proposed by Redman et al. (2002) could be explained by this mechanism. Here different endophytes may produce antioxidants that differ in their capacities to quench various types of reactive oxygen or may differ in their capacities to reach specific tissues undergoing stress.

Other researches propose a more general mechanism for stress tolerance in endophyte-infected plants. Torres et al. (2012) and Hamilton et al. (2012) proposed that enhanced oxidative stress tolerance in grasses infected by clavicipitaceous endophytes was the result of induced upregulation of plant-produced antioxidants and other stress defensive compounds due to secretion of ROS and auxins by the endophyte into plant tissues. In particular, the secretion of hydrogen peroxide into plant tissues, a known plant defense signal molecule, may be responsible for increasing the readiness of many endophyte-infected grasses to endure biotic and abiotic stresses.

Antioxidants may increase the tolerance of plants to many oxidative stresses and also increase the resistance to pathogens that use ROS to incite disease (Clarke et al. 2006). Therefore, ROS-producing endophytes may increase the hardiness of plant hosts in multiple ways. In some food crop plants, ROS-producing endophytes may increase the nutritional value of the crop by enhancing production of antioxidant nutrients. An example of such an application could be in a crop like cranberries where endophytic fungi are common in fruits and leaves (Jeffers 1991). Some of these fungi may be latent pathogens and responsible for fruit rot or other diseases, but others appear to be nonpathogenic (Oudemans et al. 1998). In a preliminary study of seven of the most common endophytes in cranberry, we identified several that secreted observable amounts of ROS in cultures. The leaf, stem, and fruit endophyte *Pestalotia vaccinnii* produced notable quantities of superoxides in potato dextrose agar cultures, while the endophytic field rot pathogen *Phyllosticta vaccinii* produced significant amounts of hydrogen peroxide in potato dextrose agar cultures as well. Several other endophytic pathogens (*Colletotrichum gloeosporioides*, *Physalospora vaccinii*, and *Strasseria geniculata*) produced weak reactions to stains for peroxides and superoxides.

An understanding of the mechanism by which endophytic microbes enhance stress tolerance of host plants remains elusive. New approaches to answer the "mechanism" question will most likely involve carefully controlled experiments combined with genome expression analyses to determine precisely what genes in the host and endophyte upregulate under particular stress conditions.

4 Nutritional Endosymbiotic Systems

Healthy plants are colonized by many different endophytes and epiphytes, both fungal and bacterial (Döbereiner 1992; James 2000; Alvarez-Loayza et al. 2011; Bacon and Hinton 2011; Stone et al. 2000; Fürnkranz et al. 2012; Taulé et al. 2012). The ability of many bacterial endophytes to fix atmospheric nitrogen implicates them as key components of Nutritional Endosymbiotic Systems (Rosenblueth and Martínez-Romero 2006; Reinhold-Hurek and Hurek 2011). There is a large body of research on "associative nitrogen fixation" that seeks to determine if plants are obtaining fixed nitrogen from endophytic diazotrophic bacteria (Döbereiner 1992; Döbereiner et al. 1994; James et al. 1994; Kloepper 1994; Hurek et al. 1988, 1994; James 2000; Mantelin and Touraine 2004; Zhang et al. 2008; Dakora et al. 2008; Magnani et al. 2010). Generally, investigators either use an assay such as acetylene reduction or isotopic nitrogen tracking to ascertain whether gaseous nitrogen is being assimilated into plants (Stewart et al. 1967; Radajewski et al. 2000). However, the question of whether nitrogen moves into plant tissues or remains associated with microbes in any substantial manner is generally not answered clearly. Other studies focus on plant growth enhancements due to the presence of specific microbes on plants (Kloepper 1994). This work is complicated by the fact that endophytes frequently produce growth regulators. Furthermore, any growth enhancements in plants may be attributed to the growth regulatory compounds rather than to the nitrogen derived from the microbes (Barazani and Friedman 1999). Recently, we have found evidence for a mechanism whereby grass seedlings obtain nutrients from seed-transmitted diazotrophic bacteria through oxidation using plant-secreted reactive oxygen (White et al. 2012a). We denominated this mechanism "oxidative nitrogen scavenging" (ONS) since the plant employs ROS (specifically H_2O_2), to degrade microbes and oxidize their constituent protein components prior to proteolysis and absorption. Our studies have centered on documenting ONS in pooid grasses, including *Poa annua*, *Poa pratensis*, *Festuca arundinacea*, *Festuca rubra*, and *Lolium perenne*. In tall fescue, we have identified two endophytic diazotrophic bacteria that may be part of a nutritional endosymbiotic system, including *Pantoea agglomerans* and an unidentified species of *Pseudomonas*, both of which are seed-transmitted and colonize-germinating seedlings. Bacteria may be transmitted on the caryopsis surface on glumes and paleas that closely adhere to caryopses. All seed collections of these grass species obtained from natural populations or from commercially available samples bear similar diazotrophic bacteria. Our experiments suggest that the bacteria proliferate in meristems of seedlings, but are degraded predominantly on seedling roots to provide organic forms of nitrogen and perhaps other nutrients that are needed for the rapid seedling growth. Roots secrete H_2O_2 onto bacterial populations on and within roots. Microscopic examination of bacteria on root hairs and other root epidermal cells has shown that the rod-shaped bacterial cells first swell to become spherical, lose their nucleic acid and protein contents, and eventually disappear from plant surfaces. In experiments using grass seedlings, we have found that proper seedling root development depends on the presence of the bacteria on roots of seedlings (unpublished). Using seeds that were rigorously surface disinfected to remove all bacteria, those that were germinated on water agarose medium produced seedlings whose roots did not develop properly when compared to roots germinated from non-surface sterilized seeds. With bacteria present, seedling roots showed proper gravitropic response with roots growing downward into the agarose medium and developed root hairs that extended into the agarose medium. Without bacteria, seedling roots frequently did not grow downward and the few roots that found their way into the medium did not produce root hairs.

In other experiments using similarly sterilized seedlings, we were able to restore proper root development by incorporating 0.1 % proteins (egg albumin, lipase, or cellulase) into the agarose medium. These simple experiments suggest that the grasses, at least in the seedling stage, require bacteria largely as a nutrient source to fuel early seedling development. The fact that proteins are sufficient enough to restore root development indicates that the effect is nutritional one rather than the result of a microbial-produced hormone that affects development.

There is evidence that ONS, or a more developed phagocytic digestive system to extract nutrients from bacteria, may be widespread in plants. Paungfoo-Lonhienne et al. (2010) demonstrated that tomato plants internalized exogenously applied bacteria into root cortical cells that were then degraded and their nutrients transported into shoot tissues. We conducted a preliminary survey of seedlings of 23 species of plants in 16 families of vascular plants for evidence of ONS. All seeds used in that study were rigorously disinfected to remove all exogenous bacteria and were subsequently germinated on sterile water agarose medium to reduce any contamination or exogenous

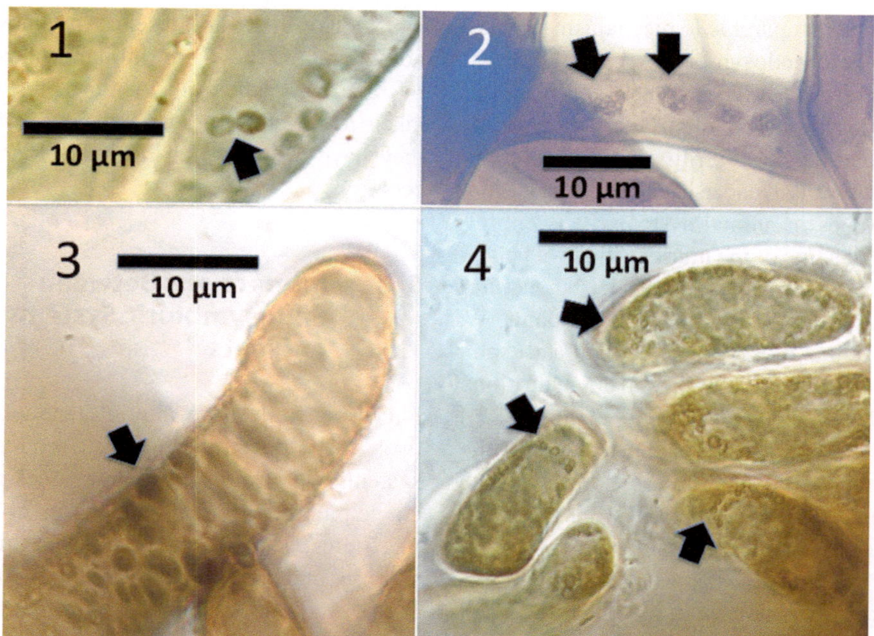

Fig. 21.1 Seedling root cells showing evidence of microbivory of bacteria (stained with diaminobenzidine tetrachloride/horseradish peroxidase to visualize H2O2 (*red-brown* color); counterstained with aniline blue/lactophenol to visualize proteins (*blue*)). (1) Root cap cell of *Yucca schottii* with dividing bacteria (*arrow*) in vesicles. (2) Root hairs of sedge *Fimbristylis cymosa* with clusters of degraded bacteria internally (*arrows*). (3) Root hair of seedling of Portia tree (*Thespesia populnea*) with oxidizing bacteria (*arrow*) on surface. (4) Root cap cells of *Moringa oleifera* with oxidizing bacteria internally (*arrows*)

nutrients. Under these low nutrient conditions, we frequently observed bacteria within vesicles in the cytoplasm of root hairs and root epidermal and cortical cells. Using ROS and protein probes (White et al. 2012a), we confirmed oxidative degradation of the intracellular bacteria, where degrading bacterial cells swelled and lost capacity to stain for protein contents (see Fig. 21.1). Although our survey is preliminary, we were able to determine that oxidative degradation of intracellular bacteria is a phenomenon that occurs in many plant species in diverse habitats. Desert plants that show this phenomenon include Agavaceae and Cactaceae. Vines in many temperate and tropical plant families (e.g., Anacardiaceae, Araceae, Araliaceae, Caprifoliaceae, Orchidaceae, Polypodiaceae, Ranunculaceae, and Vitaceae) also may rely on nutrient scavenging from bacteria to provide sufficient nutrients to fuel plant development. In vines, bacteria can be visualized in meristematic cells and are distributed intracellularly throughout tissues of plants (unpublished data). Oxidative degradation occurs generally in tendrils and stem epidermal tissues or aerial roots.

Nutritional Endosymbiotic Systems may occur where the nutrients obtained from microbes are not strictly nitrogen based, but, instead, provide other forms of nutrients that plants require. Specific examples here could include biotin, folic acid, niacin, and thiamine. When plants are grown axenically in tissue culture, these nutrients must frequently be provided exogenously. This proposed function has a parallel in animal systems where bacteria in the gut supply the host with vitamin K, vitamin B_{12}, biotin, folic acid, and pantothenate (Hooper et al. 2002). In a study of whiteflies (*Bemisia* spp.) and their bacterial endosymbionts (*Portiera* spp.), the whitefly host was hypothesized to obtain carotenoids from its endosymbionts (Sloan and Moran 2012). The transfer mechanisms of nutrients from microbes in animals to host tissues are still unknown, but they could be comparable to mechanisms in plant systems where endosymbionts are degraded to extract nutrients. Similarities between plant and animal endosymbiotic systems could provide the basis for using plant Nutritional Endosymbiotic Systems as models to understand nutrient acquisition from microbes in animal systems. It is entirely possible that some of the nutrients that we believe are produced by plants may in fact be acquired from endosymbiotic microbes within plants. We hypothesize that most plants obtain at least some of their nitrogenous nutrients from Nutritional Endosymbiotic Systems (White et al. 2012b). However, our preliminary evidence suggests that plants differ with respect to the extent that they rely on endosymbiotic microbes and the ONS process to provide nitrogen. Epiphytes, vines, and some desert species appear to heavily utilize Nutritional Endosymbiotic Systems to obtain nitrogen (unpublished data).

The study of Nutritional Endosymbiotic Systems has the potential to impact fields of agriculture, ecology, evolutionary biology, and human health. Nutritional Endosymbiotic Systems could form the foundation of the nitrogen cycle in ecosystems, such as deserts, where limited available nitrogen is present in soils (Whitford 2002). A thorough understanding of these systems could lead to applications in agriculture. For example, new strategies may emerge for cultivation of more efficient food, fuel, and fiber crops with reduced inorganic nitrogen applications. From the perspective of evolutionary biology, there are several potential impacts. The earliest land plants (e.g., *Cooksonia*, *Sawdonia*, *Rhynia*, *Horneophyton*) in the Ordovician, Silurian, and Devonian lacked root systems to absorb nutrients from soils efficiently (Taylor and Taylor 1993). It is possible that they employed Nutritional Endosymbiotic Systems involving internal oxidation and digestion of diazotrophic bacteria as a source of nitrogen and other nutrients. This possibility could explain why we see oxidation of intracellular bacteria in diverse plant families ranging from ferns to dicots. These early plants are also known to have associated with glomalean fungi that may have functioned like mycorrhizae in assisting with absorption of some nutrients (Taylor et al. 1995). However, at present, we know almost nothing about how Nutritional Endosymbiotic Systems work, and further studies must be undertaken to develop our understanding of impacts they have on plant nutrition, development, and ecology.

5 Interactions Between Endosymbiotic Systems of Plants

Plants generally have multiple endophytes, and these may constitute multiple endosymbiotic systems. For example, pooid grasses may have clavicipitaceous fungal endophytes in the shoot meristems, leaves, and culms that function in defense and perhaps provide stress tolerance. Plants may also contain diazotrophic bacterial endophytes that are oxidized on the surface and interiors of roots to provide nutrients, such as nitrogen, for growth. Thus, a particular grass plant individual may possess at least two different endosymbiotic systems, one that is defensive and the other that is nutritional. One of the interesting phenomena with regard to clavicipitaceous endophytes in grasses is that infection will frequently result in enhanced growth of plants compared to plants that are not infected. This effect has been attributed to increased photosynthetic efficiency or to growth regulator production by the fungal endophyte (Spiering et al. 2006). Because some clavicipitaceous endophytes have been documented to produce auxins, increased growth of some hosts has also been attributed to auxin-induced growth stimulation (Yue et al. 2000; Vadassery et al. 2008).

Investigators have only recently begun to examine interactions between symbiotic systems in plants. Novas et al. (2011) examined the effects of a clavicipitaceous endophyte, *Neotyphodium*, on growth of VA mycorrhizae, a nutritional symbiotic system, on roots of the host *Bromus*. These investigators reported an increase in the colonization of mycorrhizae as a result of clavicipitaceous endophyte infection. Many pooid grasses also oxidize diazotrophic bacteria in order to utilize them as a nutrient source (White et al. 2012a).

Any enhancement of oxidation of bacteria in roots by the clavicipitaceous endophytes would be expected to increase nitrogenous nutrients available to plants and increase growth. There is some indirect evidence that *Neotyphodium coenophialum* endophyte infection in tall fescue grass may alter oxidative reactions on roots of the host. Malinowski and Belesky (2006) demonstrated that tall fescue plants bearing the endophyte secreted higher levels of antioxidant phenolics from roots. The phenolics may be secreted to protect plant roots from reactive oxygen secreted by roots onto bacteria. Additionally, Lyons et al. (1990) found that organic and inorganic forms of nitrogen increased in tall fescue grasses as a result of endophyte infection. Stimulation of Nutritional Endosymbiotic Systems in grasses directly or indirectly by endophytic fungi could account for enhanced nitrogen content and increases in the efficiency of photosynthesis as a result. This is an interesting possibility that will require further experimentation in order to evaluate.

6 Microbial Consortia and Other Factors Affecting Endosymbiotic Systems

Nutritional Endosymbiotic Systems are often composed of microbial consortia. These microbes provide their plant hosts with key nutrients [such a nitrogenous compounds] that may cause an increase in plant biomass, fecundity, and crop yield. Consortia of nitrogen-fixing bacteria have been documented in a number of plant hosts such as wild rice, sugarcane, and grasses (Minamisawa et al. 2004; Miyamoto et al. 2004; Zhang et al. 2008; Taulé et al. 2012). Nitrogen-fixing endophytic bacteria typically include common diazotrophic soil species such as *Azotobacter diazotrophicus*, *Azoarcus* spp., *Pantoea agglomerans*, and *Klebsiella oxytoca*. However, recent evidence has uncovered the presence of novel endophytic anaerobic clostridia along with the presence of common diazotrophs (Miyamoto et al. 2004; Minamisawa et al. 2004). This finding stresses the importance of utilizing nontraditional microbiological isolation techniques in endophyte research. Future research challenges include the development of isolation and cultivation techniques for these novel bacteria and assessing the diversity of endophytes utilizing culture-independent methods that may detect the presence of unculturable or novel microorganisms that have been overlooked previously. Consortia of nitrogen-fixing bacteria have the potential of improving plant growth in marginal soils and serve as biofertilizers for agricultural crops, thus reducing the environmental impact of fertilizing practices.

How microbial consortia function is unknown; however, it is conceivable that a consortium of endophytic microbes may interact with hosts in order to complete an endosymbiotic system. Any given endophytic microbial consortium may potentially confer more than one benefit to its plant host. Therefore, these systems could be categorized within more than one type of endosymbiotic system. Endosymbiotic consortia may be composed of fungi, bacteria, and even viruses. It seems feasible that favorable combinations of microbial endophytes may have a synergistic effect that could be more efficient at enhancing plant growth than the individual symbionts.

Microbial consortia may interact with phytopathogenic fungi or bacteria to complete a defensive endosymbiotic system. This could alter the physiology of the pathogen and induce a nonpathogenic endophytic state, thereby avoiding disease in the plant host. *Fusarium oxysporum* is commonly documented as a wilt-causing phytopathogen in some plants or a nonpathogenic plant growth-promoting endophyte of other plants. Minerdi et al. (2008) determined that the virulence of a pathogenic strain of *F. oxysporum* can be reduced by a bacterial consortium. *Fusarium oxysporum* MSA 35 was isolated from wilt-suppressive soils and was observed microscopically to contain numerous bacterial cells attached to the hyphae. The species of bacteria in this consortium were *Serratia marcescens*, *Stenotrophomonas maltophilia*, *Achromobacter xylosoxidans*, and *Bacillus halodurans*. This consortium had numerous effects on the growth and pathogenicity of *F. oxysporum*. Among the effects observed were changes in pigmentation, sporulation, aerial hyphae production, and hyphal thickness. Further studies showed that members of this microbial consortium communicate with each other and their plant hosts through emissions of volatile organic compounds. Volatiles emitted by co-cultures of the fungus with the microbial consortium appeared to promote growth in lettuce, more so than volatiles emitted by *F. oxysporum* lacking the consortium of bacterial symbionts (Minerdi et al. 2011).

The idea that microbial consortia are able to alter the virulence of pathogens that devastate or destroy economically important crops is one that must be further explored. Understanding the mechanisms by which microbial consortia are able to decrease the virulence of specific phytopathogens may be crucial when seeking solutions for disease control. Because of potentially complex population interactions, applying this knowledge in order to implement successful biocontrol strategies for agricultural crops vulnerable to disease will be a challenge.

Recent research has suggested viruses as possible modulators of Stress Tolerance Endosymbiotic Systems. A unique endophytic system was documented by Márquez, et al. (2007) where a mycovirus infecting the fungus *Curvularia protuberata* resulted in enhanced thermal tolerance in the host plant, panic grass (*Dichanthelium lanuginosum*). The *Curvularia* thermal tolerance virus (CThTV) appeared to alter the physiology of the fungal endophyte. A further study of this three-way symbiotic system indicated that the virus induced the expression of fungal genes that are thought to be involved in thermotolerance. CThTV altered the expression of fungal genes involved in the biosynthetic pathways of various osmoprotectants such as trehalose, glycine betaine, taurine, and melanin (Morsy et al. 2010). The production of these osmoprotectants may be a strategy used by this and many other endophytes to protect themselves from environmental stress, such as increased temperature. Mycoviruses and bacteriophages are often overlooked and may be indispensable parts of successful endosymbiotic systems in plants. Viruses that alter the gene expression of microbial endophytes are likely to differ in a case-by-case basis, and thus, the genes they express may differ. It is possible that many endophytic fungi may be infected and altered by the presence of viruses. Understanding specifically how these viruses are able to alter fungal endophyte physiology could lead to the design of microbial consortia that allow their plant hosts to inhabit otherwise inhospitable environments.

Studies have uncovered a number of bacteria that are symbiotic with mycorrhiza and are commonly referred to as "mycorrhiza helper bacteria" (Bonfante and Iulia-Andra 2009). Arbuscular mycorrhizal fungi, such as *Gigaspora margarita*, have been described as a niche for many rhizobacteria, some of them being vertically transmitted endohyphal symbionts (Bianciotto and Bonfante 2002; Bianciotto et al. 2004). Some species of bacteria described as mycorrhizal symbionts are common soil bacteria such as *Pseudomonas aeruginosa* and *Burkholderia cepacia* (Sundram et al. 2011). Other symbionts may be novel species such as the endohyphal symbiont "*Candidatus Glomeribacter gigasporum*" (Bianciotto et al. 2003). Mycorrhizas are common endophytes of plant roots that may also be part of a nutritional endosymbiotic system by partnering with certain bacteria.

7 Conclusions

The plant microbiome consists of bacterial and fungal endophytes, many of which have not been identified. In this chapter, we advocate a functional view of the plant microbiome where the microbes function to enhance survival and growth of the host plants. We propose that plant endophytic microbes are critical to plant growth and development, providing nutrients, enhancing stress tolerance, and defending plants from herbivores. There is good support for this moderate functional view in the large body of research on endophytes and beneficial microbes of plants. However, we also propose that endophytes were critical to the evolution of plants. Perhaps, without endophytes land plants may never have evolved or would be very different than the plants we see today. Developing a full understanding of plant microbiomes and the endosymbiotic systems of plants may permit us to produce hardier and more resistant food, fiber, and fuel crops using reduced agrichemical inputs.

Acknowledgments We are grateful to the New Jersey Agricultural Experiment Station and Central Washington University for resources and financial support. We are also grateful to Dr. John Craighead for discussions regarding the roles of bacteria in the human microbiome.

References

Afkhami ME, Rudgers JA (2008) Symbiosis lost: imperfect vertical transmission of fungal endophytes in grasses. Am Nat 172:405–416

Ainsworth GC (1981) Introduction to the history of plant pathology. Cambridge University Press, Cambridge, 315pp

Álvarez-Loayza P, White JF Jr, Torres MS, Balslev H, Kristiansen T (2011) Light converts endosymbiotic fungus to pathogen, influencing seedling survival and niche-space filling of a common tropical tree, *Iriartea deltoidea*. PLoS ONE 6(1):e16386. doi:10.1371/journal.pone.0016386

Arnold AE, Maynard Z, Gilbert GS (2001) Fungal endophytes in dicotyledonous neotropical trees: patterns of abundance and diversity. Mycol Res 105:1502–1507

Arnold AE, Mejia LC, Kyllo D, Rojas EI, Maynard Z, Robbins N, Herre EA (2003) Fungal endophytes limit pathogen damage in a tropical tree. Proc Natl Acad Sci U S A 100:15649–15654

Austin DF (1973) The American Erycibeae (Convolvulaceae): *Maripa*, *Dicranostyles*, and *Lysiostyles*. I. Systematics. Ann Missouri Bot Gard 60:306–412

Azevedo JL, Maccheroni W, Pereira JO, de Araujo WL (2000) Endophytic microorganisms: a review on insect control and recent advances on tropical plants. Electron J Biotechnol 3:e1–e4, http://www.ejb.org/content/vol3/issue

Bacon CW, Hinton DM (2011) In planta reduction of maize seedling stalk lesions by the bacterial endophyte *Bacillus mojavensis*. Can J Microbiol 57:1–8

Barazani OZ, Friedman J (1999) Is IAA the major root growth factor secreted from plant-growth-mediating bacteria? J Chem Ecol 25:2397–2406

Bashan Y, Singh CS, Levanony H (1989) Contribution of *Azospirillum brasilense* Cd to growth of tomato seedlings is not through nitrogen fixation. Can J Bot 67:2429–2434

Bianciotto V, Bonfante P (2002) Arbuscular mycorrhizal fungi: a specialised niche for rhizospheric and endocellular bacteria. Antonie Van Leeuwenhoek 81:365–371

Bianciotto V, Lumini E, Bonfante P (2003) 'Candidatus glomeribacter gigasporum' gen. nov., sp. nov., an endosymbiont of arbuscular mycorrhizal fungi. Int J Syst Evol Microbiol 53:121–124

Bianciotto V, Genre A, Jargeat P, Lumini E, Bécard G, Bonfante P (2004) Vertical transmission of endobacteria in arbuscular mycorrhizal fungus *Gigaspora margarita* through generation of vegetative spores. Appl Environ Microbiol 70(6):3600–3608

Bills GF (1996) Isolation and analysis of endophytic fungal communities from woody plants. In: Redlin SC, Carris LM (eds) Endophytic fungi in grasses and woody plants: systematics, ecology, and evolution. American Phytopathological Society Press, St Paul, pp 31–65

Bonfante P, Iulia-Andra A (2009) Plants, mycorrhizal fungi, and bacteria: a network of interactions. Annu Rev Microbiol 63:363–383

Bonos SA, Wilson MM, Meyer WA, Funk CR (2005) Suppression of red thread in fine fescues through endophyte-mediated resistance. Online Appl Turfgrass Sci. doi:10.1094/ATS-2005-0725-01-RS

Bravo JA, Forsythe P, Chew MV, Escaravage E, Savignaca HM, Dinana TG, Bienenstock J, Cryan JF (2011) Ingestion of *Lactobacillus* strain regulates emotional behavior and central GABA receptor expression in a mouse via the vagus nerve. Proc Natl Acad Sci U S A 108:16050–16055, www.pnas.org/cgi/doi/10.1073/pnas.1102999108

Cheplick GP, Clay K (1988) Acquired chemical defences in grasses: the role of fungal endophytes. Oikos 52:309–318

Cho KM, Hong SY, Lee SM, Kim YH, Kahng GG, Lim YP, Kim H, Yun HD (2007) Endophytic bacterial communities in ginseng and their antifungal activity against pathogens. Microb Ecol 54:341–351

Clarke BB, White JF Jr, Hurley RH, Torres MS, Sun S, Huff DR (2006) Endophyte mediated suppression of dollar spot disease in fine fescue. Plant Dis 90:994–998

Clay K (1988) Fungal endophytes of grasses: a defensive mutualism between plants and fungi. Ecology 69:10–16

Clay K, Cheplick GP (1989) Effect of ergot alkaloids from fungal endophyte-infected grasses on fall armyworm (*Spodoptera frugiperda*). J Chem Ecol 15:169–182

Clay K, Schardl CL (2002) Evolutionary origins and ecological consequences of endophyte symbiosis with grasses. Am Nat 160:S99–S127

Dakora FD, Chimphango SBM, Valentine AJ, Wlmerich C, Newton WE (2008) Biological nitrogen fixation: towards poverty alleviation through sustainable agriculture. In: Proceedings of the 15th international nitrogen fixation congress and the 12th international conference of the African Association for Biological Nitrogen Fixation, Cape Town, 21–26 Jan 2007. Current plant sciences and biotechnology in agriculture, vol 42. Springer, New York

Döbereiner J (1992) History and new perspectives of diazotrophs in association with non-leguminous plants. Symbiosis 13:1–13

Döbereiner J, Baldani VLD, Olivares FL, Reis VM (1994) Endophytic diazotrophs: the key to BNF in gramineous plants. In: Hegasi NA, Fayez M, Monib M (eds) Nitrogen fixation with non-legumes. American University in Cairo Press, Cairo, pp 395–408

Dugan FM, Sitton J, Sullivan RF, White JF Jr (2002) The *Neotyphodium* endophyte of wild barley (*Hordeum brevisubulatum* subsp. *violaceum*) grows and sporulates on leaf surfaces of the host. Symbiosis 32:147–159

Faeth SH, Saikkonen K, Helander M, Sullivan TJ, Rambo JL (1999) Endophytic fungi in native populations of grasses: against conventional wisdom of the anti-herbivore mutualism and the plant diversity hypothesis. Am Zool 39:120A–121A

Feng Y, Shen D, Song W (2006) Rice endophyte *Pantoea agglomerans* YS19 promotes host plant growth and affects allocations of host photosynthates. J Appl Microbiol 100:938–945

Fürnkranz M, Lukesch B, Müller H, Huss H, Grube M, Berg G (2012) Microbial diversity inside pumpkins: microhabitat-specific communities display a high antagonistic potential against phytopathogens. Microb Ecol 63:418–428

Hamilton CE, Gundel PE, Helander M, Saikkonen K (2012) Endophytic mediation of reactive oxygen species and antioxidant activity in plants: a review. Fungal Divers 54:1–10

Heath RL, Packer L (1968) Photoperoxidation in isolated chloroplasts. I. Kinetics and stoichiometry of fatty acid peroxidation. Arch Biochem Biophys 125: 189–198

Hooper LV, Midvedt T, Gordon JI (2002) How host-microbial interactions shape the nutrient environment of the mammalian intestine. Annu Rev Nutr 22: 283–307

Huang W-Y, Cai Y-Z, Xing J, Corke JH, Sun M (2007) A potential antioxidant resource: endophytic fungi from medicinal plants. Econ Bot 61:14–30

Hurek T, Reinhold-Hurek B, Kellenberger, Grimm B, Fendrik J, Neimann EG (1988) Occurrence of effective nitrogen scavenging bacteria in the rhizosphere of kallar grass. Plant Soil 110:339–348

Hurek T, Reinhold-Hurek B, Kellenberger E, Van Montagu M (1994) Root colonization and systemic spreading of *Azoarcus* sp. strain BH72 in grasses. J Bacteriol 176:1913–1923

James EK (2000) Nitrogen fixation in endophytic and associative symbiosis. Field Crop Res 65:197–209

James EK, Reis VM, Olivares FL, Baldani JI, Dobereiner J (1994) Infection of sugar cane by the nitrogen-fixing bacterium *Acetobacter diazotrophicus*. J Exp Bot 45: 757–766

Jeffers S (1991) Seasonal incidence of fungi in symptomless cranberry leaves and fruit treated with fungicides during bloom. Phytopathology 81:636–644

Jennings DD, Ehrenshaft M, Pharr DM, Williamson JD (1998) Roles for mannitol and mannitol dehydrogenase in active oxygen-mediated plant defense. Proc Natl Acad Sci U S A 95:15129–15133

Kloepper JW (1994) Plant growth-promoting rhizobacteria (other systems). In: Okon Y (ed) *Azospirillum*/plant associations. CRC Press, Inc., Boca Raton, pp 137–166

Kuldau G, Bacon CW (2008) Clavicipitaceous endophytes: their ability to enhance grass resistance to multiple stresses. Biol Control 46:57–71

Kumar M, Yadav V, Tuteja N, Johri AK (2009) Antioxidant enzyme activities in maize plants colonized with *Piriformospora indica*. Planta 155:780–790

Latch GCM, Potter LR, Tyler BF (1987) Incidence of endophytes in seeds from collections of *Lolium* and *Festuca* species. Ann Appl Biol 111:59–64

Leistner E, Steiner U (2009) Fungal origin of ergoline alkaloids present in dicotyledonous plants (Convolvulaceae). In: Anke T, Weber D (eds) Physiology and genetics, The mycota XV. Springer, Berlin, pp 197–208

Lewis GC, White JF Jr, Bonnefont J (1993) Evaluation of grasses infected with fungal endophytes against locusts. Ann Appl Biol Test Agrochem Cultiv 14:142–143

Lucero ME, Unc A, Cooke P, Dowd S, Sun S (2011) Endophyte microbiome diversity in micropropagated *Atriplex canescens* and *Atriplex torreyi* var *griffithsii*. PLoS ONE 6(3):e17693. doi:10.1371/journal.pone.0017693

Lyons PC, Evans JJ, Bacon CW (1990) Effects of the fungal endophyte *Acremonium coenophialum* on nitrogen accumulation and metabolism in tall fescue. Plant Physiol 92:726–732

Magnani G, Didonet C, Cruz L, Picheth C, Pedrosa F, Souza E (2010) Diversity of endophytic bacteria in Brazilian sugarcane. Genet Mol Res 9:258

Malinowski DP, Belesky DP (2006) Ecological importance of *Neotyphodium* spp. grass endophytes in agroecosystems. Grassland Sci 52:1–14

Malinowski M, Belesky DP, Lewis GC (2005) Abiotic stresses in endophytic grasses. In: Roberts C, West C, Spiers D (eds) *Neotyphodium* in cool season grasses. Blackwell Publishing, Ames, pp 187–199

Mantelin S, Touraine B (2004) Plant growth-promoting bacteria and nitrate availability: impacts on root development and nitrate uptake. J Exp Bot 55:27–34

Markert A, Steffan N, Ploss K, Hellwig S, Steiner U, Drewke C, Li SM, Boland W, Leistner E (2008) Biosynthesis and accumulation of ergoline alkaloids in a mutualistic association between *Ipomoea asarifolia* (Convolvulaceae) and a clavicipitalean fungus. Plant Physiol 147:296–305

Márquez LM, Redman R, Rodríguez R, Roossinck M (2007) A virus in a fungus in a plant: three-way symbiosis required for thermal tolerance. Science 315:513–515

Matthews J, Clay K (2001) Influence of fungal endophyte infection on plant-soil feedback and community interactions. Ecology 82:500–509

Minamisawa C, Nishioka K, Miyaki T, Ye B, Miyamoto T, You M, Saito A, Saito M, Barraquio W, Teaumroong N, Sein T, Sato T (2004) Anaerobic nitrogen-fixing consortia consisting of *Clostridia* isolated from gramineous plants. Appl Environ Microbiol 70:3096–3102

Minerdi D, Moretti M, Gilardi G, Barberio C, Gullino M, Garibaldi A (2008) Bacterial ectosymbionts and virulence silencing in a *Fusarium oxysporum* strain. Environ Microbiol 10:1725–1741

Minerdi D, Bossi S, Maffei M, Gullino ML, Garibaldi A (2011) *Fusarium oxysporum* and its bacterial consortium promote lettuce growth and expansin A5 gene expression through microbial volatile organic compound (MVOC) emission. FEMS Microbiol Ecol 76:342–351

Miyamoto T, Kawahara M, Minamisawa K (2004) Novel endophytic nitrogen-fixing clostridia from the grass *Miscanthus sinensis* as revealed by terminal restriction fragment length polymorphism analysis. Appl Environ Microbiol 70:6580–6586

Moon CD, Craven KD, Leuchtmann A, Clement SL, Schardl CL (2004) Prevalence of interspecific hybrids amongst asexual fungal endophytes of grasses. Mol Ecol 13:1455–1467

Morsy M, Oswald J, He J, Tang Y, Roossinck M (2010) Teasing apart a three-way symbiosis: transcriptome analysis of *Curvularia protuberata* in response to viral infection and heat stress. Biochem Biophys Res Com 401:225–230

Moy M, Belanger F, Duncan R, Freehof A, Leary C, Meyer W, Sullivan R, White JF Jr (2000) Identification of epiphyllous mycelial nets on leaves of grasses infected by clavicipitaceous endophytes. Symbiosis 28:291–302

Novas MV, Ianonne LJ, Godeas AM, Scervino JM (2011) Evidence for leaf endophyte regulation of root symbionts: effect of *Neotyphodium* endophytes on the pre-infective state of mycorrhizal fungi. Symbiosis. doi:10.1007/s13199-011-0140-4

Ortíz-Castro R, Valencia-Cantero E, López-Bucio J (2008) Plant growth promotion by *Bacillus megaterium* involves cytokinin signaling. Plant Sig Behav 3:263–265

Oudemans PV, Caruso FL, Stretch AW (1998) Cranberry fruit rot in the Northeast: a complex disease. Plant Dis 82:1176–1184

Panaccione DG (2005) Origins and significance of ergot alkaloid diversity in fungi. FEMS Microbiol Lett 251:9–17

Panaccione DG, Cipoletti JR, Sedlock AB, Blemings KP, Schardl CL, Machado C, Seidel GE (2006) Effects of ergot alkaloids on food preference and satiety in rabbits, as assessed with gene-knockout endophytes in perennial ryegrass (*Lolium perenne*). J Agric Food Chem 54:4582–4597

Paungfoo-Lonhienne C, Rentsch D, Robatzrk S, Webb RI, Sagulenko E, Nasholm T, Schmidt S, Lonhienne TGA (2010) Turning the table: plants consume microbes as a source of nutrients. PLoS ONE 5(7):e11915. doi:10:1371/journal.pone.0011915

Porras-Alfaro A, Bayman P (2011) Hidden fungi, emergent properties: endophytes and microbiomes. Annu Rev Phytopathol 49:291–315

Puente ME, Bashan Y (1994) The desert epiphyte *Tillandsia recurvata* harbours the nitrogen-fixing bacterium *Pseudomonas stutzeri*. Can J Bot 72:406–408

Puente ME, Li C, Bashan Y (2009) Rock-degrading endophytic bacteria in cacti. Environ Exp Bot 66: 389–401

Radajewski S, Ineson P, Parekh NR, Murrell JC (2000) Stable-isotope probing as a tool in microbial ecology. Nature 403:646–649

Rasmussen S, Parsons AJ, Fraser K, Xue H, Newman JA (2008) Metabolic profiles of *Lolium perenne* are differentially affected by nitrogen supply, carbohydrate content, and fungal endophyte infection. Plant Physiol 146:1440–1453

Redman RS, Sheehan KB, Stout RG, Rodriguez RJ, Henson JM (2002) Thermotolerance generated by plant/fungal symbiosis. Science 298:1581

Redman RS, Kim YO, Woodward CJDA, Greer C, Espino L (2011) Increased fitness of rice plants to abiotic stress via habitat adapted symbiosis: a strategy for mitigating impacts of climate change. PLoS ONE 6(7):e14823. doi:10.1371/journal.pone.0014823

Reinhold-Hurek B, Hurek T (2011) Living inside plants: bacterial endophytes. Curr Opin Plant Biol 14:435–443

Richardson MD, Chapman GW, Hoveland CS, Bacon CW (1992) Sugar alcohols in endophyte-infected tall fescue. Crop Sci 32:1060–1061

Rodriguez RJ, Freeman DC, McArthur ED, Kim YO, Redman RS (2009a) Symbiotic regulation of plant growth, development and reproduction. Commun Integr Biol 2:141–143

Rodriguez RJ, Woodward C, Kim YO, Redman RS (2009b) Habitat-adapted symbiosis as a defense against abiotic and biotic stresses. In: White JF Jr, Torres MS (eds) Defensive mutualism in microbial symbiosis. CRC Press, Boca Raton, pp 335–346

Rosenblueth M, Martínez-Romero E (2006) Bacterial endophytes and their interactions with hosts. Mol Plant Microbe Interact 19:827–837

Rudgers JA, Koslow JM, Clay K (2004) Endophytic fungi alter relationships between diversity and ecosystem properties. Ecol Lett 7:42–51

Rudgers JA, Mattingly WB, Koslow JM (2005) Mutualistic fungus promotes plant invasion into diverse communities. Oecologia 144:463–471

Saikkonen K, Helander M, Ranta H, Neuvonen S, Virtanen T, Suomela J, Vuorinen P (1996) Endophyte-mediated interactions between woody plants and insect herbivores? Entomol Exp Appl 80:269–271

Saikkonen K, Helander M, Faeth SH, Schulthess F, Wilson D (1999) Endophyte-grass-herbivore interactions: the case of *Neotyphodium* endophytes in Arizona fescue populations. Oecologia 121:411–420

Saikkonen K, Helander M, Faeth SH (2004) Fungal endophytes: hitchhikers of the green world. In: Gillings M, Holmes A (eds) Plant microbiology. BIOS Scientific Publishers Limited, Oxford, pp 77–95

Schardl CL, Craven KD (2003) Interspecific hybridization in plant-associated fungi and oomycetes: a review. Mol Ecol 12:2861–2873

Schardl CL, Phillips TD (1997) Protective grass endophytes: where are they from and where are they going? Plant Dis 81:430–438

Schardl CL, Leuchtmann A, Spiering MJ (2004) Symbioses of grasses with seedborne fungal endophytes. Annu Rev Plant Biol 55:315–340

Schulz B, Boyle C (2005) The endophytic continuum. Mycol Res 109:661–686

Schulz B, Boyle C, Draeger S, Römmert A-K, Krohn K (2002) Endophytic fungi: a source of novel biologically active secondary metabolites. Mycol Res 106:996–1004

Sloan DB, Moran NA (2012) Endosymbiotic bacteria as a source of carotenoids in whiteflies. Biol Lett. doi:10.1098/rsbl.2012.0664

Spiering MJ, Moon CD, Wilkinson HH, Schardl CL (2005) Gene clusters for insecticidal loline alkaloids in the grass-endophytic fungus *Neotyphodium uncinatum*. Genetics 169:1403–1414

Spiering MJ, Greer DH, Schmid J (2006) Effects of the fungal endophyte, *Neotyphodium lolii*, on net photosynthesis and growth rates of perennial ryegrass (*Lolium perenne*) are independent of in planta endophyte concentration. Ann Bot 98:379–387

Steiner U, Ahimsa-Müller MA, Markert A, Kucht S, Groß J, Kauf N, Kuzma M, Zych M, Lamshöft M, Furmanowa M, Knoop V, Drewke C, Leistner E (2006) Molecular characterisation of a seed transmitted clavicipitaceous fungus occurring on dicotyledoneous plants (Convolvulaceae). Planta 224:533–544

Stewart WD, Fitzgerald GP, Burris RH (1967) In situ studies on N_2 fixation using the acetylene reduction technique. Proc Natl Acad Sci U S A 58:2071–2078

Stone JK, Bacon CW, White JF Jr (2000) An overview of endophytic microbes: endophytism defined. In: Bacon CW, White JF Jr (eds) Microbial endophytes. Marcel-Dekker, New York, pp 3–30

Strobel G, Daisy B (2003) Bioprospecting for microbial endophytes and their natural products. Microbiol Mol Biol Rev 67:491–502

Sullivan TJ, Rodstrom J, Vandop J, Librizzi J, Graham C (2007) Symbiont-mediated changes in *Lolium arundinaceum* inducible defenses: evidence from changes in gene expression and leaf composition. New Phytol 176:673–679

Sundram S, Meon S, Seman IA, Othman R (2011) Symbiotic interaction of endophytic bacteria with arbuscular mycorrhizal fungi and its antagonistic effect on *Ganoderma boninense*. J Microbiol 49:551–557

Suryanarayanan TS, Kumaresan V, Johnson JA (1998) Foliar fungal endophytes from two species of the mangrove *Rhizophora*. Can J Microbiol 44:1003–1006

Tadych M, Ambrose KV, Bergen MS, Belanger FC, White JF Jr (2012) Taxonomic placement of *Epichloë poae* sp. nov. and horizontal dissemination to seedlings via conidia. Fungal Divers 5:1–15

Taulé C, Mareque C, Barloco C, Hackembruch F, Reis V, Sicardi M, Battistoni F (2012) The contribution of nitrogen fixation to sugarcane (*Saccharum officinarum* L.), and the identification and characterization of part of the associated diazotrophic bacterial community. Plant Soil 356:35–49

Taylor TN, Taylor EI (1993) The biology and evolution of fossil plants. Prentice Hall, New York

Taylor TN, Remy W, Hass H, Kerp H (1995) Fossil arbuscular mycorrhizae from the early Devonian. Mycologia 87:560–573

Torres MS, White JF, Zhang X, Hinton DM, Bacon CW (2012) Endophyte-mediated adjustments in host morphology and physiology and effects on host fitness traits in grasses. Fungal Ecol 5:322–330

Vadassery J, Ritter C, Venus Y, Camehi I, Varma A et al (2008) The role of auxins and cytokinins in the mutualistic interaction between *Arabidopsis* and *Piriformospora indica*. Mol Plant Microbe Interact 21:1371–1383

Waller F, Achatz B, Baltruschat H, Fodor J, Becker K, Fischer M, Heier T, Hückelhoven R, Neumann C, Wettstein D, Franken P, Kogel KH (2005) The endophytic fungus *Piriformospora indica* reprograms barley to salt-stress tolerance, disease resistance, and higher yield. Proc Natl Acad Sci U S A 102: 13386–13391

West ER, Cother EJ, Steel CC, Ash GJ (2010) The characterization and diversity of bacterial endophytes of grapevine. Can J Microbiol 56:209–216

White JF Jr (1987) The widespread distribution of endophytes in the Poaceae. Plant Dis 71:340–342

White JF Jr (1988) Endophyte-host associations in forage grasses. XI. A proposal concerning origin and evolution. Mycologia 80:442–446

White JF Jr, Cole GT (1985) Endophyte-host associations in forage grasses. III. *In vitro* inhibition of fungi by *Acremonium coenophialum*. Mycologia 77:487–489

White JF, Torres MS (2010) Is endophyte-mediated defensive mutualism oxidative stress protection? Physiol Plant 138:440–446

White JF Jr, Morrow AC, Morgan-Jones G, Chambless DA (1991) Endophyte-host associations in forage grasses. XIV. Primary stromata formation and seed transmission in *Epichloë typhina*: developmental and regulatory aspects. Mycologia 83:72–81

White JF Jr, Glenn AE, Chandler KF (1993) Endophyte-host associations in grasses. XVIII. Moisture relations and insect herbivory of the stromal leaf of *Epichloë typhina*. Mycologia 85:195–202

White JF, Martin TI, Cabral D (1996) Endophyte-host associations in grasses. Conidia formation by *Acremonium* endophytes in the phylloplanes of *Agrostis hyemalis* and *Poa rigidifolia*. Mycologia 88:174–178

White JF, Sullivan R, Balady G, Gianfagna T, Yue Q, Meyer W, Cabral D (2001) A fungal endosymbiont of the grass *Bromus setifolius*: distribution in some Andean populations, identification and examination of beneficial properties. Symbiosis 31:241–257

White JF Jr, Crawford H, Torres MS, Mattera R, Bergen M, Irisarry I (2012a) A proposed mechanism for nitrogen acquisition by grass seedlings through oxidation of symbiotic bacteria. Symbiosis 57:161–171

White JF, Johnson H, Torres MS, Irizarry I (2012b) Nutritional endosymbiotic systems in plants: bacteria function like 'quasi-organelles' to convert atmospheric nitrogen into plant nutrients. Plant Pathol Microbiol 3:e104

Whitford WG (2002) Ecology of desert systems. Elsevier Sciences Ltd./Academic, London

Wilkinson HH, Siegel MR, Blankenship JD, Mallory AC, Bush LP, Schardl CL (2000) Contribution of fungal loline alkaloids to protection from aphids in a grass-endophyte mutualism. Mol Plant-Microbe Interact 13:1027–1033

Yue Q, Miller CJ, White JF, Richardson MD (2000) Isolation and characterization of fungal inhibitors from *Epichloë festucae*. J Agric Food Chem 48:4687–4692

Zhang YP, Nan ZB (2007) Growth and anti-oxidative systems changes in *Elymus dahuricus* is affected by *Neotyphodium* endophyte under contrasting water availability. J Agron Crop Sci 193:377–386

Zhang G, Peng G, Wang E, Yan H, Yuan Q, Zhang W, Lou X, Wu H, Tan Z (2008) Diverse endophytic nitrogen-fixing bacteria isolated from wild rice *Oryza rufipogon* and description of *Phytobacter diazotrophicus* gen. nov. sp. nov. Arch Microbiol 189:431–439

Zimmerman NB, Vitousek PM (2012) Fungal endophyte communities reflect environmental structuring across Hawaiian landscape. Proc Natl Acad Sci U S A 109: 13022–13027

22 Microbial Endophytes: Future Challenges

Charles W. Bacon and Dorothy M. Hinton

Abstract

Endophytes are represented by a diverse group of prokaryotic (bacteria or cyanobacteria) or eukaryotic (fungi or parasitic vascular plants) organisms that form lifelong associations within tissues of plants. Ecologically, these associations are viewed as mutualistic and as sources of secondary metabolites capable of serving as novel medicinals and agrichemicals. It is this area that serve to stimulate the large research investigations from all parts of the planet. The challenges as we see them are multifaceted. These include an understanding of the genetics nature of microbial endophytes, how endophytes communicate and partition themselves within hosts, how do these biotrophic organisms obtain nutrients, and are specific nutrient acquisitions key to the final effects observed? Further, are there basic difference between bacterial endophytes and fungal endophytes? What influence the host interactions to produce the desired effects, and how is the stability of the system affected. Thus, future challenges are dependent on identifying, delineating, dissecting, and defining the mechanisms whereby hosts and their symbionts accomplish this curious lifestyle. Defining these biological mechanisms will ensure the present and future successful technological applications of microbial endophytes.

1 Introduction

The emphasis in greater sustainability and an increase in public concern for hazards associated with synthetic chemical pesticides and transgenic plants have produced a resurgence of interest in the use of introduced microorganisms for biological control of plant pathogens. Most of these microorganisms are inconsistent in their performance in biological control resulting in reduced commercial

C.W. Bacon (✉) • D.M. Hinton
Toxicology and Mycotoxin Research Unit, USDA, Agriculture Research Service, Russell Research Center, 950 College Station Road, Athens, GA 30605, USA
e-mail: charles.bacon@ars.usda.gov

V.C. Verma and A.C. Gange (eds.), *Advances in Endophytic Research*, DOI 10.1007/978-81-322-1575-2_22, © Springer India 2014

development and widespread use. The major reason for this lack of performance is inadequate colonization of the target site, variation in expression of control at that site, and need for numerous applications. Most of the biocontrol organisms are either soil or surface dwellers and have very little affinity to plants as specific colonizers as evidenced by ineffective controls of disease following repeated applications (see Thomashow and Weller 1996; Hallmann et al. 1997; Hallmann 2001 for review). However, we are concerned with here a unique group of microbial organisms that form endophytic associations within plants, and several of these have been discussed in other chapters in this book. Microbial endophytes actively colonize below and aboveground host tissues and establish long-term associations, actually lifelong natural associations, without imposing any obvious harm to the host. Since the work demonstrating the production of ruminant deterrents by clavicipitalean endophytes of pasture grasses (Bacon et al. 1977), much emphasis has been placed on these and many other microbial endophytes. Still after several decades of active research involving an international cadre of research scientists, we have not learned the essentials necessary to predict nor maintain associations, but have developed some important hypotheses. Of consequence is the central dogma that these associations are of merit and play an important role in plant defense. Thus, the hypothesis of defensive mutualism serves to drive the thinking of so many endophytic systems. There are however objections to this hypothesis, although poorly supported and not replaced by another rational for existence. Based on limited information, there is no doubt that microbial endophytes will have an important role in the improvements of crops, as pharmaceutical agents, and contributing to the ecological success of several non-crop plants. The utilization of these organisms is anticipated to be unlimited as most of these have not yet been exploited. Future exploitation of these organisms are highly dependent on our understanding of the apparent roles they play within the symbioses of such a diversity of hosts as well as their continued discovery from a diversity of habitats. However, the successful utilization of endophytes is dependent upon deciphering any generalized inter- and intra-interactions with hosts although we envision that there also may be highly specific or strain interactions. The future challenges are deciphering the many roles each play within the symbioses and acceptance and successful application of each to a particular environmental situation.

What are the future challenges? The answers to these questions are intended to stimulate research in directions of basic research on such an obviously applied problem. A major premise forming the basis for microbial endophytes is that they mechanistically operate similar within a host. This is probably not the case, and as such the challenges lie in dissecting out that which is fundamental from that which produces specific and highly desirable ecological attributes. Many attempts have been made at surrogate transformations of endophytic systems with the hope of obtaining specific material for specific areas. However, success using such transformations is not always achieved as some endophytes are mysterious and appear not to effect a change in the host or even worst become pathogenic. Neither, the manner in which specific endophytes of vast genetic diversity effect host fitness traits is known nor can we predict the long-term stability of these effects within a population to affect an ecological successful planting. Are these traits environmentally triggered, or are they manifested regardless of the environment? Following is a list of questions that we feel are discussion points from which major challenges in endophyte research should be explored and answered before we can understand the nature of the relationship and reap the full benefits of microbial endophytes. However, not all questions are presented here, but the major ones are posed for guidance in providing our current status for understanding the science of "endophytology" that can serve as a guide for meeting the challenges for the present and future applications of microbial endophytes.

2 Major Challenges

2.1 Are the Diversities of Habitats Important for Uses of Endophytes?

Endophytic associations are ubiquitous and are represented by an even greater number of microbionts that have developed this strategy including algae, fungi, bacteria, and viruses. The ubiquitous nature is illustrated by some surveys conducted for endophytic fungi (Germida et al. 1998; Schulz et al. 1998; Petrini and Petrini 1989; Petrini 1986, 1990; Hoff et al. 2004) and bacteria as well as the curious associations among microbes with other organisms (Wrede et al. 2012) (and papers cited therein). Endophytes are further subdivided into functional roles as the helpful scheme presented by Rodriguez et al. (Rodriguez et al. 2009). It is important to be cognizant that endophytic associations can involve a third microbial component, but the extent of this multiple mutualistic interactions is unknown. Endophytes are both obligate and facultative. In most instances the obligate endophytism includes the very rare and novel taxa, none of which are found as free-living organisms. However, the facultative associations include microbes that consist of strains or species of otherwise pathogenic or saprophytic organisms some of which are capable of free living as saprotrophs even if it means at the expense of the senescing host tissue. Endophytic organisms are associated with the entire spectrum of plants that include monocots and dicots, either herbaceous or woody, and terrestrial or aquatic. They are found in environmental extremes including the great deserts, arctic tundra, oceans, and sulfur hot springs, as well as lush tropical forest (Suryanarayanan et al. 1998; Imada et al. 2007; Stierle and Stierle 2005; Suryanarayanan and Kumaresan 2000). In terms of evolution, these environmental variables are considered to be inducers of endophytic associations particularly as it relates to microbial symbionts (Klitgord and Segre 2010; Schardl and Phillips 1997; Schardl and Moon 2003; Scott and Schardl 1993; Saikkonen et al. 2001). Due to this vast array of ecosystems and endophytic associations, the successful utilization of endophytic plants is expected to be highly dependent on the interaction environmental variables from which they were derived. While speculative, a complex environment might predict the problems and or limit the nature of the uses of endophytes. The challenges facing such a diverse assemblage of endophytic systems are to find a common physiological or genetic relationship so that each and every endophyte can be utilized, and its performance predicted. Further, will this common relationship reside in both bacterial and fungal endophytes?

2.2 What Are the Difficulties Encountered in the Uses of Endophytes?

Regardless of the many uses of endophytes, the two major categories can be divided into in vitro uses and in vivo uses. Most successful applications of microbial endophytes are those that seek pharmacological metabolites produced during in vitro fermentations. In vitro, the isolated endophyte is used for the production of one of many pharmacological agents conducted under laboratory fermentations that include novel antibiotics, immunosuppressants, antimycotics, and anticancer drugs. The many metabolites that are produced by endophytes are reviewed, and some of these have reference to specific novel compounds leading to important drug discovers (Findlay et al. 1995; Li et al. 2006; Mei and Flinn 2010; Priti et al. 2009; Southcott and Johnson 1997; Strobel 2002; Yu et al. 2010; Strobel and Daisy 2003; Strobel et al. 2004). In vitro, while demonstrating a potential for specific metabolic production, the optimal production of most of these in culture have yet to be discovered. There is marked variation in amounts of specific metabolite produced by strains from each microbial endophyte. Media compositions and genetic of the organisms have been shown to favor strains of a species which

Table 22.1 Total surfactin production by strains of *Bacillus mojavensis* cultured in nutrient medium and Pharmamedia and tested for antagonism to *F. verticillioides* measured after 14 days on nutrient agar (Bacon et al. 2012)

		Total surfactin (μg/ml)	
Strains	Location	Nutrient broth	Pharmamedia
RRC 101	RRC	2.55	17.12
RRC 101fa	RRC	2.89	19.65
RRC 112	RRC	0.00	0.73
RRC 112fa	RRC	0.00	0.63
RRC 111	RRC	0.00	2.31
RRC 113	RRC	0.00	2.63
RRC 114	RRC	0.00	2.18
ATCC 51516	Mojave	3.39	22.55
NRRL B-14698 [T]	Mojave	58.42	42.14
NRRL B-14699	Mojave	0.00	0.21
NRRL B-14700	Mojave	2.41	15.22
NRRL B-14701	Mojave	0.00	1.18
NRRL B-14702	Mojave	5.03	22.94
NRRL B-14703	Gobi	0.00	2.42
NRRL B-14704	Gobi	0.00	2.06
NRRL B-14705	Gobi	0.00	37.21
NRRL B-14706	Gobi	0.00	10.11
NRRL B-14707	Gobi	0.72	0.29
NRRL B-14708	Gobi	6.28	89.30
NRRL B-14709	Gobi	9.85	30.27
NRRL B-14710	Gobi	0.00	0.10
NRRL B-14711	Gobi	1.71	27.16
NRRL B-14712	Gobi	10.05	17.55

can have no effect on strains within that same species (Table 22.1) (Bacon et al. 2012; Bacon 1985, 1988). Those interested in endophytes for the production of pharmacological agents conduct wide screens for conditions required for increase production. In additions to cultural modifications, there are also varied genetic or molecular modifications of microbial endophytes that can serve the purpose on increasing the production of specific compounds. However, the challenge here is finding the specific genes and other constructs that can be effective in general to market such a modification. Some progress has been made and includes the in vitro production of alkaloids, terpenoids, polyketones, peptides, isocumarines, complex sugars, quinols, and phenols. Of monumental importance is the isolation and demonstration of the ovarian and breast diterpenoid anticancer agent paclitaxel or Taxol produced by the endophytic fungus *Taxomyces andreanae* initially isolated from the bark of the western yew tree, *Taxus brevifolia*, although subsequently established as being derived from the endophyte in this tree and several additional symbiotic yew species (Bashyal et al. 1999; Strobel et al. 1996, 1999; Lin et al. 2003; Liu et al. 2006). Success is far from being complete as most of the pharmacological agents consist of very novel and complex structural groups representing various but several metabolic pathways that complicate such studies. For example, Taxol is derived from both the mevalonate and non-mevalonate pathways (Soliman et al. 2011), and very little is known about the essential genes encoding the enzymes leading to the pathways. However, strain improvement and fermentation engineering have resulted in an increase in the yield of this important anticancer agent (Zhou et al. 2010).

The in vivo or *in planta* uses of microbial endophytes present the greater challenges for success. This is due to a variety of complicated issues that address the mechanisms of actions responsible for the host specificity, if any, leading to the elicitation and metabolite production that will protect the host from the environmental challenges. For each endophyte there are methods of inoculating the hosts, including some that are seed transmitted resulting in infected seedling upon germination. A desirable goal for further development of biocontrol strategies using endophytes is the genetic modifications of strains for specific locations and cultivars of plants with specific agronomic traits that will be maintained hopefully throughout the growing season. Due to the ease of genetically modifying microbial endophytes, one use involves the molecular modifications of endophytes and their reconstruction into natural or at least native hosts for altered expression. Such surrogately transformed plants have been accomplished with success (Schardl 1994; Tsai et al. 1992; Murray et al. 1992; Young et al. 2005). However, details of several other microbial endophytes have only just been initiated, and in this regard, the work dealing with the Gram-positive bacteria is behind those detailing transformation of

Gram-negative bacteria. Surrogate transformations have worked in instances where the modifications abolished the *in planta* changes, which in this case was the removal of ergot alkaloids from desirable strains of *Neotyphodium* species of grass endophytes. In yet another instance, gene deletion of one class of toxins leads to the production and accumulation of an entirely different class of toxins. In theory the ecological stamina of the surrogately transformed hosts may not be applicable to all environments. Further, mutation in at least fungal endophytes can alter the phenotypes of grass hosts. Additional approaches utilize nonnative fungal endophytes such as strains that do not produce cattle toxins resulting in improved versions of native grasses (Parish et al. 2003a, b). The resulting limited ecotypes are however tailor made for resistance to biotic pests (Parish et al. 2003b).

With the exception of toxic metabolites from fungal endophytic associations, very few metabolites demonstrated from in vitro studies have been shown to be produced under *in planta* or in vivo uses. However, strains indicated as being negative under cultural condition may not only be positive in the *in planta* situation but also very efficient in metabolite production. Very limited information is available concerning effects of soil culture requirements on *in planta* production. Microbial endophytes are biotrophic but in the broader sense heterotrophic and depend on carbon, nitrogen, and energy provided from its hosts. There are therefore series of interactions responsible for the final synthesis of complex secondary compounds that provide for the long-term survival strategies of microbial endophytes such as host defense. Certainly abiotic factors imposed on the symbioses are also contributors since they relate to the host survival. Studies dealing with maintenance and establishment of the symbioses are relatively recent and the information presented thus far is rather scarce (Tanaka et al. 2005, 2006, 2007; Bostock 2005; Mundy et al. 2006; Taylor and McAinsh 2004). Control mechanisms between the two symbionts are unclear, and the interaction with environmental variable may influence both or only one of the symbionts. Cross talk and other communications are considered salient features between the two although studies indicating the evidence and mechanism of action are varied for both fungal and bacterial endophytes (Tanaka et al. 2005, 2006, 2007; Bostock 2005; Mundy et al. 2006; Rasmussen et al. 2007; Taylor and McAinsh 2004). There are some indications that carbon dioxide and nitrogen concentrations affect the content of endophytic hyphae and produce a considerable influence on the production of nitrogen-based secondary compounds such as ergot alkaloids and other mycotoxins (Lyons et al. 1990; Arechavaleta et al. 1992; Hunt et al. 2005; Draper et al. 2011; Rasmussen et al. 2012). Bacterial influence on hosts may be recognized by production of a defensive metabolite phenotypically, however, there is very little information concerning their ecological dependence and affects on host.

2.3 What Are the Problems Associated with Increased Production of Metabolites?

Fermentation parameters are inadequate both for the growth of the microbial endophyte and the duration of product accumulation. Most novel endophytes are notorious for slow growth and rapid decline under culture conditions. Genetic stability of some species occurs more frequently than others. Other endophytes are solid producers of desirable metabolites but rapidly decline although strict cultural protocols are followed. Increased metabolite production can be obtained by one of several mutational approaches. One approach is heterologous production which is used primarily in the increased production of pharmacological therapeutics. The transfer of multiple copies of genes or gene clusters from an endophyte into a nonhost that is amendable to culture can result in the production of compounds in higher amounts due to the stability of the foreign host and its ease of culture (Wenzel and Muller 2005; Zhang et al. 2011). Several hosts that are available for heterologous expression include *E. coli*, *Streptomyces* sp., and species of yeast. However, not all hosts are suitable for heterologous expression, and occasionally due to host metabolism only some intermediates of the desired product are produced. Trial and error might fulfill acceptable expressions utilizing new and modified techniques (Wenzel and Muller 2005; Zhang et al. 2011). Heterologous expression as a tool for increasing the biological activity of endophytes used for *in planta* biological control has not been explored to any extent, although such genetic modifications might fall under the concerns of regulatory agencies, e.g., GMO uses.

2.4 What Are the Legal Concerns and Impediments to Uses of Endophyte-Enhanced Plants?

The intended uses of endophytically infected plants for conservation grasses forage grass improvement, and disease protection might also involve some concern from human health that can also relate to various regulatory agencies. The use of microbial endophytes have not received regulatory attention from none of the three US regulatory agencies that are responsible for genetically modified crops, and these include the Environmental Protection Agency (EPA), the Food and Drug Administration, and the US Department of Agriculture. Since each of these agencies regulates various aspects of transgenic crops from several perspectives, the uses of endophytically enhanced plants are without regulation although there are perhaps self constraints dealing with their use. The uses of endophytes do not take on the environmental concerns as genetic engineering or recombinant DNA as the host genetic materials is not modified. Further, in most instances the host plant is naturally infected by microbes although not necessarily by the native endophyte. The concept as used in this review is one of endophyte technology as opposed to biotechnology which implies to most, including regulatory agencies as technology resulting from DNA manipulations. Thus the uses of endophytes do not have the same public concerns as GMO and biopharming of agronomically important plants such as corn, rice, and soybeans. The uses of native and nonnative endophytes in other countries are apparently equally accepted as in the USA.

Currently, there is concern from several US state agencies for the certification process of seed sold for various endophyte enhancements. This relates to validation that the seed contain known levels of viable endophytes. The cost of such seed depends on validation of its percentage viable endophyte certification for that seed lot, but endophyte viability is difficult to establish temporally. This is one of the biggest impediments to the use of endophyte-infected seed, especially those intended for turf uses. This is due to the lack of an immediate viability test for seed at the time certification labels are placed on seed. Demonstration of endophyte presence in seed is procedure and is outlined in several regulatory procedures. Determination of endophyte viability in seed is done usually by certain seedling grow-out test that document viability, but this takes from 4–6 week post germination, which means that the seed originally tested is now 4–6 weeks older, invalidating the percentage life infection status, which depending on test seed lot storage condition may be considerably less. Perhaps the various molecular techniques may solve this problem. The lack of an endophyte viability test includes a large portion of the forage industry. Until such tests are developed, it is known that endophytically infected grass seed remains relatively constant under refrigeration at 4 °C or slightly higher. Bacterial endophytes that are added to the seed coat do not have this as a problem, especially since most of the species are spore forming *Bacillus* species that are added as spores either in water or coating. The *Bacillus* spore maintains high viability on seed for several months at room temperature and is compatible with most seed coatings.

2.5 Additional Questions

The endophytic organisms used in patents include bacteria, fungi, and viruses, although the majority includes bacteria. What are the patented uses of endophytes? From 1976 to the present, the reduced number of patents listed at Patent Storm, a patent listing site (http://www.patentstorm.us/) (2013), indicates that currently there were well over 600 patents for numerous uses (Table 22.2).

Table 22.2 Selected US Patents for bacterial and fungal endophytes from a patent search site (Patent Storm, 1/29/2013: http://www.patentstorm.us/)

Patent number	Title	Issue date
7037879	Pest control method for grass family plants using endophytic bacteria	05/02/2006
6815591	Enhancing endophyte in grass	11/09/2004
7465855	Nontoxic endophytes, plants injected therewith, and methods for injecting plants	12/16/2008
7084331	Rice containing endophytic bacteria and method of producing it	08/01/2006
7642424	Tall fescue endophyte E34	01/05/2010
7892813	Fungal endophytes	02/22/2011
7976857	Grass endophytes	07/12/2011
5914107	Method of introducing an endophytic fungus into rough bluegrass, *Poa trivialis*, and *Poa compressa*	06/22/1999
6072107	Ryegrass endophytes	06/06/2000
7259004	Endophytic streptomycetes from higher plants with biological activity	08/21/2007
8101551	Production and use of endophytes as novel inoculants for promoting enhanced plant vigor, health, growth, and yield-reducing environmental stress	01/24/2012
6548745	Italian rye grass and a method of introducing endophytic fungi into an Italian rye grass	04/15/2003
20080229441	Fungal endophytes of *Elymus canadensis*	09/18/2008
20090105076	Production and use of endophytes as novel inoculants for enhanced plant vigor, health, growth, and yield	04/23/2009
20090181447	Grass endophytes	07/16/2009
5994117	Use of *Bacillus subtilis* as an endophyte for the control of diseases caused by fungi	11/30/1999
6335188	Endophyte ergot alkaloid synthetic compounds, compounds which encode therefore, and related methods	01/01/2002
20110173727	Endophyte-enhanced seedlings with increased pest tolerance	07/14/2011
20120198590	Antifungal metabolites from fungal endophytes of *Pinus strobus*	08/02/2012
20110262401	Grass endophytes	10/27/2011
20080022420	Be9301a tall fescue with endophytes	01/24/2008
20110162116	Method for growing plants and ROS content	06/30/2011
20110289627	Modified cry3a toxins and nucleic acid sequences coding therefore	11/24/2011
20120149571	Inoculants including *Bacillus* bacteria for inducing production of volatile organic compounds in plants	06/14/2012
20120165513	Processes for isolation and purification of enfumafungin, a novel antifungal compound produced by an endophytic *Hormonema* species	06/28/2012
20120210464	Insecticidal proteins	08/16/2012
20120260372	Transgenic plants expressing modified cry3abacteria, in particular *Bacillus thuringiensis* or E. coli	10/11/2012
20120270776	Novel pesticide toxins	10/25/2012
20050090395	Biological control deciduous trees with new strains of *Chondrostereum purpureum* isolates	04/28/2005
20050095283	Compositions and methods for topically treating diseases from *Taxus brevifolia* (Pacific yew), *Taxomyces andreanae*, and endophytic fungus of the Pacific yew	05/05/2005
20050215764	Biological polymer from *Taxus brevifolia* (Pacific yew) and *Taxomyces andreanae* and endophytic fungus of the Pacific yew	09/29/2005
20050249667	Process for treating a biological organism from dried bark of *Taxus brevifolia* (Pacific yew) and *Taxomyces andreanae* and endophytic fungus of the Pacific yew	11/10/2005
20060085870	Modified Cry3A toxins as bacteria	04/20/2006
20060147371	Water-soluble compound paclitaxel obtained from *Taxus brevifolia* (Pacific yew) and *Taxomyces andreanae* and endophytic fungus of the Pacific yew	07/06/2006
20060275887	Mycobacteria compositions and methods of use in bioremediation	12/07/2006

(continued)

Table 22.2 (continued)

Patent number	Title	Issue date
20070026506	Method for the production of Taxol and/or taxanes from cultures of hazel cells	02/01/2007
20070240237	Expression in use of novel pesticidal toxins	10/11/2007
6515016	Composition and methods of paclitaxel which was obtained from *Taxomyces andreanae* and an endophytic fungus of the Pacific yew	02/04/2003
20050142162	Taxus brevifolia (Pacific yew) and *Taxomyces andreanae* and endophytic fungus of the Pacific yew	06/30/2005
20070207183	Zein-coated medical device obtained from endophytic fungus of the Pacific yew	09/06/2007
7393678	Inoculants for enhancing plant growth from*Klebsiella pneumoniae* inoculants	07/01/2008
6599930	Coniosetin and derivatives Cryptocin, from the endophytic fungus *Cryptosporiopsis* cf. *quercina*	07/29/2003
6069299	Fungus and insect control with chitinolytic enzymes	05/30/2000
5558997	Monoclonal antibodies to *Mycosphaerella* species	09/24/1996
5731173	Fructosyltransferase enzyme, method for its production, and DNA encoding the enzyme	03/24/1998
6759397	Ginsenoside chemotherapy, Taxol, and taxane production by *Taxomyces andreanae*	07/06/2004

Added to these uses are also those intended to increase plant biomass for biofuels or bioenergy. This suggests that the challenges for endophytes are being met although not as conveniently as anticipated. Due to the economic benefits from bioprospecting, the in vitro uses of endophytes far outnumber the *in planta* uses. The preponderances of patents have been issued for in vitro uses of endophytic microbes indicating the large economic benefits derived from bioprospecting for new and novel antibiotics and other rare medicinals. However, there is a considerable financial outlay for developing and using novel medicinals, whereas the expense and time for developing an in vivo system involve considerably less time and money. The numbers of patents suggest in both cases that the introduction and acceptance of novel endophytes for either in vitro or in vivo uses are well worth the effort. There are additional minor challenges, and these have particular relevance to the *in planta* uses of endophytes. Do endophytes offer multiple protections to hosts and is the distribution a factor in any observed beneficial effects? What are the natural distributions of endophytes in plants? The quest for specific endophytic systems rest on the primary use, and this in terms dictates the distribution in plants. What are bases for host recognition and for genotypic specificities for endophyte and hosts, and is this necessary for mutualistic expression?

Successful uses of microbial endophytes depend on demonstrating either positive *in planta* responses or production of pharmacological compounds in vitro. However, both depend on the successful isolation of the endophyte and its identity. Both of these can be difficult since some endophytic organisms fail to grow in culture and cannot be isolated which makes identity difficult. Several new molecular approaches have been developed that now allow for the determination of those taxa that are difficult to isolate, culture, and identify. Use of molecular procedures have allowed for successful studies including isolation and or identification of multiple individuals within a large population of endophytes as well as their biochemical assessments. Such methods are also useful for ecological and evolutionary characterizations of endophytes (Draper et al. 2011; Rasmussen et al. 2009; Matsumura et al. 2003; Felitti et al. 2006; Bailly et al. 2007; Duang et al. 2006; Guo et al. 2000; Handelman 2004; Pirttila et al. 2000). Thus, it is anticipated that many more endophytic systems will be identified, analyzed, and utilized. Further, these techniques should also prove useful in determining *in planta* that compounds produced within the association are those produced by the fungus without modification by the plant obviating a concern for the genetic role of the host in the observed effect.

3 Conclusion

The increased emphasis in sustainability by the public and the concern for pollution, as well as hazards associated with highly toxic synthetic chemical pesticides and transgenic plants, have generated a large international political interest in the environment. Alternative strategies such as the use of introduced microorganisms for biological control have gained a worldwide interest, and endophytic organisms are at the forefront. Most of these microorganisms are bacteria, followed by fungi, and the use of either as biocontrol agents can result in a highly desired product suitable for widespread uses. Most of the biocontrol endophytes are seed borne, therefore the systems are viewed as self-propagating and host contained agents with lasting effect during the seasons or seasons following their applications. The major reasons for any performance problem are inadequate colonization of the target site, variation in expression of control at that site, and the need for numerous applications, but continued research should alleviate such problems. Endophytic organisms maybe seed borne and require only one seed application from which a biological association with its host is established. For example, bacterial endophytes are rapidly becoming a distinct and important class of biocontrol organisms as indicated by the recent increase in publications and patents, which reflect the interest, and ease of application for their benefits to agriculture and technology. Yet to be explored are the multiple interactions involving three or more symbionts and the varied roles the members of archaea contribute to any mutualistic associations with bacteria, vertebrates, and invertebrates. Thus, we are at the beginning of this exciting area of science, and future investigations should provide the fundamental basis for endophytic mutualists, as well as initiating additional searches for other natural mutualistic endophytes that will place the technology of microbial endophytes on a firm scientific basis.

References

Arechavaleta M, Bacon CW, Plattner RD et al (1992) Accumulation of ergopeptide alkaloids in symbiotic tall fescue grown under deficits of soil water and nitrogen fertilizer. Appl Environ Microbiol 58:857–861

Bacon CW (1985) A chemically defined medium for the growth and synthesis of ergot alkaloids by the species of *Balansia*. Mycologia 77:418–423

Bacon CW (1988) Procedure for isolating the endophyte from tall fescue and screening isolates for ergot alkaloids. Appl Environ Microbiol 54:2615–2618

Bacon CW, Porter JK, Robbins JD et al (1977) *Epichloe typhina* from toxic tall fescue grasses. Appl Environ Microbiol 34:576–581

Bacon CW, Hinton DM, Mitchell TR et al (2012) Characterization of endophytic strains of *Bacillus Mojavensis* and their production of surfactin isomers. Biol Control 62:1–9

Bailly J, Fraissinet-Tachet L, Verner M-C (2007) Soil eukaryotic functional diversity, a metatranscriptomic approach. ISME J 1:632–642

Bashyal B, Li JY, Strobel G et al (1999) Seimatoantlerium nepalense, an endophytic taxol producing coelomycete from himalayan yew (*Taxus Wallachiana*). Mycotaxon 72:33–42

Bostock RM (2005) Signal crosstalk and induced resistance: straddling the line between cost and benefit. Annu Rev Phytopathol 43:545–580

Draper J, Rasmussen S, Zubair H (2011) Metabolite analysis and metabolomics in the study of biotrophic interactions between plants and microbes. In: Biology of plant metabolomics, Blackwell Publishing, Oxford. Ann Plant Rev 43:1–24

Duang LM, Jeewon R, Lumyoung S et al (2006) DGGE coupled with ribosomal DNA phylogenies reveal uncharacterized fungal phylotypes on living leaves of *Magnolia Liliifera*. Fungal Divers 23:121–138

Felitti S, Shields K, Ramsperger M et al (2006) Transcriptome analysis of *Neotyphodium* and *Epichloe* grass endophytes. Fungal Genet Biol 43:465–475

Findlay JA, Buthelezi S, Lavoie R et al (1995) Bioactive isocoumarins and related metabolites from conifer endophytes. J Nat Prod 58:1759–1766

Germida JJ, Siciliano SD, De Freitas JR (1998) Diversity of root-associated bacteria associated with field-grown canola (*Brassica napus*) and wheat (*Triticum aestivum*). FEMS Microbiol Ecol 26:43–50

Guo LD, Hyde KD, Liew ECY (2000) Detection and taxonomic placement of endophytic fungi within frond tissues of *Livistona chinensis* based on rDNA sequences. Mol Phylogenet Evol 20:1–13

Hallmann J (2001) Plant interactions with endophytic bacteria. In: Jeger MJ, Spence NJ (eds) Biotic interactions in plant-pathogen associations. CABI Publishing, New York, pp 87–119

Hallmann J, Quadt-Hallmann A, Mahaffee WF et al (1997) Bacterial endophytes in agricultural crops. Can J Microbiol 43:895–914

Handelman J (2004) Metagenomics: application of genomics to uncultured microorganisms. Microbiol Mol Biol Rev 68:669–685

Hoff JA, Klopfenstein NB, McDonald GI (2004) Fungal endophytes in woody roots of Douglas-fir (*Pseudotsuga menziesii*) and Ponderosa pine (*Pinus ponderosa*). For Pathol 34:255–271

Hunt MG, Rasmussen S, Newton PCD et al (2005) Near-term impacts of elevated CO_2, nitrogen and fungal endophyte-infection on *Lolium perenne* L. growth, chemical composition and alkaloid production. Plant Cell Environ 28:1345–1354

Imada C, Koseki N, Kmata M et al (2007) Isolation and characterization of antibacterial substances produced by marine actinomycetes in the presence of seawater. Actinomycetologica 21:27–31

Klitgord N, Segre D (2010) Environments that induce synthetic microbial ecosystems. PLoS Comput Biol 6:e1001002

Li QY, Zu YG, Shi RZ et al (2006) Review camptothecin: current perspectives. Curr Med Chem 13:2021–2039

Lin FC, Liu XH, Wang KK et al (2003) Recent research and prospect on taxol and its producing fungi. Appl Microbiol Biotechnol 86:1701–1717

Liu K, Ding X, Chen CL (2006) Isolation and characterization of endophytic taxol-producing fungi from *Taxus chinensis*. J Ind Microbiol Biotechnol 43:534–538

Lyons PC, Evans JJ, Bacon CW (1990) Effects of the fungal endophyte *Acremonium coenophialum* on nitrogen accumulation and metabolism in tall fescue. Plant Physiol 92:726–732

Matsumura H, Reich S, Ito A et al (2003) Gene expression analysis of plant host-pathogen interactions by SuperSAGE. Proc Natl Acad Sci U S A 100: 15718–15723

Mei C, Flinn BS (2010) The use of beneficial microbial endophytes for plant biomass and stress tolerance improvement. Recent Patents Biotech 4:81–95

Mundy J, Nielsen HB, Brodersen P (2006) Crosstalk. Trends Plant Sci 11:63–64

Murray FR, Latch GCM, Scott DB (1992) Surrogate transformation of perennial ryegrass *Lolium perenne*, using genetically modified *Acremonium* endophyte. Mol Gen Genet 233:1–9

Parish JA, McCann MA, Watson RH et al (2003a) Use of nonergot alkaloid-producing endophytes for alleviating tall fescue toxicosis in sheep. J Anim Sci 81:1316–1322

Parish JA, McCann MA, Watson RH et al (2003b) Use of nonergot alkaloid-producing endophytes for alleviating tall fescue toxicosis in stocker cattle. J Anim Sci 81:2856–2868

PatentStorm (2013) Endophytes. http://www.patenstorm.us/patents/6060051.html. Accessed 29 Jan 2013

Petrini O (1986) Taxonomy of endophytic fungi of aerial plant tissues. In: Fokkema NJ, Van Den Heuvel J (eds) Microbiology of the phyllosphere. Cambridge University Press, Cambridge, pp 175–187

Petrini O (1990) Endophytic fungi in British ericaceae: a preliminary study. Trans Br Mycol Soc 83:510–512

Petrini LE, Petrini O (1989) Recovery of endophytes of *Abies balsamea* from needles and galls of *Paradiplosis tumifex*. Phytoprotection 70:97–103

Pirttila AM, Laukkanen H, Pospiech H et al (2000) Detection of intracellular bacteria in the buds of Scotch pine (*Pinus sylvestris* L.) by in situ hybridization. Appl Environ Microbiol 66:3073–3077

Priti V, Ramesha BT, Sing S et al (2009) Opinion: how promising are endophytic fungi as alternative sources of plant secondary metabolites? Curr Sci 97:477–478

Rasmussen S, Parsons AJ, Liu Q, Xue H et al (2007) High nutrient supply and carbohydrate content reduce endophyte and alkaloid concentration. In: Popay AJ, Thom ER (eds) Proceedings of the 16th international symposium on fungal endophytes of grasses. New Zealand Grassland Research Association, Dunedin, pp 135–138

Rasmussen S, Parson AJ, Newman JA (2009) Metabolomics analysis of the lolium perenne-Neotyphodium *lolii* symbiosis: more than just alkaloids? Phytochem Rev 8:535–550

Rasmussen S, Parsons AJ, Fraser K et al (2012) High nitrogen supply and carbohydrate content reduce fungal endophyte and alkaloid concentration in *Lolium perenne*. Plant Physiol 146:1440–1453

Rodriguez RJ, White JF Jr, Arnold AE et al (2009) Fungal endophytes: diversity and functional roles. New Phytol 182:314–330

Saikkonen K, Wali P, Helander M et al (2001) Evolution of endophyte-plant symbioses. Trends Plant Sci 9:275–280

Schardl CL (1994) Molecular and genetic methodologies and transformation of grass endophytes. In: Bacon CW, White JF Jr (eds) Biotechnology of endophytic fungi of grasses. CRC Press, Boca Raton, pp 151–165

Schardl CL, Moon CD (2003) Processes of species evolution in *Ephchloe/Neotyphodium* endophytes of grasses. In: White JF Jr, Bacon CW, Hywel-Jones NL, Spatafora JW (eds) Clavicipitalean fungi. Marcel Dekker, New York, pp 273–310

Schardl CL, Phillips TD (1997) Protective grass endophytes: where are they from and where are they going. Plant Dis 81:430–438

Schulz B, Guske S, Dammann U (1998) Endophyte host interactions II. Defining symbiosis of the endophyte-host interaction. Symbiosis 25:213–227

Scott B, Schardl C (1993) Fungal symbionts of grasses: evolutionary insights and agricultural potential. Trends Microbiol 1:196–200

Soliman SSM, Tsao R, Raizada MN (2011) Chemical inhibitors suggests endophytic fungal paclitaxel is derived from both mevalonate and non-mevalonate-like pathways. J Nat Prod 74:2497–2504

Southcott KA, Johnson JA (1997) Isolation of endophytes from two species of palm, from Bermuda. Can J Microbiol 43:789–792

Stierle AA, Stierle DB (2005) Bioprospecting in the Berkeley pit: bioactive metabolites from acid mine waste extremophiles. Stud Nat Prod Chem 32:1123–1175

Strobel GA (2002) Rainforest endophytes and bioactive compounds. Crit Rev Biotechnol 22:315–333

Strobel G, Daisy B (2003) Bioprospecting for microbial endophytes and their natural products. Microbiol Mol Biol Rev 67:491–502

Strobel G, Yang XS, Sears J et al (1996) Taxol from *Pestalotiopsis microspora*, an endophytic fungus of *Taxus wallachiana*. Microbiology 142:435–440

Strobel GA, Miller RV, Martinez-Miller C et al (1999) Cryptocandin, a potent antimycotic from the endophytic fungus *Cryptosporiopsis* cf. *quercina*. Microbiology 145:1919–1926

Strobel GA, Daisy B, Castillo U et al (2004) Natural products from endophytic microorganisms. J Nat Prod 67:268

Suryanarayanan TS, Kumaresan V (2000) Endophytic fungi of some halophytes from an estuarine mangrove forest. Mycol Res 104:1465–1467

Suryanarayanan TS, Kumaresan V, Johnson JA (1998) Foliar fungal endophytes from two species of the Mangrove *Rhizophora*. Can J Microbiol 44:1003–1006

Tanaka A, Tapper B, Popay AJ et al (2005) A symbiosis expressed non-ribosomal peptide synthetase from a mutualistic fungal endophyte of perennial ryegrass confers protection to the symbiotum from insect herbivory. Mol Microbiol 57:1036–1050

Tanaka A, Christensen MJ, Takemoto D et al (2006) Reactive oxygen species play a role in regulating a fungus-perennial ryegrass mutualistic interaction. Plant Cell 18:1052–1066

Tanaka A, Christensen MJ, Takemoto D et al (2007) Endophyte production of reactive oxygen species is critical for maintaining the mutualistic symbiotic interaction between *Epichloe festucae* and Pooid grasses. In: Popay AJ, Thom ER (eds) Proceedings of the 6th international symposium on fungal endophytes of grasses, New Zealand Grassland Association, Dunedin, pp 185–188

Taylor JE, McAinsh MR (2004) Signalling crosstalk in plants: emerging issues. J Exp Bot 55:147–149

Thomashow LS, Weller DM (1996) Current concepts in the use of introduced bacteria for biological disease control: Mechanisms and antifungal metabolites. In: Stacy G, Keen NT (eds) Plant-microbe-interactions. Chapman and Hall, New York, pp 187–235

Tsai H-F, Siegel MR, Schardl CL (1992) Transformation of *Acremonium coenophialum*, a protective fungal symbiont of the grass *Festuca arundinacea*. Curr Genet 22:399–406

Wenzel S, Muller R (2005) Recent developments towards the heterologous expression of complex bacterial natural product biosynthetic pathways. Curr Opin Biotechnol 16:594–606

Wrede C, Dreier A, Kokoschka S et al (2012) Archaea in symbioses. Archaea 2012

Young CA, Bryant MK, Christensen MJ et al (2005) Molecular cloning and genetic analysis of a symbiosis-expressed gene cluster for lolitrem biosynthesis from a mutualistic endophyte of perennial ryegrass. Mol Genet Genomics 274:13–29

Yu HZC, Guo L, Li W et al (2010) Recent developments and future prospects of antimicrobial metabolites produced by endophytes. Microbiol Res 165:437–449

Zhang HR, Boghigian BA, Armando JE et al (2011) Methods and options for the heterologous production of complex natural products. Nat Prod Rep 28:125–151

Zhou XZH, Liu L, Lin J (2010) A review: recent advances and future prospects of taxol-producing endophytic fungi. Appl Microbiol Biotechnol 86:1701–1717

About the Editors

Dr. Vijay Chandra Verma joined Department of Botany, Banaras Hindu University, India, for his Ph.D., which he obtained in 2009. During his Ph.D., Dr. Verma explored the endophytic microbes associated with neem plant and their ecological roles for getting inexplicable host-endophyte interactions. He has also investigated the potential of endophytes in production of host-specific compounds and reported for the first time ever that azadirachtin and piperine are the two most important molecules that can be exclusively sourced from their respective host and can also be produced by fungal endophytes of the same host plants. He visited Technical University of Dortmund as DAAD fellow and recently has been recognized as Fast Track Young Scientist by the Department of Science and Technology, India. He has published more than 25 publications in the journals of international repute and has edited one book. Dr. Verma is also serving as associate editor for Universal Journal of Microbiology and Biochemistry and as voluntary reviewer for various journals. Further having research interest in host-endophyte interactions and natural product chemistry, he is also interested in the fields of nanobiotechnology and medical parasitology.

V.C. Verma and A.C. Gange (eds.), *Advances in Endophytic Research*,
DOI 10.1007/978-81-322-1575-2, © Springer India 2014

Prof. Alan Christopher Gange obtained his Ph.D. from University of London, UK, in 1985. In 1992, he joined the Royal Holloway University of London as lecturer and became professor of microbial ecology in 2007 and is currently serving as head of the department, School of Biological Sciences, Royal Holloway University of London. He has over 30 years of research and teaching experience. He supervises more than 16 Ph.D.s and 10 postdocs, and currently 9 Ph.D.s and a postdoc are working in his group. He has published more than 100 peer-reviewed papers in highly reputed international journals including ***Science***. His research interest is to study the multitrophic interactions which affect the diversity and structure of plant communities. He is specifically interested in how nonpathogenic fungi in plants affect the insects. Prof. Gange is holding several responsible positions in many professional bodies of international repute including treasurer, British Ecological Society; ex-NERC peer review college member; and member of British Ecological Society, Ecological Society of America. He is currently a fellow of Royal Entomological Society of London. He is also serving editorial boards of many reputed journals, editor for ***Journal of Turfgrass Research***, and ***Antenna***, associate editor for ***Insect Conservation and Diversity***, and regional editor for ***Soil Biology and Biochemistry***.